Arctic Ocean
Norway
Sweden
Finland
Russia
1 Estonia
2 Latvia
3 Lithuania
Denmark
The Neth.
Belarus
Poland
Germany
Belg.
Lux
Lich.
Aust
Hung
France
Switz.
San Ma.
Ukraine
Moldova
Rom
Bulg
Georgia
Kazakhstan
Mongolia
Andorra
Spain
Monaco
Vat. C.
Italy
Alb
Greece
Malta
Turkey
Armenia
Azerbaijan
Uzbekistan
Turkmenistan
Kyrgystan
Tajikistan
North Korea
South Korea
Japan
China
Tunisia
Cyprus
Lebanon
Israel
Syria
Iraq
Iran
Afghanistan
Pakistan
Morocco
Algeria
Libya
Egypt
Jordan
Kuwait
Bahrain
Saudi Arabia
Qatar
United Arab Emirates
Oman
Yemen
Nepal
Bhutan
Bangladesh
India
Hong Kong (U.K.)
China (Taiwan)
Laos
Pacific
Ocean
Mali
Niger
Chad
Sudan
Eritrea
Myanmar (Burma)
Thailand
Vietnam
Philippines
Burkina Faso
Nigeria
Djibouti
Ethiopia
Somalia
Cambodia (Kampuchea)
Sri Lanka
Brunei
Malaysia
Singapore
Palau
Federated States of Micronesia
Mashall Islands
Ivory Coast
Ghana
Cameroon
Cent. Af. Rep.
Uganda
Togo
Benin
São Tome and Principe
Kenya
(Maldives)
Indonesia
Kiribati
Nauru
Equatorial Guinea
Gabon
Congo
Zaire
Rwanda
Burundi
Cabinda
Tanzania
Seychelles
Malawi
Solomon Islands
Tuvalu
Papua New Guinea
Angola
Comoros
Indian Ocean
Zambia
Zimbabwe
Mozambique
Madagascar
Namibia
Botswana
Mauritius
Vanuatu
Fiji
Swaziland
Lesotho
South Africa
Australia
New Zealand
0°
20°
40°
60°
80°
100°
120°
140°
160°
180°
Bier

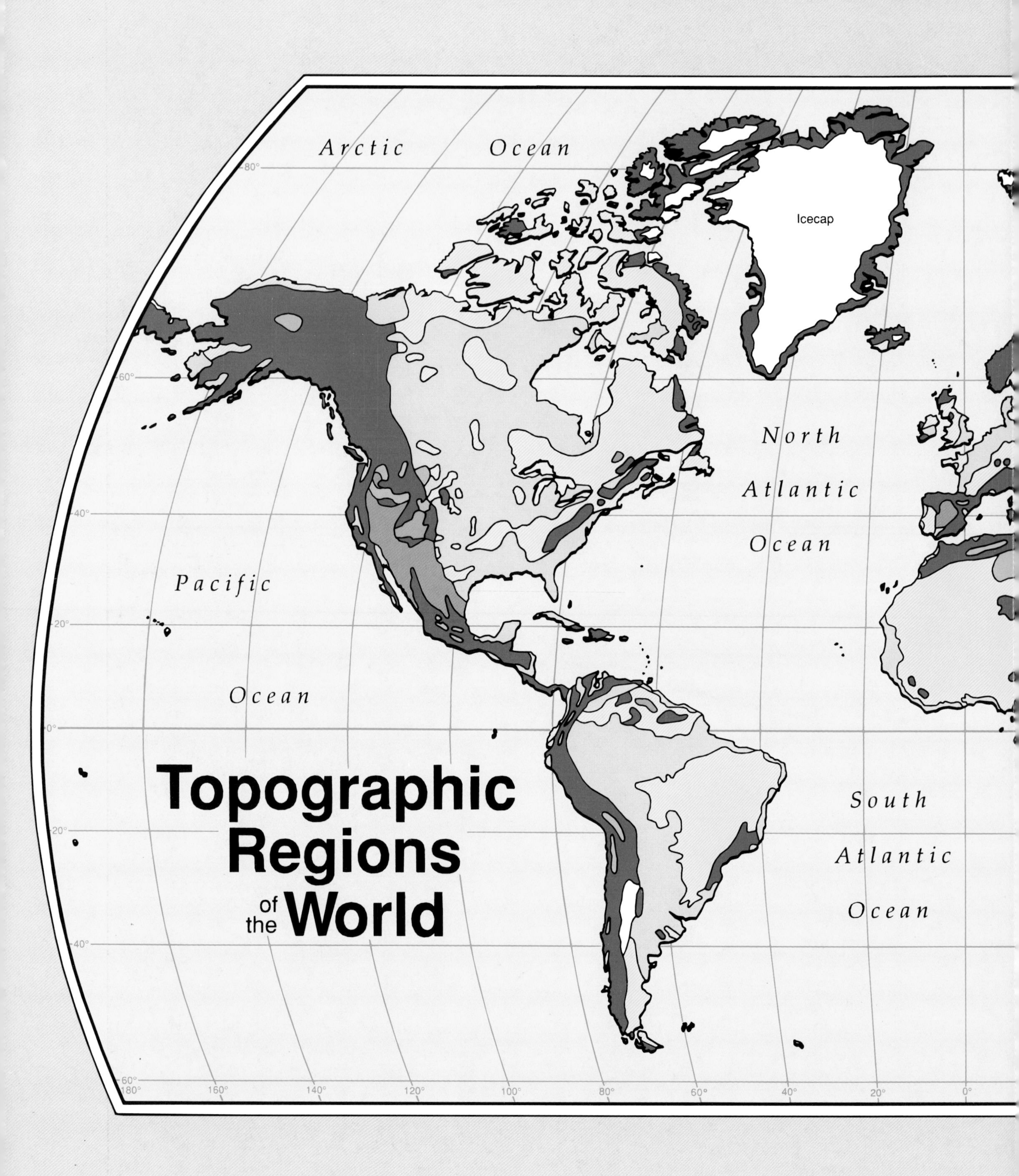

Arctic Ocean
Icecap
North Atlantic Ocean
Pacific Ocean
South Atlantic Ocean
Topographic Regions of the World
80°
60°
40°
20°
0°
20°
40°
60°
180°
160°
140°
120°
100°
80°
60°
40°
20°
0°

Countries of the World
Brooks and Roberts Modified
Van der Grinten Projection
Arctic Ocean
Pacific Ocean
North Atlantic Ocean
Greenland
Alaska (U.S.A.)
Canada
Ireland
4 Slovenia
5 Croatia
6 Bosnia-
7 Yugoslavia
8 Macedonia
10 Slovakia
United States
Bermuda (U.K.)
Hawaii (U.S.A.)
Cuba
The Bahamas
Haiti
Dominican Republic
Puerto Rico (U.S.A.)
Anguilla
Antigua
St. Lucia
St. Vincent & The Grenadines
Barbados
Trinidad & Tobago
Dominica
Grenada
Jamaica
Mexico
Belize
Honduras
Guatemala
El Salvador
Nicaragua
Costa Rica
Panama
Venezuela
Guyana
Suriname
French Guiana (Fr.)
Colombia
Ecuador
Cape Verde
Senegal
Gambia
Guinea Bissau
Guinea
Sierra Leone
Kiribati
Western Samoa
Niue
Tonga
Peru
Brazil
Bolivia
Paraguay
Chile
Argentina
Uruguay
Falkland Islands
80°
60°
40°
20°
0°
180°
160°
140°
120°
100°
80°
60°
40°
20°

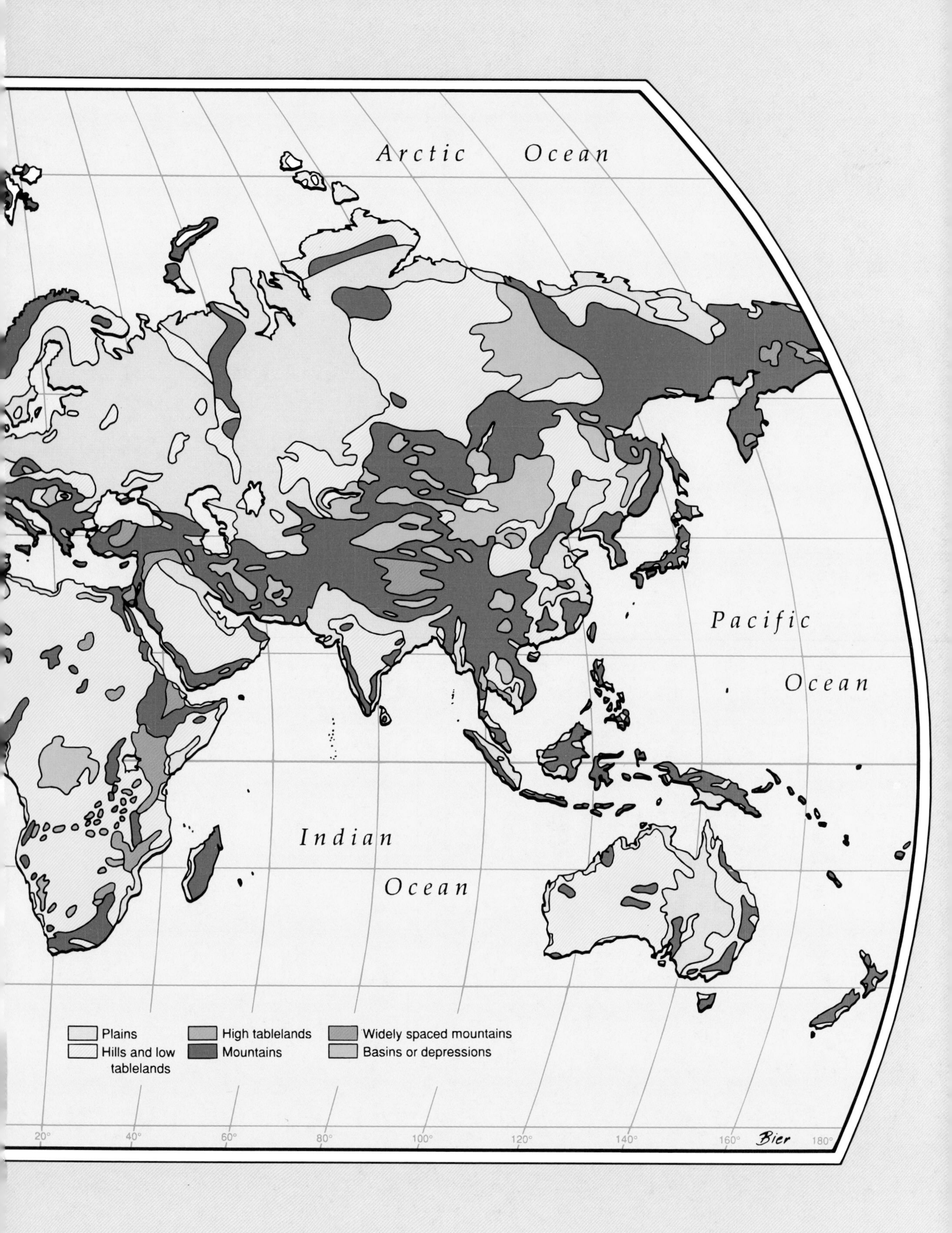
Arctic Ocean
Pacific Ocean
Indian Ocean
Plains
Hills and low tablelands
High tablelands
Mountains
Widely spaced mountains
Basins or depressions
20°
40°
60°
80°
100°
120°
140°
160°
180°
Bier

Introduction to Geography

Introduction to Geography

Seventh Edition

Arthur Getis
San Diego State University

Judith Getis

Jerome D. Fellmann
University of Illinois, Urbana-Champaign

Boston Burr Ridge, IL Dubuque, IA Madison, WI New York San Francisco St. Louis
Bangkok Bogotá Caracas Lisbon London Madrid
Mexico City Milan New Delhi Seoul Singapore Sydney Taipei Toronto

McGraw-Hill Higher Education

A Division of The ***McGraw-Hill*** *Companies*

INTRODUCTION TO GEOGRAPHY, SEVENTH EDITION

This book is printed on recycled, acid-free paper containing 10% postconsumer waste.

1 2 3 4 5 6 7 8 9 0 QPD/QPD 0 9 8 7 6 5 4 3 2 1 0

ISBN 0–697–38506–X

Vice president and editorial director: *Kevin Kane*
Publisher: *Edward Bartell*
Sponsoring editor: *Daryl Buflodt*
Developmental editor: *Lu Ann Weiss*
Marketing manager: *Lisa Gottschalk*
Editing associate: *Joyce Watters*
Senior production supervisor: *Mary Haas*
Design manager: *Stuart Paterson*
Photo research coordinator: *John Leland*
Senior supplement coordinator: *David Welsh*
Compositor: *Precision Graphics*
Typeface: *10/12 Palatino*
Printer: *Quebecor-Dubuque*

Cover/interior design: *Elise Lansdon*
Cover image: *FPG International*
Photo research: *Toni Michaels/PhotoFind*

Library of Congress Cataloging-in-Publication Data

Getis, Arthur, 1934–
Introduction to geography / Arthur Getis, Judith Getis, Jerome D.
Fellmann.—7th ed.
cm.
Includes bibliographical references and index.
ISBN 0–697–38506–X
1. Geography. I. Getis, Judith, 1938– . II. Fellmann, Jerome
Donald, 1926– . III. Title. IV. Title: Geography.
G128.G495 2000
910—dc21 98–55195
CIP

www.mhhe.com

Brief Contents

CONTENTS

PART TWO

THE CULTURE–ENVIRONMENT TRADITION 188

CHAPTER 6 POPULATION GEOGRAPHY 191

CHAPTER 7 CULTURAL GEOGRAPHY 229

PART FOUR

THE AREA ANALYSIS TRADITION 474

CHAPTER 13 THE REGIONAL CONCEPT 476

PREFACE

As did its predecessors, this seventh edition of *Introduction to Geography* seeks to introduce college students to the breadth and excitement of the field of geography. Like previous editions, its content is organized around the major research traditions of the discipline, a structure that adopters of earlier editions have found to be attractive to students and convenient and flexible for instructors.

Although the framework of presentation has been retained in this present edition, we have added and deleted materials to reflect new research findings in the different topical areas of geography and the spatial consequences of continuing changes in world political and economic circumstances. In addition to necessary chapter revisions and updating of facts, analyses, and viewpoints mandated by current events, we have made every effort to incorporate in this revision many of the helpful content and organizational suggestions offered by colleagues and users.

Inevitably, of course, in a climate of rapid end-of-century alterations in established economic, political, social, population, and environmental structures and relationships, reality outpaces textbook analysis. The time lags between world events and book publication and, later, between publication and actual class assignment inevitably mean that, at best, some of the text's content will be out of date and, at worst, some may be glaringly wrong at the time of student use. When events outpace publication, a mutually supportive partnership between geography textbook authors and classroom instructors is essential. We—and most importantly the students—must rely on the instructor to provide the currency of information and geographic interpretation essential to correct a text overtaken by circumstances.

PURPOSE

We recognize that many students will have only a single college course and textbook in geography. Our purpose for those students is to convey concisely and clearly the nature of the field, its intellectual challenges, and the logical interconnections of its parts. Even if they take no further work in geography, we are satisfied that they will have come into contact with the richness and breadth of our discipline and have at their command new insights and understandings for their present and future role as informed adults. Other students may have the opportunity and interest to pursue further work in geography. For them, we believe, this text will make apparent the content and scope of the subfields of geography, emphasize its unifying themes, and provide the foundation for further work in their areas of interest.

APPROACH

The approach we take allows the major research traditions of geography to dictate our principal themes. Chapter 1 introduces students to the four organizing traditions that have emerged through the long history of geographical thought and writing: earth science, culture–environment, location, and area analysis. Each of the four parts of this book centers on one of these geographic perspectives. Within each part (except that on area analysis) are chapters devoted to the subfields of geography, each placed with the tradition to which we think it belongs. Thus, the study of weather and climate is part of the earth science tradition; population geography is considered under the culture–environment tradition; and urban geography is included with the locational perspective.

Of course, our assignment of a topic may not seem appropriate to all users since each tradition contains many emphases and themes. Some subfields could logically be attached to more than one of the recognized traditions. The rationale for our clustering of chapters is given in the brief introductions to each part of the text. The tradition of area analysis—of regional geography—is presented in a single final chapter that draws on the preceding traditions and themes and is integrated with them by cross-references.

This revision reflects topical and regional changes since the last edition. We have replaced or revised many of the maps and diagrams, updated tables, and secured many new photographs. The frequently lengthy captions accompanying all illustrations convey additional information and explanation and serve as extensions of the text—not just identification or documentation of the figure. Chapter opening vignettes capture the reader's interest and attention in preparation for the subject matter that follows, while boxed inserts throughout the text further develop points or ideas discussed in each chapter. In short, every effort has been made to gain and retain student attention—the essential first step in the learning process.

Increasingly for today's students, that learning process is electronically based. In recognition of the growing reliance we are all placing on Internet and World Wide Web sources of information, we have included in each chapter a boxed discussion and preliminary guide to "websites" that either themselves provide data or serve as guides to other home page sources of data related to the contents of the chapter. Because of constantly changing home page addresses and continual addition and deletion of individual sites, we do not pretend that the references given in those boxed presentations are exhaustive, represent the best sites available on the given topics, or accurately report latest addresses. We hope, however, they will be useful starting points for student exploration and for the modifications, corrections, and additions that instructors will be able to supply. The authors welcome suggested additions, deletions, or adjustments to the boxed discussions and lists; those received and reviewed will be made available to others through this book's home page maintained by the publisher and referenced in the "On-Line" box of Chapter 1.

Chapter 1 in the larger sense prepares the student for the later substantive chapters. It introduces the field of geography as a whole, noting its breadth of interests and the unifying questions, themes, and concepts that structure all geographic inquiry. It also outlines the organization of the book and explains the several "traditions" forming its framework.

Important to that framework is the final chapter of the book devoted to the area analysis tradition. The case studies and examples that Chapter 13, "The Regional Concept," contains illustrate the regional geographic application of the systematic themes developed by the earlier chapters. Regional understanding has always been an important motivation and justification of geography as a discipline; Chapter 13 is designed to introduce students to the diversity of regional geographic exposition. It may be read either as a separate chapter or in conjunction with the earlier material. That is, each systematic chapter contains a reference to the section of Chapter 13 where a relevant regional geographic example is to be found. That referenced case study can then be incorporated to demonstrate the relationships of regional and systematic geography, to show the "real world" application of geographic understandings, and to provide a springboard for further case studies as class or instructor interest may dictate. In addition, the regional studies may serve as models for independent student reports by applying to specific cases the insights and techniques of analysis developed in the separate substantive chapters.

Flexibility

A useful textbook must be flexible enough in its organization to permit an instructor to adapt it to the time and subject matter constraints of a particular course. Although designed with a one-quarter or one-semester course in mind, this text may be used in a full year introduction to geography when employed as a point of departure for special topics and amplifications introduced by the instructor or when supplemented by additional readings and class projects. Moreover, the chapters are reasonably self-contained and need not be assigned in the sequence here presented. The "traditions" structure may be dropped and the chapters rearranged to suit the emphases and sequences preferred by the instructor or found to be of greatest interest to the students. The format of the course should properly reflect the joint contribution of instructor and book rather than be dictated by the book alone.

Learning Aids

Learning aids at the conclusion of each chapter include a Summary, a list of Key Words introduced in or essential to that chapter, For Review and Consideration questions, and a limited Selected References listing of important recent or classic considerations of the chapter's subject matter. We have tried to include both relatively widely available recent titles, many containing additional extensive bibliographies, and more specialized articles and monographs useful to students who are motivated by interest or assignment to delve more deeply into particular subfields of geography.

At the end of the book we have placed a comprehensive Glossary of terms and, as a special Appendix, a modified version of the *1998 World Population Data Sheet* of the Population Reference Bureau. In addition to basic demographic data and projections for countries, regions, and continents, the *Data Sheet* includes selected economic and social statistics helpful in national and regional comparisons. Although inevitably dated and subject to change, the appendix data nonetheless will provide for some years a wealth of useful comparative information for student projects, regional and topical analyses, and study of world patterns.

Our Ancillary Package

To assist the instructor, an **on-line** *Instructor's Manual* highlights the main ideas of each chapter, offers topics for class discussions, and provides approximately 70 suggested test questions for each chapter. McGraw-Hill also makes available sets of slides and overhead transparencies reproducing many of the maps, drawings, and photographs in the text; computerized testing materials (*MicroTest* for PC or Mac) for instructors; and an **on-line** *Study Guide* for students. The *Student Study Guide* will be updated to feature web questions to enhance use of the text's boxed readings.

In addition, instructors may order the text packaged with the *Nystrom Desk Atlas, Rand McNally Atlas of World Geography,* or the *Base Map Collection* at significant savings. Also available for packaging are the *Student Atlas of World Politics,* selected *Annual Editions©*, and Dushkin Atlas©.

The text can be packaged with the Rand McNally New Millennium CD-ROM and the Dorling Kindersley Eye Witness World Atlas CD-ROM. These CD-ROMs provide an abundance of current information by linking to the web.

Annual Editions Online: Geography is an online version of Dushkin's successful *Annual Editions: Geography*. Included on this site are 15–25 selected articles from the popular press, a topic guide, section overviews, annotated Table of Contents, test items, on-line assessment and quizzing (built into the site), an on-line search engine to search for additional articles, Web links to support the articles and an Instructor's Resource Guide. Access to this material is free with the purchase of a new textbook. Simply go to **http://www.dushkin.com/aeonline/** and enter the passcode supplied with your textbook.

Be sure to visit *Introduction to Geography's* Web site at: **http://www.mhhe.com/earthsci/geography/getis7e** where you can find on-line quizzing, regional and embassy links, PowerPoint lecture outlines and more.

ACKNOWLEDGMENTS

A number of reviewers greatly improved the content of this and earlier editions of *Introduction to Geography* by their critical comments and suggestions. Although we could not act on every helpful suggestion or adopt every useful observation, all were carefully and gratefully considered. In addition to those acknowledgments of assistance detailed in previous editions, we note with appreciation the thoughtful assistance recently rendered by:

Jeffrey D. Allender, University of Central Arkansas
Gregory E. Faiers, University of Pittsburgh at Johnstown
Gary L. Fowler, University of Illinois at Chicago
Emily B. Good, Northeast Illinois University
Greg R. Harris, Furman University
Gordon Hopper, University of Toronto at Mississauga Erindale College
Erik C. Howenstine, Northeastern Illinois University
Rob B. Kent, University of Akron
Howard Kittleson, Riverland Community College
Lisle S. Mitchell, University of South Carolina
James Penn, Southeastern Louisiana University
Harun Rasid, Lakehead University
Gerald L. Reynolds, University of Central Arkansas
Melissa Tollinger, East Carolina University
Thomas A. Wikle, Oklahoma State University
Leon Yacher, Southern Connecticut State University
David Zurick, Eastern Kentucky University

Sixth Edition
Tanya Allison, Montgomery College
Gary Belcher, Kentucky Christian College
Thomas Krabacher, California State University-Sacramento
Gary Brown, Montgomery College
David Dalton, College of the Ozarks
John Milbauer, Northeastern State University
Toshi Ikagawa, Eastern Kentucky University
John Bailey, Western Montana College
Lauren Scott, San Diego State University

Introduction to Geography **Focus Group Members**
Jeffrey Allender, University of Central Arkansas
Don Hagen, Northwest Missouri State University
Cary Komoto, University of Wisconsin—Rice Lake
William Laatsch, University of Wisconsin—Green Bay
Raoul Miller, University of Minnesota, Duluth
Erik Prout, Louisiana State University
Lallie Scott, Northeastern State University
Hubert B. Stroud, Arkansas State University

We gratefully express appreciation to these and unnamed others for their help and contributions and specifically absolve them of responsibility for decisions on content and for any errors of fact or interpretation that users may detect.

We acknowledge with pleasure the assistance rendered by Laura Martin Makey of San Diego Sate University in the preparation of the Internet discussions for several chapters of the text. We are also indebted to W. D. Brooks and C. E. Roberts, Jr., of Indiana State University for the projection used for many of the world maps in this book: a modified van der Grinten. Most of the maps, graphs, and charts in this edition still reflect the cartographic and design skills of James A. Bier, our close collaborator for all previous editions of *Introduction to Geography*. Although other demands on his time prevent his continued association with the text, its authors will always remain deeply grateful for his past invaluable contributions and continuing advice and personal friendship.

Finally, we note with deep appreciation and admiration the efforts of the publisher's "book team," separately named on the copyright page, who collectively shepherded this revision to completion. We are grateful for their highly professional interest, guidance, and support.

Arthur Getis
Judith Getis
Jerome D. Fellmann

LEARNING SYSTEMS

KEEPING UP WITH THE WORLD

This well-established text has been keeping students current with world trends for many years, and the seventh edition continues that tradition and purpose. While ***Introduction to Geography*** is still organized around the four themes of geography—physical, cultural, locational, and regional—it has also been extensively revised and updated with new text, additional photography, and revised maps reflecting new information and analysis and changing current events.

Revisions to ***Introduction to Geography*** include:

- Substantially rewritten and expanded text. New or revised sections include *The Nature of Geography; Evolution of the Discipline; Basic Geographic Concepts;* and others.
- Redrawn world pattern maps and selected graphs and charts (using current data).
- Extensively revised and expanded sections on tertiary and quaternary service activities.
- Boxed articles and discussions on *The Kyoto Protocol; A Population Implosion?; The Matter of Race;* and *Gender and Migration.*
- Updated information on the latest happenings in Eastern Europe.

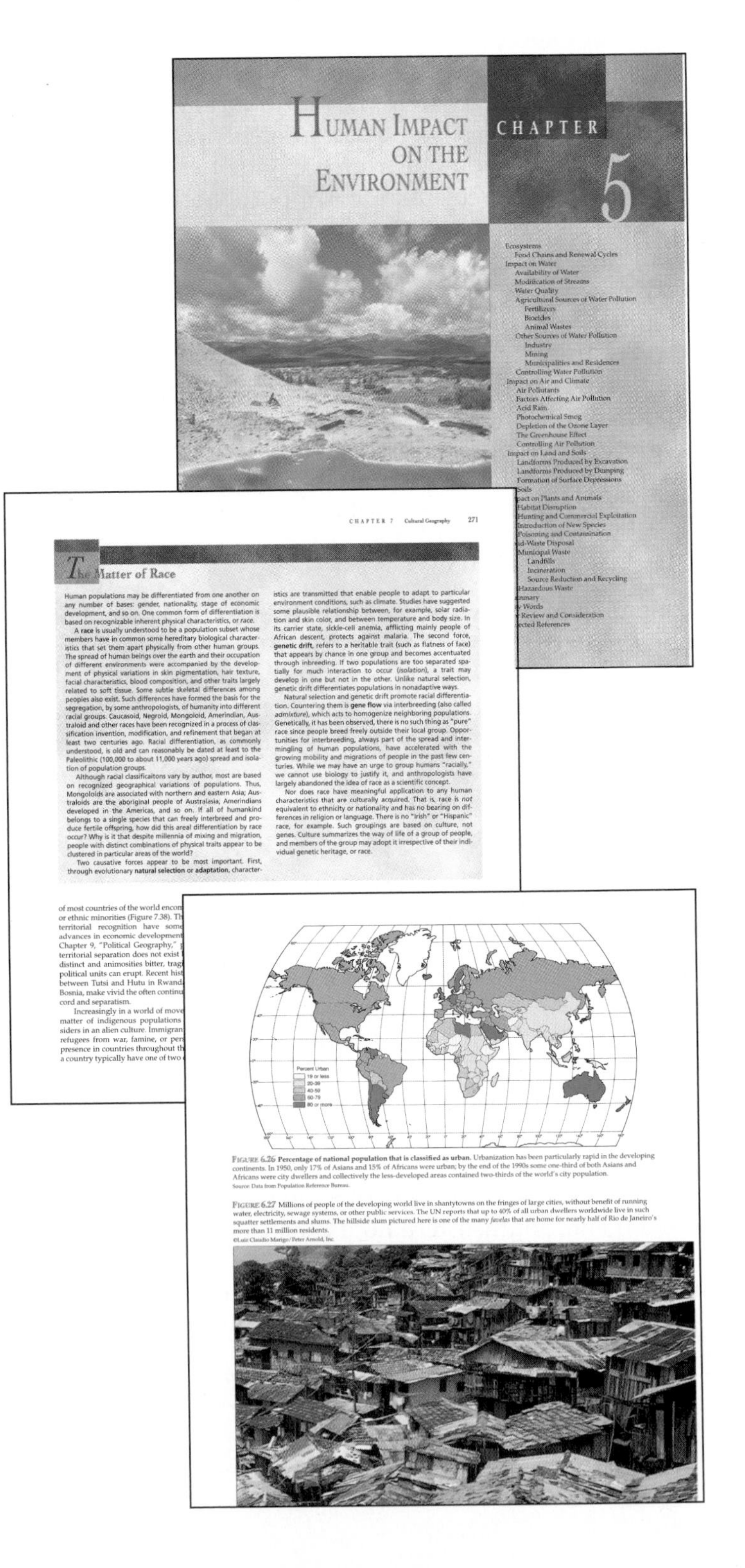

HUMAN IMPACT ON THE ENVIRONMENT

CHAPTER 5

Ecosystems
Food Chains and Renewal Cycles
Impact on Water
Availability of Water
Modification of Streams
Water Quality
Agricultural Sources of Water Pollution
Fertilizers
Biocides
Animal Wastes
Other Sources of Water Pollution
Industry
Mining
Municipalities and Residences
Controlling Water Pollution
Impact on Air and Climate
Air Pollutants
Factors Affecting Air Pollution
Acid Rain
Photochemical Smog
Depletion of the Ozone Layer
The Greenhouse Effect
Controlling Air Pollution
Impact on Land and Soils
Landforms Produced by Excavation
Landforms Produced by Dumping
Formation of Surface Depressions
Soils
…pact on Plants and Animals
Habitat Disruption
Hunting and Commercial Exploitation
Introduction of New Species
Poisoning and Contamination
…id-Waste Disposal
Municipal Waste
Landfills
Incineration
Source Reduction and Recycling
Hazardous Waste
…mmary
…y Words
…Review and Consideration
…ected References

CHAPTER 7 Cultural Geography 271

The Matter of Race

Human populations may be differentiated from one another on any number of bases: gender, nationality, stage of economic development, and so on. One common form of differentiation is based on recognizable inherent physical characteristics, or *race.*

A **race** is usually understood to be a population subset whose members have in common some hereditary biological characteristics that set them apart physically from other human groups. The spread of human beings over the earth and their occupation of different environments were accompanied by the development of physical variations in skin pigmentation, hair texture, facial characteristics, blood composition, and other traits largely related to soft tissue. Some subtle skeletal differences among peoples also exist. Such differences have formed the basis for the segregation, by some anthropologists, of humanity into different racial groups. Caucasoid, Negroid, Mongoloid, Amerindian, Australoid and other races have been recognized in a process of classification invention, modification, and refinement that began at least two centuries ago. Racial differentiation, as commonly understood, is old and can reasonably be dated at least to the Paleolithic (100,000 to about 11,000 years ago) spread and isolation of population groups.

Although racial classificaitons vary by author, most are based on recognized geographical variations of populations. Thus, Mongoloids are associated with northern and eastern Asia; Australoids are the aboriginal people of Australasia; Amerindians developed in the Americas, and so on. If all of humankind belongs to a single species that can freely interbreed and produce fertile offspring, how did this areal differentiation by race occur? Why is it that despite millennia of mixing and migration, people with distinct combinations of physical traits appear to be clustered in particular areas of the world?

Two causative forces appear to be most important. First, through evolutionary **natural selection** or **adaptation**, characteristics are transmitted that enable people to adapt to particular environment conditions, such as climate. Studies have suggested some plausible relationship between, for example, solar radiation and skin color, and between temperature and body size. In its carrier state, sickle-cell anemia, afflicting mainly people of African descent, protects against malaria. The second force, **genetic drift**, refers to a heritable trait (such as flatness of face) that appears by chance in one group and becomes accentuated through inbreeding. If two populations are too separated spatially for much interaction to occur (*isolation*), a trait may develop in one but not in the other. Unlike natural selection, genetic drift differentiates populations in nonadaptive ways.

Natural selection and genetic drift promote racial differentiation. Countering them is **gene flow** via interbreeding (also called *admixture*), which acts to homogenize neighboring populations. Genetically, it has been observed, there is no such thing as "pure" race since people breed freely outside their local group. Opportunities for interbreeding, always part of the spread and intermingling of human populations, have accelerated with the growing mobility and migrations of people in the past few centuries. While we may have an urge to group humans "racially," we cannot use biology to justify it, and anthropologists have largely abandoned the idea of race as a scientific concept.

Nor does race have meaningful application to any human characteristics that are culturally acquired. That is, race is *not* equivalent to ethnicity or nationality and has no bearing on differences in religion or language. There is no "Irish" or "Hispanic" race, for example. Such groupings are based on culture, not genes. Culture summarizes the way of life of a group of people, and members of the group may adopt it irrespective of their individual genetic heritage, or race.

of most countries of the world encon
or ethnic minorities (Figure 7.38). Th
territorial recognition have some
advances in economic development
Chapter 9, "Political Geography,"
territorial separation does not exist
distinct and animosities bitter, trag
political units can erupt. Recent hist
between Tutsi and Hutu in Rwand
Bosnia, make vivid the often continu
cord and separatism.
Increasingly in a world of move
matter of indigenous populations
siders in an alien culture. Immigran
refugees from war, famine, or pers
presence in countries throughout th
a country typically have one of two

FIGURE 6.26 **Percentage of national population that is classified as urban.** Urbanization has been particularly rapid in the developing continents. In 1950, only 17% of Asians and 15% of Africans were urban; by the end of the 1990s some one-third of both Asians and Africans were city dwellers and collectively the less-developed areas contained two-thirds of the world's city population.
Source: Data from Population Reference Bureau.

FIGURE 6.27 Millions of people of the developing world live in shantytowns on the fringes of large cities, without benefit of running water, electricity, sewage systems, or other public services. The UN reports that up to 40% of all urban dwellers worldwide live in such squatter settlements and slums. The hillside slum pictured here is one of the many *favelas* that are home for nearly half of Rio de Janeiro's more than 11 million residents.
©Luiz Claudio Marigo/Peter Arnold, Inc.

OUTSTANDING FEATURES YOU'LL FIND . . .

Chapter Opening Vignettes

Interesting, real-world anecdotes are presented at the beginning of each section.

The Gauda's[1] son is eighteen months old. Every morning, a boy employed by the Gauda carries the Gauda's son through the streets of Gopalpur. The Gauda's son is clean; his clothing is elegant. When he is carried along the street, the old women stop their ceaseless grinding and pounding of grain and gather around. If the child wants something to play with, he is given it. If he cries, there is consternation. If he plays with another boy, watchful adults make sure that the other boy does nothing to annoy the Gauda's son.

Shielded by servants, protected and comforted by virtually everyone in the village, the Gauda's son soon learns that tears and rage will produce anything he wants. At the same time, he begins to learn that the same superiority which gives him license to direct others and to demand their services places him in a state of danger. The green mangoes eaten by all of the other children in the village will give him a fever; coarse and chewy substances are likely to give him a stomachache. While other children clothe themselves in mud and dirt, he finds himself constantly being washed. As a Brahmin [a religious leader], he is taught to avoid all forms of pollution and to carry out complicated daily rituals of bathing, eating, sleeping, and all other normal processes of life.

In time, the Gauda's son will enter school. He will sit motionless for hours, memorizing long passages from Sanskrit holy books and long poems in English and Urdu. He will learn to perform the rituals that are the duty of every Brahmin. He will bathe daily in the cold water of the private family well, reciting prayers and following a strict procedure. The gods in his house are major deities who must be worshipped every day, at length and with great care.[2]

The Gauda and his family are not Americans, as references and allusions in the preceding paragraphs make plain. The careful reader would infer, correctly, that they are Indians—and if one were to read more of the book from which this excerpt is taken, it would become evident that the Gauda lives in a village in southern India. The class structure, the religion, the language, the food, and other strands of the fabric of life mentioned in the passage place the Gauda, his family, and his village in a specific time and place. They bind the people of the region together as sharers of a common culture and set them off from those of other areas with different cultural heritages. The 6 billion people who were the subject of Chapter 6 are of a single human family, but it is a family differentiated into many branches, each characterized by a distinctive *culture.*

[1]Gauda = village headman
[2]Excerpt from *Gopalpur: A South Indian Village*, by Alan R. Beals, copyright © 1962 by Holt, Rinehart and Winston, Inc., and renewed 1990 by Alan R. Beals, reprinted by permission of the publisher.

230

To some writers in newspapers and the popular press, "culture" means the arts (literature, painting, music, etc.). To a social scientist, **culture** is the specialized behavioral patterns, understandings, and adaptations that summarize the way of life of a group of people. In this broader sense, culture is as much a part of the regional differentiation of the earth as are topography, climate, and other aspects of the physical environment. The visible and invisible evidences of culture—buildings and farming patterns, language and political organization—are elements in the spatial diversity that invites and is subject to geographic inquiry. Cultural differences in area result in human landscapes with variations as subtle as the differing "feel" of urban Paris, Moscow, and New York or as obvious as the sharp contrasts of rural Zimbabwe and the Prairie Provinces of Canada (Figure 7.1).

Since such differences exist, cultural geography exists, and one branch of cultural geography addresses a whole range of "why?" and "what?" and "how?" questions. Why, since humankind constitutes a single species, are cultures so varied? What are the most pronounced ways in which cultures and culture regions are distinguished? What were the origins of the different culture regions we now observe? How, from whatever limited areas in which single culture traits and amalgams developed, were they diffused over a wider portion of the globe? Why do cultural contrasts between recognizably distinct groups persist even in such presumed "melting pot" societies as that of the United States or in the outwardly homogeneous, long-established countries of Europe? These and similar questions are the concerns of the present chapter and, in part, of Chapters 8 and 9.

COMPONENTS OF CULTURE

Culture is transmitted within a society to succeeding generations by imitation, instruction, and example. It is learned, not biological, and has nothing to do with instinct or with genes. As members of a social group, individuals acquire integrated sets of behavioral patterns, environmental and social perceptions, and knowledge of existing technologies. Of necessity we learn the culture in which we are born and reared. But we need not—indeed, cannot—learn its totality. Age, sex, status, or occupation dictate the aspects of the cultural whole in which we become fully indoctrinated.

A culture, that is, despite overall generalized and identifying characteristics and even an outward appearance of uniformity, displays a social structure—a framework of roles and interrelationships of individuals and established groups. Each individual learns and adheres to the rules and conventions not only of the culture as a whole but, importantly, of those specific to the subgroup to which he or she belongs. And that subgroup may have its own recognized social structure.

Culture is a complexly interlocked web of behaviors and attitudes. Realistically, its full and diverse content cannot be

"On-line" Boxes

There are a lot of interesting geography-related Internet sites out there. These boxes contain some of the best geography sites on the Web. Additionally, every text is packaged with an internet password providing access to Dushkin Publishing Group's on-line articles from the popular press. **http://www.dushkin.com/aeonline/**

CHAPTER 10 Economic Geography 355

On-Line ECONOMIC GEOGRAPHY

The range of subject matter in economic geography—and in this chapter—is broad and the number World Wide Web sites touching on some or all of it is great. Primary and secondary activities, domestic and international trade, developed and developing economies, economic theory, and the like, all are part of a vast web library of information. More than for most other subjects of this book, economic geography investigation will involve "surfing" the web, following leads, links, and interconnections in pursuit of the particular topics of interest to you. A few suggestions for getting started—tied to the main chapter topics—are offered here; nearly all provide extensive cross-references and links to related websites.

Primary Activities **Agriculture** sites worth checking include the *Food and Agriculture Organization of the United Nations*

http://www.fao.org/

and the *United States Department of Agriculture*

http://www.usda.gov/

Both provide guides to their own activities and agencies and links to other, related sites, as does the *Agriculture Canada* page.

http://www.agr.ca/

The *Iowa State University's Agriculture WWW Sites* page gives extensive links to agriculture-related sites on-line and is a good starting point for those with specific crop and livestock interests.

http//www.ag.iastate.edu/other.html

Other primary resources are considered on both international and purely domestic sites, including *Resources for the Future*

http://www.rff.org

and the U.S. Department of Interior's *Minerals Information* page.

http://minerals.er.usgs.gov/minerals

The *World Resources Institute* maintains websites for its areas of work in economics, forests, energy, sustainable agriculture, resources, and environmental management. Look for them through the Institute's home page.

http://www.wri.org/

Trade *The World Trade Organization*, established in 1995, is the principal agency of the world's multilateral trading system. Its home page includes access to documents discussing international conferences and agreements, reviewing its publications, and summarizing the current state of world trade.

http://www.wto.org/

Two other useful websites of broader topical interest, both yielding research results and statistical data for worldwide and national economies, are well worth checking. *The World Bank* is a leading source for country studies, research, and statistics covering all aspects of economic development and world trade; its site provides access to the contents of the bank's publications including its *World Development Report*, and to its research areas.

http://www.worldbank.org/

The Paris-based *Organization for Economic Cooperation and Development* site gives access to comparative statistics from 25 economically advanced countries and to the documents produced by the OECD itself.

www.oecd.org/

Secondary and Tertiary Activities The *U.S. Department of Commerce* is charged with promoting American business, manufacturing, and trade. Its home page connects with the websites of its constituent agencies.

http://www.doc.gov/

The *Bureau of Labor Statistics* page contains economic data, including unemployment rates, worker productivity, employment surveys and statistical summaries.

http://stats.bls.gov/blshome.html

A valuable Canadian counterpart is *Industry Canada*, with links to its own sectors, to other related branches of government, and to pertinent industry pages.

http://info.ic.gc.ca/

General "Economic" Sites *Econolink* advertises itself as featuring "the best web sites that have anything to do with economics." Its listings are selective with brief descriptions of site content; many referenced sites themselves have web links. Although more purely "economic" than "economic geographic," the page will help connect you to topics of particular interest to you.

http://www.progress.org/econolink

A much differently based and distributed form of extensive subsistence agriculture is found in all of the warm, moist, low-latitude areas of the world. There, many people engage in a kind of nomadic farming. Through clearing and use, the soils of those areas lose many of their nutrients (as soil chemicals are dissolved and removed by surface and groundwater or nutrients are removed from the land in the vegetables picked and eaten), and farmers cultivating them need to move on after harvesting several crops. In a sense, they rotate fields rather than crops to maintain productivity. This type of **shifting cultivation** has a number of names, the most common of which are *swidden* (an English localism for "burned clearing") and *slash-and-burn*.

End-of-Chapter Pedogogy

Reviewing material is enhanced with summaries, key word lists, and review and critical thinking questions.

CHAPTER 11 The Geography of Natural Resources 429

erate. The clearance of tropical forests is a matter of global, not just local, concern.

The growing demand for resources, induced by population increases and economic development, strains the earth's supply of raw materials. We began this chapter by presenting two significantly different views of the future. Because the shape of the future will be determined by the way the present generation of people thinks and acts, the wise and careful management of natural resources of all types is essential for all the world's economies, regardless of their stage of development.

Key Words

aquaculture 419	oil shale 399
biomass 403	perpetual resource 388
coal gasification 399	photovoltaic cell (PV) 405
coal liquefaction 399	potentially renewable resources 389
energy 391	proved reserves 390
estuarine zone 420	renewable resource 388
geothermal energy 405	resource 388
hydropower 404	solar energy 405
liquefied natural gas (LNG) 397	synthetic fuel 399
nonrenewable resource 389	tar sand 400
nuclear fission 401	tragedy of the commons 418
nuclear fusion 403	

For Review and Consideration

1. What is the basic distinction between a *renewable* and a *nonrenewable* resource? Why do estimates of *proved*

8. What, in general, are the leading mining countries? What role do developing countries play in the production of critical raw materials? How have producing countries reacted to the threatened scarcity of copper?
9. Since food resources are considered renewable, why is there concern about their exploitation? Discuss three ways of increasing food production. What problems do they pose?
10. What is the *estuarine zone*? What role does it play in the maintenance of marine life? In what ways is the ecology of the zone being assaulted?
11. What vital ecological functions do forests perform? Where are the tropical rain forests located, and what concerns are raised by their destruction?
12. Discuss three ways of reducing demands on resources.

Selected References

Bender, William, and Margaret Smith. *Population, Food, and Nutrition*. Washington, D.C.: Population Reference Bureau, Inc. 1997.

Castillon, David A. *Conservation of Natural Resources: A Management Approach*. 2d ed. Dubuque, Iowa: Wm. C. Brown, 1997.

Coull, James R. *World Fisheries Resources*. New York: Routledge, 1993.

Cutter, Susan, Hilary L. Renwick, and William H. Renwick. *Exploitation, Conservation, Preservation: A Geographic Perspective on Natural Resource Use*. 2d ed. New York: John Wiley & Sons, 1991.

Dower, Roger, et al. *A Sustainable Future for the United States*. Washington, D.C.: World Resources Institute, 1996.

Elliott, David. *Energy, Society and Environment*. New York: Routledge, 1997.

Energy for Tomorrow's World. The World Energy Council. New York: St. Martin's Press, 1993.

...ld Food Problem. 2d ed. Cambridge, Mass.: ...96.

...rgy. Philadelphia: Saunders College Pub-

...eral Resources and the Environment. New ...blishing Co., 1994.

...hapman. *Environmental Resources*. New ... Sons, 1996.

... *Conserving the World's Biological Diversity*. ...esources Institute, 1990.

...urce Conservation and Management. Bel-...worth Publishing Co., 1990.

...ura Tangley. *Trees of Life: Saving Tropical* ...logical Wealth. Washington, D.C.: World ..., 1991.

...orld Nuclear Power. New York: Routledge,

... *Rainforests*. New York: Routledge, 1992.

...esources: Allocation, Economics, Policy. 2d ed. ...ge, 1990.

...Air and Water: Resources and Environment in ...y. London: Edward Arnold, 1991.

FIGURE 11.44 **Aluminum cans await recycling.** It takes only 5% as much electricity to make aluminum from scrap as from raw materials. In other words, manufacturers can make 20 cans out of recycled material with the same energy it takes to make a single can out of new material. Another good reason to recycle is that an aluminum can thrown away in the trash will still be in the landfill 100 years from now.
©Ann Duncan/Tom Stack & Associates

throwing away an aluminum soft drink container wastes as much energy as filling the can half full of gasoline and pouring it on the ground.

Finally, the *substitution* of other energy sources for gas and oil, and the substitution of other materials for nonfuel minerals in short supply can be more ... the technology to exploit them econom... oped, coal and oil shale could supply ... the future. In addition, the renewable en... as biomass, solar, and geothermal pow... nite in their amount and variety. While ... source is likely to be as important as oi... they could make a significant contributi...

Summary

Our economic and material well-being depend on our use of natural resources. Renewable resources like soil and plants are those that can be regenerated in nature as fast as or faster than societies exploit them, although even renewable resources can be depleted if the rate of use exceeds that of regeneration. Nonrenewable resources—the fossil fuels and nonfuel minerals—are generated so slowly that they are thought of as existing in finite amounts. The proved reserves of a resource are the amounts that have been identified and that can be extracted profitably.

The Industrial Revolution was characterized by a shift from societal dependence on renewable resources to those derived from nonrenewable minerals, chiefly fossil fuels. The industrially advanced countries depend for about 40% of their commercial energy on crude oil, a scarce and unevenly distributed resource whose known reserves are likely to be exhausted before the middle of the next century. Only two regions, Russia and the Middle East, have large reserves of natural gas, another resource likely to approach depletion by the mid-21st century. Synthetic petroleum and gas can be produced from a number of sources, but at the present time all synfuels have high economic and environmental costs. Although nuclear power plants produce less than one-third of the world's electricity as a whole, the pattern is one of uneven dependence. Some countries receive more than half of their electricity from such plants, while others have none.

Renewable natural resources are more widely and evenly distributed than the nonrenewable ones. Wood and other forms of biomass are the primary source of energy for more than half of the world's people. Hydropower is a major source of electricity in many countries. Other renewable resources, including geothermal energy and wind power, make a more localized and limited contribution to energy needs.

The earth's crust contains a variety of nonfuel mineral resources from which people fashion metals, glass, stone, and other products. All are nonrenewable. Some exist in vast amounts, others in relatively small quantities; some are widely distributed, others concentrated in just a few locations.

About one-tenth of the earth's surface is used for the intensive production of food. Although food production has increased as fast as world population, more people than ever before are malnourished. Methods of expanding food

Boxed Articles and Discussions

Stay on top of current topics (such as "A Population Implosion?"). Use these readings as catalysts for class discussions. *Geography and Public Policy* boxed readings highlight important or controversial issues and encourage students to reflect on these topics, forming educated opinions.

208 PART TWO The Culture-Environment Tradition

A Population Implosion?

For much of the last half of the 20th century, demographers and economists have focused on a "population explosion" and its implied threat of a world with too many people and too few resources of food and minerals to sustain them. By the end of the century, those fears for some observers were being replaced by a new prediction of a world with too few rather than too many people.

That possibility is suggested by two related trends. The first became apparent by 1970 when it was noted that the total fertility rates (TFRs) of 19 countries, almost all of them in Europe, had fallen below the **replacement level**—the level of fertility at which populations replace themselves—of 2.1. Simultaneously, Europe's population pyramid began to become noticeably distorted, with a smaller proportion of young and a growing share of middle-aged and retired-age inhabitants. The decrease in native working-age cohorts had already, by 1970, encouraged the influx of non-European "guest workers" whose labor was needed to maintain economic growth and to sustain the generous security provisions guaranteed to what was becoming the oldest population of any continent.

Many countries of Western and Eastern Europe sought to reverse their birth rate declines by adopting pronatalist policies. The communist states of the East rewarded pregnancies and births with generous family allowances, free medical and hospital care, extended maternity leaves, and child care. France, Italy, the Scandinavian countries, and others gave similar bonuses or awards for first, second, and later births. Despite those inducements, however, reproduction rates continued to fall in Western states and, after the dissolution of communism in the East, to plummet there, too. Despite continuing large-scale in-migration of non-Europeans during the decade, the continent's total population remained essentially constant between 1990 and 1999. "In demographic terms," France's prime minister remarked, "Europe is vanishing."

Europe's experience soon was echoed in other societies of advanced economic development on all continents. By 1995, the United States, Canada, Australia, New Zealand, Japan, Taiwan, South Korea, Singapore, and other older and newly industrializing countries (NICs) registered fertility rates below the replacement level. As they have for Europe, simple projections foretold their aging and declining population. Japan's current slight increase, for example, will become a decrease in 2006 when its population will be older than Europe's; Taiwan forecasts negative growth by 2035.

The second trend indicating to many that world population numbers might stabilize and even decline during the lifetimes of today's college cohort is a simple extension of the first: TFRs are being reduced to or below the replacement levels in countries at all stages of economic development in all parts of the world. While only 18% of total world population in 1975 lived in countries with a fertility rate below replacement level, nearly one-half did so by the end of the century. By 2015, demographers estimate, nearly half the world's countries and over two-thirds of its population will show TFRs below 2.1 children per woman. Exceptions to the trend are and still will be found in Africa, especially sub-Saharan Africa, and in some areas of South, Central, and West Asia; but even in those regions fertility rates have been decreasing in recent years. "Powerful globalizing forces [are] at work pushing towards fertility reduction everywhere," was a 1997 observation from the French National Institute of Demographic Studies.

That conclusion is plausibly supported by assumptions of the United Nation's 1996 "low variant" world population projection. Noting that total fertility rates for more developed regions have already fallen to about 1.5 from 2.0 at start of the 1990s, the UN conjectured a further drop to about 1.4 by 2006. For the less developed regions, the rate dropped from 4.0 to 3.3 during the 1990s alone; the low variant model projects it will decline to about 2.0 in 2020 and 1.6 in 2050. Should those low variant assumptions prove valid, global depopulation could commence in less than 40 years. Between 2040 and 2050, the projection reports, world population would fall by about 85 million (roughly the amount it is increasing each year at the end of the 1990s), and shrink further by about 25% with each successive generation.

If the UN low variant scenario is realized in whole or in part—it is currently rejected by most demographers—a much different worldwide demographic and economic future is promised than that prophesied so recently by "population explosion" forecasts and by demographers' majority estimate of continuing growth to 11 billion in A.D. 2200. Declining rather than increasing pressure on world food and mineral resources would be in our future along with shrinking rather than expanding world, regional, and national economies. Even the achievement of **zero population growth**, a condition for individual countries when births plus immigration equal deaths plus emigration, has social and economic consequences not always perceived by its advocates. These inevitably include an increasing proportion of older citizens, fewer young people, a rise in the median age of the population, and a growing old-age dependency ratio with ever-increasing pension and social services costs borne by a shrinking labor force. Actual population decline, now the common European condition, would only exaggerate those consequences on a worldwide basis.

and antibiotic-resistant diseases, pesticide resistance of disease-carrying insects, and such new scourges of both the less developed and more developed countries as AIDS (acquired immune deficiency syndrome) cast doubt on the finality of that "ultimate" stage (see "Our Delicate State of Health").

Even the resurgence of old and emergence of new scourges such as malaria, tuberculosis, and AIDS are unlikely to have serious foreseeable demographic consequences. The United Nations, for example, has estimated that in a hypothetical worst case—that is, if all of Africa

CHAPTER 1

INTRODUCTION

The rapid conversion of its native forests to farm land is one reflection of growing population pressures in the East African country of Kenya.
©Frans Lanting/Minden Pictures

Government inspectors had issued the first of two warnings about the dam nearly a year before, but no repairs had been made. Fortunately, workers spotted the cracks early, and when the dam broke, letting loose a 6-meter (20-foot) wave, the villages had been cleared and all train and highway traffic stopped. What poured through the breach was not just water, but a thick deadly brine, the impounded discharge of a major fertilizer plant. On its rush to the river, the brine swept away railroad tracks, ripped up roads, smashed through a village, inundated 2000 hectares (5000 acres) of the nation's richest farmland, and spilled into the purest river left in the western part of the country. Worse than the immediate physical destruction were the aftereffects. Two thousand tons of fish were destroyed along with all of the vegetation that could support their replacement. Water supplies to two major and many minor cities were cut off as the formerly pure river became "brinier than the saltiest seawater," and a million tons of salt were deposited in a layer 10.5 meters (35 feet) thick at the bottom of a major reservoir 480 kilometers (300 miles) downstream.

The accident happened in 1983 in the western part of Ukraine and affected the Dniester River from near the Polish border all the way to the Black Sea. The details of its date and location are less important than the lessons contained in the event. The wall of brine, the destruction of farmland and fish, the salt layer at the lake bottom, and the scramble for alternate water supplies are dramatic evidence of the frequently unbearable pressures humans place on their environment. Ever-accumulating radioactive waste, deadly manufactured chemicals, accelerated erosion through unwise forestry and farming practices, the creation of deserts through overgrazing, the poisoning of soils by salts from faulty irrigation and of groundwater through deep-well injection of the liquid garbage of modern industry—all are examples of unsustainable human demands on natural systems. The social and economic actions of humans occur within the context of the environment and have environmental consequences too serious to ignore.

The interaction of human and environmental systems works both ways. People can inflict irreparable damage on the environment and the environment can exact a frightening toll from societies that inappropriately exploit it or even innocently occupy it.

November is the cyclone season in the Bay of Bengal. At the northern end of the bay lie the islands and the lowlands of the Ganges delta, a vast fertile land, mostly below 9 meters (30 feet) in elevation, made up of old mud, new mud, and marsh. Densely settled by desperately land-needy people, this delta area is home to the majority of the population of Bangladesh.

Early in November of 1970, a low-pressure weather system moved across the Malay Peninsula of Southeast Asia and gained strength in the Bay of Bengal, generating winds of nearly 240 kilometers (150 miles) per hour. As it moved northward, the storm sucked up and drew along with it a high wall of water. On the night of November 12, with a full moon and highest tides, the cyclone and its battering ram of water slammed into the islands and the deltaic mainland. When it had passed, some of the richest rice fields in Asia were gray with the salt that ruined them, islands totally covered with paddies were left as giant sand dunes, and an estimated 500,000 people had perished.

Should the tragedy—with its potential for annual recurrence—be called the result of the blind forces of nature, or should it be seen as the logical outcome of a state of overpopulation that forced human settlement on lands more wisely left as the realm of river and sea? (Figure 1.1)

As a discipline, geography does not attempt to make value judgments about such questions. Geography does claim, however, to be a valid and revealing approach to contemporary questions of political, economic, social, and ecological concern. Humans and environment in interaction; the distribution of natural phenomena affecting human use of the earth; the cultural patterns of settlement and exploitation of the physical world—these are the themes of that encompassing discipline called *geography.*

The Nature of Geography

Geography is often referred to as the *spatial* science, that is, the discipline concerned with the use of earth space. In fact, *geography* literally means "description of the earth," but that task is really the responsibility of nearly all the sciences. Geography might better be defined as the study of spatial variation, of how—and why—things differ from place to place on the surface of the earth. It is, further, the study of how observable spatial patterns evolved through time. If things were everywhere the same, if there were no spatial variation, the kind of human curiosity that we call "geographic" simply would not exist. Without the certain conviction that in some interesting and important way landscapes, peoples, and opportunities differ from place to place, there would be no discipline of geography.

But we do not have to deal in such abstract terms. You consciously or subconsciously display geographic awareness in your daily life. You are where you are, doing what you are doing, because of locational choices you faced and spatial decisions you made. You cannot be here reading this book and simultaneously be somewhere else—working, perhaps, or at the gym. And should you now want to go to work or take an exercise break, the time involved in going from here to there (wherever "there" is) is time not available

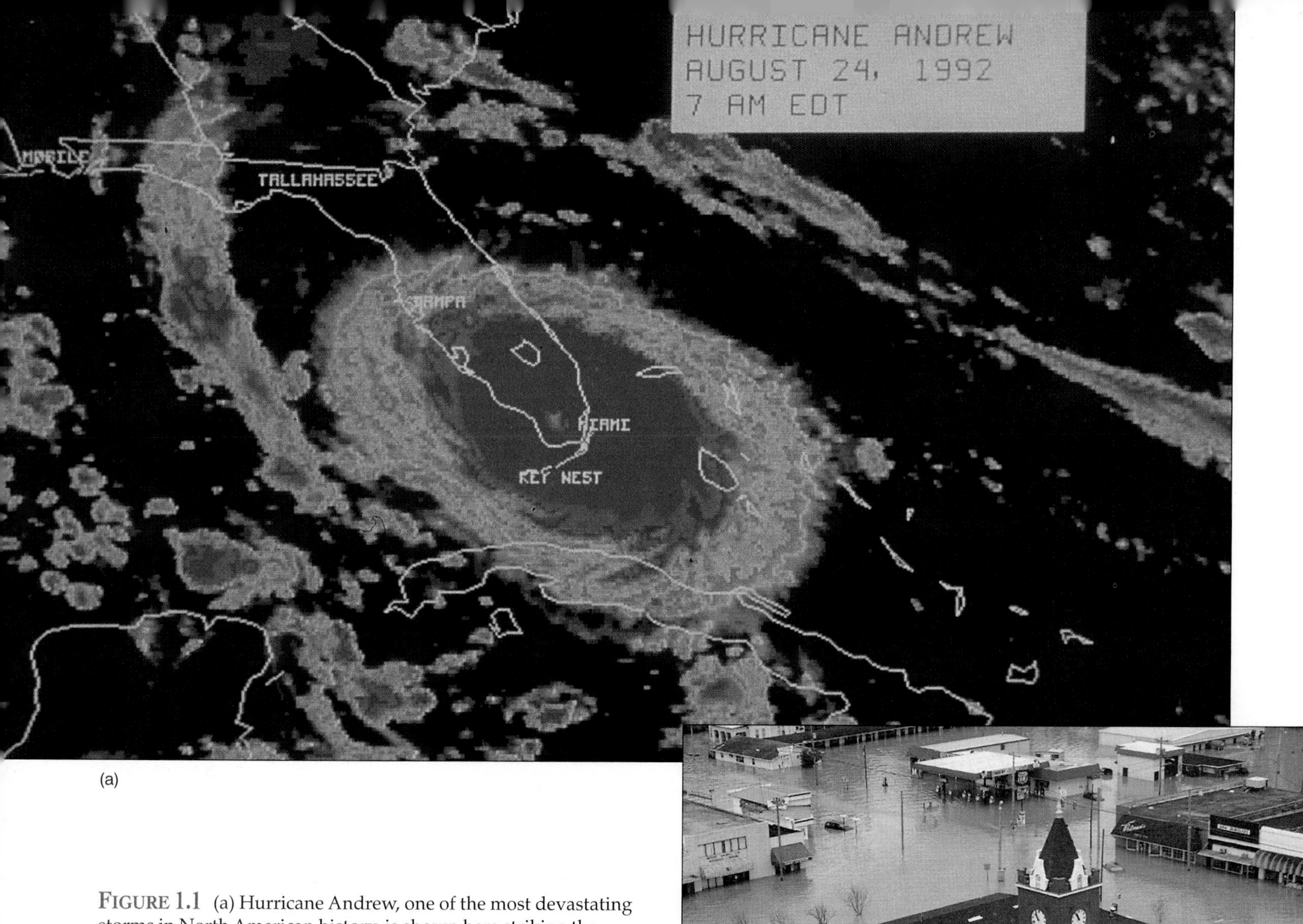

(a)

(b)

FIGURE 1.1 (a) Hurricane Andrew, one of the most devastating storms in North American history, is shown here striking the Florida coast on August 24, 1992. Winds of 264 kilometers per hour (164 mph) and a storm surge of water leveled the communities of Homestead and Florida City and destroyed Homestead Air Force Base in southern Dade County. In September, 1996, Hurricane Fran crashed into the Carolinas, flooding coastal towns and lowland areas northward into Pennsylvania, killing at least 17 persons and causing an estimated $1 billion in damage. Its onslaught and (b) the flooding of the valley town of Elba, Alabama, in early March, 1998, are further grim reminders that intensive development of low-lying coastal and river flood plain areas is an open invitation to potential catastrophic loss of life and property. (See also Figure 8.6.)
Source: (a) NOAA; (b) Associated Press.

for other activities in other locations. Of course, the act of going implies knowing where you are now, where "there" is in relation to "here," and the paths or routes you can take to cover the distance.

These are simple examples of the observation that "space matters" in a very personal way. You cannot avoid the implications of geography in your everyday affairs. Your understanding of your hometown, your neighborhood, or your college campus is essentially a geographic understanding. It is based on your awareness of where things are, of their spatial relationships, and of the varying content of the different areas and places you frequent. You carry out your routine activities in particular places and move on your daily rounds within defined geographic space, following logical paths of connection between different locations. At the same time, those activities and movements are necessarily affected by the physical environment—the terrain features, the weather or climatic conditions, and the like—in which they take place.

Just as geography matters in your personal life, so it matters on the larger stage as well. Decisions made by corporations about the locations of manufacturing plants or warehouses in relation to transportation routes and markets are spatially rooted. So, too, are those made by shopping center developers and locators of parks and grade schools. At an even grander scale, judgments about the projection of national power or the claim and recognition of "spheres of influence and interest" among rival countries are related to the implications of distance and area.

Geography, therefore, is about space and the content of space. We think of and respond to places from the standpoint not only of where they are but, rather more importantly, of what they contain or what we think they contain. Reference to a place or an area usually calls up images about its physical nature or what people do there and often suggests, without conscious thought, how those physical things and activities are related. "Bangladesh," "farming," and "disastrous flooding" or "Colorado," "mountains," and "skiing" might be simple examples. The content of area, that is, has both physical and cultural aspects, and geography is always concerned with understanding both (Figure 1.2).

Evolution of the Discipline

Geography's combination of interests was apparent even in the work of the early Greek geographers who first gave structure to the discipline. Geography's name was reputedly coined by the Greek scientist Eratosthenes over 2200 years ago from the words *geo,* "the earth," and *graphein,* "to write." From the beginning, that writing focused both on the physical structure of the earth and on the nature and activities of the people who inhabited the different lands of the known world. To Strabo (*c.* 64 B.C.–A.D. 20) the task of geography was to "describe the several parts of the inhabited world . . . to write the assessment of the countries of the world [and] to treat the differences between countries." Even earlier, Herodotus (*c.* 484–425 B.C.) had found it necessary to devote much of his book to the lands, peoples, economies, and customs of the various parts of the Persian Empire as necessary background to an understanding of the causes and course of the Persian wars.

Greek (and, later, Roman) geographers measured the earth, devised the global grid of parallels and meridians (marking latitudes and longitudes), and drew upon that grid surprisingly sophisticated maps of their known world (Figure 1.3). They explored the apparent latitudinal variations in climate and described in numerous works the familiar Mediterranean basin and the more remote, partly rumored lands of northern Europe, Asia, and equatorial Africa. Employing nearly modern concepts, they described

FIGURE 1.2 Ski development at Whistler Mountain, British Columbia, Canada clearly shows the interaction of physical environment and human activity. Climate and terrain have made specialized human use possible. Human exploitation has placed a cultural landscape on the natural environment, thereby altering it.
©Karl Weatherly/Corbis Media

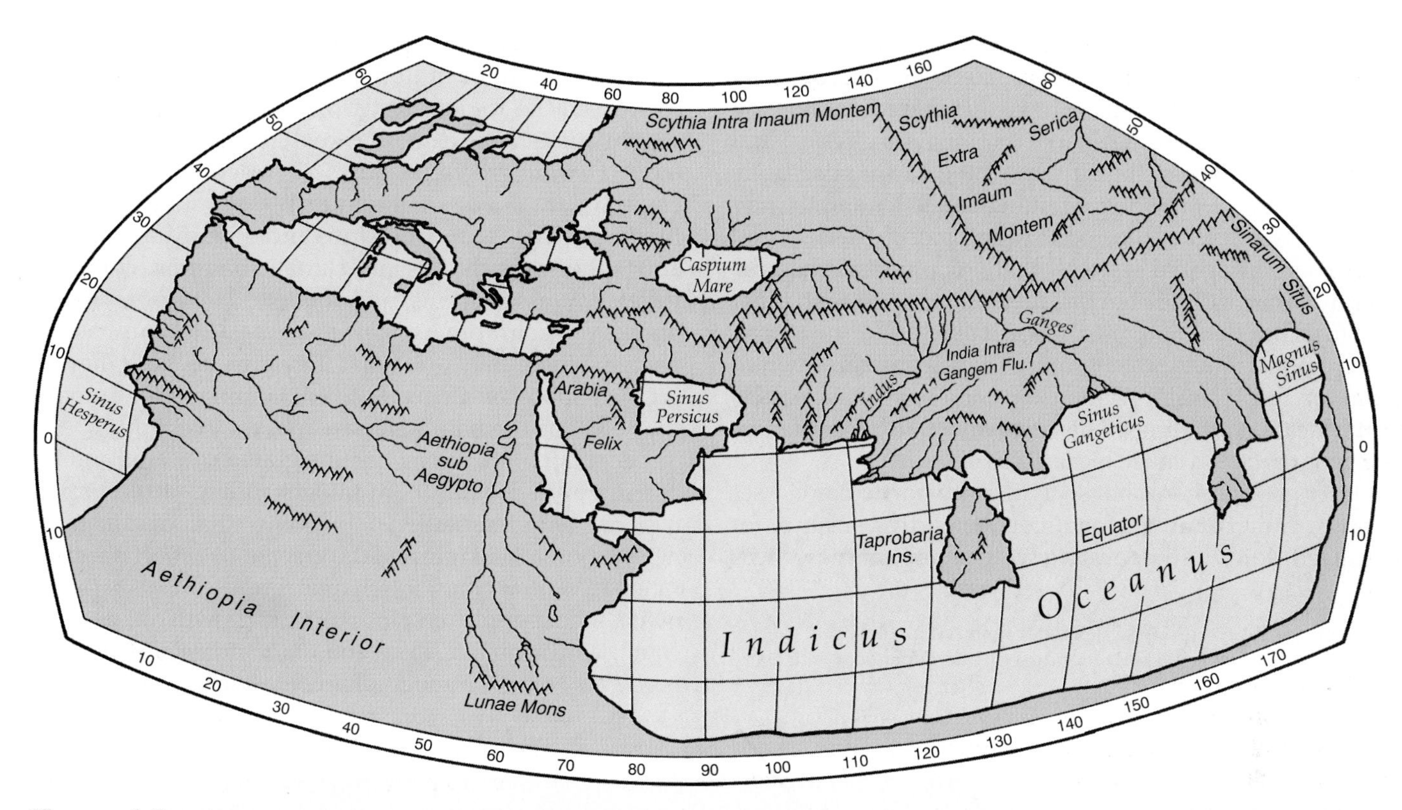

FIGURE 1.3 World map of the 2nd century A.D. Greco-Egyptian geographer-astronomer Ptolemy. Ptolemy (Claudius Ptolemaeus) adopted a previously developed map grid of latitude and longitude based on the division of the circle into 360°, permitting a precise mathematical location for every recorded place. Unfortunately, errors of assumption and measurement rendered both the map and its accompanying six-volume gazetteer inaccurate. Many variants of Ptolemy's map were published in the 15th and 16th centuries. The version shown here summarizes the extent and content of the original.

river systems, explored cycles of erosion and patterns of deposition, cited the dangers of deforestation, described areal variations in the natural landscape, and noted the consequences of environmental abuse. Against that physical backdrop, they focused their attention on what humans did in home and distant areas—how they lived; what their distinctive similarities and differences were in language, religion, and custom; and how they used, altered, and perhaps destroyed the lands they inhabited. Strabo, indeed, cautioned against the assumption that the nature and actions of humans were determined by the physical environment they inhabited. He observed that humans were the active elements in a human–environmental partnership.

The interests guiding the early Greek and Roman geographers were and are enduring and universal. The ancient Chinese, for example, were as involved in geography as an explanatory viewpoint as were Westerners, though there was no exchange between them. Further, as Christian Europe entered its Dark and Middle Ages between A.D. 800 and 1400 and lost its knowledge of Greek and Roman geographical work, Muslim scholars—who retained that knowledge—undertook to describe and analyze their known world in its physical, cultural, and regional variation.

Modern geography had its origins in the surge of scholarly inquiry that, beginning in the 17th century, gave rise to many of the traditional academic disciplines we know today. In its European rebirth, geography from the outset was recognized—as it always had been—as a broadly based integrative study. Patterns and processes of the physical landscape were early interests, as was concern with humans as part of the earth's variation from place to place. The rapid development of geology, botany, zoology, climatology, and other natural sciences by the end of the 18th century strengthened regional geographic investigation and increased scholarly and popular awareness of the intricate interconnections of things in space and between places. By that same time, accurate determination of latitude and longitude and scientific mapping of the earth made assignment of place information more reliable and comprehensive. During the 19th century, national censuses, trade statistics, and ethnographic studies gave firmer foundation to human geographic investigation. By the end of the 19th century, geography had become a distinctive and respected discipline in universities throughout Europe and in other regions of the world where European academic examples were followed. The proliferation of professional geographers and geography programs resulted in the development of a whole series of increasingly specialized disciplinary subdivisions, many represented by separate chapters of this book.

Geography's specialized subfields are not isolated from one another; rather, they are closely interrelated. Geography in all its subdivisions is characterized by three dominating interests. The first is in the spatial variation of physical and human phenomena on the surface of the earth; geography examines relationships between human societies and the natural environments that they occupy and modify. The second is a focus on the spatial systems[1] that link physical phenomena and human activities in one area of the earth with other areas. Together, these interests lead to a third enduring theme, that of regional analysis: geography studies human–environmental (or "ecological") relationships and spatial systems in specific locational settings. This areal orientation pursued by some geographers is called *regional geography.*

Other geographers choose to identify particular classes of things, rather than segments of the earth's surface, for specialized study. These *systematic geographers* may focus their attention on one or a few related aspects of the physical environment or of human populations and societies. In each case, the topic selected for study is examined in its interrelationships with other spatial systems and areal patterns. *Physical geography* directs its attention to the natural environmental side of the human–environment structure. Its concerns are with landforms and their distribution, with atmospheric conditions and climatic patterns, with soils or vegetation associations, and the like. The other systematic branch of geography is *human geography.* Its emphasis is on people: where they are, what they are like, how they interact over space, and what kinds of landscapes of human use they erect on the natural landscapes they occupy.

A grasp of the broad, yet integrated concerns and topics of geography is vital to an understanding of the important national and international problems that dominate daily news reports. Acid rain and the greenhouse effect, the deterioration of central cities, international trade imbalances, inadequate food supply and population growth in developing countries, turmoil in Africa and the Middle East—all of these problems occur in a geographic context, and geography helps

[1]A "system" is simply a group of elements organized in a way that every element is to some degree directly or indirectly interdependent with every other element.

Careers in Geography

Geography admirably serves the objectives of a liberal education. It can make us better informed citizens, more able to understand the important issues facing our communities, our country, and our world and better prepared to contribute solutions.

Can it, as well, be a pathway to employment for those who wish to specialize in the discipline? The answer is "yes," in a number of different types of jobs. One broad cluster is concerned with supporting the field itself through teaching and research. Teaching opportunities exist at all levels, from elementary to university postgraduate. Teachers with some training in geography are in increasing demand in elementary and high schools in the United States, reflecting geography's inclusion as a core subject in the federally adopted *Goals 2000: Educate America Act* (Public Law 103-227) and the national determination to create a geographically literate society (see "The National Geography Standards," p. 17). At the college level, specialized teaching and research in all branches of geography have long been established, and geographically trained scholars are prominently associated with urban, community, and environmental studies, regional science, locational economics, and other interdisciplinary programs.

Because of the breadth and diversity of the field, training in geography involves the acquisition of skills and approaches applicable to a wide variety of jobs outside the academic world. Modern geography is both a physical and social science and fosters a wealth of technical skills. The employment possibilities it presents are as many and varied as are the agencies and enterprises dealing with the natural environment and human activities and with the acquisition and analysis of spatial data.

Many professional geographers work in government, either at the state or local level or in a variety of federal agencies and international organizations. Although many positions do not carry a geography title, physical geographers serve as water and other natural resource analysts, weather and climate experts, soil scientists, and the like. An area of recent high demand is for environmental managers and technicians. Geographers who have specialized in environmental studies find jobs in both public and private agencies. Their work may include assessing the environmental impact of proposed development projects on such things as air and water quality and endangered species, as well as preparing the environmental impact statements required before construction can begin.

Human geographers work in many different roles in the public sector. Jobs include data acquisition and analysis in health care, transportation, population studies, economic development, and international economics. Many geography graduates find positions as planners in local and state governmental agencies concerned with housing and community development, park and recreation planning, and urban and regional planning. They map and analyze land use plans and transportation systems, monitor urban land development, make informed recommendations about the location of public facilities, and engage in basic social science research.

Most of these same specializations are found in the private sector. Geographic training is ideal for such tasks as business planning and market analysis; factory, store, and shopping center site selection; community and economic development programs for banks, public utilities, and railroads; and similar applications. Publishers of maps, atlases, news and travel magazines, and the like, employ geographers as writers, editors, and mapmakers.

to explain them. To be geographically illiterate is to deny oneself not only the ability to comprehend local and world problems but also the opportunity to contribute meaningfully to the development of policies for dealing with them.

Very importantly, too, an understanding of the broad disciplinary subdivisions examined in this book may suggest the great diversity of job opportunities awaiting those who pursue college training in geography. While geographic literacy is indispensable to general citizen understanding of the world we live in and the problems it faces, equally importantly geographic training can help open the way to wonderfully rewarding and diversified careers (see "Careers in Geography").

Some Basic Geographic Concepts

The topics included within the broad field of geography are diverse. That very diversity, however, emphasizes the reality that all geographers—whatever their particular topical or regional interests—are united by the similar questions they ask and the common set of concepts they employ to consider their answers. Of either a physical or cultural phenomenon, they will inquire: What is it? Where is it? How did it come to be what and where it is? Where is it in relation to other physical or cultural realities that affect it or are affected by it? How is it part of a functioning whole? How does its location affect people's lives and the content of the area in which it is found?

These and similar questions are spatial in focus and systems analytical in approach and are derived from enduring central themes in geography. In answering them, geographers draw upon a common store of concepts, terms, and methods of study that together form the basic structure and vocabulary of geography. Collectively, geographers believe that recognizing spatial patterns is the essential starting point for understanding how people live on and shape the earth's surface. That understanding is not just the task and interest of the professional geographer; it should be, as well, part of the mental framework of all informed persons.

Geographers use the word *spatial* as an essential modifier in framing their questions and forming their concepts.

The combination of traditional, broad-based liberal arts perspective with the technical skills required in geographic research and analysis gives geography graduates a competitive edge in the current labor market. These field-based skills include familiarity with geographic information systems (GIS, explained in Chapter 2), cartography and computer mapping, remote sensing and photogrammetry, and competence in data analysis and problem solving. In particular, students with expertise in GIS, who are knowledgeable about data sources, hardware, and software, are finding they have ready access to employment opportunities. The following table, based on the booklet "Careers in Geography,"[1] summarizes some of the professional opportunities open to students who have specialized in one (or more) of the various subfields of geography.

Geographic Field of Concentration	Employment Opportunities
Cartography and geographic information systems	Cartographer for federal government (agencies such as Defense Mapping Agency, U.S. Geological Survey, or Environmental Protection Agency) or private sector (e.g., Environmental Systems Research Institute, ERDAS, Intergraph, or Bentley); map librarian; GIS specialist for planners, land developers, real estate agencies, utility companies, local government; remote-sensing analyst; surveyor
Physical geography	Weather forecaster; outdoor guide; coastal zone manager; hydrologist; soil conservation/agricultural extension agent
Environmental studies	Environmental manager; forestry technician; park ranger; hazardous waste planner
Cultural geography	Community developer; Peace Corps volunteer; health care analyst
Economic geography	Site selection analyst for business and industry; market researcher; traffic/route delivery manager; real estate agent/broker/appraiser; economic development researcher
Urban and regional planning	Urban and community planner; transportation planner; housing, park, and recreation planner; health services planner
Regional geography	Area specialist for federal government; international business representative; travel agent; travel writer
Geographic education or general geography	Elementary/secondary school teacher; college professor; overseas teacher

[1] "Careers in Geography," by Richard G. Boehm. Washington, D.C.: National Geographic Society, 1996. Previously published by Peterson's Guides, Inc.

Geography, they say, is a *spatial science.* It is concerned with *spatial behavior* of people, with the *spatial relationships* that are observed between places on the earth's surface, and with the *spatial processes* that create or maintain those behaviors and relationships. The word *spatial* comes, of course, from *space,* and to geographers it always carries the idea of the way things are distributed, the way movements occur, and the way processes operate over the whole or a part of the surface of the earth. The geographer's space, then, is earth space, the surface area occupied or available to be occupied by humans. Spatial phenomena have locations on that surface, and spatial interactions occur between places, things, and people within the earth area available to them. The need to understand those relationships, interactions, and processes helps frame the questions that geographers ask.

Those questions have their starting point in basic observations about the location and nature of places and about how places are similar to or different from one another. Such observations, though simply stated, are profoundly important to our comprehension of the world we occupy.

- Places have location, direction, and distance with respect to other places.
- A place has size; it may be large or small. Scale is important.
- A place has both physical structure and cultural content.
- The characteristics of places develop and change over time.
- The elements of places are interrelated with other places.
- Places may be generalized into regions of similarities and differences.

These are basic notions understandable to everyone. They also are the means by which geographers express fundamental observations about the earth spaces they examine and put those observations into a common framework of reference. Each of the concepts is worth further discussion, for they are not quite as simple as they at first seem.

Location, Direction, and Distance

Location, direction, and *distance* are everyday ways of assessing the space around us and identifying our position in relation to other things and places of interest. They are also essential in understanding the processes of spatial interaction that figure so importantly in the study of both physical and human geography.

Location

The location of places and things is the starting point of all geographic study as well as of our personal movements and spatial actions in everyday life. We think of and refer to location in at least two different senses, *absolute* and *relative.*

Absolute location is the identification of place by some precise and accepted system of coordinates; it therefore is sometimes called *mathematical location.* We have several such accepted systems of pinpointing positions. One of them is the global grid of parallels and meridians—that is, latitude and longitude (discussed in Chapter 2, page 22). With it the absolute location of any point on the earth can be accurately described by reference to its degrees, minutes, and seconds of *latitude* and *longitude* (Figure 2.4).

Other coordinate systems are also in use. Survey systems such as the township, range, and section description of property in much of the United States give mathematical locations on a regional level, while street address precisely defines a building according to the reference system of an individual town. Absolute location is unique to each described place, is independent of any other characteristic or observation about that place, and has obvious value in the legal description of places, in measuring the distance separating places, or in finding directions between places on the earth's surface.

When geographers—or real estate agents—remark that "location matters," however, their reference is usually not to absolute but to **relative location**—the position of a place or thing in relation to that of other places or things (Figure 1.4). Relative location expresses spatial interconnection and interdependence. On an immediate and personal level, we think of the location of the school library not in terms of its street address or room number but where it is relative to our classrooms, or the cafeteria, or some other reference point. On the larger scene, relative location tells us that peo-

FIGURE 1.4 **The reality of *relative location* on the globe may be** strikingly different from the impressions we form from flat maps. The position of Russia with respect to North America when observed from a polar perspective emphasizes that relative location properly viewed is important to our understanding of spatial relationships and interactions between the two world areas.

ple, things, and places exist not in a spatial vacuum but in a world of physical and cultural characteristics that differ from place to place.

New York City, for example, may in absolute terms be described as located at (approximately) latitude 40°43′ N (read as 40 degrees, 43 minutes north) and longitude 73°58′ W. We have a better understanding of the *meaning* of its location, however, when reference is made to its spatial relationships: to the continental interior through the Hudson–Mohawk lowland corridor or to its position on the eastern seaboard of the United States. Within the city, we gain understanding of the locational significance of Central Park or the Lower East Side not solely by reference to the street addresses or city blocks they occupy, but by their spatial and functional relationships to the total land use, activity, and population patterns of New York City.

In view of these different ways of looking at location, geographers make a distinction between the *site* and the *situation* of a place. **Site,** an absolute location concept, refers to the physical and cultural characteristics and attributes of the place itself. It is more than mathematical location, for it tells us something about the internal features of that place. **Situation,** on the other hand, refers to the external relations of the locale. It is an expression of relative location with particular reference to items of significance to the place in question. Site and situation in the city context are further examined in Chapter 12.

Direction

Direction is a second universal spatial concept. Like location, it has more than one meaning and can be expressed in absolute or relative terms. **Absolute direction** is based on the cardinal points of north, south, east, and west. These appear uniformly and independently in all cultures, derived from the obvious "givens" of nature: the rising and setting of the sun for east and west, the sky location of the noontime sun and of certain fixed stars for north and south.

We also commonly use **relative** or *relational* **directions.** In the United States we go "out West," "back East," or "down South"; we worry about conflict in the "Near East" or economic competition from the "Far Eastern countries." These directional references are culturally based and locationally variable, despite their reference to cardinal compass points. The Near and the Far East locate parts of Asia from the European perspective; they are retained in the Americas by custom and usage, even though one would normally travel westward across the Pacific, for example, to reach the "Far East" from California, British Columbia, or Chile. For many Americans, "back East" and "out West" are reflections of the migration paths of earlier generations for whom home was in the eastern part of the country, to which they might look back. "Up North" and "down South" reflect our accepted custom of putting north at the top and south at the bottom of our maps.

Distance

Distance joins location and direction as a commonly understood term that has dual meanings for geographers. Like its two companion spatial concepts, distance may be viewed in both an absolute and a relative sense.

Absolute distance refers to the spatial separation between two points on the earth's surface measured by some accepted standard unit such as miles or kilometers for widely separated locales, feet or meters for more closely spaced points. **Relative distance** transforms those linear measurements into other units more meaningful for the space relationship at question.

To know that two competing malls are about equidistant in miles from your residence is perhaps less important in planning your shopping trip than is knowing that because of street conditions or traffic congestion one is 5 minutes and the other 15 minutes away (Figure 1.5). Most people, in fact, think of time distance rather than linear distance in their daily activities; downtown is 20 minutes by bus, the library is a 5-minute walk. In some instances, money rather than time may be the distance transformation. An urban destination might be estimated to be a $10 cab ride away, information that may affect either the decision to make the trip at all or the choice of travel mode to get there.

A *psychological* transformation of linear distance is also frequent. The solitary late-night walk back to the car through an unfamiliar or dangerous neighborhood seems far longer than a daytime stroll of the same distance through familiar and friendly territory. A first-time trip to a

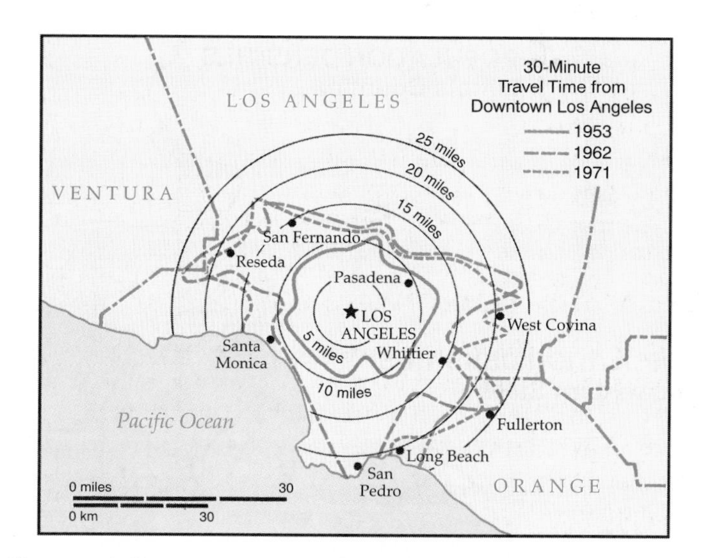

FIGURE 1.5 **Lines of equal travel time** (*isochrones*) mark off different linear distances from a given starting point, depending on the condition of the route and terrain and changes in the roads and traffic flows over time. On this map, the areas within 30 minutes' travel time from downtown Los Angeles are recorded for the period 1953 to 1971.

Redrawn by permission from Howard J. Nelson and William A. V. Clark, *The Los Angeles Metropolitan Experience,* page 49, Association of American Geographers, 1976.

new destination frequently seems much longer than the return trip over the same path. Nonlinear distance and spatial interaction are further considered in Chapter 8.

Size and Scale

When we say that a place may be large or small, we speak both of the nature of the place itself and of the generalizations that can be made about it. In either instance, geographers are concerned with **scale,** though we may use that term in different ways. We can, for example, study a problem such as population or landforms at the local scale or on a global scale. Here, the reference is purely to the size of unit studied. More technically, scale tells us the relationship between the size of an area on a map and the actual size of the mapped area on the surface of the earth. In this sense, as Chapter 2 makes clear, scale is a feature of every map and is essential to recognizing what is shown on that map.

In both senses of the word, scale implies the degree of generalization represented (Figure 1.6). Geographic inquiry may be broad or narrow; it occurs at many different size-scales. Climate may be an object of study, but research and generalization focused on climates of the world will differ in degree and kind from study of the microclimates of a city. Awareness of scale is very important. In geographic work concepts, relationships, and understandings that have meaning at one scale may not be applicable at another.

For example, the study of world agricultural patterns may refer to global climatic regimes, cultural food preferences, levels of economic development, and patterns of world trade. These large-scale relationships are of little concern in the study of crop patterns within single counties of the United States, where topography, soil and drainage conditions, farm size, ownership, and capitalization, or even personal management preferences may be of greater explanatory significance.

Physical and Cultural Attributes

All places have individual physical and cultural attributes distinguishing them from other places and giving them character, potential, and meaning. Geographers are concerned with identifying and analyzing the details of those attributes and, particularly, with recognizing the interrelationship between the physical and cultural components of area: the human–environmental interface.

The physical characteristics of a place refer to such natural aspects as its climate, soil, water supplies, mineral resources, terrain features, and the like. These **natural landscape** attributes provide the setting within which human action occurs. They help shape—but do not dictate—how people live. The resource base, for example, is physically determined, though how resources are perceived and utilized is culturally conditioned.

Environmental circumstances directly affect agricultural potential and reliability; indirectly, they may affect such matters as employment patterns, trade flows, population distributions, national diets, and so on. The physical environment simultaneously presents advantages and disadvantages with which humans must deal. Thus, the danger of cyclones in the Bay of Bengal must be balanced

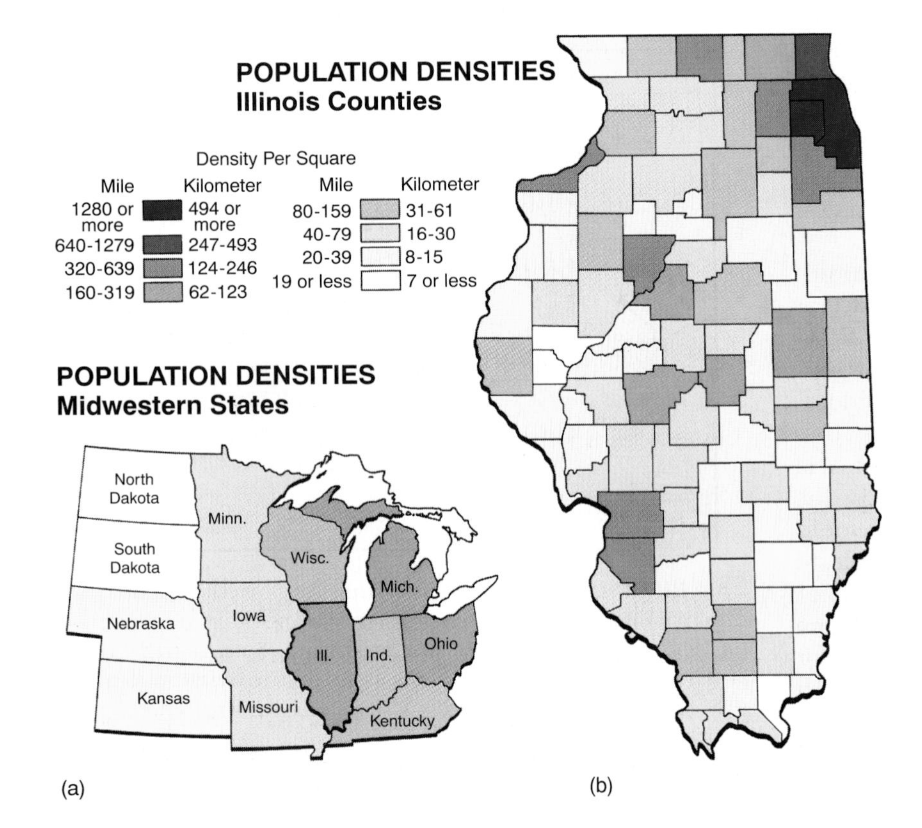

FIGURE 1.6 **Population density and map scale.** "Truth" depends on one's scale of inquiry. Map (a) reveals that the maximum population density of Midwestern states is no more than 123 people per square kilometer (319 per sq mi). From map (b), however, we see that population densities in two Illinois counties exceed 494 people per square kilometer (1280 per sq mi). If we were to reduce our scale of inquiry even further, examining individual city blocks in Chicago, we would find densities as high as 2000 people per square kilometer (5200 per sq mi). Scale matters!

On-Line Introduction

The Internet is a vast network of computers electronically joining together millions of people and many thousands of organizations and institutions from all over the world. In recent years the Internet has become both a major means of personal communication (e-mail) and in its multimedia component, the World Wide Web or WWW, an important tool for academic research and general information gathering. The World Wide Web lets users navigate through a graphic interface, combining hypertext and hypermedia to link documents, images, video clips, and sound files. Most colleges and universities (and now even K–12 schools) have Internet access in their libraries and computer labs. In some cases, Internet access is even available in classrooms and has become an integral part of the teaching and learning environment.

There are numerous Internet resources for geography, some that duplicate information found in traditional sources such as books and journals, and some that supplement more familiar resources. There are also "websites" that present information in ways not possible with traditional tools, and, hence, are unique to the Internet.

Reflecting ongoing changes in our learning environment and recognizing the importance of the Internet as an instructional and research medium, we include in each chapter of this text a special boxed section (like this one) discussing World Wide Web sites that may themselves be data sources of value in expanding topics of the chapter or, instead, be linked listings of other websites concerned with those chapter topics.

Due to the dynamic nature of the Internet, some of the online addresses listed in this book may have changed or possibly no longer exist when you try to consult them. In most instances, cross-references and directions to replacement addresses are cited at the old website locations for your guidance. And new and useful sites are constantly being developed. You are invited to report new sources and addresses you have found valuable and wish to share with other readers. Check our home page for directions for leaving an e-mail message with your information and suggestions.

http://www.mhhe.com/earthsci/geography/getis/

For those interested in professional geographic associations and activities, the home page of the Association of American Geographers is a good starting point. It offers information about the association itself, its publications, regional divisions and "specialty groups" (complete with e-mail addresses), and provides links to such other organizations as the Canadian Association of Geographers, the National Council for Geographic Education, and the National Geographic Society.

http://www.aag.org/

The *Resources* page of the National Geographic Society is well worth your visit. It provides access to "Information Central," "Glad You Asked," and the resources of the "NGS Library" among other connections.

http://www.nationalgeographic.com/resources/

Or visit National Geographic's *Geoguide* to their "digital field-trips."

http://www.nationalgeographic.com/resources/education/geoguide

Are you interested in furthering your geography education or in learning about geography programs worldwide? Two good starting points are: the Ryerson University geography department list

http://www. geo.ryerson.ca/~gta/html/geograph.html

and a similar list maintained by the University of Texas.

http://www.utexas.edu/depts/grg/virtdept/resources/depts/depts.htm

Among the best general guides to geography resources on the Internet are those of the State University of New York at Buffalo

http://ncgia.geog.buffalo.edu/GIAL/netgeog.html

and of the University of Wisconsin-Stevens Point.

http://www.uwsp.edu/acaddept/geog/resour.htm

Michigan State University maintains a useful guide to *Geography-related Servers* that is well-worth noting.

http://www.ssc.msu.edu/~geo/geoglinks.html

Finally, a visit to the *Geography* site at About.com (the former MiningCo.com) is well worth while. It features annotated net links to selected geographical resources, on-line maps, data, and weekly articles, along with a chat room and bulletin board.

http://geography.About.com

against the agricultural bounty derived from the region's favorable terrain, soil, temperature, and moisture conditions. Physical environmental patterns and processes are explored in Chapters 3 and 4 of this book.

At the same time, by occupying a given place, people modify its environmental conditions. The existence of the United States Environmental Protection Agency (and its counterparts elsewhere) is a reminder that humans are the active and frequently harmful agents in the continuing interplay between the cultural and physical worlds (Figure 1.7). Virtually every human activity leaves its imprint on the earth's soil, water, vegetation, animal life, and other

FIGURE 1.7 Sites (and sights) such as this devastation of ruptured barrels and petrochemical contamination near Texas City, Texas, are all-too-frequent reminders of the adverse environmental impacts of humans and their waste products. Many of those impacts are more subtle, hidden in the form of soil erosion, water pollution, increased stream sedimentation, plant and animal extinction, deforestation, and the like.
©W. Frerck/Odyssey

FIGURE 1.8 This Landsat image reveals contrasting cultural landscapes along the Mexico–California border. Move your eyes from the Salton Sea (the dark patch at the top of the image) southward to the agricultural land extending to the edge of the picture. Notice how the regularity of the fields and the bright colors (representing growing vegetation) give way to a marked break, where irregularly shaped fields and less prosperous agriculture are evident. Above the break is the Imperial Valley of California; below the border is Mexico.
NASA.

resources, and on the atmosphere common to all earth space, as Chapter 5 makes clear.

The visible imprint of that human activity is called the **cultural landscape.** It, too, exists at different scales and at different levels of visibility. Contrasts in agricultural practices and land use between Mexico and southern California are evident in Figure 1.8, while the signs, structures, and people of Los Angeles' Chinatown leave a smaller, more confined imprint within the larger cultural landscape of the metropolitan area itself.

The physical and human characteristics of places are the keys to understanding both the simple and the complex interactions and interconnections between people and the environments they occupy and modify. Those interconnections and modifications are not static or permanent, but are subject to continual change.

Attributes of Place Are Always Changing

The physical environment surrounding us seems eternal and unchanging but, of course, it is not. In the framework of geologic time, change is both continuous and pronounced. Islands form and disappear; mountains rise and are worn low to swampy plains; vast continental glaciers form, move, and melt away, and sea levels fall and rise in response. Geologic time is long, but the forces that give shape to the land are timeless and relentless.

Even within the short period of time since the most recent retreat of continental glaciers—some 10,000 or 11,000 years ago—the environments occupied by humans have been subject to change. Glacial retreat itself marked a period of climatic alteration, extending the area habitable by humans to include vast reaches of northern Eurasia and North America formerly covered by thousands of feet of ice. With moderating climatic conditions came associated changes in vegetation and fauna. On the global scale, these were natural environmental changes; humans were as yet too few in numbers and too limited in technology to alter materially the course of physical events. On the regional scale, however, even early human societies exerted an impact on the environments they occupied. Fire was used to clear forest undergrowth, to maintain or extend grassland for grazing animals and to drive them in the hunt, and, later, to clear openings for rudimentary agriculture.

With the dawn of civilizations and the invention and spread of agricultural technologies, humans accelerated their management and alteration of the now no longer "natural" environment. Even the classical Greeks noted how the landscape they occupied differed—for the worse—from its former condition. With growing numbers of people and particularly with industrialization and the spread of European exploitative technologies throughout the world, the pace of change in the content of area accelerated. The built landscape—the product of human effort—increasingly replaced the natural landscape. Each new settlement or city, each agricultural assault on forests, each new mine, dam, or factory changed the content of regions and altered the temporarily established spatial interconnections between humans and the environment.

Characteristics of places today are the result of constantly changing past conditions. They are the forerunners of differing human–environmental balances yet to be struck. Geographers are concerned with place at given moments of time. But to understand fully the nature and development of places, to appreciate the significance of their relative locations, and to understand the interplay of their physical and cultural characteristics, geographers must view places as the present result of past operation of distinctive physical and cultural processes (Figure 1.9).

You will recall that one of the questions geographers ask about a place or thing is: How did it come to be what and where it is? This is an inquiry about process and about becoming. The forces and events shaping the physical and explaining the cultural environment of places today are an important focus of geography and are the topics of most of the chapters of this book. To understand them is to appreciate the changing nature of the spatial order of our contemporary world.

Interrelations between Places

The concepts of relative location and distance that we earlier introduced lead directly to another fundamental spatial reality: Places are interrelated with other places in structured and comprehensible ways. In describing the processes and patterns of that **spatial interaction,** geographers add *accessibility* and *connectivity* to the ideas of location and distance.

A basic law of geography tells us that in a spatial sense everything is related to everything else but that relationships are stronger when things are near one another. Our observation, therefore, is that interaction between places diminishes in intensity and frequency as distance between them increases—a statement of the idea of **distance decay,** which we explore in Chapter 8.

Consideration of distance implies assessment of **accessibility.** How easy or difficult is it to overcome the "friction of distance"? That is, how easy or difficult is it to surmount the barrier of the time and space separation of places? Distance isolated North America from Europe until the development of ships (and aircraft) that reduced the effective distance between the continents. All parts of the ancient and medieval city were accessible by walking; they were "pedestrian cities," a status lost as cities expanded in area and population with industrialization. Accessibility between city districts could only be maintained by the development of public transit systems whose fixed lines of travel increased ease of movement between connected points and reduced it between areas not on the transit lines themselves.

(a)

(b)

FIGURE 1.9 **The process of change in a cultural landscape.** (a) Before the advent of the freeway, this portion of suburban Long Island, New York, was largely devoted to agriculture. (b) The construction of the freeway and cloverleaf interchange ramps altered nearby land uses to replace farming with housing developments and new commercial and light industrial activities.
U.S. Geological Survey.

Accessibility therefore suggests the idea of **connectivity,** a broader concept implying all the tangible and intangible ways in which places are connected: by physical telephone lines, street and road systems, pipelines and sewers; by unrestrained walking across open countryside; by radio and TV broadcasts beamed outward uniformly from a central source; and in nature even by movements of wind systems and flows of ocean currents. Where routes are fixed and flow is channelized, *networks*—the patterns of routes connecting sets of places—determine the efficiency of movement and the connectedness of points.

There is, inevitably, interchange between connected places. **Spatial diffusion** is the process of dispersion of an idea or a thing (a new consumer product or a new song, for example) from a center of origin to more distant points. The rate and extent of that diffusion are affected, again, by the distance separating the origin of the new idea or technology and other places where it is eventually adopted. Diffusion rates are also affected by such factors as population densities, means of communication, obvious advantages of the innovation, and importance or prestige of the originating node. Further discussion of spatial diffusion is found in Chapter 8.

Geographers study of the dynamics of spatial relationships. Movement, connection, and interaction are part of the social and economic processes that give character to places and regions (Figure 1.10). Geography's study of those relationships recognizes that spatial interaction is not just an awkward necessity but a fundamental organizing principle of the physical and social environment.

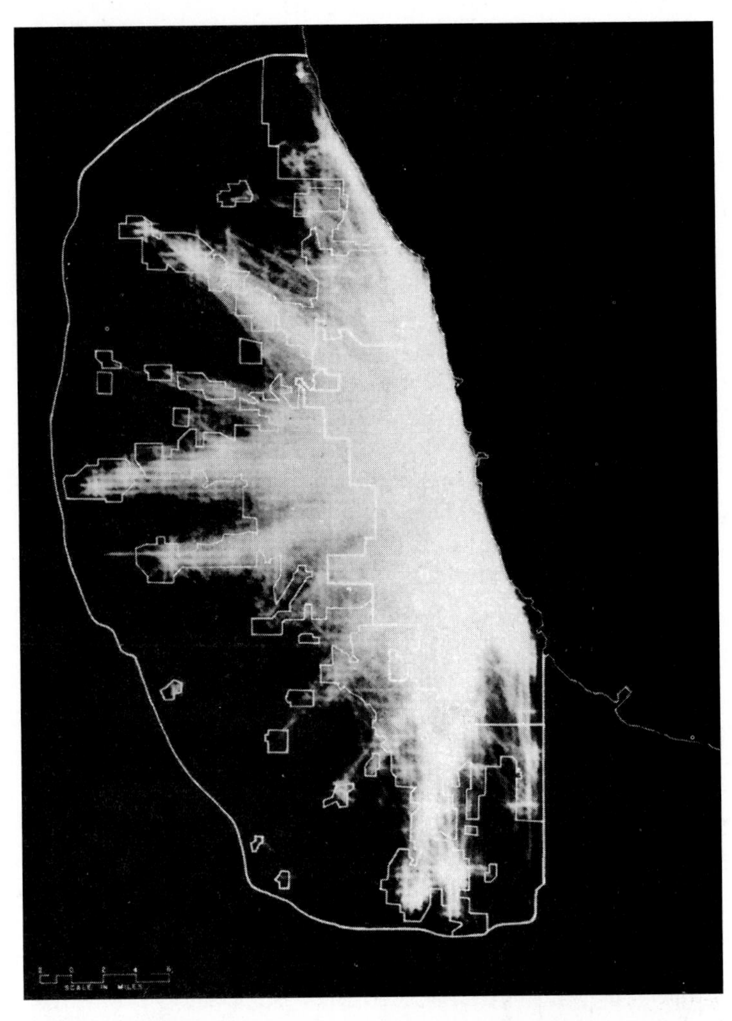

FIGURE 1.10 **The routes of 5 million automobile trips** made each day in Chicago during the late 1950s are recorded on this light-display map. The boundaries of the region of interaction that they created are clearly marked. Those boundaries (and the dynamic region they defined) were subject to change as residential neighborhoods expanded or developed, as population relocations occurred, and road patterns altered over time.

Chicago Area Transportation Study, *Final Report, 1959.* Vol. I, p. 44, Fig. 22: "Desire Lines of Internal Automobile Driver Trips."

Place Similarity and Regions

The distinctive characteristics of places—physical, cultural, locational—immediately suggest to us two geographically important ideas. The first is that no two places on the surface of the earth can be *exactly* the same. Not only do they have different absolute locations, but—as in the features of the human face—the precise mix of physical and cultural characteristics of place is never exactly duplicated. Since geography is a spatial science, the inevitable uniqueness of place would seem to impose impossible problems of generalizing spatial information.

That this is not the case results from the second important idea: the physical and cultural features of an area and the dynamic interconnections of people and places show patterns of spatial similarity. Often the similarities are striking enough for us to conclude that spatial regularities exist. They permit us to recognize and define **regions,** earth areas that display significant elements of uniformity. Places are, therefore, both unlike and like other places, creating patterns of areal differences and coherent spatial similarity.

The problems of the historian and the geographer are similar. Each must generalize about items of study that are essentially unique. The historian creates arbitrary but meaningful and useful historical periods for reference and study. The "Roaring Twenties" or the "Victorian Era" are shorthand summary names for specific time spans, internally quite complex and varied, but significantly distinct from what went before or followed after. The region is the geographer's equivalent of the historian's epoch: a device that segregates into component parts the complex reality of the earth's surface. In both the time and space need for generalization, attention is focused on key unifying elements or similarities of the era or area selected for study. In both the historical and geographical cases, the names assigned to those times and places serve to identify the time span or region and to convey between speaker and listener a complex set of interrelated attributes.

Regions are not "given" in nature any more than "eras" are given in the course of human events. Regions are devised; they are spatial summaries designed to bring order to the infinite diversity of the earth's surface. At their root, they are based on the recognition and mapping of **spatial distributions**—the territorial occurrence of environmental, human, or organizational features selected for study. As many spatial distributions exist as there are imaginable physical, cultural, or connectivity elements of area to examine (Figure 1.11). Those that are selected for study, however, are those that contribute to the understanding of a specific topic or problem.

Although there are as many individual regions as the objectives of spatial study and understanding demand, two generalized *types* of regions are recognized. A **formal region** is one of essential uniformity in one or a limited number of related physical or cultural features. Your home state is a formal political region within which uniformity of law and administration is found. The Columbia Plateau or the tropical rain forest are areas of uniform physical characteristics making up formal natural regions (Figure 1.12).

A **functional region,** in contrast, may be visualized as a spatial system. Its parts are interdependent, and throughout its extent the functional region operates as a dynamic, organizational unit. A functional region has unity not in the sense of static content but in the manner of its operational connectivity. It has a node or core area surrounded by the

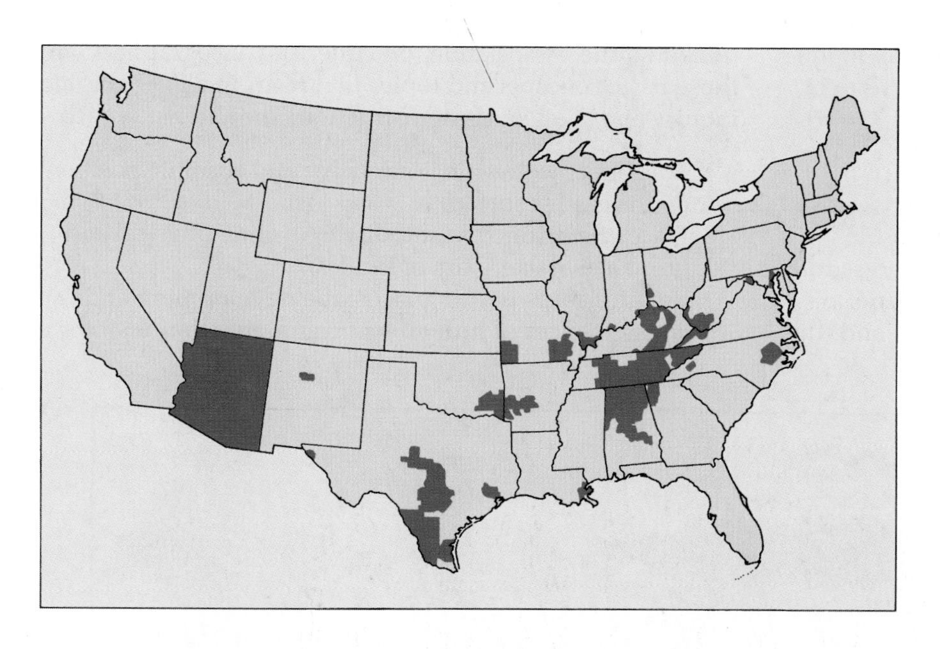

FIGURE 1.11 **Pulmonary tuberculosis mortality rates for white males, 1965–1971.** All spatial data may be mapped. As this example of the distribution of pulmonary tuberculosis demonstrates, mapped distributions frequently reveal regional patterns that invite analysis. Areas emphasized had above average mortality rates. The question is, Why?

Thomas J. Mason, et al., *An Atlas of Mortality from Selected Diseases.* NIH Publication #81-2397, May 1981. National Institutes of Health, Bethesda, Maryland.

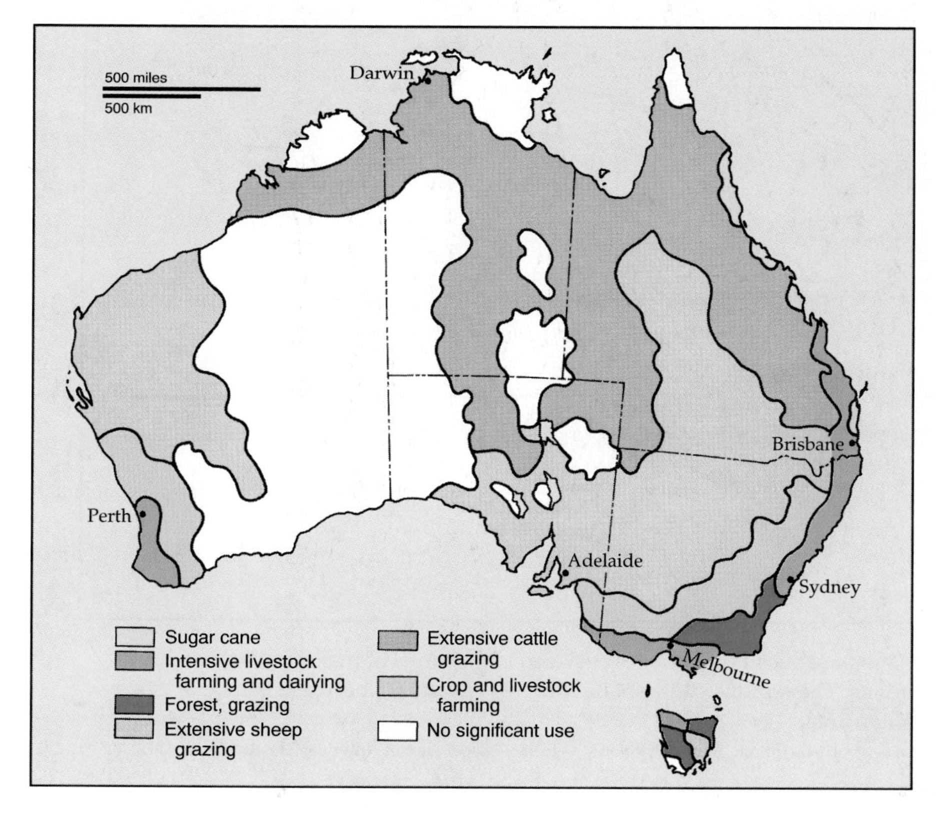

FIGURE 1.12 This generalized **land use map of Australia** is composed of *formal regions* whose internal economic characteristics show essential uniformities setting them off from adjacent territories of different condition or use.

total region defined by the type of control exerted. Trade areas of towns, national "spheres of influence," and the territories subordinate to the financial, administrative, wholesaling, or retailing centrality exercised by such regional capitals as Chicago, Atlanta, or Minneapolis are cases in point (Figure 1.13). Further examples of formal and functional regionalism will be encountered incidentally in the following chapters and make up the total content of Chapter 13.

As you read the chapters of this book, notice how many different examples of regions are presented in map form and discussed in the text. Note, too, how those depictions and discussions vary between, primarily, formal and functional regions as the subjects and purposes of the examples change. Chapter 13 contains additional special regional studies illuminating specific topics that are the subjects of Chapters 3 to 12.

Geography's Themes and Standards

The basic geographic concepts discussed in the preceding sections of this chapter reflect—and are statements of—both the "fundamental themes in geography" and the "National Geography Standards" that together have helped organize and structure the study of geography over the past several years at all grade and college levels. Both the "themes" and "standards" focus on the development of geographic literacy. The former represent an instructional approach keyed to identification and instruction in the knowledge, skills, and perspectives students should gain from a structured program in geographic education. The latter—"standards"—codify the essential subject matter, skills, and perspectives of geography essential to the mental equipment of all educated adults.

The *five fundamental themes* as summarized by a joint committee of the National Council for Geographic Education and the Association of American Geographers are those basic concepts and topics that recur in all geographic inquiry and at all levels of instruction. They are:

- Location: the meaning of relative and absolute position on the earth's surface;
- Place: the distinctive and distinguishing physical and human characteristics of locales;
- Relationships within places: the development and consequences of human–environment relationships;

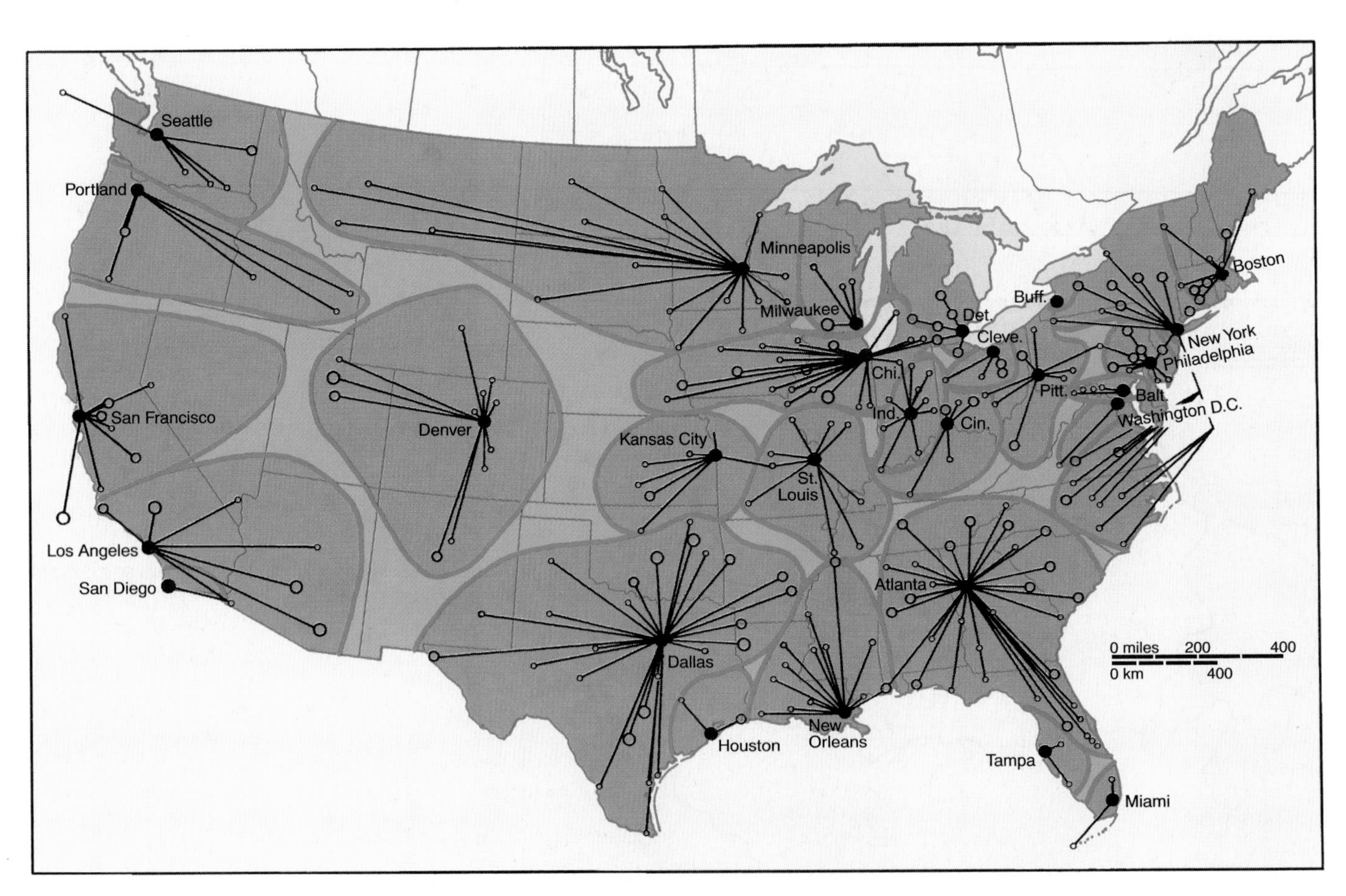

FIGURE 1.13 The *functional regions* shown on this map are based on linkages between large banks of major central cities and the "correspondent" banks they serve in smaller towns. The regions suggest one form of *connectivity* between principal cities and locales beyond their own immediate metropolitan area.

Redrawn by permission from *Annals of the Association of American Geographers,* John R. Borchert, vol. 62, page 358, Association of American Geographers, 1972.

- Movement: patterns and change in human spatial interaction on the earth;
- Regions: how they form and change.

The National Geography Standards were detailed in 1994 as part of the nationally adopted *Goals 2000: Educate America Act* (see "The National Standards"). Designed specifically as guidelines to the essential geographic literacy to be acquired by students who have gone through the U.S. public school system, the "standards" address the same conviction underlying this present *Introduction to Geography*—that being literate in geography is a necessary part of the mental framework of all informed persons.

Organization of This Book

The breadth of geographic interest and subject matter, the variety of questions that focus geographic inquiry, and the diversity of concepts and terms geographers employ require a simple, logical organization of topics for presentation to students new to the field. Despite its outward appearance of complex diversity, geography should be seen to have a broad consistency of purpose achieved through the recognition of a limited number of distinct but closely related "traditions." William D. Pattison, who suggested this unifying viewpoint, and J. Lewis Robinson (among others) who accepted and expanded Pattison's reasoning, found that four traditions were logical and inclusive ways of clustering geographic inquiry. While not all geographic work is confined by the separate traditions, one or more of them is implicit in most geographical studies. The unifying categories—the four traditions within which geographers work—are:

1. the earth science tradition;
2. the culture–environment tradition;
3. the locational (or spatial) tradition;
4. the area analysis (or regional) tradition.

The mutual interdependence of the four traditions is suggested by Figure 1.14, as are their common individual and collective ties to the full range of research and study

The National Standards

The inclusion of Geography in the *Goals 2000* national education program reflects the conviction that a grasp of the skills and understandings of geography are essential in an American educational system "tailored to the needs of productive and responsible citizenship in the global economy." Along with the "basic observations" just reviewed, the National Geography Standards 1994 help frame the kinds of understanding we will seek in the following pages and suggest the purpose and benefit of further study of geography.

The 18 Geography Standards tell us:
The geographically informed person knows and understands:

The World in Spatial Terms

1. How to use maps and other geographic tools and technologies to acquire, process, and report information from a spatial perspective.
2. How to use mental maps to organize information about people, places, and environments in a spatial context.
3. How to analyze the spatial organization of people, places, and environments on Earth's surface.

Places and Regions

4. The physical and human characteristics of places.
5. That people create regions to interpret Earth's complexity.
6. How culture and experience influence people's perceptions of places and regions.

Physical Systems

7. The physical processes that shape the patterns of Earth's surface.
8. The characteristics and spatial distribution of ecosystems on Earth's surface.

Human Systems

9. The characteristics, distribution, and migration of human populations on Earth's surface.
10. The characteristics, distribution, and complexity of Earth's cultural mosaics.
11. The patterns and networks of economic interdependence on Earth's surface.
12. The processes, patterns, and functions of human settlement.
13. How the forces of cooperation and conflict among people influence the division and control of Earth's surface.

Environment and Society

14. How human actions modify the physical environment.
15. How physical systems affect human systems.
16. The changes that occur in the meaning, use, distribution, and importance of resources.

The Uses of Geography

17. How to apply geography to interpret the past.
18. How to apply geography to interpret the present and plan for the future.

Source: *Geography for Life: National Geography Standards* 1994 (Washington, D.C.: National Geographic Research and Exploration, 1994).

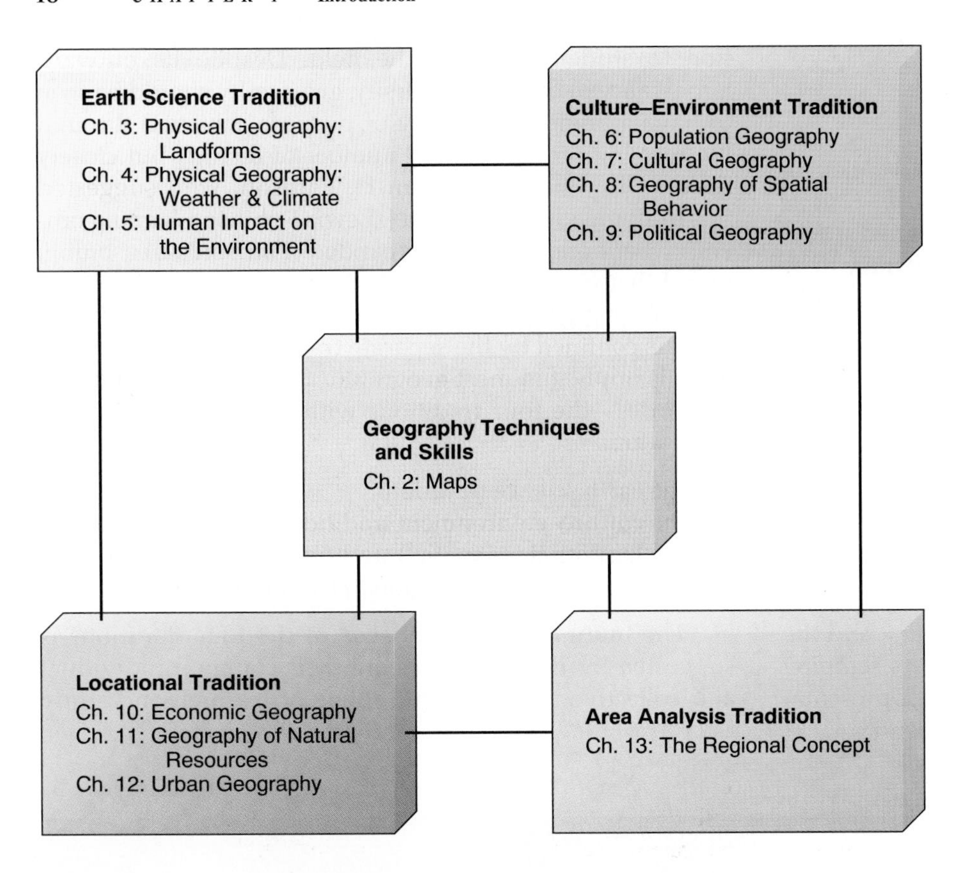

FIGURE 1.14 The four traditions of geography do not stand alone. Rather, each is interconnected with the others and all together depend on unifying research skills and tools. As the diagram indicates, the chapters of this text are grouped by reference to the "traditions" to help you recognize the broad divisions of geography as well as its underlying and unitary nature. To avoid diagram clutter, lines connecting each box to every other have not been drawn but should be understood.

techniques that geographers employ: maps, of course, but also remote sensing, statistical tools, geographic information systems, and other spatial analytical techniques.

We have employed the four traditions as the device for clustering the chapters of this book (from Chapter 3 onward), hoping they will help you recognize the unitary nature of geography while appreciating the diversity of topics studied by geographers. They are, in a sense, introduced by Chapter 2 which recognizes maps (and related tools and techniques) as essential, distinctive, and unifying tools of geographers.

The **earth science tradition** is that branch of the discipline that addresses itself to the earth as the habitat of humans. It is the tradition that in ancient Greece represented the roots of geography, the description of the physical structure of the earth and of the natural processes that give it detailed form. In modern terms, it is the vital environmental half of the study of human–environment systems, which together constitute geography's subject matter. The earth science tradition prepares the physical geographer to understand the earth as the common heritage of humankind and to find solutions to the increasingly complex web of pressures placed on the earth by its expanding, demanding human occupants. Consideration of the elements of the earth science tradition constitutes Part I (Chapters 3 to 5) of this text.

Part II (Chapters 6 to 9) details some of the content of the **culture–environment tradition.** Within this theme of geography, consideration of the earth as a purely physical entity gives way to a primary interest in how people *perceive* the environments they occupy. Its focus is on culture. The landscapes that are explored and the spatial patterns that are central are those that are cultural in origin and expression. People in their numbers, distributions, and diversity; in their patterns of social and political organization; and in their spatial perceptions and behaviors, are the orienting concepts of the culture–environment tradition. The theme is distinctive in its thrust, but tied to the earth science tradition because populations exist, cultures emerge, and behaviors occur within the context of the physical realities and patternings of the earth's surface.

The **locational tradition**—or, as it is sometimes called, the *spatial tradition*—is the subject of the chapters of Part III (Chapters 10 to 12). It is a tradition that underlies all of geographic inquiry. As Robinson suggested, if we can agree that geology is rocks, that history is time, and that sociology is people, then we can assert that geography is earth space. The locational tradition is primarily concerned with the distribution of cultural phenomena or physical items of significance to human occupance of the earth.

Part, but by no means all, of the locational tradition is concerned with distributional patterns. More central are scale, movement, and areal relationships. Map, statistical, geometrical, and systems analysis research are among the techniques employed by geographers working within the locational tradition. Irrespective, however, of the analytical tools used or the sets of phenomena studied—economic activities, resource distributions, city systems, or others—the underlying theme is the geometry, or the distribution, of the

TABLE 1.1

Regional Studies Contained in Chapter 13

Topic	Page Nos. in Ch. 13	Chapter Reference	Reference Page
Landforms as regions	479–480	Ch. 3: Physical Geography: Landforms	59
Dynamic regions in weather and climate	480–482	Ch. 4: Physical Geography: Weather and Climate	101
Ecosystems as regions	482–484	Ch. 5: Human Impact on the Environment	146
Population as regional focus	484–485	Ch. 6: Population Geography	191
Language as region	485–486	Ch. 7: Cultural Geography	229
Mental regions	486–487	Ch. 8: Geography of Spatial Behavior	278
Political regions	487–489	Ch. 9: Political Geography	308
Economic regions	489–490	Ch. 10: Economic Geography	348
Natural resource regions	490–492	Ch. 11: The Geography of Natural Resources	387
Urban regions	492–494	Ch. 12: Urban Geography	431

phenomenon discussed and the flows and interconnections that unite it to related physical and cultural occurrences.

The **area analysis tradition** is considered in Chapter 13, which makes up Part IV of our *Introduction to Geography.* Again, the roots of this tradition may be traced to antiquity. Strabo's *Geography* was addressed to the leaders of Augustan Rome as a summary of the nature of places in their separate characters and conditions—knowledge deemed vital to the guardians of an empire. Imperial concerns may long since have vanished, but the study of regions and the recognition of their spatial uniformities and differences remain. Such uniformities and differences, of course, grow out of the structure of human–environment systems and interrelations that are the study of geography. To illustrate the role and diversity of regional studies in geography, much of Chapter 13 comprises special regional investigations and models illuminating specific topics that are the themes of topical Chapters 3 to 12 of this book. References to those special examples are given in Table 1.1 and repeated in the various chapter sections to which they relate.

The identification of the four traditions of geography is not only an organizational convenience, but is also a recognition that within that diversity of subject matter called geography, unity of interest is ever preserved. The traditions, though recognizably distinctive, are intertwined and overlapping. We hope their use as organizing themes—and their further identification in short introductions to the separate sections of this book—will help you grasp the unity in diversity that is the essence of geographic study.

KEY WORDS

absolute direction 9
absolute distance 9
absolute location 8
accessibility 13
area analysis tradition 19
connectivity 14
cultural landscape 12
culture–environment tradition 18
distance decay 13
earth science tradition 18
formal (uniform) region 15
functional (nodal) region 15
locational tradition 18
natural landscape 10
region 14
relative direction 9
relative distance 9
relative location 8
scale 10
site 9
situation 9
spatial diffusion 14
spatial distribution 15
spatial interaction 13

FOR REVIEW AND CONSIDERATION

1. In what two meanings and for what different purposes do we refer to *location?* When geographers say "location matters," what aspect of location commands their interest?
2. What does the term *cultural landscape* imply? Is the nature of the cultural landscape dictated by the physical environment?
3. What kinds of distance transformations are suggested by the term *relative distance?* How is the concept of *psychological distance* related to relative distance?
4. How are the ideas of *distance, accessibility,* and *connectivity* related to processes of *spatial interaction?*
5. Why do geographers concern themselves with *regions?* How are *formal* and *functional* regions different in concept and definition?
6. What are the *four traditions* of geography? Do they represent unifying or divided approaches to geographic understanding?

SELECTED REFERENCES

Abler, Ronald F., Melvin G. Marcus, and Judy M. Olson, eds. *Geography's Inner Worlds: Pervasive Themes in Contemporary American Geography.* New Brunswick, N.J.: Rutgers University Press, 1992.

Demko, George J., with Jerome Agel, and Eugene Boe. *Why in the World: Adventures in Geography.* New York: Anchor Books/Doubleday & Co., Inc., 1992.

Fenneman, Nevin M. "The Circumference of Geography." *Annals of the Association of American Geographers* 9 (1919):3–11.

Holt-Jensen, Arild. *Geography: Its History and Concepts.* 2d ed. Totowa, N.J.: Barnes and Noble, 1988.

Lanegran, David A., and Risa Palm. *An Invitation to Geography.* 2d ed. New York: McGraw-Hill, 1978.

Livingstone, David N. *The Geographical Tradition.* Cambridge, Mass.: Blackwell, 1992.

Marshall, Bruce, ed. *The Real World: Understanding the Modern World Through the New Geography.* Boston: Houghton Mifflin Co., 1991.

Martin, Geoffrey J., and Preston E. James. *All Possible Worlds: A History of Geographical Ideas.* 3d ed. New York: John Wiley & Sons, Inc., 1993.

Massey, Doreen. "Introduction: Geography Matters," in Doreen Massey and John Allen, eds. *Geography Matters! A Reader.* Pp. 1–11. New York: Cambridge University Press, 1984.

McDonald, James R. "The Region: Its Conception, Design and Limitations." *Annals of the Association of American Geographers* 56 (1966):516–528.

Morrill, Richard L. "The Nature, Unity and Value of Geography." *Professional Geographer* 35, no. 1 (Feb. 1983):1–9.

National Research Council. Rediscovering Geography Committee. *Rediscovering Geography: New Relevance for Science and Society.* Washington, D.C.: National Academy Press, 1997.

Pattison, William D. "The Four Traditions of Geography." *Journal of Geography* 63 (1964):211–216.

Robinson, J. Lewis. "A New Look at the Four Traditions of Geography." *Journal of Geography* 75 (1976):520–530.

Rogers, Ali, Heather Viles, and Andrew Goudie. *The Student's Companion to Geography.* Cambridge, MA.: Basil Blackwell, Inc., 1992.

Wheeler, James O. "Notes on the Rise of the Area Studies Tradition in U.S. Geography, 1910–1929." *Professional Geographer* 38 (1986):53–61.

White, Gilbert F. "Geographers in a Perilously Changing World." *Annals of the Association of American Geographers* 75 (1985): 10–15.

Wood, Tim F. "Thinking in Geography." *Geography* 72 (1987): 289–299.

Don't forget about Dushkin's *Annual Editions Online: Geography* at http://dushkin.com/aeonline/. See preface for details.

CHAPTER 2

MAPS

SPOT satellite image of two of the Galápagos Islands, Isabela and Fernandina.
Photo Researchers, Inc.

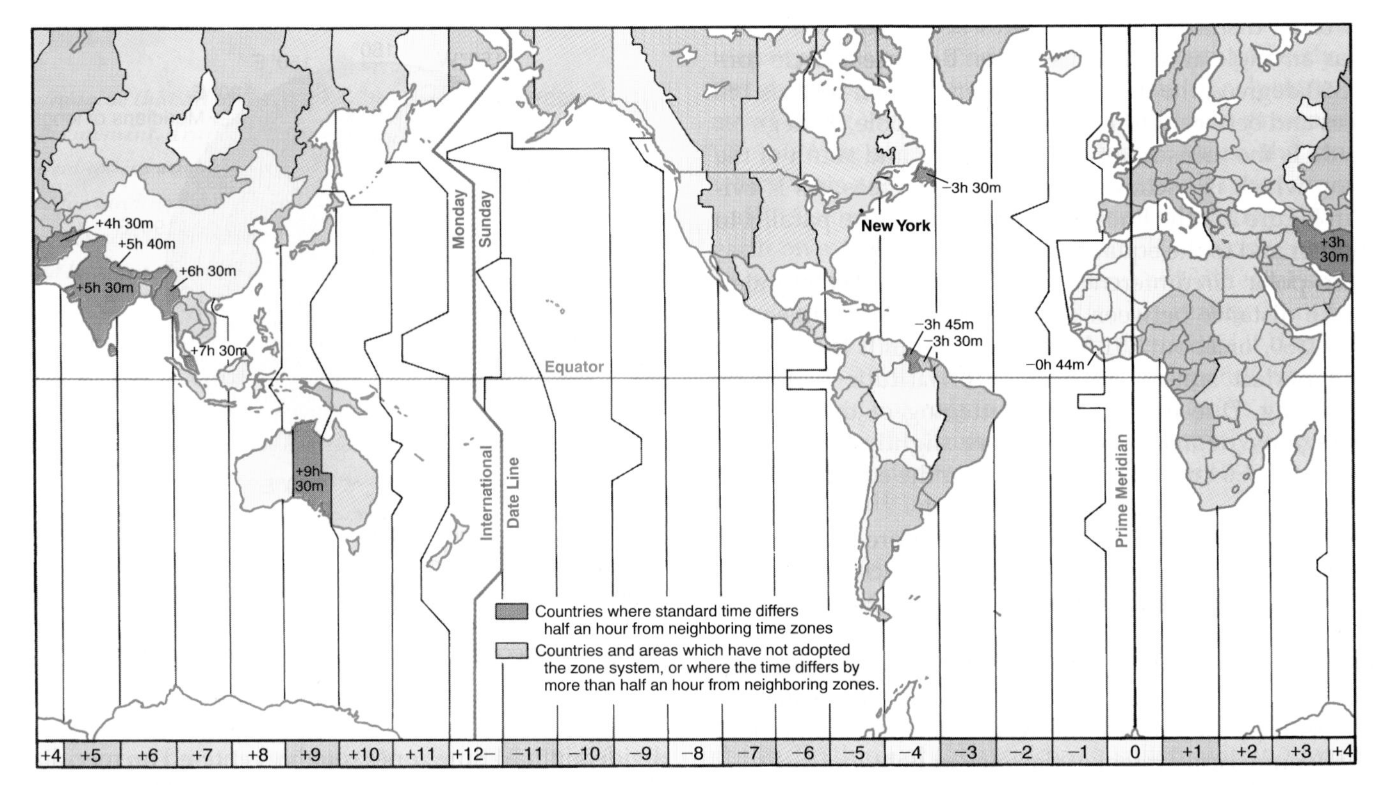

FIGURE 2.3 **World time zones.** Each time zone is about 15° wide, but variations occur to accommodate political boundaries. The figures at the bottom of the map represent the time difference in hours when it is 12 noon in the time zone centered on Greenwich, England. New York is in column –5, so the time there is 7 A.M. when it is noon at Greenwich. Modifications to the universal system of time zones are numerous. Thus, Iceland operates on the same time as Britain, although it is a time zone away. Spain, entirely within the boundaries of the GMT zone, sets its clocks at +1 hour, whereas Portugal conforms to GMT. China straddles five time zones, but the whole country operates on Beijing time (+ 8 hours). In South America, Chile (in the –5 hour zone) uses the –4 hour designation, while Argentina uses the –3 hour zone instead of the –4 hour zone to which it is better suited.

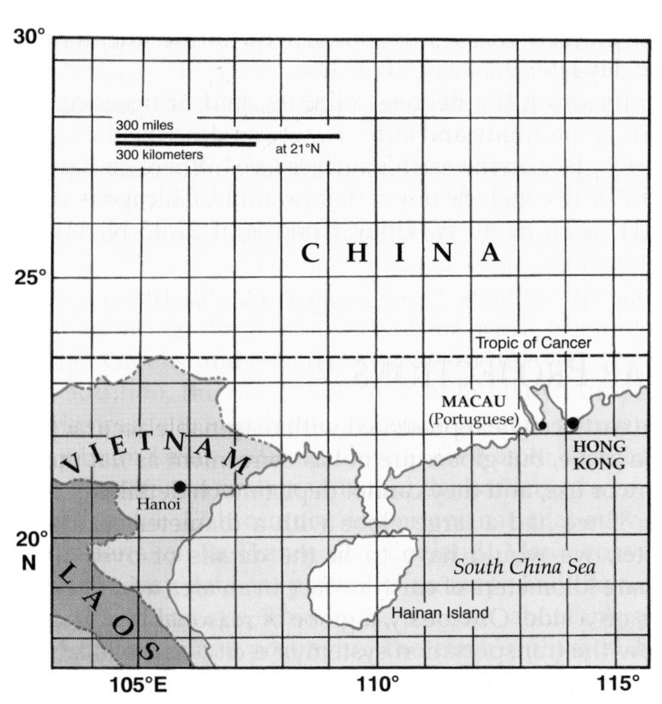

FIGURE 2.4 The latitude and longitude of Hong Kong are 22°15′N, 114°10′E. What are the coordinates of Hanoi?

intact. So, in drawing a map, the relationships between points are inevitably distorted in some way. The term **map projection** designates the way the curved surface of the globe is represented on a flat map.

Properties of Map Projections

Because no projection can be entirely accurate, the serious map user needs to know in what respects a particular map correctly reproduces earth features, and in what respect it distorts them. The four main properties of maps—area, shape, distance, and direction—are distorted in different ways and to different degrees by various projections.

Area

Some projections enable the cartographer to represent the areas of regions in correct or constant proportion to earth reality. That means that any square inch on the map represents an identical number of square miles (or of similar units) anywhere else on the map. As a result, the shape of the portrayed area is inevitably distorted. A square on the earth, for example, may become a rectangle on the map, but that rectangle has the correct area. Such projections are called **equal-area** or **equivalent.**

WHERE ON EARTH ARE YOU?

Do you know where you are? Exactly where you are with reference to the globe grid of latitude and longitude? A handheld GPS receiver can tell you your position anywhere on earth.

GPS stands for **Global Positioning System.** The technology uses a string of Department of Defense satellites that orbit some 20,000 kilometers (12,500 mi) above the earth. Each satellite carries four atomic clocks so accurate that they might gain or lose an average of one second in 30,000 years. As they orbit, the satellites continuously transmit their positions, time signals, and other data. The satellites are arranged so that at least four are above the horizon at any time, all over the earth, available for simultaneous measurement. A GPS receiver records the positions of a number of the satellites simultaneously, then determines its latitude, longitude, altitude, and the time.

Positions calculated from GPS signals are accurate to within about 15 meters (49 ft). If two receivers are used, accuracy increases to 3 meters (10 ft). And if a receiver is used in conjunction with a fixed transmitter on earth whose position is known, positions can be measured within centimeters.

GPS technology was originally designed for military applications, particularly naval and aerial navigation. Its success was evident in the 1991 Gulf War with Iraq, when U.S. troops used the devices to find their way across the Saudi Arabian desert. Now these waterproof, lightweight receivers are also being used in a variety of civilian applications, including geodetic control surveying, monitoring geologic fault lines, firefighting, and oil, gas, and mineral exploration. As the receivers become even smaller, lighter, and less expensive, increasing numbers of sailors, hikers, and other noncommercial users are likely to adopt them to find out exactly where they are at any given time.

Several automobile manufacturers recently have made in-car navigation systems an option in their new cars. The systems tap GPS signals to monitor the car's exact location, comparing it to a computerized atlas stored on a compact disk in the trunk. The car's location, constantly updated, appears on a computer screen mounted on the dashboard. The navigation systems enable motorists to find out exactly where they are and how to reach their destination. For example, the driver can give a street address or the name of a movie theater, hospital, or other building, and the computer will calculate the best route and display it on the screen. Some systems use an electronic voice to provide directions.

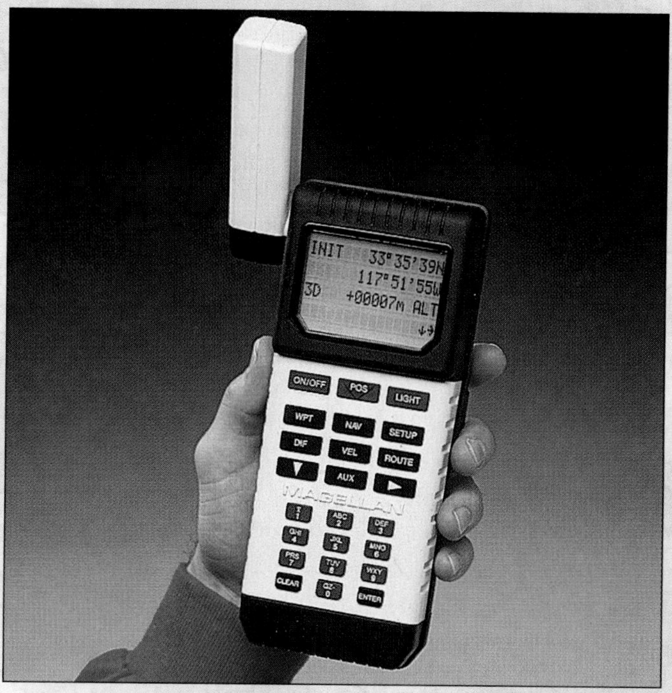

This **GPS receiver** displays latitude, longitude, and altitude on a small liquid crystal display screen.
Courtesy of Magellan Systems Corporation.

Shape

Although no projection can provide correct shapes for large areas, some accurately portray the shapes of small areas by preserving correct angular relationships. That is, an angle on the globe is rendered correctly on the map. Maps that have true shapes for small areas are based on **conformal projections.** Parallels and meridians always intersect at right angles on such maps, as they do on a globe, but the size of areas is distorted. *A map cannot be both equivalent and conformal.*

Distance

Distance relationships are nearly always distorted on a map, but some projections do maintain true distances in one direction or along certain selected lines. Others, called **equidistant projections,** show true distance in all directions, but only from one or two central points (see, for example, Figure 2.9.)

A planar equidistant map centered on Detroit, for example, shows the correct distance between Detroit and the cities of Boston, Los Angeles, and any other point on the map. But it does *not* show the correct distance between Los Angeles and Boston. *A map cannot be both equidistant and equal-area.*

Direction

As is true of distances, directions between all points cannot be shown without distortion. **Azimuthal projections** do exist, however, that enable the map user to correctly measure the directions from a single point to any other point. Directions or azimuths from points other than the central point to other points are not accurate. An azimuthal projection may also be equivalent, conformal, or equidistant.

Types of Projections

While all projections can be described mathematically, some can be thought of as being constructed by geometrical techniques rather than by mathematical formulas. In geometrical projections, the grid system is transferred from the globe to a geometrical figure, such as a cylinder or a cone, which, in turn, can be cut and then spread out flat without any stretching or tearing. The surfaces of cylinders, cones, and planes are said to be **developable surfaces:** the first two because they can be cut and laid flat without distortion, a plane because it is flat at the outset.

The selection of the surface to be developed or of the specific mathematical formula to be employed is determined by the properties of the globe grid that one elects to retain. **Globe properties** are as follows:

1. all meridians are equal in length; each is one-half the length of the equator;
2. all meridians converge at the poles and are true north-south lines;
3. all lines of latitude (parallels) are parallel to the equator and to each other;
4. parallels decrease in length as one nears the poles;
5. meridians and parallels intersect at right angles;
6. the scale on the surface of the globe is the same everywhere in all directions.

Two types of circles appear on the earth's spherical grid. A **great circle** is formed on the surface of a sphere by a plane that passes through the center of the sphere. Thus, the equator is a great circle, and each meridian is half of a great circle. Every great circle bisects the earth, dividing it equally into hemispheres. An arc segment of the great circle joining them is the shortest distance between any two points on the earth's surface. A **small circle** is the line created by the intersection of a spherical surface with a plane that does *not* pass through its center. Except for the equator, all parallels of latitude are small circles. Different projections will represent great and small circles in different ways.

Cylindrical Projections

The **Mercator projection** (Figure 2.5) is one of the most commonly used (and misused) **cylindrical projections.**

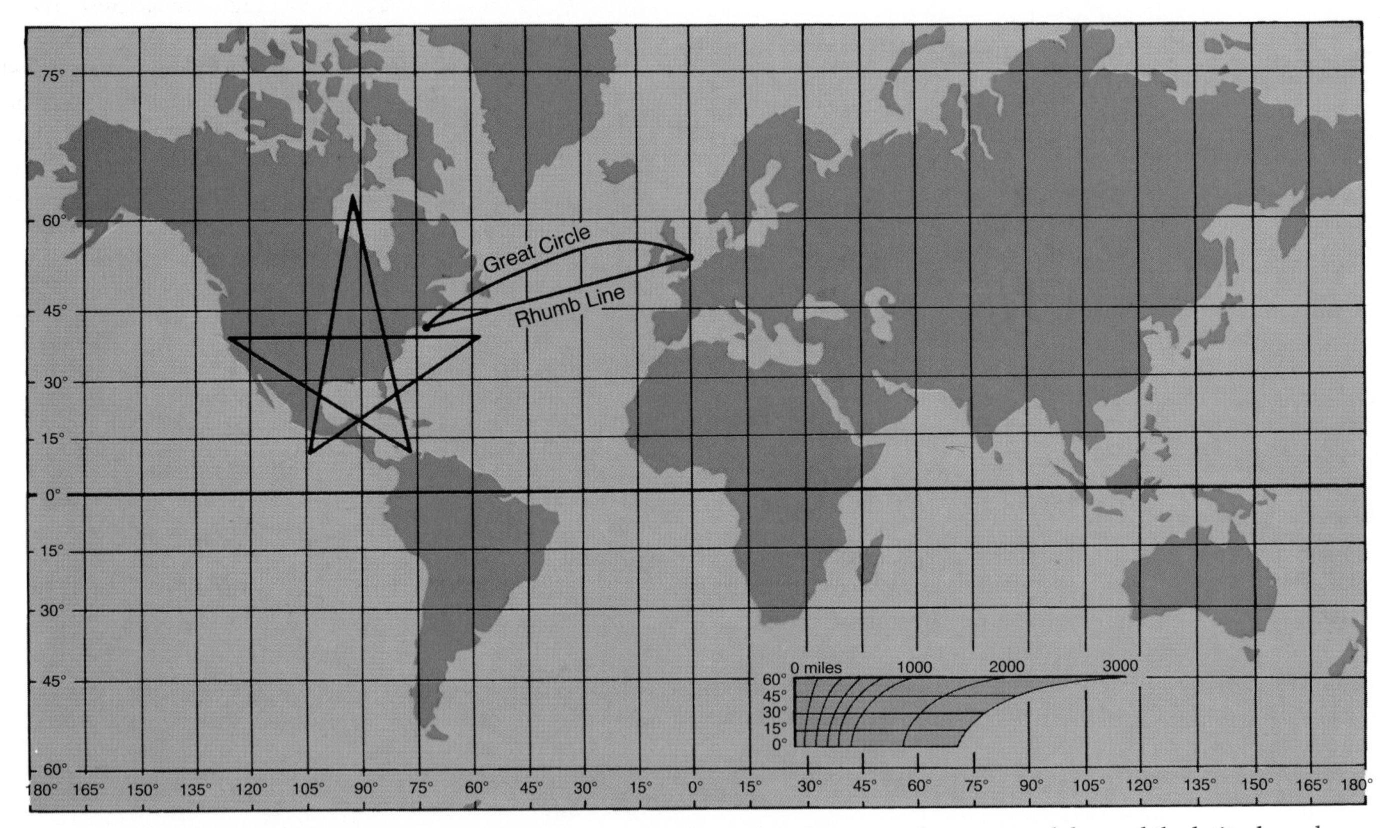

FIGURE 2.5 **Distortion on the Mercator projection.** A perfect five-pointed star was drawn on a globe, and the latitude and longitude of the points of the star were transferred to the Mercator map shown above. The manner in which the star is distorted reflects the way the projection distorts land areas. Mercator maps are usually accompanied by a diagram showing the varying scale of distance at different latitudes. The most significant property of the Mercator projection is that it is the only one on which any straight line is a line of constant compass bearing, or *rhumb line.* With the exception of the equator, great circles appear as curved lines on a Mercator projection. Although a rhumb line usually is not the shortest distance between two points, navigators can draw a series of straight lines between the starting point and destination that will approximate the great circle route.

Named for its inventor, it has been mathematically adjusted to serve navigational purposes. Suppose we roll a piece of paper around a globe so that it touches the sphere at the equator. Instead of the paper being the same height as the globe, however, it extends far beyond the poles. If we place a light source at the center of the globe, the light will project a shadow map upon the cylinder of paper. When we unroll the cylinder, the map will appear as shown in Figure 2.5.

Note the variance between the grid we have just projected and the true properties of the globe grid. The grid lines cross each other at right angles, as they do on the globe, and they are all straight north-south or east-west lines. But the meridians do not converge at the poles as they do on a globe. Instead, they are equally spaced, parallel, vertical lines. Because the meridians are equally far apart, the parallels of latitude have all become the same length. The Mercator map balances the spreading of meridians toward the poles by spacing the latitude lines farther and farther apart. Mathematical tables show the cartographer how to space the parallels to balance the spreading of meridians so that the result is a conformal projection. The shapes of small areas are true, and even large regions are fairly accurately represented.

Note that the Mercator map in Figure 2.5 stops at 75°N and 60°S. A very large map would be needed to show the polar regions, and the sizes of those regions would be enormously exaggerated. The poles themselves can never be shown on a Mercator projection tangent at the equator. In fact, the enlargement of areas with increasing latitude is so great that a Mercator map should not be published without a scale of miles like that shown.

The Mercator projection has often been misused in atlases and classrooms as a general-purpose world map. Many schoolchildren have grown up convinced that the state of Alaska is nearly the size of all the lower 48 states put together. To check the distortion of this projection, compare the relative sizes of Greenland (840,000 sq. mi.) and Mexico (756,066 sq. mi.) on Figure 2.5 with those on Figure 2.6, an equal-area projection.

The proper use of the Mercator is for navigation. In fact, it is the standard projection used by navigators because of a peculiarly useful property: a straight line drawn anywhere on the map is a line of constant compass bearing. If such a line, called a **rhumb line,** is followed, a ship's or plane's compass will show that the course is always at a constant angle with respect to geographic north. On no other projection is a rhumb line both straight and true as a direction.

Conic Projections

Of the three developable geometric forms—cylinder, cone, and plane—the cone is the closest in form to one-half of a globe. **Conic projections,** therefore, are widely used to depict hemispheres or smaller parts of the earth.

A useful projection in this category, and the easiest to understand, is the *simple conic* projection. Imagine that a cone is laid over half the globe, as in Figure 2.7, tangent to the globe at the 30th parallel. Distances are true only along the tangent circle, which is called the *standard parallel.* When

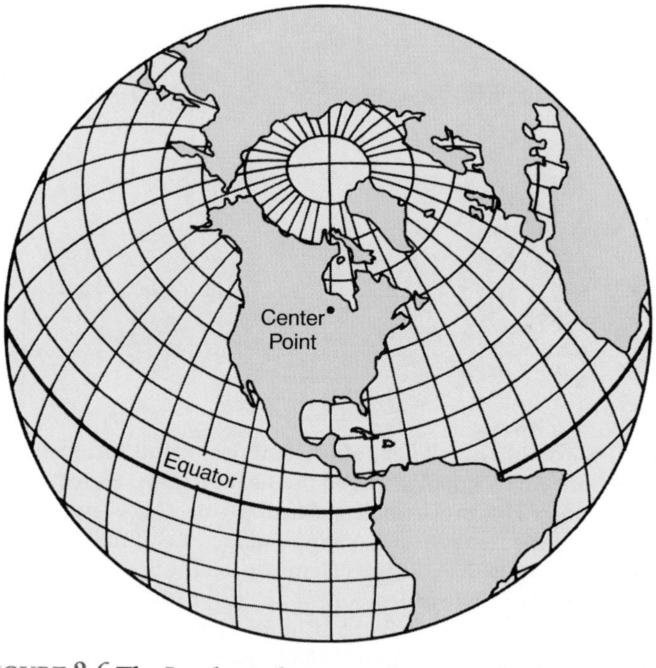

FIGURE 2.6 **The Lambert planar equal-area projection** is mathematically derived to display the property of equivalence, which means that areas on the map are shown in true proportion to the same areas on the earth. Directions are true only from the center point. The projection is well suited to areas extending equally in all directions from the center point.

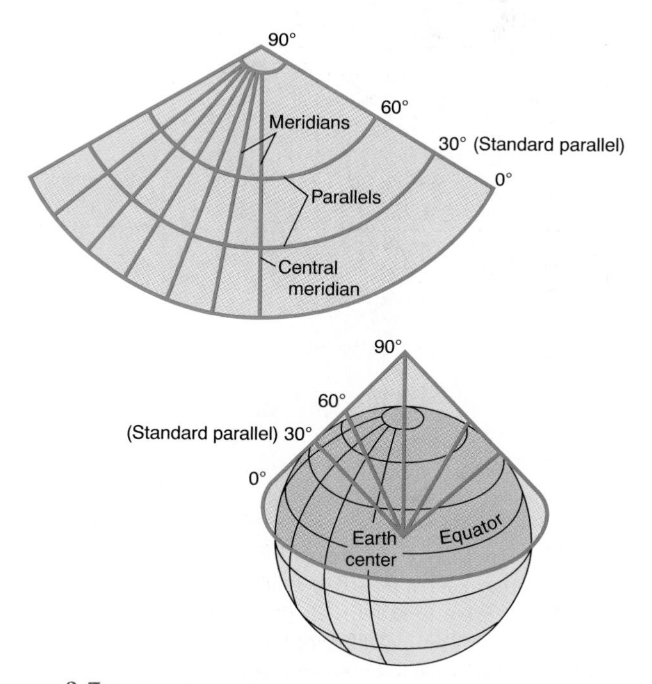

FIGURE 2.7 **A simple conic projection with one standard parallel.** Most conics are adjusted so that the parallels are spaced evenly along the central meridian.

the cone is developed, of course, the standard parallel becomes an arc of a circle, and all other parallels become arcs of concentric circles. With a central light source, the parallels become increasingly farther apart as they approach the pole, and distortion is accordingly exaggerated.

One can lessen the amount of distortion by shortening the length of the central meridian, spacing the parallels of latitude at equal distances on that meridian, and making the 90th parallel (the pole) an arc rather than a point. Most of the conic projections in general use employ such mathematical adjustments. When more than one standard parallel is used, a *polyconic* projection results.

Conic projections can be adjusted to achieve desired qualities; hence, they are widely used. They are particularly suited for showing areas in the midlatitudes that have a greater east-west than north-south extent. Both area and shape can be represented by conic projections without serious distortion (Figure 2.8). Many official map series, such as the topographic sheets and the world aeronautical charts of the U.S. Geological Survey (USGS) and the Aerospace Center of the Defense Mapping Agency, use types of conic projections. They are also used in many atlases.

Planar Projections

Planar projections are constructed by placing a plane surface tangent to the globe at a single point (Figure 2.9).

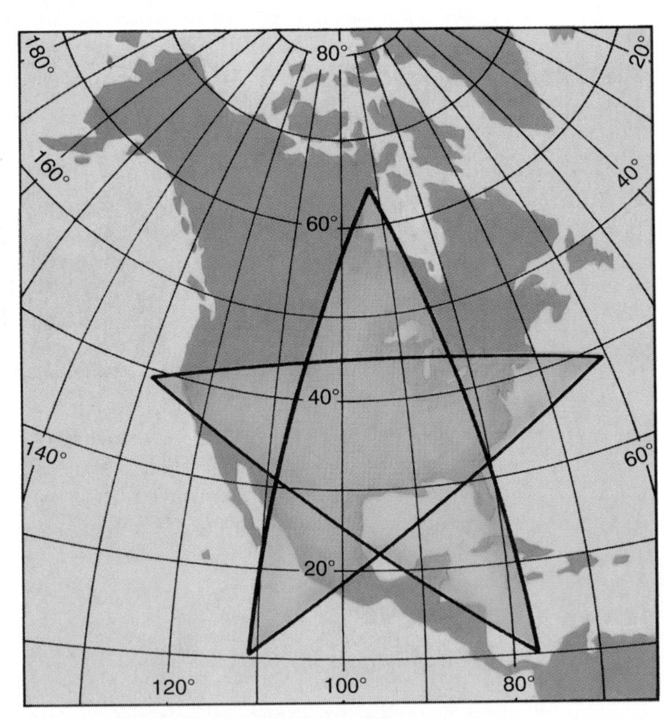

FIGURE 2.8 **The polyconic projection.** The map is produced by bringing together east-west strips from a series of cones, each tangent at a different parallel. This projection differs from the simple conic in that the parallels of latitude are not arcs of concentric circles and the meridians are curved rather than straight lines. While neither equivalent nor conformal, the projection portrays shape well. Note how closely the star resembles a perfect five-pointed star.

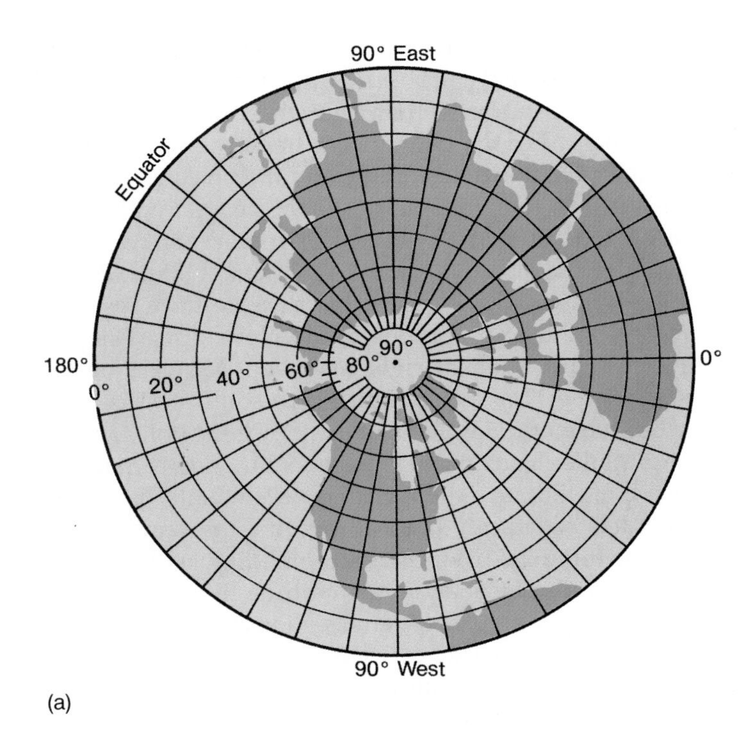

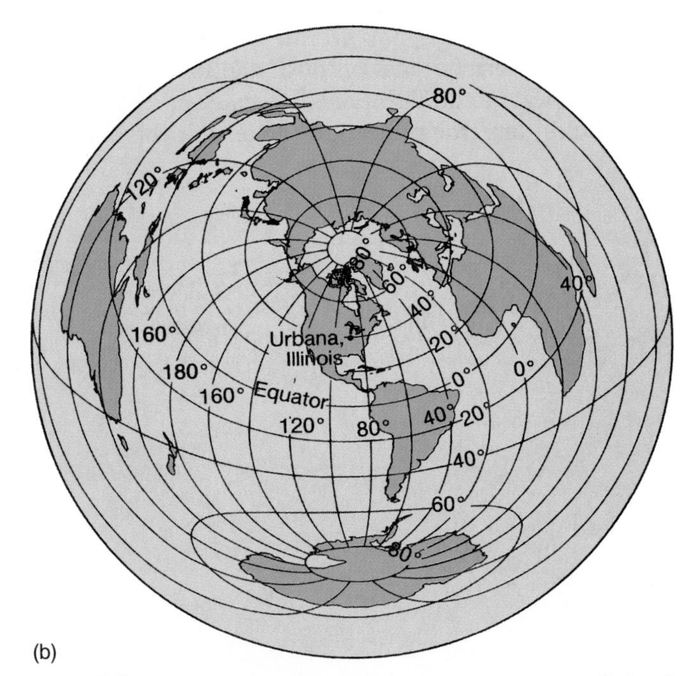

FIGURE 2.9 (a) **The planar equidistant projection.** Parallels of latitude are circles equally spaced on the meridians, which are straight lines. This projection is particularly useful because distances from the center to any other point are true. If the grid is extended to show the Southern Hemisphere, the South Pole is represented as a circle instead of a point. (b) **A planar equidistant projection centered on Urbana, Illinois.** The scale of miles applies only to distances from Urbana or on a line through it. The scale on the rim of the map, representing the antipode of Urbana, is infinitely stretched.

Although the plane may touch the globe anywhere the cartographer wishes, the polar case with the plane centered on either the North or the South Pole is easiest to visualize (Figure 2.9a).

This *equidistant* projection is useful because it can be centered anywhere, facilitating the correct measurement of distances from that point to all others. For this reason it is often used to show air navigation routes that originate from a single place. When the plane is centered on places other than the poles, the meridians and the parallels become curiously curved, as is evident in Figure 2.9b.

Because they are particularly well suited for showing the arrangement of polar landmasses, planar maps are commonly used in atlases. Depending on the particular projection used, true shape, equal area, or some compromise between them can be depicted. In addition, one of the planar projections is widely used for navigation and telecommunications. The *gnomonic* projection, shown in Figure 2.10, is the only one on which all great circles (or parts thereof) appear as straight lines. Because great circles are the shortest distances between two points, navigators need only connect the points with a straight line to find the shortest route.

Other Projections

Projections can be developed mathematically to show the world or a portion of it in any shape that is desired: ovals, hearts, trapezoids, stars, and so on. One often-used projection is *Goode's Homolosine.* Usually shown in its interrupted form, as in Figure 2.11, it is actually a combination of two different projections. This equal-area projection also represents shapes well.

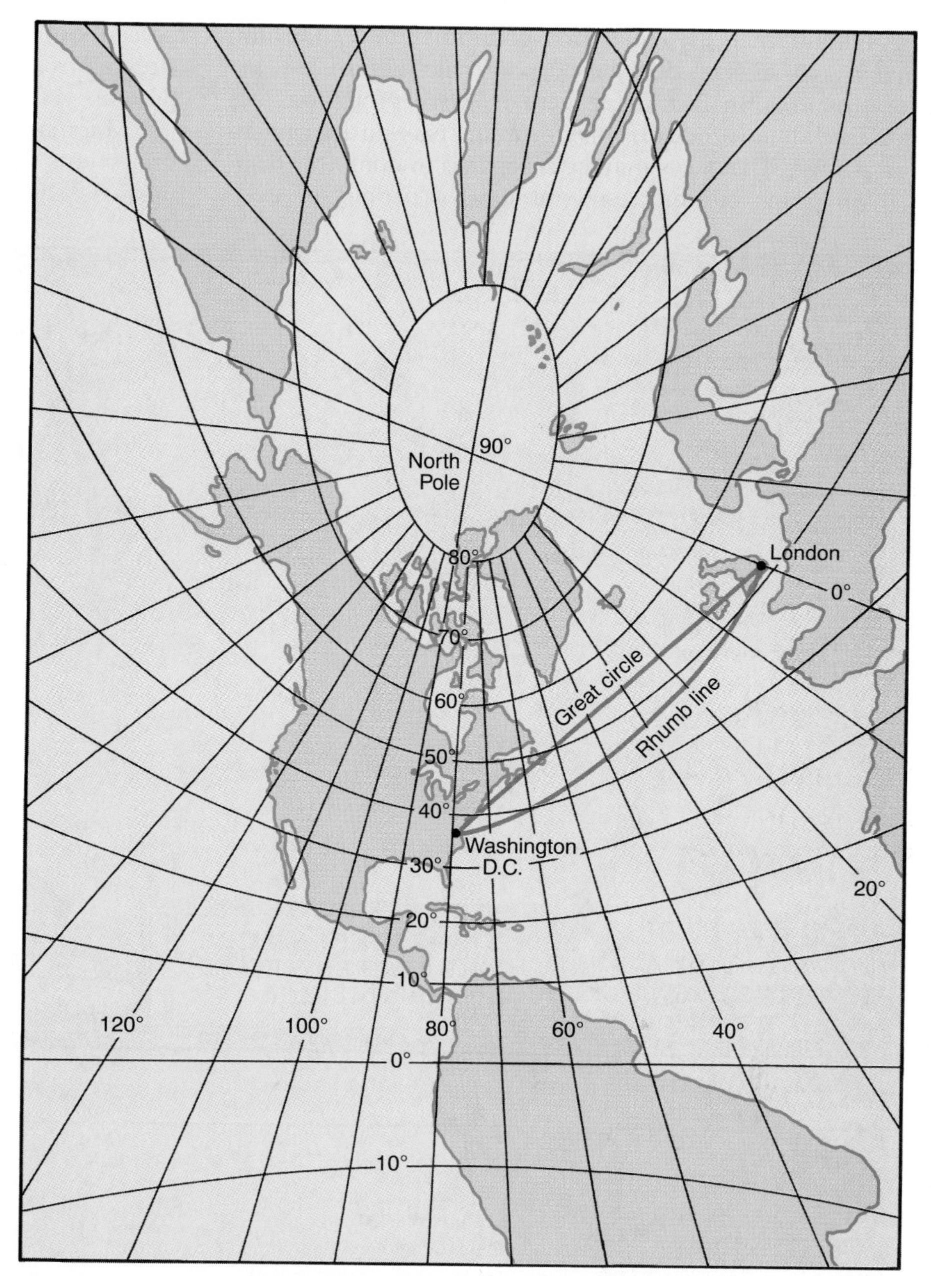

FIGURE 2.10 **The gnomonic projection** is the only one on which all great circles appear as straight lines. Rhumb lines are curved. In this sense, it is the opposite of the Mercator projection, on which rhumb lines are straight and great circles are curved (see Figure 2.5). Note that the distortion of shapes and areas increases away from the center point. The map is not conformal, equal-area, or equidistant.

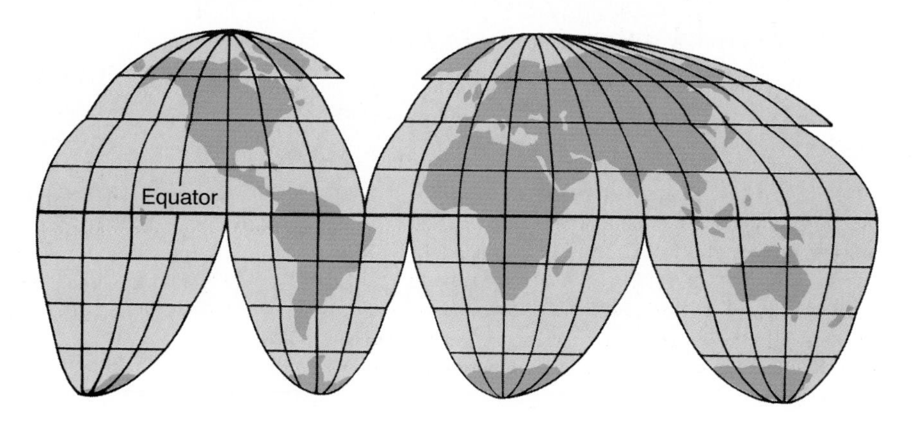

FIGURE 2.11 **Goode's Homolosine projection** is a combination of two different projections. This projection can also interrupt the continents to display the ocean areas intact.

Not all maps are equal-area, conformal, or equidistant; many, such as the widely used polyconic (Figure 2.8), are compromises. In fact, some very effective projections are non-Euclidian in origin, transforming space in unconventional ways. Distances may be measured in nonlinear fashion (in terms of time, cost, number of people, or even perception), and maps that show relative space may be constructed from these data. Two examples of such transformations are shown in Figure 2.12.

Mapmakers must be conscious of the properties of the projections they use, selecting the one that best suits their purposes. If the map shows only a small area, the choice of a

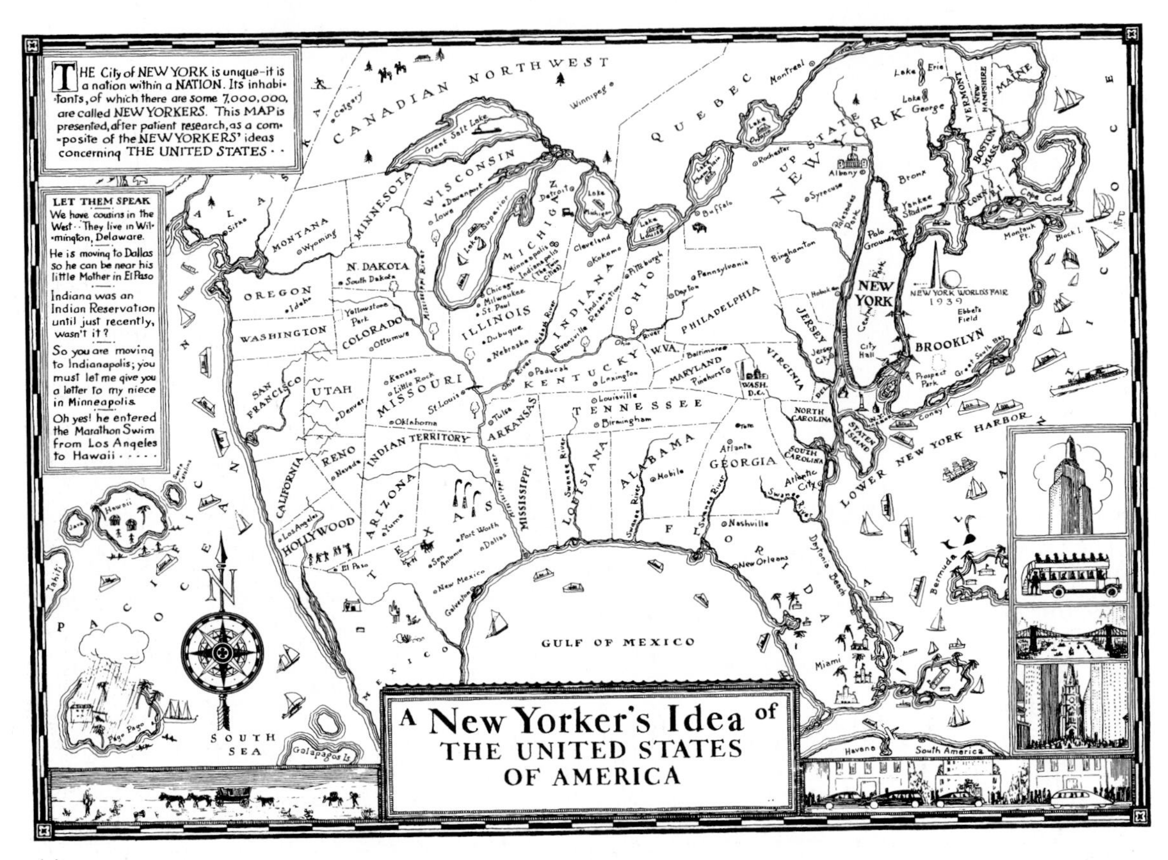

(a)

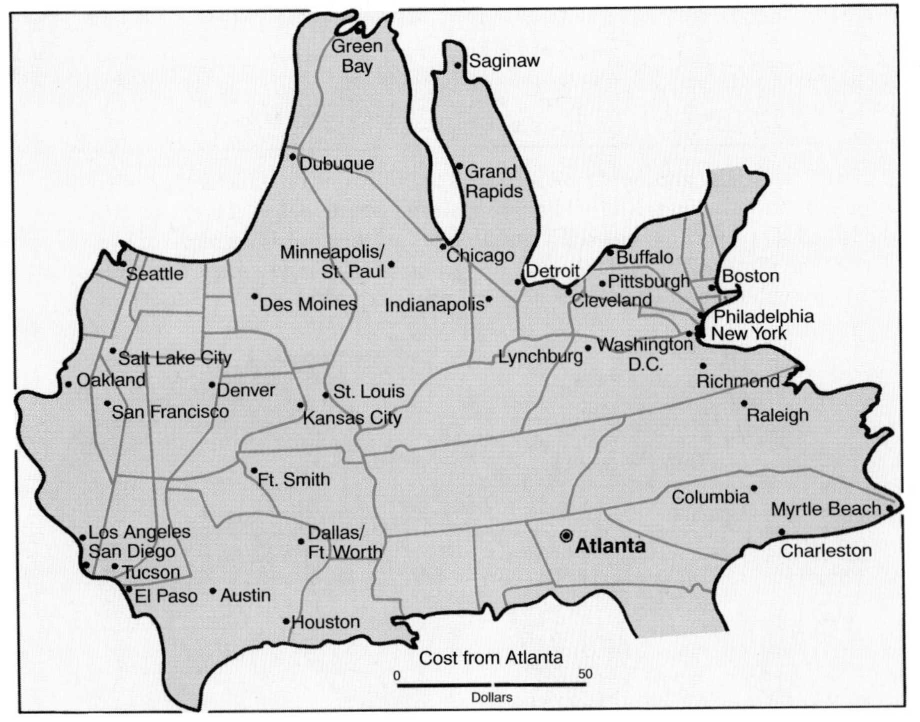

(b)

FIGURE 2.12 (a) **A New Yorker's idea of the United States of America** portrays the country not as it actually is, but as the artist conceives it. Think of it as a map based on an unknown projection. Distance, area, shape, and direction are all distorted, and a "psychological" rather than a literally true map is achieved. Such mental maps are discussed in Chapter 8. (b) **Airline cost distance from Atlanta, Georgia.** The linear scale is based upon one-way coach airfares on American Airlines, September 1987. Note how the transformation distorts shape and results in such anomalies as placing San Francisco in Nevada.

projection is not critical—virtually any can be used. The choice becomes more important when the area to be shown extends over a considerable longitude and latitude; then, the selection of a projection depends upon the purpose of the map. Navigation maps usually employ a Mercator, true direction, or equidistant projection. If numerical data are being mapped, the relative sizes of the areas involved should be correct, so that one of the many equal-area projections is likely to be used (see "The Peters Projection"). Display maps usually employ conformal projections, although compromise and equal-area projections are also common. Most atlases indicate which projection has been used for

Geography and Public Policy

The Peters Projection

Although map projections might appear to be a rather sterile, uninteresting topic, the projection shown here is creating significant controversy. Developed and promoted by Arno Peters, a German historian, the **Peters projection** purports to reflect concern for the problems of the Third World by providing a less European-centered representation of the world. In contrast to the Mercator projection, which greatly distorts areas, the Peters projection is an equal-area map; therefore, it is better suited than the Mercator map for comparing distributions. Peters claims that his map shows the densely populated parts of the earth and the countries of the Third World in proper proportion to one another. He has persuaded a number of agencies with special interest in the Third World, including the World Council of Churches and several United Nations organizations (UNESCO, UNDP, and UNICEF), to adopt the map.

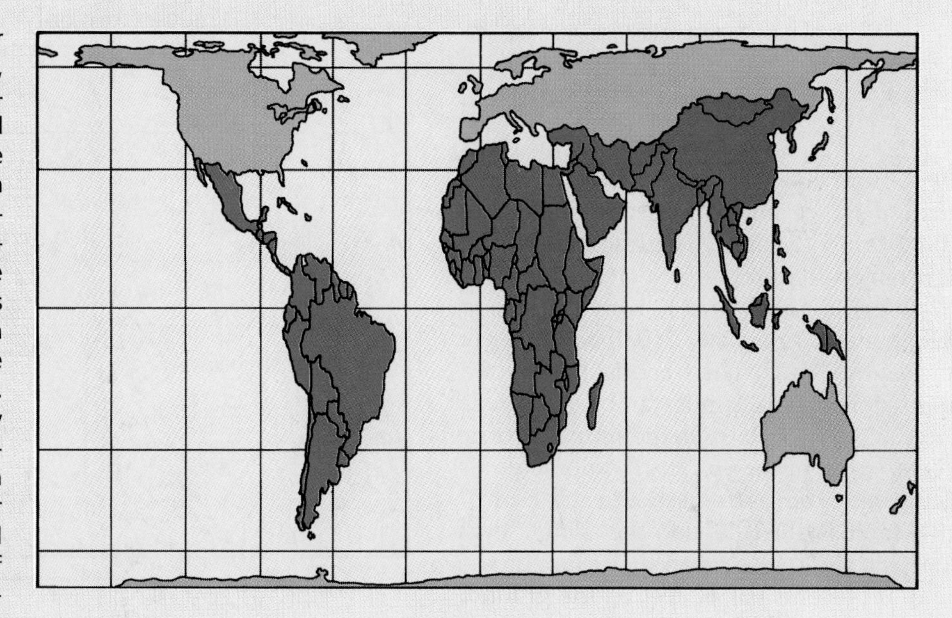

But the Peters projection is as inappropriate as the Mercator map, argue two geographers (P. Porter and P. Voxland, "Distortion in Maps: The Peters Projection and Other Devilments," *Focus,* Summer 1986). Characterizing Peters's cartography as "perverse and wrongheaded," they criticize his projection on several grounds.

First, it badly distorts shapes in the tropics and at high latitudes. Although Peters claims to support the Third World, his map preserves shapes more accurately for the developed world. Second, distances and directions cannot be measured except under very limited conditions. Third, Peters claims his projection will meet every cartographic need, which is simply incorrect. Finally, the projection is not, in fact, new but is a modification of an equal-area projection developed by James Gall in 1885.

Questions to Consider

1. Since many projections yield an equal-area world map with less distortion of shapes than does the Peters projection, what purpose is served by using the latter? Why do you think the Peters map has been adopted by organizations interested in the Third World?
2. The publisher of the Peters map, Friendship Press, asserts that the map shows "fairness to all people by setting forth all countries in their true size and location." Do you agree? Why or why not?
3. Most people assume that a map represents reality; they do not look at maps critically. If the Peters map were the only world map that people studied, what impression would they have of the countries of the world? How would their views differ from those of people familiar only with Mercator's map?

each map, thus informing the map reader of the properties of the maps and their distortions.

Selection of the map grid, determined by the projection, is the first task of the mapmaker. A second decision involves the scale at which the map is to be drawn.

SCALE

The **scale** of a map is the ratio between the measurement of something on the map and the corresponding measurement on the earth. Scale is typically represented in one of three ways: verbally, graphically, or numerically as a representative fraction. As the name implies, a *verbal* scale is given in words, such as "1 inch to 1 mile." A *graphic* scale of 1 inch to 1 mile is shown below.

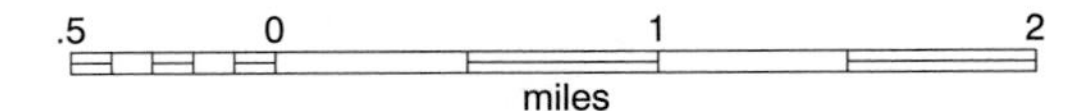

A *representative fraction* (RF) scale gives two numbers, the first representing the map distance and the second indicating the ground distance. The fraction may be written in a number of ways. There are 5280 feet in 1 mile and 12 inches in 1 foot; 5280 times 12 equals 63,360, the number of inches in 1 mile. The fractional scale of a map at 1 inch to 1 mile can be written as 1:63,360 or 1/63,360. On the simpler metric scale, 1 centimeter to 1 kilometer is 1:100,000. The units used in each part of the fractional scale are the same, thus 1:63,360 could also mean that 1 foot on the map represents 63,360 feet on the ground, or 12 miles—which is, of course, the same as 1 inch represents 1 mile. Numerical scales are the most accurate of all scale statements and can be understood in any language. Figure 2.14 is based on a fractional scale of 1:24,000.

The map scale, or ratio between the map dimensions and those of reality, can range from very large to very small. A *large-scale map*, such as a plan of a city, shows an area in considerable detail. That is, the ratio of map to ground distance is relatively large, for example 1:600 (1 inch on the map represents 600 inches or 50 feet on the ground) or 1:24,000. At this scale, features such as buildings and highways can be drawn to scale. Figure 2.14 is an example of a large-scale map. *Small-scale maps*, such as those of countries or continents, have a much smaller ratio. Buildings, roads, and other small features cannot be drawn to scale and must be magnified and represented by symbols to be seen. Figures 2.3 and 2.4 are small-scale maps. Although no rigid numerical limits differentiate large-scale from small-scale maps, most cartographers would consider large-scale maps to have a ratio of 1:50,000 or less, and maps with ratios of 1:500,000 or more to be small-scale.

Each of the four maps in Figure 2.13 is drawn at a different scale. Although each is centered on Boston, notice how scale affects both the area that can be shown in a square that is 2 inches on a side and the amount of detail that can be depicted. On map (a), at a scale of 1:25,000, about 2.6 inches represent 1 mile, so that the 2-inch square shows less than 1 square mile. At this scale, one can identify individual buildings, highways, rivers, and other landscape features. Map (d), drawn to a scale of 1 to 1 million (1:1,000,000 or 1 inch represents almost 16 miles), shows an area of almost 1000 square miles. Now, only major features, such as main highways and the location of cities, can be shown, and even the symbols

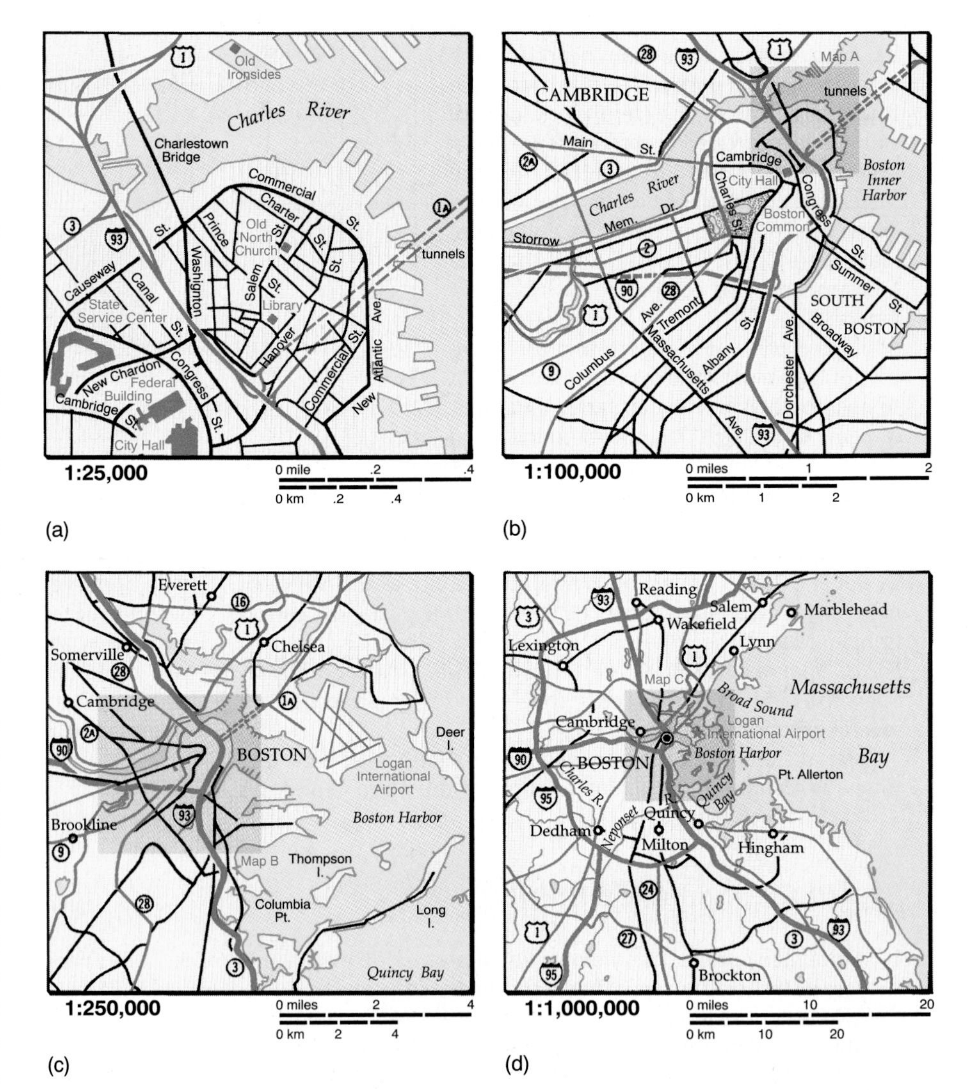

FIGURE 2.13 **The effect of scale on area and detail.** The larger the scale, the greater the number and kinds of features that can be included because the map shows a more restricted area than does a map at a smaller scale.

used for that purpose occupy more space on the map than would the features depicted if they were drawn true to scale.

Small-scale maps like (c) and (d) in Figure 2.13 are said to be very *generalized.* They give a general idea of the relative locations of major features but do not permit accurate measurement. They show significantly less detail than do large-scale maps and typically smooth out such features as coastlines, rivers, and highways.

Topographic Maps and Terrain Representation

When we speak of the topography of an area, we refer to its terrain. **Topographic maps** portray the surface features of relatively small areas, often with great accuracy (Figure 2.14). They not only show the elevations of landforms, streams, and other water bodies, but may also display features that people have added to the natural landscape. These might include transportation routes, buildings, and such land uses as orchards, vineyards, and cemeteries. Boundaries of all kinds, from state boundaries to field or airport limits, may also be depicted on topographic maps.

The U.S. Geological Survey (USGS), the chief federal agency for topographic mapping in this country, produces several map series, each on a standard scale. Complete topographic coverage of the United States is available at scales of 1:250,000 and 1:100,000. Maps are also available at various other scales. A single map in one of these series is called a *quadrangle.* Topographic quadrangles at the scale of 1:24,000 exist for the entire area of the 48 contiguous states and Hawaii, a feat that requires more than 54,000 maps. Each map covers a rectangular area that is 7.5 minutes of latitude by 7.5 minutes of longitude. As is evident from Figure 2.14, these 7.5-minute quadrangle maps provide detailed information about the natural and cultural features of an area.

Because of Alaska's large size and sparse population, the primary scale for mapping that state is 1:63,360 (1 inch represents 1 mile). The Alaska quadrangle series consists of 2700 maps.

In Canada, the responsibility for national mapping lies with Survey and Mapping and Remote Sensing, Natural Resources (NRCAN). Maps at a scale of 1:250,000 are available for the entire country; the more heavily populated southern part of Canada is covered by 1:50,000-scale maps. Provincial mapping agencies produce detailed maps at even larger scales.

The USGS produces a sheet listing the symbols it employs on topographic maps (Figure 2.15), and some older maps provide legends on the reverse side. Note that in the case of running water, separate symbols are used in the legend and on the map to depict perennial (permanent) streams, intermittent streams, and springs; the location and size of rapids and falls are also indicated. There are three different symbols for dams and two more for types of bridges. On maps of cities, where it would be impossible to locate every building separately, the built-up area is denoted by special tints, and only streets and public buildings are shown.

Cartographers use a number of techniques to represent the three-dimensional surface of the earth on a two-dimensional map. The easiest way to show relief is to use numbers called *spot heights* to indicate the elevation of selected points. A *bench mark* is a particular type of spot height (see "Geodetic Control Data"). The principal device used to show elevation on topographic maps, however, is the **contour line,** along which all points are of equal elevation above a datum plane, usually mean sea level. Contours are imaginary lines, perhaps best thought of as the outlines that would occur if a series of parallel, equally spaced horizontal slices were made through a vertical feature. Figure 2.16 shows the relationship of contour lines to elevation for an imaginary island.

The *contour interval* is the vertical spacing between contour lines, and it is normally stated on the map. The more irregular the surface, usually, the greater is the number of contour lines that will need to be drawn; the steeper the slope, the closer are the contour lines rendering that slope. Contour intervals of 10 and 20 feet are often used, though in relatively flat areas the interval may be only 5 feet. In mountainous areas, the spacing between contours is greater: 50 feet, 100 feet, or more. On Figure 2.14, for example, the contour interval is 50 feet. Normally, contour intervals are constant on any single map, although small-scale maps like those of continents often have a variable contour interval—showing, perhaps, contour lines at 500, 1000, 2000, 3000, 5000, 7000, and 10,000 feet. If the intervals were not variable, rugged areas would be clogged by too many contour lines.

Contour lines are the most accurate method of representing terrain, giving the map reader information about the elevation of any place on the map and the size, shape, and slope of all relief features. They are not truly pictorial, however. Most map readers find it difficult to visualize the landscape from contour lines. To heighten the graphic effect of a topographic map, contours are sometimes supplemented by the use of **shaded relief.** This method of representing the three-dimensional quality of an area is illustrated by Figure 2.17. An imaginary light source, usually in the northwest, can be thought of as illuminating a model of the area, simulating the appearance of sunlight and shadows and creating the illusion of three-dimensional topography. Portions that are in the shadow are darkened on the map.

The tremendous amount of information on topographic maps makes them useful to engineers, regional planners, land use analysts, and developers, as well as to hikers and casual users. Given such a wealth of information, the experienced map reader can make deductions about both the physical character of the area and the economic and cultural use of the land.

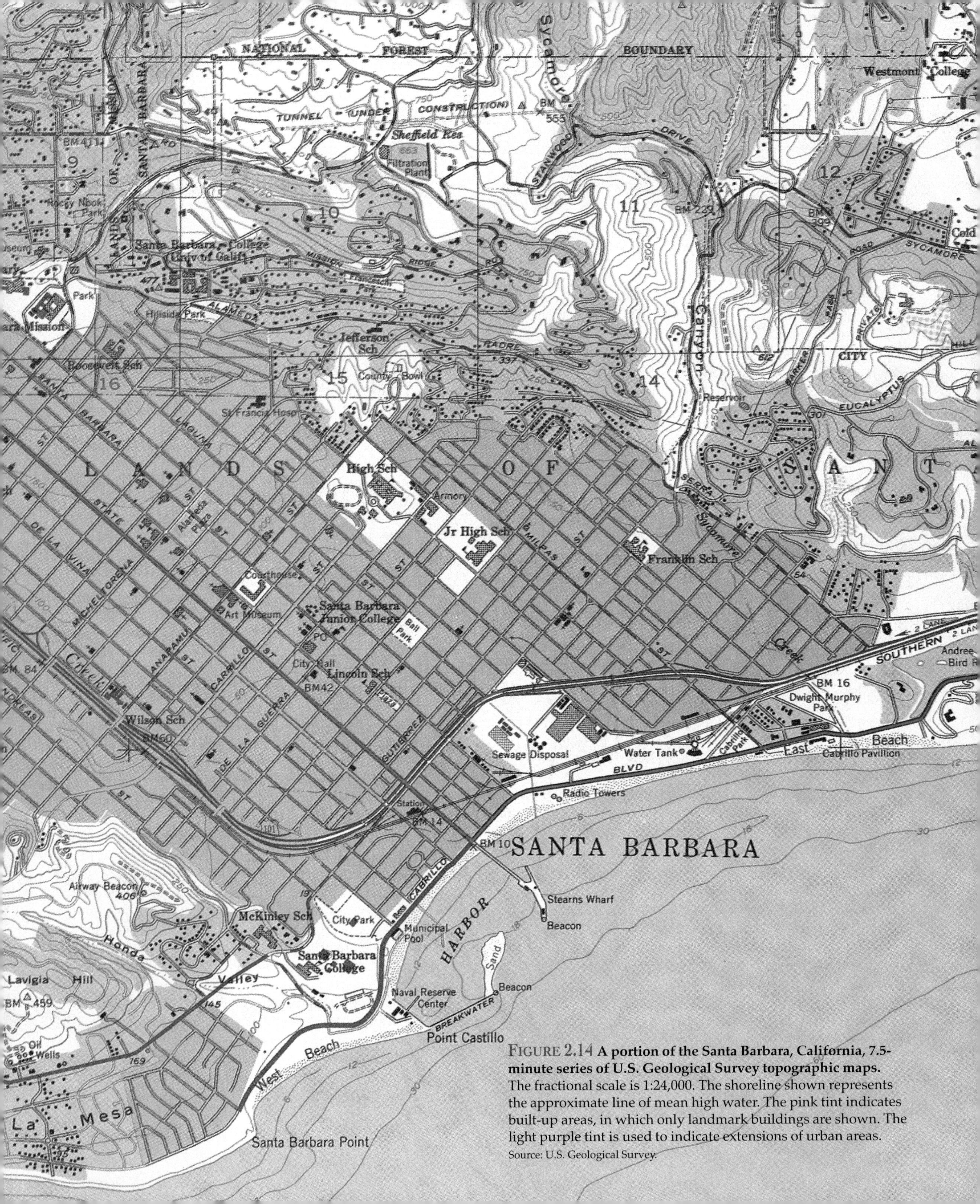

FIGURE 2.14 **A portion of the Santa Barbara, California, 7.5-minute series of U.S. Geological Survey topographic maps.** The fractional scale is 1:24,000. The shoreline shown represents the approximate line of mean high water. The pink tint indicates built-up areas, in which only landmark buildings are shown. The light purple tint is used to indicate extensions of urban areas. Source: U.S. Geological Survey.

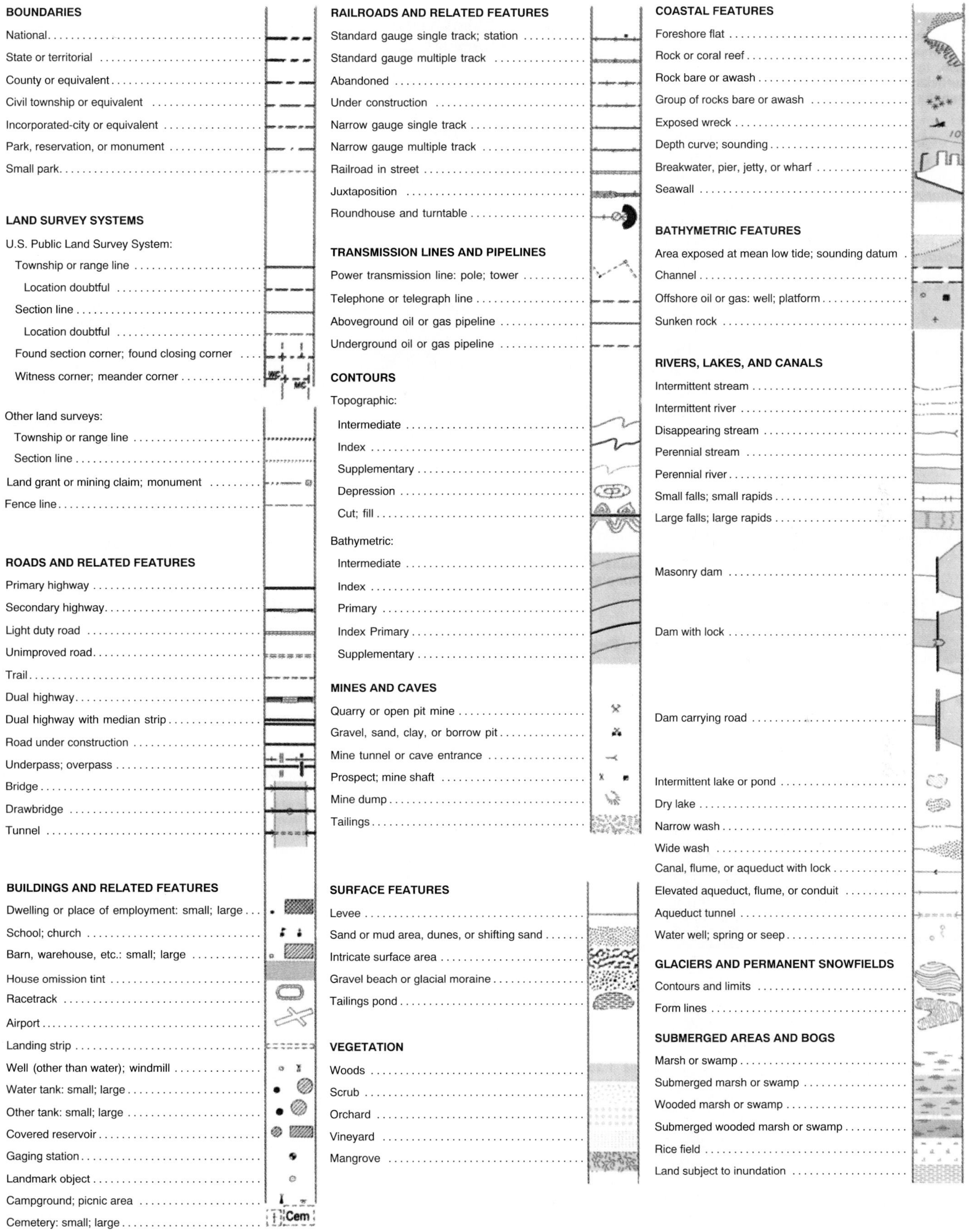

BOUNDARIES
National
State or territorial
County or equivalent
Civil township or equivalent
Incorporated-city or equivalent
Park, reservation, or monument
Small park

LAND SURVEY SYSTEMS
U.S. Public Land Survey System:
Township or range line
Location doubtful
Section line
Location doubtful
Found section corner; found closing corner
Witness corner; meander corner

Other land surveys:
Township or range line
Section line
Land grant or mining claim; monument
Fence line

ROADS AND RELATED FEATURES
Primary highway
Secondary highway
Light duty road
Unimproved road
Trail
Dual highway
Dual highway with median strip
Road under construction
Underpass; overpass
Bridge
Drawbridge
Tunnel

BUILDINGS AND RELATED FEATURES
Dwelling or place of employment: small; large
School; church
Barn, warehouse, etc.: small; large
House omission tint
Racetrack
Airport
Landing strip
Well (other than water); windmill
Water tank: small; large
Other tank: small; large
Covered reservoir
Gaging station
Landmark object
Campground; picnic area
Cemetery: small; large

RAILROADS AND RELATED FEATURES
Standard gauge single track; station
Standard gauge multiple track
Abandoned
Under construction
Narrow gauge single track
Narrow gauge multiple track
Railroad in street
Juxtaposition
Roundhouse and turntable

TRANSMISSION LINES AND PIPELINES
Power transmission line: pole; tower
Telephone or telegraph line
Aboveground oil or gas pipeline
Underground oil or gas pipeline

CONTOURS
Topographic:
Intermediate
Index
Supplementary
Depression
Cut; fill

Bathymetric:
Intermediate
Index
Primary
Index Primary
Supplementary

MINES AND CAVES
Quarry or open pit mine
Gravel, sand, clay, or borrow pit
Mine tunnel or cave entrance
Prospect; mine shaft
Mine dump
Tailings

SURFACE FEATURES
Levee
Sand or mud area, dunes, or shifting sand
Intricate surface area
Gravel beach or glacial moraine
Tailings pond

VEGETATION
Woods
Scrub
Orchard
Vineyard
Mangrove

COASTAL FEATURES
Foreshore flat
Rock or coral reef
Rock bare or awash
Group of rocks bare or awash
Exposed wreck
Depth curve; sounding
Breakwater, pier, jetty, or wharf
Seawall

BATHYMETRIC FEATURES
Area exposed at mean low tide; sounding datum
Channel
Offshore oil or gas: well; platform
Sunken rock

RIVERS, LAKES, AND CANALS
Intermittent stream
Intermittent river
Disappearing stream
Perennial stream
Perennial river
Small falls; small rapids
Large falls; large rapids
Masonry dam
Dam with lock
Dam carrying road
Intermittent lake or pond
Dry lake
Narrow wash
Wide wash
Canal, flume, or aqueduct with lock
Elevated aqueduct, flume, or conduit
Aqueduct tunnel
Water well; spring or seep

GLACIERS AND PERMANENT SNOWFIELDS
Contours and limits
Form lines

SUBMERGED AREAS AND BOGS
Marsh or swamp
Submerged marsh or swamp
Wooded marsh or swamp
Submerged wooded marsh or swamp
Rice field
Land subject to inundation

FIGURE 2.15 **Standard U.S. Geological Survey topographic map symbols.**
Source: U.S. Geological Survey.

Geodetic Control Data

The horizontal position of a place, specified in terms of latitude and longitude, constitutes only two-thirds of the information needed to locate it in three-dimensional space. Also needed is a vertical control point defining elevation, usually specified in terms of altitude above sea level. Together, the horizontal and vertical positions comprise *geodetic control data.* A network of more than 1 million of these precisely known points covers the entire United States.

Each point is indicated by a bronze marker fixed in the ground. You may have seen some of the vertical markers, called *bench marks,* on mountaintops, hilltops, or even on city sidewalks. Every U.S. Geological Survey (USGS) map shows the markers in the area covered by the map, and the USGS maintains Geodetic Control Lists containing the description, location, and elevation of each marker. A bench mark is indicated on the map by the letters BM, a small x, and the elevation.

These lists were revised in 1987 when, after 12 years' effort, federal scientists completed the recalculation of the precise location of some 250,000 bench marks across the country. In using a satellite locating system for the first national resurvey of control points since 1927, the National Oceanic and Atmospheric Administration (NOAA) found, for example, that New York's Empire State Building is 36.7 meters (120.4 ft) northeast of where it formerly officially stood. The Washington Monument has been moved northeast by 28.8 meters (94.5 ft), the dome of California's state capitol in Sacramento has been relocated 91.7 meters (300.9 ft) southwest, and Seattle's Space Needle now has a position 93 meters (305.1 ft) west and 20 meters (65.6 ft) south of where maps now show it. The satellite survey provides much more accurate locations than did the old system of land measurement of distances and angles. The result is more accurate maps and more precise navigation.

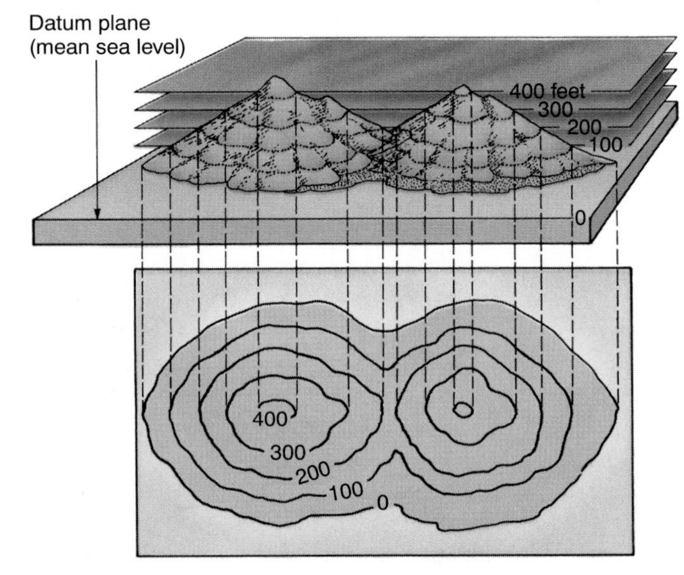

FIGURE 2.16 **Contours drawn for an imaginary island.** The intersection of the landform by a plane held parallel to sea level is a contour representing the height of the plane above sea level.

Patterns and Symbols on Maps

The study of the spatial pattern of things, whether people, cows, or traffic flows, is the essence of the locational tradition in geography, a subject explored in Part III of this book. Maps are used to record the location of these phenomena, and different kinds of techniques are employed to depict their presence or numbers at specific points, in given areas, or along lines.

Point Symbols

A topographic map, using symbols, shows the location at points of many kinds of things, such as churches, schools, and cemeteries. Each symbol counts as one occurrence. Sometimes, however, our interest is in showing the variation in the number of things that exist at several points; for example, the population of selected cities, the tonnage handled at certain terminals, or the number of passengers at given airports.

There are two chief means of symbolizing such distributions, as Figure 2.18 indicates. One method is to choose a

FIGURE 2.17 **A shaded-relief map of a portion of the Swiss Alps near Interlaken.** The scale is 1:50,000. Notice how effectively shading portrays a three-dimensional surface.

Reproduced by permission of the Swiss Federal Office of Topography, March 9, 1995.

symbol, usually a dot, to represent a given quantity of the mapped item (such as 50 people) and to repeat that symbol as many times as necessary (Figure 2.18a). Such a map is easily understood because the dots give the map reader a visual impression of the pattern. Sometimes pictorial symbols, for example, human figures or oil barrels, are used instead of dots.

If the range of the data is great, the cartographer may find it inconvenient to use a repeated symbol. For example, if one port handles 50 or 100 times as much tonnage as another, that many more dots would have to be placed on the map. To circumvent this problem, the cartographer can choose a second method and use proportional symbols. The size of the symbol is varied according to the quantities represented. Thus, if bars are used, they can be shorter or longer as necessary. If squares or circles are used, the *area* of the symbol ordinarily is proportional to the quantity shown (Figure 2.18b).

There are occasions, however, when the range of the data is so great that even circles or squares would take up too much room on the map. In such cases, three-dimensional symbols, usually spheres or cubes, are used, and their *volume* is proportional to the data. Unfortunately, many people fail to perceive the added dimension implicit in volume,

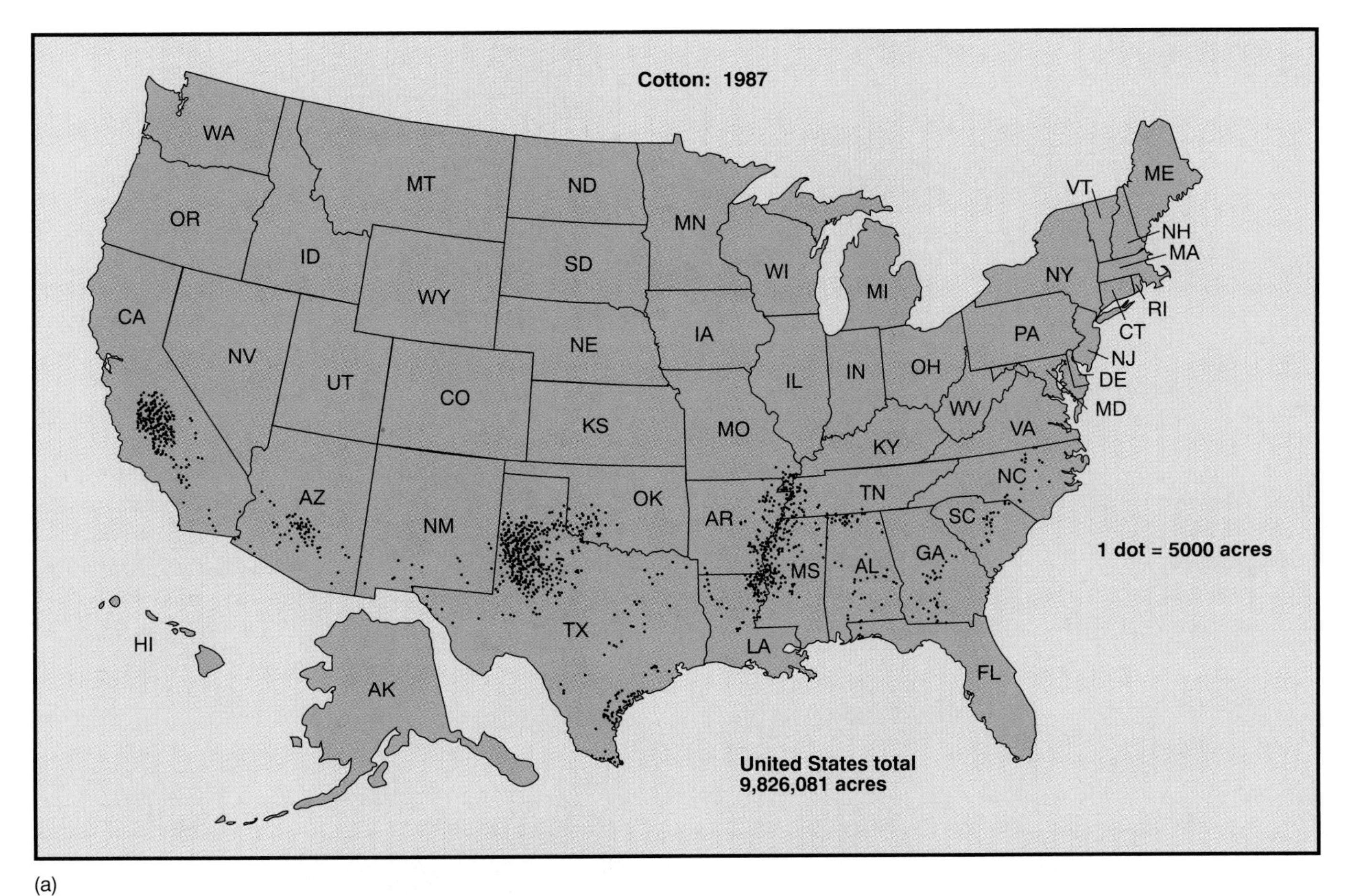

(a)

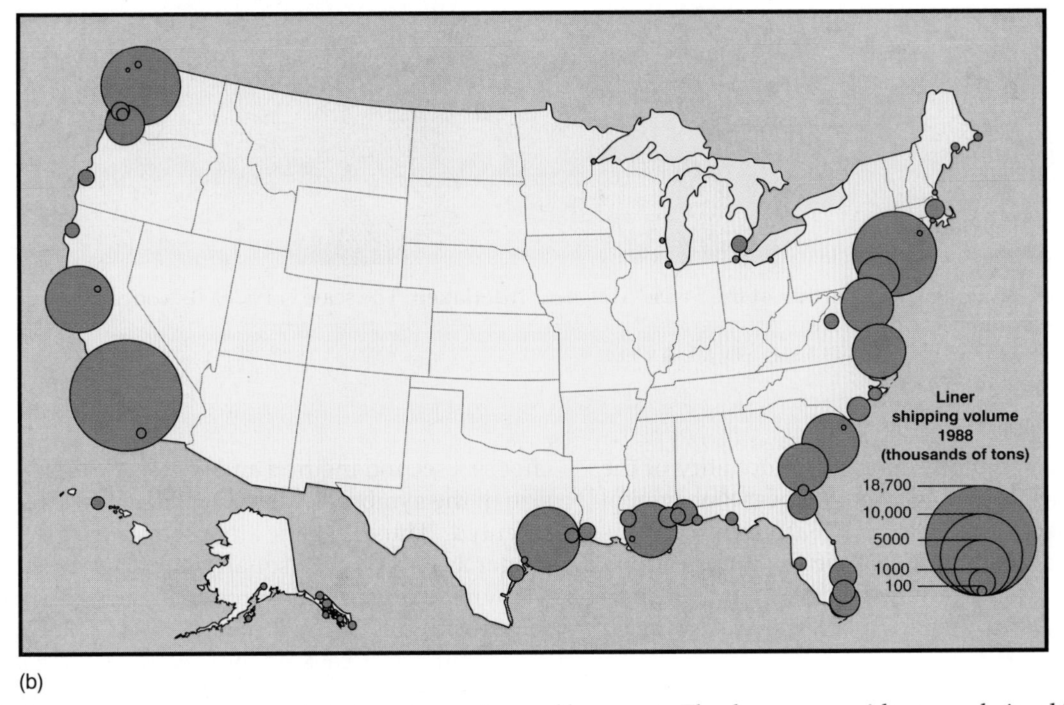

(b)

FIGURE 2.18 (a) **Cotton acreage in 1987.** Dots have been allocated by county. The dot map provides a good visual impression of the distribution and relative density of a phenomenon—in this case, how the amount of cotton grown varies from place to place. (b) **Liner shipping volume for U.S. ports in 1988.** On this map, the area of the circle is proportional to the total shipping volume of a port: imports, exports, and in-transit shipments. The scale in the bottom right corner aids the reader in interpreting the map.

(a) From *1987 Census of Agriculture,* Vol. 2, Subject Series, Part 1, *Agricultural Atlas of the United States,* page 159. U.S. Department of Commerce, Bureau of the Census, June 1990.

and most cartographers do not recommend the use of such symbols (Figure 2.19).

Area Symbols

Maps showing distribution by area fall into two general categories: those showing differences in kind and those showing differences in quantity. Atlases contain numerous examples of the first category, such as patterns of religions, languages, political entities, vegetation, or types of rock. Normally, different colors or patterns are used for different areas, as shown in Figure 2.20.

Maps that show how the *amount* of a phenomenon varies from area to area are called **choropleth maps.** The term is derived from the Greek words *choros* (place) and *pleth* (magnitude or value). The quantities shown may be absolute numbers (e.g., the population of counties) or derived values such as percentages, ratios, rates, and densities (e.g., population density by county). The data are grouped into a limited number of classes, each represented by a distinctive color, shade, or pattern. Figure 2.21 is an example of a choropleth map. In this case, the areal units are states. Other commonly used subdivisions are counties, townships, cities, and census divisions.

Three main problems characterize maps that show the distribution of a phenomenon in an area:

1. they give the impression of uniformity to areas that may actually contain significant variations;
2. boundaries attain unrealistic precision and significance, implying abrupt changes between areas when, in reality, the changes may be gradual;
3. unless colors are chosen wisely, some areas may look more important than others.

An interesting variation of area maps is the **cartogram,** a map simplified to present a single idea in a diagrammatic way. Figures 2.12 and 2.22 are examples of cartograms. Depending on the idea that the cartographer wishes to convey, the sizes and shapes of areas may be altered, distances and directions may be distorted, and contiguity may or may not be preserved.

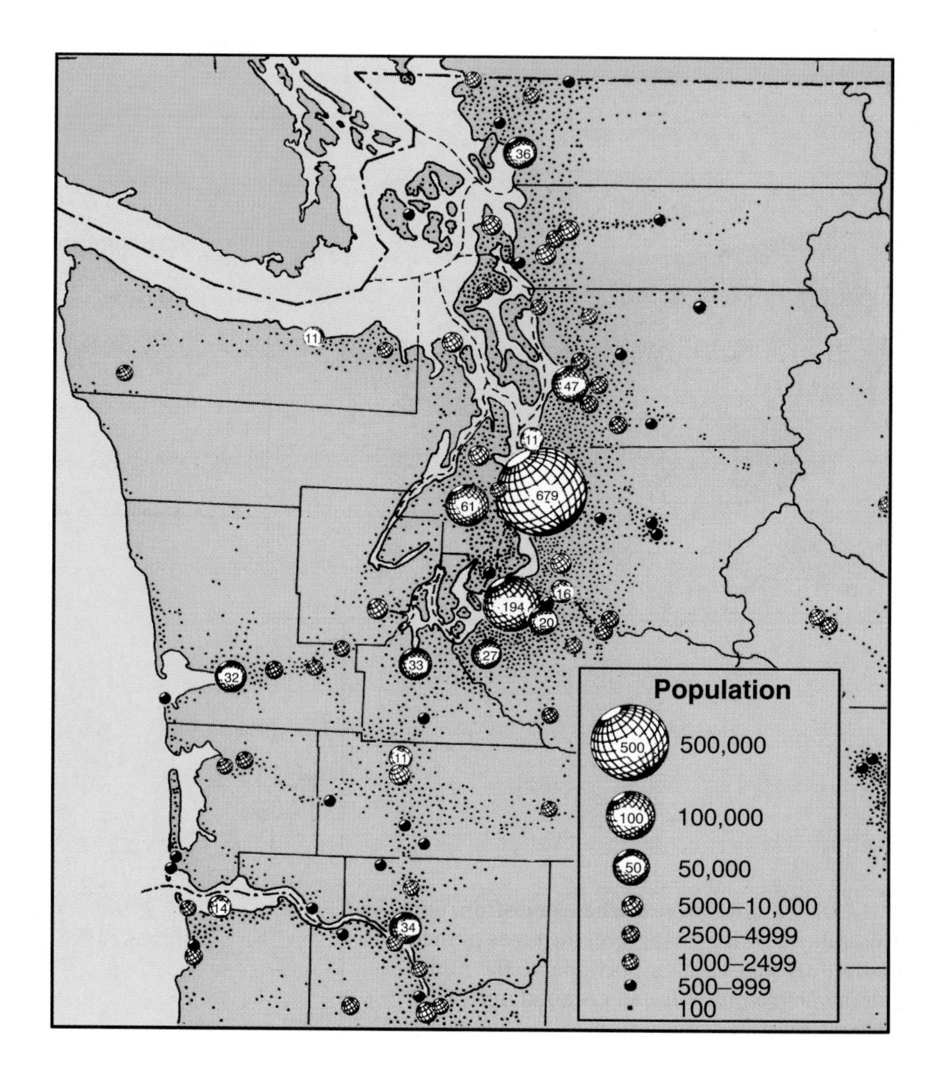

FIGURE 2.19 **The distribution of population in western Washington in 1950.** The cartographer used proportional spheres to represent large urban population concentrations. Although proportional volume symbols enable the mapmaker to represent a wide range of data, the symbols often are misinterpreted.

FIGURE 2.20 **Language regions of Africa.** Maps such as this one may give the false impression of uniformity within a given area, for example, that Bantu and no other languages are spoken over much of Africa. Such maps are intended to represent only the predominant language in an area.

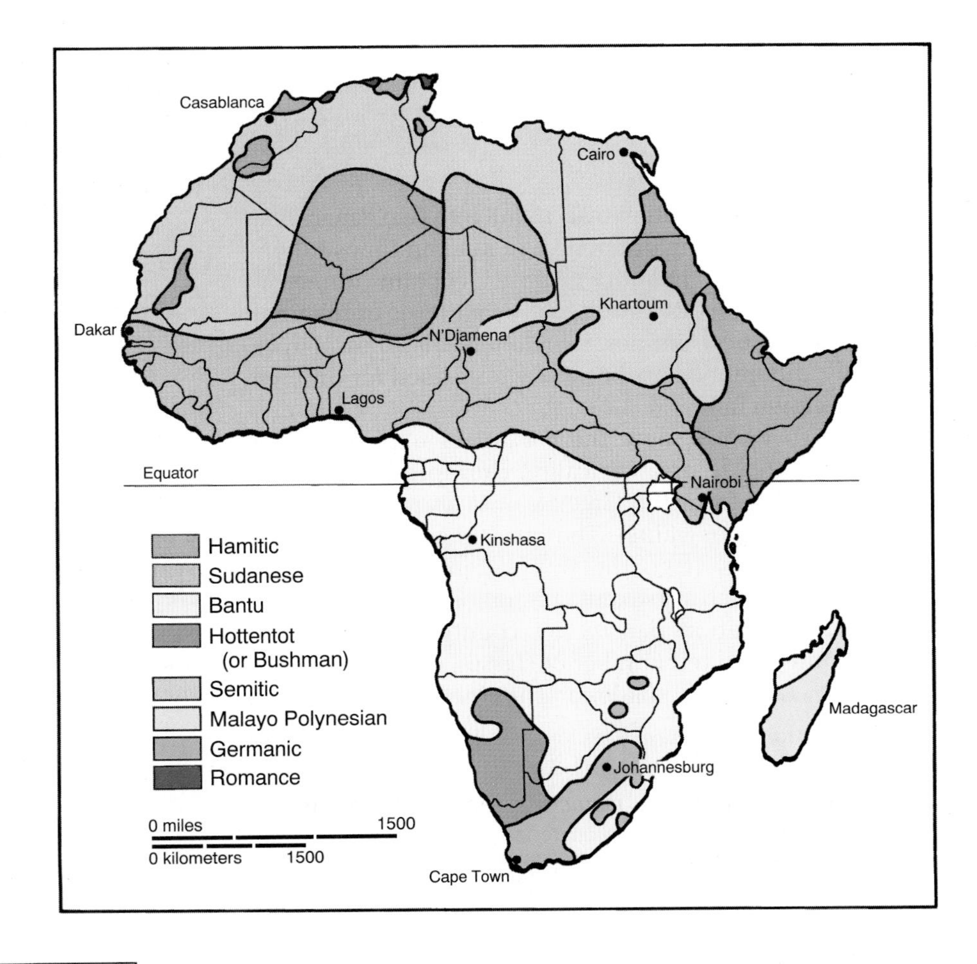

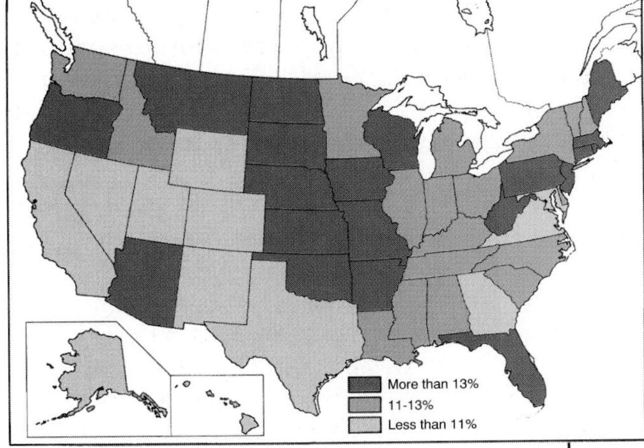

FIGURE 2.21 **Choropleth map showing percentage of population 65 years of age and over by state, 1990.** Quantitative variation by area is more easily visualized in map than in tabular form.

Source: U.S. Bureau of the Census.

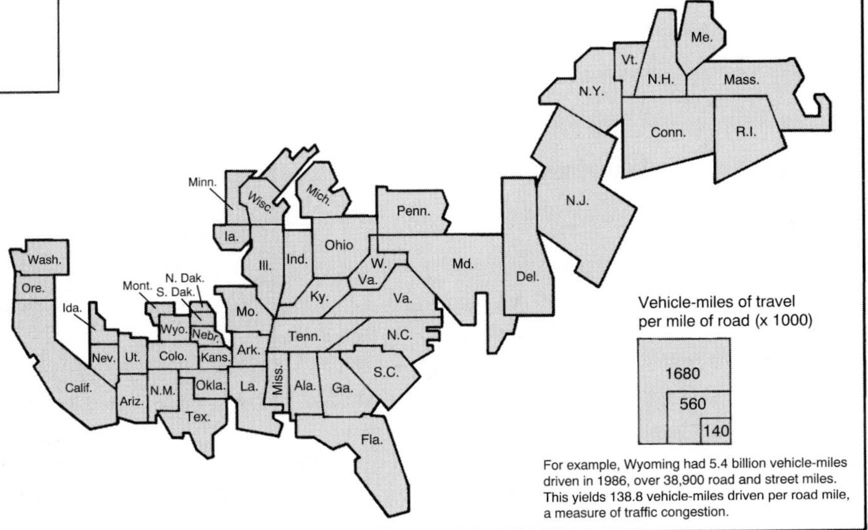

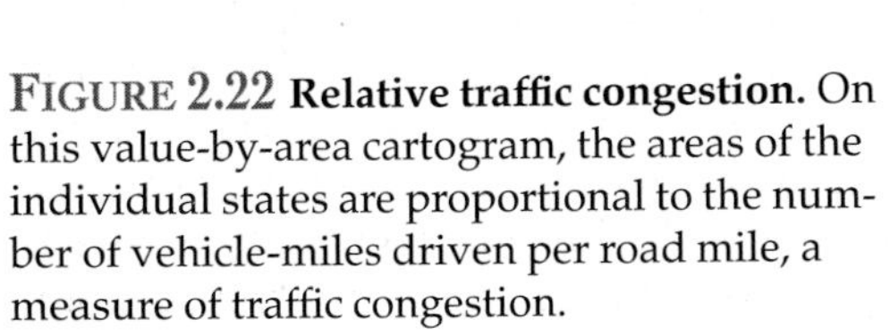

FIGURE 2.22 **Relative traffic congestion.** On this value-by-area cartogram, the areas of the individual states are proportional to the number of vehicle-miles driven per road mile, a measure of traffic congestion.

Cartogram designed by Bernard J. van Hamond.

Line Symbols

Some lines on maps do not have numerical significance. The lines representing rivers, political boundaries, roads, and railroads, for example, are not quantitative. They are indicated on maps by such standardized symbols as the ones shown here and in Figure 2.15.

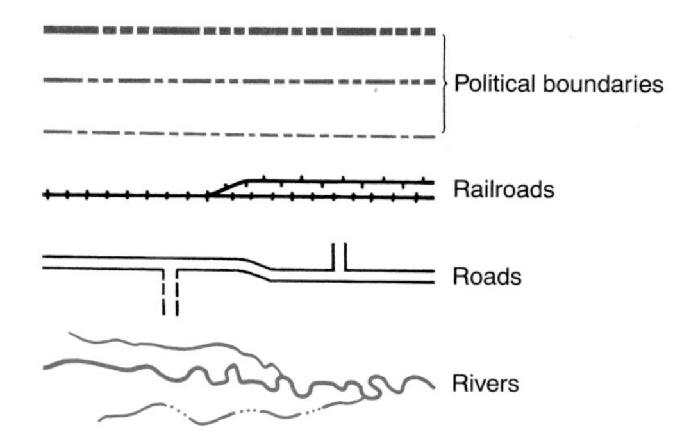

Often, however, lines on maps do denote specific numerical values. Contour lines that connect points of equal elevation above mean sea level are a kind of **isoline,** or line of constant value. Other examples of isolines are *isohyets* (equal rainfall), *isotherms* (equal temperature), and *isobars* (equal barometric pressure). The implications of isolines as regional boundaries are discussed in Chapter 13.

Flow-line maps are used to portray traffic or commodity flows along a given route, usually a waterway, highway, or railway. The location of the route taken, the direction of movement, and the amount of traffic can all be depicted. The amount shown may be either the total or a per mile figure. In Figure 2.23, the width of the flow line is proportional to the number of interstate migrants in the United States. Another flow-line map appears in Figure 11.6.

Remote Sensing

When topographic maps were first developed, it was necessary to obtain the data for them through fieldwork, a slow and tedious process, which involved relating a given point on the earth's surface to other points by measuring its distance, direction, and altitude. The technological developments that have taken place in aerial photography since the 1930s have made it possible to speed up production and greatly increase the land area represented on topographic maps. Aerial photography is only one of a number of remote sensing technqiues now employed.

Remote sensing is a relatively new term, but the process it describes—detecting the nature of an object from a distance—has been taking place for well over a century. Soon after the development of the camera, photographs were made from balloons and kites. Even carrier pigeons wearing miniature cameras that took exposures automatically at set intervals were used to take aerial photographs of Paris. The airplane, first used for mapping in the 1930s, provided a platform for the camera and the photographer so that it was possible to take photographs from planned positions.

Aerial Photography

Although there is now a variety of sensing devices, aerial photography employing cameras with returned film remains a widely used remote sensing technique. Mapping from the air has certain obvious advantages over surveying from the ground, the most evident being the bird's-eye view that the cartographer obtains. Using stereoscopic devices, the cartographer can determine the exact slope and size of features, such as mountains, rivers, and coastlines. Areas that are otherwise hard to survey, such as mountains and deserts, can be mapped easily from the air. Furthermore, millions of square miles can be surveyed in a very short time. Aerial photographs must, of course, be interpreted by using such clues as the size, shape, tone, and color of the recorded objects before maps can be made from them. Maps based on aerial photographs can be made quickly and revised easily so that they are kept up-to-date. With aerial photography, the earth can be mapped more accurately, more completely, and more rapidly than ever before.

In 1975, the U.S. Department of the Interior instituted the National Mapping Program to improve the collection and analysis of cartographic data and to prepare maps that would assist decision makers who deal with resource and environmental problems. One of the first goals of the program was to achieve complete coverage of the country with orthophotographic imagery for all areas not already mapped at the scale of 1:24,000.

An **orthophotomap** is a topographic map on which natural and cultural features are portrayed by color-enhanced photographic images with certain map symbols (contours, boundaries, names) added as needed for interpretation. The prefix *ortho* (from the Greek word *orthos,* meaning "correct") is used here because the aerial photographs have been rectified, or corrected, to minimize distortion, giving the orthophoto the geometric characteristics of a map. Note in Figure 2.24 that in contrast to the conventional topographic map, the photograph is the chief means of representing information. Orthophotomaps have a variety of uses, aiding forest management, soil surveys, geological investigations, flood hazard and pollution studies, and city planning.

Standard photographic film detects reflected energy within the visible portion of the electromagnetic spectrum (Figure 2.25). Although they are invisible, *near-infrared* wavelengths can be recorded on special sensitized infrared film. Discerning and recording objects that are not visible to the human eye, infrared film has proved particularly useful for the classification of vegetation and hydrographic

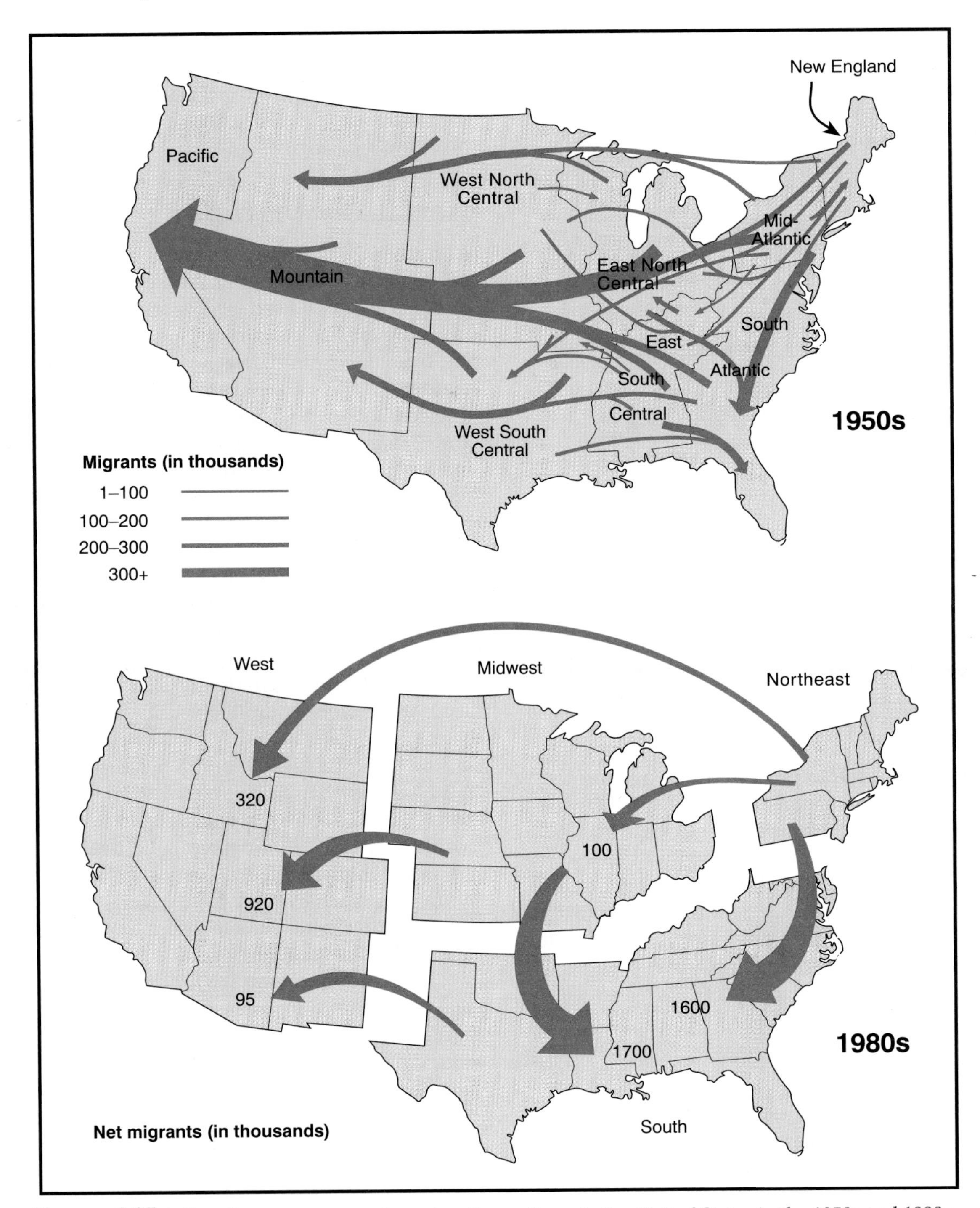

FIGURE 2.23 **A flow-line map** contrasting migration patterns in the United States in the 1950s and 1980s.

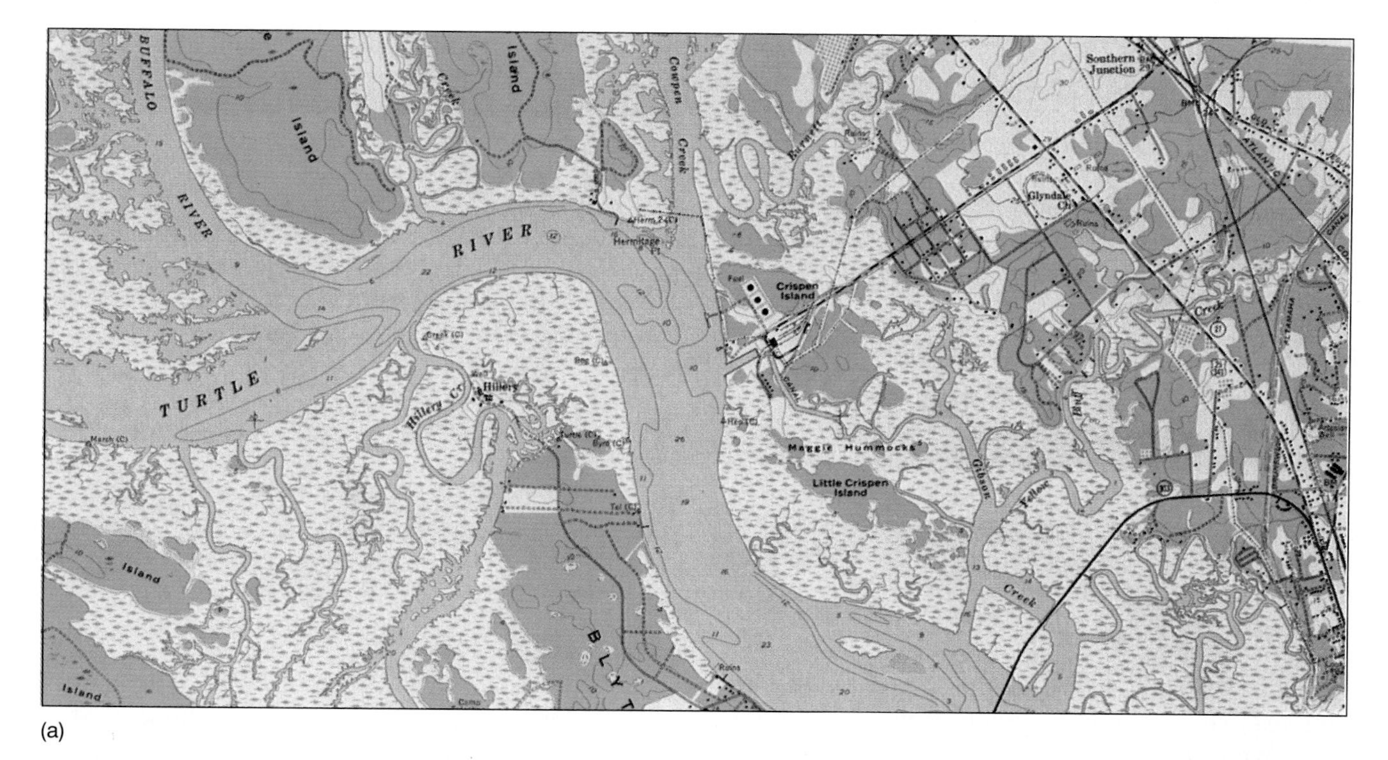

(a)

(b)

FIGURE 2.24 **Topographic map (a) and orthophotomap (b) of a portion of the Brunswick West quadrangle in southeastern Georgia.** Because aerial photographs show subtle topographic detail in areas of very low relief, orthophotomaps are especially well suited for portraying marshlands and coastal zones. Orthophoto data were also used in constructing Figure 2.32.

Source: U.S. Geological Survey.

features. Color-infrared photography yields what are called **false-color images,** "false" because the film does not produce an image that appears natural. For example, leaves of healthy vegetation have a high infrared reflectance and are recorded as red on color-infrared film, while unhealthy or dormant vegetation appears as blue, green, or gray. Clear water appears as black, but sediment-laden water is light blue.

Nonphotographic Imagery

For wavelengths longer than 1.2 micrometers (a micrometer is 1 one-millionth of a meter) on the electromagnetic spectrum, sensing devices other than photographic film must be used. **Thermal scanners,** which sense the energy emitted by objects on earth, are used to produce images of thermal radiation (Figure 2.26). That is, they record the longwave radiation (which is proportional to surface temperature) emitted by water bodies, clouds, and vegetation as well as by buildings and other structures. Unlike conventional photography, thermal sensing can be employed during nighttime as well as daytime, giving it military applications. It is widely used for studying various aspects of water resources, such as ocean currents, water pollution, surface energy budgets, and irrigation scheduling.

Operating in a different band of the electromagnetic spectrum, **radar** systems can also be used during the day or night. They transmit pulses of energy toward objects and sense the energy reflected back. The data are used to create images, such as that shown in Figure 2.27, which was produced by radar equipment mounted on an airplane. Because radar can penetrate clouds and vegetation, as well as darkness, it is particularly useful for monitoring the locations of airplanes, ships, and storm systems and for detecting variation in soil moisture conditions.

Radar imagery has even proved to be valuable for archaeological study. Radar signals have penetrated beneath desert sands to discover ancient ruins along the Silk Road of western China, reveal the course of the Nile River in antiquity, and pinpoint the location of the buried lost city of Ubar in Oman. Wealthy and beautiful, and the center of the frankincense trade, Ubar had disappeared and been covered by sands more than 2000 years ago. Radar signals can also penetrate the foliage of dense forests to reveal the underlying topography. Using radar maps as their guide, archaeologists recently discovered remnants of previously unknown temples and waterways, and evidence of an earlier culture, in Angkor, the ancient capital of the Khmer empire in what is now Cambodia.

Satellite Imagery

In the last 30 years, both manned and unmanned spacecraft have supplemented the airplane as the vehicle for imaging the terrain. Concurrently, many steps have been taken to automate mapping, including the use of electronic mapping techniques, automatic plotting devices, and automatic data processing. Many images are now taken either from continuously orbiting satellites, such as those in the U.S. Landsat series and the French SPOT series, or from manned spacecraft flights, such as those of the Apollo and Gemini missions. Among the advantages of satellites are the speed of coverage and the fact that views of large regions can be obtained.

In addition, because they are equipped to record and report back to the earth information from parts of the electromagnetic spectrum that are outside the range of human eyesight, these satellites enable us to map the invisible. A number of agencies in the United States, Japan, and Russia have launched satellites specifically to monitor the weather. Data obtained by satellites have greatly improved weather forecasting and the tracking of major storm systems and in the process, saved countless lives. The satellites are one of the sources of the weather maps shown daily on television and in newspapers.

The **Landsat satellites,** first launched in 1972, take about 1 hour and 40 minutes to orbit the earth and can provide repetitive coverage of almost the entire globe every 18 days. Rather than recording data photographically, the

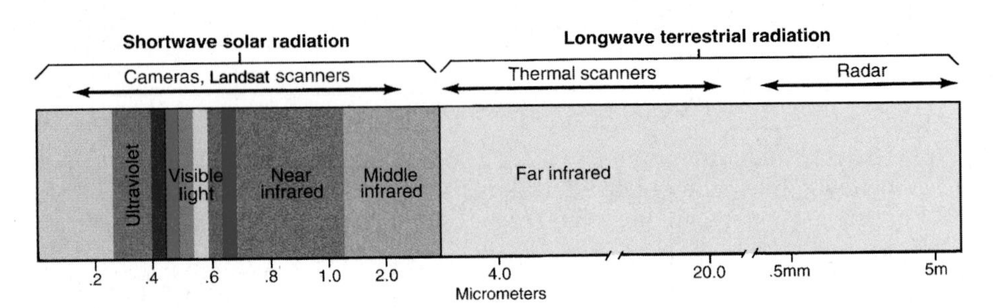

FIGURE 2.25 **Wavelengths of the electromagnetic spectrum in micrometers.** One micrometer equals 1 one-millionth of a meter. Sunlight is made up of different wavelengths. The human eye is sensitive to only some of these wavelengths, the ones we see in the colors of the rainbow. Although invisible, near-infrared wavelengths can be recorded on special sensitized film and by scanners on satellites. The scanners measure reflected light in both the visible and near-infrared portions of the spectrum. Wavelengths longer than 4.0 micrometers characterize terrestrial radiation.

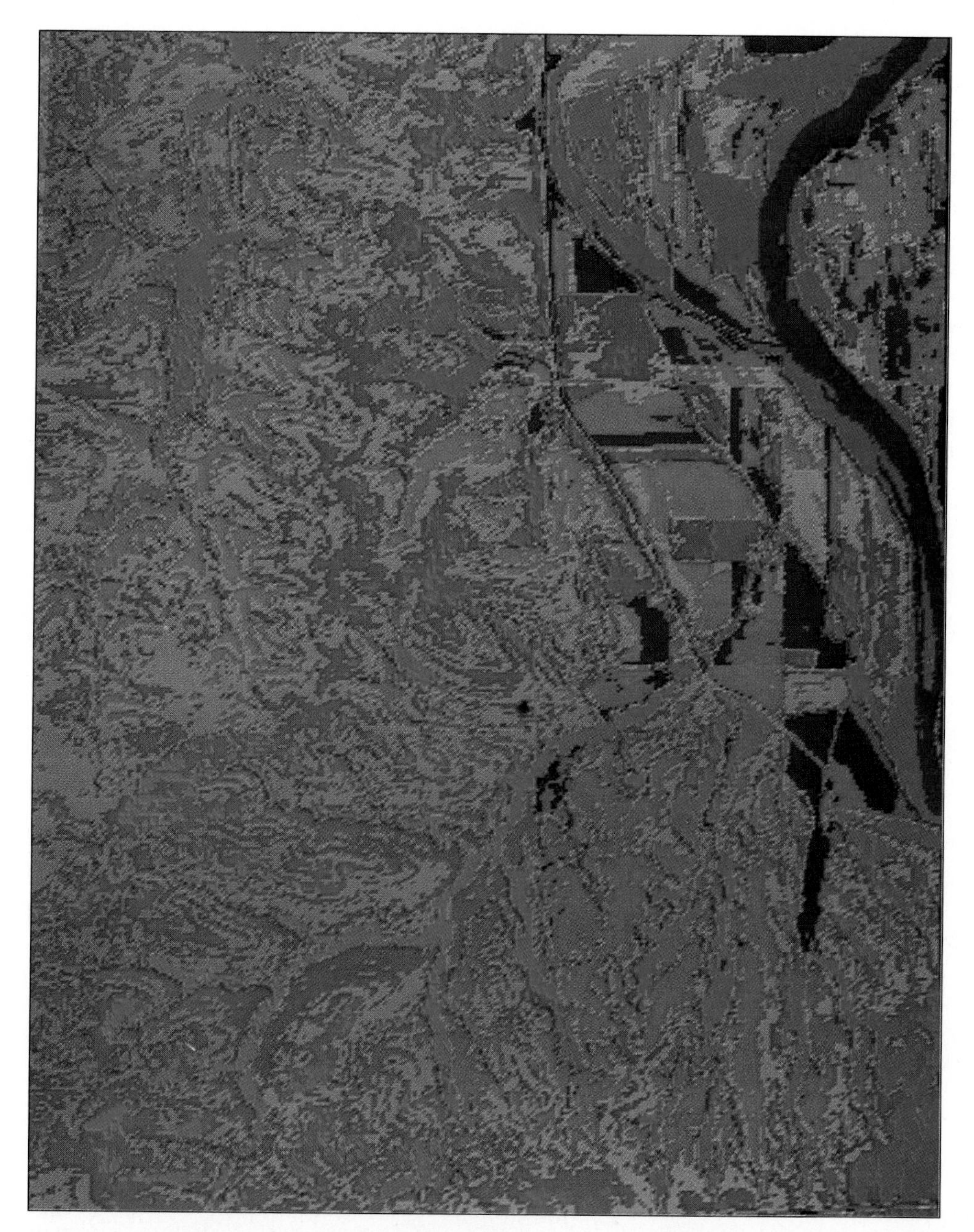

FIGURE 2.26 **A thermal radiation image of an area south of Manhattan, Kansas, June 6, 1987.** The warm (red, orange) colors represent areas with the highest surface temperatures—the bare fields in the upper right corner of the image and the road. The higher elevations in the sandstone region on the left-hand side of the image are warmer than the lowlands and woodlands at lower elevations.

CESAR Laboratory, San Diego State University, courtesy of Allen Hope.

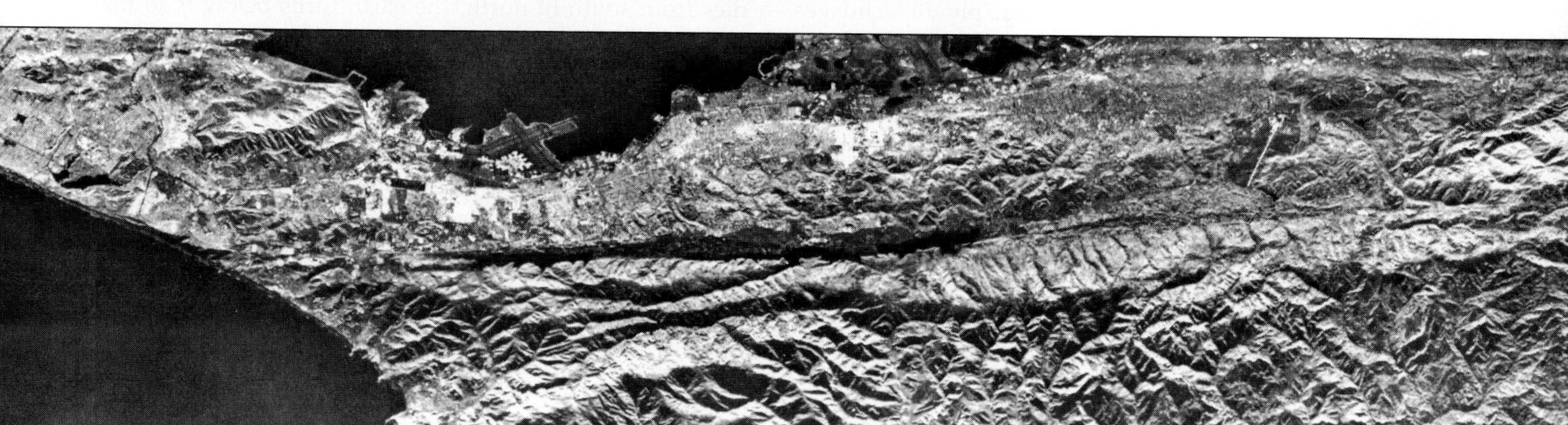

FIGURE 2.27 **Radar image of the southern portion of the San Francisco peninsula.** The San Andreas fault is clearly visible.

NASA.

On-Line MAPS

Maps are extremely well suited to the image-intensive nature of the World Wide Web (WWW). Maps of the world, various regions, and countries are all available on-line. In addition, most map companies and GIS providers have sites, as do government agencies.

A good place to start browsing is with one of the index sites, all of them with the majority of linkage references in common.

Map-Related Web Sites is compiled and updated by the Perry-Castañeda Library Map Collection at the University of Texas at Austin. It is one of the most comprehensive map collections on the Internet.

http://www.lib.utexas.edu/Libs/PCL/Map_collection/map_sites/map_sites.html

Among the several useful *Infomine: Internet Resource Collections* of the University of California Library is "Maps and GIS."

http://logic17.ucr.edu/

George Mason University hosts *Cartography Resources: Commercial Mapping* as part of the WWW Virtual Library. This searchable website also provides access to related lists on remote sensing and GIS.

http://geog.gmu.edu/gess/jwc/cartogrefs.html

The *ProsperNet* "Maps" home page contains a brief but varied set of map agency and map collection links.

Many websites offer instructional material about specific cartographic projections, problems, properties, or techniques. In addition to those cited here, others will be located as you pursue the leads and links suggested in the index sites noted above.

Again, the Perry-Castañeda Library of the University of Texas, Austin, has a useful site to visit. Using the same URL address given above, select "Cartographic References." Each of the listed "Map Projections" has a brief history of the projection, its features, use, and appearance.

http://www.lib.utexas.edu/Libs/PCL/Map_collection/map_sites/map_sites.html

Hunter College Map Projections home page contains information relating to map projections. Also found: a number of projection identification "games" in the Globes and Amusements section and many links to various sources of maps with unique projections.

http://everest.hunter.cuny.edu/mp/

The Peters Projection site is dedicated to the projection created by Arno Peters to address some of the problems of existing maps. There are comparisons of the Peters and Mercator projections and a discussion of how the maps we use influence the way we view the world.

http://www.webcom.com/~bright/petermap.html

Mapquest by GeoSystems Global Corporation is an on-line interactive map service that includes a brief history of cartography and the "Map Room" Map Skills department with exercises to help you learn the grid system, scale, and map projections.

http://www.geosys.com/

In its *The Geographer's Craft* program, the Department of Geography at the University of Texas, Austin, has a detailed "Overview" discussion of map projections. That home page also

Landsat satellite relays electronic signals to receiving stations, where computers convert them into photolike images that can be adjusted to fit special map projections. Composite images can be made by combining information from different wavelengths of light energy.

Landsat carries scanning instruments that pick up sunlight reflected by foliage, water, rocks, and other objects. One sensor, the multispectral scanner (MSS), covers the visible and near-infrared range, from 0.4 to 1.1 micrometers. The other sensor, the higher resolution thematic mapper (TM), has 7 wavebands, several in the thermal ranges up to 11.7 micrometers.

Landsat's cameras are capable of resolving objects about 30 meters (100 ft) apart. Even sharper images are yielded by the French *SPOT* (Satellite Probatoire d'Observation de la Terre) *satellite* launched in 1986. Its sensors can show objects that are less than 10 meters (33 ft) apart and can also produce three-dimensional pictures. Like Landsat, the SPOT satellite is in a polar orbit, which means that as it flies from south to north, the earth turns below it so that each orbit covers a strip of surface adjacent to the previous one. SPOT images the earth at the same local time on consecutive passes and repeats its pattern of successive ground tracks at 26-day intervals.

A Landsat image of southern California appears in Figure 2.28. Others appear in Figures 1.8 and 3.23. Analyses of Landsat images have practical applications in agriculture and forest inventory, land use classification, identification of geologic structures and associated mineral deposits, and monitoring of natural disasters.

Mapping is only one of the applications of remote sensing, which has also proven to be an effective method of conducting resource surveys and monitoring the natural environment. Geologists have found remote sensing to be particularly useful in conducting resource surveys in desert and remote areas. For example, information about vegetation

On-Line *CONTINUED*

provides access to related courses on coordinate systems, geodetic datums, and global positioning systems.

http://www.utexas.edu/depts/grg/gcraft/notes/mapproj/mapproj.html

How Far Is It? is an interactive program that determines the latitude and longitude of a place and calculates the distance between any two cities in the United States.

http://www.indo.com/distance

The U.S. Geological Survey has a number of *Fact Sheets* to assist in map reading. Among the topics discussed are *Map Scales, Finding Your Way with Map and Compass,* and *What Do Maps Show?*

http://info.er.usgs.gov/fact-sheets/map-scales/index.html

http://info.er.usgs.gov/fact-sheets/finding-your-way/finding-your-way.html

http://www.usgs.gov/education/teacher/what-do-maps-show/?/index.html

Geographic information systems and remote sensing techniques are of increasing interest to student and professional geographers. Index sites that can help you get acquainted with the Internet resources on those topics include the following. Many others will be encountered as you investigate.

Geographic Information Systems WWW Resource is an alphabetically arranged list of GIS-related sites maintained by the University of Edinburgh. Of special note is the *ARC/INFO Tutorial,* a step-by-step guide to basic ARC/INFO commands that create, edit, and produce geographic data.

http://www.geo.ed.ac.uk/home/giswww.html

A worldwide lisiting of "GIS and Remote Sensing" sites is a major component of Utrecht University's *Nice Geography Sites.*

http://www.frw.ruu.nl:80/nicegeo.html#gis

The WWW Virtual Library: Remote Sensing Organizations is a complete listing of remote sensing organizations throughout the world, along with its companion site, *Other Information.*

http://www.vtt.fi/aut/ava/rs/virtual/organizations.html

http://www.vtt.fi/aut/ava/rs/virtual/other.html

Iowa State University's *Global Positioning System Resources* provides an extensive overview of GPS technology, with links to other GPS sites.

http://www.cnde.iastatre.edu/gps.html

Various governmental agencies and private companies maintain home pages, only two of which are noted here.

USGS National Mapping Information, a home page of the U.S. Geological Survey, provides accurate and up-to-date cartographic data for the United States. Its "Educational Resources" option gives access to earth science educational material for teachers and (K–12) students.

http://mapping.usgs.gov/

Canada's *National Atlas Information Service* is responsible for developing and maintaining an authoritative synthesis of the geography of Canada.

http://ellesmere.ccm.emr.ca/

or folding patterns of rocks can be used to help identify likely sites for mineral or oil prospecting. Remote sensing imagery has been used to monitor a variety of environmental phenomena, including water pollution, the effects of acid rain, and rain forest destruction. As noted earlier, weather satellites can monitor frontal systems and are a valuable contribution to worldwide weather forecasting. Because remotely sensed images can be used to calculate such factors as biomass production and rates of transpiration and photosynthesis, they are invaluable for modeling relationships between the atmosphere and the earth's surface.

Geographic Information Systems

A major development in cartography is the use of computers to assist in making maps. Within the last 25 years, computers have become an integral part of almost every stage of the cartographic process, from the collection and recording of data to the production and revision of maps. Although the initial cost of equipment is high, the investment is repaid in the more efficient and more accurate production and revision of maps.

Computers are at the heart of what has come to be known as a **geographic information system (GIS)**. A GIS is a computer-based set of procedures for assembling, storing, manipulating, analyzing, and displaying geographically referenced information. Any data that can be located spatially can be entered into a GIS. The five major components of a GIS are:

1. a data input component that converts maps and other data from their existing form into digital or computer-readable form;
2. a data management component used to store and retrieve data;

(a)

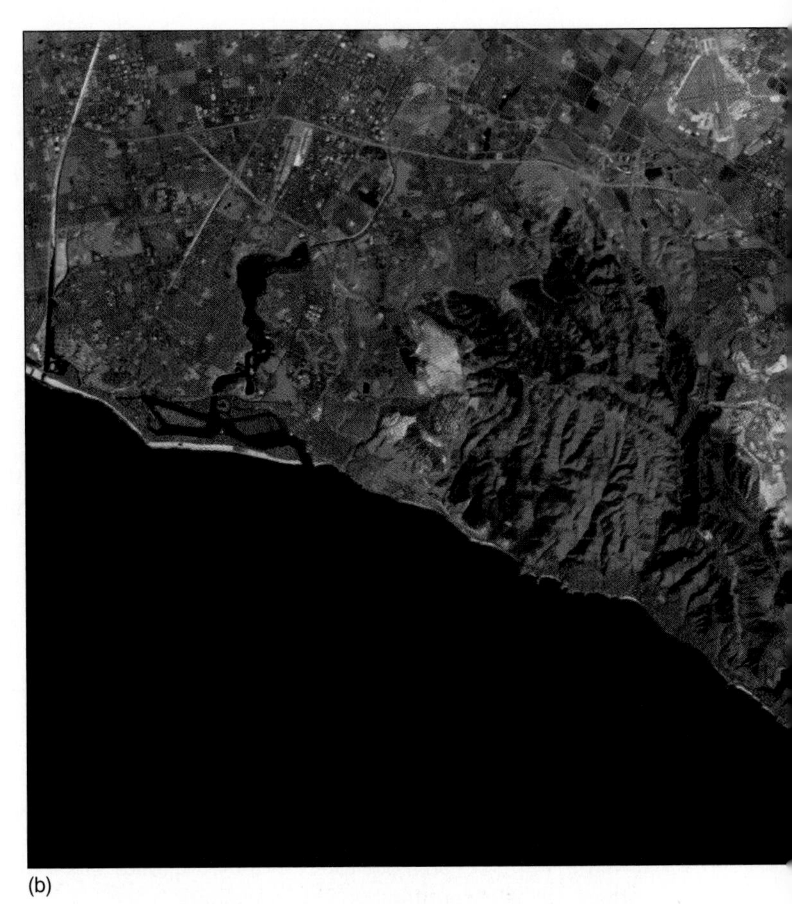

(b)

FIGURE 2.28 **Landsat images of Laguna Beach, California.** (a) This image was taken on October 10, 1992. Healthy vegetation on the San Joaquin Hills (right of center) is shown in green, urban areas are shown in shades of blue and light purple, and the Pacific Ocean is dark blue. The white areas in the lower left corner are clouds. (b) This Landsat image was taken on October 29, 1993, two days after a firestorm swept across the hills, destroying more than 400 houses and businesses. The burned area is in magenta.

Landsat imagery courtesy of EOSAT, Lanham, Maryland.

3. data manipulation functions that allow data from disparate sources to be used simultaneously;
4. analysis functions that enable the extraction of useful information from the data;
5. a data output component that makes it possible to visualize maps and tables on the computer monitor or as hard copy (such as paper).

The first step in developing a GIS is to create the **geographic database** (a digital record of geographic information) from such sources as maps, field surveys, aerial photographs, satellite imagery, and so on. Geographic information is of two types: locational (spatial) data and nonlocational (attribute) data. *Spatial* data describe the location of, connections, and relationships among point, line, and area features. The second type of geographic information is descriptive. It gives the characteristics of the point, line, and area features in terms of certain *attributes,* which may be either qualitative (e.g., the types and names of roads in a given area) or quantitative (e.g., the widths of the roads). Place names are one kind of attribute (see Cambodia or Kampuchea? Burma or Myanmar?). Attributes are stored in the computer in tabular form as sets of numbers and characteristics to be accessed when they are needed to create maps or perform analyses. Usually, the tables are related to each other through some similar information found in several tables (a relational database).

Spatial data typically are stored in a GIS in one of two ways (Figure 2.29). In a *raster* format, a map is subdivided into rows of tiny uniform cells, generally square in shape. The cells may be as small as 0.001 square inch. The location of each cell or *pixel* (picture element) is defined by its row and column numbers, and the location of a feature is defined by the positions of the cells it occupies. The value stored for each cell indicates the type of subject or condition found at that location, with the condition pertaining to the entire cell.

The second type of storage utilizes a *vector* format. Vector digital data are recorded as the coordinates of points, lines (a series of point coordinates), or areas (shapes bounded by lines). As indicated by Figure 2.29b, vector data files look more like traditional hand-drawn maps than do

CAMBODIA OR KAMPUCHEA? BURMA OR MYANMAR? CHANGES IN PLACE NAMES ON MAPS

We can virtually guarantee that by the time you are reading this textbook, it will be out of date in at least one respect—some boundaries and place names will have changed during the time between completion of the manuscript and its publication. All boundaries and many place names reflect political conditions. As they change, so, often, do the names associated with them.

The end of European colonialism in Africa, for example, resulted in the formation of many new countries, most of which eventually adopted African names. Thus, Southern Rhodesia renamed itself Zimbabwe, while Northern Rhodesia became Zambia. The Republic of Upper Volta was renamed Burkina Faso, and the Belgian Congo became Zaire until 1997, when a new government renamed the country the Democratic Republic of the Congo.

Other name changes result from the formation or dissolution of federations. In 1990, the Yemen Arab Republic and the People's Democratic Republic of Yemen merged to become the Republic of Yemen, and East and West Germany united to form the Federal Republic of Germany. The dissolution of the Union of Soviet Socialist Republics (USSR) in 1990–91 gave impetus to wholesale name changes. The fifteen former Soviet Socialist Republics each declared themselves independent republics. Slight changes in spelling or terminology occurred as, for example, Byelorussia was renamed Belarus, Moldavia became Moldova, and the Ukraine became simply Ukraine. The fall from favor of former political idols was reflected in changes in the names of many cities. For example, Leningrad took back its original name of St. Petersburg, and Gorky became Nizhni Novgorod once again.

Confusion occurs if name changes aren't accepted by other countries. The United States, for example, has not officially recognized the changes of Cambodia to Kampuchea, Burma to Myanmar, and its capital from Rangoon to Yangon. Note that both old and new names of those countries appear on the map inside the front cover of this book.

Changes in place names are of particular interest to map publishers such as the National Geographic Society, Rand McNally, and the various federal agencies that produce maps, which strive to keep them from becoming obsolete. The *Board on Geographic Names (BGN)*, in the Department of the Interior, formulates place-name policy and provides the official names to be applied to places both in the United States and abroad by federal mapping agencies. Most large private map producers also follow the board's recommendations. The BGN has its counterparts in other countries, and the United Nations Group of Experts on Geographical Names is attempting to standardize the treatment of place names.

Finally, for place names within the United States, the USGS produces the *Geographic Names Information System,* the country's official database for place names. Containing more than two million entries, it is a geographic database in digital form intended for use in computer-assisted cartography. It provides information for all known places, geographic features, and areas in the United States that are identified on maps by a proper name.

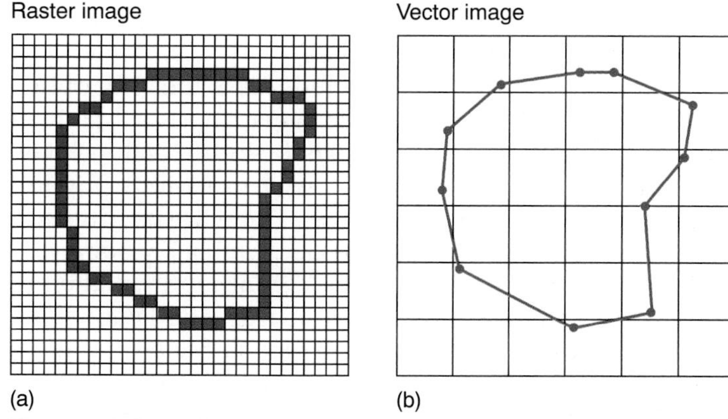

FIGURE 2.29 **Raster and vector versions of the same image.** A raster image is presented as a set of individual cells. The smaller the cells, the more the raster image will resemble a vector image.
Redrawn with permission from John Campbell, *Map Use and Analysis.* 3d ed., Fig. 19.1, p. 298. Copyright 1998 WCB/Mc-Graw Hill, Dubuque, Iowa.

raster data files, because vector digitizing preserves the form of point, line, and area features.

No matter which format is used to represent the spatial component of geographic information, the data must be entered into a GIS. One commonly used method is manual digitizing (Figure 2.30a). Using a special digitizing table, a person retraces the map with a handheld cursor that records the position of objects on the map. After digitizing, attribute codes are added to identify what each digitized symbol or line represents. A faster technique is to use an *optical scanner,* which consists of a scanning head and a large cylindrical drum on which the map is placed (Figure 2.30b). The scanning head takes readings as the drum rotates. When one rotation is completed, the head moves a fraction of a millimeter and repeats the process. The scanner produces a digital image.

Once geographic information is in the computer in digital form, the data can be manipulated, analyzed, and displayed with a speed and precision not otherwise possible.

(a)

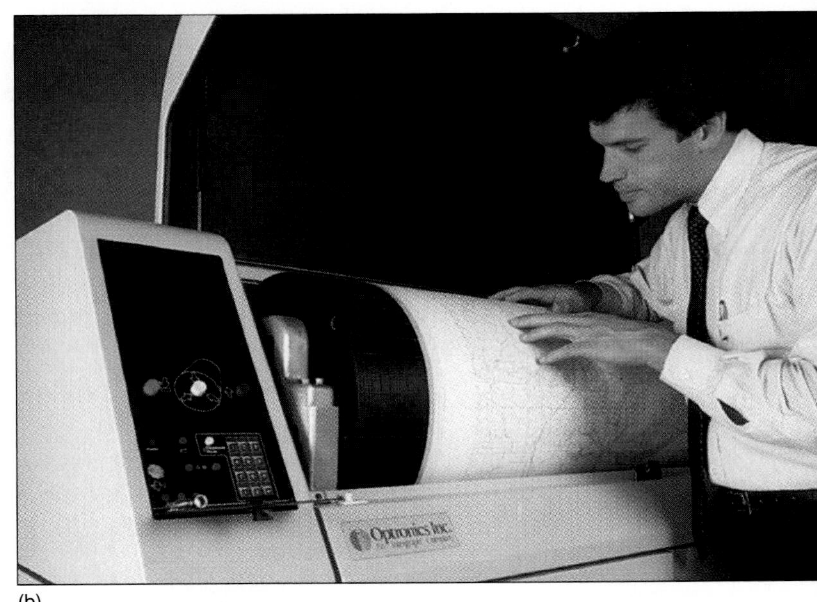

(b)

FIGURE 2.30 **Map digitization.** (a) Maps can be converted to digital form by using a handheld computer mouse to trace each feature. (b) Electronic scanning devices automatically convert map information into digital form.

(a) Courtesy of Environmental Systems Research Institute, Inc.

(b) Courtesy of Intergraph Corporation.

Because computers can process millions of facts in seconds, they are particularly useful for researchers who need to analyze many variables simultaneously. The development of geographic information systems has deemphasized the use of the map to store information and enabled researchers to concentrate on using the map for analyzing and communicating spatial information. With the appropriate software, a computer operator can display any combination of data, showing the relationships among variables almost instantly (Figure 2.31). In this sense, a GIS allows an operator to ask interesting questions and to generate maps that were virtually impossible to create 20 years ago.

Suppose that a state needed to select a site for a radioactive waste disposal facility. With a database drawn from maps of earthquake intensity, oil and gas field locations, sand and gravel resources, aquifers, and other distributions, the suitability of sites could be evaluated. Areas prone to earthquakes or subsidence (sinking or settling) could be immediately excluded from consideration, while those with such characteristics as simple geologic structure, surficial materials of low permeability, and deep aquifers would be more favorably assessed.

GIS operations can produce several types of output: displays on a computer monitor, listings of data, or hard copy. When a map is to be produced, the operator can quickly call up the desired data. Geographic information systems are particularly useful for revising existing maps because outdated data—for example, population sizes—can be modified or replaced easily. Hundreds of software packages are now available for map production, particularly for commonly used projections. They enable the cartographer to produce plots of boundaries, coastlines, and grid systems at any desired scale.

A number of bureaus and agencies in the federal government prepare data for use in geographic information systems. The USGS has completed digital coverage for transportation and hydrographic features found on its 1:100,000-scale maps and for most information found on its 1:2,000,000-scale maps. Currently, it is digitizing categories of data from its 7.5-minute topographic quadrangles and hopes to complete the National Digital Cartographic Data Base by the year 2000. It will include all of the information that now appears on the agency's maps.

The USGS helped the Census Bureau develop a GIS that incorporates census data as well as various physical and cultural landscape features. The Census Bureau's TIGER (Topologically Integrated Geographic Encoding and Referencing system) files are the digital representation of the United States used in the 1990 census. The TIGER database identifies every street in the United States and gives attribute information such as census tract, block numbers, and ZIP codes. A process called *file matching* enables users to integrate TIGER files with other information and thus allows for analysis using the full range of GIS functions.

Many bureaus and agencies at the local, regional, and state levels are especially attracted by the usefulness of GIS.

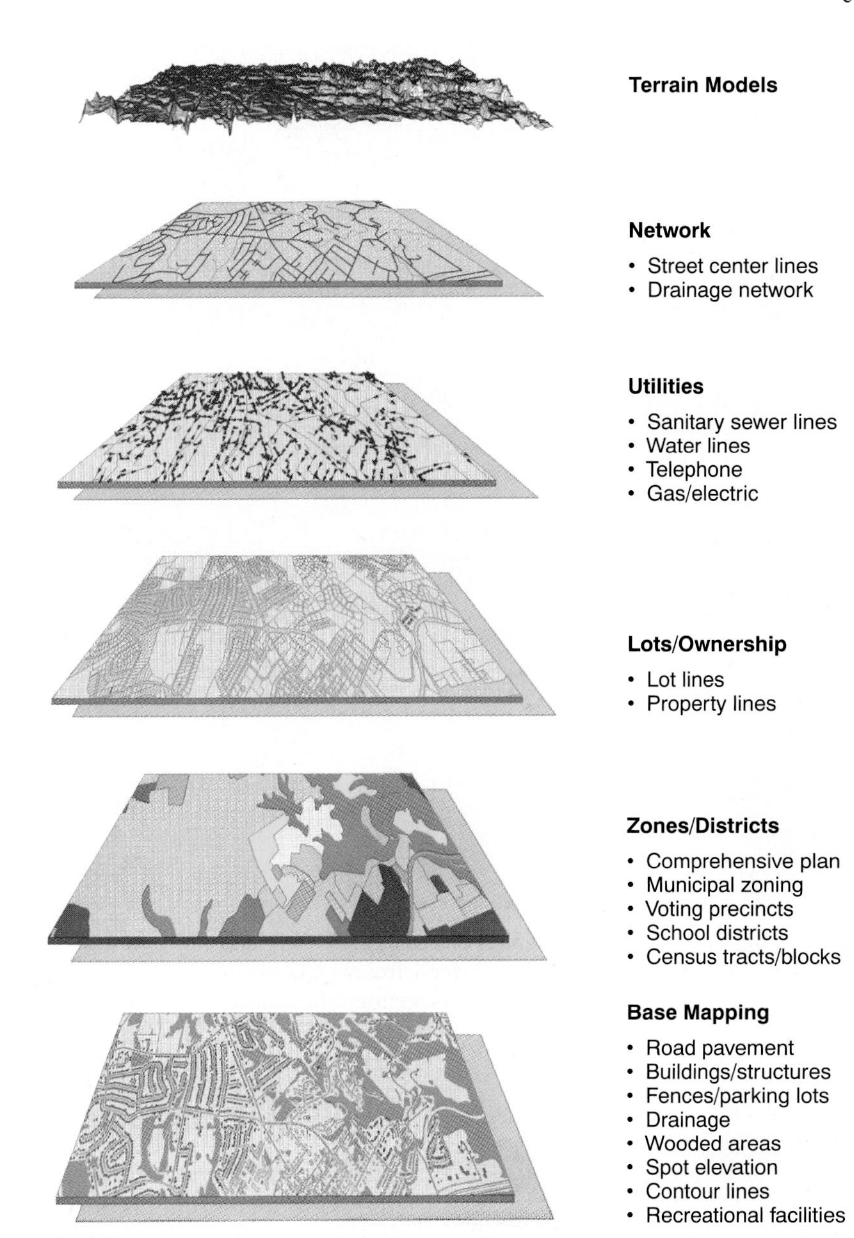

FIGURE 2.31 **Information layering** is the essence of a GIS. Map information that has been converted to digital data is stored in the computer in different data "layers." A GIS enables the user to combine just the layers that are desired to produce a composite map and to analyze how those different variables relate to each other.

These include such departments as highway and traffic control, public utilities, and planning. The database for a city GIS might contain, among others, political boundaries, census tracts, a building inventory, addresses, streets, traffic lights, sewer lines, and gas mains (Figure 2.32).

A large number of companies in the private sector are also using computerized mapmaking systems. Among others, oil and gas companies, restaurant chains, soft-drink bottlers, and car rental companies rely on GIS systems to perform such diverse tasks as identifying drilling sites, picking locations for new franchises, analyzing sales territories, and calculating optimal driving routes. Some of the many uses of GIS are illustrated in the chapters that follow. Note, for example, their application to earthquake monitoring (Chapter 3) and environmental degradation (Chapter 5).

FIGURE 2.32 **Downtown Omaha, Nebraska.** Omaha is one of many cities that is developing a comprehensive GIS that will consolidate spatial data collected by the various departments and agencies of the local government. The data will include street names and addresses, zoning maps, tax assessment files, and public works facilities. This map was prepared from topographic and orthophoto data.
Courtesy of Analytical Surveys, Inc., Colorado Springs, Colorado.

SUMMARY

In this chapter, we have not attempted to discuss all aspects of the field of cartography. Mapmaking and map design, systems of land survey, map compilation, and techniques of map reproduction are among the topics that have been deliberately omitted. Our intent has been to introduce those aspects of map study that will aid in map reading and interpretation and to hint at new techniques of map creation and of the design and purpose of geographic information systems.

Maps are among the oldest and most basic means of communication. They are as indispensable to the geographer as words, photographs, or quantitative techniques of analysis. Geographers are not unique in their dependence on maps. People involved in the analysis and solution of many of the problems facing the world also rely on them. Environmental protection, control of pollution, conservation of natural resources, and land use planning are just a few of the issues that call for the accurate representation of elements on the earth's surface.

The roots of modern mapping lie in the 17th century. Three key developments that made it both desirable and possible to map the surface features of the earth were the rediscovery of the works of earth scientists of ancient Greece, transoceanic voyages, and the invention of the printing press.

The grid system of latitude and longitude is used to locate points on the earth's surface. Latitude is the measure of distance north and south of the equator, while longitude is the angular distance east or west of the prime meridian. Both latitude and longitude are given in degrees, which for greater precision can be subdivided into minutes and seconds.

All systems of representing the curved earth on a flat map distort one or more earth features. Any given projection will distort area, shape, distance, and/or direction. Cartographers select the projection that best suits their purpose and may elect to use an equal-area, conformal, or equidistant projection, or one that shows correctly the directions from a single point to all others. Many useful projections have none of these qualities, however.

Among the most accurate and most useful large-scale maps are the topographic quadrangles produced by a country's chief mapping agency. They contain a wealth of information about both the physical and cultural landscape and are used for a variety of purposes.

In recent years, remote sensing techniques have resulted in the faster and more accurate mapping of the earth. Both aerial photography and satellite imagery can discern reflected energy in the visible and near-infrared portions of the electromagnetic spectrum. Applications of remote sensing include mapping, monitoring the environment, and conducting resource surveys. The need to store, process, and retrieve the vast amounts of data generated by remote sensing has spurred the development of geographic information systems.

As you read the remainder of this book, note the many different uses of maps. For example, notice in Chapter 3 how important maps are to your understanding of the theory of continental drift; in Chapter 7, how maps aid geographers in identifying cultural regions; and in Chapter 8, how behavioral geographers use maps to record people's perceptions of space.

Key Words

azimuthal projection 25
cartogram 39
choropleth map 39
conformal projection 25
conic projection 27
contour line 33
cylindrical projection 26
developable surface 26
equal-area projection 24
equidistant projection 25
equivalent projection 24
false-color image 44
geographic database 48
geographic information system (GIS) 47
Global Positioning System (GPS) 25
globe properties 26
great circle 26
grid system 22
International Date Line 23
isoline 41
Landsat satellite 44
latitude 23
longitude 23
map projection 24
Mercator projection 26
orthophotomap 41
Peters projection 31
planar projection 28
prime meridian 23
radar 44
remote sensing 41
rhumb line 27
scale 32
shaded relief 33
small circle 26
thermal scanner 44
topographic map 33

For Review and Consideration

1. What important map and globe reference purpose does the *prime meridian* serve? Is the prime or any other meridian determined in nature or devised by humans? How is the prime meridian designated or recognized?
2. What happens to the length of a degree of longitude as one approaches the poles? What happens to a degree of latitude between the equator and the poles?
3. From a world atlas, determine in degrees and minutes the locations of New York City; Moscow, Russia; Sydney, Australia; and your hometown.
4. List at least five properties of the globe grid. Examine the projections used in Figures 2.5, 2.8, and 2.9a. In what ways do each of these projections adhere to or deviate from globe grid properties?
5. Briefly make clear the differences in properties and purposes of *conformal, equivalent,* and *equidistant* projections. Give one or two examples of the kinds of map information that would best be presented on each type of projection. Give one or two examples of how misunderstandings might result from data presented on an inappropriate projection.
6. In what different ways may *map scale* be presented? Convert the following map scales into their verbal equivalents.
   1:1,000,000 1:63,360 1:12,000
7. What is the purpose of a *contour line?* What is the *contour interval* on Figure 2.16? What landscape feature is implied by closely spaced contours?
8. What kinds of data acquisition are suggested by the term *remote sensing?* Describe some ways in which different portions of the spectrum can be sensed. To what uses are remotely sensed images put?
9. What are the basic components of a *geographic information system?* How is spatial information recorded in a *geographic database?* What are some of the uses of computerized mapmaking systems?
10. The table below gives the gross domestic product per capita of selected European countries (in U.S. dollars, 1995). On outline maps or as cartograms, represent the data in two different ways.

Austria	$19,000	Luxembourg	24,800
Belgium	19,500	Netherlands	19,500
Denmark	21,700	Norway	24,500
France	20,200	Portugal	11,000
Germany	17,900	Spain	14,300
Greece	9,500	Sweden	20,100
Ireland	15,400	Switzerland	22,400
Italy	18,700	United Kingdom	19,500

Selected References

American Cartographic Association. Committee on Map Projections. *Choosing a World Map: Attributes, Distortions, Classes, Aspects*. Special Publication No. 2. Falls Church, Va.: American Congress on Surveying and Mapping, 1988.

—. *Matching the Map Projection to the Need*. Special Publication No. 3. Falls Church, Va.: American Congress on Surveying and Mapping, 1991.

—. *Which Map Is Best? Projections for World Maps*. Special Publication No. 1. Falls Church, Va.: American Congress on Surveying and Mapping, 1986.

Brown, Lloyd A. *The Story of Maps*. Boston: Little, Brown, 1949; reprint ed., New York: Dover Publications, 1977.

Campbell, James. *Introduction to Remote Sensing*. 2d ed. New York: Guilford Publications, Inc., 1996.

Campbell, John. *Map Use and Analysis*. 3d ed. Dubuque, Iowa: WCB/McGraw-Hill, 1998.

Dent, Borden C. *Cartography: Thematic Map Design*. 4th ed. Dubuque, Iowa: Wm. C. Brown Publishers, 1996.

Leick, Alfred. *GPS Satellite Surveying*. 2d ed. New York: John Wiley & Sons, Inc., 1995.

Lillesand, Thomas M., and Ralph W. Kiefer. *Remote Sensing and Image Interpretation*. 3d ed. New York: John Wiley & Sons, Inc., 1994.

MacEachren, Alan M. *How Maps Work*. New York: Guilford Publications, Inc., 1995.

Miller, Victor, and Mary Westerbrook. *Interpretation of Topographic Maps*. Columbus, Ohio: Charles E. Merrill, 1989.

Monmonier, Mark. *Drawing the Line: Tales of Maps and Cartocontroversy*. New York: Henry Holt, 1995.

—. *How to Lie with Maps*. 2d ed. Chicago: Univ. of Chicago Press, 1996.

Muehrcke, Phillip C., and Julianna O. Muehrcke. *Map Use: Reading, Analysis, Interpretation*. 3d ed. Madison, Wis.: JP Publications, 1992.

Robinson, Arthur H., et al. *Elements of Cartography*. 6th ed. New York: John Wiley & Sons, Inc., 1995.

Stillwell, H. Daniel. "Global Distortion: Is It Time to Retire the Mercator Projection?" *Mercator's World*, Vol. 2, no. 1 (Sept./Oct. 1997):54–59.

Thrower, Norman J. W. *Maps and Civilization: Cartography in Culture and Society*. Chicago: Univ. of Chicago Press, 1996.

Tyner, Judith. *Introduction to Thematic Cartography*. Englewood Cliffs, N.J.: Prentice Hall, 1992.

Wilford, John Noble. "Revolutions in Mapping." *National Geographic*, Vol. 193, no. 2 (February 1998):6–39.

Wood, Dennis. *The Power of Maps*. New York: Guilford Press, 1992.

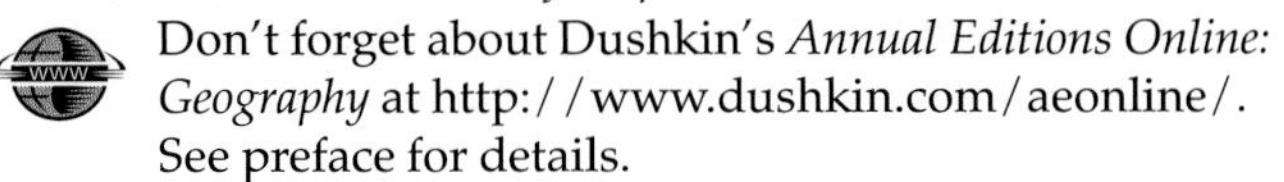

Don't forget about Dushkin's *Annual Editions Online: Geography* at http://www.dushkin.com/aeonline/. See preface for details.

PART ONE

THE EARTH SCIENCE TRADITION

For nearly a month the mountain had rumbled, emitting puffs of steam and flashes of fire. Within the past week, on its slopes and near its base, deaths had been recorded from floods, mud slides, and falling rock. A little after 8:00 on the morning of May 8, 1902, the climax came for volcanic Mount Pelée and the thriving port of Saint Pierre on the island of Martinique. To the roar of one of the biggest explosions the world has ever known and to the clanging of church bells aroused in swaying steeples, a fireball of gargantuan size burst forth from the upper slope of the volcano. Lava, ash, steam, and superheated air engulfed the town, and 29,933 people met their death.

More selective, but for those victimized just as deadly, was the sudden "change in weather" that struck central Illinois on December 20, 1836. Within an hour, preceded by winds gusting to 70 miles per hour, the temperature plunged from 40°F to –30°F. On a walk to the post office through slushy snow, Mr. Lathrop of Jacksonville, Illinois, found, just as he passed the Female Academy, that "the cold wave struck me, and as I drew my feet up the ice would form on my boots until I made a track . . . more like that of [an elephant] than a No. 7 boot." Two young salesmen were found frozen to death along with their horses; one "was partly in a kneeling position, with a tinderbox in one hand, a flint in the other, with both eyes open as though attempting to light the tinder in the box." Others died, too—inside horses that had been disemboweled and used as makeshift shelters; in fields, woods, and on roads both a short distance and an eternity from the travelers' destinations.

Fortunately, few human–environmental encounters are so tragic as these. Rather, the physical world in all its spatial variation provides the constant and generally passive background against which the human drama is played. It is to that background that physical geographers, acting within the earth science tradition of the discipline, direct their attention. Their primary concern is with the natural rather than the cultural landscape and with the interplay between the encompassing physical world and the activities of humans.

The interest of those working within the earth science tradition of geography, therefore, is not necessarily in the physical sciences as an end in themselves but in physical processes as they create landscapes and environments of significance to humankind. The objective is not solely to trace the physical and chemical reactions that have produced a piece of igneous rock or to describe how a glacier has scoured the rock; it is the relationship of the rock to people. Geographers are interested in what the rock can tell us about the evolution of the earth as our home or in the significance of certain types of rock for the distribution of mineral resources and fertile soil.

The questions that physical geographers ask do not usually deal with the catastrophic, though catastrophe is an occasional element of human–earth-surface relationships. Rather they ask questions that go to the root of understanding the earth as the home and the workplace of humankind. What does the earth as our home look like? How have its components been formed? How are they changing, and what changes are likely to occur in the future through natural or human causes? How are environmental features, in their spatially distinct combinations, related to past, present, and prospective human use of the earth? These are some of the basic questions that geographers who follow the earth science tradition attempt to answer.

The three chapters in Part I of our review of geography deal, in turn, with landforms, weather and climate, and human impact on the environment. Although the subject matter of each clearly identifies it with the fundamental earth science background of the discipline, a brief word about the content and purpose of each chapter is appropriate.

The treatment of so vast a field as landforms (Chapter 3) must be highly selective within an introductory text. The aim here is simply to summarize the great processes by which landforms are created and to depict the general classes of features resulting from those processes without becoming overly involved in scientific reasoning and technical terms.

In Chapter 4, "Physical Geography: Weather and Climate," the major elements of the atmosphere—temperature, precipitation, and air pressure—are discussed, and their regional generalities and associations in patterns of world climates are presented. The study of weather and climate adds coherence to our understanding of the earth as the home of people. Frequent reference will be made to the

—*Continued next page*

Eroded lava cliffs along the Na Pali coast, Kauai, Hawaii.
Frans Lanting/Minden Pictures

Continued from previous page—

climatic background of human activities in later sections of our study.

Part I concludes with a look at the imprint humans have made upon the earth's surface and resources. Chapter 5, "Human Impact on the Environment," explores some of the implications of a basic geographic observation: In the interaction between humans and the physical environment, humans are the active and frequently destructive agents of change. Human exploitation of the earth results in both intended and unanticipated changes to the natural environment.

The focus of Part I, therefore, is on the earth as the environment and habitat of humans. Both the diversity and reciprocal nature of human–environmental interactions are recurring themes not only of the following three chapters but in all geographic inquiry.

Physical Geography: Landforms

CHAPTER

3

Mt. Edith Cavell, Jasper National Park, Alberta, Canada.
© Richard During/Tony Stone Images

Although too early for sunbathers and snorkelers, the Hawaiian Islands will have a new island to add to their collection, which contains such scenic beauties as Oahu, Maui, and Kauai. It is Loihi, one-half mile below sea level, just 17 miles from the big island of Hawaii. Because the speed of its ascent must be measured in geologic time, it probably will not appear above the water surface for another million or so years. It is a good example, however, of the ceaseless changes that take place on the earth's surface. A stationary hot spot that occasionally spews forth lava lies under the constantly westward-moving Pacific plate, giving rise to the chain of Hawaiian islands. As the westernmost of the islands erode and sink below sea level, new islands arise at the eastern end. In Loihi's most recent explosion in 1996, scientists feared that a tidal wave would be set off at the surface that could devastate the islands, including Honolulu and Waikiki Beach. Fortunately, this was not the case.

Humans on their trip through life come into contact with the ever changing, active, moving physical environment. Most of the time, we are able to live comfortably with the changes, but when a freeway is torn apart by an earthquake, or floodwaters force us to abandon our homes, we suddenly realize that we spend a good portion of our lives trying to adapt to the challenges the physical environment has for us.

For the geographer, things just will not stand still—not only little things, like icebergs or new islands rising out of the sea, or big ones, like exploding volcanoes changing their shape and form, but also monstrous things such as continents that wander about like nomads, and ocean basins that expand, contract, and split in the middle like worn-out coats. It is a fascinating story, this changing home of humans.

Geologic time is long, but the forces that give shape to the land are timeless and constant. Processes of creation and destruction are continually at work to fashion the seemingly eternal structure upon which humankind lives and works. Two types of forces interact to produce those infinite local variations in the surface of the earth called *landforms:* (1) forces that push, move, and raise the earth's surface; and (2) forces that scour, wash, and wear down the surface. Mountains rise and are then worn away. The eroded materials—soil, sand, pebbles, rocks—are transported to new locations and help to create new landforms. How long these processes have worked, how they work, and their effects are the subject of this chapter.

Much of the research needed to create the story of landforms results from the work of geomorphologists. *Geomorphology,* a branch of the fields of geology and physical geography, is the study of the origin, characteristics, and development of landforms. Geomorphology emphasizes the study of the various processes that influence the erosion, transportation, and deposition of materials. A modern thrust is in the area of the interrelationships between plant and animal life and landforms.

In a single chapter, we can only begin to explore the many and varied contributions of geomorphologists. After discussing the context within which landform change takes place, we consider the forces that are building up the earth's surface and then review the forces wearing it down. Since most earth surface-changing processes occur over long periods of time, amounting to millions of years, perspective can be gained by developing a sense of what is meant by the time span within which the drama of earth change takes place.

Geologic Time

The earth was formed about 4.7 billion years ago. When we think of a person who lives to be 100 years old as having had a long life, it becomes clear that the earth is incredibly old indeed. Because our usual concept of time is dwarfed when we speak of billions of years, it is useful to compare the age of the earth with something more familiar.

Imagine that the height of the World Trade Center in New York City represents the age of the earth. The twin towers are 110 stories, or 412 meters (1353 ft), tall. In relative terms, even the thickness of a piece of paper laid on the rooftop would be too great to represent an average person's lifetime. Of the total building height, only 4.7 stories represent the 200 million years that have elapsed since the present ocean basins began to form. The first hominids, or humanlike creatures, made their appearance on earth about 15 million years ago, or the height equivalent of one-third of a story. Earth history is so long, and involves so many major geologic events, that scientists have divided it into a series of recognizable, distinctive stages. These stages are depicted in Figure 3.1.

At this moment, the landforms on which we live are ever so slightly being created and destroyed. The processes involved have been in operation for so long that any given location most likely was the site of ocean and land at a number of different times in its past. Many of the landscape features on earth today can be traced back several hundred million years. The processes responsible for building up and tearing down those features are occurring simultaneously, but usually at different rates.

In the last 40 years, scientists have developed a useful framework within which one can best study our constantly changing physical environment. Their work is based on the early 20th-century geological studies of Alfred Wegener, who proposed the theory of **continental drift.** He believed that all landmasses were once united in one super continent and that over many millions of years the continents broke away from each other, slowly drifting to their current positions. Although Wegener's theory was initially

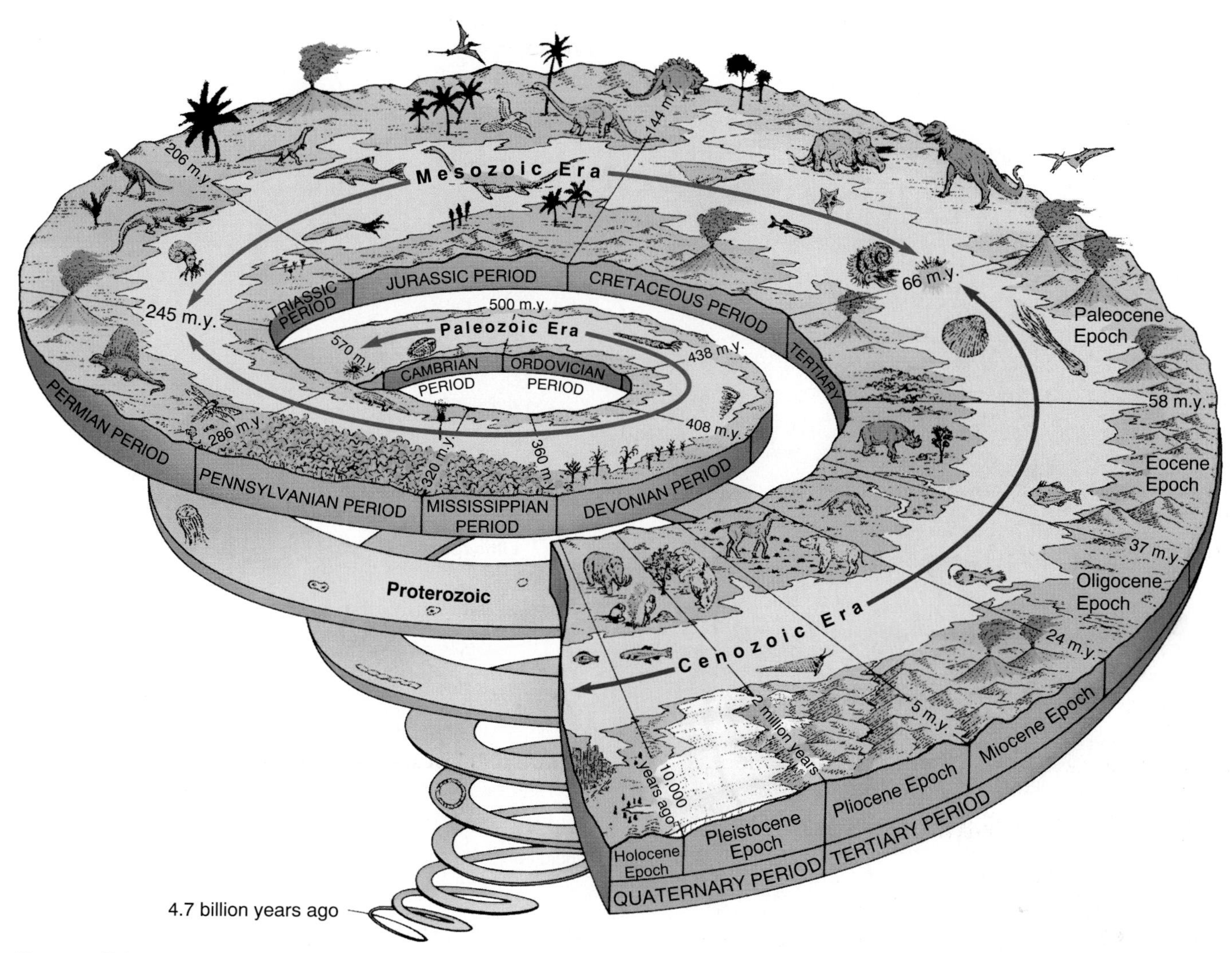

FIGURE 3.1 **A diagrammatic history of the earth.** The sketch depicts some of the known characteristics of the named geologic periods.

Source: After U.S. Geological Survey publication, Geologic Time.

rejected outright, new evidence and new ways of rethinking old knowledge have led to wide acceptance in recent years by earth scientists of the idea of moving continents. Wegener's ideas were a forerunner of the broader **plate tectonics** theory, which encompasses continental drift, seafloor spreading, the movement of lithospheric plates, and the deformation of the earth's crust. The word tectonics comes from the Greek word for builder, *tekton.*

Movements of the Continents

The landforms mapped by cartographers are only the surface features of a thin cover of rock, the earth's *crust.* Above the core and the lower mantle of the earth is a partially molten plastic layer called the **asthenosphere** (Figure 3.2). The asthenosphere supports a thin but strong solid shell of rocks called the **lithosphere,** of which the outer, lighter portion is the earth's crust. The crust consists of one set of rocks found below the oceans and another set that makes up the continents.

The lithosphere is broken into about twelve large, rigid plates, each of which, according to the theory of plate tectonics, slides or drifts very slowly over the heavy semimolten asthenosphere. A single plate may contain both oceanic and continental crust. Figure 3.3 shows that the North American plate, for example, contains the northwest Atlantic Ocean and most, but not all, of North America. The peninsula of Mexico (Baja California) and part of California are on the Pacific plate.

Scientists are not certain why lithospheric plates move. One reasonable theory suggests that heat and heated material from the earth's interior rise by convection into particular crustal zones of weakness. These zones are sources for the divergence of the plates. The cooled materials then sink downward in subduction zones (discussed later). In this way, the plates are thought to be set in motion. Strong evidence indicates that about 200 million years ago, the entire

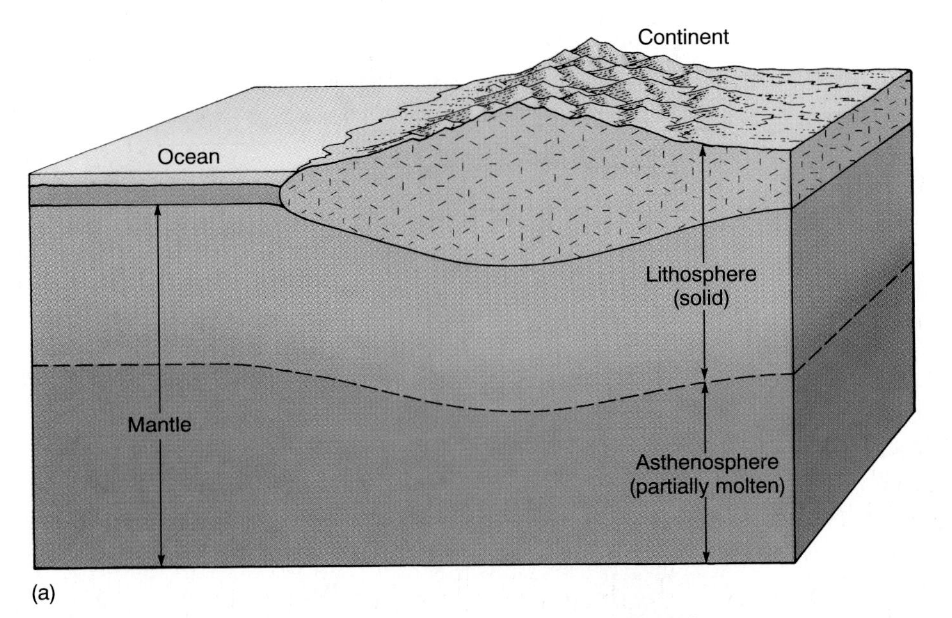

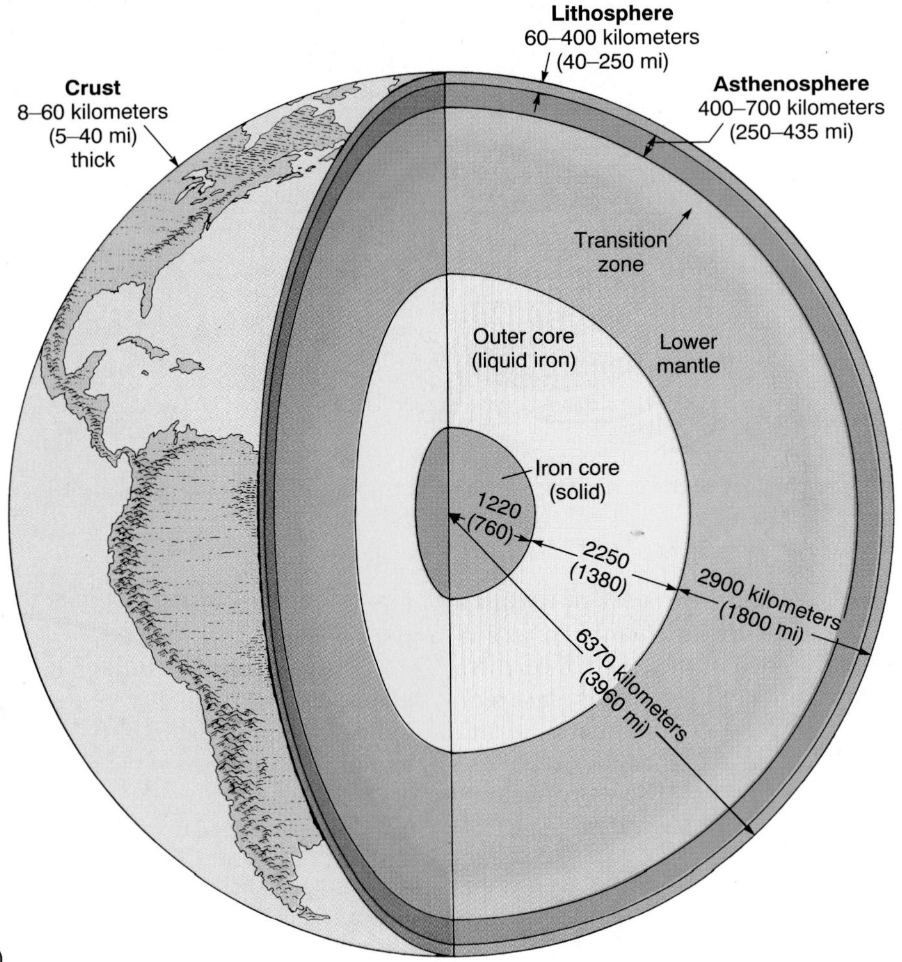

FIGURE 3.2 (a) **The outer zones of the earth** (not to scale). The *lithosphere* includes the crust and the uppermost mantle. The *asthenosphere* lies entirely within the upper mantle. The distance between the highest point on earth—Mt. Everest in Nepal (8848 m or 29,028 ft)—and the lowest point on the ocean floor—Mariana Trench near the Western Pacific island of Guam (10,924 m or 35,840 ft)—is 19.8 km (12.3 mi). (b) The very thin crust of the earth overlies a layered planetary interior. Zonation of the earth occurred early in its history. Radioactive heating melted the original homogeneous planet. A dense iron core settled to the center; a surface and the remnant lower mantle, overlain by a transition zone and the asthenosphere, formed between them. Escaping gases eventually created the atmosphere and the oceans.

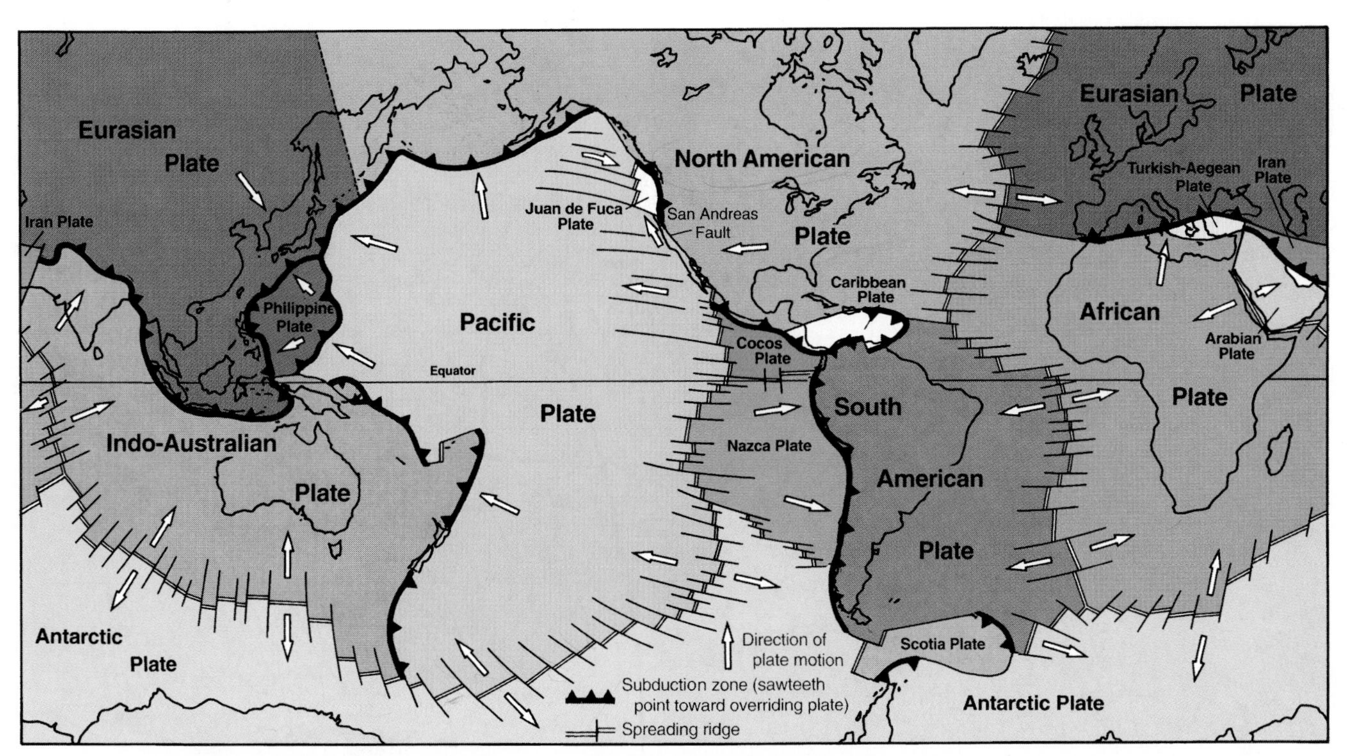

FIGURE 3.3 The large lithospheric plates move as separate entities and collide. Assuming the African plate to be stationary, relative plate movements are shown by arrows. Seafloor spreading, the triggering mechanism, takes place along the axes of the ridges. Recent research indicates that the single Indo-Australian plate shown here may in fact be two plates with a unique, broad boundary between them. Deep-sea trenches, mountain ranges, volcanoes, and earthquakes occur along plate collision lines.
Source: After W. Hamilton, U.S. Geological Survey.

continental crust was connected in one super continent, which Wegener named **Pangaea** ("all earth"). Pangaea was broken into plates as the seafloor began to spread, in which the major force came from the widening of what is now the Atlantic Ocean (Figure 3.4).

Materials from the asthenosphere have been rising along the Atlantic Ocean fracture and, as a result, the seafloor has continued to spread. The Atlantic Ocean is now 6920 kilometers (4300 mi) wide at the equator. If it widens by a bit less than 2.5 centimeters (1 in.) per year, as scientists have estimated, one could calculate that the separation of the continents did, in fact, begin about 200 million years ago. Notice on Figure 3.5 how the ridge line that makes up the axis of the ocean runs parallel to the eastern coast of North and South America and the western coast of Europe and Africa.

According to plate tectonics theory, collisions occurred as the lithospheric plates moved. The pressure exerted at the intersections of plates resulted in earthquakes, which over periods of many years combined to change the shape and features of the landforms. Figure 3.6 shows the location of near-surface earthquakes for a recent time period. Comparison with Figure 3.3 illustrates that the areas of greatest earthquake activity are at plate boundaries.

The famous San Andreas Fault in California is part of a long fracture separating two lithospheric plates, the North American and the Pacific. Earthquakes occur along **faults** (sharp breaks or fractures in rock along which there is slippage) when the tension and the compression at the junction become so great that only an earth movement can release the pressure. The San Andreas case is called a *transform* fault, which occurs when one plate slips past another in a horizontal motion. Because the Atlantic Ocean is still widening at the rate of about 2.5 centimeters (1 in.) per year, earthquakes must occur from time to time to relieve the stress along the tension zone in the mid-Atlantic and along other fracture lines, such as the San Andreas Fault (see "Damage from Earthquakes").

Despite the availability of scientific knowledge about earthquake zones, the general disregard for this danger is a difficult cultural phenomenon with which to deal. Every year there are hundreds and sometimes thousands of casualties resulting from inadequate preparation for earthquakes. In some well-populated areas, the chances that damaging earthquakes will occur are very great. The distribution of earthquakes shown in Figure 3.6 implies the potential dangers to densely settled areas of Japan, the Philippines, parts of Southeast Asia, and the western rim of the Americas.

Movement of the lithospheric plates results in the formation of deep-sea trenches and continental-scale mountain ranges, as well as in the occurrence of earthquakes. The continental crust is made up of lighter rocks than is the oceanic crust. Where plates with different types of crust at

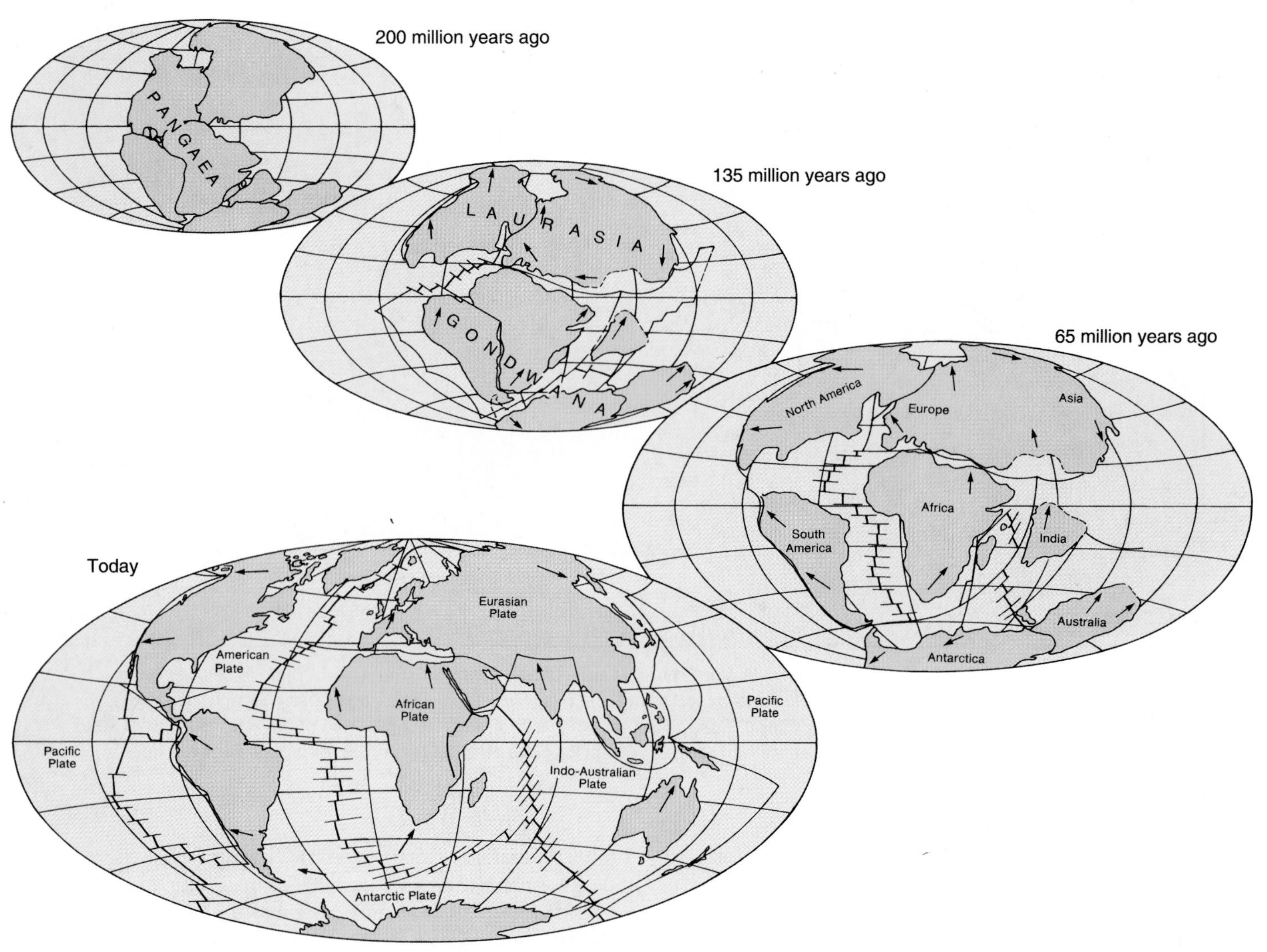

FIGURE 3.4 **The drifting of the continents.** Two hundred million years ago the continents were connected as one large landmass. After they split apart, the continents moved to their present positions. Notice how India broke away from Antarctica and collided with the Eurasian landmass. The Himalayas were formed at the zone of contact.
American Petroleum Institute.

their edges push against each other, there is a tendency for the denser oceanic crust to be forced down into the asthenosphere. This causes long and deep trenches to form below the ocean. This type of collision is termed **subduction** (Figure 3.7). The edge of the overriding continental plate is uplifted to form a mountain chain that runs close to and parallel with the offshore trench. The subduction zones of the world are shown in Figure 3.3.

Most of the Pacific Ocean is underlain by a plate that, like the others, is constantly pushing and being pushed. The continental crust on adjacent plates is being forced to rise and fracture, making an active earthquake and volcano zone of the Pacific Ocean rim (sometimes called the "ring of fire"). In recent years, major earthquakes and volcanic activity have occurred in Colombia, Mexico, Central America, the Pacific Northwest of the United States, Alaska, Hawaii, Armenia, Turkey, China, and the Philippines. Figure 3.8 shows the location of volcanoes known to have erupted at some time in the past. The tremendous explosion that rocked Mount St. Helens in the state of Washington in 1980 is an example of continuing volcanic activity along the Pacific rim. Many scientists who believed that a damaging earthquake would occur along the San Andreas Fault were not surprised by the October 17, 1989, earthquake whose epicenter on the fault was 64 kilometers (40 mi) south of San Francisco.

Plate intersections are not the only locations susceptible to readjustments in the lithosphere. As lithospheric plates have moved, the earth's crust has been cracked or broken in virtually thousands of places. Some breaks are weakened to the point that they allow molten material from the asthenosphere to find its way to the surface. The molten

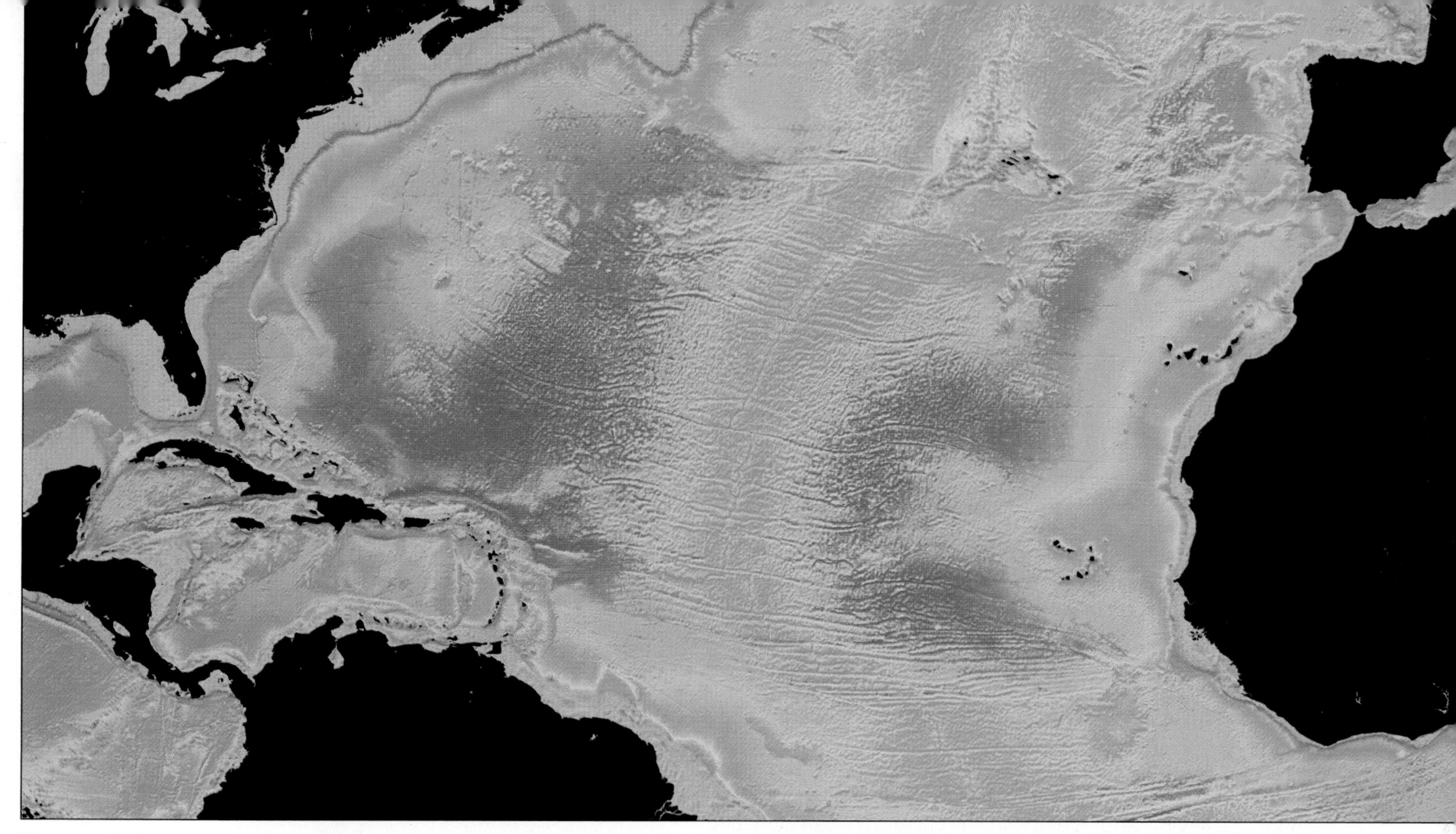

FIGURE 3.5 An accurate map of the seafloor created by the National Oceanic and Atmospheric Administration using gravity measurements taken from satellite readings. The configuration of the seafloor is evidence of the dynamic processes shaping continents and ocean basins.

Source: David T. Sandwell, 1995. Scripps Institution of Oceanography.

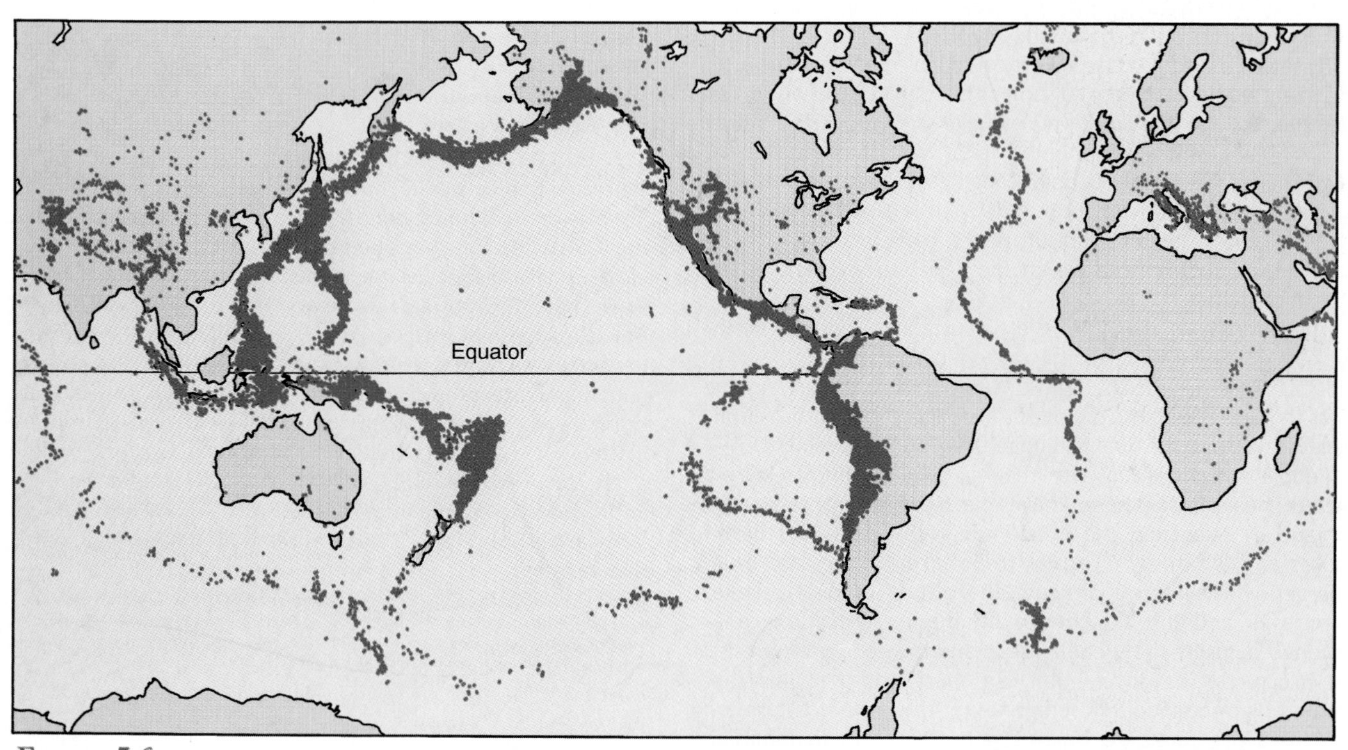

FIGURE 3.6 This map of the epicenters of earthquakes over a seven-year period also reveals lithospheric plate margins. Compare it with Figure 3.3.

Reprinted by permission from Bulletin of Seismological Society of America, 1969.

DAMAGE FROM EARTHQUAKES

For several days the weather in southern California had been almost perfect. Warm days, cool evenings, and a nearly full moon lulled people into a sense of complacency. But 14.5 kilometers (9 mi) beneath the Los Angeles district of Northridge, the inexorable motion of two tectonic plates was about to release a tremendous amount of energy toward the earth's surface. At 4:31 A.M. on January 17, 1994, a violent earthquake struck the San Fernando valley; its epicenter was in Northridge, 32 kilometers (20 mi) northwest of downtown Los Angeles. Measured at 6.8 on the Richter scale, the quake was felt as far away as Las Vegas, Nevada, 440 kilometers (275 mi) to the east.

The earthquake damaged more than 40,000 structures, 1600 of them so badly that they had to be torn down. There were more than 60 casualties, of which 16 people died when an apartment complex collapsed, and hundreds more were injured. Virtually every building on the Northridge campus of California State University was damaged. Broken water mains flooded streets, while ruptured gas pipelines spewed fireballs into the air and ignited fires in a number of districts. The quake caused widespread damage to streets and highways across the San Fernando valley. The overpasses of three heavily traveled freeways buckled and collapsed; their repair was expected to take at least one year. To add to people's woes, the earthquake and its hundreds of aftershocks (some as high as 5.5 on the Richter scale) sent dust rising above the mountains, leading to an outbreak of valley fever. Technically known as coccidioidomycosis, this flu-like illness is caused by inhaling a fungus that lives in soil and becomes airborne when dust is forced into the air.

As severe as the damage was, it could have been much worse, and likely would have been in some other parts of the world. For example, the 6.4 magnitude earthquake that struck the Indian state of Maharashtra in September 1993 was estimated to have killed more than 30,000 people. Tens of thousands more were injured, and hundreds of thousands were left homeless. Fewer than 100 people died as a result of the more powerful Northridge quake. To what can we attribute such an extreme difference in casualties?

The amount of damage an area sustains during an earthquake is governed in part by factors over which people have no control. These include the type of soil and rock underlying the areas that are struck, the depth of the quake, the way the seismic waves travel, and the quake's impact on ground movement.

People *can* affect, however, the way they respond to known hazards; heavy losses are not inevitable. Governments can curtail damage by educating the public about the nature of earthquakes and the safety measures to be taken in the case of an emergency. They can coordinate the efforts of disaster-relief agencies. In addition, they can pass and enforce building codes designed to insure that people live in and travel on quake-resistant structures. The government can enact land-use plans that prohibit placing schools, hospitals, and other large structures on or near active faults, or on such sensitive earth materials as

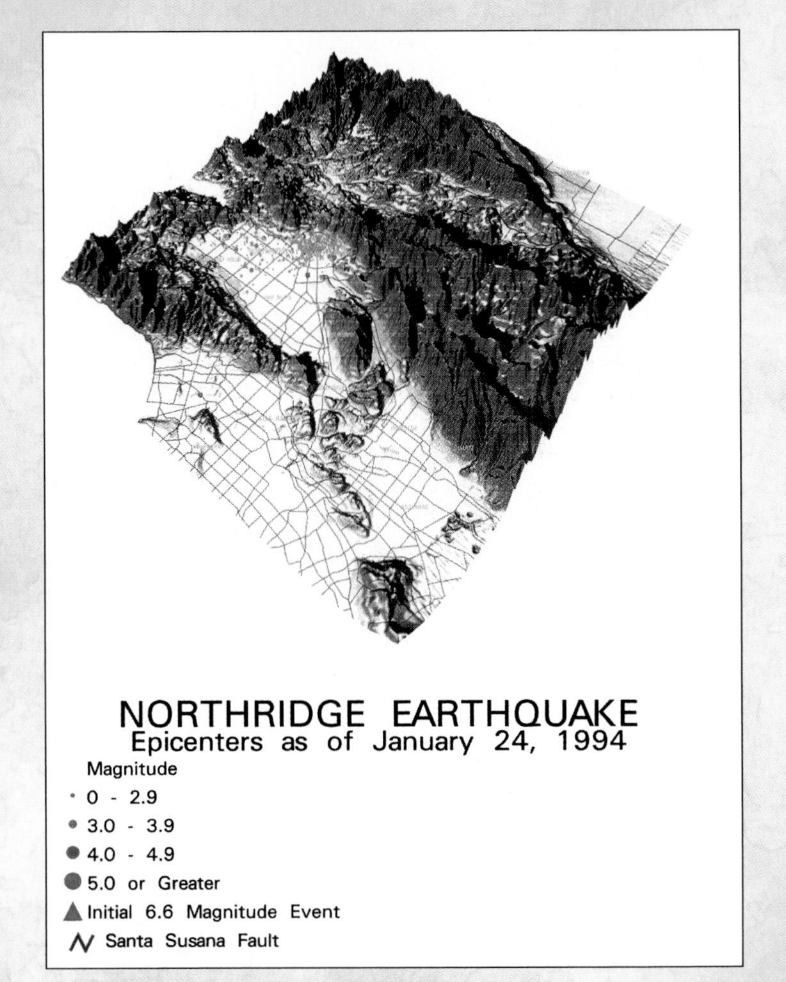

California Institute of Technology seismologists working with the software technology of the Earth Systems Research Institute (ESRI) and data supplied by the U.S. Geological Survey and Thomas Brothers Maps were able to produce the map shown here. Elevation data allowed the computer to create a three-dimensional surface of the San Fernando Valley area of Los Angeles. Draped over that surface were a shaded relief map, major streets, and aftershock epicenters. Note that the original epicenter in Northridge is well south of the hundreds of aftershocks that occurred at the northern end of the valley and in the nearby Santa Susana Mountains. (The software employed to make the map is called ARC/INFO Rev. 7.0. The hardware is Sun SPARCstation and the plotter is a CalComp Electrostatic.)

Source: Graphic image courtesy of Environmental Systems Research Institute, Inc., and California Institute of Technology Seismological Laboratory, Pasadena. Portions of this image are copyrighted by Thomas Bros. Maps and reproduced with permission granted by Thomas Bros. Maps.

water-saturated silts or unconsolidated clay deposits that are likely to be areas of high damage.

California began a major renovation of its highway system after the 1971 Sylmar earthquake in the Los Angeles basin. More than 1200 bridges were strengthened in the years that followed. The 1989 Loma Prieta quake in the San Francisco area gave impetus to another major effort to reinforce overpasses, freeway columns, and other highway structures. There is little doubt that many more of these would have collapsed during the Northridge quake had they not been reinforced and modified.

The earthquake that struck Kobe, Japan, on January 17, 1995—the first anniversary of the Northridge quake—shows that preparedness cannot guarantee safety or prevent severe damage. After the 1923 earthquake that leveled Tokyo and claimed more than 140,000 lives, the Japanese government enacted strict building codes to make buildings, dams, bridges, and other structures as seismically safe as possible. Steps were taken to strengthen school buildings, construct embankments against quake-generated tidal waves, and improve evacuation roads. Despite these measures, the Kobe quake killed more than 5000 people and damaged or destroyed more than 50,000 buildings. Why?

There are several reasons. First, much of Kobe is built on soft alluvial soil; the sediments liquefy and settle after a quake. Second, western Japan had not suffered a major quake for nearly 50 years, so many of the older buildings and highways had never been retrofitted with stronger supports. Almost all of those that were built to the latest earthquake-resistance standards emerged from the quake unscathed. Finally, the government's disaster-relief policies were woefully inadequate. Kobe had virtually no stockpiles of emergency supplies of water, food, blankets, and first-aid kits. Also lacking were search-and-rescue teams of engineers and doctors, search dogs, and fiber-optic cameras to look for survivors inside the rubble. Roads were clogged with traffic, making it difficult for soldiers and firefighters to reach the city.

This is one of seven bridges that collapsed during the Northridge earthquake; 233 others were damaged.
The San Diego Union-Tribune, photo by Don Kohlbauer.

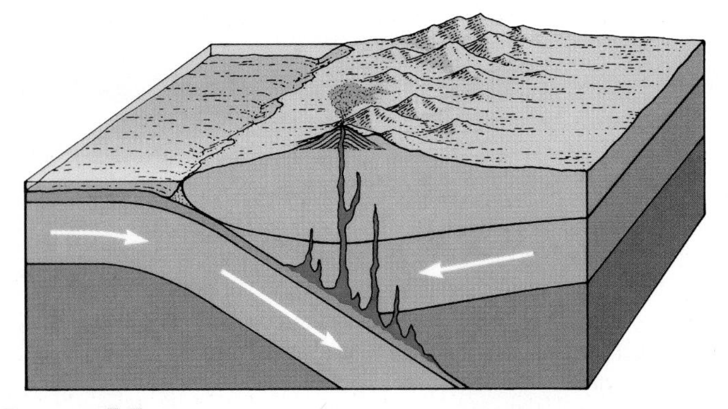

FIGURE 3.7 **The process of subduction.** When lithospheric plates collide, the heavier oceanic crust is usually forced beneath the lighter continental material. See Figure 3.3 for the subduction zones of the world.

material may explode out of a volcano or ooze out of cracks. Later in this chapter, when we discuss the earth-building forces, we will return to the discussion of volcanic activity. First, however, it is necessary to describe the materials that make up the earth's surface.

EARTH MATERIALS

The rocks of the earth's crust vary according to mineral composition. Rocks are made up of particles that contain various combinations of such common elements as oxygen, silicon, aluminum, iron, and calcium, together with

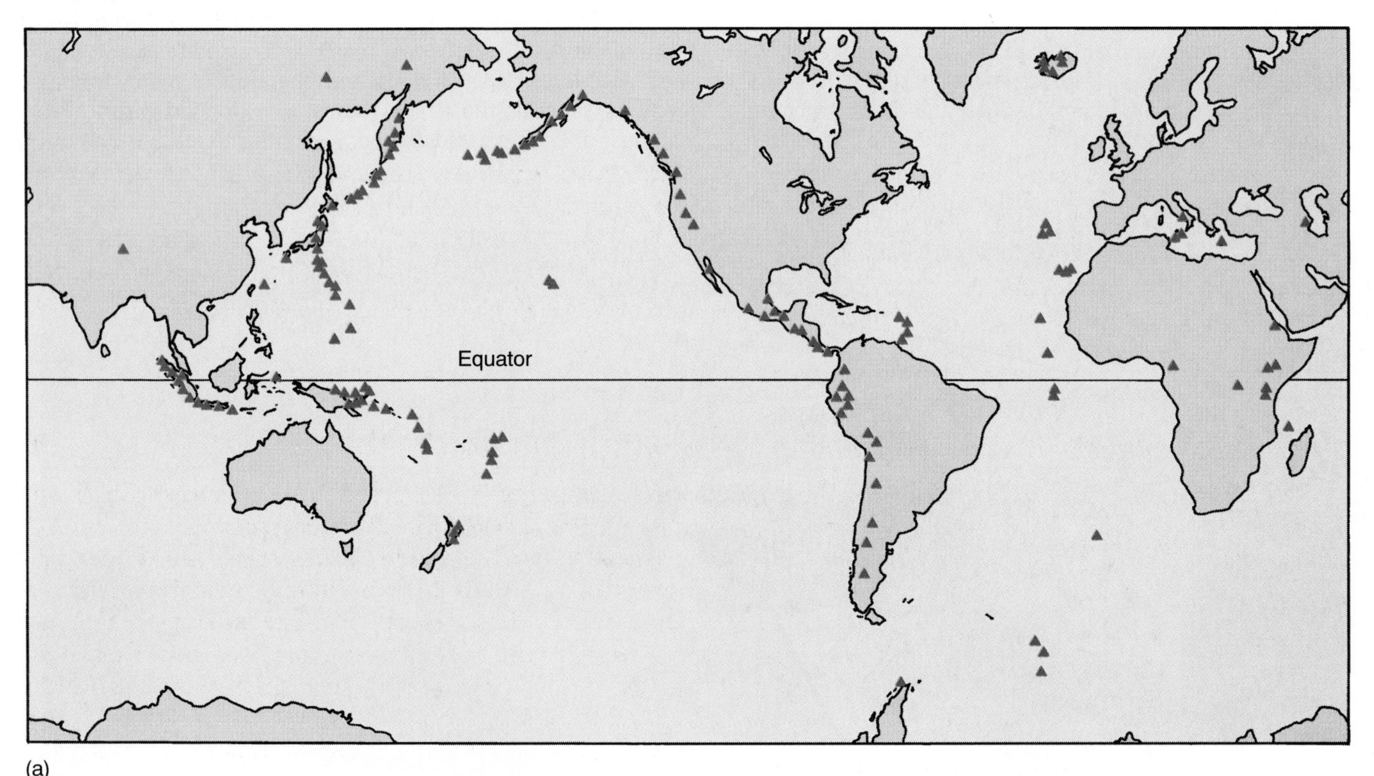

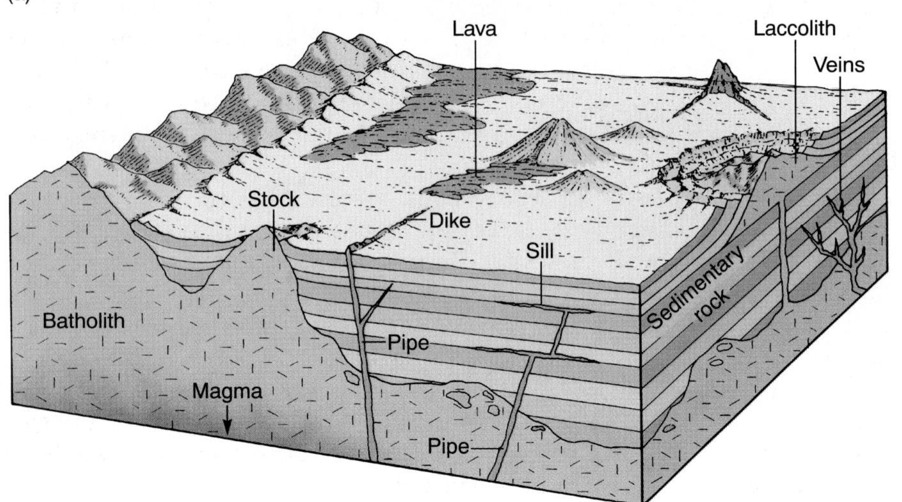

FIGURE 3.8 (a) **The distribution of volcanoes.** Note the association of volcanic activity with plate boundaries as shown on Figure 3.3. (b) **Extrusive and intrusive forms of volcanism.** Lava and *ejecta* (ash and cinders) are extrusions of rock material onto the earth's surface in the form of cones or horizontal flows. *Batholiths* and *laccoliths* are irregular masses of crystalline rock that have cooled slowly below the earth's surface (intrusions), and in this diagram they have become surface features because of the erosion of overlying material.

Sources: Map plotted from the Environmental Data and Information Service, NOAA.

less abundant elements. A particular chemical combination that has a hardness, density, and definite crystal structure of its own is called a **mineral.** Some well-known minerals are quartz, feldspar, and silica. Depending on the nature of the minerals that form them, rocks may be hard or soft, dense or open, one color or another, or chemically stable or not. While some rocks resist decomposition, others are very easily broken down. Among the more common varieties of rock are granites, basalts, limestones, sandstones, and slates.

Although one can classify rocks according to their physical properties, the more common approach is to classify them by the way they evolved. The three main groups of rocks are igneous, sedimentary, and metamorphic.

Igneous Rocks

Igneous rocks are formed by the cooling and hardening of earth material. Weaknesses in the crust give molten material from the asthenosphere an opportunity to find its way into or onto the crust. When the molten material cools, it hardens and becomes rock. The name for underground molten material is *magma;* above ground it is *lava. Intrusive* igneous rocks are formed below ground level by the hardening of magma, while *extrusive* igneous rocks are created above ground level by the hardening of lava (Figure 3.9).

Depending on the speed of cooling of lava and magma, different minerals form. Because it is not exposed to the coolness of the air, magma hardens slowly, allowing silicon and oxygen to unite and form quartz, a hard, dense mineral. With other components, grains of quartz combine to form the intrusive igneous rock called *granite.*

The lava that oozes out onto the earth's surface and makes up a large part of the ocean basins contains a considerable amount of sodium or calcium aluminosilicates. These form the mineral called *feldspar* and together with several other minerals make up the extrusive igneous rock called *basalt,* the most common rock on earth. If, instead of oozing, the lava erupts from a volcano crater, it may cool very rapidly. Some of the rocks formed in this manner contain air spaces and are light and angular, such as *pumice.* Some may be dense, even glassy, as is *obsidian.* The glassiness occurs when lava meets standing water and cools suddenly.

Sedimentary Rocks

Sedimentary rocks are composed of particles of gravel, sand, silt, and clay that were eroded from already existing rocks. Surface waters carry the sediment to oceans, marshes, lakes, or tidal basins. Compression of these materials by the weight of additional deposits on top of them, and a cementing process brought on by the chemical action of water and certain minerals, causes sedimentary rock to form.

Sedimentary rocks evolve under water in horizontal beds called *strata* (Figure 3.10). Usually one type of sediment collects in a given area. If the particles are large—for instance, the size of gravel—a gravelly rock called *conglomerate* forms. Sand particles are the ingredient for *sandstone,* while silt and clay form *shale* or *siltstone.*

Sedimentary rocks also derive from organic material, such as coral, shells, and marine skeletons. These materials settle into beds in shallow seas and congeal, forming *limestone.* If the organic material is mainly vegetation, it can develop into a sedimentary rock called *coal. Petroleum* is also a biological product, formed during the millions of years of burial by chemical reactions that transform some of the organic material into liquid and gaseous compounds. The oil and gas are light, therefore they ooze through the pores of the surrounding rock to places where dense rocks block their upward movement.

Sedimentary rocks vary considerably in color (from coal black to chalk white), hardness, density, and resistance to chemical decomposition. Large parts of the continents contain sedimentary rocks. For example, nearly the entire eastern half of the United States is overlain with these rocks. Such formations indicate that in the geologic past, seas covered even larger proportions of the earth than they do today.

Metamorphic Rocks

Metamorphic rocks are formed from igneous and sedimentary rocks by earth forces that generate heat, pressure, or chemical reaction. The word *metamorphic* means "changed shape." The internal earth forces that cause the movement and collision of lithospheric plates may be so great that heat and pressure change the mineral structure of a rock, forming new rocks. For example, under great pressure, shale, a sedimentary rock, becomes *slate,* a rock with different properties. Limestone, under certain conditions, may become *marble,* and granite may become *gneiss* (pronounced nice). Materials metamorphosed at great depths and exposed only after overlying surfaces have been slowly eroded away are among the oldest rocks known on earth. Like igneous and sedimentary rocks, however, their formation is a continuing process.

Rocks are the constituent ingredients of most landforms. The strength or weakness, permeability, and chemical content control the way rocks respond to the forces that shape and reshape them. Two principal processes alter rocks: (1) the tectonic forces that tend to build landforms up; and (2) the gradational processes that wear landforms down. All rocks are part of the *rock cycle* through which old rocks are continually transformed into new ones by these processes. No rocks have been preserved unaltered throughout the earth's history.

FIGURE 3.9 **Various rock types.** (a) Shale (sedimentary), (b) limestone (sedimentary), (c) gneiss (metamorphic), (d) basalt (igneous), (e) sandstone (sedimentary), (f) marble (metamorphic), (g) slate (metamorphic), (h) quartzite (metamorphic).
a–h: Hubbard Scientific Company.

Tectonic Forces

The earth's crust is altered by the constant forces resulting from plate movement. *Tectonic* (generated from within the earth) forces shaping and reshaping the earth's crust are of two types, either diastrophic or volcanic. **Diastrophism** is the great pressure acting on the plates that deforms the surface by folding, twisting, warping, breaking, or compressing rock. **Volcanism** is the force that transports heated material to or toward the surface of the earth. When particular places on the continents are under pressure, the changes that take place can be as simple as the bowing or cracking of rock or as dramatic as lava exploding from the crater and sides of Mount St. Helens.

Diastrophism

In the process of plate tectonics, pressures build in various parts of the earth's crust, and slowly, over thousands of years, the crust is transformed (see "Mount Everest: The Jewel in the Crown," page 74). By studying rock formations, geologists are able to trace the history of the development of a region. Over geologic time, most continental areas have been subjected to both tectonic and gradational activity—building up and tearing down. They usually have a complex history of broad warping, folding, faulting, and leveling. Some flat plains in existence today may hide a history of great mountain development in the past.

Broad Warping

Great forces resulting, perhaps, from the movement of continents may bow an entire continent. Also, the changing weight of a large region, perhaps due to melting continental glaciers, may result in the **warping** of the surface. For example, the down-warping of the eastern United States is evident in the many irregularly shaped stream estuaries. As the coastal area warped downward, the sea has advanced, forming estuaries and underwater canyons.

Folding

When the pressure caused by moving continents is great, layers of rock are forced to buckle. The result may be a

(e) (f) (g) (h)

warping or bending effect, and a ridge or series of parallel ridges or **folds** may develop. If the stress is pronounced, great wavelike folds form (Figure 3.11). The folds can be thrust upward many thousands of feet and laterally for many miles. The folded ridges of the eastern United States are, at present, low parallel mountains—300 to 900 meters (1000 to 3000 ft) above sea level—but the rock evidence suggests that the tops of the present mountains were once the valleys between 9100-meter (30,000-ft) crests (Figure 3.12).

Faulting

A fault is a break or fracture in rock. The stress causing a fault results in displacement of the earth's crust along the fracture zone. Figure 3.13 depicts examples of fault types. There may be uplift on one side of the fault or downthrust on the other. In some cases, a steep slope known as a fault *escarpment,* which may be several hundred miles long, is formed. The stress can push one side up and over the other side, or a separation away from the fault may cause the sinking of land, creating a *rift valley* (Figure 3.14).

Many faults are merely cracks (called *joints*) with little noticeable movement along them. In other cases, however, mountains such as the Sierra Nevadas of California have risen as the result of faulting. Sometimes, the movement has been horizontal along the surface rather than upward or downward. The San Andreas transform fault, mentioned earlier and pictured in Figure 3.15, is such a case.

Earthquakes

Whenever movement occurs along a fault, or at some other point of weakness, an earthquake results. The greater the movement, the greater the magnitude of the earthquake (see "Scaling Earthquakes"). Tension builds in rock as tectonic forces are applied, and when a critical point is finally reached, an earthquake occurs and tension is reduced.

The earthquake that occurred in Alaska on Good Friday in 1964 was one of the strongest known, with a magnitude of 8.2 on the Richter scale. Although the stress point of that earthquake was below ground 121 kilometers (75 mi) away from Anchorage, vibrations called *seismic waves* caused earth movement in the weak clays underneath the

FIGURE 3.10 The sedimentary rocks of the Grand Canyon in Arizona are evident in this photograph.
©Robert N. Wallen.

MOUNT EVEREST: THE JEWEL IN THE CROWN

The fastest growing mountain range on earth happens also to contain the world's tallest set of mountains. Nearly in the center of the Himalayas stands the earth's highest peak, Mount Everest. Currently, Mount Everest is measured as standing 8848 meters (29,028 ft) above sea level. Recent measurements, however, indicate that Mount Everest and many of the other peaks, such as K-2, are growing at about 1 centimeter (one-half inch) a year.

As the mountains build ever higher, their great weight softens underlying materials, causing the mountains to slip back. In other words, there are two forces at work. The force building the mountains is that of the Indian plate moving northward, crashing into the Eurasian plate. The Himalayas, at the edge of the Eurasian plate, react to the great force by pushing higher and higher. But apparently, a second force that works against the first keeps mountains on earth from rising to heights of 15,000 or 18,000 meters (roughly 50,000 or 60,000 ft). One may think of these great mountains as being in a kind of equilibrium—the bigger they get, the heavier they get, and the more likely they are to sag.

The battle between the two plates began about 45 million years ago. Usually, the aggressor plate is forced under the more stationary plate (subduction). In this case, however, the rocks of the two plates are similar in weight and density. Thus, subduction has not taken place, and what normally would have been a modest crinkle on the earth's surface turned into the tallest, most rugged mountain range on earth. Particularly interesting is one view of the Himalayas. The Indian plate contains the relatively low elevation Indian subcontinent and, as a result, the view of the Himalayas from the plains of northern India is one of the great sights on earth. No wonder Mount Everest is the jewel in the mountain-climber's crown.

© Dinodia Picture Agency, Bombay.

city. Sections of Anchorage slid downhill, and part of the business district dropped 3 meters (10 ft).

Table 3.1 indicates the characteristic effects associated with earthquakes of different magnitudes. Figure 3.16 shows various types of earthquake-induced damage.

If an earthquake occurs below an ocean, the movement can cause a **tsunami,** a large destructive sea wave (see "The Tsunami"). Though not noticeable on the open sea, a tsunami may become 9 or more meters (30 ft) high as it approaches land, sometimes thousands of kilometers from the earthquake site. The islands of Hawaii now have a tsunami warning system, developed following the devastation in Hilo in 1946.

Earthquakes occur daily in hundreds of places throughout the world. Most are slight and only noticeable on *seismographs,* instruments that record seismic waves. But from time

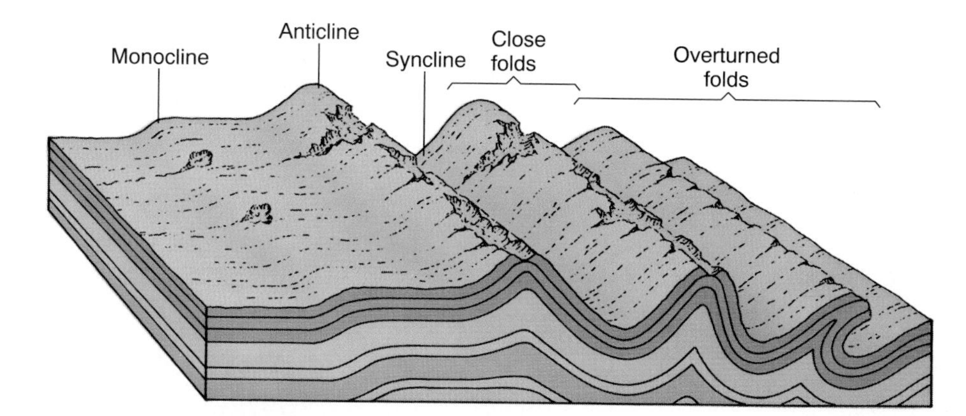

FIGURE 3.11 Diagram of stylized forms of folding. Degrees of folding vary from slight undulations of strata with little departure from the horizontal to highly compressed or overturned beds.

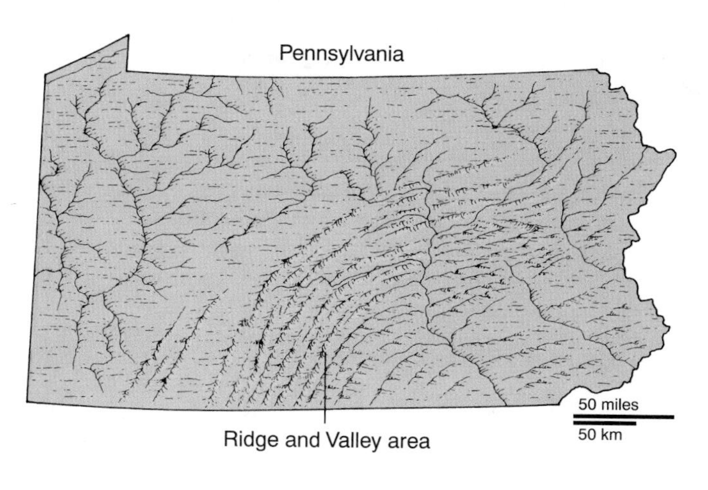

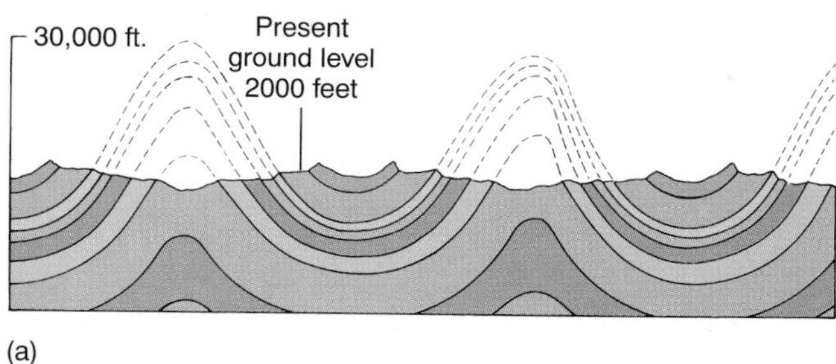

FIGURE 3.12 (a) The Ridge and Valley region of Pennsylvania, now eroded to hill lands, is the relic of 9100-meter (30,000-ft) folds that were reduced to form *synclinal* (downarched) hills and *anticlinal* (uparched) valleys. The rock in the original troughs, having been compressed, was less susceptible to erosion. (b) A syncline in Maryland on the border of Pennsylvania.

(b) ©Mark C. Burnett/Photo Researchers, Inc.

to time, large-scale earthquakes occur, such as those in Mexico in 1985 (4200 deaths) and China in 1976 (242,000 deaths). Most earthquakes take place on the Pacific rim (Figure 3.6), where stress from the outward-moving lithospheric plates is greatest. The Aleutian Islands of Alaska, Japan, Central America, and Indonesia experience a number of moderately severe earthquakes each year.

Volcanism

The second tectonic force is volcanism. The most likely places through which molten materials can move toward the surface are at the intersections of plates. However, other fault-weakened zones are also subject to volcanic activity (compare Figures 3.7 and 3.8).

If sufficient internal pressure forces the magma upward, weaknesses in the crust, or faults, enable molten materials to reach the surface. The material ejected onto the earth's surface may arrive as a great explosion forming a steep-sided cone termed a *strato* or *composite volcano* (Figure 3.17). The eruption may also be without explosions, forming a gently sloping *shield volcano*.

The major volcanic belt of the world coincides with the major earthquake and fault zones. This is the zone of convergence between two plates. A second zone of volcanic activity is at diverging plate boundaries, such as in the center of the

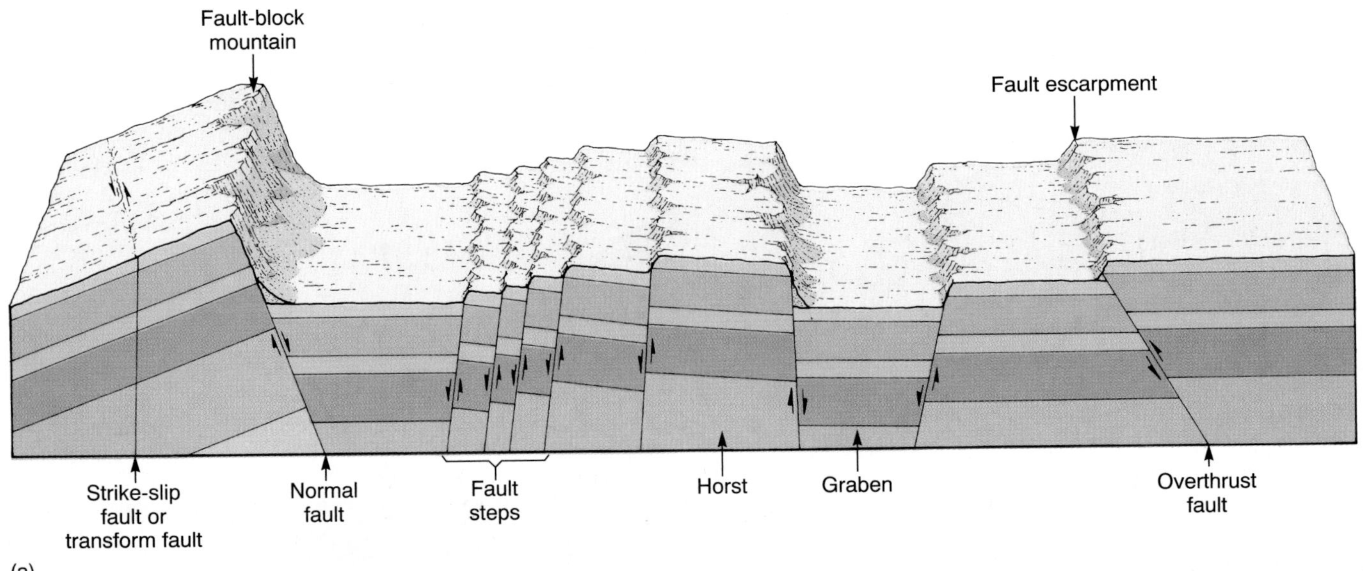

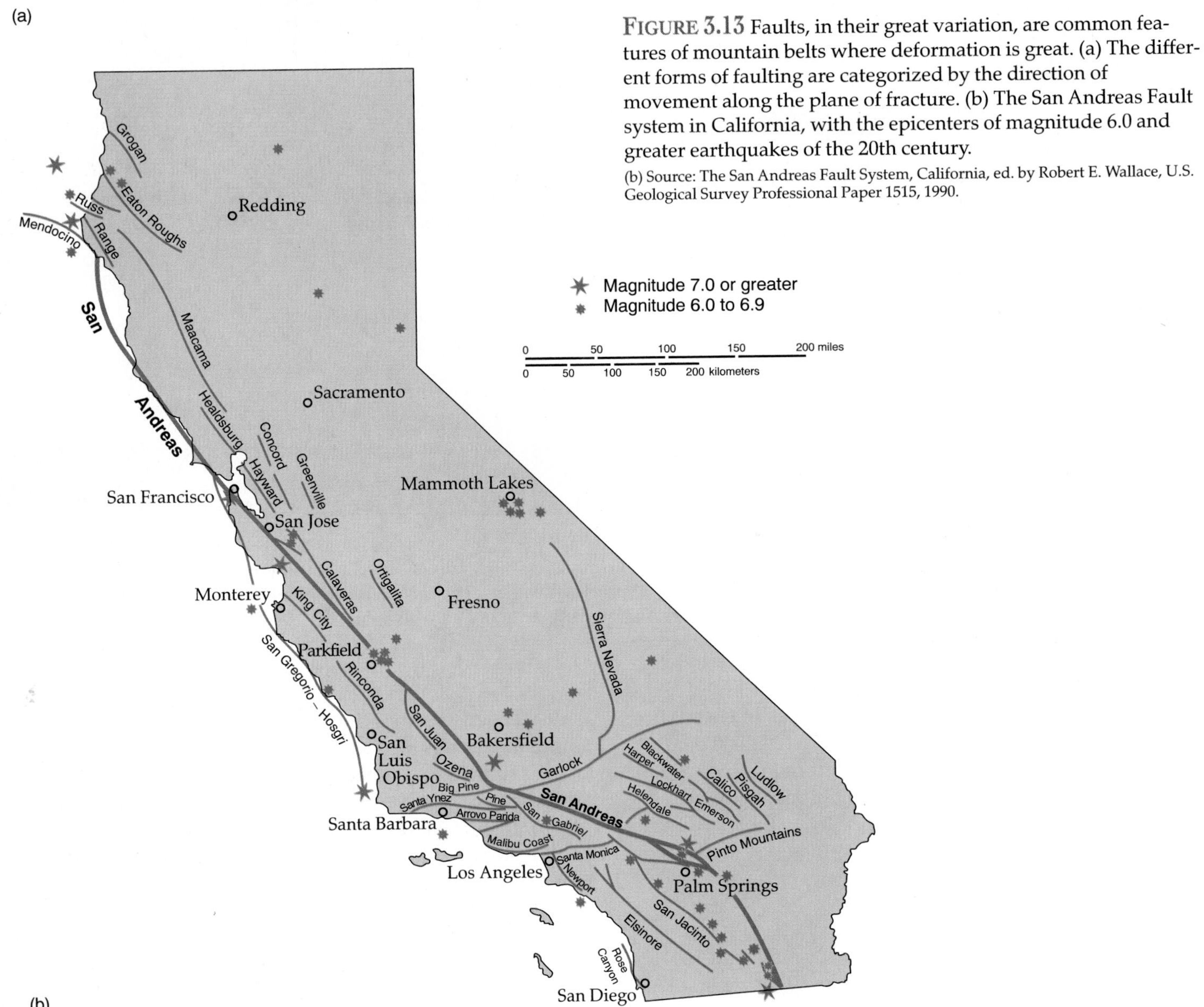

FIGURE 3.13 Faults, in their great variation, are common features of mountain belts where deformation is great. (a) The different forms of faulting are categorized by the direction of movement along the plane of fracture. (b) The San Andreas Fault system in California, with the epicenters of magnitude 6.0 and greater earthquakes of the 20th century.

(b) Source: The San Andreas Fault System, California, ed. by Robert E. Wallace, U.S. Geological Survey Professional Paper 1515, 1990.

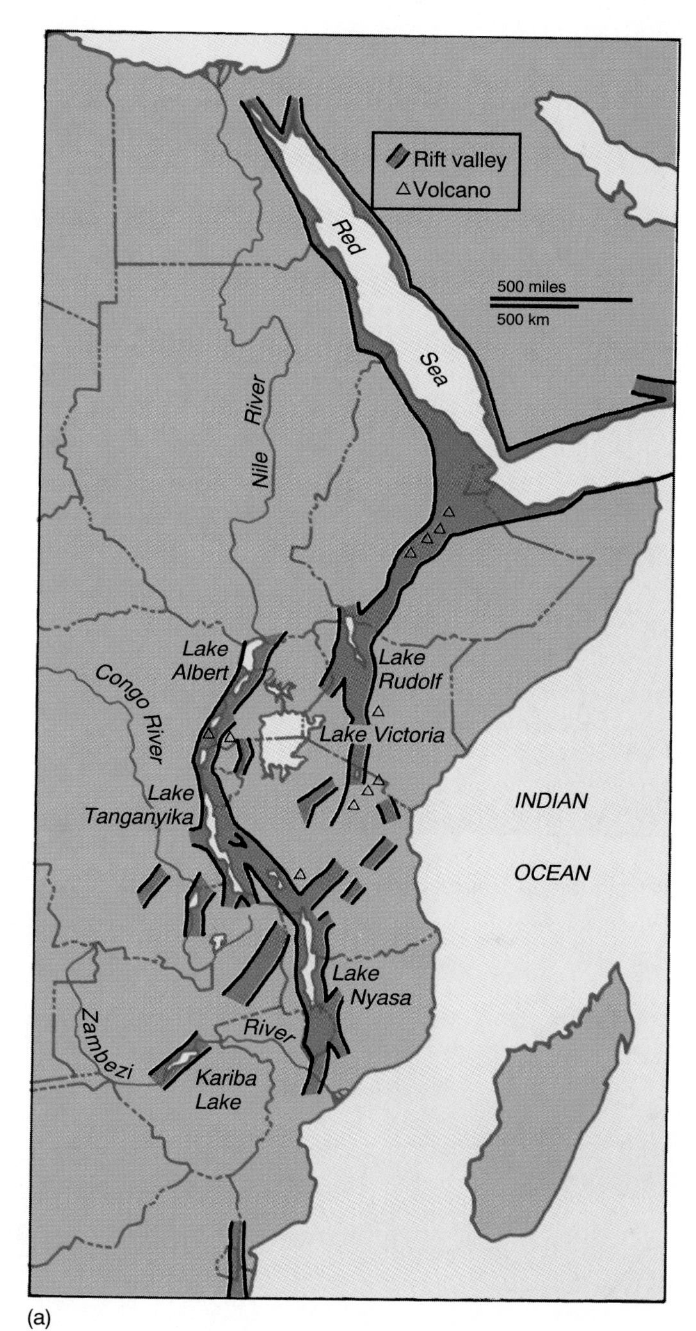

(a)

(b)

FIGURE 3.14 (a) Great fractures in the earth's crust resulted in the creation, through subsidence, of an extensive *rift valley* system in East Africa. The parallel faults, some reaching more than 610 meters (2000 ft) below sea level, are bordered by steep walls of the adjacent plateau, which rises to 1500 meters (5000 ft) above sea level and from which the structure dropped. (b) In Tanzania, the same tectonic forces have created a rift valley, one edge of which is shown here.

(b) ©N. Myers/Bruce Coleman, Inc.

FIGURE 3.15 View along the San Andreas Fault. A *transform fault*, the San Andreas marks a part of the slipping boundary between the Pacific and the North American plates. The inset map shows the relative southward movement of the American plate; dislocation averages about 1 centimeter (0.4 in) per year.

(photo) ©Kevin Shafer/Tom Stack & Associates.

FIGURE 3.16 Two views of earthquake damage. (a) A portion of the top deck of Interstate 880 collapsed when a major earthquake struck the San Francisco Bay area on October 17, 1989. Cars on the lower deck were crushed. (b) On January 17, 1995, an earthquake measured at 7.2 on the Richter scale devastated large parts of Kobe, Japan. The quake killed more than 5000 people, damaged or destroyed more than 50,000 buildings, and left hundreds of thousands of people temporarily homeless. Although the quake lasted just 20 seconds, it toppled buildings and elevated highways with seeming ease. Shown here is part of the Hanshin Expressway, which connects Osaka and Kobe. Huge fires, fed by ruptured gas mains, raged out of control for hours. Because the quake disabled much of the city's water system, firefighters were unable to contain the flames.

(a) AP/Wide World Photos.

(b) Reuters/Bettmann.

(a)

(b)

SCALING EARTHQUAKES

In 1935, C. F. Richter devised a scale of earthquake *magnitude*. An earthquake is really a form of energy expressed as wave motion passing through the surface layer of the earth. Radiating in all directions from the earthquake focus, seismic waves gradually dissipate their energy at increasing distances from the *epicenter* (the point on the earth's surface directly above the focus). On the Richter scale, the amount of energy released during an earthquake is estimated by measurement of the ground motion that occurs. Seismographs record earthquake waves, and by comparison of wave heights, the relative strength of quakes can be determined. Although Richter scale numbers run from 0 to 9, there is no absolute upper limit to earthquake severity. Presumably, nature could outdo the magnitude of the most intense earthquakes recorded so far, which reached 8.5–8.6.

Because magnitude, as opposed to intensity, can be measured accurately, the Richter scale has been widely adopted. Nevertheless, it is still only an approximation of the amount of energy released in an earthquake. In addition, the height of the seismic waves can be affected by the rock materials under the seismographic station, and some seismologists believe that the Richter scale underestimates the magnitude of major tremors.

In recent years, seismologists have used a measure called *moment magnitude*. This measure accounts for the movement of the earth's surface during an earthquake. If the movement (slip) is great, compared to the size (length) of the fault, then the moment magnitude will be great. A small slip on a large fault is considered a minor earthquake. The Northridge earthquake had similar Richter and moment values, but the disastrous quake in Alaska in 1964 had an 8.2 Richter value and a 9.0 moment magnitude.

TABLE 3.1
RICHTER SCALE OF EARTHQUAKE MAGNITUDE

Magnitude[a]	Characteristic Effects of Earthquakes Occurring Near the Earth's Surface[b]
0	Not felt
1	Not felt
2	Not felt
3	Felt by some
4	Windows rattle
5	Windows break
6	Poorly constructed buildings destroyed; others damaged
7	Widespread damage; steel bends
8	Nearly total damage
9	Total destruction

[a]Since the Richter scale is logarithmic, each increment of a whole number signifies a tenfold increase in magnitude. Thus, a magnitude 4 earthquake produces a registered effect upon the seismograph 10 times greater than a magnitude 3 earthquake.

[b]The damage levels of earthquakes are presented in terms of the consequences felt or seen in populated areas; the recorded seismic wave heights remain the same whether or not structures on the surface are damaged. The actual impact of earthquakes on humans varies not only with the severity of the quake and with such secondary effects as tsunamis or landslides, but also with the density of population and the quality of the buildings in the affected areas.

Atlantic Ocean. Molten material can either flow smoothly out of a crater or be shot into the air with great force. Some relatively quiet volcanoes have long gentle slopes indicative of smooth flow, while explosive volcanoes have steep sides. Steam and gases are constantly escaping from the nearly 300 active volcanoes in the world today.

When pressure builds, a crater can become a boiling cauldron with steam, gas, lava, and ash billowing out (Figure 3.18). In the case of Mount St. Helens in 1980, a large bulge had formed on the north slope of the mountain. An earthquake preceded an explosion in this bulging area, shooting debris into the air, completely devastating an area of about 400 square kilometers (150 sq mi), causing about 10 centimeters (4 in.) of ash to rain down on most of Washington and parts of Idaho and Montana, and reducing the elevation of the mountain by more than 300 meters (1000 ft).

In many cases, the forces beneath the crust are not intense enough to allow magma to reach the surface. In these instances, magma hardens into a variety of underground formations of igneous rock that barely affect surface landform features. However, gradational forces may erode overlying rock so that igneous rock, which is usually hard and resists erosion, becomes a surface feature. The Palisades, a rocky ridge facing New York City from the west, is this type of landform.

On other occasions, a weakness below the earth's surface may allow the growth of a mass of magma but denies exit to the surface because of firm overlying rock. Through the pressure it exerts, however, the magmatic intrusion may still buckle, bubble, or break the surface rocks. In addition, domes of considerable size may develop, such as the Black Hills of South Dakota (see Figures 13.4 and 13.5).

Evidence from the past shows that lava has sometimes flowed through fissures or fractures without forming volcanoes. These oozing lava flows have covered large areas to great depths. The Deccan Plateau of India and the Columbia Plateau of the Pacific Northwest in the United States are examples of this type of process (Figure 3.19).

GRADATIONAL PROCESSES

Gradational processes are responsible for the reduction of land surface. If a land surface where a mountain once stood is now a low, flat plain, gradational processes have been at

The Tsunami

A tsunami follows any submarine earthquake that causes fissures or cracks in the earth's surface. Water rushes in to fill the depression caused by the falling away of the ocean bottom. The water then moves outward, building in momentum and rhythm, as swells of tremendous power. The waves that hit Hilo, Hawaii, following the April 1, 1946, earthquake off of Dutch Harbor, Alaska, were moving at approximately 640 kilometers per hour (400 mph) with a crest-to-crest spacing of some 130 kilometers (80 mi).

The long swells of a tsunami are largely unnoticed in the open ocean. Only when the wave trough scrapes sea bottom in shallow coastal areas does the water pile up into precipitous peaks. The seismic sea waves at Hilo on the exposed northeast part of the island of Hawaii were estimated at between 14 and 30 meters (45 and 100 ft). The water smashed into the city, deposited more than 4 meters (14 ft) of silt in its harbor, left fish stranded in palm trees, created millions of dollars in damage, and caused 173 deaths—many of these casualties were people who had gone to the shore to see the giant waves arrive.

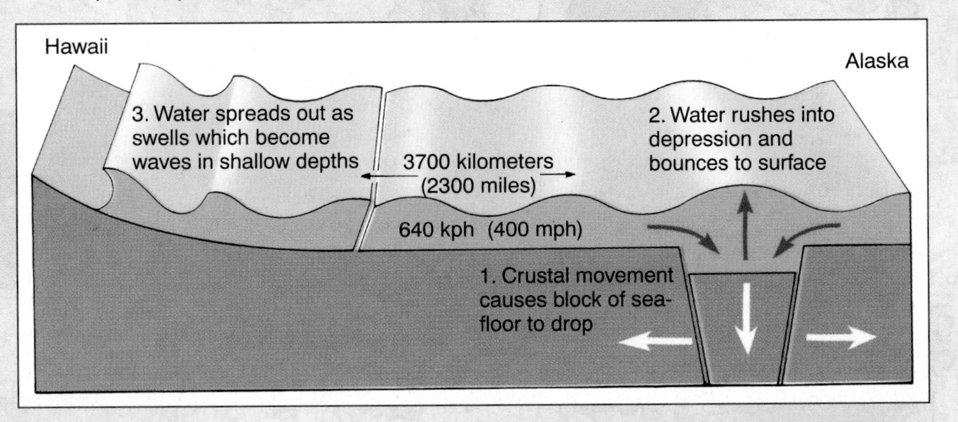

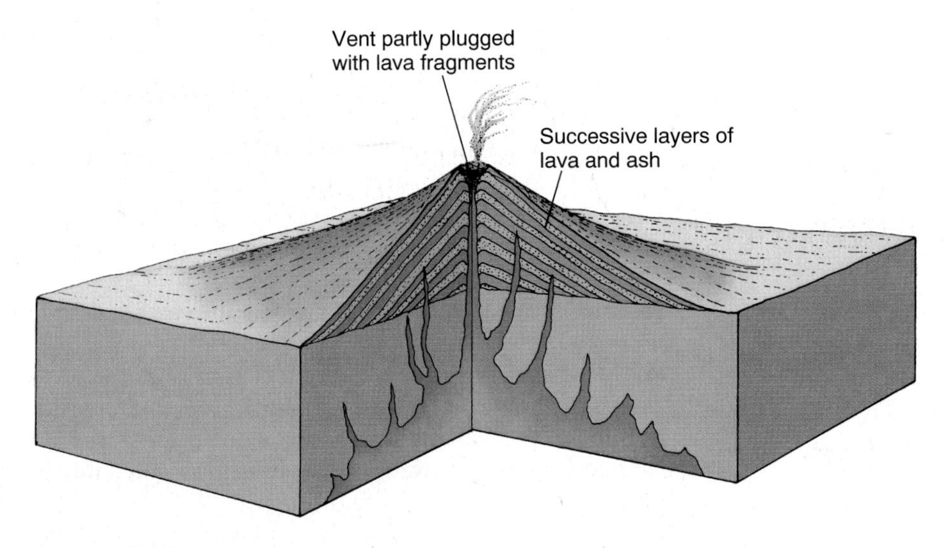

FIGURE 3.17 Sudden decompression of gases contained within lavas results in explosions of rock material to form ashes and cinders. Composite volcanoes, such as the one diagrammed, are composed of alternate layers of solidified lava and ash and cinders.

work. The worn, scraped, or blown away material is deposited in new places and, as a result, new landforms are created. In terms of geologic time, the Rocky Mountains are a recent phenomenon; gradational processes are active there, just as on all land surfaces, but they have not yet had time to reduce these huge mountains.

Three kinds of gradational processes occur: *weathering, gravity transfer,* and *erosion.* Both mechanical and chemical weathering processes play a role in preparing bits of rock for the creation of soils and for movement to new sites by means of gravity or erosion. The force of gravity transfers any loosened, higher-lying material, and the agents of running water, moving ice, wind, waves, and currents erode and carry these loose materials to other areas, where landforms are created or changed.

Mechanical Weathering

Mechanical weathering is the physical disintegration of earth materials at or near the surface. A number of processes cause mechanical weathering; the three most important are frost action, the development of salt crystals, and root action.

If water that soaks into a rock (between particles or along joints) freezes, ice crystals grow and exert pressure on the rock. When the process is repeated—freezing, thawing, freezing, thawing, and so on—the rock begins to disintegrate. Salt crystals act similarly in dry climates, where groundwater is drawn to the surface by *capillary* action (water rising because of surface tension). This

(a)

FIGURE 3.18 (a) Mount St. Helens, Washington, before it erupted. (b) The eruption of Mount St. Helens on May 18, 1980, blew off 4 cubic kilometers (1.5 cubic miles) of its crest, lowering the summit by nearly 400 meters (1300 ft). (c) The force of the explosion flattened trees in a 400-square-kilometer (150-sq-mi) area.

(a) David Muench/Tony Stone Images.

(b), (c) U.S. Geological Survey.

(b)

(c)

FIGURE 3.19 Fluid basaltic lavas created the Columbia Plateau, covering an area of 130,000 square kilometers (50,000 sq mi). Some individual flows were more than 100 meters (300 ft) thick and spread up to 60 kilometers (40 mi) from their original fissures.
(photo) ©Linda J. Moore.

action is similar to the process in plants whereby liquid plant nutrients move upward through the stem and leaf system. Evaporation leaves behind salt crystals that help disintegrate rocks. Roots of trees and other plants may also find their way into rock joints and, as they grow, they break and disintegrate rock. These are all mechanical processes because they are physical in nature and do not alter the chemical composition of the material upon which they act.

Chemical Weathering

A number of **chemical weathering** processes cause rock to decompose rather than disintegrate. In other words, rocks separate into component parts by chemical reaction rather than fragmenting. The three most important processes are oxidation, hydrolysis, and carbonation. Because each of these depends on the availability of water, less chemical weathering occurs in dry and cold areas than in moist and warm ones. Chemical reactions are accelerated in the presence of moisture and heat.

Oxidation occurs when oxygen combines with rock minerals, such as iron, to form oxides. As a result, some rock areas in contact with oxygen begin to decompose. Decomposition also results when water comes into contact with certain rock minerals, such as aluminosilicates. The chemical change that occurs is called *hydrolysis.* When carbon dioxide gas from the atmosphere dissolves in water, a weak carbonic acid forms. The action of the acid, called *carbonation,* is particularly evident on limestone because the calcium bicarbonate salt created in the process readily dissolves and is removed by ground and surface water.

Weathering, either mechanical or chemical, does not itself create distinctive landforms. Nevertheless, it acts to prepare rock particles for erosion and for the creation of soil. After the weathering process decomposes rock, the force of gravity and the erosional agents of running water, wind, and moving ice are able to carry the weathered material to new locations.

Soils

Mechanical and chemical weathering create soil. *Soil* is the thin layer of fine material containing organic matter, air, water, and weathered rock materials that rests on bedrock. When digging down into the soil, one can recognize a series of different-colored layers. Soil scientists identify three sections, including:

1. the top, or the *A-horizon,* which is usually the darkest. It may contain much organic matter, such as decayed plant leaves, twigs, and animal remains, as well as clays and sand grains;
2. the second level, or the *B-horizon,* which is much lighter in color. It is made up largely of clay with small and large bits of minerals and lesser amounts of organic matter;
3. the lowest level, or the *C-horizon,* which is the upper, broken surface of bedrock mixed with clay and only small amounts of organic matter. The older the soil gets, and the warmer and wetter the climate becomes, the deeper and more discernible the

C-horizon will be. Soil age and climate are factors that enable us to better distinguish these three horizons (Figure 3.20).

Soil type is much more a function of the climate of the region in which it occurs than it is of the kind of bedrock below it. Temperature and rainfall act on minerals, in conjunction with the decaying of overriding vegetation, to form soils. The process of soil formation takes many hundreds and sometimes thousands of years. However, dust storms and poor conservation practices by farmers and land managers can deplete an area's crucial A-horizon in just a few years. The topic of soils is discussed in more detail in Chapter 4.

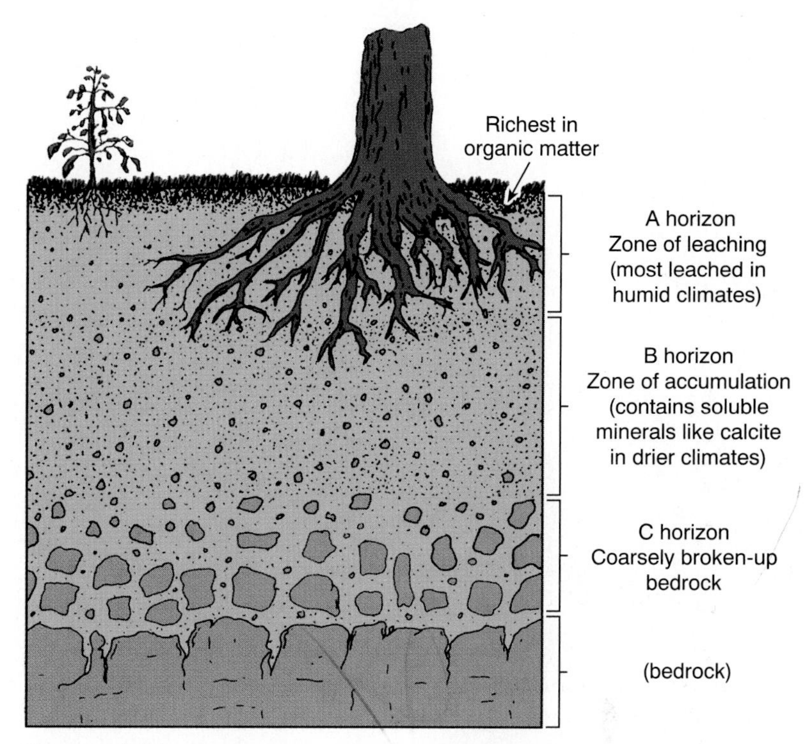

FIGURE 3.20 A representative soil profile showing the various soil horizons.

Gravity Transfer

The force of **gravity transfer**—that is, the attraction of the earth's mass for bodies at or near its surface—is constantly pulling on all materials. Small particles or huge boulders, if not held back by bedrock or other stable material, will fall down slopes. Spectacular acts of gravity include avalanches and landslides. More widespread, but less noticeable, are movements such as soil-creep and the flow of mud down hillsides.

Especially in dry areas, a common but very dramatic landform created by the accumulation of rock particles at the base of hills and mountains is the *talus slope,* pictured in Figure 3.21. As pebbles, particles of rock, or even larger stones break away from exposed bedrock on a mountainside because of weathering, they fall and accumulate, producing large conelike landforms. The larger rocks travel farther than the fine-grained sand particles, which remain near the top of the slope.

Erosional Agents and Deposition

Erosional agents, such as wind and water, carve already existing landforms into new shapes. Fast-moving agents carry debris, and slow-moving agents drop it. The material that has been worn, scraped, or blown away is deposited in new places, and new landforms are created. Each erosional agent is associated with a distinctive set of landforms.

Running Water

Running water is the most important erosional agent. Water, whether flowing across land surfaces or in stream channels, plays an enormous role in wearing down and building up landforms.

Running water's ability to erode depends upon several factors: (1) the amount of precipitation; (2) the length and steepness of the slope; and (3) the kind of rock and vegetative cover. Steeper slopes and faster flowing water results, of course, in more rapid erosion. Vegetative cover sometimes slows the flow of water. When this vegetation is reduced, perhaps because of farming or livestock grazing, erosion can be severe, as shown in Figure 3.22.

Even the impact force of precipitation—heavy rain or hail—can cause erosion. After hard rain dislodges soil, the surface becomes more compact; therefore, further precipitation fails to penetrate the soil. The result is that more water, prevented from seeping into the ground, becomes available for surface erosion. Soil and rock particles in the water are carried to streams, leaving behind gullies and small stream channels.

Both the force of water and the particles contained in the stream are agents of erosion. The particles act as abrasives, scouring the surface over which they move. Abrasion, or wearing away, takes place when particles strike against stream channel walls and along the streambed. Because of the force of the current, large particles, such as gravel, slide along the streambed, grinding rock on the way.

Floods and rapidly moving water are responsible for dramatic changes in channel size and configuration, sometimes forming new channels. In cities where paved surfaces cover soil that would otherwise have absorbed or held water, runoff is accentuated so that nearby rivers and streams rapidly increase in size and velocity after heavy rains. Oftentimes, flash floods and severe erosion result.

Small particles, such as clay and silt, are suspended in water and constitute—together with material dissolved in the water or dragged along the bottom—the *load* of a stream. Rapidly moving floodwaters carry huge loads. As high water or floodwater recedes, and stream velocity

(a)

FIGURE 3.21 **Examples of mechanical weathering.** (a) Rockfall from this butte has created a pronounced talus slope. (b) Creeping soil has caused trees to tilt.
(a) ©Robert N. Wallen.
(b) ©Visuals Unlimited.

decreases, sediment contained within the stream no longer remains suspended, and particles begin to settle. Heavy, coarse materials drop the quickest; finer particles are carried farther. The decline in velocity and the resulting deposition are especially pronounced and abrupt when streams meet slowly moving water in bays, oceans, and lakes. Silt and sand accumulate at the intersections, creating *deltas,* as pictured in Figure 3.23.

A great river, such as the Chang Jiang (Yangtze) in China, has a large, growing delta, but less prominent deltas exist at the mouths of many streams. Until the recent completion of the Aswan Dam, the huge delta of the Nile River had been growing. Now much of the silt is being dropped in Lake Nasser behind the dam.

In plains adjacent to streams, land is sometimes built up by the deposition of *stream load.* If the deposited material is rich, it may be a welcome and necessary part of farming activities, such as that historically known in Egypt along the Nile. Should the deposition be composed of sterile sands and boulders, however, formerly fertile bottomland may be destroyed. By drowning crops or inundating inhabited areas, the floods themselves, of course, may cause great human and financial loss. More than 900,000 lives were lost in the floods of the Huang He (Yellow River) of China in 1887.

Stream Landscapes

A landscape is in a particular state of balance between the uplift of land and its erosion. Rapid uplift is not followed by nicely ordered stages of erosion. Recall that uplift and erosion take place simultaneously. At a given location one force may be greater than the other at a given time, but as yet there is no way to predict the "next" stage of landscape evolution.

Perhaps the most important factor differentiating the effect of streams on landforms is whether the recent climate

(b)

FIGURE 3.22 **Gullying** can result from poor farming techniques, including overgrazing by livestock or many years of continuous row crops. Surface runoff removes topsoil easily when vegetation is too thin to protect it.
©Grant Heilman/Grant Heilman Photography, Inc.

(for example, the last several million years) has tended to be humid or arid.

Stream Landscapes in Humid Areas

Perhaps weak surface material or a depression in rock allows the development of a stream channel. In its downhill run in mountainous regions, a stream may flow over precipices, forming *falls* in the process. The steep downhill gradient allows streams to flow rapidly, cutting narrow V-shaped channels in the rock (Figure 3.24a). Under these conditions, the erosional process is greatly accelerated. Over time, the stream may have worn away sufficient rock for the falls to become rapids, and the stream channel becomes incised below the height of the surrounding landforms. This is evident in the upper reaches of the Delaware, Connecticut, and Tennessee rivers.

In humid areas, the effect of stream erosion is to round landforms (Figure 3.24b). Streams flowing down moderate gradients tend to carve valleys that are wider than those in mountainous areas. Surrounding hills become rounded, and valleys, called **floodplains,** flatten. Streams work to widen the floodplain. Their courses meander, constantly carving out new erosional channels. The channels left behind as new ones are cut become *oxbow*-shaped lakes, hundreds of which are found in the Mississippi River floodplain (Figure 3.24c).

In nearly flat floodplains, the highest elevations may be the banks of rivers, where *natural levees* are formed by the filtering of silt at river edges during floods. Flood waters that breach levees are particularly disastrous. Damaging everything in their path, flood waters fill the floodplain as they equalize its elevation with that of the swollen river. The U.S. Army Corps of Engineers has augmented the natural levees in particularly susceptible areas such as the banks of the lower Mississippi River.

Stream Landscapes in Arid Areas

A distinction must be made between the results of stream erosion in humid regions as opposed to arid areas. The lack of vegetation in arid regions greatly increases the erosional force of running water. Water originating in mountainous areas sometimes never reaches the sea if the channel runs through a desert. In fact, stream channels may be empty except during rainy periods, when water rushes down hillsides to collect and form temporary lakes called *playas.* In the process, **alluvium** (sand and mud) builds up in the lakes and at lower elevations, and *alluvial fans* are formed along hillsides (Figure 3.25). The fan is produced by the scattering of silt, sand, and gravel outward as the stream reaches lowlands at the base of the slope it traverses. If the process has been particularly longstanding, alluvial deposits may bury the eroded mountain masses. In desert regions in Nevada, Arizona, and California, it is not unusual to observe partially buried mountains poking through alluvium.

Because streams in arid areas have only a temporary existence, their erosional power is less certain than that of the freely flowing streams of humid areas. In some

FIGURE 3.23 **Landsat image of the delta of the Mississippi River.** Notice the ongoing deposition of silt and the effect that both river and gulf currents have on the movement of the silt.
NASA.

instances, they barely mark the landscape; in other cases, swiftly moving water may carve deep, straight-sided *arroyos.* Often, water may rush onto an alluvial plain in a complicated pattern resembling a multistrand braid, leaving in its wake an alluvial fan. The channels resulting from this rush of water are called *washes.* The erosional power of unrestricted running water in arid regions is dramatically illustrated by the steep-walled configuration of *buttes* and *mesas* (large buttes), such as those in Mitchell Mesa, Texas, shown in Figure 3.26.

Groundwater

Some of the water supplied by rain and snow sinks underground into pores and cracks in rocks and soil, not in the form of an underground pond or lake, but simply as very wet subsurface material. When it accumulates, a zone of saturation forms. As indicated in Figure 3.27, the upper level of this zone is the **water table;** below it, the soils and rocks are saturated with water. Groundwater moves constantly, but very slowly (only centimeters a day). Most remains underground, seeking the lowest level. When the surface of the land dips below the water table, however, ponds, lakes, and marshes form. Some water finds its way to the surface by capillary action in the ground or in vegetation. Groundwater, particularly when charged with carbon dioxide, dissolves soluble materials by a chemical process called *solution.*

Although groundwater tends to decompose many types of rocks, its effect on limestone is most spectacular. Many of the great caves of the world have been created by the underground movement of water through limestone regions. Water sinking through the overlying rock leaves carbonate deposits as it drips. The deposits hang from cave roofs (*stalactites*) and build upward from cave floors (*stalagmites*). In some areas, the uneven effect of groundwater erosion on limestone leaves a landscape pockmarked by a series of *sinkholes,* surface depressions in an area of plentiful caverns.

Karst topography refers to a large limestone region marked by sinkholes, caverns, and underground streams, as shown in Figure 3.28. Central Florida, a karst area, has suffered considerable damage from the creation and widening of sinkholes. This type of topography gets its name from a region on the Adriatic Sea at the Italy-Yugoslavia border. The Mammoth Cave region of Kentucky, another karst area, has many kilometers of interconnected limestone caves.

(a)

(b)

(c)

FIGURE 3.24 (a) V-shaped valley of a rapidly downcutting stream, the Yellowstone River in Wyoming. (b) A meandering stream, the Okovango River, in Botswana. (c) Oxbow-shaped lakes adjoining a meandering stream in Alaska.
(a) ©Robert N. Wallen.
(b) ©Frans Lanting/Minden Pictures.
(c) ©B. Anthony Stewart/National Geographic Image Collection.

On-Line Physical Geography: Landforms

The best first stop on the web for physical geography resources is the index maintained by Prof. James E. Burt at the University of Wisconsin-Madison. Included are images, data sets, and data libraries. Entries are arranged by subject, followed by a list of some popular information libraries.

http://feature.geography.wisc.edu/phys.htm

Also at the University of Wisconsin is an excellent archive of images and movies.

http://www.ssec.wisc.edu/

A prime federal government source of earth science information is the U.S. Geological Survey (USGS).

http://www.usgs.gov/

Some sites of particular relevance to the study of landforms are noted below. *This Dynamic Planet* contains text and figures depicting volcanoes, earthquakes, impact craters, and plate tectonics.

http://pubs.usgs.gov/pdf/planet.html

The *Cascades Volcano Observatory Home Page* is maintained by the USGS via the Cascades Volcano Observatory. It has a comprehensive list of volcanoes, offers information on volcanically induced geologic and hydrologic hazards, and contains a photo archive of volcanoes and volcanic phenomena. It includes links to other components of the USGS Volcano Hazards Program, such as the Alaska and Hawaii Volcano Observatory and the international Volcano Disaster Assistance Program.

http://vulcan.wr.usgs.gov/home.html

Earthquake Information from the USGS-Geologic Division offers a wealth of earthquake data. Information on recent earthquakes as well as historical events is available, along with predictive intensity maps, research on crustal deformation, and a tutorial on how the epicenter of a quake is calculated.

http://quake.wr.usgs.gov

Coasts in Crisis, a USGS publication, describes the ever-changing nature of coastlines, and both the natural and human causes of coastal change.

http://pubs.usgs.gov/circular/c1075

Michigan Tech Volcanoes Page contains information about current global volcanic activity, research in remote sensing of volcanoes and their eruptive products, hazard mitigation, links to government agencies and research institutions, and even some volcano humor.

http://www.geo.mtu.edu/volcanoes/

NWS-San Francisco Bay Area Natural Hazards provides addresses to links pertaining to tsunamis, earthquakes, tornadoes, and hurricanes, furnished by the National Weather Service.

http://www.nws.mbay.net/hazards.html

Virtual Earthquake, created by California State University/Los Angeles, is a simulation that enables students to locate the epicenter of an earthquake and determine its magnitude on the Richter scale. Students can choose from four geographic areas for their simulation. Type

http://vflylab.calstatela.edu/edesktop/VirtApps/VirtualEarthQuake/VQuakeIntro.html

The Jet Propulsion Laboratory at Caltech University maintains several interesting websites. *Space Radar Images of Earth* are available for categories such as geology, oceans, rivers, glaciers, and volcanoes.

http://www.jpl.nasa.gov/radar/sircxsar/

Also of interest are the *Global Time Series.* Measurements made with the Global Positioning System are sensitive enough to detect motion of the earth's tectonic plates. Click on a map of the world to see time series of latitude, longitude, and height. Horizontal velocities, mostly due to motion of the earth's tectonic plates, are represented on the maps by arrows extending from each site.

http://sideshow.jpl.nasa.gov/mbh/series.html

Hydrology-Related Internet Resources, or *HydroWeb,* a large list of links related to the hydrologic sciences, is maintained at the Pacific Northwest National Laboratory's Environmental Technology Division (ETD).

http://etd.pnl.gov:2080/hydroweb.html

U.S. Water News Homepage is an electronic version of a water news publication, covering issues of water supply, water quality, water policy and legislation, as well as other relevant topics.

http://www.uswaternews.com/homepage.html

Glaciers

Another way erosion and deposition occur is through the effect of glaciers. Although they are much less extensive today, glaciers covered a large part of the earth's land area as recently as 8000 to 15,000 years ago during the *Pleistocene* geologic epoch (see Figure 3.1). Many landforms were created by the erosional or depositional effects of glaciers.

Glaciers form only in very cold places with short or nonexistent summers, where annual snowfall exceeds annual snowmelt and evaporation. The weight of the snow causes it to compact at the base and form ice. When the snowfall reaches a thickness of about 100 meters (328 ft), ice at the bottom becomes plasticlike and begins to move slowly. A **glacier,** then, is a large body of ice moving slowly

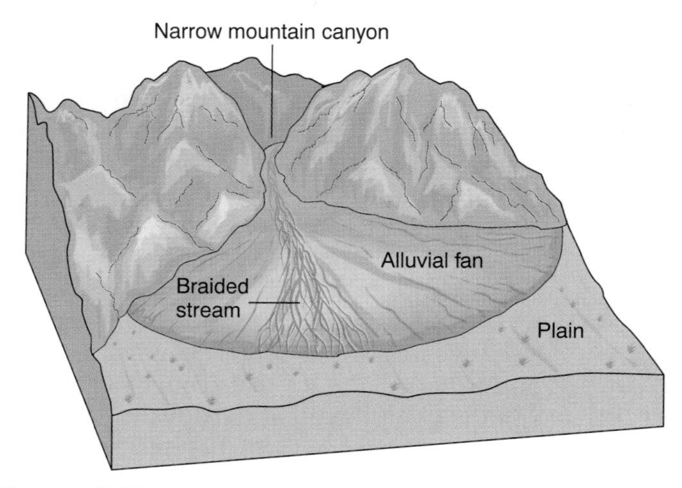

FIGURE 3.25 **Alluvial fans** are built where the velocity of streams is reduced as they flow out upon the more level land at the base of the mountain slope. The abrupt change in slope and velocity greatly reduces the stream's capacity to carry its load of coarse material. Deposition occurs, choking the stream channel and diverting the flow of water. With the canyon mouth fixing the head of the alluvial fan, the stream sweeps back and forth, building and extending a broad area of deposition.
Redrawn from Charles C. Plummer and David McGeary, *Physical Geology*, 8th ed.

FIGURE 3.26 **Canyonlands National Park, Utah.** The resistant caprock of the mesa protects softer, underlying strata from downward erosion. Where the caprock is removed, lateral erosion lowers the surface, leaving the mesa as an extensive and pronounced relic of the former higher-lying landscape.

FIGURE 3.27 The groundwater table generally follows surface contours but in subdued fashion. Water flows slowly through the saturated rock, emerging at earth depressions that are lower than the level of the water table. During a drought, the table is lowered and the stream channel becomes dry.

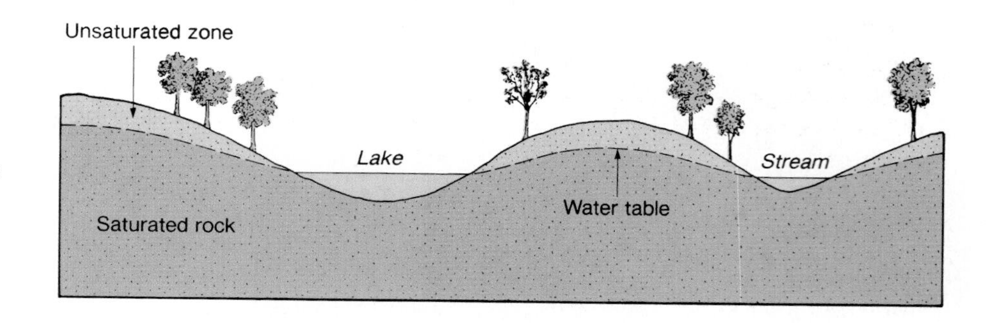

down a slope or spreading outward on a land surface (Figure 3.29). Some glaciers appear to be stationary simply because the melting and evaporation at the glacier's edge equals the speed of the ice advance. Glaciers can, however, move as much as a meter per day.

Most theories of glacial formation concern earth climatic cooling. Perhaps some combination of the following theories explains the evolution of glaciers. The first theory attributes the ice ages to periods when there may have been excessive amounts of volcanic dust in the atmosphere. The argument is that the dust, by reducing the amount of solar energy reaching the earth, effectively lowered temperatures at the surface. A second theory attributes the ice ages to known changes in the shape, tilt, and seasonal positions of the earth's orbit around the sun over the last half-million years. Such changes alter the amount of solar radiation received by the earth and its distribution over the earth. A recent theory suggests that when large continental plates drift over polar regions, temperatures on earth become more extreme and, as a result, induce the development of glaciers. This theory, of course, cannot explain the most recent ice ages.

Today, continental-size glaciers exist only on Antarctica and Greenland, but mountain glaciers are found in many parts of the world. About 10% of the earth's land area is under ice. During the most recent advance of ice, the continental ice of Greenland was part of an enormous glacier that covered nearly all of Canada (Figure 3.30) and the northernmost portions of the United States and Eurasia. The giant glacier reached thicknesses of 3000 meters (10,000 ft) (the depth in Greenland today), enveloping entire mountain systems.

The weight of glaciers breaks up underlying rock and prepares the rock for transportation by the moving mass of ice. Consequently, glaciers alter landforms by weathering and erosion. Glaciers scour the land as they move, leaving surface scratches on rocks that remain. Much of eastern Canada has been scoured by glaciers that left little soil but many ice-gouged lakes and streams. The erosional forms created by glacial scourings have a variety of names. A *glacial trough* is a deep, U-shaped valley visible only after the glacier has

receded. If the valley is below sea level today, as in Norway or British Columbia, *fiords,* or arms of the sea, are formed. Some of the landforms created by scouring are shown in Figure 3.29. Figure 3.31 shows *tarns,* small lakes in the hollowed-out depressions of a *cirque.* Cirques are formed by ice erosion at the head of a glacial valley.

Glaciers create landforms when they deposit the debris they have transported. These deposits, called *glacial till,* consist of rocks, pebbles, and silt. As the great tongues of ice move forward, debris accumulates in parts of the glacier. The ice that scours valley walls and the ice at the tip of the advancing tongue are particularly filled with debris (Figure 3.32). As a glacier melts, it leaves behind hills of glacial till, called **moraines.**

FIGURE 3.28 Limestone erodes easily in the presence of water. (a) *Karst* topography, such as that shown here, occurs in humid areas where limestone in flat beds is at the surface. (b) This satellite photo of central Florida shows the many round lakes formed in the sinkholes of a karst landscape.
(b) NASA.

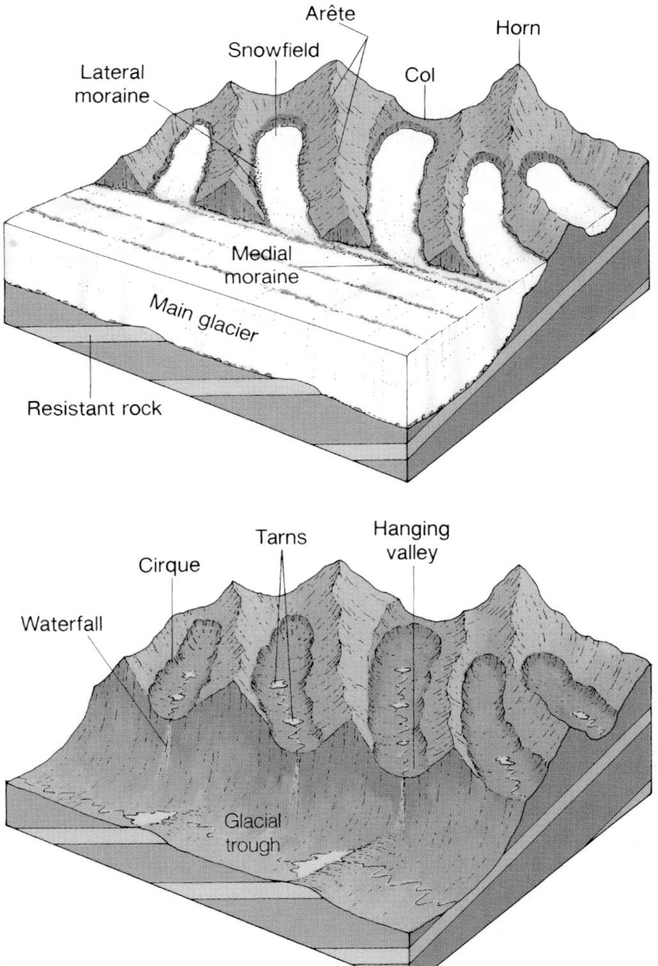

FIGURE 3.29 **The evolution of alpine glacial landforms.** Frost shattering and ice movement carve *cirques,* the irregular bottoms of which may contain lakes (*tarns*) after glacial melt. Where cirque walls adjoin from opposite sides, knife-like ridges called *arêtes* are formed, interrupted by overeroded passes or *cols.* The intersection of three or more arêtes creates a pointed peak, or *horn.* Rock debris falling from cirque walls is carried along by the moving ice. *Lateral moraines* form between the ice and the valley walls; *medial moraines* mark the union of such debris where two valley glaciers join.

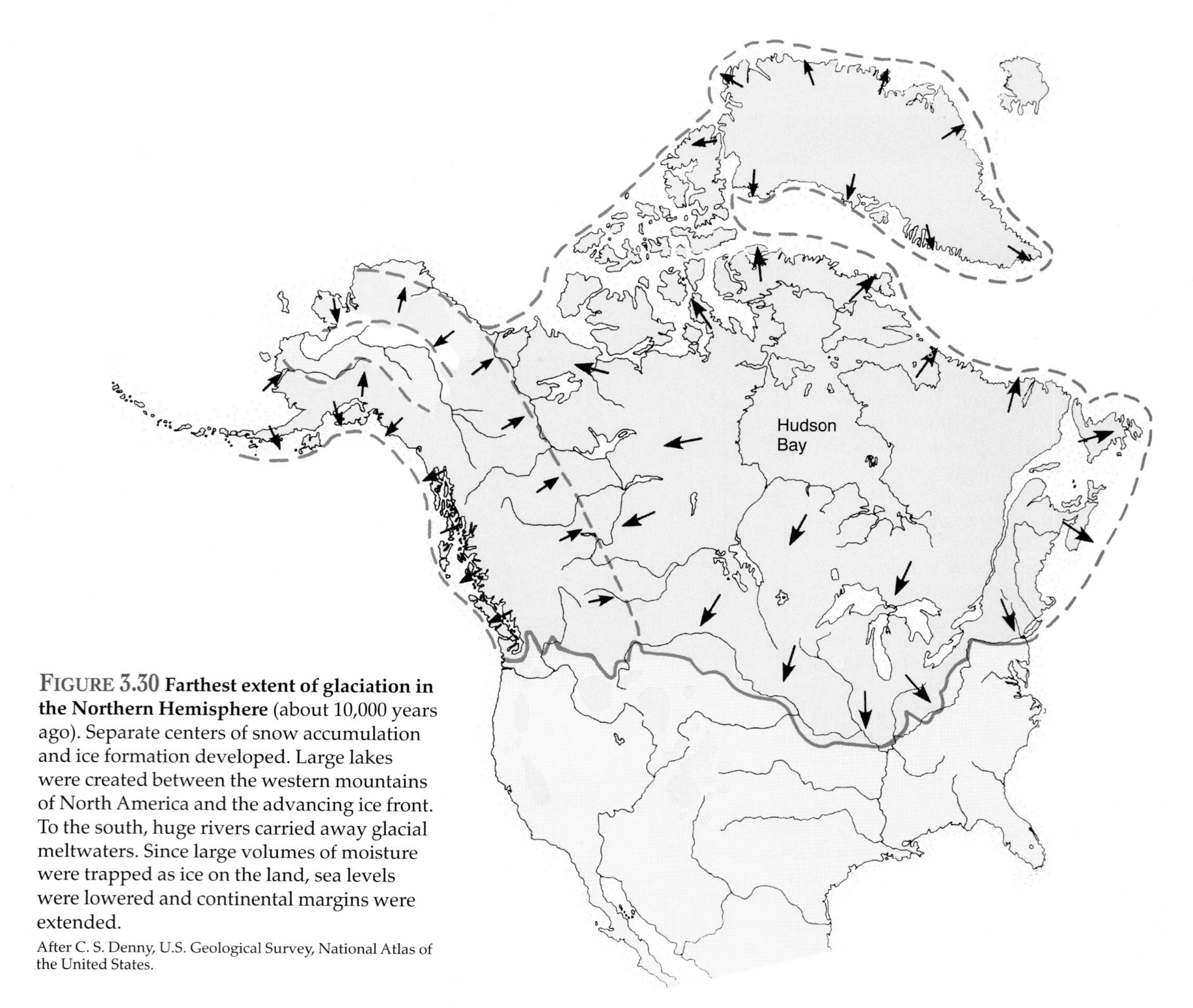

FIGURE 3.30 **Farthest extent of glaciation in the Northern Hemisphere** (about 10,000 years ago). Separate centers of snow accumulation and ice formation developed. Large lakes were created between the western mountains of North America and the advancing ice front. To the south, huge rivers carried away glacial meltwaters. Since large volumes of moisture were trapped as ice on the land, sea levels were lowered and continental margins were extended.

After C. S. Denny, U.S. Geological Survey, National Atlas of the United States.

FIGURE 3.31 **Tarns in glacial cirques.** The glacially formed lakes shown here are in the Glacier Peak Wilderness Area, Washington.

PERMAFROST

In 1577, on his second voyage to the New World in search of the Northwest Passage, Sir Martin Frobisher reported finding ground in the far north that was frozen to depths of "four or five fathoms, even in summer" and that the frozen condition "so combineth the stones together that scarcely instruments with great force can unknit them." The permanently frozen ground, now termed *permafrost,* underlies perhaps a fifth of the earth's land surface. In the lands surrounding the Arctic Ocean, its maximum thickness is about 600 meters (2000 ft).

For almost 300 years after Frobisher's discovery, little attention was paid to this frost phenomenon. But in the 19th and 20th centuries, during the building of the Trans-Siberian railroad, the construction of buildings associated with the discovery of gold in Alaska and the Yukon Territory, and the development of the oil pipeline that now connects Prudhoe Bay in northern Alaska with Valdez in southern Alaska, attention has focused on the unique nature of permafrost.

These Alaskan railroad tracks have been warped by the effect of melting permafrost.
O. J. Ferrains, U.S. Geological Survey/L. A. Yehle photo.

Uncontrolled construction activities lead to the thawing of permafrost. This, in turn, produces unstable ground susceptible to soil movement and landslides, ground subsidence, and frost heaving. Scientists have found that successful use requires that the least possible disturbance of the frozen ground occur. The Alaska pipeline was built above the earth's surface in order to allow the passage of migrating wildlife and to reduce the likelihood that the relatively warm oil would disturb the permafrost and thus destroy the pipeline. It was necessary to preserve the insulating value of the ground surface as much as possible, thus the vegetation mat was not removed. Instead, a coarse gravel fill was added along the surface below the pipeline.

Many other landforms have been formed by glaciers. The most important is the *outwash plain,* a gently sloping area in front of a melting glacier. The melting along a broad front sends thousands of small streams running out from the glacier in braided fashion, streams that deposit neatly stratified glacial till. Outwash plains, which are essentially great alluvial fans, cover a wide area and provide new, rich parent-material for soil formation. Most of the midwestern part of the United States owes some of its soil fertility to relatively recent glacial deposition.

Before the end of the most recent ice age, at least three previous advances occurred during the million years of the Pleistocene period. Firm evidence is not available on whether we have emerged from the cycle of ice advance and retreat. Factors concerning the earth's changing temperature, which are discussed in Chapter 4, must be considered before assessing the likelihood of a new ice advance. For the first half of this century, the world's glaciers were melting faster than they were building up. Current trends are not clear, although there is fear that the greenhouse effect (discussed in Chapter 5) is warming the earth and will cause the seas to rise.

Waves and Currents

While glacial action is intermittent in earth history, the breaking of ocean waves on continental coasts and islands is unceasing and causes considerable change in coastal landforms. As waves reach shallow water close to shore, they are forced to heighten until a breaker is formed, as shown in Figure 3.33. The uprush of water not only carries sand for deposition but also erodes the landforms at the coast, while the backwash carries the eroded material away. This type of action results in different kinds of landforms, depending on conditions.

If land at the coast is well above sea level, the wave action causes cliffs to form. Cliffs then erode at a rate dependent on the rock's resistance to the constant assault of salt water. During storms, a great deal of power is released by the forward thrust of waves, and much weathering and erosion take place. Landslides are a hazard during coastal storms, and they occur particularly in areas where weak sedimentary rock or glacial till exists.

Beaches are formed by the deposition of sand grains contained in the water. The sand originates from the vast

(a)

FIGURE 3.32 (a) Arêtes and lateral and medial moraines. (b) Receding ice sheets create various landforms. Moraines form at the retreating edge, while eskers and drumlins are formed as the glacier melts. Streams carry silt to form outwash plains.
(a) © Steve McCutcheon / Alaska Pictorial Service.

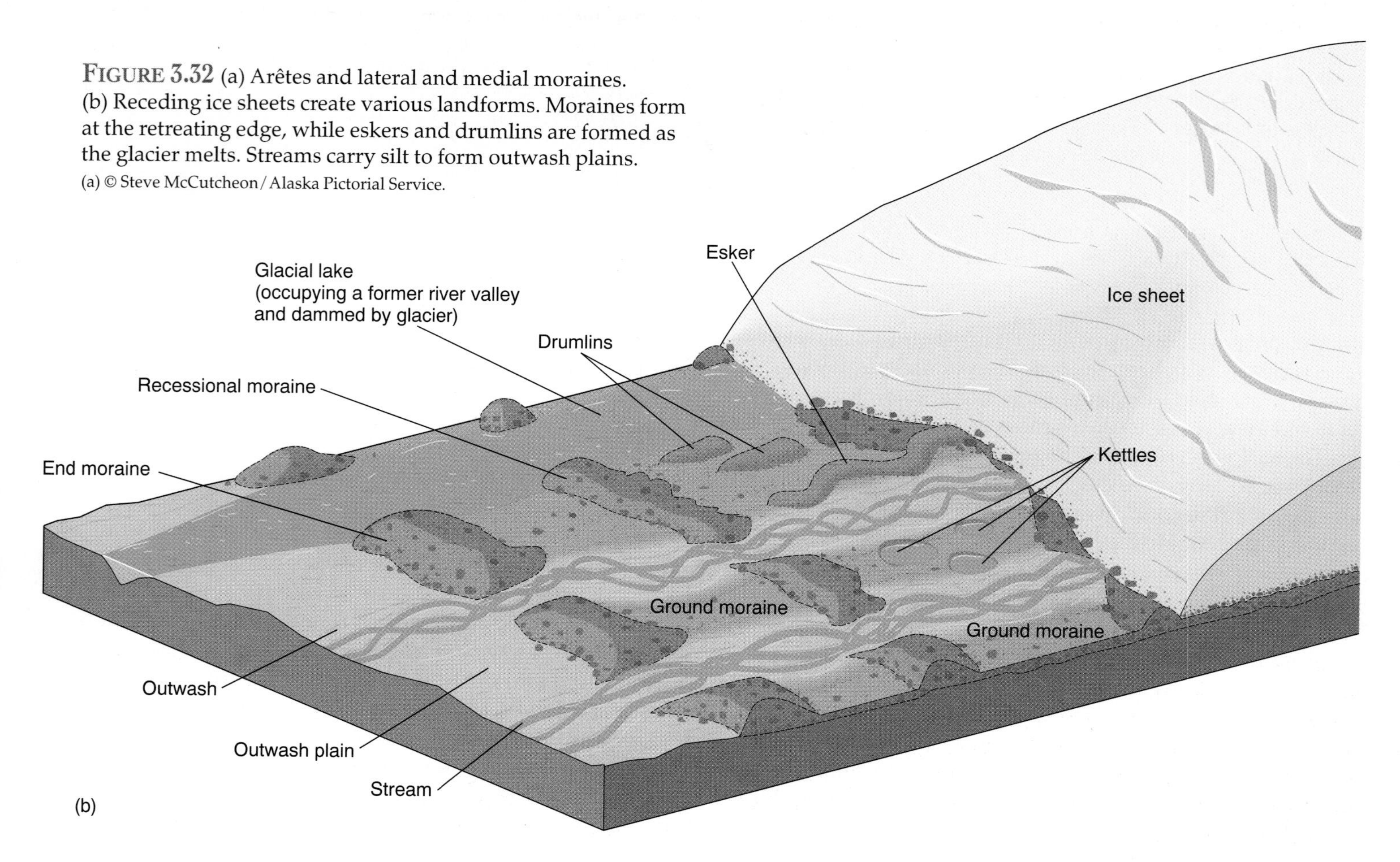

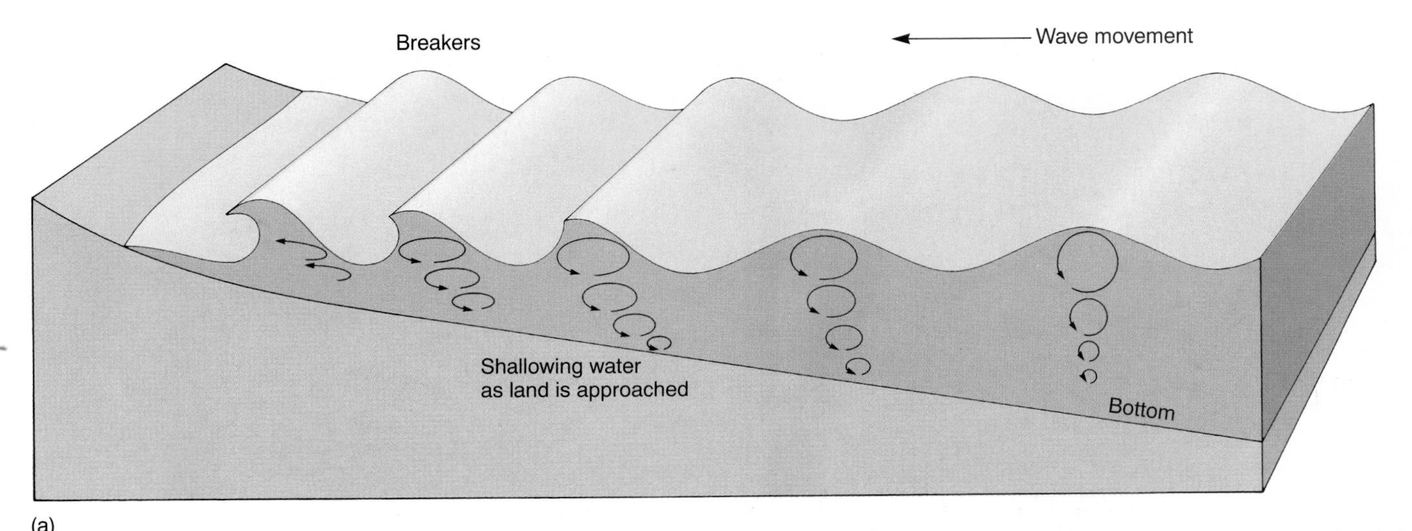

FIGURE 3.33 **Formation of waves and breakers.** (a) As the offshore swell approaches the gently sloping beach bottom, sharp-crested waves form, build up to a steep wall of water, and break forward in plunging surf. (b) Evenly spaced breakers form as successive waves touch bottom along a regularly sloping shore.
(b) ©Carla W. Montgomery.

amount of coastal erosion and streams (Figure 3.34). *Longshore currents,* which move roughly parallel to the shore, transport the sand, forming beaches and *sandspits.* A more sheltered area increases the chances of a beach being built.

The backwash of waves, however, takes sand away from beaches if no longshore current exists. As a result, *sandbars* develop a short distance away from the shoreline. If sandbars become large enough, they eventually close off the shore, creating a new coastline that encloses *lagoons* or *inlets. Salt marshes* very often develop in and around these areas. Figure 3.35 shows one kind of area partially carved by waves.

Coral reefs, made not from sand but from coral organisms growing in shallow tropical water, are formed by the secretion of lime in the presence of warm water and sunlight. Reefs, consisting of millions of colorful skeletons, develop short distances offshore. Off the coast of northeastern Australia lies the most famous coral reef, the Great Barrier Reef. *Atolls,* found in the South Pacific, are reefs formed in shallow water around a volcano that has since been covered or nearly covered by water.

Wind

Unlike in humid areas, where the effect of wind is confined mainly to sandy beach areas, in dry climates, wind is a powerful agent causing weathering, erosion, and deposition. Limited vegetation in dry areas leaves exposed particles of sand, clay, and silt subject to movement by wind. Thus, many of the sculptured features found in dry areas result from mechanical weathering, that is, from the abrasive action of sand and dust particles as they are blown against rock surfaces. Sand and dust storms occurring in a drought-stricken farm area may make it unusable for agriculture. Inhabitants of Oklahoma, Texas, and Colorado suffered greatly in the 1930s when their farmlands became the "Dust Bowl" of the United States.

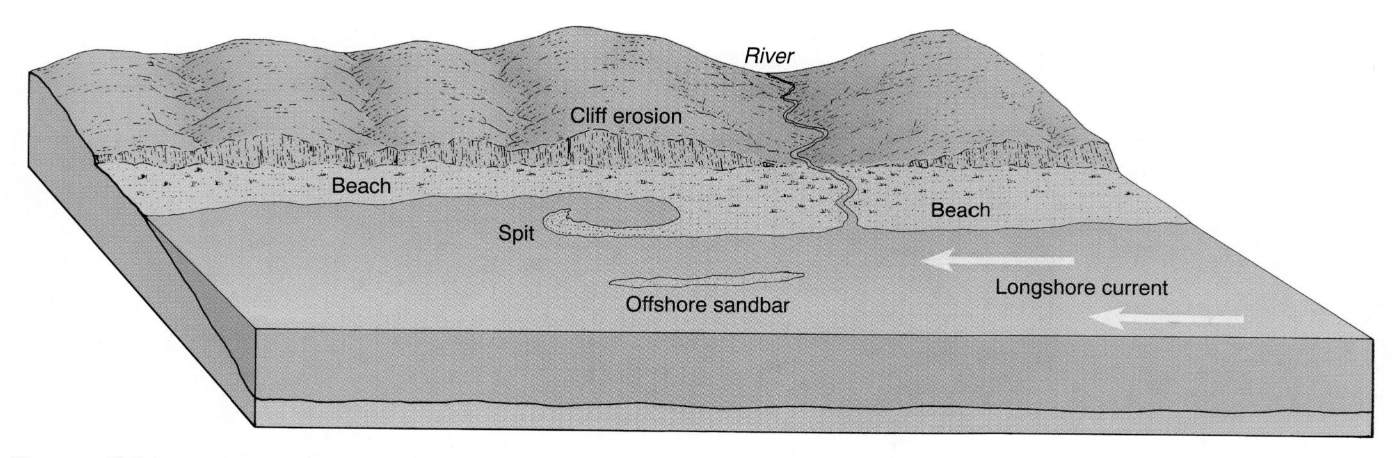

FIGURE 3.34 The cliffs behind the shore are eroded by waves during storms and high water. Sediment from the cliff and river forms the beach deposit; the longshore current moves some sediment downcurrent to form a *spit*. Offshore *sandbars* are created from material removed from the beach and deposited by retreating waves. Generally not all the features shown here would occur in a single setting.

FIGURE 3.35 Sea cliffs, headlands, embayments, and offshore erosional remnants are typical of cliffed shorelines such as this one along the coast of Oregon. Beaches occur only near the mouths of rivers or in embayments.

FIGURE 3.36 The prevailing wind from the left has given these transverse dunes a characteristic gentle windward slope and a steep, irregular leeward slope.
Courtesy of James A. Bier

Several types of landforms are produced by wind-driven sand. Figure 3.36 depicts one of these. Although sandy deserts are much less common than gravelly deserts, their characteristic landforms are better known. Most of the Sahara, the Gobi, and the western U.S. deserts are covered with rocks, pebbles, and gravel and not sand. Each also has a small portion (and the Saudi Arabian Desert has a large area) covered with sand blown by wind into a series of waves or *dunes*. Unless vegetation stabilizes them, the dunes move as sand is blown from their windward faces onto and over their crests. One of the most distinctive sand desert dunes is the crescent-shaped *barchan*. Along seacoasts and inland lakeshores, in both wet and dry climates, wind can create sand ridges that reach a height of 90 meters (300 ft). Sometimes, coastal communities and farmlands are threatened or destroyed by moving sand (see "Beaches on the Brink").

Another kind of wind-deposited material, silty in texture and pale yellow or buff in color, is called **loess.** Encountered usually in midlatitude westerly wind belts, it covers extensive areas in the United States (Figure 3.37), central Europe, central Asia, and Argentina. It has its greatest development in northern China, where loess covers hundreds of thousands of square miles, often to depths of more than 30 meters (100 ft). The wind-borne origin of loess is confirmed by its typical occurrence downwind from extensive desert areas, though major deposits are assumed

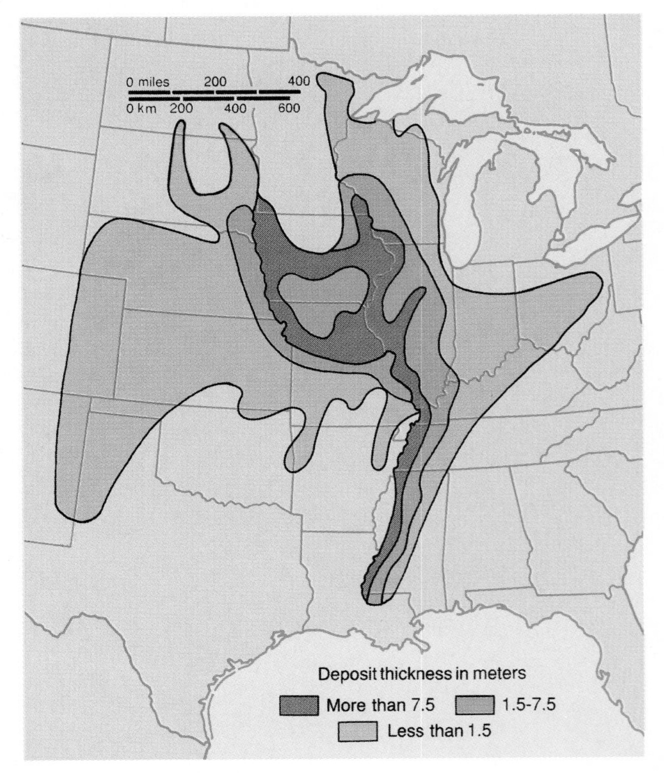

FIGURE 3.37 Location of windblown silt deposits, including *loess*, in the United States. The thicker layers, found in the upper Mississippi valley area, are associated with the wind movement of glacial debris. Farther west, in the Great Plains, wind-deposited materials are sandy in texture, not loessial.
Adapted from *Geology of Soils* by Charles B. Hunt, copyright 1972 W. H. Freeman and Company. Reprinted by permission of the author.

Geography and Public Policy

Beaches on the Brink

Headlines such as "Beaches on the Brink," "Storm-Lashed Cape Is a Fragile Environment," "Fighting the Development Tide," and "Jersey Looks for Money to Restore the Shore," signal a growing concern with the condition of coastlines. In addition, they raise the central question of how we can utilize coastlines without, at the same time, destroying them.

Because many of the world's people live or vacation on coasts, and because such areas are often densely populated, coastal processes have a considerable impact on humans. Nature's forces are continually shaping and reshaping the coasts; they are dynamic environments, always in a state of flux. Some processes are dramatic and induce rapid change: tropical cyclones (hurricanes and typhoons), tsunamis, and floods can wreak havoc, take thousands of lives, and cause millions of dollars in damage in a matter of hours. A less hazardous process is beach erosion, although it tends to magnify the effects of storms.

Some beach erosion is caused by natural processes, both marine and land. Waves carry huge loads of suspended sand, and longshore currents constantly move sand along the shoreline. The weathering and erosion of sea cliffs produce sediment, and rivers carry silt from mountains to beaches.

Human activities affect both erosion and deposition, however. Dams, for example, decrease the flow of sand to the water's edge by trapping sediment upstream. We fill marshes, construct dikes, bulldoze dunes or remove their natural stabilizing vegetation. We hasten erosion when we build roads, houses,

An eroded cliffside in Bodega Bay, California.
©AP/Dan Krauss/Wide World Photos.

—Continued on page 98

Continued from page 97

and other structures on cliffs and dunes, or when we plant trees and lawns on top of them.

Especially vulnerable to erosion are barrier islands, narrow strips of sand parallel to the mainland. Under natural conditions, they are not stable places; their ends typically migrate, and during storms, waves can wash right over them. Some barrier islands are highly developed and densely populated, including two good examples, Atlantic City and Miami Beach.

Once hotels and condominiums, railroads, and highways have been built along the waterfront, people attempt to protect their investments by preventing beaches from eroding. Structures such as jetties, groins, and breakwaters are designed to trap sand and thus retard erosion, but they are not uniformly successful. While some are locally beneficial by forming new areas of deposition, they almost invariably accelerate erosion in adjacent areas. The efforts of one community are often negated by those of nearby towns and cities, underscoring the need for agencies to coordinate efforts and develop comprehensive land use plans for an extensive length of shoreline.

An alternative to erecting artificial structures is beach replenishment: adding sand to beaches to replace that lost by erosion. This provides recreational opportunities; at the same time, it buffers property from damage by storms. Sand may be dredged from harbors or pumped from offshore sandbars. The disadvantages of this technique are that dredging disrupts marine organisms, sand of the right texture may be hard to obtain, and replenished beaches can be short-lived. For example, more than $5 million was spent replenishing the beach at Ocean City, New Jersey, in 1982. The beach disappeared within three months, after a series of northeasters hit the area.

The expense of maintaining coastal zones raises two basic questions: who benefits and who should pay. Some people contend that the interests of those who own coastal property are not compatible with the public interest, and it is unwise to expend large amounts of public funds to protect the property of only a few. People argue that shorefront businesses and homeowners are the chief beneficiaries of shore protection measures, that they often deny the public access to the beaches in front of their property, and, therefore, they should pay for the majority of the cost of maintaining the shoreline. At present, however, this is rarely the case; costs are typically shared by communities, states, and the federal government. Indeed, 51 federal programs subsidize coastal development and redevelopment. The largest is the National Flood Insurance Program, which offers low-cost insurance to homeowners in flood-prone areas. People have used the protection of that insurance to build anywhere, even in high-risk coastal areas.

Questions to Consider

1. From 1994 to the present, the Army Corps of Engineers has spent millions of dollars annually to pump sand from the ocean bottom onto eroded beaches in New Jersey. This ongoing project is based on the assumption that additional replenishment will be required every five or six years as beaches erode. Currently, the federal government pays 65% of the cost, the state government 25%, and local governments 10%. New Jersey Senator Frank Lautenberg contends that beach replenishment is critical for the region's future. "People's lives and property are at stake. Jersey's beaches bring crucial tourist dollars to the state." But James Tripp of the Environmental Defense Fund argues that beach rebuilding is simply throwing taxpayers' money into the ocean. "Pumping all the sand in the world is not going to save the day." Do you think the beach replenishment project is a wise use of taxpayers' money? Does the federal government have an obligation to protect or rebuild storm-damaged beaches? Why or why not?
2. Coastal erosion is not a problem for beaches, only for people who want to use them. Do you believe that we should learn to live with erosion, not building in the coastal zone unless we are prepared to consider our structures temporary and expendable? Should communities adopt zoning plans that prohibit building on undeveloped lands within, say, 50 meters (164 ft) of the shore?
3. Should the federal government curtail programs that provide inexpensive storm insurance for oceanfront houses and businesses, as well as speedy grants and loans for storm repairs not covered by insurance? Should people be allowed to rebuild storm-damaged buildings even if they are vulnerable to future damage? Why or why not?
4. Coastal erosion will become more serious if the current rise in sea level—about 1 inch every 12 years—continues or even increases as global warming causes sea water to expand or the polar ice caps to melt. Many of the world's major cities would be threatened by this rise in sea level. What are some of those cities? How might they protect themselves?

to have resulted from wind erosion of nonvegetated sediment deposited by meltwater from retreating glaciers. Because rich soils usually form from loess deposits, if climatic circumstances are appropriate, these areas are among the most productive agricultural lands in the world.

Landform Regions

Every piece of land not covered by buildings and other structures contains clues as to how it has changed over time. Geomorphologists interpret these clues, studying such things as earth materials and soils, the availability of water, drainage patterns, evidences of erosion, and glacial history. The scale of analysis may be as small as a stream or as large as a *landform region,* a large section of the earth's surface where a great deal of homogeneity occurs among the types of landforms that characterize it.

A map on the inside front cover of this book shows, in a general way, the kinds of landform regions found in different parts of the world. Note how the mountain belts generally coincide with plate boundaries not found beneath the sea (see Figure 3.3) and with the earthquake-prone areas (see Figure 3.6). Vast plains exist in North and South America, Europe, Asia, and Australia. Many of these regions were created under former seas and appeared as land when seas contracted. These, and the smaller plains areas, are the drainage basins for some of the great rivers of the world, such as the Mississippi-Missouri, Amazon, Volga, Nile, Ganges, and Tigris-Euphrates. The valleys carved by these rivers and the silt deposited by them are among the most agriculturally productive areas in the world. The plateau regions are many and varied. The African plateau region is the largest. Much of the African landscape is characterized by low mountains and hills whose base is about 700 meters (2300 ft) above sea level. Generally quiet from the standpoint of tectonic activity, Africa is largely made up of geologically ancient continental blocks that have been in an advanced stage of erosion for millions of years.

Humans affect and are affected by the landscape, landforms, moving continents and earthquakes; but, except at times of natural disaster, these elements of the physical world are, for most of us, quiet, accepted background. More immediate in affecting our lives and fortunes are the great patterns of climate. Climate helps define the limits of the economically possible with present levels of technology, the daily changes of weather that affect the success of picnics and crop yield alike, and the patterns of vegetation and soils. We turn our attention to these elements of the natural environment in Chapter 4.

Summary

In the most recent 200 million of the earth's 4.7 billion years, continental plates broke away from Pangaea and drifted on the asthenosphere to their present positions. Rocks, the materials that constitute the earth's surface, are classified as igneous, sedimentary, and metamorphic. At or near plate intersections, tectonic activity is particularly in evidence in two forms. Diastrophism, such as faulting, results in earthquakes and, on occasion, tsunamis. Volcanism moves molten materials toward the earth's surface.

The building up of the earth's surface is balanced by three gradational processes—weathering, gravity transfer, and erosion. Weathering, both mechanical and chemical, prepares materials for transport by disintegrating rocks. It is also instrumental in the development of soils. Talus slopes and soil-creep are examples of the effect of gravitational transfer. The erosional agents of running water, groundwater, glaciers, waves and currents, and wind move materials to new locations. Examples of landforms created by the collection of eroded materials are alluvial fans, deltas, natural levees, moraines, and sand dunes.

Key Words

For Review and Consideration

1. What evidence makes the theory of *plate tectonics* plausible?
2. What is the meaning and name of the process that occurs when two plates collide?
3. In what ways may rocks be classified? List three classes of rocks according to their origin. In what ways can they be distinguished from one another?
4. Explain what is meant by *gradation* and *volcanism.*
5. What is meant by *folding, joint,* and *faulting?*
6. Draw a diagram indicating the varieties of ways *faults* may occur.
7. With what earth movements are earthquakes associated? What is a *tsunami* and how does it develop?
8. What is the distinction between *mechanical* and *chemical weathering?* Is weathering responsible for landform creation? In what ways do glaciers engage in mechanical weathering?

9. Explain the origin of the various landforms one usually finds in desert environments.
10. How do *glaciers* form? What landscape characteristics are associated with glacial erosion? With glacial deposition?
11. What landform features can you identify on Figure 3.15? What were the agents of their creation?
12. How are alluvial fans, deltas, natural levees, and moraines formed?
13. How is groundwater erosion differentiated from surface water erosion?
14. How are the processes that bring about change due to waves and currents related to the processes that bring about change by the force of wind?
15. What processes account for the landform features of the area in which you live?

SELECTED REFERENCES

Ahnert, Frank. *Introduction to Geomorphology.* London: Edward Arnold Publishers, Ltd., 1998.

Benn, Douglas I., and David J. A. Evans. *Glaciers and Glaciation.* New York: John Wiley & Sons, Inc., 1997.

Bradshaw, Michael, and Ruth Weaver. *Foundations of Physical Geography.* Dubuque, Iowa: Wm. C. Brown Publishers, 1995.

Briggs, David, Peter Smithson, Kenneth Addison, and Ken Atkinson. *Fundamentals of the Physical Environment.* 2d ed. New York: Routledge, 1997.

Chorley, Richard J. *Water, Earth, and Man.* London: Methuen Ltd., 1969.

de Blij, H. J., and Peter O. Muller. *Physical Geography of the Global Environment.* New York: John Wiley & Sons, Inc., 1995.

French, Hugh M. *The Periglacial Environment.* 2d ed. Harlow, Essex, England: Addison Wesley Longman, 1996.

Goudie, Andrew. *The Changing Earth: Rates of Geomorphological Processes.* Oxford: Blackwell Publishers, 1995.

Knighton, David. *Fluvial Forms and Processes.* New York: John Wiley & Sons, Inc., 1995.

Lancaster, Nicholas. *The Geomorphology of Desert Dunes.* New York: Routledge, 1995.

Livingston, Ian, and Andrew Warren. *Aeolian Geomorphology: An Introduction.* Harlow, Essex, England: Addison Wesley Longman, 1996.

Ollier, Cliff. *Volcanoes.* New York: Basil Blackwell, Inc., 1988.

Progress in Physical Geography. London: Edward Arnold Publishers, Ltd. Various issues.

Rolls, David, and Will J. Bland. *Weathering: An Introduction to Basic Principles.* London: Edward Arnold Publishers, Ltd., 1998.

Ross, David. *Introduction to Oceanography.* Harlow, Essex, England: Addison Wesley Longman, 1995.

Strahler, Alan H., and Arthur N. Strahler. *Introducing Physical Geography.* New York: John Wiley & Sons, Inc., 1994.

Thorn, Colin E. *An Introduction to Theoretical Geomorphology.* London: Unwin Hyman, 1988.

Trenhaile, Alan S. *Coastal Dynamics and Landforms.* Oxford: Oxford University Press, 1997.

Viles, Heather, and Tom Spencer. *Coastal Problems: Geomorphology, Ecology and Society at the Coast.* New York: John Wiley & Sons, Inc., 1995.

Wallen, Robert N. *Introduction to Physical Geography.* Dubuque, Iowa: Wm. C. Brown Publishers, 1992.

Don't forget about Dushkin's *Annual Editions Online: Geography* at http://www.dushkin.com/aeonline/. See preface for details.

PHYSICAL GEOGRAPHY: WEATHER AND CLIMATE

CHAPTER 4

Thunderstorm in the prairies.

It started as a small disturbance in the Atlantic Ocean, east of the Caribbean islands in early July. But as is often the case in the summer hurricane season, the disturbance soon became a tropical storm, and then, after its winds were clocked at greater than 125 kilometers per hour (74 mph), it was designated Bertha, the second full-blown hurricane of the 1996 season. Fed by warm tropical waters, the hurricane advanced westward (Figure 4.1).

In the tourist town of Charlotte Amalie, on the island of St. Thomas in the Virgin Islands, cleanup, especially the replacement of roofs blown off by Hurricane Marilyn in 1995, was proceeding slowly. The blue tarps used to protect the houses began to dot the sky like so many flying carpets as Bertha began to demonstrate its fury.

Bertha did not stop after creating havoc in the Virgin Islands, Puerto Rico, and the Bahamas. High winds downed trees, and damaged houses. Wind-driven high water flooded coastal regions; roads were washed out. Six people were killed. The hurricane plowed into coastal North Carolina at the height of the summer vacation season. A million people left the low-lying coastal areas from Florida to Virginia for fear that waves, winds, and rising water would put their lives in jeopardy. Fortunately, early warning bulletins from Miami's Hurricane Warning Center gave people time to escape. Still, Hurricane Bertha caused great disruption and devastation in coastal North Carolina communities hit by 169-kilometer-per-hour (105-mph) winds, heavy rain, and seas that rose 2.4 meters (8 ft) above high tide. Before it spun itself out, coastal areas from North Carolina to Nova Scotia, Canada, experienced beach erosion and heavy rains.

The power of hurricanes is concentrated in a narrow path. Whether such meteorological events occur in Asia or North America, they do great damage. The lives of all in the paths of these storms are affected. Tropical storms are one extreme type of weather phenomenon. Most people are "weather watchers"—they watch television forecasts with great interest and plan their lives around weather events. In this chapter, we review the subsection of physical geography concerned with weather and climate. It deals with normal, patterned phenomena from which such abnormalities as Hurricane Bertha occasionally emerge.

A weather forecaster describes current conditions for a limited region, such as a metropolitan area, and predicts future weather conditions. If the elements that make up the **weather,** like temperature, wind, and precipitation, are recorded at specified moments in time, such as every hour, an inventory of weather conditions can be developed. By finding trends in data that have been gathered over an extended period of time, we can speak about typical conditions. These characteristic circumstances describe the **climate** of a region. Weather is a moment's view of the lower atmosphere, while climate is a description of typical weather conditions in an area or at a place over a period of time. Geographers analyze the differences in weather and climate from place to place in order to understand how climatic elements affect human occupance of the earth.

In geography, we are particularly interested in the physical environment that surrounds us. That is why the **troposphere,** the lowest layer of the earth's atmosphere, attracts our attention. This layer, extending about 10 kilometers (6 mi) above the ground, contains virtually all of the air, clouds, and precipitation of the earth.

In this chapter, we try to answer the questions usually raised regarding characteristics of the lower atmosphere. By discussing these answers from the viewpoint of averages or average variations, we attempt to give a view of the earth's climatic differences, a view held to be very important for understanding the way people use the land. Climate is a key to understanding, in a broad way, the distribution of world population. People have great difficulty living in areas that are, on average, very cold, very hot, very dry, or very wet. They are also negatively affected by huge storms or flooding. In this chapter, we first discuss the elements that constitute weather conditions, and then describe the various climates of the earth.

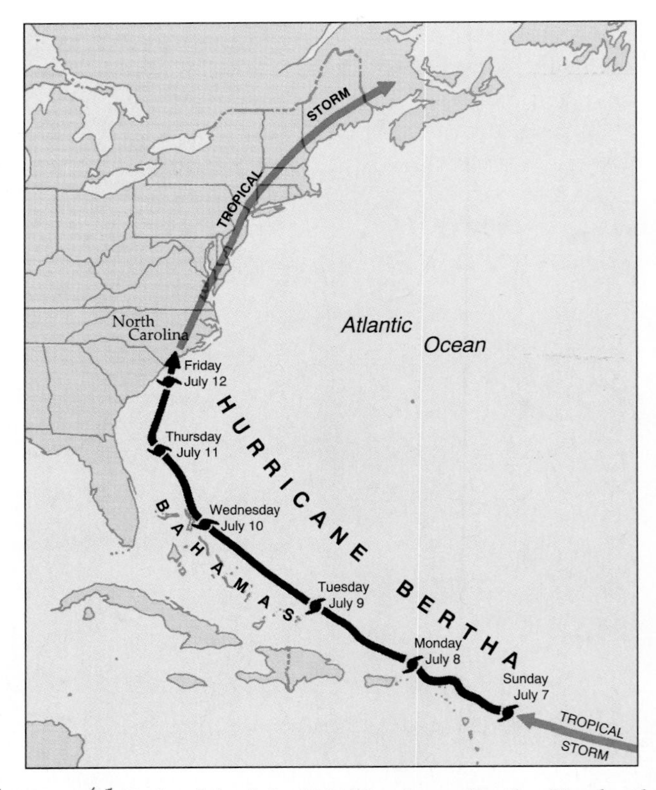

FIGURE 4.1 **Path of the July 1996 Hurricane Bertha.** Hardest hit were the Bahamas and the North Carolina coast, but a great deal of rain and wind affected areas as far north as Maine.

Air Temperature

Perhaps the most fundamental question about weather is: Why do temperatures vary from place to place? The answer to this question requires the discussion of a number of concepts to help focus on the way heat accumulates on the earth's surface.

Energy from the sun, called *solar energy,* is transformed into heat, primarily at the earth's surface and secondarily in the atmosphere. Not every part of the earth or its overlying atmosphere receives the same amount of solar energy. At any given place, the amount of incoming solar radiation, or **insolation,** available depends on the intensity and duration of radiation from the sun. These are determined by both the angle at which the sun's rays strike the earth and the number of daylight hours. These two fundamental factors, plus the following five modifying variables, determine the temperature at any given location:

1. the amount of water vapor in the air;
2. the degree of cloud cover (or cover in general);
3. the nature of the surface of the earth;
4. the elevation above sea level;
5. the degree of air movement.

Let us look at these factors briefly.

Earth Inclination

The earth, as Figure 4.2 indicates, does not spin on an axis that is perpendicular to a line connecting the center of the earth and the sun. Rather, the earth's axis is tilted about 23.5° away from the perpendicular; every 24 hours the earth rotates once on that axis, as shown in Figure 4.3. While rotating, the earth is slowly revolving around the sun in a nearly circular annual orbit. The axis of the earth, that is, the imaginary line connecting the North Pole to the South Pole, always remains in the same position. In other words, there is parallelism of the earth's axis during rotation and revolution (Figure 4.4). If the earth were not tilted from the perpendicular, the solar energy received *at a given latitude* would not vary during the course of the year. The rays of the sun would directly strike the equator, and as distance away from the equator became greater, rays would strike the earth at ever-increasing angles, therefore diminishing the intensity of the energy and giving climates a latitudinal standardization (Figures 4.5 and 4.6).

Because of the inclination, however, the location of highest incidence of incoming solar energy varies during the course of the year. When the Northern Hemisphere is tilted directly toward the sun, the vertical rays are felt as far north as 23.5°N latitude. This position of the earth occurs about June 21, the summer *solstice* for the Northern

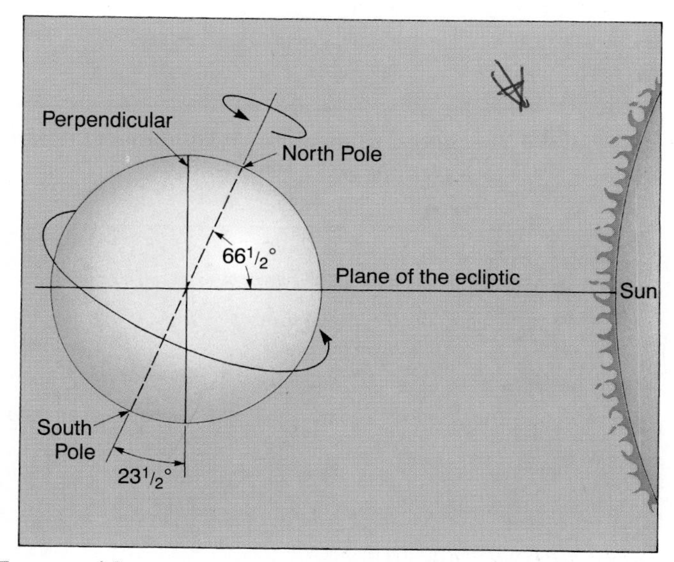

FIGURE 4.2 The earth spins on an axis tilted about 23.5° from the perpendicular or 66.5° from the *plane of the ecliptic,* an imaginary plane that contains the lines connecting the center of the earth at all times of the year with the center of the sun.

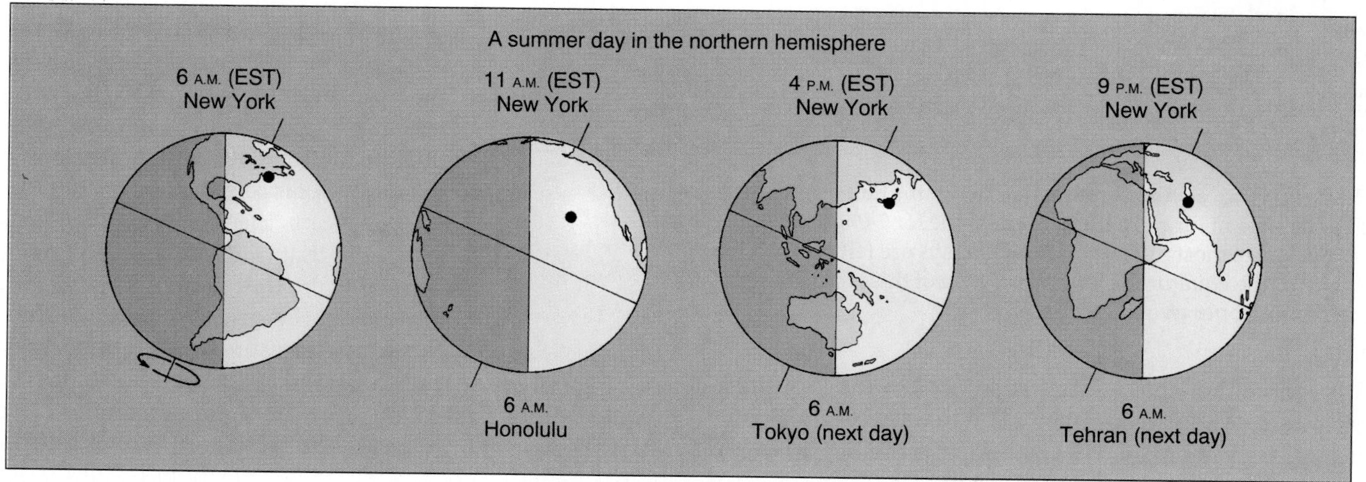

FIGURE 4.3 **The process of the 24-hour rotation of the earth on its axis.**

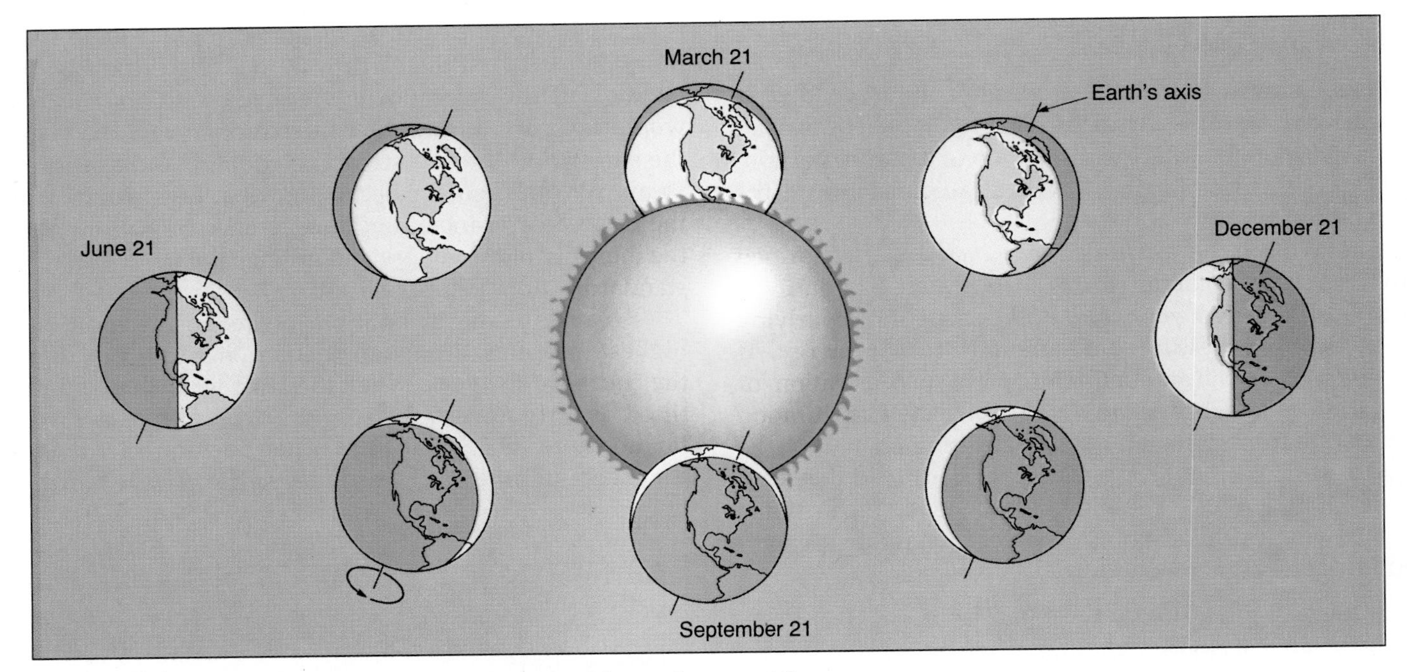

FIGURE 4.4 **The process of the yearly revolution of the earth around the sun.**

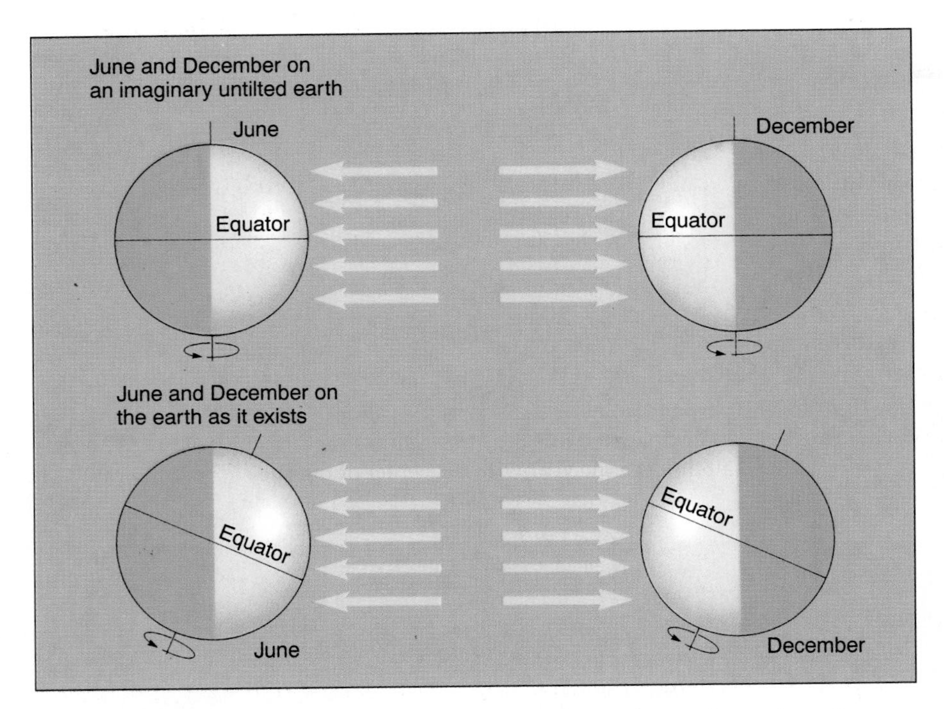

FIGURE 4.5 Notice on the bottom two diagrams that as the earth revolves, the north polar area in June is bathed in sunshine for 24 hours, while the south polar area is dark. The most intense of the sun's rays are felt north of the equator in June and south of the equator in December. None of this is true in the untilted examples shown on the upper two diagrams.

Hemisphere and the winter solstice for the Southern Hemisphere. About December 21, when the vertical rays of the sun strike near 23.5°S latitude, it is the beginning of summer in the Southern Hemisphere and the onset of winter in the Northern Hemisphere. During the rest of the year, the position of the earth relative to the sun results in direct rays migrating from about 23.5°N to 23.5°S and back again. On about March 21 and September 21 (the spring and autumn *equinoxes*), the vertical rays of the sun strike the equator.

The tilt of the earth also means that the length of days and nights varies during the year. One-half of the earth is always illuminated by the sun, but only at the equator is it light for 12 hours each day of the year. As distance away from the equator becomes greater, the hours of daylight or darkness increase, depending on whether the direct rays of the sun are north or south of the equator. In the summer, daylight increases to the maximum of 24 hours in the summer polar region, and during the same period, nighttime finally reaches 24 hours in length in the other polar region.

Because of the 24-hour daylight, it would seem that much solar energy should be available in the summer polar region, but this is not the case. The angle of the sun is so narrow (the sun is low in the sky) that solar energy is spread over a wide surface. By contrast, the combination of relatively long days and sun angles close to 90° makes an enormous amount of energy available to areas in the neighborhood of 15° to 30° north and south latitude during each hemisphere's summer.

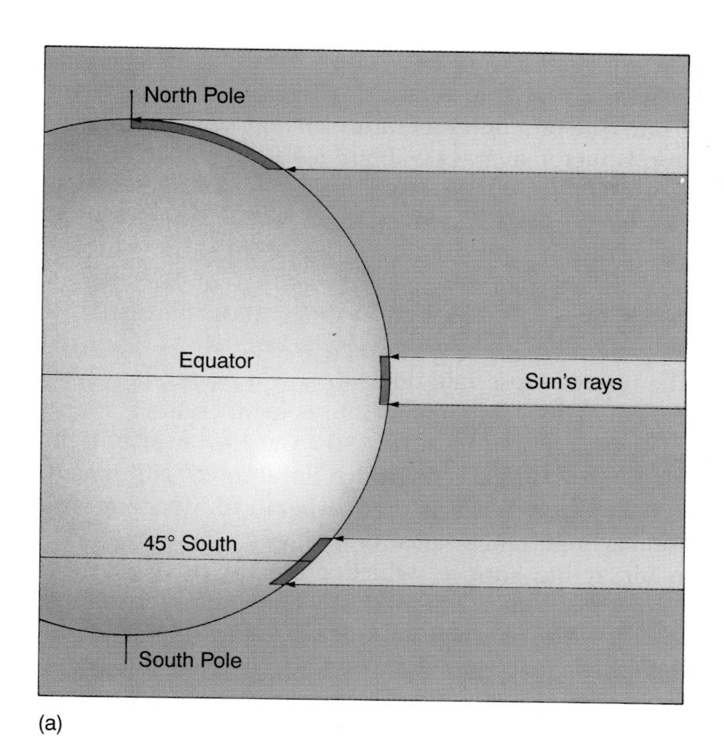

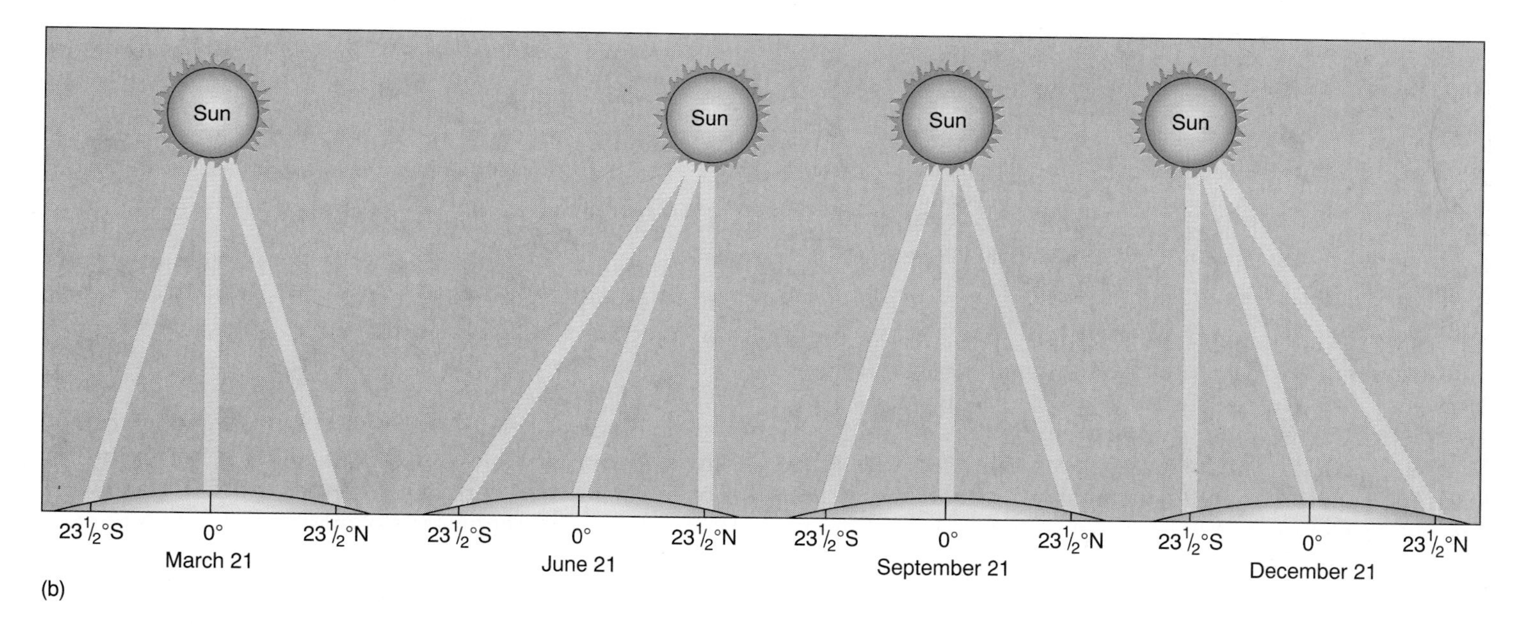

FIGURE 4.6 (a) Three equal, imaginary rays from the sun are shown striking the earth at different latitudes at the time of the equinox. As distance away from the equator increases, the rays become more diffused, showing how the sun's intensity is diluted in the high latitudes. (b) Imaginary rays from the sun at spring and autumn equinoxes and summer and winter solstices.

Reflection and Reradiation

Much of the potentially receivable insolation is, in fact, sent back to outer space or diffused in the troposphere in a process known as **reflection**. Clouds, which are dense concentrations of suspended, tiny water or ice particles, reflect a great deal of energy. Light-colored surfaces, especially snow cover, also serve to reflect large amounts of solar energy.

Energy is lost through reradiation as well as reflection. In the **reradiation** process, the earth acts as a communicator of energy. As indicated in Figure 4.7, the shortwave energy that is absorbed into the land and water is returned to the atmosphere in the form of longwave terrestrial radiation. On a clear night, when no clouds can block or diffuse movement, temperatures continually decrease, as the earth reradiates as heat the energy it has received and stored during the course of the day.

Some kinds of earth surface material, especially water, store solar energy more effectively than others. Because water is transparent, solar rays can penetrate a great distance below its surface. If water currents are present, heat is distributed even more effectively. On the other hand, land surfaces are opaque, so all of the energy received from the sun is concentrated at the surface. Land, having more heat available at the surface, reradiates its energy faster than water. Air is heated by the process of reradiation from the earth and not directly by energy from the sun passing through it. Thus, because land heats and cools more rapidly than water, hot and cold temperature extremes recorded on earth occur on land and not the sea.

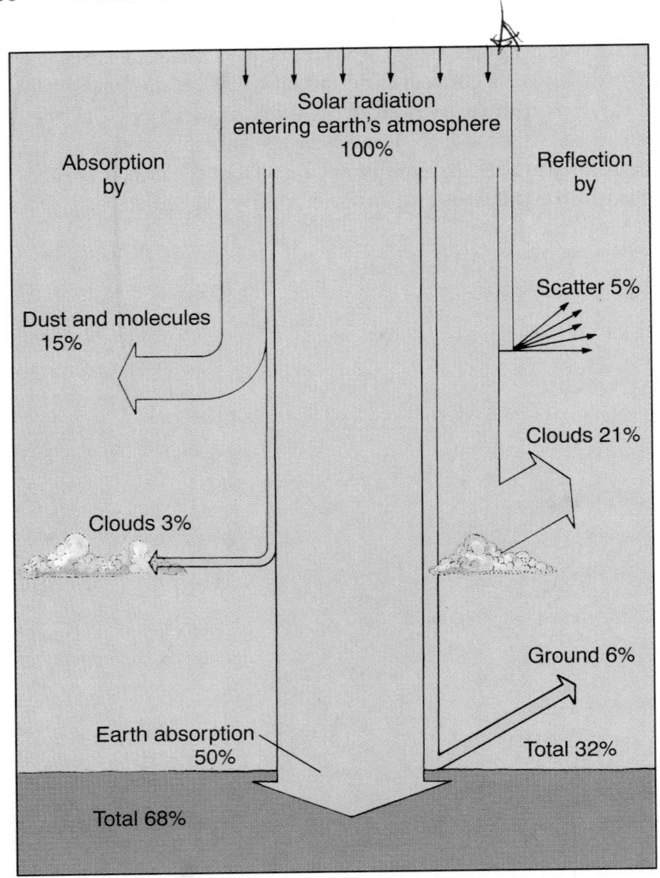

FIGURE 4.7 Consider the incoming solar radiation as 100%. The portion that is absorbed into the earth (50%) is eventually released to the atmosphere and then reradiated into space. Notice that the outgoing radiation is equal to 100%, showing that there is an energy balance on the earth.

Temperatures are moderated by the presence of large bodies of water near land areas. Note in Figure 4.8 that coastal areas have lower summer temperatures and higher winter temperatures than those places at the same distance from the equator, excluding seacoasts. Land areas affected by the moderating influences of water are considered *marine* environments; those areas not affected by nearby water are *continental* environments.

Temperatures vary in a cyclical way from day to day. In the course of a day, as incoming solar energy exceeds energy lost through reflection and reradiation, temperatures begin to rise. The ground stores some heat, and temperatures continue to rise until the angle of the sun becomes so narrow that energy received no longer exceeds that lost by the reflection and reradiation processes. Not all of the heat loss occurs during the night, but long nights appreciably deplete stored energy.

The Lapse Rate

We may think that as we move vertically away from the earth toward the sun, temperatures would increase. However, this is not true within the troposphere. The earth absorbs and reradiates heat; therefore, temperatures are usually warmest at the earth's surface and lower as elevation increases. Note on Figure 4.9 that this temperature **lapse rate** (the rate of temperature change with altitude in the troposphere) averages about 6.4°C per 1000 meters (3.5°F per 1000 ft). For example, the difference in elevation between Denver and Pikes Peak is about 2700 meters (9000 ft), which normally results in a 17°C (32°F) difference in temperature. Jet planes flying at an altitude of 9100 meters (30,000 ft) are moving through air that is about 56°C (100°F) colder than ground temperatures.

The normal lapse rate does not always hold, however. Rapid reradiation sometimes causes temperatures to be higher above the earth's surface than at the surface itself. This particular condition, in which air at lower altitudes is cooler than air aloft, is called a **temperature inversion.** An inversion is important because of its effect on air movement. Warm air at the surface, which normally rises, may be blocked by the even warmer air of a temperature inversion (Figure 4.10). Thus surface air is trapped; if it is filled with automobile exhaust emissions or smoke, a serious smog condition may develop (see "The Donora Tragedy"). Because of the configuration of nearby mountains, Los Angeles, pictured in Figure 4.11, often experiences temperature inversions, causing sunlight to be reduced to a dull haze.

The effect of air movement on temperature is made clear in the following section on air pressure and winds.

Air Pressure and Winds

A second fundamental question about weather and climate concerns **air pressure:** How do differences in air pressure from place to place affect weather conditions? The answer to this question first requires that we explain why differences in air pressure occur.

Air is a gaseous substance whose weight affects air pressure. If it were possible to carve out 16.39 cubic centimeters (1 cu in.) of air at the earth's surface and weigh it, along with all the other cubic centimeters of air above it, under normal conditions the total weight would equal approximately 6.67 kilograms (14.7 lbs) of air as measured at sea level. Actually, this is not very heavy when you consider the dimensions of the column of air: 2.54 centimeters by 2.54 centimeters (1 in. by 1 in.) by about 9.7 kilometers (6 mi), or about 6.2 cubic meters (220 cu ft). The weight of air 4.8 kilometers (3 mi) above the earth's surface, however, is considerably less than 6.67 kilograms (14.7 lbs) because there is correspondingly less air above it. Thus, it is clear that air is heavier and air pressure is higher closer to the earth's surface. It is a physical law that for equal amounts of cold and hot air, the cold air is heavier. This law exemplifies why hot-air balloons, filled with lighter air, can rise into the atmosphere. A cold morning is characterized by relatively heavy air, but as afternoon temperatures rise, air becomes lighter.

As early as the 17th century, it was discovered that when a column of mercury is contained within a tube, normal atmospheric pressure at sea level is sufficient to balance the

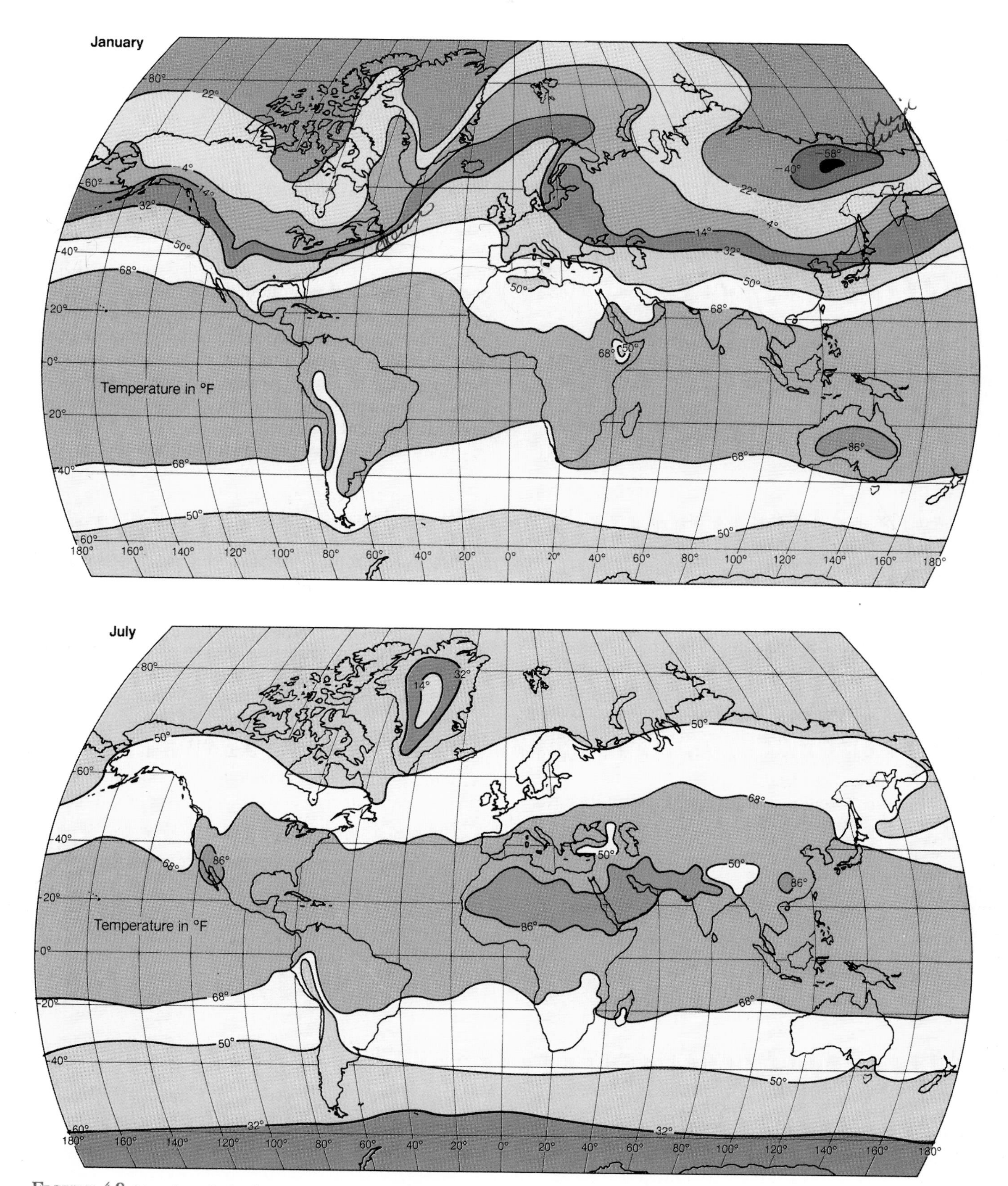

FIGURE 4.8 At a given latitude, water areas are warmer than land areas in winter and cooler in summer. *Isotherms* are lines of equal temperature.

weight of the mercury column to a height of 76 centimeters (29.92 in.). *Barometers* of varying types are used to record changes in air pressure. Barometric readings in inches of mercury are a normal part, along with recorded temperatures, of every weather report. Since air pressure at a given location changes as surface heating lightens air, barometers record a drop in atmospheric pressure when air is heating, and a rise in pressure when air is cooling.

In order to visualize the effect of air movements on weather, it is useful to think of air as a liquid made up of

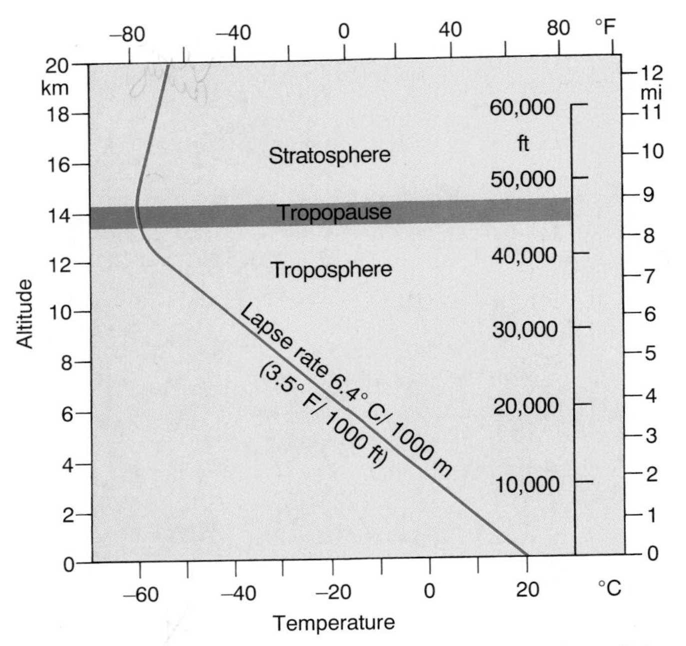

FIGURE 4.9 The temperature lapse rate under typical conditions.

two fluids with different densities (representing light air and heavy air); for example, water and gasoline. If the fluids are put into a tank at the same time, the lighter liquid will move to the top, representing the vertical motion of air. The heavier liquid spreads out horizontally along the bottom of the tank, becoming the same thickness everywhere. This flow represents the horizontal movement of the air or wind. Air attempts to achieve an equilibrium by evening out pressure imbalances that result from the heating and cooling processes previously discussed. Air races from heavy (cold) air locations to light (warm) air locations. Thus, the greater the differences in air pressure between places, the greater the wind.

Pressure Gradient Force

Because of differences in the nature of the earth's surface—water, snow cover, dark green forests, cities, and so on—and the other previously mentioned factors that affect energy receipt and retention, zones of high and low air pressure develop. Sometimes these high- and low-pressure zones cover entire continents, but usually they are considerably smaller—several hundred miles wide—and within these regions, small differences are noted over short distances. When pressure differences exist between areas, a **pressure gradient** is formed.

In order to balance pressure differences that have developed, air from the heavier high-pressure areas flows to low-pressure zones. Heavy air stays close to the earth's surface as it moves, producing winds, and helping to speed the upward movement of warm air. The velocity, or speed, of the wind is in direct proportion to pressure differences. As depicted in Figure 4.12, winds are caused by pressure differences that induce airflow from points of high pressure to points of low pressure. If distances between high- and low-pressure zones are short, pressure gradients are steep and wind velocities are great. The equalization process results in more gentle air movements when zones of different pressures are far apart.

The Convection System

A room's temperature is lower near the floor than the ceiling because warm air rises and cool air descends. The circulatory

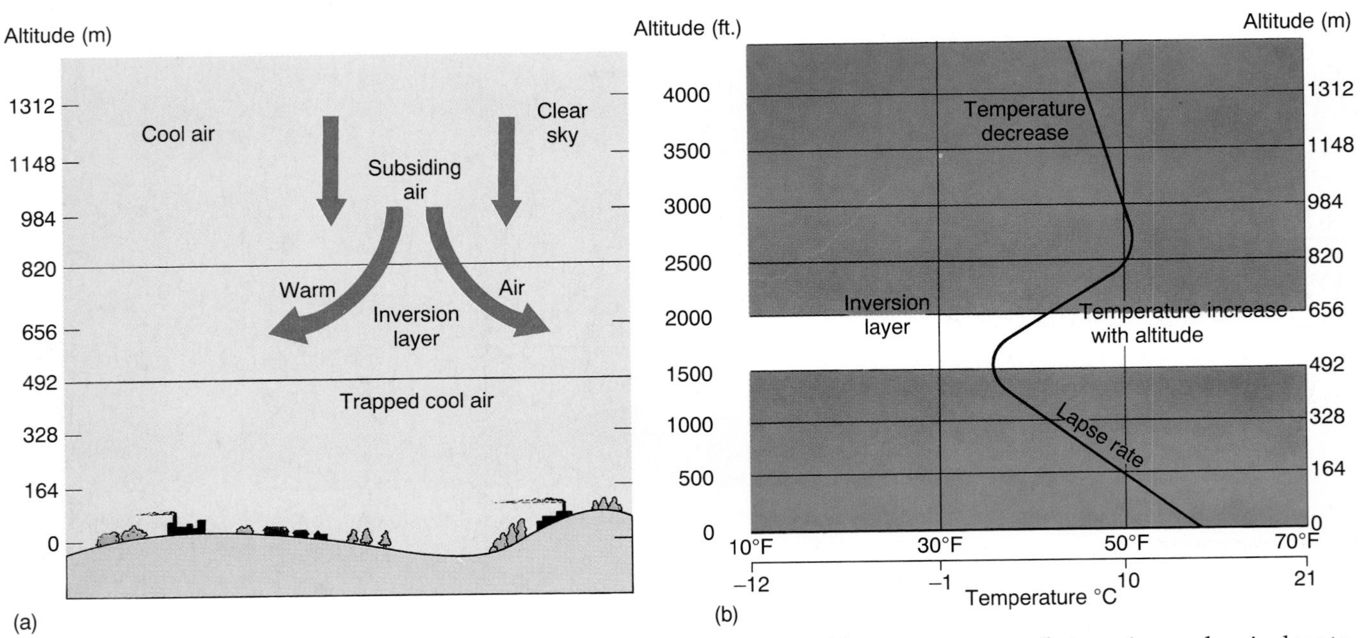

FIGURE 4.10 **Temperature inversion.** (a) A layer of warm, subsiding air acts like a cap, temporarily trapping cooler air close to the ground. (b) Note that air temperature decreases with distance from the ground until the warm inversion layer is reached, at which point the temperature increases.

THE DONORA TRAGEDY

A heavy fog settled over the valley town of Donora, Pennsylvania, in late October 1948. Stagnant, moisture-filled air was trapped in the valley by surrounding hills and by a temperature inversion that held cooler air, gradually filling with smoke and fumes from the town's zinc works, against the ground under a lid of lighter, warmer, upper air. For five days the smog increased in concentration; the sulfur dioxide emitted from the zinc works continually converted to deadly sulfur trioxide by contact with the air.

Both old and young, with and without past histories of respiratory problems, reported to doctors and hospitals a difficulty in breathing and unbearable chest pains. Before the rains washed the air clean nearly a week after the smog buildup, 20 were dead and hundreds of others hospitalized. A normally harmless, water-saturated inversion had been converted to deadly poison by a tragic union of natural weather processes and human activity.

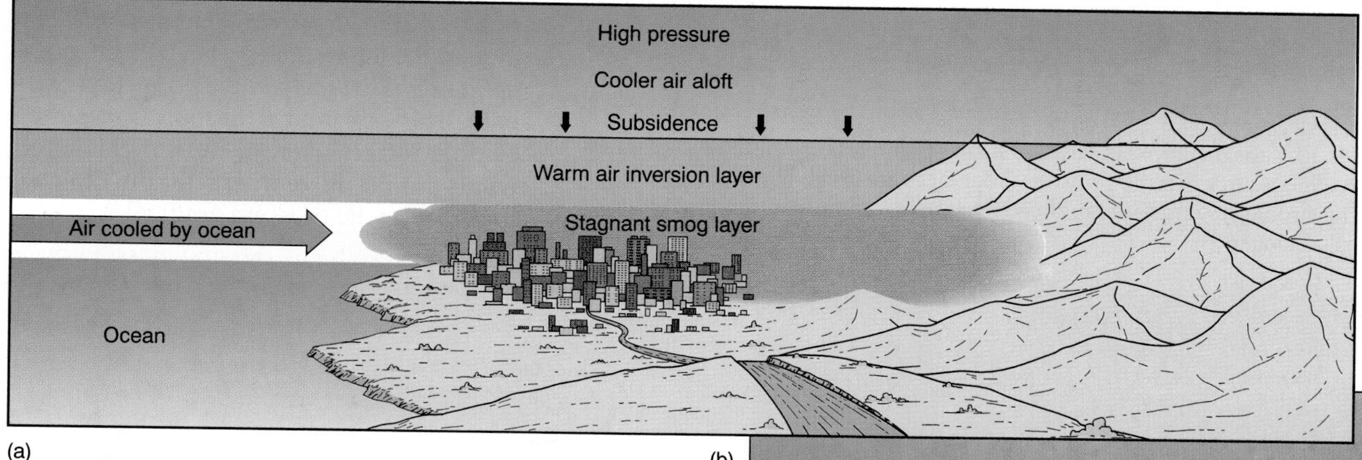

(a)

(b)

FIGURE 4.11 **Smog in the Los Angeles area.** (a) Below the inversion layer, stagnant air holds increasing amounts of pollutants, caused mainly by automobile exhausts. (b) Afternoon smog in Los Angeles.

(b) ©Ken Biggs/Tony Stone Images.

motion of descending cool air and ascending warm air is known as **convection.** A convectional wind system results from the flow of air that replaces warm, rising air and the rapid movement of replacement air.

Land and Sea Breezes

A good example of a convectional system is **land** and **sea breezes** (Figure 4.13a). Close to a large body of water, the differential daytime heating between land and water is great. As a result, warmer air over the land rises vertically, only to be replaced by cooler air from over the sea. At night, just the opposite occurs; the water is warmer than the land, which has reradiated much of its heat, and the result is a land breeze toward the sea. These two winds make seashore locations in warm climates particularly comfortable.

Mountain and Valley Breezes

Gravitational force causes the heavy cool air that accumulates over snow in mountainous areas to descend into lower valley locations, as suggested in Figure 4.13b. Consequently, valleys can become much colder than the slopes, and a temperature inversion occurs. Slopes are the preferred sites for agriculture in mountainous regions because cold air from **mountain breezes** can cause freezing conditions in the valleys. In densely settled narrow valleys where industry is concentrated, air pollution can become particularly dangerous. Mountain breezes usually occur during the night; **valley breezes**—caused by warm air moving up slopes in mountainous regions—are usually a daytime phenomenon. The canyons of southern California are the scenes of strong mountain and valley breezes. In addition, during the dry season, they become dangerous areas for the spread of brush and forest fires.

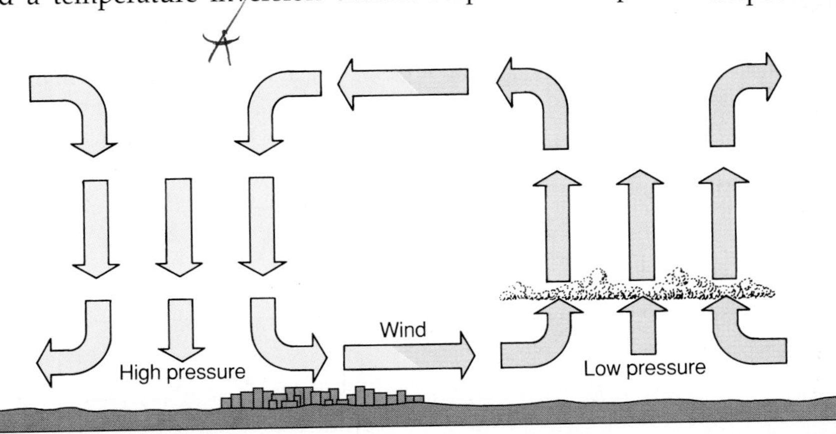

FIGURE 4.12 If the distance between centers of high pressure and low pressure is short, the wind is strong. If the pressure differences are great, winds are again strong. The strongest winds are produced by extreme pressure differences over very short distances.

Coriolis Effect

In the process of moving from high to low pressure, wind appears to veer toward the right in the Northern Hemisphere and toward the left in the Southern Hemisphere, no matter what the compass direction of the path. This apparent deflection is called the **Coriolis effect.** Were it not for this effect, winds would move in exactly the direction specified by the pressure gradient.

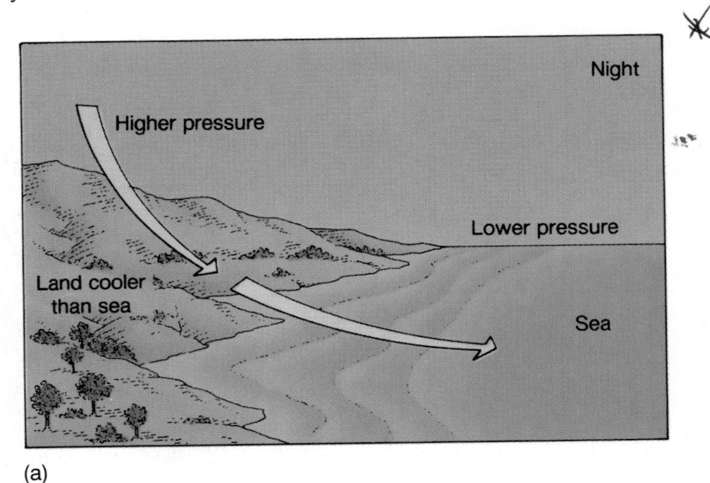

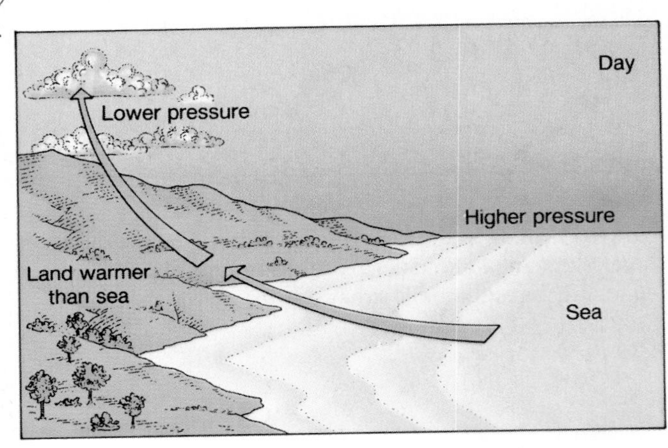

(a)

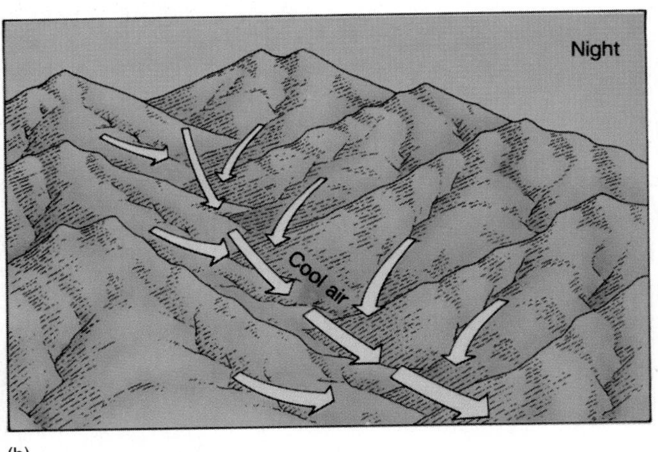

(b)

FIGURE 4.13 **Convectional wind effects due to differential heating and cooling.** (a) Land and sea breezes. (b) Mountain and valley breezes.

To illustrate the impact of the Coriolis effect upon winds, a familiar example may be helpful. Imagine a line of ice skaters holding hands while skating in a circle, with one skater nearest the center of the circle. This skater turns slowly, while the outermost skater must skate very rapidly in order to keep the line straight. In a similar way, the equatorial regions of the earth are rotating at a much faster rate than the areas around the poles.

Next, suppose that the center skater threw a ball directly toward the skater at the end of the line. By the time the ball arrived, it would pass behind the outside skater. If the skaters are going in a counterclockwise direction—as the earth appears to be moving viewed from the position of the North Pole—the ball appears to the person at the North Pole to pass to the right of the outside skater. If the skaters are going in a clockwise direction—as the earth appears to be moving viewed from the South Pole—the deflection is to the left. Because air (like the ball) is not firmly attached to the earth, it, too, will appear to be deflected. The air maintains its direction of movement, but the earth's surface moves out from under it. Since the position of the air is measured relative to the earth's surface, the air appears to have diverged from its straight path.

The Coriolis effect and the pressure gradient force produce spirals rather than simple straight-line patterns of wind, as indicated in Figure 4.14. The spiral of wind is the basic form of the many storms that are so important to the earth's air circulation system. These storm patterns are discussed later in this chapter.

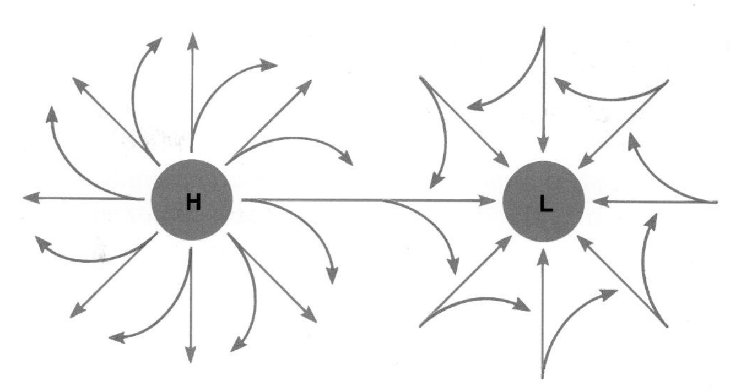

FIGURE 4.14 **The effect of the Coriolis force on flowing air.** The straight arrows indicate the paths that winds would follow flowing out of an area of high (H) pressure or into one of low (L) pressure were they to follow the paths dictated by pressure differentials. The curved arrows represent the apparent deflecting effect of the Coriolis force. Wind direction—indicated on the diagram by selected curved arrows—is always given by the direction *from* which the wind is coming.

The Frictional Effect

Wind movement is slowed by the frictional drag of the earth's surface. The effect is strongest at the surface and declines until it becomes ineffective at about 1524 meters (about one mile) above the surface. Not only is wind speed decreased, but wind direction is changed. Instead of following a course exactly dictated by the pressure gradient force or by the Coriolis effect, the **frictional effect** causes wind to follow an intermediate path.

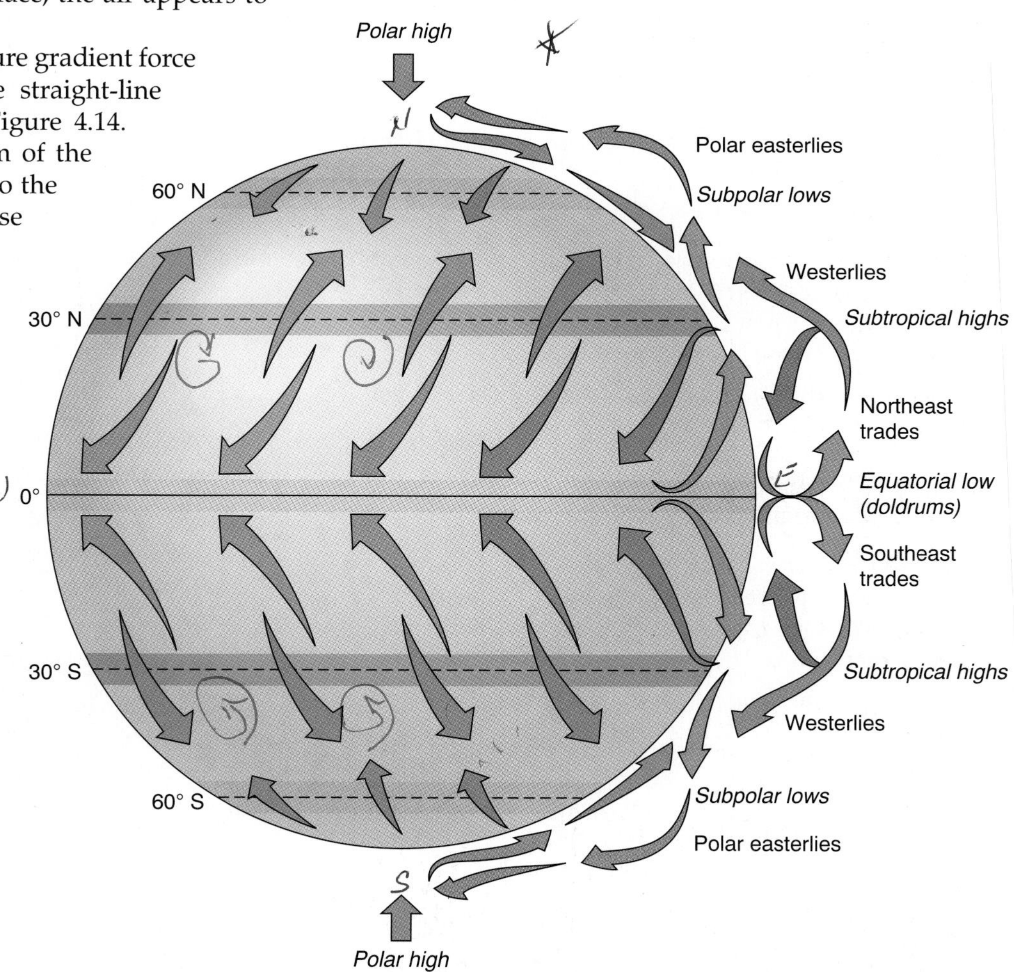

FIGURE 4.15 **The planetary wind and pressure belts as they would develop on an earth of homogeneous surface.** The high and low pressure belts represent surface pressure conditions; the wind belts are prevailing surface wind movements responding to pressure gradients and the Coriolis effect. Contrasts between land and water areas on the earth's surface, particularly evident in the Northern Hemisphere, create complex distortions of this simplified pattern.

Wind Belts

Equatorial areas of the earth are zones of low pressure. Intense solar heating in these areas is responsible for a convectional effect. Note in Figure 4.15 how the warm air rises and tends to move

On-Line Physical Geography: Weather and Climate

WeatherNet, sponsored by The Weather Underground at the University of Michigan, calls itself "the Internet's premier source of weather information." It provides access to thousands of forecasts, images, and the Net's largest collection of weather links. Also available are the latest short- and medium-range computer model graphics, tabular forecast output, and National Weather Service discussions. There are pictures of weather conditions at more than 120 places in North America, surface and upper-air analyses, including temperature maps, regional weather plots, and jet stream maps.

http://cirrus.sprl.umich.edu/wxnet/

The *NASA-Goddard Climate and Radiation Branch* site provides information about ongoing research projects. Although some of the reports are rather technical in nature, there are excellent explanations of topics such as tropospheric aerosols, cloud radiative processes, climate analysis, climate theory and modeling, and the use of satellite remote sensing.

http://climate.gsfc.nasa.gov/

The Weather Channel, a popular cable television channel, now maintains its own pages on the web. The site provides daily U.S. and international weather forecasts, maps, and helpful weather information for teachers.

http://www.weather.com/weather

Storm Chasers Homepage is designed for the storm chaser to get the latest weather information from the National Weather Service (NWS), and provides the NWS a place to receive chaser reports, inquiries, and information from chasers.

http://www.stormchaser.niu.edu/chaser/chaser2.html

Tornado Project Online provides information about tornadoes, their intensity, and common tornado myths.

http://www.tornadoproject.com/index.html

The *Woods Hole Oceanographic Institute,* one of the country's leading centers for oceanographic research, maintains this comprehensive resource covering a wide variety of oceanographic information.

http://www.whoi.edu/

National Oceanic and Atmospheric Administration is the general page for NOAA information and links to all major NOAA sites, a few of which are noted in the next column. The mission of NOAA is to describe and predict changes in the earth's environment, and conserve and manage wisely the nation's coastal and marine resources to ensure sustainable economic opportunities.

http://www.noaa.gov/

National Weather Service Welcome Page is the government source for weather information. There are links to the NOAA climatic and historic data records.

http://www.nws.noaa.gov/index.html

The *Storm Prediction Center* of the National Severe Storms Laboratory within NOAA monitors and forecasts severe weather for the continental United States. The site has the latest information about severe weather and a large historical archive of information back to 1950.

http://www.nssl.noaa.gov/~spc/

The *Tropical Prediction Center,* located in Miami, Florida, maintains a home page. Its branch, the National Hurricane Center (NHC), tracks tropical cyclones over the Atlantic, Caribbean, Gulf of Mexico, and the Eastern Pacific from May through November.

http://www.nhc.noaa.gov/

The *NOAA Western Region* page has extensive climatological and weather data for the Western United States. It has a text-only option for faster loading (no images).

http://nimbo.wrh.noaa.gov/

NOAA/PMEL/TAO El Niño Theme Page, another NOAA site, describes the regional and global consequences of the El Niño phenomenon, as well as discussing prediction methods and models.

http://www.pmel.noaa.gov/toga-tao/el-nino/nino-home.html

Tracking El Niño—NOVA Online Adventure, by the Public Broadcasting System (PBS) enables users to discover the effect of El Niño on global weather. Satellite images enhance the textual descriptions.

http://www.pbs.org/wgbh/nova/elnino/

A Comprehensive Bibliography on the El Niño Phenomenon is maintained by the Center for Ocean-Atmospheric Prediction Studies (COAPS) at Florida State University.

http://www.coaps.fsu.edu/lib/biblio/enso-bib-intro.html

away from the *equatorial low* pressure in both a northerly and southerly direction. As equatorial air rises, it cools and eventually becomes heavy. The light air near the surface cannot support the cool, heavy air. The heavy air falls, forming zones of high pressure. These areas of *subtropical high* pressure are located at about 30°N and 30°S of the equator.

When this cooled air reaches the earth's surface, it, too, moves in both a northerly and southerly direction. The Coriolis effect, however, modifies wind direction and creates, in the Northern Hemisphere, belts of winds called the *northeast trades* in the tropics and the *westerlies* (really the southwesterlies) in the midlatitudes. The names refer to the direction from which the winds come. Most of the United States lies within the belt of westerlies; that is, the air usually moves across the country from southwest to northeast. A series of ascending air cells also exists over the oceans to the north of the westerlies called the *subpolar low*. These areas tend to be cool and rainy. The *polar easterlies* connect the subpolar low areas to the *polar high*.

The general planetary air-circulation pattern is modified by local wind conditions. It should be clear that these belts move in unison as the vertical rays of the sun change position. For example, equatorial low conditions are evident in the area just north of the equator during the Northern Hemisphere summer and just south of the equator during the Southern Hemisphere summer. Recent evidence suggests that sun spot activity, which varies in a cycle averaging about 11 years, affects the surface wind direction within the wind belts. Air circulation will be discussed in more detail in the section on movement of air masses.

The strongest flows of upper air winds, 9 to 12 kilometers (30,000 to 40,000 ft), are the **jet streams.** These air streams, moving at 160 to 320 kilometers per hour (100 to 200 mph), from west to east in both the Northern and Southern Hemispheres, circle the earth in an undulating pattern, first north then south as they move westward. There are three to six undulations at any one time in the Northern Hemisphere, but the waves are not always continuous. These undulations, or waves, control the flow of air masses on the earth's surface. More stable undulations are likely to create similar day-to-day weather conditions. These waves tend to separate cold polar air from warm tropical air. When a wave dips far to the south in the Northern Hemisphere, cold air is brought equatorward and warm air moves poleward, bringing severe weather changes to the midlatitudes. The jet stream is more pronounced in the winter than in the summer.

Nowhere does the manner in which the seasonal shift takes place have such a profound effect on humanity as in the densely populated areas of southern and eastern Asia. The wind, which comes from the southwest during summer in India, reaches the landmass after picking up a great deal of moisture over the warm Indian Ocean. As it crosses the coast mountains and foothills of the Himalayas, the monsoon rains begin. A **monsoon** wind is one that changes direction seasonally. The summer monsoon wind brings heavy showers over most of South Asia.

In the southern and eastern parts of Asia, the farm economy, and particularly the rice crop, is totally dependent on summer monsoon rainwater. If, for any of several possible reasons, the wind shift is late or the rainfall is significantly more or less than optimum, crop failure may result. The undue prolongation of the summer monsoon rains in 1978 caused disastrous flooding, crop failure, and the loss of lives in eastern India and southeast Asia.

The transition to dry northerly monsoon winter winds occurs gradually across the region, first becoming noticeable in the north in September. By January, most of the subcontinent is dry. Then, beginning in March in southern areas, the yearly cycle repeats itself.

Ocean Currents

Surface ocean currents correspond roughly to wind direction patterns because the winds of the world set ocean currents in motion. In addition, just as differences in air pressure cause wind movements, so do differences in the density of water cause water movement. When water evaporates, residues of salt and other minerals that will not evaporate are left behind, making water denser. High-density water exists in areas of high pressure, where descending dry air readily picks up moisture. In areas of low pressure, where rainfall is plentiful, ocean water is low in density. Wind direction (including the Coriolis effect) and the differences in density cause water to move in wide paths from one part of the ocean to another.

There is an important difference between surface air movements and surface water movements. Landmasses are barriers to water movement, deflecting currents and sometimes forcing them to move in a direction opposite to the main current. Air, on the other hand, moves freely over both land and water.

The shape of an ocean basin also has an important effect on ocean current patterns. For example, the north Pacific current, which moves from west to east, strikes the western coast of Canada and the United States. The current is then forced to move both north and south, although the major movement is the cold ocean current that moves south along the California coast. In the Atlantic Ocean, however, as Figure 4.16 indicates, the current is deflected in a northeasterly direction by the shape of the coast (Nova Scotia and Newfoundland jut far into the Atlantic). It then moves freely across the ocean, past the British Isles and Norway, finally reaching the extreme northwest coast of Russia. This massive movement of warm water to northerly lands, called the **North Atlantic drift,** has enormous significance to inhabitants of those areas. Without it, northern Europe would be much colder.

Ocean currents affect not only the temperature, but also the precipitation on land areas adjacent to the ocean. A cold ocean current near land causes the air just above the water to be cold while the air above is warm. There is very little opportunity for convection, thus denying moisture to nearby land. Coastal deserts of the world usually border cold ocean currents. On the other hand, warm ocean currents—such as those off the coast of India—bring moisture to the adjacent land area, especially when prevailing winds are landward (see "El Niño," pp. 116-117).

The earlier question about ways in which differences in air pressure affect weather conditions is now answered, in terms of warm and cool air movement over various surfaces at different times of the year and different times of the day. A more complete answer regarding the causes of different types of weather conditions, however, requires an explanation of the susceptibility of places to receive precipitation, because rainfall and wind patterns are highly related.

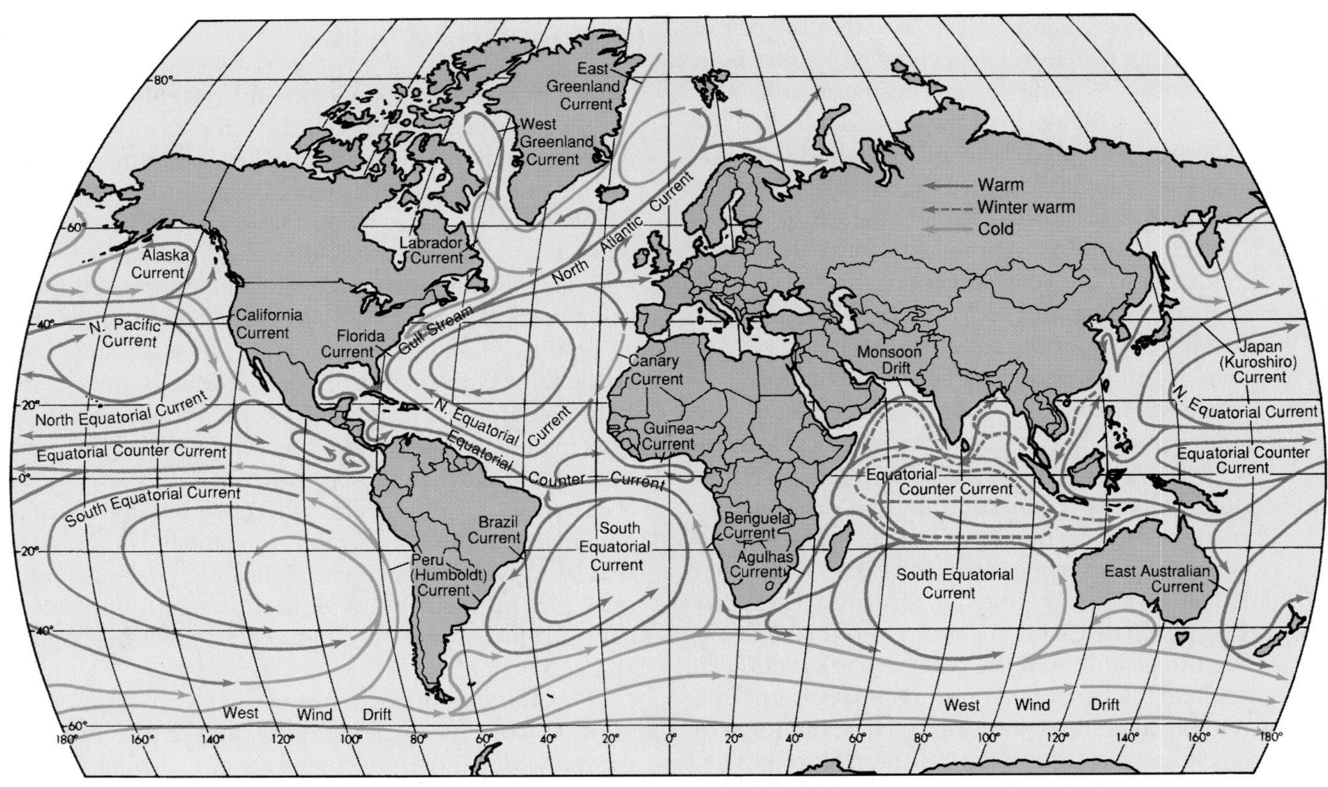

FIGURE 4.16 **The principal surface ocean currents of the world.** Notice how the warm waters of the Gulf of Mexico, the Caribbean, and the tropical Atlantic Ocean drift to northern Europe.

Moisture in the Atmosphere

Air contains water vapor (what we feel as humidity), which is the source of all precipitation. **Precipitation** is any form of water particles—rain, sleet, snow, or hail—that fall from the atmosphere and reach the earth's surface. Ascending air can expand easily because less pressure is on it. When heat from the lower air spreads through a larger volume in the troposphere, the mass of air becomes cooler. Cool air is less able to hold water vapor than warm air (Figure 4.17).

Air is said to be *saturated* when it contains so much water vapor that it condenses (changes from a gas to a liquid) and forms droplets if fine particles, called *condensation nuclei,* are present. These particles, mostly dust, pollen, smoke, and salt crystals, nearly always exist. At first, the tiny water droplets are usually too light to fall. As so many droplets coalesce into larger drops, and become too heavy to remain suspended in air, they fall as rain. When temperatures below the freezing point cause water vapor to form ice crystals instead of water droplets, then snow is created (Figure 4.18).

Large numbers of rain droplets or ice crystals form clouds, which are supported by slight upward movements of air. The form and altitude of clouds depend on the amount of water vapor in the air, the temperature, and wind movement. Descending air in high-pressure zones usually yields cloudless skies. Whenever warm, moist air rises, clouds form. The most dramatic cloud formation is probably the *cumulonimbus,* pictured in Figure 4.19. This is the anvil-head cloud that

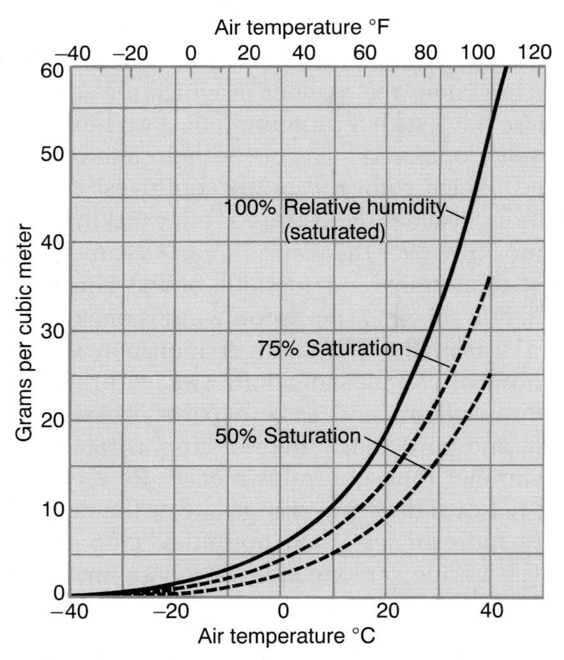

FIGURE 4.17 **The water-carrying capacity of air and relative humidity.** The actual water in the air (water vapor) divided by the water-carrying capacity (×100) equals the relative humidity. The solid line represents the maximum water-carrying capacity of air at different temperatures.

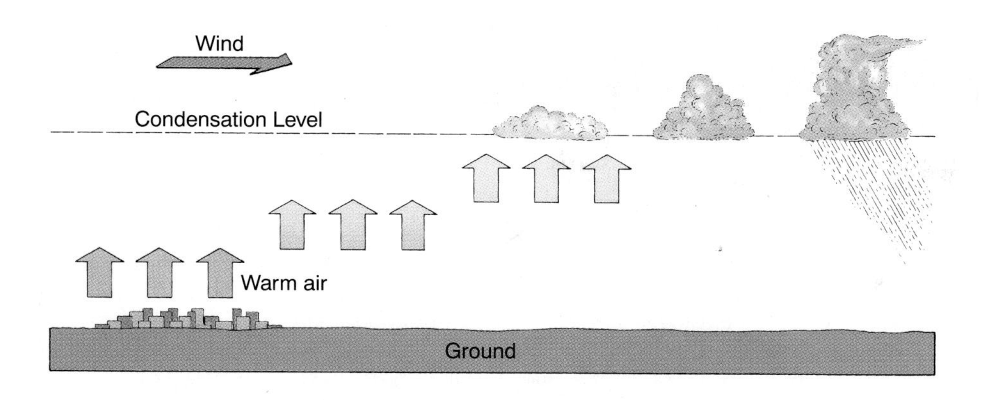

FIGURE 4.18 As warm air rises, it cools. As it cools, its water vapor condenses and clouds form. If the water content of the air is greater than the air's capacity to retain moisture, some form of precipitation is likely to occur.

(a)

(b)

(c)

(d)

FIGURE 4.19 **Cloud types:** (a) fair weather cumulus, (b) cumulonimbus, (c) stratus, (d) cirrus.

(a) ©Doug Armand/Tony Stone Images.
(b) ©Barry Wilson.
(c) A. Copley/VU.
(d) ©Steve McCutcheon/Alaska Pictorial Service.

El Niño

El Niño is a term coined years ago by fishermen who noticed that the normally cool waters off the coasts of Ecuador and Peru were considerably warmer every three or four years around Christmastime, hence the name El Niño, Spanish for "the child," referring to the infant Jesus. The fish catch was significantly reduced during these periods. If fishermen were able to identify the scientific associations that present-day oceanographers and climatologists make, they would have recognized a host of other effects that follow from El Niño.

During the winter of 1997–1998, an unusually severe El Niño caused enormous damage and hundreds of deaths. The west coast of the United States, especially California, was inundated with rainfall, amounts double, triple, and even quadruple normal. For the November to March winter period, San Francisco received 40.25 inches of rain—normal is 16.39. The 15 inches in February 1998 was the most for that month in the 150 years of record keeping in San Francisco. The resort city of Acapulco, Mexico, was badly battered by torrential rains and high, wind-blown tides. Parts of South America, especially Ecuador, Peru, and Chile, were ravaged by floods and mudslides, while droughts and fires scorched eastern South America, Australia, and parts of Asia, especially Indonesia. A stronger than usual southern branch of the jet stream generated by El Niño spawned dozens of tornadoes that killed more than one hundred people in Alabama, Georgia, and Florida.

During periods of El Niño, winds that usually blow from east to west over the central Pacific Ocean, from the cold ocean current to the warm waters of East Asia, slow or even reverse. This phenomenon occurs every two to seven years, but with different degrees of intensity. For example, an El Niño occurred in 1986–87 and again in 1991–92, but the modest warm-water buildup did not cause extraordinary circumstances, while the 1982–83 and 1997–98 events were among the most extreme on record. The cold-water peak between El Niño occurrences is called **La Niña.** The last major La Niña occurred in 1988, a year marked by drought for large portions of North America.

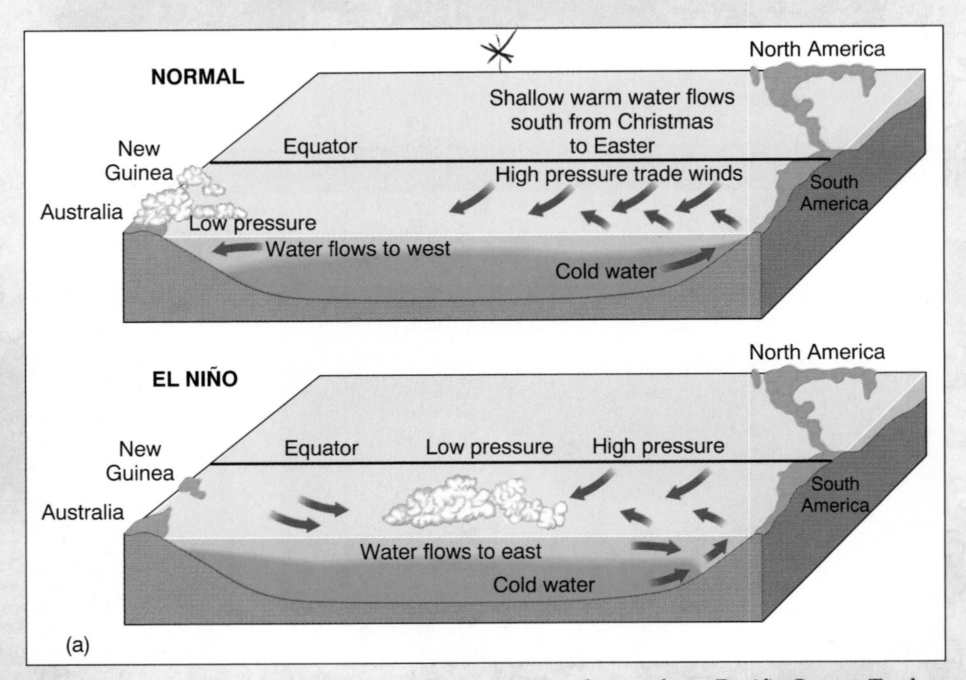

(a) The top diagram shows normal circumstances in the southern Pacific Ocean. Trade winds blow warm surface water westward and allow cold water to come to the surface along the South American coast. The bottom diagram shows that during El Niño, winds from near Australia blow warm water eastward to the coast of South America. (b) Sea surface temperatures in degrees Celsius are shown for La Niña, normal, and El Niño conditions. Notice how the extent of the warm water (red, orange colors) changes, particularly in the eastern Pacific Ocean.

Source: (a) From Michael Bradshaw and Ruth Weaver, *Physical Geography: An Introduction to Earth Environments* (St. Louis: Mosby, 1993), p. 211.

(b) Courtesy of Richard W. Reynolds/NOAA.

often accompanies heavy rain. Low, gray *stratus* clouds appear more often in cooler seasons than in warmer months. The very high, wispy *cirrus* clouds that may appear in all seasons are made entirely of ice crystals.

The amount of water vapor in the air compared to the amount when condensation would begin at a given temperature is called the air's **relative humidity.** As air gets warmer, the amount of water vapor it can contain increases. If the relative humidity is 100%, the air is completely saturated with water vapor. A relative humidity value of 60% on a hot day means the air is extremely humid and very uncomfortable. A 60% reading on a cold day, however, indicates that although the air contains relatively large amounts of water vapor, it holds, in absolute terms, much less vapor than on a hot, muggy day. This example demonstrates that relative humidity is meaningful only if we keep air temperature in mind.

Dew on the ground in the morning means that nighttime temperatures dropped to the level at which condensation took place (see Figure 4.17). The critical temperature for condensation is called the **dew point.** Foggy or cloudy conditions

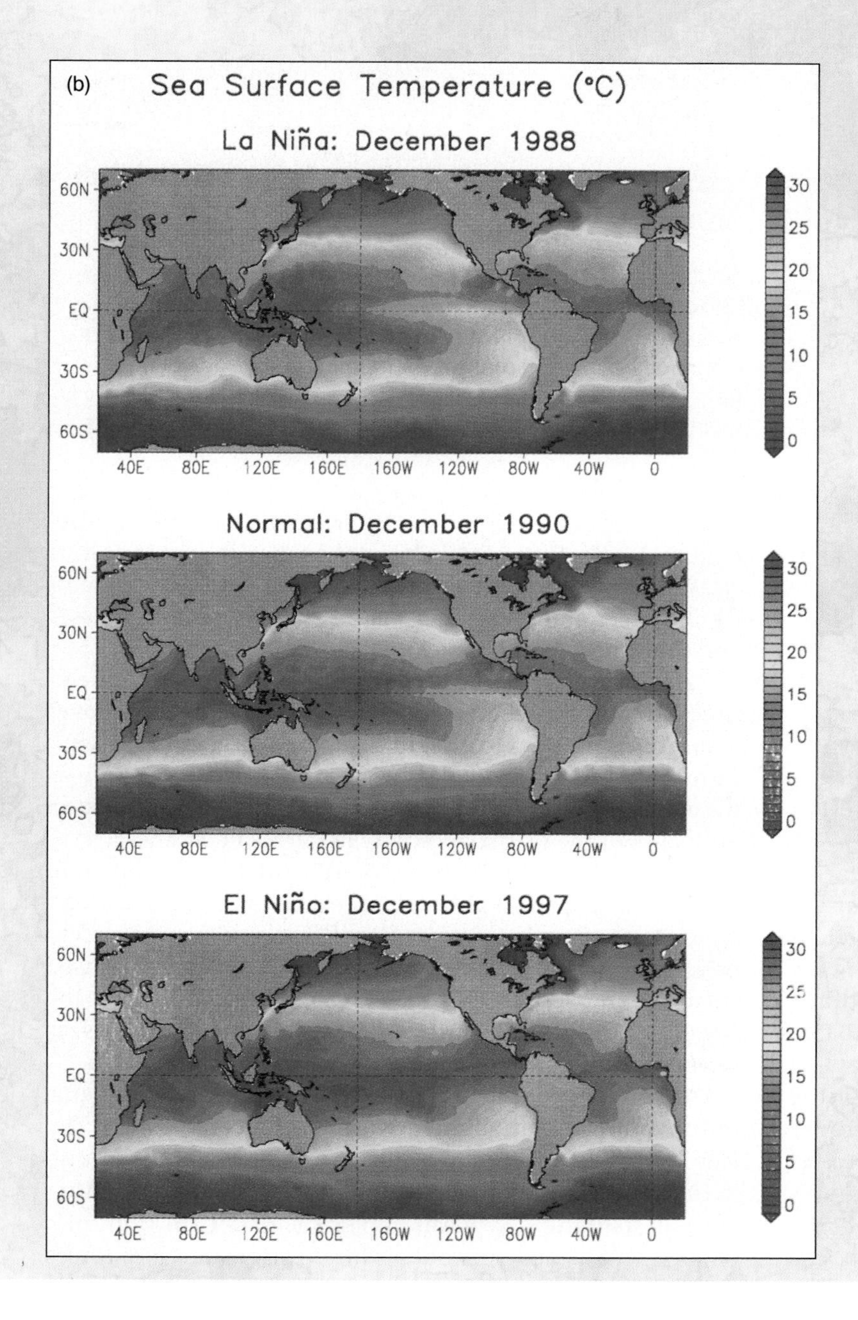

The El Niño condition is an example of the interaction of atmospheric pressure and ocean temperature. The atmospheric and oceanic states encourage each other. Under normal conditions, the contrast in temperatures across the ocean helps drive winds that, in turn, keep pushing water to the west, maintaining the contrast in water temperature. But when a condition called the *southern oscillation* occurs there is warming in the eastern Pacific, enhancing the usual temperature contrasts between the equator and the earth's poles. Atmospheric pressure rises near Australia, the wind falters, and El Niño is created off the coast of South America. The greater the temperature disparity combined with moisture available from the Pacific Ocean, the more severe the weather.

on the earth's surface imply that the dew point has been reached and that relative humidity is valued at 100%.

Types of Precipitation

When large masses of air rise, precipitation may take place in one of three types: (1) convectional; (2) orographic; or (3) cyclonic or frontal.

The first type, **convectional precipitation,** results from rising, heated, moisture-laden air. As air rises, it cools. When its dew point is reached, condensation and precipitation occur, as Figure 4.20 shows. This process is typical of summer storms or showers in tropical and continental climates. Usually the ground is heated during the morning and early afternoon. Warm air that accumulates begins to rise, first forming cumulus clouds and then cumulonimbus clouds. Finally, lightning, thunder, and heavy rainfall occur, which may affect each part of the ground for only a brief period since the storm is moving. It is common for these convectional storms to occur in late afternoon or early evening.

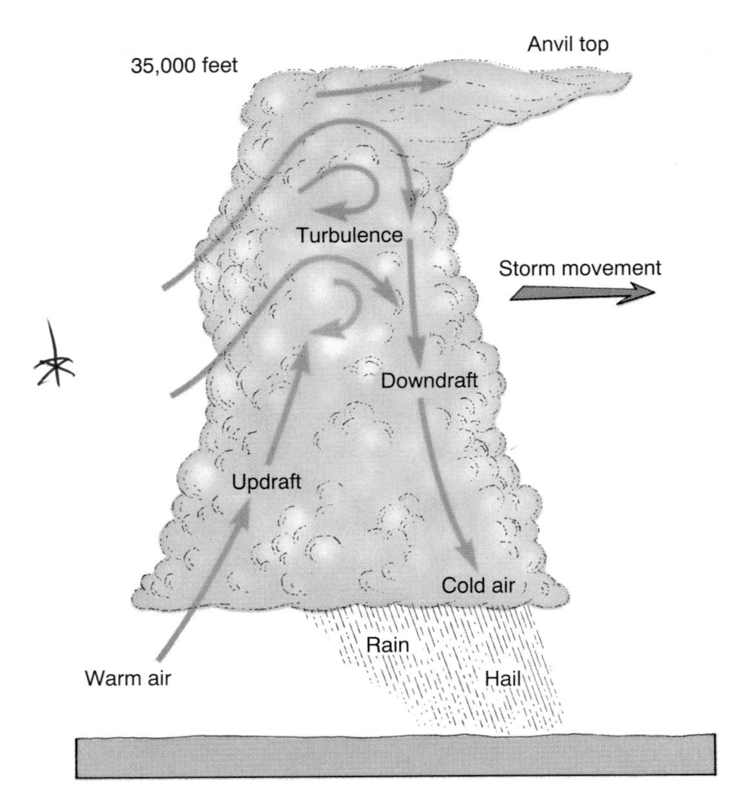

FIGURE 4.20 When warm air laden with moisture rises, a cumulonimbus cloud may develop and convectional precipitation occur. The turbulence within the system creates a downdraft of cold upper-altitude air.

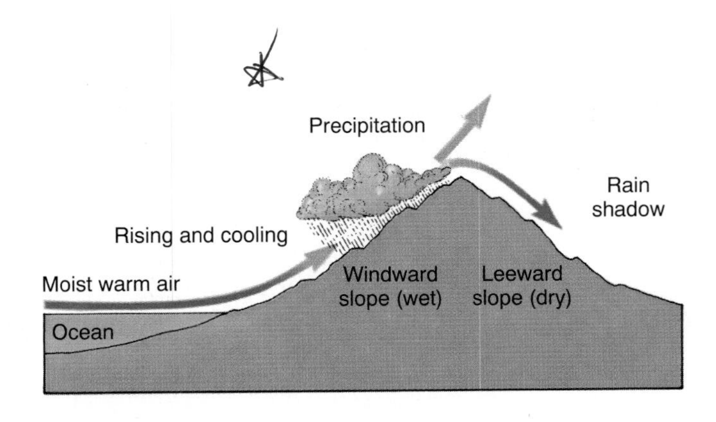

FIGURE 4.21 **Orographic precipitation.** Surface winds may be raised to higher elevations by hills or mountains lying in their paths. If such orographically lifted air is sufficiently cooled, precipitation occurs. Descending air on the leeward side of the upland barrier becomes warmer, its capacity to retain moisture is increased, and water absorption rather than release takes place.

If quickly rising air currents violently circulate air within a cloud, ice crystals may form near the top of the cloud. When these ice crystals are large enough to fall, a new updraft can force them back up, enlarging the pieces of ice. This process may occur repeatedly until the updrafts can no longer sustain the ice pieces as they fall to the ground in the form of hail.

Orographic precipitation, the second type and depicted in Figure 4.21, occurs as warm air is forced to rise because hills or mountains block moisture-laden winds. This type of precipitation is typical in areas where mountains and hills are close to oceans or large lakes. Saturated air from over the water blows onshore, rising as the land rises. Again, the processes of cooling, condensation, and precipitation take place. The *windward* side—the side exposed to the prevailing wind—of the hills and mountains receives a great deal of precipitation. The opposite side, called the *leeward,* and the adjoining regions downwind are very often dry. The air that passes over the mountains or hills descends and warms. As we have seen, descending air does not produce precipitation, and warming air absorbs moisture from surfaces it passes over. A graphic depiction of the great differences in rainfall over very short distances is shown on the map of the state of Washington in Figure 4.22.

Cyclonic or **frontal precipitation,** the third type, is common to the midlatitudes, where cool and warm air masses meet. Although it is less frequent, this type of precipitation also occurs in the tropics as the originator of hurricanes and typhoons. In order to understand cyclonic or frontal precipitation, first visualize the nature of air masses and the way cyclones develop.

Air masses are large bodies of air with similar temperature and humidity characteristics throughout; they form over a **source region.** Source regions include large areas of uniform surface and relatively consistent temperatures, such as the cold land areas of northern Canada, the north central part of Russia, or the warm tropical water areas in oceans close to the equator. Source regions for North America are shown in Figure 4.23. During a period of a few days or a week, an air mass may form in a source region. For example, in the fall in northern Canada, when snow has already covered the vast subarctic landscape, cold, heavy, dry air develops over the frozen land surface. Further discussion on air masses as regional entities may be found in Chapter 13.

When this continental polar air mass is large, it begins to move toward the lighter, warmer air to the south. The leading edge of the tongue of air is called a **front.** The front, in this case, separates the cold, dry air from whatever other air is in its path. If a warm, moist air mass is in front of a polar air mass, heavier cold air hugs the ground and forces lighter air above it upward. The rising moist air condenses, and frontal precipitation occurs. On the other hand, the movement of warm air over cold air pushes the cold air back, again causing precipitation. In the first case, when cold air moves toward warm air, cumulonimbus clouds form and precipitation is brief and heavy. As the front passes, temperatures drop appreciably, the sky clears, and air becomes noticeably drier. In the second case, when warm air moves over cold air, steel-gray nimbostratus (*nimbo* means "rain") clouds form and precipitation is steady and long-lasting. As the front passes, warm, muggy air becomes characteristic of the area. Figure 4.24 summarizes the movement of fronts.

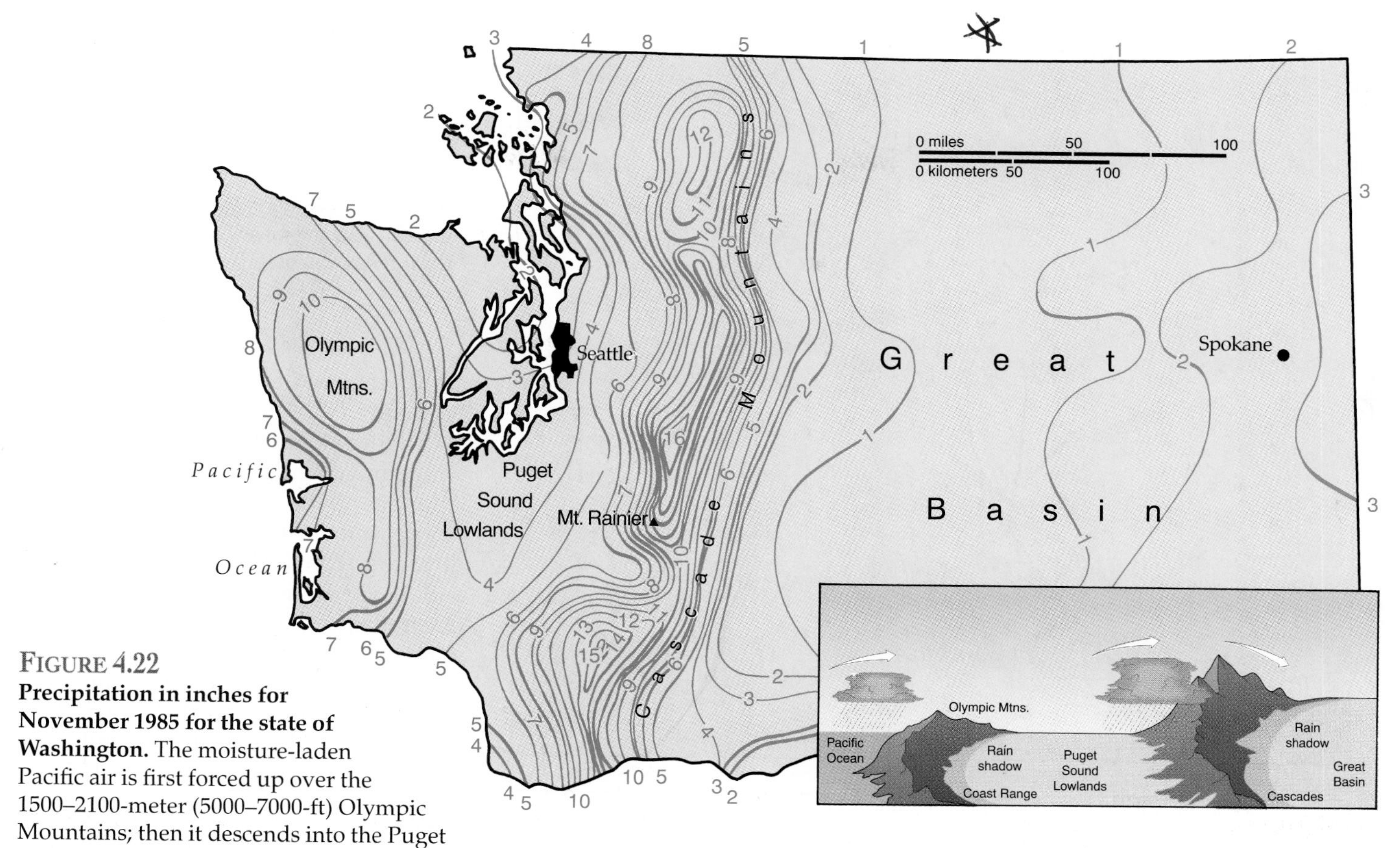

FIGURE 4.22 **Precipitation in inches for November 1985 for the state of Washington.** The moisture-laden Pacific air is first forced up over the 1500–2100-meter (5000–7000-ft) Olympic Mountains; then it descends into the Puget Sound lowlands; then it goes up over the 2700–4300-meter (9000–14,000-ft) Cascade Mountains and finally down into the Great Basin of eastern Washington.

Inset: From Robert N. Wallen, *Introduction to Physical Geography.* Copyright © 1993. McGraw-Hill Company, Inc., Dubuque, Iowa. All Rights Reserved. Reprinted by permission.

Storms

Two air masses coming into contact (a front) creates the possibility of storms developing. If the contrasts in temperature and humidity are sufficiently great, or if wind directions of the two touching masses are opposite, a wave might develop in the front as shown in Figure 4.25. Once established, the waves may enlarge and become wedgelike. One wedge represents cooler air moving along the surface, while the other wedge signifies warmer air moving up and over the cold air. In both wedges, warm air rises and a low-pressure center forms. Considerable precipitation is accompanied, in the Northern Hemisphere, by counterclockwise winds around the low-pressure area. A large system of air circulation centered on a region of low atmospheric pressure is called a midlatitude **cyclone,** which can develop into a storm.

A cyclone may be a weak storm or one of great intensity. The tropical storm is likely to begin over warm waters, far enough from the equator for a significant Coriolis effect to occur. Here, a wave that is unassociated with a front may form. If a wedge develops because conditions are just right, an intense tropical storm may grow that is fed by energy embodied in the rising moist, warm air. This storm is called

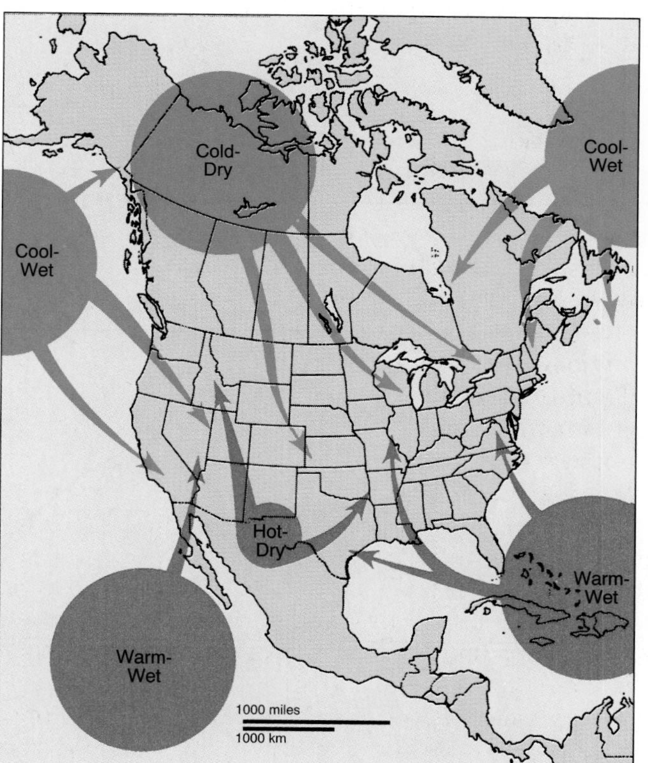

FIGURE 4.23 **Source regions for air masses in North America.** The United States and Canada, lying between major contrasting air-mass source regions, are subject to numerous storms and weather changes. See also Figure 13.6.

From T. McKnight, *Physical Geography: A Landscape Appreciation,* 4e, c1993. Adapted by permission of Prentice-Hall, Englewood Cliffs, New Jersey.

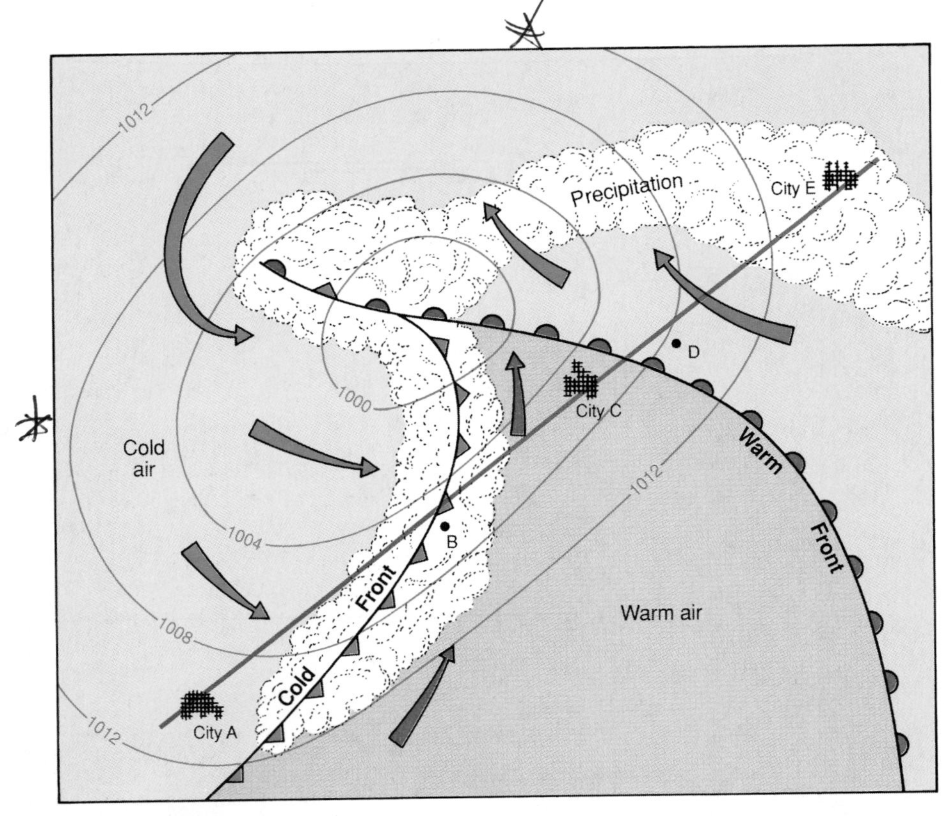

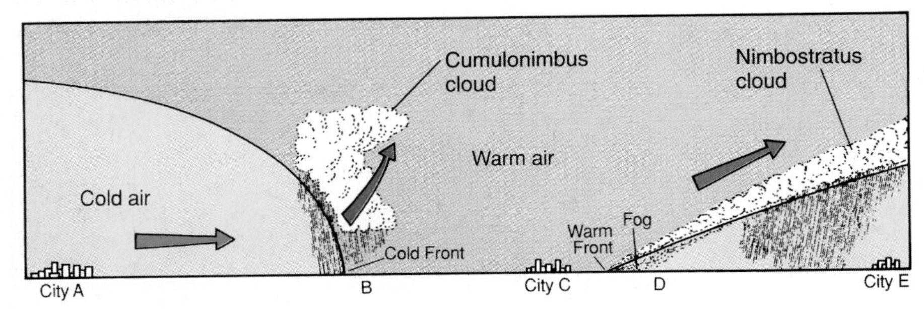

FIGURE 4.24 In this diagram, the cold front has recently passed over city A and is heading in the direction of B. The warm front is moving away from B and toward city C. The wind direction is shown by arrows and the air pressure by isobars, lines of equal atmospheric pressure. The isobars indicate that the lowest pressure is found at the intersection of the warm and cold fronts.

a **hurricane** in the Atlantic region and a **typhoon** in the Pacific area.

Figure 4.26 shows the paths and wind movements of hurricanes in the Atlantic region. The winds of these storms move in a counterclockwise direction, converging near the center, and rising in several concentric belts. Great damage is caused by the high winds (greater than 74 miles per hour) and the surge of ocean water into coastal lowlands. At the hurricane's center, called the *eye,* air descends and results in gentle breezes and relatively clear skies. Over land, these storms lose their warm-water energy source and subside quickly. If they move farther into colder northern waters, they are pushed or blocked by other air masses and also lose their energy source and abate. Table 4.1 describes increasingly devastating hurricanes.

The *New York Times* reported on January 8, 1996, that "A monstrous, crippling blizzard that experts said would make history attacked much of the East yesterday with blinding snow that was expected to become two feet deep before ending today." A **blizzard** is the occurrence of heavy snow and high winds. The "Blizzard of '96" was created when a disturbance at 9150 meters (30,000 ft) in the Rocky Mountains caused the jet stream to veer northeast along the East Coast. At the same time, a typical low-pressure system along the Gulf of Mexico began to move north, drawing an additional stream of moist

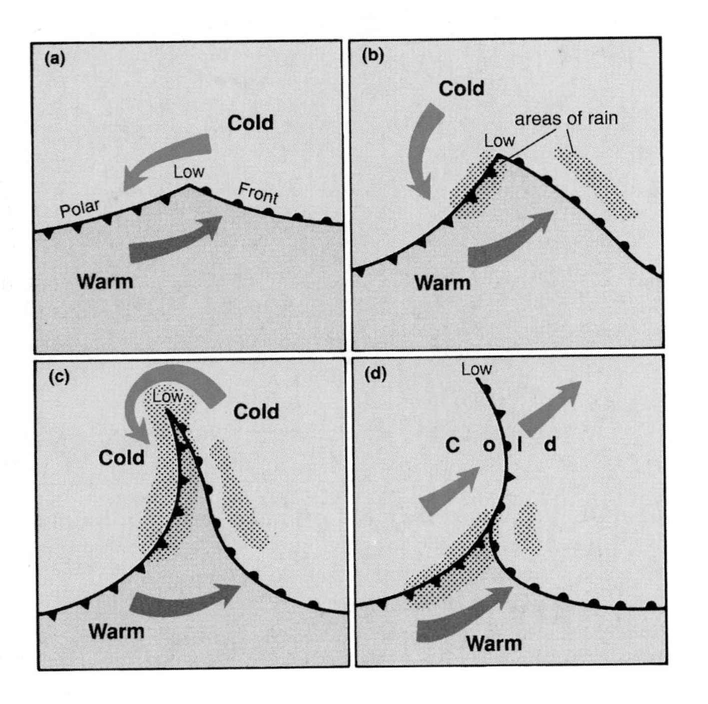

FIGURE 4.25 When wedges of warm and cold air develop along a low-pressure trough in the midlatitudes, the possibility of cyclonic storm formation occurs. (a) A wave begins to form along the polar front. (b) In the Northern Hemisphere, cold air begins to turn in a southerly direction, while warm air moves north. The meeting lines of these unlike air masses are called *fronts.* (c) Cold air, generally moving faster than warm air, begins to overtake the warm air, forcing it to rise, and, in the process, the storm deepens. (d) Eventually, two sections of cold air join; the warm air forms a pocket overhead, removing it from its energy and moisture source. The cyclonic storm dissipates as the polar front is reestablished.

TABLE 4.1

THE FORCE OF HURRICANES

CATEGORY	BAROMETRIC PRESSURE (INCHES)	WIND (MPH)	POTENTIAL DAMAGE
1	Over 28.94	74–95	Damage mainly to trees, shrubbery, unanchored mobile homes; storm surge damage for all categories
2	28.50–28.94	96–110	Some trees blown down; major damage to exposed mobile homes; some damage to roofs
3	27.91–28.49	111–130	Trees stripped of foliage, large trees blown down; mobile homes destroyed; some structural damage to small buildings
4	27.17–27.90	131–155	All signs blown down; extensive damage to windows, doors, and roofs; flooding inland as far as six miles; major damage to lower floors of structures near shore
5	Less than 27.17	Over 155	Severe damage to windows, doors, and roofs; small buildings overturned and blown away; major damage to structures less than 4.6 meters (15 ft) above sea level within 458 meters (500 yds) of shore

air from the Atlantic Ocean. In the meantime, a high-pressure system was moving south with Arctic air from Canada. These movements converged on the northeast United States. The result was nearly a meter (2 to 3 ft) of crippling snow that brought cities such as Washington, D.C., Baltimore, Philadelphia, New York, and Boston to a standstill (Figure 4.27a).

The most violent of all storms is the **tornado.** It is also the smallest storm (Figure 4.27b), typically measuring less than 30 meters (100 ft) in diameter. Tornadoes are spawned in the huge cumulonimbus clouds that sometimes travel in advance of a cold front along a squall line. During the spring or fall, when adjacent air masses differ the most, the central United States is prone to many of these funnel-shaped killer clouds. Although winds are estimated to reach 800 kilometers per hour (500 mph), these storms are small and usually travel on the ground for less than a mile, so only limited areas are affected, but the areas affected are devastated (see Figure 4.27c).

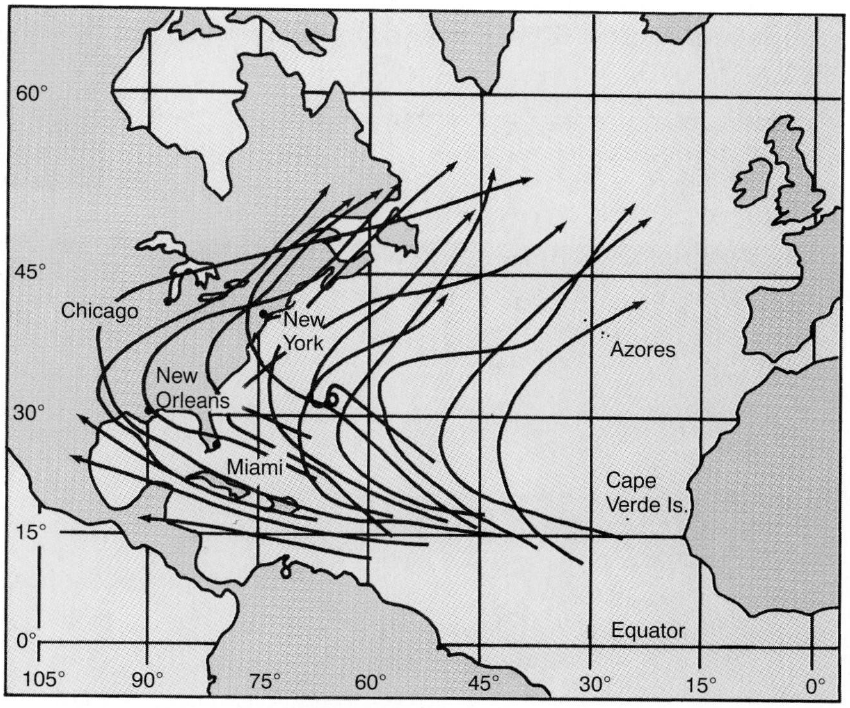

FIGURE 4.26 **Tracks of typical hurricanes.** In the United States, the most vulnerable areas are the Gulf Coast of Florida; Cape Hatteras, North Carolina; Long Island, New York; and Cape Cod, Massachusetts.

Source: After U.S. Navy Oceanographic Office.

CLIMATE

We have traced some of the causes of weather changes that occur as air from high-pressure zones flows toward low-pressure areas, fronts pass and waves develop, dew points are reached, and sea breezes arise. Some parts of the world experience these changes more rapidly and more often than do other parts.

Day-to-day weather conditions can be explained by the principles we have developed. However, the understanding of the effect of weather elements—temperature, precipitation, and air pressure and winds—cannot be comprehended unless a person is conscious of the earth's surface features. Weather forecasters in each location on earth must deal with weather elements in the context of their local environment.

The complexities of daily weather conditions may be summarized by statements about climate. The climate of an

(a)

(b)

FIGURE 4.27 Storms. (a) Blizzards bring transportation systems of cities to a halt. (b) In the United States, tornadoes occur most frequently in the central and south central parts of the country (especially in Oklahoma, Kansas, and the Texas Panhandle), where cold polar air very often meets warm, moist Gulf air. (c) Extensive damage caused by a tornado in Ogden, Illinois, on April 19, 1996.

(a) Leland Bobbe/Tony Stone Images.
(b) © Sheila Beaugher/Liaison International.
(c) John Dixon/Champaign Urbana *News Gazette*

(c)

area is a generalization based on daily and seasonal weather conditions. Are summers warm on the average? Is heavy snow in the winter likely? Are winds normally from the southeast? Are climatic averages typical of daily weather conditions, or are the day-to-day or week-to-week variations so great that one should speak of average variations rather than just averages? These are the questions we must ask in order to form an intelligent description of the differences in conditions from place to place.

The two most important elements that differentiate weather conditions are temperature and precipitation. While air pressure is also an important weather element, differences in air pressure are hardly noticeable without the use of a barometer. Thus, we may regard warm, moderate, cold, or very cold temperatures as characteristic of a place or region. In addition, high, moderate, and low precipitation are good indicators of the degree of humidity or aridity in a place or region. We shall take these two scales, define the terms more precisely, and map the areas of the world having various combinations of temperatures and precipitation.

Because extreme seasonal changes do occur, two global climatic maps are shown in Figure 4.28; one for (a) winters and one for (b) summers. Maps could have been developed for each of the four seasons, or for the 12 months of the year. Rather, these two maps give a good, though brief, description of climatic differences. Keep in mind that a summer map of the world is a combination of Northern Hemisphere climates from June 21 to September 21 and Southern Hemisphere climates from December 21 to March 21, because the seasons are reversed in the two hemispheres.

Figure 4.29 depicts the various climates of the world and is based on the type of information presented in Figure 4.28. In rugged mountain areas, climate at any one location may differ appreciably from nearby areas, depending on such factors as elevation, latitude, southward- or northward-facing slopes, exposure to moisture-laden winds, and so on. Figure 4.29 represents just one of a number of similar, more complex climate schemes that are used today. Best known is the Köppen system, developed in 1918, which is based on natural vegetation in addition to temperature and precipitation criteria.

The Tropical Climates

Tropical climates are generally associated with earth areas lying between the northernmost and southernmost lines of the sun's vertical rays—the *Tropic of Cancer* and the *Tropic of Capricorn.* The location of tropical climates is shown in Figure 4.30.

The numbers in the section headings that follow refer to the keys on Figure 4.28. The first number represents typical winter conditions, and the second represents summer conditions. Each number represents idealized conditions. Table 4.2 serves as a guide for comparing the different climates.

Tropical Rain Forest (1;1)

The areas that straddle the equator are generally located within the equatorial low-pressure zone. These regions are called **tropical rain forest.** They are warm, wet climates in both the winter and summer (Figure 4.31). Rainfall usually comes from daily convectional thunderstorms and although most days are sunny and hot, by afternoon, cumulonimbus clouds form and convectional rain falls.

Tropical rain forests are typically filled with natural vegetation, which is still present but declining rapidly because of intentional fires in large areas of the Amazon Basin of South America and the Zaire River Basin of Africa. Tall, dense forests of broadleaf trees and heavy vines predominate. Among the hundreds of species of trees found in tropical rain forests, both dark woods and light woods, as well as spongy softwoods like balsa, and hardwoods like teak and mahogany exist (Figure 4.32). Rain forests also extend away from the equator along coasts where prevailing winds supply a constant source of moisture to coastal uplands. In addition, the orographic effect provides sufficient precipitation for heavy vegetation to develop in these forests.

The typical soils of these regions are *oxisols.* As a result of rapid weathering, most soil nutrients that are necessary for cultivated plants are absent. Only with large inputs of fertilizers can the soils be made to sustain continued agricultural use.

Savanna (3;1)

As the sun's vertical rays extend farther from the equator in the summer, the equatorial low-pressure zone follows the sun's path. Thus, areas to the north and south of the rain forest are wet in the summer months, although still hot, but are dry the remainder of the year because the moist equatorial low has been replaced by the dry air of subtropical highs. These areas are known as **savanna** lands because of the kind of natural vegetation that grows here.

The natural vegetation of savannas resembles a form of scrub forest; however, these areas are now recognized as a grassland with widely dispersed trees. The natural tendency toward a more forested cover has been reduced by the periodic clearing by burning that local agriculturalists and hunters engage in. Savannas sometimes seem to have been purposely designed because of their parklike appearance, as indicated in Figure 4.33. The east African region of Kenya and Tanzania contains well-known grasslands and fire-resisting species of trees, where large animals like giraffes, lions, and elephants roam. The *campos* and *llanos* of South America are other huge savanna areas.

The wetter portions of the savanna lands are often underlain by *ultisols,* soils that develop in warm, wet-dry regions beneath forest vegetation (Figure 4.33b). These soils are weak in nutrients for cultivated plants, but respond favorably when lime and fertilizers are applied. In drier parts of the savanna, the characteristic soils are *vertisols,*

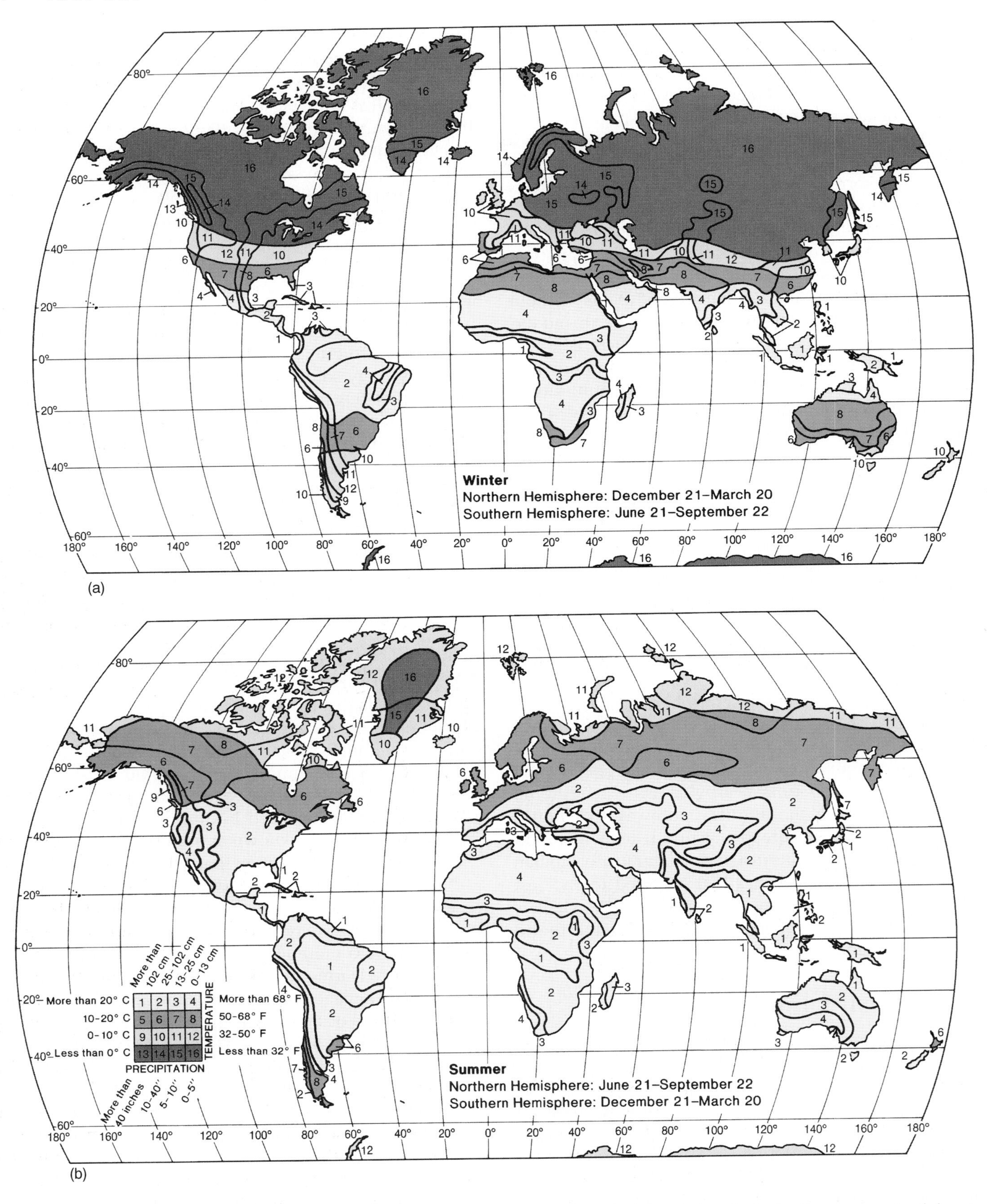

FIGURE 4.28 These maps combine temperature and precipitation data to display seasonal variations in the basic components of climate. In reading the maps, remember that the winter climate shown in (a) represents the winter season for both the Northern and Southern Hemispheres. This means that December, January, and February data were used for the north latitudes and June, July, and August data were used for the south latitudinal parts of the world. The result is a winter view of the world, one of greatly varying temperatures and small amounts of precipitation. (b) The summer view shows a world of nearly universally warm temperatures and large amounts of precipitation.

which form under grasses in warm climates. When it rains, the surface becomes plasticlike and some of the soil slides into the cracks formed during the dry season. Consequently, vertisols are difficult to till and are used productively as grazing land.

Generally, breaks between different climates are hard to differentiate and distinguish. Instead, zones of transition are apparent. These transitional zones are typical of plains and plateaus—and not so gradual in mountainous regions. Less dense forests exist between the thick tropical rain forest and the savanna.

A special case in Asia needs mentioning, however. When summer **monsoon** winds carry water-laden air to the mainland, a significant increase in rainfall on the hills, mountains, and adjacent plains occurs. The amount of precipitation in the savanna region is not as abundant—notice the pattern of precipitation on Figure 4.34. As a result, vegetation is much denser even though the winters are dry. Jungle growth and large forests are the natural vegetation. Much of this vegetation, however, has ceased to exist because people have been using the land for rice and tea production for many generations.

Hot Deserts (7;4)

On the poleward side of the savannas, grasses begin to shorten, and desert shrubs become evident. This is where we approach the belt of subtropical high pressure that brings considerable sunshine, hot summer weather, and very little precipitation. Note the minute amount of rainfall shown in Figure 4.35. The precipitation that does fall is convectional, but sporadic. As conditions become drier, fewer and fewer drought-resistant shrubs appear and, in some areas, only gravelly and sandy deserts exist, as suggested on Figure 4.36.

The great, hot deserts of the world, such as the Sahara, the Arabian, the Australian, and the Kalahari, are all the products of high-pressure zones (see Figure 4.45). Often, the driest parts of these deserts are along the western coasts, where cold ocean currents are found, as Figure 4.28 indicates. The soils, called *aridisols,* respond well to cultivation when irrigation is made available. Earlier mention was made of the relationship between cold ocean currents and deserts.

The Humid Midlatitude Climates

Figure 4.37 shows the location of several climate types that are all humid, that is, not having desert conditions in the winter, summer, or both. In addition, winter temperatures well below those of the tropical climates are characteristic of the humid midlatitudes. These climate types would be neatly defined, paralleling the lines of latitude, were it not for mountain ranges, warm or cold ocean currents, and particularly, land-water configurations. These factors cause the greatest variations in the middle latitudes.

Mediterranean Climate (6;3)

Midlatitude winds generally blow from the west in both the Northern and Southern Hemispheres, and a significant amount of the precipitation is produced from frontal systems. Thus, it is important to know if the water is cold or warm near land areas. Several climatic zones are noticeable in the middle latitudes, all marked by warm summer temperatures except those in areas cooled by westerly winds from the ocean.

To the poleward side of the hot deserts, a transition zone occurs between the subtropical high and the moist westerlies zones. Here, cyclonic storms bring rainfall only in the winter, when the westerlies shift toward the equator. Summers are dry and hot as the subtropical highs shift slightly poleward (Figure 4.38). Winters are not cold. These conditions describe the **Mediterranean climate,** which is often found on the western coasts of continents in the middle latitudes. Southern California, the Mediterranean area itself, western Australia, the tip of South Africa, and central Chile in South America are characterized by this type of climate. In these areas, where precipitation is sufficient, shrubs and small deciduous trees such as the scrub oak grow (Figure 4.39).

The Mediterranean climate area—long and densely settled in southern Europe, the Near East, and North Africa—has more moisture and a greater variety of vegetation and soil types than the desert. Clear, dry air predominates, winters are relatively short and mild, and plants and flowers grow year-round. Even though summers are hot, nights are usually cool and clear. Much of the vegetation in this area is now in the form of crops.

Marine West Coast Climate (10;6)

Closer to the poles, but still within the westerly wind belt, are areas of **marine west coast climate.** Here, cyclonic storms play a relatively large role. In the winter, more rainfall and cooler temperatures prevail than in the Mediterranean zones. Compare the patterns in Figures 4.38 and 4.40. In the transitional zone just poleward of the Mediterranean climate, little rainfall occurs during the summer. Closer to the poles, however, rainfall increases appreciably in the summer and even more so in the winter. Marine winds from the west moderate both summer and winter temperatures. Thus, summers are pleasantly cool, and winters, though cold, do not normally produce freezing temperatures.

This climate affects relatively small land areas in all but one region. Because northern Europe contains no great mountain belt to thwart the west-to-east flow of moist air, the marine west coast climate stretches well across the continent to Poland. In Poland, cyclonic storms originating in the Arctic regions are noticeable. Northern Europe's moderate climate owes its existence to few mountains and a relatively warm ocean current whose influence is felt for nearly 1600 kilometers (1000 mi), from Ireland to central Europe.

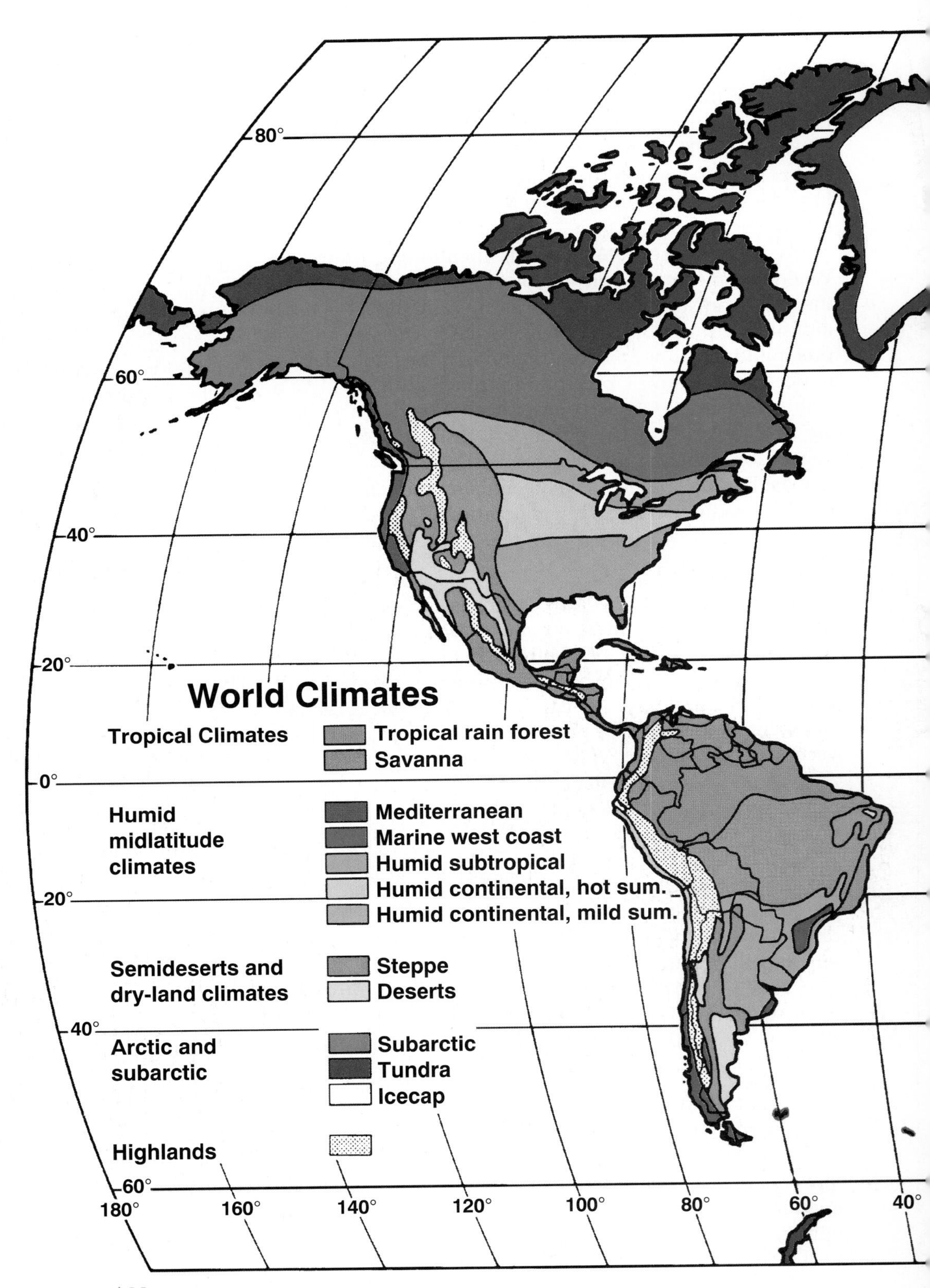

FIGURE 4.29 Climates of the world.

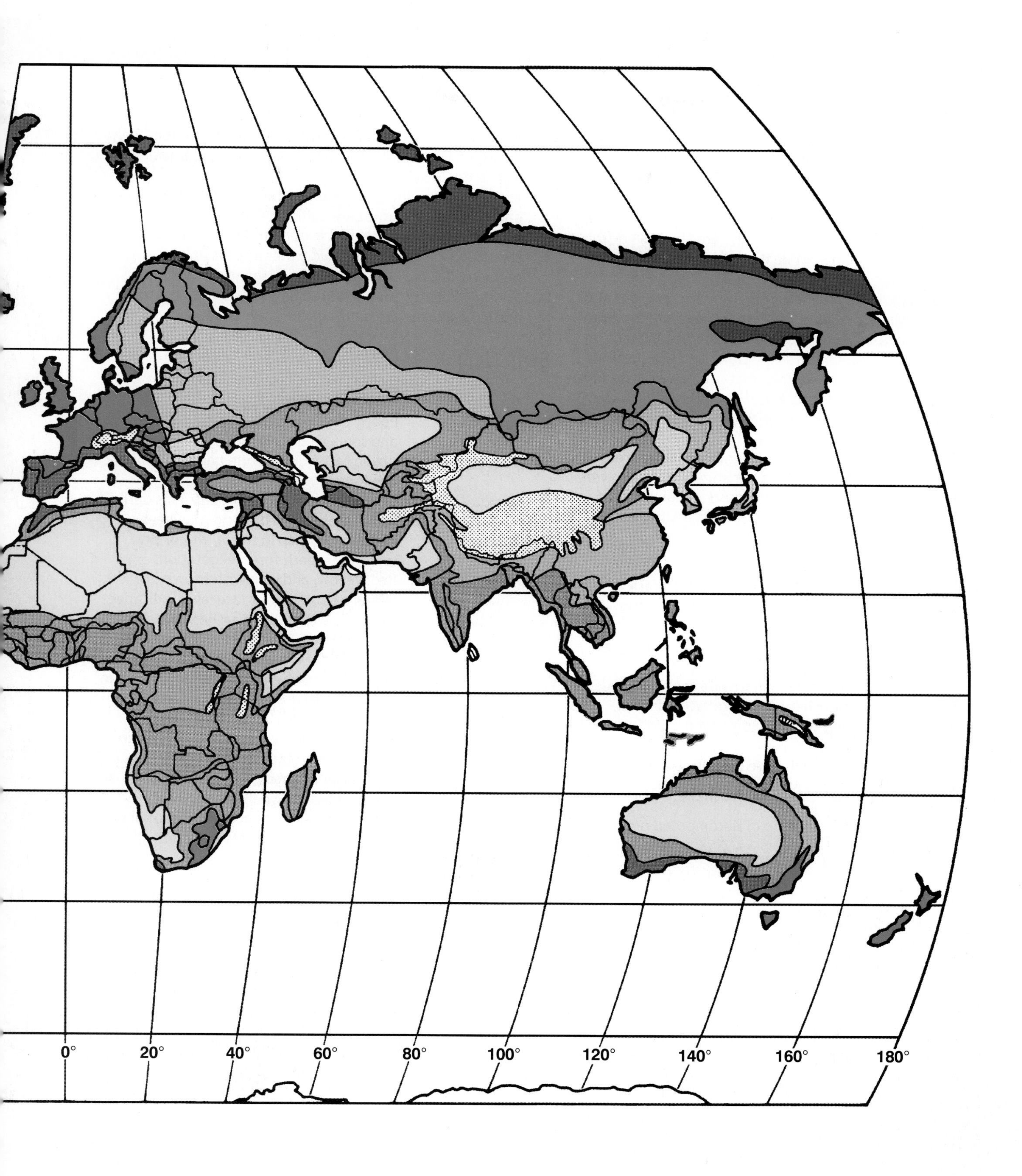
0°
20°
40°
60°
80°
100°
120°
140°
160°
180°

SOIL TYPES

The previous chapter mentioned that the gradational processes of mechanical and chemical weathering create soils. Recall, too, that soil type is much more a product of the climate of the region in which it occurs than it is of the kind of bedrock below it. Higher temperatures and more abundant moisture accelerate weathering processes. The deepest soils are found in the warmest wettest environments, while the soil is only thinly developed in arctic regions.

Soil is made up of not only slowly disintegrating rock, but also more rapidly decaying, overlying vegetation. Vegetation, more than bedrock, influences the type of soil that evolves. Water that drains through soil carries chemicals in solution and tiny particles. A large amount of rainfall in warm climates has a devastating effect on the supply of plant nutrients. When this occurs, even though soil is deep, it loses rich nutrients to the ever-present groundwater.

Although broad classes of soil types extend over large areas of the earth's surface, the individual characteristics of soil in a given place may be markedly different over very short distances. Variation is due to the following soil-forming factors:

- the chemical composition of rock that provides the basic material (the *geologic* factor);
- temperature and moisture conditions (the *climatic* factor);
- the slope of land and, thus, the drainage conditions (the *topographic* factor);
- organisms in the soil, such as ants, worms, algae, fungi, and bacteria (the *living organisms* factor);
- overlying plants and plant roots that provide the link between soil nutrients and plants (the *vegetation* factor);
- the necessary time to bring these factors together to form soils (the *chronologic* factor);
- in addition, humans and animals have greatly altered soils.

Soils are different in many ways, varying in such characteristics as:

1. *humus content.* Humus is made up of decomposed organic matter;
2. *color.* Dark colors usually indicate high humus content; light colors indicate highly leached soils in wet areas, and alkaline soils in dry areas;
3. *texture.* Sand is the largest soil particle type, followed by silt, and then clay. The most agriculturally productive soil, called *loam,* is an equal combination of all three textures;
4. *structure.* The capacity to hold water and air, as well as permeability, are both significant to the structure of soils;
5. *acidity/alkalinity.* Soils balanced between being very acidic and very alkaline are most agriculturally productive.

SOIL CLASSIFICATION

Soil Type	Brief Description
Inceptisols	Poorly developed; form in cold climates and some river valleys.
Spodosols	Acidic; lack humus; form beneath coniferous forests.
Alfisols	Gray-brown; rich in plant nutrients derived from fallen leaves of deciduous trees.
Ultisols	Reddish; develop in warm, wet or dry regions beneath forest vegetation.
Oxisols	Red, yellow, and yellowish-brown just below a darkened surface layer indicative of extreme chemical weathering.
Mollisols	Dark brown to black; extremely rich in nutrients; form beneath grasses.
Vertisols	High in clay content; form under grasses in climates with pronounced wet and dry periods.
Entisols	Poorly developed; thin and sandy; often found in mountain environments.
Aridisols	Light in color; sandy and desertlike; related to saline or alkaline soils.
Histosols	Form in the arctic as peat; consist of plant remains that accumulate in water.

The orographic effect from mountains in areas like the northwestern United States, western Canada, and southern Chile produces enormous amounts of precipitation, often in the form of snow on the windward side. Vast coniferous forests—needle-leaf trees, such as pines, spruces, and firs—cover the mountains' lower elevations. Because the mountains prevent moist air from continuing to the leeward side, midlatitude deserts are found to the east of these marine west coast areas.

The main soils of marine west coast areas are *spodosols.* They are strongly acidic and low in plant nutrients, a result of the acidic humus that develops from the fallen needles of coniferous trees. Fertilizers must be used for farming in order to neutralize the acidity.

Humid Subtropical Climate (6;2)

On the eastern coasts of continents, the transition is from the equatorial climate to the **humid subtropical climate.** Convectional summer showers and winter cyclonic storms are the sources of precipitation. As illustrated in Figure 4.41, this climate is characterized by hot, moist summers and moderate, moist winters. In the fall, on occasion, hurricanes that develop in tropical waters strike the coastal areas.

The generally even distribution of rainfall allows for the presence of deciduous forests containing hardwood trees like oak and maple, whose leaves turn orange and red before falling in autumn. In addition, conifers become mixed with deciduous trees as a second-growth forest.

In Chapter 2, the three horizons of soil were described. Clearly, it would take a very detailed system to identify each of the various combinations of soil horizon thicknesses, soil factors, and soil characteristics. The table and map present the most general classification system. Keep in mind that each of the names provided represents a wide range of soil characteristics. These names are used later in this chapter to describe typical, general soil types found in the various world climates.

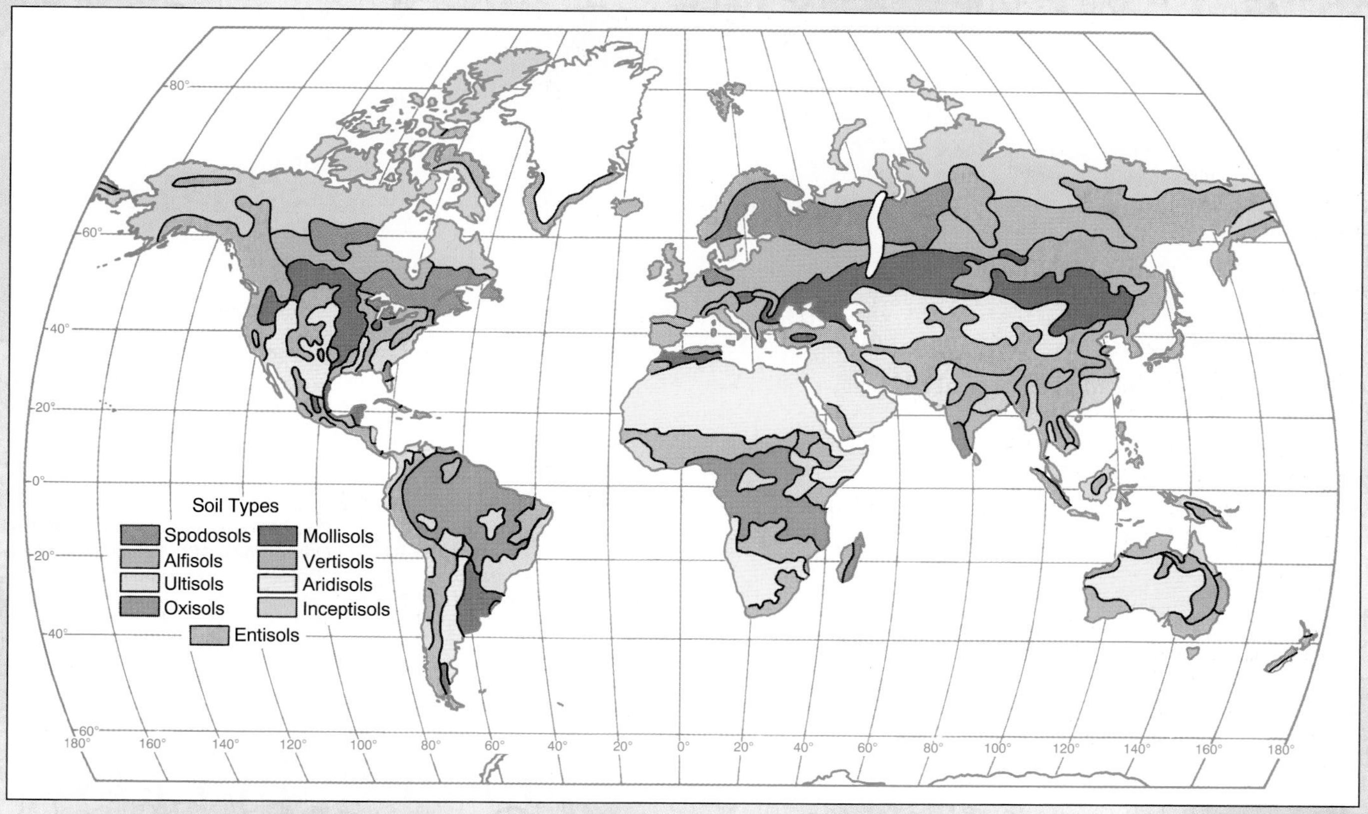

Source: T. L. McKnight, *Physical Geography: A Landscape Appreciation,* 2d ed., Prentice-Hall, Inc., 1987.

The poleward transition to the continental climates is accompanied by increasingly colder winters and shorter summers. In this direction as well, cyclonic storms become more responsible for rainfall than convectional showers. The region can no longer be characterized as humid subtropical; rather it is described as *humid continental* (see next section). Southern Brazil, the southeastern United States, and southern China all have a humid subtropical climate.

Beneath the deciduous forests of both the humid subtropical and the humid continental climates are *alfisols.* The A-horizon is generally gray-brown in color, and these soils are usually rich in plant nutrients. Derived from the strongly basic humus created by fallen leaves of deciduous trees, alfisols retain moisture during the hot days of summer, allowing for productive agriculture.

Humid Continental Climate (10;2)

Air masses that originate close to the poles and drift toward the equator, and other air masses that drift toward the poles from the tropics produce frontal precipitation. Whenever warmer air or marine air blocks cold continental air masses, or vice versa, frontal storms develop. The climates that these air masses influence are described as **humid continental.** Figures 4.42 and 4.43 show the range and the dominance of winter temperatures within this climatic type.

The continental climate may be contrasted to marine west coast climates in that the former has prevailing

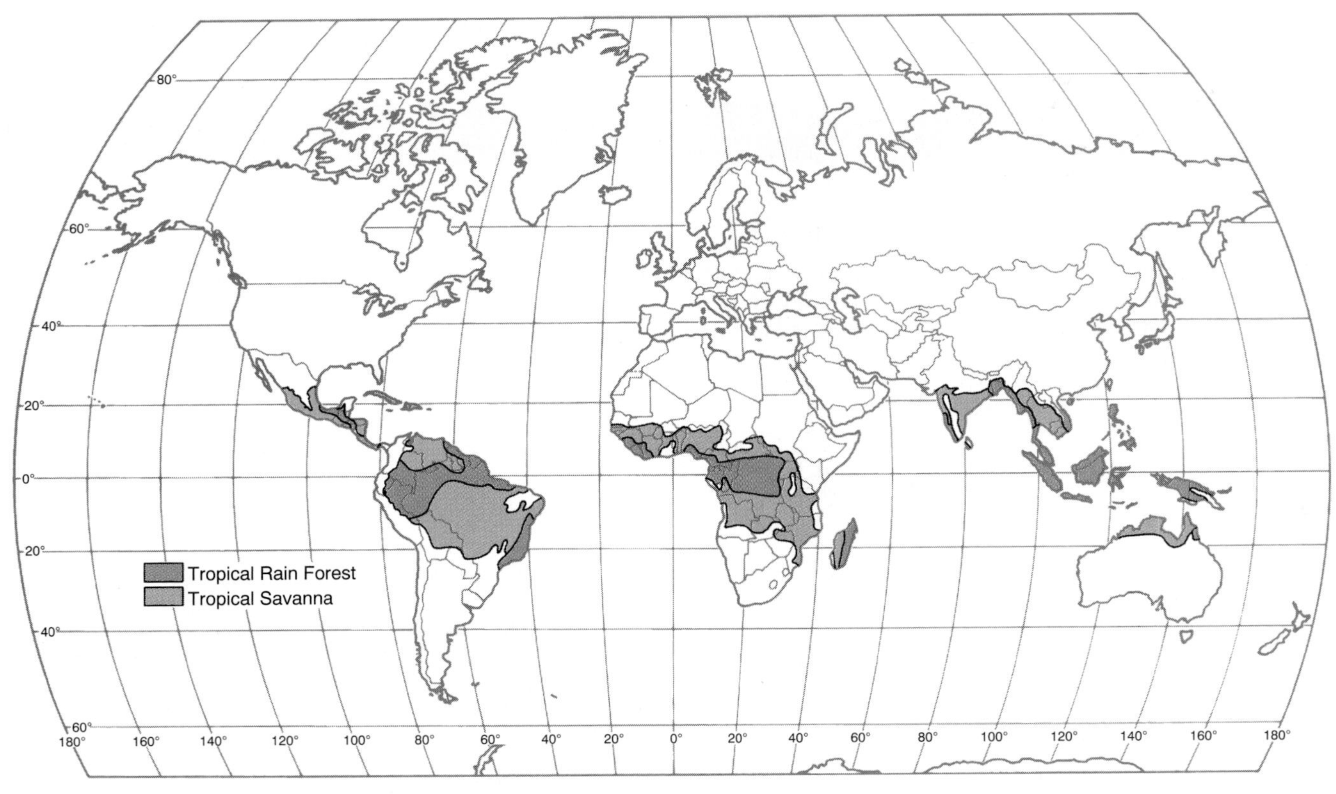

FIGURE 4.30 The location of tropical climates.

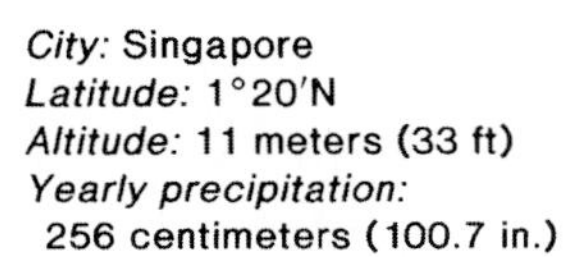

Climate designation: 1;1
Climate name: Tropical Rain Forest
Other cities with similar climates:
Colombo (2;2), Panama City (1;1),
Jakarta (1;1), Lagos (1;2)

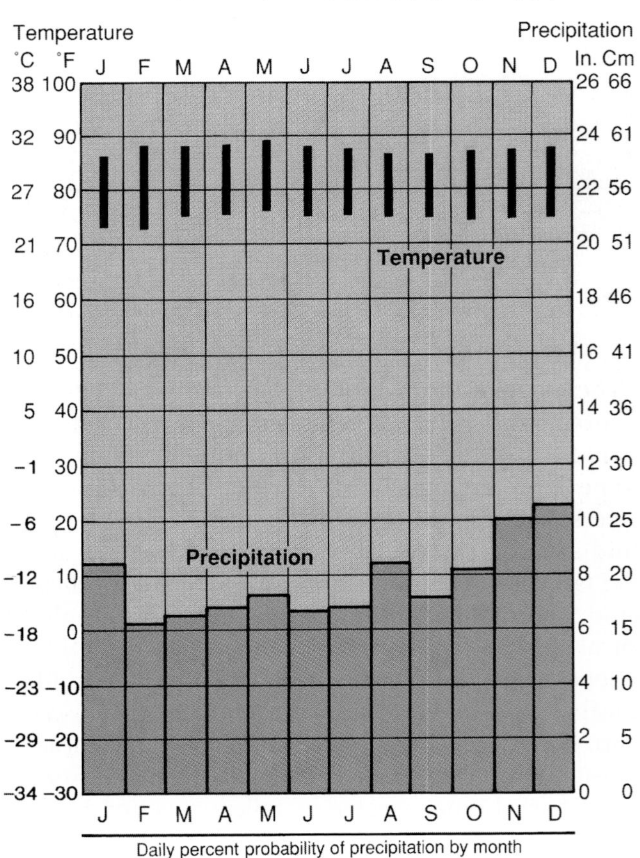

FIGURE 4.31 This and succeeding climate charts (*climagraphs*) show average daily high and low temperatures for each month, the average precipitation for each month, and the probability of precipitation on any particular day in a designated month. For Singapore, the average daily high temperature in August is 30.5°C (87°F), the low is 24°C (75°F). The rainfall for the month, on average, is 21 centimeters (8.4 in.), and on a given day in August, there is a 42% chance of rainfall.

TABLE 4.2

Climate Characteristics

Climate Type	Temperature and Precipitation	Soil, Vegetation, Wild Life
Tropical		
Tropical rain forest (1;1)	Constant high temperatures Rainfall heavy all year; convectional High amount of cloud cover High humidity	Dense, many species of trees Jungle where light penetrates Many small animals, insects Oxisols
Savanna (3;1)	High temperatures Rainfall heavy in summer high-sun period; convectional Dry in winter low-sun period Monsoon: highest temperature just before rainy season	Forests to grassland depending on rainfall amount Large animals Ultisols, vertisols, and ardisols
Hot deserts (7;4)	Extremely high temperatures in summer; warm winters Very little rainfall Low humidity	Shrubs in gravelly or sandy environments Reptiles Aridisols
Humid Midlatitude		
Mediterranean (6;3)	Warm to hot summers Mild to cool winters Dry summers Frontal precipitation in winter Generally low humidity	Chapparal vegetation (scrub oak trees and bushes) Alfisols, aridisols
Marine west coast (10;6)	Westerly winds year-around Mild summers Cool to cold winters Low rainfall in summer Frontal rainfall in winter	Vast coniferous forests in orographic regions Deciduous forest in plains Spodosols
Humid subtropical (6;2)	Hot summers Mild winters Convectional showers in summer Frontal precipitation in winter	Deciduous forests Coniferous forests, especially in sandy soils Mainly alfisols
Humid continental (10;2)	Hot to mild summers Cool to very cold winters Convectional showers in summer Frontal snowfall in winter	Coniferous forests Spodosols
Semidesert and Dryland		
Steppe (10;4) and Desert (4;4)	Warm to hot summers Cold winters Some convectional rainfall in summer Some frontal snowfall in winter	Grass and desert shrubs Mollisols in grasslands Aridisols in deserts
Arctic and Subarctic		
Subarctic (16;7), Tundra (16;16) (16;12), and Icecap (16;16)	Cool to cold, short summers Extremely cold winters Dry climates with some summer and winter precipitation	Stunted coniferous forests to mosses and lichens to permanent ice Inceptisols
Highlands	Great variety of conditions based on elevation, prevailing winds, sun or non-sun facing slopes, latitude, valley or non-valley, ruggedness	

Natural Vegetation

Each climate is typified by a particular mixture of **natural vegetation,** that is, the plant life that would exist in each area if humans did not directly interfere with the growth process. Natural vegetation, little of which remains today in areas of human settlement, has close interrelationships with soils, landforms, groundwater, elevation, and other features of habitat, including animals. The tie between *biomes*—the major structural subdivisions of assemblages of plants and animals—and climate is particularly close. Indeed, the earliest maps of world climatic zones were based not upon statistical variation in recorded temperatures, pressures, precipitation, and other measures of the atmosphere, but upon observed variation in vegetative regions. The discussion of climates, therefore, benefits from the visualization of regional differences, by reference to the type of mature plant communities that, in nature, developed within those climates.

The accompanying map shows the general pattern of the earth's natural vegetation regions. In the hotter parts of the world, where rainfall is heavy and well-scattered throughout the year, the vegetation type is *tropical rain forest. Forests,* in general, are made up of trees growing closely together, creating a continuous and overlapping leaf canopy. In the tropics, the forest consists of hundreds of tree species in any small area. Because the canopy blocks the sun's rays, only sparse undergrowth exists. When tropical rainfall is seasonal, *savanna* vegetation occurs, characterized by a low grassland with occasional patches of forests or individual trees. The high evaporation rate denies the savanna region sufficient moisture for dense vegetation.

Mediterranean or *chaparral* vegetation is found in the hot summer and mild, damp winter midlatitudes characterizing California, Australia, Chile, South Africa, and the Mediterranean Sea regions. This type of vegetation consists mainly of shrubs and trees of limited size, like the live oak. Together, they form a low, dense vegetation that is green during the wet season and brown during the dry season. Most dryland areas support some vegetation. *Semidesert* and *desert* vegetation is made up of dwarf trees, shrubs, and various types of cactus, though in gravelly and sandy areas, virtually no plants exist.

In temperate parts of the world with modest year-round rainfall, such as central North America, southern South America, and south-central Asia, the most prevalent type of vegetation is *prairie* or *steppe.* These are extensive grasslands, usually growing on high humus content soils. When rainfall is higher in temperate areas, natural vegetation turns to *deciduous woodlands.* These types of trees, such as oak, elm, and sycamore, lose their leaves during the cold season.

Beyond temperate zones, in northerly regions that have mild summers and very cold winters, *coniferous forests* are common. Evaporation rates are low. Usually, only a few species of trees predominate, such as pines and spruces. Still farther north, forests yield to *tundra* vegetation, a complex mix of very low-growing shrubs, mosses, lichens, and grasses.

At various elevations in mountain areas, the possibility of seeing many of these vegetation types is great. For example, moving from east to west in tropical Peru, one is initially in the Amazon River's tropical rain forests, and then in the deciduous forests of

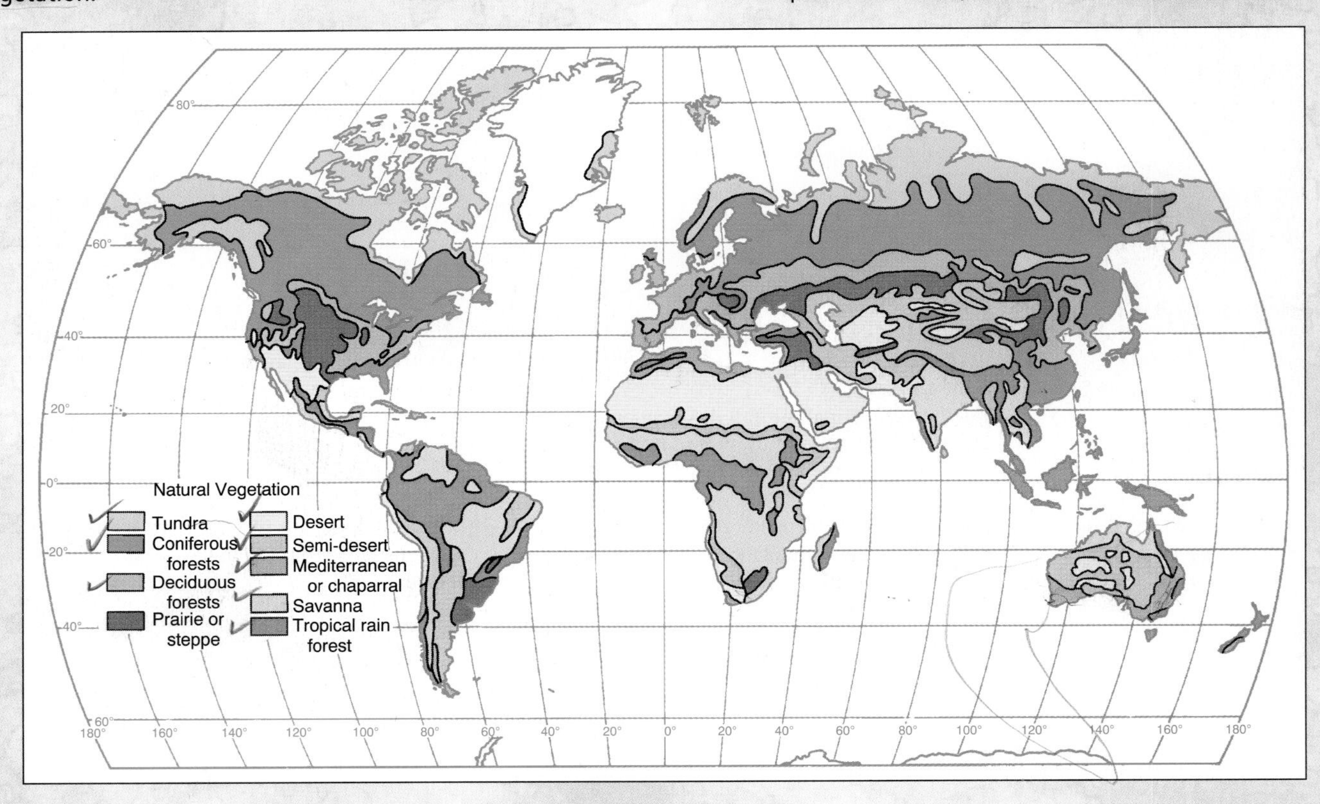

the low Andes Mountains. Then, upon entering valleys in the high Andes, coniferous forests appear, followed by mosses and rocks above the treeline. Descending onto the western slopes, grasslands give way to chaparral areas on the low slopes. Finally, semidesert and desert vegetation appears at the base of the mountains close to the Pacific Ocean.

This map of the western portion of the conterminous United States illustrates the power of the latest technology in bringing important information on vegetation into a usable format. The map was produced for aiding in the assessment of wildfire risk. It clearly shows how vegetation types correspond closely to climates. Compare this map to the world vegetation map. Note that patterns of vegetation on the U.S. map are far more complicated than the general case shown on the world map.

The map was produced using satellite imagery to identify the various electromagnetic wavelengths emitted by the vegetation for millions of pixels. Each pixel represents a very small portion of the surface. The resulting numerical data were classified and given colors to represent the different types of vegetation. After gathering all of the data, it was then important to transfer all of this information onto a suitable map base. The map base came from the U.S. Geological Survey and contained elevation and hydrographic information. All of the data were then transferred electronically onto an appropriate map projection, the Albers Equal-Area Conic projection.

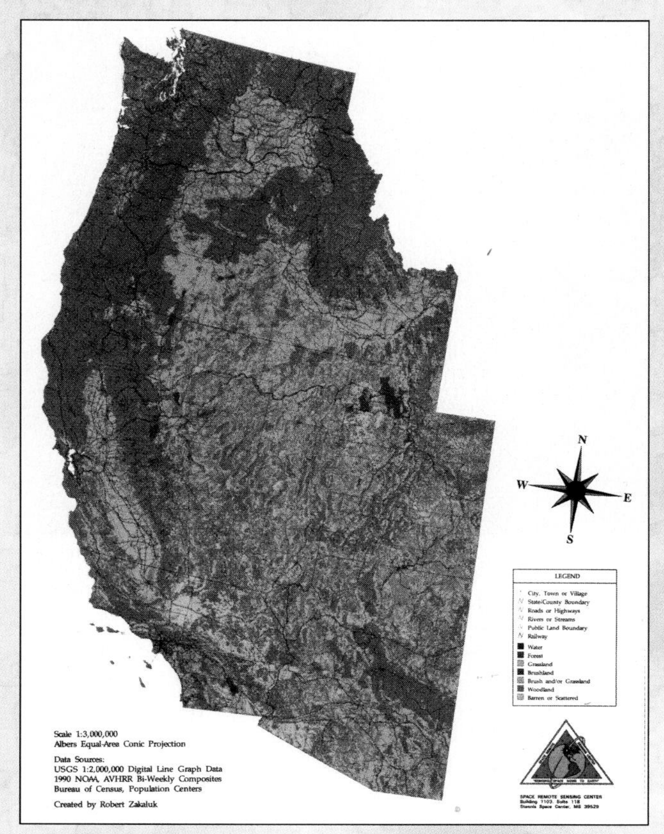

Source: Space Remote Sensing Center/Stennis Space Center, MS.

FIGURE 4.32 **Tropical rain forest.** The vegetation is characterized by tall, broadleaf, hardwood trees and vines.
© Lawrence Naylor/Photo Researchers, Inc.

(a)

(b)

FIGURE 4.33 The parklike landscapes of grasses and trees characteristic of the (a) drier and (b) wetter tropical savanna.
(a) © Aubrey Lang/Valan Photos.
(b) © Thomas J. Bassett, Dept. of Geography, University of Illinois.

City: Yangôn, Myanmar
Latitude: 16°46′N
Altitude: 5.5 meters (18 ft)
Yearly precipitation: 252 centimeters (99.2 in.)

Climate designation: 3;1
Climate name: Savanna (monsoon type)
Other cities with similar climates: Bombay (4;1), Calcutta (3;1), Miami (3;1)

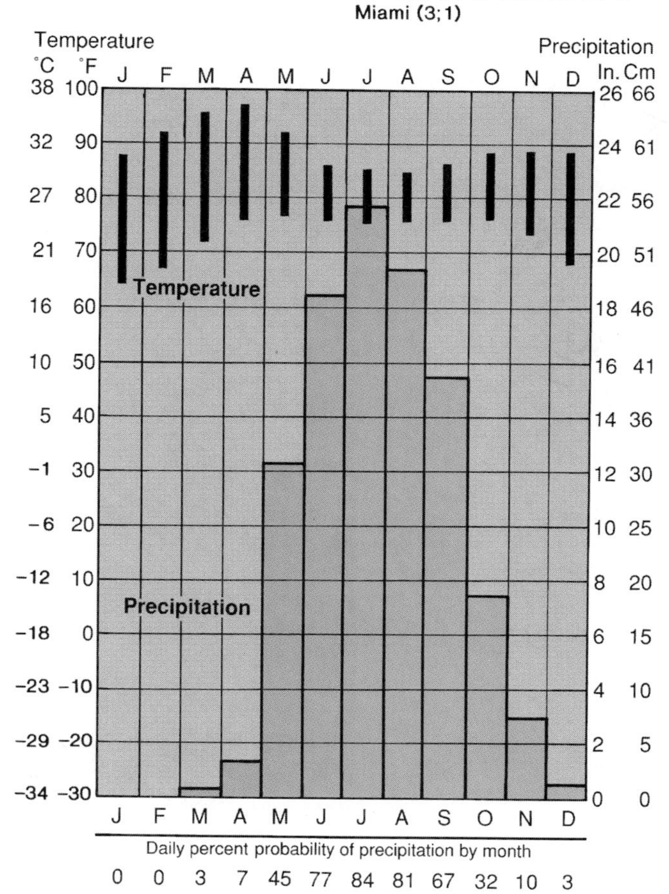

FIGURE 4.34 **Climagraph for Yangôn, Myanmar.** (See Figure 4.31 for explanation.)

City: Cairo, Egypt
Latitude: 29°52′N
Altitude: 116 meters (381 ft)
Yearly precipitation: 2 centimeters (0.7 in.)

Climate designation: 7;4
Climate name: Hot Desert
Other cities with similar climates: Mecca (4;4), Karachi (8;4)

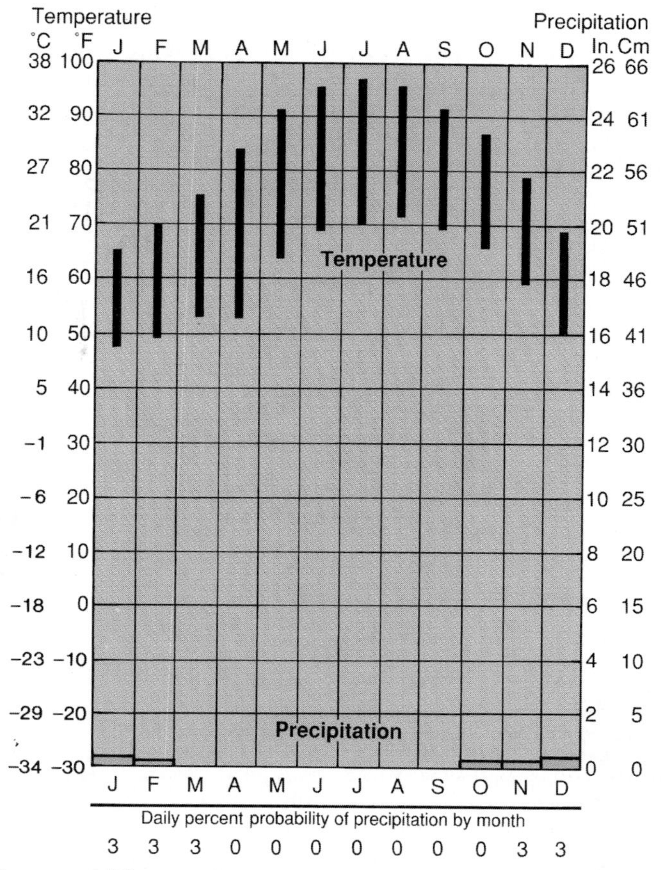

FIGURE 4.35 **Climagraph for Cairo, Egypt.** (See Figure 4.31 for explanation.)

FIGURE 4.36 **Death Valley, California.** Devoid of stabilizing vegetation, desert sands are constantly rearranged in complex dune formations.

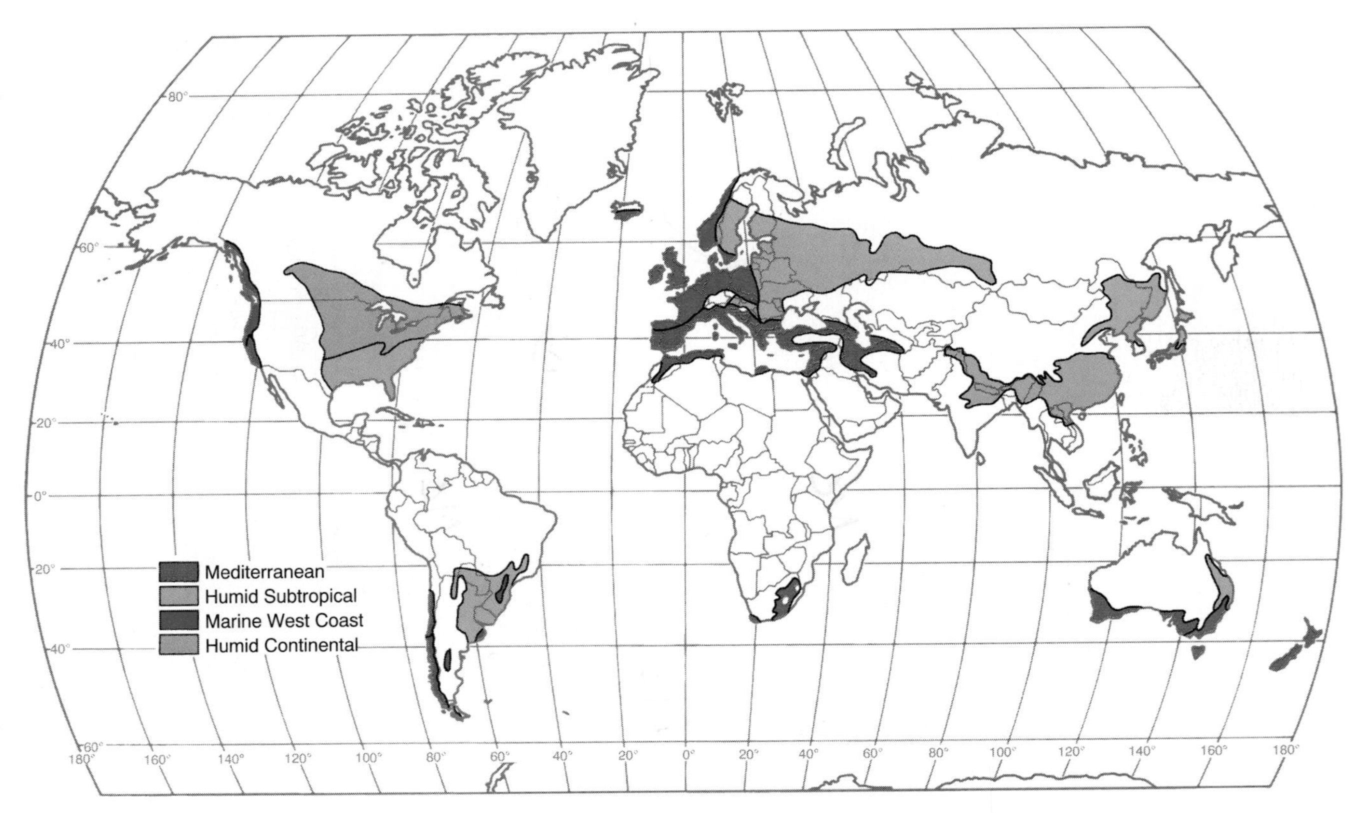

FIGURE 4.37 **The location of humid midlatitude climates.**

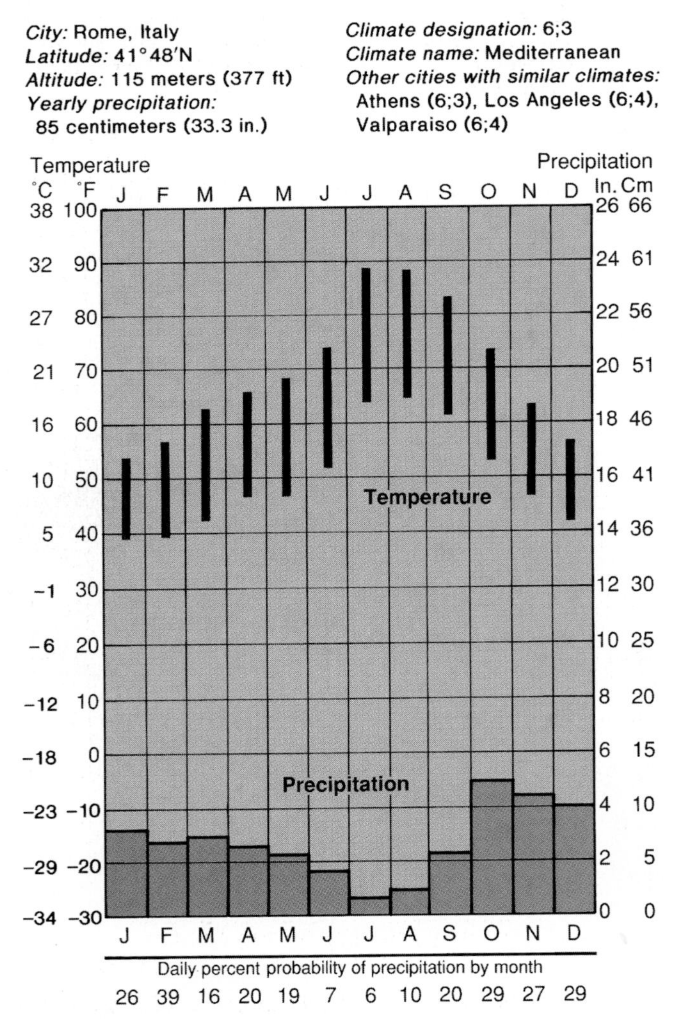

FIGURE 4.38 **Climagraph for Rome, Italy.** (See Figure 4.31 for explanation.)

winds from the land, the latter from the sea. Coniferous forests become more plentiful in the direction of the poles, until temperatures become so low that trees are denied an adequate growing season and therefore their growth is stunted (Figure 4.44). Along with coniferous forests are the infertile spodosols. The transition from the very cold air masses of the winter to the occasional convectional storms of the summer means that four distinct seasons are apparent.

Three huge areas of the world are characterized by a humid continental climate: (1) the northern and central United States and southern Canada; (2) most of the European portion of Russia; and (3) northern China. Because there are no land areas at a comparable latitude in the Southern Hemisphere, this climate is not represented there. In fact, the only non-mountain cold climate in the Southern Hemisphere is the polar climate of Antarctica.

Midlatitude Semideserts and Dryland Climates (10;4)

The location of these climates is shown in Figure 4.45. In the interior of continents where mountains block west winds, or in lands far from the reaches of moist tropical air, extensive regions of *semidesert* conditions appear.

Figure 4.46 illustrates typical temperature and precipitation patterns in these areas. Occasionally, a summer convectional storm or a frontal system with some moisture occurs. These moderately dry lands are called **steppes.** The

FIGURE 4.39 **Vegetation typical of an area with a Mediterranean climate.** Trees such as scrub oak are short and scattered.

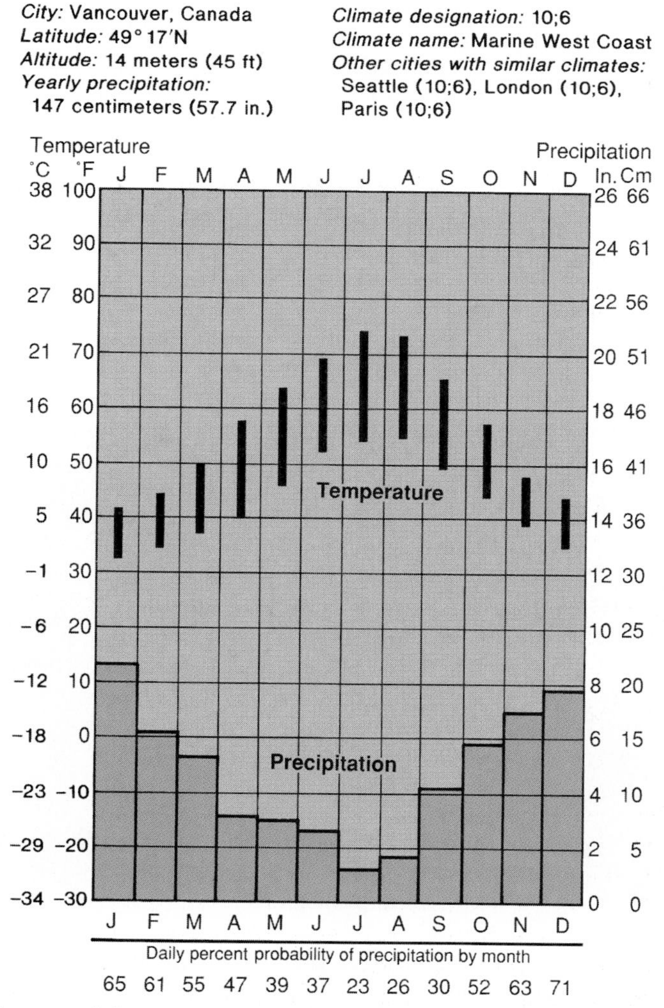

FIGURE 4.40 **Climagraph for Vancouver, Canada.** (See Figure 4.31 for explanation.)

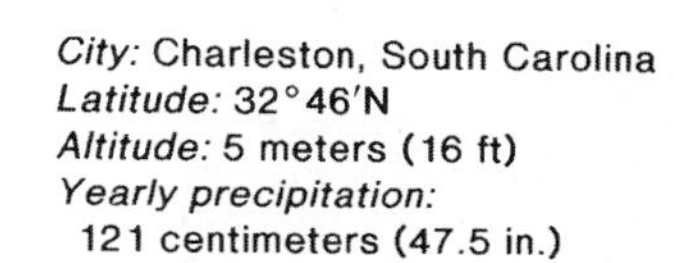

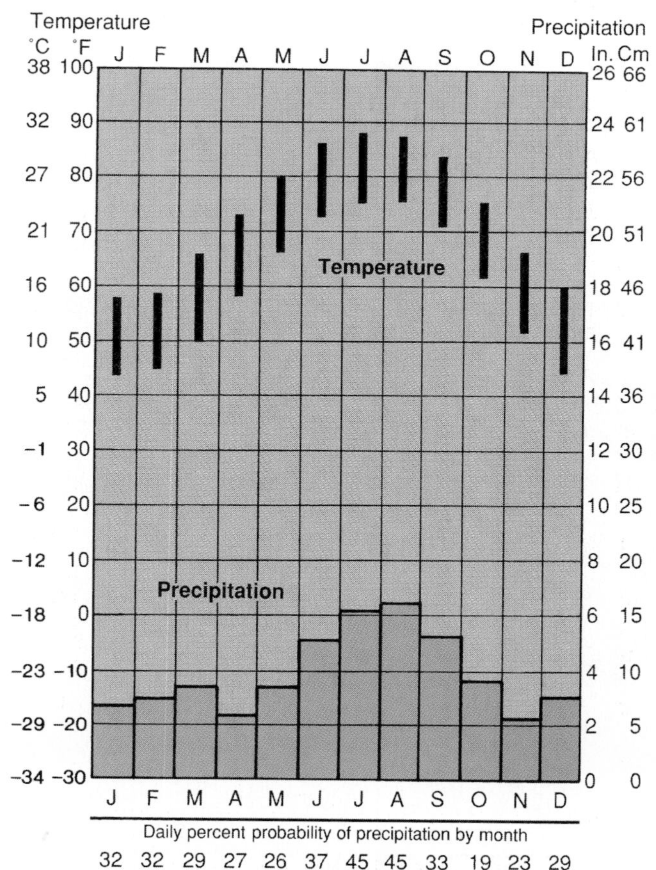

FIGURE 4.41 **Climagraph for Charleston, South Carolina.** (See Figure 4.31 for explanation.)

City: Chicago, Illinois
Latitude: 41°52′N
Altitude: 181 meters (595 ft)
Yearly precipitation: 85 centimeters (33.3 in.)
Climate designation: 14;2
Climate name: Humid Continental (warm summer)
Other cities with similar climates: New York (10;2), Berlin (11;2), Warsaw (15;2)

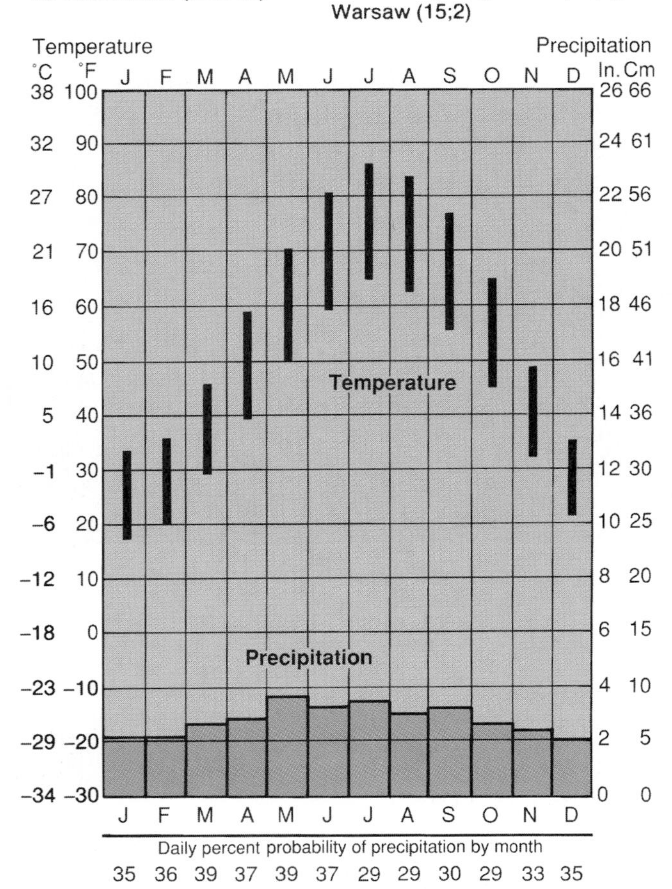

FIGURE 4.42 **Climagraph for Chicago, Illinois.** (See Figure 4.31 for explanation.)

City: Moscow, Russia
Latitude: 55°46′N
Altitude: 154 meters (505 ft)
Yearly precipitation: 55 centimeters (21.8 in.)
Climate designation: 15;6
Climate name: Humid Continental (cool summer)
Other cities with similar climates: Montreal (14;6), Winnipeg (15;6), St. Petersburg (15;6)

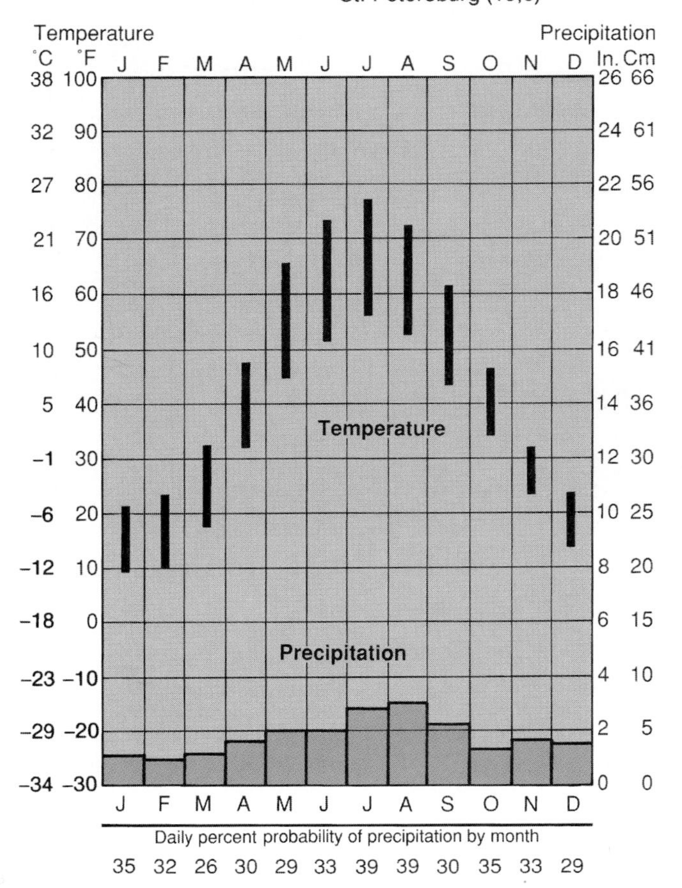

FIGURE 4.43 **Climagraph for Moscow, Russia.** (See Figure 4.31 for explanation.)

FIGURE 4.44 In the extensive region of east central Canada and the area around Moscow, Russia, the summers are long and warm enough to support a dense coniferous forest. Farther north, growth is less luxuriant.

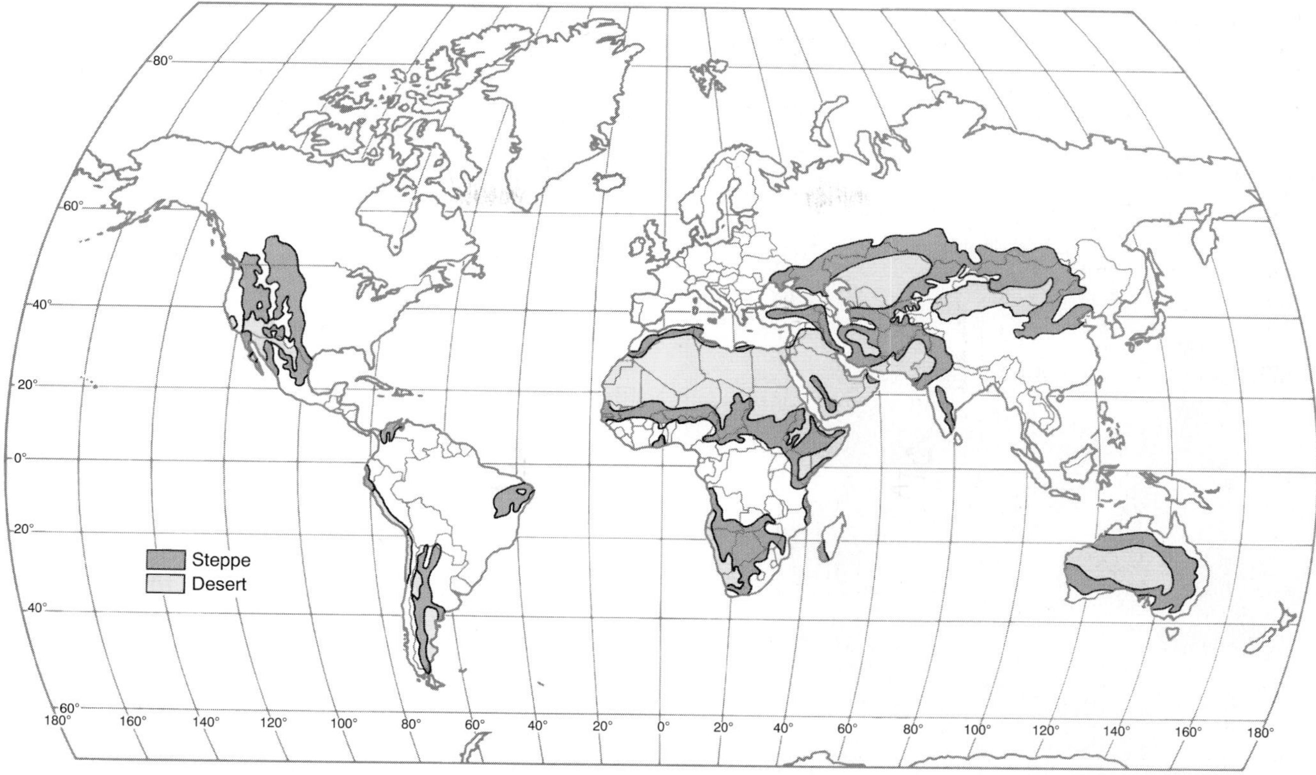

FIGURE 4.45 **The location of steppe and desert climates.**

natural vegetation is grass, although desert shrubs, pictured in Figure 4.47, are found in drier portions of the steppes. Rain is not plentiful, but soils are rich because the grasses return nutrients to the soil. The soils, called *mollisols,* have a dark brown to black A-horizon and are among the most naturally fertile soils in the world. As a result, humans have made the steppes of the United States, Canada, Ukraine, and China some of the most productive agricultural regions on earth. The steppes are also known for their hot, dry summers and biting winter winds that sometimes bring blizzards.

Subarctic and Arctic Climates (16;7)

Toward northern areas and into the interior parts of the North American and Eurasian landmasses, increasingly colder temperatures prevail (Figures 4.48 and 4.49). Trees become stunted, and eventually only mosses and other cool-weather plants of the type shown in Figure 4.50 will grow.

The word **tundra** is often used to describe the northern boundary zone beyond these treed subarctic regions. Because long, cold winters predominate, the ground is frozen most of the year. A few cool summer months, with an abundant supply of mosquitoes, break up the monotony of extreme cold. Although very cold temperatures characterize the tundra, snowfall is not very abundant. Strong easterly winds blow snow, which, combined with ice fogs and little winter sunlight, contributes to a very bleak climate.

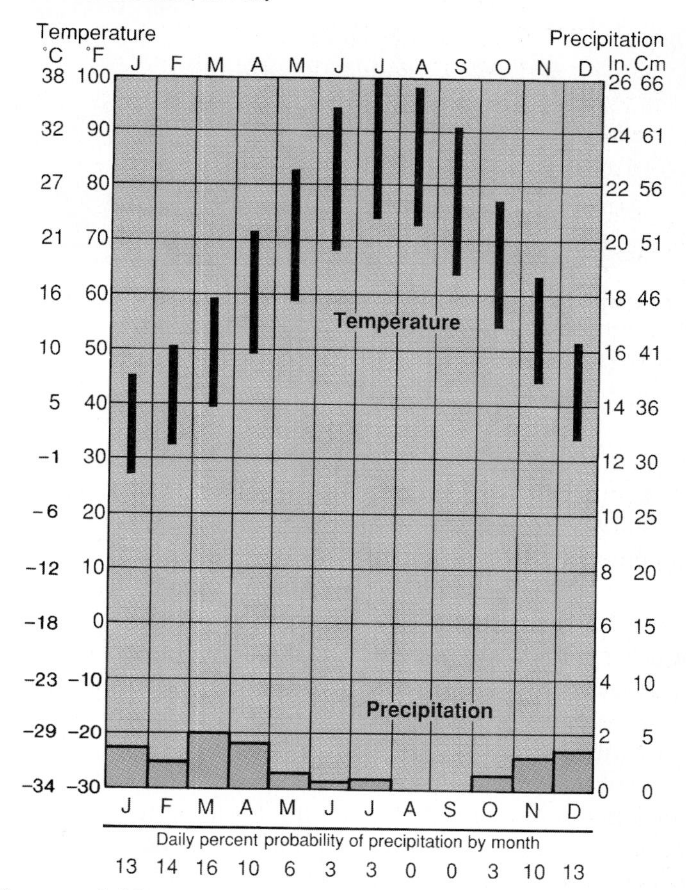

FIGURE 4.46 **Climagraph for Tehran, Iran.** (See Figure 4.31 for explanation.)

FIGURE 4.47 **Desert shrubs** in the midlatitude drylands of northern Mexico.
© William E. Ferguson.

Alaska, northern Canada, and northern Russia are covered with either the stunted trees of the subarctic climate, or by the bleak, treeless expanse of the tundra. Antarctica and Greenland, however, are icy deserts.

Soils in these vast arctic regions are varied. Perhaps the most characteristic soils are *histosols.* They tend to be peat or muck, consisting of plant remains that accumulate in water. In forested areas, they tend to resemble spodosols, which form in areas of poor drainage both in the arctic and the midlatitudes.

These thumbnail sketches of climatic conditions throughout the world give us the basic patterns of large regions. On any given day, conditions may be quite different from those discussed or mapped in this chapter. However, the physical climatological processes, in general, are what concern us. We can deepen our understanding of climates by applying our knowledge of the elements of weather.

Climatic Change

We have stressed that climates are only averages of, perhaps, greatly varying day-to-day conditions. Figure 4.51 illustrates the global variation in yearly precipitation. Temperatures are less changeable than precipitation on a year-to-year basis, but they too vary. How can we account for these variations? Scientists in research stations all around the world are investigating this question. The data they use range from daily temperature and precipitation records to calculations concerning the position of the earth in relation to the sun. Because day-to-day records for most places date back only 50 to 100 years, scientists look for additional information in rock formations, the chemical composition of earth materials, and astronomical changes.

It is a fact that a glacier covered most of Canada and approximately one-third of the United States about 20,000 years ago. In addition, at three other times in the last 1 million years the same situation occurred. These happenings have raised questions about periodic changes in climate (see "Our Inconstant Climates"). By analyzing layers of fossil microorganisms on the ocean floor, scientists have determined that during the last 400,000 years, climatic cycles of 100,000 years' duration have transpired. Some scientists believe this 100,000-year cycle corresponds to cyclical changes in the earth's orbit around the sun, and that such changes are responsible for the succession of the ice ages. When the orbit is nearly circular, the earth experiences relatively cold temperatures. When it is elliptical, as it is now, the earth is exposed to more total solar radiation and thus experiences warmer temperatures.

Another cycle of 42,000 years corresponds to the tilt of the earth relative to the orbital plane. The tilt varies from 22.1° to 24.5°. A low tilt position—that is, a more perpendicular position of the earth—is accompanied by periods of colder climate. Cooler summers are thought to be critical in the formation of ice sheets.

Even though the last ice age ended about 11,000 years ago, and the earth is now in one of its warmest periods, scientific evidence suggests that in 3000 years, substantially more ice will exist on the earth. Climate can change more abruptly than this time scale suggests, however. Relatively small changes in upper-air wind movements can significantly change a climate in decades. Today, polar air is more dominant than tropical air. Changes in precipitation amounts or in the reliability of precipitation, as well as temperature changes, can have enormous impact on agriculture and patterns of human settlement. For example, India has recently experienced monsoon

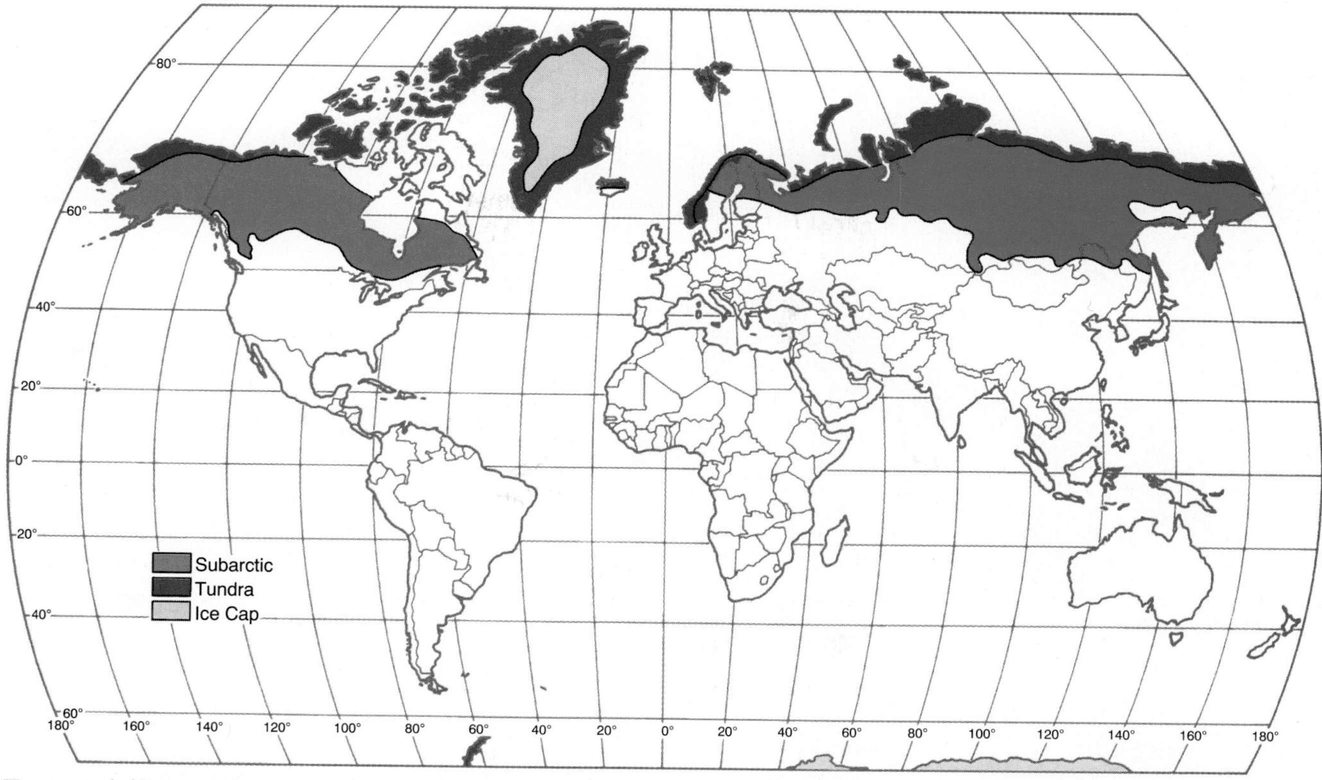

FIGURE 4.48 **The location of arctic and subarctic climates.**

winds that have fluctuated greatly in intensity. And in north Africa, summer rains have periodically failed for a number of years.

Great volcanic eruptions may also produce climatic change because dust particles spread around the world and filter out the sun's rays, to some extent. For three years after the 1883 volcanic eruption of Krakatoa near Java, a decline in temperatures throughout the world was noticeable. The July 1991 volcanic eruption of Mount Pinatubo in the Philippines spewed millions of tons of sulfur particles into the upper atmosphere. As the particles slowly spread over much of the planet, they blocked some of the sunlight that normally would have reached the earth's surface. Between the summers of 1991 and 1992, the cooling effect lowered average global temperatures by about 0.5°C.

Finally, many scientists are concerned about the effect of human activities on climatic change. Evidence shows that industrial and automobile emissions are changing the gaseous mixture in the atmosphere. These changes are among the topics discussed in Chapter 5, dealing with environmental concerns.

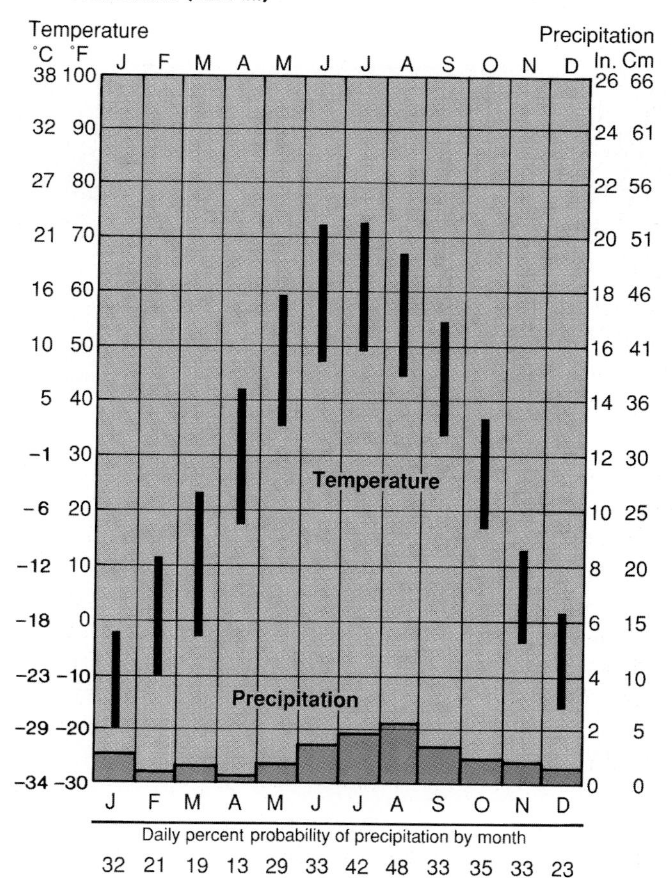

FIGURE 4.49 **Climagraph for Fairbanks, Alaska.** (See Figure 4.31 for explanation.)

FIGURE 4.50 **Tundra vegetation** in the Ruby Range of the Northwest Territories, Canada.

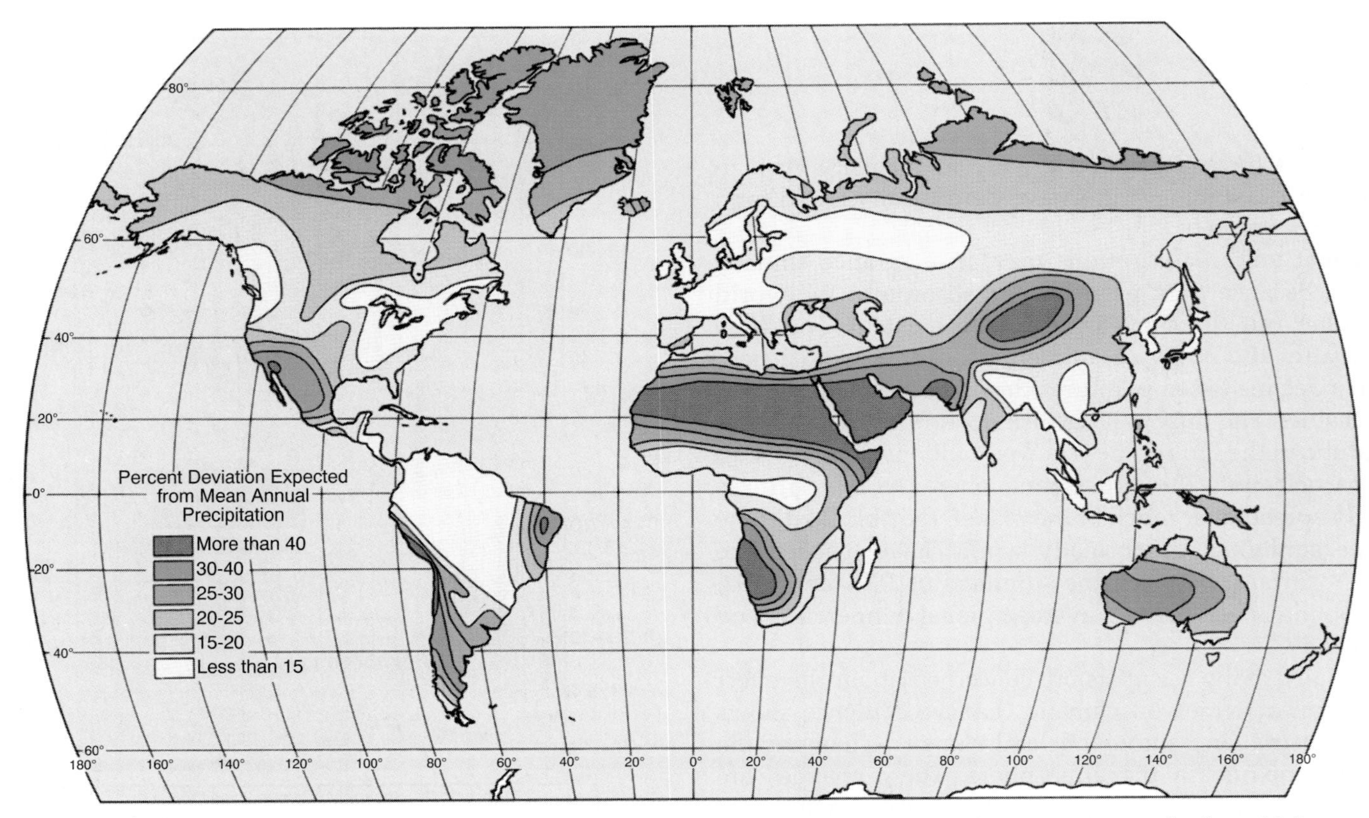

FIGURE 4.51 **The world pattern of precipitation variability.** Note that regions of low total precipitation tend to have high variability. In general, the lower the amount of long-term average annual precipitation, the lower the probability that the average will be recorded in any single year.

Geography *and* Public Policy

Our Inconstant Climates

As we have explained, different climatic conditions prevail in different parts of the world at any given time. Significant variations in climate have also occurred over time. For example, approximately 65 million years ago, at the end of the Cretaceous period, a sudden cooling of the earth's climate occurred. This cool-down is thought to have caused the extinction of some 75% of all existing plant and animal species, including most dinosaurs. To take another example, cycles of ice sheet formation and breakup occurred at least five times during the last ice age, which lasted 100,000 years and ended only 11,000 years ago.

By analyzing tree growth rings, plant pollen, lake floor sediments, historical documents, and other evidence, climatologists have identified two major climatic periods in the last millennium: a medieval warm period and a "little ice age." Between about A.D. 900 and 1300, during the medieval warm period, temperatures were as warm or warmer than they are now. Settlement and farming expanded northward and to higher altitudes, the Vikings colonized Iceland and Greenland, and vineyards flourished in Britain. Between the 15th and 19th centuries, a little ice age descended on the Northern Hemisphere. Arctic ice expanded, glaciers advanced, and drier areas of the earth were desiccated. Crop failures were common. Iceland lost one-quarter of its population between 1753 and 1759, and 1816 was the "year without a summer" in New England, when snow fell in June and frost came in July.

A pronounced warming trend began about 1880. Overall global temperatures have been rising slowly since that time. At least in the Northern Hemisphere, temperatures have been above average for all of the 20th century. Nine of the warmest years on record occurred during the 1980s and 1990s. The earth's average surface temperature rose to 16.9°C (62.45°F) in 1997, the warmest year in a record that goes back to 1856. Does this mean we are now experiencing another warm period? If so, how long is it likely to last?

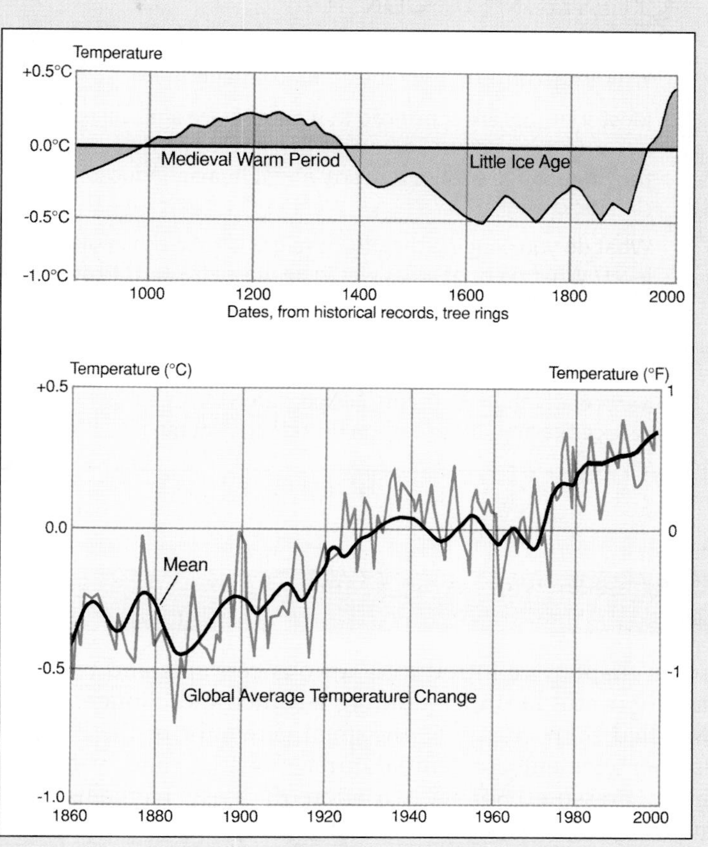

Sources: (top) J. T. Houghton, et al., *Climate Change: The IPCC Assessment,* Cambridge University Press, 1990. (bottom) NOAA.

It is not clear whether these above-average temperatures are part of a steady warming trend that simply represents a natural variation in climate, or if these temperatures can be attributed to emissions of industrial gases that trap heat in the atmosphere. Many scientists believe that human activities are contributing to global warming through what is known as the greenhouse effect, a topic discussed in Chapter 5. The greenhouse effect theoretically explains that the accumulation of carbon dioxide and other gases in the atmosphere, which is the result of burning fossil fuels, prevents heat from radiating out into space. A long-term increase of just a few degrees in the earth's average temperature would have a variety of effects, including a change in precipitation patterns and a rise in sea level as glaciers and polar ice caps thaw.

Continued on page 144

Continued from page 143

Questions to Consider

1. Why is it so difficult to predict climatic change?
2. Most scientists are convinced that another ice age will occur, although they do not agree on when it will start. In view of this, does it make sense to worry about human-induced climatic changes?
3. What do you imagine the effects might be of a rise in sea level? What types of areas would be most affected? How could people adapt to such a rise?
4. Speaking at the July 1996 United Nations Climate Change Convention, scientists and UN officials urged governments, particularly those of industrialized countries, to agree to reduce the amount of carbon dioxide they pump into the atmosphere. "We will be successful if we get agreement on further commitments that might be made in the process of controlling climate change," said the conference president. Critics of that view argue that it is too early to determine what causes warming. The Global Climate Coalition, which includes some of the largest industries in the United States, contends that lower emission targets could bind the United States "to economic and regulatory obligations that could have serious impacts on American industry and its job holders for the next 40 years." Do you think the United States should agree to reduce emissions that might contribute to global warming? Why or why not?

Summary

In this chapter we introduced various concepts and terms that are useful in understanding weather and climate. We identified solar energy as the great generator of the main weather elements of temperature, moisture, and atmospheric pressure. Spatial variation in these elements is caused by two elements: the earth's broad physical characteristics, such as greater solar radiation at the equator than at the poles; and local physical characteristics, such as the effect of water bodies or mountains on local weather conditions.

Climate regions help us simplify the complexities that arise from such special conditions as monsoon winds in Asia or cold ocean currents off the western coast of South America. In just a few descriptive sentences, it is possible to identify the essence of the wide variations in weather conditions that may exist at a place over the course of a year. When a person says that Seattle is in a region of marine west coast climate, for example, images of the average daily conditions communicate not only the facts of weather but also the reasons for conditions as they are. In addition, knowledge of climate tells us about the conditions under which one carries out life's daily tasks.

Weather and climate are among the building blocks that help us understand more clearly the role of humans on the surface of the earth. Although the following chapters focus mainly on the characteristics of human cultural landscapes, one should keep in mind that the physical landscape significantly affects human behavior.

Key Words

air mass 118
air pressure 106
blizzard 120
climate 102
convection 109
convectional precipitation 117
Coriolis effect 110
cyclone 119
cyclonic or frontal precipitation 118
dew point 116
El Niño 116
frictional effect 111
front 118
humid continental climate 129
humid subtropical climate 128
hurricane 120
insolation 103
jet stream 113
land breeze 109
La Niña 116
lapse rate 106
marine west coast climate 125
Mediterranean climate 125
monsoon 113; 125
mountain breeze 110
natural vegetation 132
North Atlantic drift 113
orographic precipitation 118
precipitation 114
pressure gradient 108
reflection 105
relative humidity 116
reradiation 105
savanna 123
sea breeze 109
source region 118
steppe 136
temperature inversion 106
tornado 121
tropical rain forest 123
troposphere 102
tundra 139
typhoon 120
valley breeze 110
weather 102

FOR REVIEW AND CONSIDERATION

1. What is the difference between *weather* and *climate?*
2. What determines the amount of *insolation* received at a given point? Does all potentially receivable solar energy actually reach the earth? If not, why?
3. How is the atmosphere heated? What is the *lapse rate;* and what does it indicate about the atmospheric heat source? Describe a *temperature inversion.*
4. What is the relationship between atmospheric pressure and surface temperature? What is a *pressure gradient,* and of what concern is it in weather forecasting?
5. In what ways do land and water areas respond differently to equal insolation? How are these responses related to atmospheric temperatures and pressures?
6. Draw and label a diagram of the planetary wind and pressure system. Account for the occurrence and character of each wind and pressure belt. Why are the belts latitudinally ordered?
7. What is *relative humidity?* How is it affected by changes in air temperatures? What is the *dew point?*
8. What are the three types of large-scale *precipitation?* How does each occur?
9. What are *air masses?* What is a *front?* Describe the development of a cyclonic storm, showing how it relates to air masses and fronts.
10. What factors were chiefly responsible for today's weather?
11. Summarize the distinguishing temperature, moisture, vegetation, and soil characteristics of each type of climate.
12. What is the climate at Tokyo, London, São Paulo, Leningrad, and Bangkok?

SELECTED REFERENCES

Aguado, Edward, and James E. Burt. *Understanding Weather and Climate.* Upper Saddle River, N.J.: Prentice-Hall, 1999.

Akin, Wallace E. *Global Patterns: Climate, Vegetation, and Soil.* Norman, Okla.: University of Oklahoma, 1991.

Bair, Frank E., ed. *The Weather Almanac.* 6th ed. Detroit: Gale Research, Inc., 1992.

Barry, R. G. *Mountain Weather and Climate.* New York: Methuen, 1981.

Barry, R. G., and R. J. Chorley. *Atmosphere, Weather, and Climate.* 7th ed. New York: Methuen, 1998.

Ellis, Steve, and Tony Mellor. *Soils and Environment.* New York: Routledge, 1995.

Glantz, Michael H. *Currents of Change: El Niño's Impact on Climate and Society.* Cambridge: Cambridge University Press, 1996.

Gordon, Adrian, Warwick Grace, Peter Schwerdtfeger, and Roland Byron-Scott. *Dynamic Meteorology: A Basic Course.* London: Edward Arnold Publishers, Ltd., 1998.

Hoyt, Douglas V., and Kenneth H. Shatten. *The Role of the Sun in Climate Change.* Oxford: Oxford University Press, 1997.

Linacre, Edward. *Climate Data and Resources.* New York: Routledge, 1992.

McGregor, Glenn R., and Simon Nieuwolt. *Tropical Climatology: An Introduction to the Climates of the Low Latitudes.* New York: John Wiley & Sons, 1998.

Moore, David M., ed. *Plant Life.* New York: Oxford University Press, 1990. The Illustrated Encyclopedia of World Geography.

Philander, S. George. *Is the Temperature Rising? The Uncertain Science of Global Warming.* Princeton: Princeton University Press, 1998.

Pielke, Roger A. *The Hurricane.* New York: Routledge, 1990.

Schneider, Stephen H., ed. *Encyclopedia of Climate and Weather.* Oxford: Oxford University Press, 1996.

Simmons, I. G. *Biogeographical Processes.* London: Allen & Unwin, 1982.

Strahler, Alan H., and Arthur N. Strahler. *Introducing Physical Geography.* New York: John Wiley & Sons, 1994.

The Weather Channel (television) provides information on and explanations of current weather conditions throughout the world.

Thompson, Russell D., and Allen Perry, eds. *Applied Climatology.* New York: Routledge, 1997.

Weatherwise. Issued six times a year by Weatherwise, Inc., 230 Nassau St., Princeton, N.J., 08540.

Webster, P. J. "Monsoons," *Scientific American.* (August 1981):109–118.

Wells, Neil. *The Atmosphere and the Ocean: A Physical Introduction.* 2d ed. New York: John Wiley & Sons, 1997.

Don't forget about Dushkin's *Annual Editions Online: Geography* at http://www.dushkin.com/aeonline/. See preface for details.

Human Impact on the Environment

Chapter 5

Polluted water at the bottom of a slag heap, Leadville, Colorado.
©David Hiser/Tony Stone Images.

Times Beach, Missouri, used to be a tight-knit, working-class community along the Meramec River southwest of St. Louis. Now the people are gone. Empty houses and stores, abandoned cars, and rusting refrigerators stand silent. "Thanks for coming," says the sign on the Easy Living Laundromat to nonexistent customers. Wildflowers bloom on overgrown lawns. Flies buzz, squirrels chase each other up trees, and an occasional coyote prowls the streets. Were they allowed, visitors might think they were seeing an uncannily realistic Hollywood stage set. Instead, the ghost town is a symbol of what happens when hazardous wastes contaminate a community.

The trouble began in the summer of 1971, when Times Beach hired a man to spread oil on its 16 kilometers (10 mi) of dirt roads to keep down the dust. Unfortunately, his chief occupation was removing used oil and other waste from a downstate chemical factory, and it was a mixture of these products that he sprayed all over town that summer and the next. The tens of thousands of gallons of purple sludge were contaminated with dioxin, the most toxic substance ever manufactured.

Effects of the dioxin exposure appeared almost immediately in two riding arenas that had been sprayed. Within days, hundreds of birds and small animals died, kittens were stillborn, and many horses became ill and died. Then residents began reporting a variety of medical disorders, including miscarriages, seizure disorders, liver impairment, and kidney cancer. Not until 1982 did the Environmental Protection Agency (EPA), alerted to high levels of dioxin in wastes stored at the chemical plant, make a thorough investigation of Times Beach. They found levels of the dioxin compound TCDD as high as 300 parts per billion in some of the soil samples; 1 part per billion is the maximum concentration deemed safe. The EPA purchased every piece of property in town and ordered its evacuation (Figure 5.1).

The story of Times Beach is one of many that could have been selected to illustrate how people can affect the quality of the water, air, and soil on which their existence depends.

Terrestrial features and ocean basins, elements of weather and characteristics of climate, flora, and fauna comprise the building blocks of that complex mosaic called the *environment,* or the totality of things that in any way may affect an organism. Plants and animals, landforms, soils and nutrients, weather and climate all comprise an organism's environment. The study of how organisms interact with one another and with their physical environment is called **ecology.** It is critically important in understanding environmental problems, which usually arise from a disturbance of the natural systems that make up our world.

Humans exist within a natural environment that they have modified by their individual and collective actions. Forests have been cleared, grasslands plowed, dams built, and cities constructed. On the natural environment, then, has been erected a cultural environment, modifying, altering, or destroying the balance of nature that existed before human impact was expressed. This chapter is concerned with the interrelation between humans and the natural environment that they have so greatly altered.

Since the beginning of agriculture, humans have changed the face of the earth, have distorted delicate balances and interplays of nature, and, in the process, have both enhanced and endangered the societies and the economies that they have erected. The essentials of the natural balance and the ways in which humans have altered it are not only our topics here but are also matters of social concern that rank among the principal domestic and international issues of our times. As we shall see, the fuels we consume, the raw materials we use, the products we create, and the wastes we discard all contribute to the harmful alteration of the **biosphere,** the thin film of air, water, and earth within which we live.

Ecosystems

The biosphere is composed of three interrelated parts:

1. the *troposphere,* some 9.5–11.25 kilometers (6–7 mi) thick;
2. the *hydrosphere,* including surface and subsurface waters in oceans, streams, lakes, glaciers, or groundwater—much of it locked in ice or earth and not immediately available for use; and

FIGURE 5.1 **Times Beach, Missouri.** Most new highway maps no longer mark the location of this former community. Roads into the town are blocked, and security personnel guard a barricade on its east edge.

3. the upper reaches of the earth's *crust,* a few thousand feet at most, containing the soils that support plant life, the minerals that plants and animals require to exist, and the fossil fuels and ores that humans exploit.

The biosphere is an intricately interlocked system, containing all that is needed for life, all that is available for living things to use, and, presumably, all that ever will be available. The ingredients of the biosphere must be, and are, constantly recycled and renewed in nature. Plants purify the air, the air helps to purify the water, the water and the minerals are used by plants and animals and are returned for reuse.

The biosphere, therefore, consists of two intertwined components: (1) a nonliving outside (solar) energy source and requisite chemicals; and (2) a living world of plants and animals. In turn, the biosphere may be subdivided into specific **ecosystems,** self-sustaining units that consist of all the organisms (plants and animals) and physical features (air, water, soil, and chemicals) existing together in a particular area. The most important principle concerning all ecosystems is that everything is interconnected. Any intrusion or interruption in the balance that has been naturally achieved inevitably results in undesirable effects elsewhere in the system. Each organism occupies a specific *niche,* or place, within an ecosystem. In the energy exchange system, each organism plays a definite role; individual organisms survive because of other organisms that also live in that environment. The problem lies not in recognizing the niches but in anticipating the chain of causation and the readjustments of the system consequent on disturbing the occupants of a particular niche.

Food Chains and Renewal Cycles

Life depends on the energy and nutrients flowing through an ecosystem. The transfer of energy and materials from one organism to another is one link in a **food chain,** defined as a sequence of organisms, such as green plants, herbivores, and carnivores, through which energy and materials move within an ecosystem (Figure 5.2). Most food chains have three or four links, although some have only two—for example, when human beings eat rice. Because the ecosystem in nature is in a continuous cycle of integrated operation, there is no start or end to a food chain. There are, simply, nutritional transfer stages in which each lower level in the food chain transfers part of its contained energy to the next higher-level consumer.

The *decomposers* pictured in Figure 5.2 are essential in maintaining food chains and the cycle of life. They cause the disintegration of organic matter—animal carcasses and droppings, dead vegetation, waste paper, and so on. In the process of decomposition, the chemical nature of the material is changed, and the nutrients contained within it become available for reuse by plants or animals. *Nutrients,* the minerals and other elements that organisms need for growth, are never destroyed; they keep moving from living to nonliving things and back again. Our bodies contain nutrients that were once part of other organisms, perhaps a hare, a hawk, or an oak tree.

The idea of a cycle is important in furthering our understanding of the natural renewal of the elements essential to life, which include carbon, oxygen, hydrogen, and phosphorus. Because the supply of these materials is fixed, they must be continuously recycled through food chains. Figure 5.3 shows one important cycle, the phosphorus cycle.

Phosphorus is an element essential to the survival of many species of plants and animals. It is an important component of DNA and of bones and teeth. Plants absorb phosphate from the soil, and other organisms acquire it by eating plants. It is returned to the soil through animal excretions or through the decay of plants and animals. When phosphate that has been used as an agricultural fertilizer is carried to rivers and eventually to the sea through runoff, only a small percentage is returned to the land through fish catches and the excretions (guano) of sea birds, which eat fish. Most of it remains in deep ocean waters until geologic processes cause uplifting of the ocean floor.

Ecosystems change constantly whether people are present or not, but humans have affected them more than has any other species. The impact of humans on ecosystems was small at first, with low population size, energy consumption, and technological levels. It has increased so rapidly and pervasively as to present us with widely recognized and varied ecological crises. Some of the effects of humans on the natural environment are the topic of the remainder of this chapter.

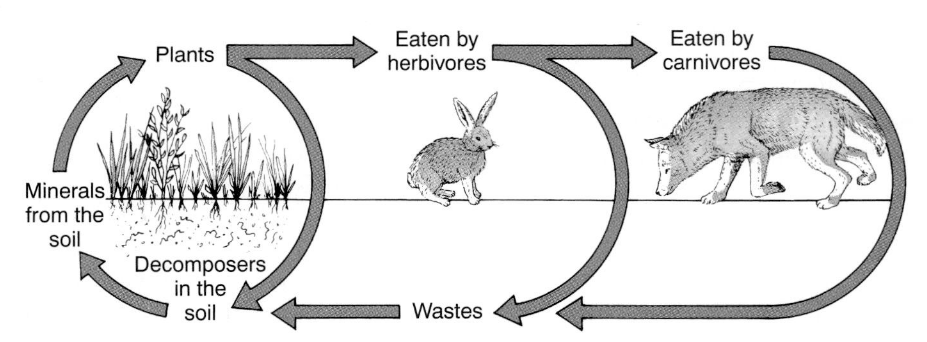

FIGURE 5.2 The supply of food in an ecosystem is a hierarchy of "who eats what"—a hierarchy that creates a food chain. In this simplified example, green plants are the *producers* (autotrophs), using nutrients and energy from the sun to make their own food. Herbivorous rabbits (*primary consumers*) feed directly on the plants, and carnivorous foxes (*secondary consumers*) feed on the rabbits. A food chain is one thread in a complex *food web,* all the feeding relationships that exist in an ecosystem. For example, a mouse might feed on the plants shown here, and then be eaten by a hawk, another food chain in this food web.

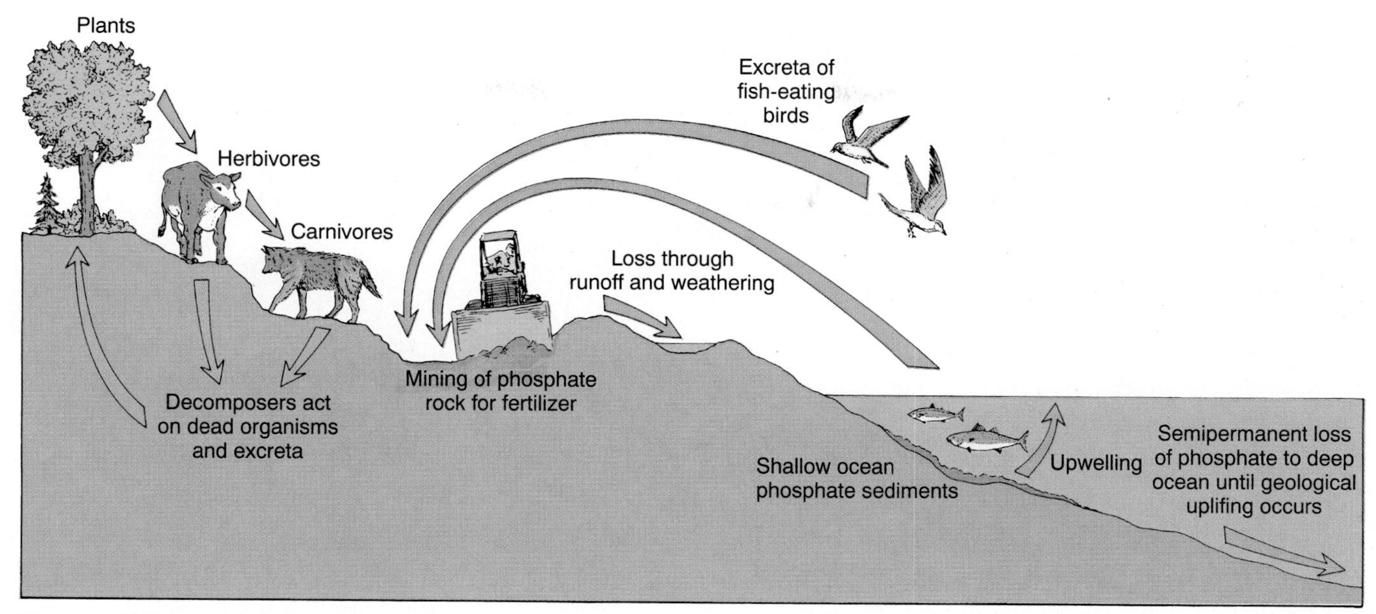

FIGURE 5.3 **The phosphorus cycle.** The natural phosphorus cycle is exceedingly slow and has been upset by human activity. Currently, phosphorus is washing into the sea faster than it can be returned to the land, primarily because it is being mined for use as a fertilizer.

IMPACT ON WATER

The supply of water is constant. The system by which it continuously circulates through the biosphere is called the **hydrologic cycle** (see Figure 5.4). In that cycle, water may change form and composition, but under natural environmental circumstances, it is purified in the recycling process and is again made available with appropriate properties to the ecosystems of the earth. *Evaporation* and *transpiration* (the emission of water vapor from plants) are the mechanisms by which water is redistributed. Water vapor collects in clouds, condenses, and then falls again to the earth. There it is reevaporated and retranspired, only to fall once more as precipitation.

People's dependence on water has long led to efforts to control its supply. Such manipulation has altered the quantity and quality of water in rivers and streams.

Availability of Water

Globally, fresh water is abundant. Enough rain and snow fall on the continents each year to cover the earth's total land area with 83 centimeters (33 in.) of water. It is usually reckoned that the volume of fresh water annually renewed by the hydrologic cycle could meet the needs of a world population 5 to 10 times its present size.

Yet in many parts of the world, water supplies are inadequate and dwindling. The problem is not with the global amount of water but with its distribution (the average amount of precipitation an area receives) and reliability (the variability of precipitation from year to year). Regional water sufficiency is also a function of the size of the population using the water and the demands it places on the resource. For the world as a whole, irrigated agriculture accounts for nearly three-quarters (73%) of freshwater use; in the poorest countries, the proportion is 90% (Figure 5.5). Industry uses about one-fifth (21%), and households and municipalities account for the remainder. Since 1940, there has been more than a fourfold increase in withdrawals of fresh water from streams, lakes, aquifers, and other sources. (An *aquifer* is a zone of water-saturated sands and gravels beneath the earth's surface; the water it contains is called *groundwater,* in contrast to surface waters such as rivers and lakes.)

"Scarcity" is the word increasingly used to describe water supplies in parts of both the developed and developing world. Insufficient water for irrigation periodically endangers crops and threatens famine; permanent streams have become intermittent in flow; lakes are shrinking; and from throughout the world come reports of rapidly falling water tables and wells that have gone dry. According to the World Bank, chronic water shortages that threaten to limit food production, economic development, sanitation, and environmental protection already plague 80 countries. Ten countries in North Africa and the Middle East actually run a *water deficit:* they consume more than their annual renewable supply, usually by pumping groundwater faster than it is renewed by rainfall.

National data can mask water scarcity problems at the local level. A number of countries have major crop-producing

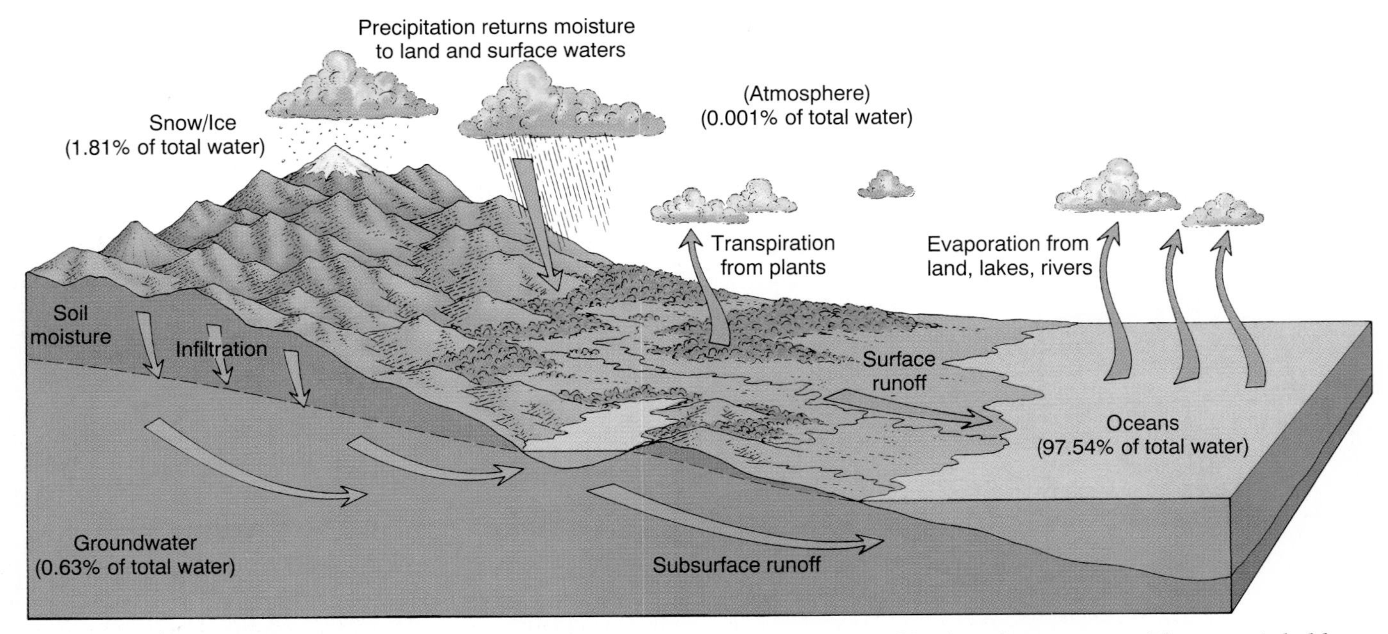

FIGURE 5.4 **The hydrologic cycle.** The sun provides energy for the evaporation of fresh and ocean water. The water is held as vapor until the air becomes supersaturated. Atmospheric moisture is returned to the earth's surface as solid or liquid precipitation to complete the cycle. Because precipitation is not uniformly distributed, moisture is not necessarily returned to areas in the same quantity as it has evaporated from them. The continents receive more water than they lose. The excess returns to the seas as surface water or groundwater. A global water balance, however, is always maintained.

FIGURE 5.5 **Irrigation in the Lake Argyle project of Western Australia.** Since 1950, irrigation has been extended to between 5 and 6 million additional hectares (12 to 15 million acres) annually. Irrigation agriculture now produces about one-third of the world's harvest from about 17% of its cropland. Usually, much more water is transported and applied to fields than crops actually require, using unnecessary amounts of a scarce resource. Much of the water is lost to the regional supply through evaporation and transpiration; often less than half of the water withdrawn for irrigation is returned to streams or aquifers for further use. "Drip irrigation," which delivers water directly to plant roots through small perforated tubes laid across the field, is one method of reducing water consumption.
Robert Frerck/Odyssey Productions/Chicago.

regions where groundwater overpumping and aquifer depletion have led to serious water shortages and restricted supplies. Two such states are China and India, home to more than 35% of the world's population. Irrigated farmland and urban and industrial growth are depleting water supplies in northern China. For much of the year, the Huang He (Yellow River) runs dry in its lower reaches before arriving at the Yellow Sea. Nearly 600 Chinese cities, mostly in the north, already have acute water shortages, and the water table beneath Beijing has dropped 37 meters (121 ft) over the last 40 years. In several states in India, including Haryana and Punjab, the country's breadbasket, water use exceeds the sustainable yield of aquifers, causing water tables to fall and wells to dry up.

Examples of unsustainable water use can also be found in the United States. Groundwater supplies are being depleted faster than they can be renewed in Arizona, New Mexico, and the agriculturally rich San Joaquin and Central Valleys of California. Demands upon the only significant source of surface water in the southwestern United States, the Colorado River, are so great that virtually no water from the stream now reaches the ocean. Aqueducts and irrigation canals siphon off its water for use in seven western states and northern Mexico. The country's largest underground water reserve, the Ogallala aquifer, which underlies nearly 20% of all U.S. irrigated land, is drying up (Figure 5.6). Stretching from South Dakota to west Texas, the aquifer supports nearly half of the country's cattle industry, a fourth of its cotton crop, and a great deal of its corn and wheat. More than 150,000 wells now puncture the aquifer, pumping water for irrigation, industry, and domestic use. The water table is falling faster than the aquifer can be replenished by nature. In some areas, the wells no longer yield enough to permit irrigation, and farmed land is decreasing; in others, water levels have fallen so far that it is uneconomical to pump it to the surface for any use.

Modification of Streams

To prevent flooding, to regulate the water supply for agriculture and urban settlements, or to generate power, people have for thousands of years manipulated rivers by constructing dams, canals, and reservoirs. Although they generally have achieved their purposes, these structures can have unintended environmental consequences (see "Blueprint for Disaster: Stream Diversion and the Aral Sea"). These include reduction in the sediment load downstream, followed by a reduction in the amount of the nutrients available for crops and fish; an increase in the salinity of the soil; and subsidence (discussed later in this chapter).

Channelization, another method of modifying river flow, is the construction of embankments and dikes and the straightening, widening, and/or deepening of channels to control floodwaters or to improve navigation. Many of the great rivers of the world, including the Nile and the Huang He, are lined by embankment systems. Like dams, these systems can have unforeseen consequences. They reduce the natural storage of floodwaters, can aggravate flood peaks downstream, and can cause excessive erosion.

Until 1960, the Kissimmee River in Florida twisted and turned as it travelled through a floodplain between Lake Kissimmee and Lake Okeechobee (Figure 5.7). The habitat supported hundreds of species of birds, reptiles, mammals, and fish. At the request of ranchers and farmers who had moved onto the wetlands and were disturbed by the tendency of the river to flood, the Army Corps of Engineers dredged the stream, turning 166 kilometers (103 mi) of meandering river into a dirt-lined canal only 90 kilometers (56 mi) long. After the completion of the canal in 1971, the wetlands disappeared, alien plant species moved in, fish populations declined drastically, and 90% of the waterfowl, including several endangered species, disappeared.

Channelization and dam construction are deliberate attempts to modify river regimes, but other types of human action also affect river flow. Urbanization, for example, has significant hydrologic impacts, including a lowering of the water table, pollution, and increased flood runoff. Likewise, the removal of forest cover increases runoff, promotes flash floods, lowers the water table, and hastens erosion. Nevertheless, the primary adverse human impact on water is felt in the area of water quality. People withdraw water from lakes, rivers, or underground deposits to use for drinking, bathing, agriculture, industry, and many other purposes. Although the water that is withdrawn returns to the water cycle, it is not always returned in the same condition as it was at the time of withdrawal. Water, like other segments of the ecosystem, is subject to serious problems of pollution.

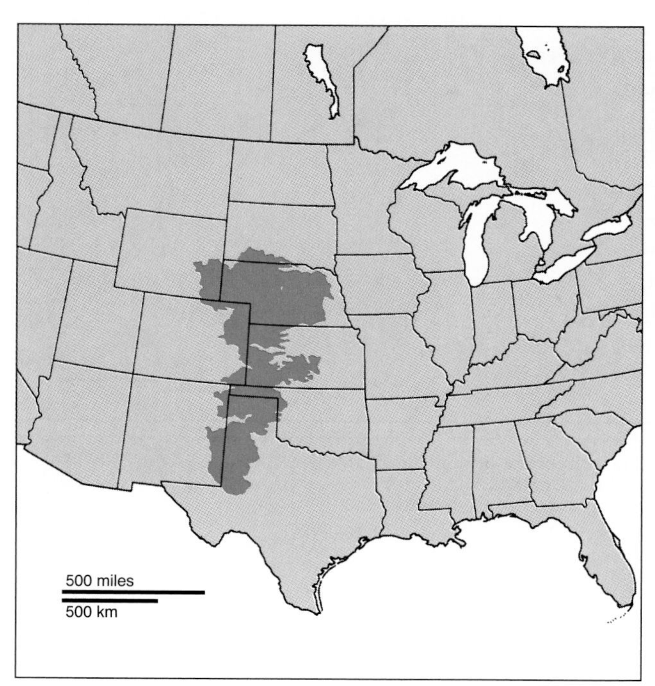

FIGURE 5.6 **The giant Ogallala aquifer,** the largest underground water supply in the United States, provides almost one-third of all groundwater used for irrigation in the country. Approximately 300 gallons of water in the field are needed to grow enough wheat for one loaf of bread, 2400 gallons to produce one pound of beef—figures that help explain why the aquifer is being depleted faster than it can be replenished by nature.

Water Quality

As a general definition, **environmental pollution** by humans means the introduction into the biosphere of wastes that, because of their volume, their composition, or both, cannot be readily disposed of by natural recycling processes. In the case of water, the central idea is that pollution exists when water composition has been so modified by the presence of one or more substances that either it cannot be used for a specific purpose or it is less suitable for that use than it was in its natural state. Pollution is brought about by the discharge into water of substances that cause unfavorable changes in its chemical or physical nature or in the quantity and quality of the organisms living in the water. Pollution is a relative term. Water that is not suitable

Blueprint for Disaster: Stream Diversion and the Aral Sea

It used to be the fourth largest lake in the world, covering an area of 69,900 square kilometers (27,000 sq mi), larger than the state of West Virginia. Now, the Aral Sea is only the sixth largest lake, and it may cease to exist by the year 2010, being converted into a number of lakes instead. The level of the sea has dropped 16 meters (48 ft) since 1960, and the volume of the water is only one-fourth of what it used to be.

The shrinkage of the lake is just one consequence of the former Soviet Union diverting nearly all the water from the lake's primary sources, the Amu Darya and the Syr Darya, to irrigate the agricultural fields of Central Asia. These are some of the others:

- As the shoreline has receded, it has left behind 26,000 square kilometers (10,000 sq mi) of salty desert waste. The high concentrations of salt, fertilizers, and pesticides are slowly turning the Aral into a dead sea. Because the water is too salty for most fish species to endure, the commercial fishing industry has already collapsed. Once a fishing port and popular resort, Müynoq is now 48 kilometers (30 mi) from the sea.
- In addition, wind storms whip up salty grit and toxic chemicals from the dried-up seabed and deposit it on cropland hundreds of kilometers away, reducing the soil's fertility. Thus, the agricultural crops (cotton, rice, fruit, and vegetables) for which the Aral Sea was sacrificed are themselves at risk.
- Forests and wetlands upon which the region's animal life depend have been decimated. Three-quarters of the animal species in the basin have vanished.
- Finally, the rivers that feed the Aral Sea have become sluggish sewers, contaminated by industrial and agricultural wastes and sewage. Although the rivers teem with viruses causing dysentery, typhoid, and other diseases, millions of people depend on them for drinking water. Over 65,000 cases of hepatitis have been reported in the region in the last 15 years, and child mortality rates are among the highest in the world. The newly independent states of Central Asia are in a poor position to reduce irrigation or undertake major health improvement projects in the Aral region.

The drastic consequences of river diversion in the Aral Sea basin illustrate why it has been called one of the earth's greatest environmental tragedies.

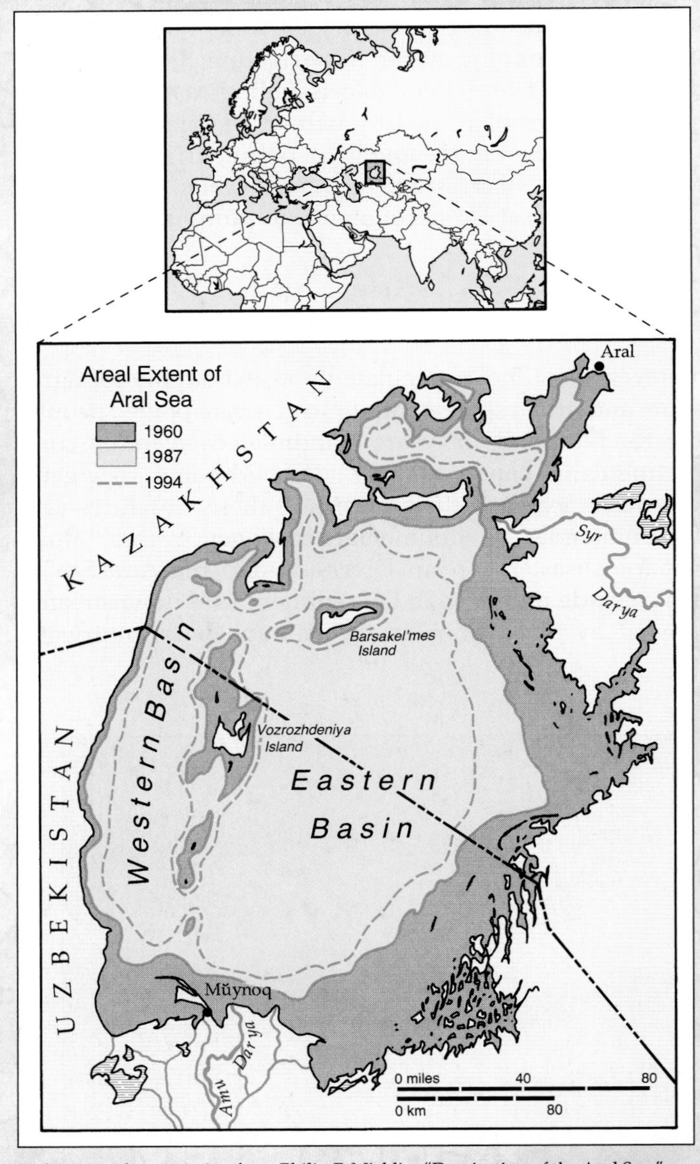

Redrawn with permission from Philip P. Micklin, "Dessication of the Aral Sea," *Science*, Vol. 242, p. 1170, 2 September 1988. Copyright ©1988 American Association for the Advancement of Science.

for drinking may be completely satisfactory for cleaning streets. Water that is too polluted for fish may provide an acceptable environment for certain water plants.

Human activity is not the only cause of water pollution. Leaves that fall from trees and decay, animal wastes, oil seepages, and other natural phenomena may affect water quality. There are natural processes, however, to take care of such pollution. Organisms in water are able to degrade, assimilate, and disperse such substances in the amounts in which they naturally occur. Only in rare instances do natural pollutants overwhelm the cleansing abilities of the recipient waters. What is happening now is that the quantities of wastes discharged by humans often exceed the ability of a given body of water to purify itself. In addition, humans are introducing pollutants, such as metals or inorganic substances, that take a very long time to break down or cannot be broken down at all by natural mechanisms.

(a)

(b)

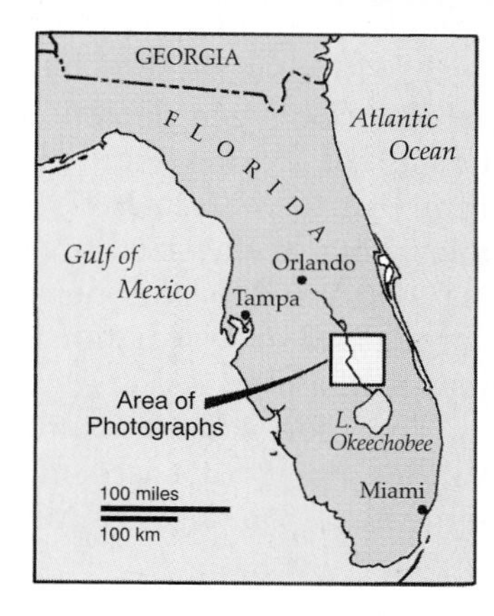

FIGURE 5.7 **The Kissimmee River** in Florida (a) before and (b) after the Army Corps of Engineers turned it into a straight canal. The habitat once supported thousands of fish, waterfowl, and such wading birds as the wood stork, snowy egret, and great blue heron. The channelization contributed to the deterioration of Lake Okeechobee and the Everglades. In 1994, the federal government and the state of Florida embarked on a 15-year project to return the Kissimmee to a meandering path. In addition, the federal Water Resources Development Act of 1996 called for completion of a comprehensive plan to restore and protect the Everglades and authorized spending for the Army Corps of Engineers to undertake projects to carry out that goal.

(a), (b) Courtesy of South Florida Water Management District.

As long as there are people on earth, there will be pollution. Thus, the problem is one not of eliminating pollution, but of controlling it. Such control can be a matter of life or death. According to United Nations figures, between 5 and 10 million people (mostly children) perish each year as a result of drinking contaminated water. They die from diarrhea and water-borne, parasitic diseases such as schistosomiasis.

The four major contributors to water pollution are agriculture, industry, mining, and municipalities and residences. Table 5.1 shows the kinds of water pollutants associated with each source. It is helpful to distinguish between "point" and "nonpoint" sources of pollution. As the name implies, *point sources* enter the environment at specific sites, such as a sewage treatment facility or an industrial discharge pipe. *Nonpoint sources* are more diffuse and, therefore, more difficult to control; examples include runoff from agricultural fields and road salts.

TABLE 5.1

SOURCES AND TYPES OF MAJOR WATER POLLUTANTS

SOURCE	TYPE OF POLLUTANT
Agriculture	Fertilizers (principally nitrogen and phosphorous); biocides; animal wastes; sediments
Industry	Synthetic organic chemicals, including polychlorinated biphenyls (PCBs) and dioxin; toxic inorganic chemicals, including heavy metals; radioactive materials; heat discharge
Mining	Acids; chlorides; heavy metals
Municipalities and residences	Nutrients; organic materials; heavy metals; toxic chemicals; chlorides; sewage

Agricultural Sources of Water Pollution

On a worldwide basis, agriculture probably contributes more to water pollution than does any other single activity. In the United States, agriculture is estimated to be responsible for about two-thirds of stream pollution. Agricultural runoff carries three main types of pollutants: fertilizers, biocides, and animal wastes.

Fertilizers

Agriculture is a chief contributor of *excess nutrients* to water bodies. Pollution occurs when nitrates and phosphates that have been used in fertilizers and that are present in animal manure drain into streams and rivers, eventually accumulating in ponds, lakes, and estuaries. The nutrients hasten the process of **eutrophication,** or the enrichment of waters by nutrients. Eutrophication occurs naturally when nutrients in the surrounding area are washed into the water, but when the sources of enrichment are artificial, as is true of commercial fertilizers, the body of water may become overloaded with nutrients. Algae and other plants are stimulated to grow abundantly, blocking the sunlight that other organisms need. When they die and decompose, the level of dissolved oxygen in the water decreases. Fish and plants that cannot tolerate the poorly oxygenated water are eliminated.

Scientists have estimated that as many as one-third of the medium- and large-size lakes in the United States have been affected by *accelerated* eutrophication. In addition, high levels of nutrients from agricultural runoff in the Mississippi River watershed have helped create a "dead zone" (an area of oxygen depletion) off the coast of Louisiana in the Gulf of Mexico (Figure 5.8). The size of the zone varies from year to year; in 1997, it covered roughly 7000 square miles, an area the size of New Jersey. The zone reaches its peak during the summer months, as the water grows warmer and solar radiation increases, causing algal populations to bloom.

The Food and Agriculture Organization of the United Nations reports that accelerated eutrophication has left 90% of the Black Sea with critically low oxygen levels, causing a precipitous decline in the total fish catch. The Baltic Sea, too, has shown increasing symptoms of eutrophication in the last 20 years; the algal blooms are thought to be at least partially responsible for declining fish catches.

FIGURE 5.8 The "dead zone" in the Gulf of Mexico, 1997. Nutrients from agricultural runoff are thought to be a critical factor in creating this area of oxygen depletion. The nutrients feed algal populations that bloom during the summer. After the algae die and sink, they decompose, depleting the oxygen near the ocean floor. The phenomenon puts the Gulf's fishing industry at risk.

Biocides

The herbicides and pesticides used in agriculture are another source of the chemical pollution of water bodies. Runoff from farms where such *biocides* have been applied contaminates both ground and surface waters. One of the problems connected with the use of biocides is that the long-term effects of such usage are not always immediately known. DDT, for example, was used for many years before people discovered its effect on birds, fish, and water plant life. Another problem is that thousands of these products, containing more than 600 active ingredients, are now in wide use, yet very few have been reviewed for safety by the Environmental Protection Agency (EPA).

Biocide contamination of groundwater exists in at least 34 states. An EPA survey found biocides in about 10% of all community water systems, with 1% containing potentially unsafe concentrations. The situation is often more serious in farm areas where residents depend on wells for their drinking water. Surveys in Minnesota and Iowa, for example, indicate that 30–60% of private wells may be tainted by runoff from farm herbicides and pesticides.

Animal Wastes

A final agricultural source of chemical pollution is animal wastes, especially in countries where animals are raised intensively. This is a problem both in feedlots, where animals are crowded together at maximum densities to be fattened before slaughter, and on the factory-like farms where hog and poultry production is increasingly concentrated. These farms and large feedlots, such as the one pictured in Figure 5.9, produce vast quantities of manure—as much raw sewage as a middle-sized city, but often without sewage treatment facilities. The main method for disposing of the manure is to spread it as a fertilizer on agricultural crops.

The water pollution that occurs from spreading manure on land is suspected by some to be responsible for recent outbreaks of the microorganism *Pfiesteria piscicida* in the Chesapeake Bay. These single-celled organisms, which live in both fresh and salt water, proliferate and become toxic when exposed to high levels of nitrogen and phosphorous, by-products of animal waste. The manure comes from the large poultry and hog farms of Maryland, Virginia, and North Carolina. Some scientists believe that excessive amounts of nutrients from the manure wash into bay tributaries, helping to breed *Pfiesteria* in its toxic form. Others think the *Pfiesteria* outbreaks are due to the fungicides that farmers spray on their crops. Whatever the cause, there is no doubt that *Pfiesteria* have killed millions of fish. There is some evidence that the organism also preys on human

FIGURE 5.9 **The Montfort Beef Company feedlot near Greeley, Colorado.** It is estimated that animal wastes in the United States total about 1.5 billion tons per year, with feedlots generating about half the total. If not treated properly, the manure pollutes both soil and water with infectious agents and excess nutrients. The sanitary disposal of organic wastes generated by such concentrations of animals is a problem only recently addressed by environmental protection agencies.

beings; scores of people have found open sores on their bodies, felt faint, and/or complained of severe headaches and memory loss after coming into contact with the organism or with water contaminated by dead fish.

Other Sources of Water Pollution

As is evident from Table 5.1, agriculture is only one of the human activities that contribute to water pollution. Other sources are industry, mining, municipalities, and residences.

Industry

In developed countries, industry probably contributes as much to contamination of the water supply as does agriculture. In the United States, about half of the water used daily is used by industry. Many industries discharge organic and inorganic wastes into bodies of water. These may be acids, highly toxic minerals, such as mercury or arsenic, or, in the case of petroleum refineries, toxic organic chemicals. The nuclear power industry has caused some water pollution when radioactive material has seeped from the tanks in which the wastes have been buried, either at sea or underground.

Such pollution can have a variety of effects. Organisms not adapted to living in contaminated water may die; the water may become unsuitable for domestic use or irrigation; or the wastes may reenter the food chain, with deleterious effects on humans. One of the most notorious pollution cases, which focused international attention on the dangers of industrial pollution, occurred in the village of Minamata in southwest Japan four decades ago. A chemical plant that used mercury chloride in its manufacturing process discharged the waste mercury into Minamata Bay, where it settled into the mud. Fish that fed on organisms in the mud absorbed the mercury and concentrated it; the fish were in turn eaten by humans. Over 700 people died, and at least 9000 others suffered deformity or other permanent disability.

A similar contamination of the water supply with mercury is occurring today in the Amazon River and its tributaries. Because mercury attaches itself to gold, an estimated half-million Brazilian prospectors use the toxic liquid to separate gold from the accompanying mud. Each year, hundreds of tons of mercury are poured in or right next to the rivers, poisoning the water and the fish that swim in it. Because it can take decades for concentrations to reach toxic levels, mercury pollution of streams is like a delayed-action time bomb. The effects of mercury poisoning in the Amazon Basin may not be known for many years.

Among the pollutants that have been discharged into the water supply in the United States are **polychlorinated biphenyls (PCBs)**, a family of related chemicals used as lubricants in pipelines and in a wide variety of electrical devices, paints, and plastics. During the manufacturing process, companies have dumped PCBs into rivers, from which they have entered the food chain. Several states have banned commercial fishing in lakes and rivers where fish have higher levels of PCBs than are considered safe. Although not all of the effects of PCBs on human health are known, they have been linked to birth defects, damage to the immune system, liver disease, and cancer. In 1977, the Environmental Protection Agency banned the direct discharge of PCBs into U.S. waters, but immense quantities of the chemicals remain in water bodies.

The petroleum industry is a significant contributor to the chemical pollution of water. Oceans are becoming increasingly contaminated by oil. Although massive oil spills like that from the tanker *Exxon Valdez* in 1989 command public attention, smaller spills routinely dump millions of gallons of oil into American waters each year. Over half the oil normally comes from oil tankers and barges, usually because of ruptures in accidents (see "GPS, GIS, and the Tampa Bay Oil Spill"). Much of the rest comes from refineries, the discharge of tank flushings and ballast from tanker holds, and seepage from offshore drilling platforms. The Gulf of Mexico, the site of extensive offshore drilling, is among the most seriously polluted major bodies of water in the world.

Acid precipitation (usually called acid rain), a by-product of emissions from factories, power plants, and automobiles, has affected the water quality and ecology of thousands of lakes and streams in the world. Because the precipitation is caused by pollutants in the air, it is discussed later in this chapter.

Many industrial processes, as well as electric power production, require the use of water as a coolant. **Thermal pollution** occurs when water that has been heated is returned to the environment and has adverse effects on the plants and animals in the water body. If the heated wastewaters are significantly warmer than the waters into which they are discharged, they can disrupt the growth, reproduction, and migration of fish populations. Many plants and fish cannot survive changes of even a few degrees in water temperature. They either die or migrate. The species that depend on them for food must also either die or migrate. Thus the food chain has been disrupted. In addition, the higher the temperature of the water, the less oxygen it contains, which means that only lower-order plants and animals can survive.

Mining

Surface mining for coal, iron, copper, gold, and other substances contributes to contamination of the water supply through the wastes it generates. Rainwater reacts with the wastes, and dissolved minerals seep into nearby water bodies. The exact chemical changes produced depend on the composition of the coal or ore slag heaps, and the reaction of the minerals with sediments or river water. In addition to altering the quality of the water, the contaminants have secondary effects on plant and animal life. Each year, for example, thousands of animals and migratory birds die in such western states as Arizona, Nevada, and California after drinking cyanide-laced waters at gold mines. The toxic water is poured over mounds of crushed rock to leach the gold and then settles into ponds and lakes that attract wildlife.

Municipalities and Residences

A host of pollutants derives from the activities associated with urbanization. The use of detergents has increased the phosphorus content of rivers, and salt (used for deicing roads) increases the chloride content of runoff. Water runoff from urban areas contains contaminants from garbage, animal droppings, litter, vehicle drippings, and the like. Because the sources of pollution are so varied, the water supply in any single area is often affected by diverse contaminants. This diversity complicates the problem of controlling water quality.

Contaminated drinking-water wells have been found in more than half of the states. Hundreds of wells in the New York metropolitan area have been closed in recent years because of chemical contamination, and thousands more may be closed in coming years. Chemicals have reached groundwater by seeping into aquifers from landfills, ruptured gasoline and fuel-oil storage tanks, septic tanks, and fields sprayed with pesticides and herbicides. The pollution of aquifers is particularly troublesome because, unlike surface waters, groundwater has a low capacity for purifying itself; it can remain contaminated for centuries.

Sewage can also be a major water pollutant, depending on how well it is treated before being discharged. This is not simply an environmental concern; it directly affects human health. Raw, untreated human waste contains viruses responsible for dysentery, polio, hepatitis, spinal meningitis, and other diseases.

Although municipal wastewater treatment is increasing in the most developed countries, more than 90% of sewage in the developing world is discharged directly into streams, lakes, and coastal waters without treatment of any kind. Fully 70% of total surface waters in India are polluted, in large part because only about 200 of its more than 3000 cities have full or partial sewage collection and treatment facilities. Of Taiwan's 20 million people, only 600,000 are served by sewers. Raw sewage from 3.6 million inhabitants of Hong Kong flows into Victoria Harbor.

Although sewage-treatment plants are common in the United States, only half of the American population lives in communities with facilities that meet the minimum goals set by the federal Clean Water Act. Aged sewage systems in

GPS, GIS, AND THE TAMPA BAY OIL SPILL

Tampa Bay, Florida, is one of the busiest ports in the United States. Early in the morning of August 10, 1993, three ships collided a few kilometers from the Sunshine Skyway Bridge near Saint Petersburg. One vessel, carrying 8 million gallons of jet fuel, burst into flames and burned for more than 14 hours. Another leaked nearly 400,000 gallons of number 6 heating oil.

Officials in charge of response and cleanup needed maps showing the bay's natural resources displayed in conjunction with the changing extent of the spill. The natural resources include turtle nesting sites, artificial reefs, mangroves, marshes, and seagrass beds. Throughout the spill, the staff of the Florida Marine Research Institute used Global Positioning System (GPS) receivers from helicopters to record the locations of the vessels and the changing perimeter of the spill. The GPS files were imported into the Marine Resources GIS to produce the necessary maps. The time between data gathering and finished map output was only three and one-half hours; maps more than five hours old were considered out-of-date.

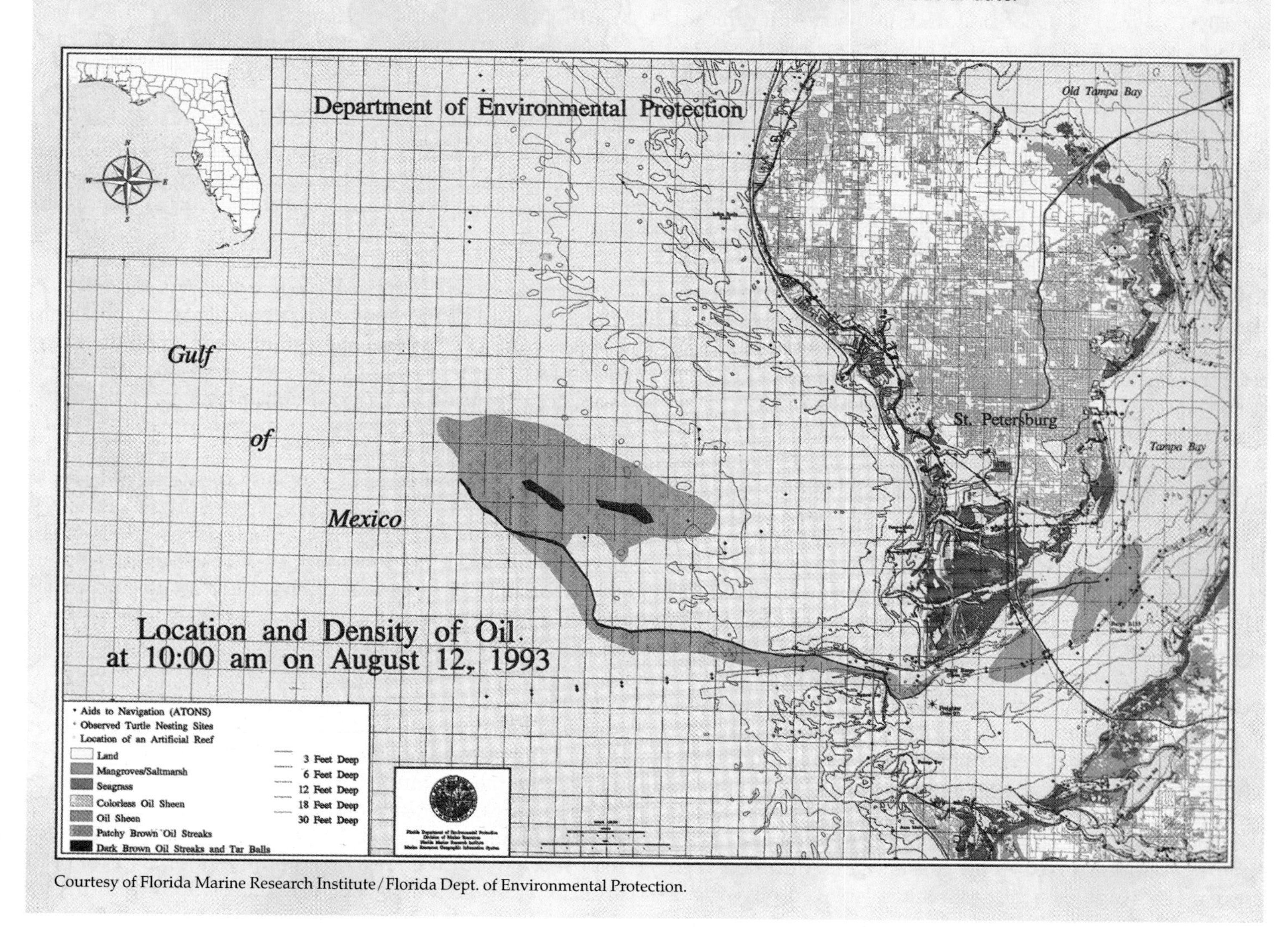

Courtesy of Florida Marine Research Institute/Florida Dept. of Environmental Protection.

1100 cities still discharge poorly treated sewage into streams, lakes, and oceans. When the sewer system of Dade County, Florida, ruptures, as it does periodically, millions of gallons of raw sewage pour into the Miami River, which empties into Biscayne Bay in downtown Miami.

In many communities, special problems arise after heavy rains, when storm water filled with animal wastes, street debris, and lawn chemicals floods the sewers. As treatment plants become overloaded, both runoff and raw sewage are diverted into rivers, bays, and oceans. New York City alone has more than 500 storm water outlets that overflow in heavy rains, pouring some 65 billion gallons of untreated sewage (about 10% of the city's total sewage) into the Hudson River and Long Island Sound each year. But

1200 other cities in the East, Midwest, and Northwest also have sewage networks that are combined with the overflow systems for storm water.

Controlling Water Pollution

In recent years, concern over increased levels of pollution has brought about major improvements in the quality of some surface waters, both in the United States and abroad. The U.S. federal government in 1972 took the lead in regulating water pollution with the enactment of the Clean Water Act. Its objective was "to restore and maintain the chemical, physical, and biological integrity of the nation's waters." Congress established uniform nationwide controls for each category of major polluting industry and directed the government to pay most of the cost of new sewage-treatment plants. Since 1972, such plants have been built to serve over 80 million Americans, and industries have spent billions of dollars to comply with the Clean Water Act by reducing organic waste discharges.

The gains have been impressive. Many rivers and lakes that were ecologically dead or dying are now thriving. Once dumping grounds for all kinds of human and industrial waste, the Hudson, Potomac, Cuyahoga, and Trinity Rivers are cleaner, more inviting, and more productive than before, and they now support fishing, swimming, and recreational boating. Similarly, Seattle's Lake Washington and the Great Lakes are healthier than they were two decades ago. Recently, authorities have announced ambitious plans to clean up the waters of Chesapeake Bay, the country's largest estuary, and to undo much of the damage that has been inflicted on Florida's Everglades by improving the water quality of the Kissimmee River and Lake Okeechobee.

Environmental awareness in other countries has also prompted legislation and action. For example, the river Thames in southern England, which had become seriously contaminated by the dumping of sewage and industrial wastes, is now cleaner than it has been in centuries. The enforcement of stringent pollution-control standards has halted the downward trend in quality. Algae and seaweed, fish and wildfowl have returned to the river in abundance.

Even the Mediterranean Sea is on its way to gradual recovery. When the 18 countries that border the sea signed the Convention for the Protection of the Mediterranean Sea against Pollution in 1976, all coastal cities dumped their untreated sewage into the sea, tankers spewed oily wastes into it, and tons upon tons of phosphorus, detergents, lead, and other substances contaminated the waters. Now, many cities have built or are building sewage-treatment plants, ships are prohibited from indiscriminate dumping, and some national governments are beginning to enforce control of pollution from land-based sources.

Such gains should not mislead us. While some of the most severe problems have been attacked, serious pollution still plagues about one-fourth of American rivers and streams, lakes and reservoirs. The solution to water pollution lies in the effective treatment of municipal and industrial wastes; the regulation of chemical runoff from agriculture, mining, and forestry; and the development of less-polluting technologies. Although pollution-control projects are expensive, the long-term costs of pollution are even higher.

Impact on Air and Climate

The **troposphere,** the thin layer of air just above the earth's surface, contains all the air that we breathe. Every day thousands of tons of pollutants are discharged into the air by cars and incinerators, factories and airplanes. Air is polluted when it contains substances in sufficient concentrations to have a harmful effect on living things.

Air Pollutants

Truly clean air has probably never existed. Just as there are natural sources of water pollution, so are there substances that pollute the air without the aid of humans. Ash from volcanic eruptions, marsh gases, smoke from forest fires, and wind-blown dust are natural sources of air pollution.

Normally these pollutants are of low volume and are widely dispersed throughout the atmosphere. On occasion, a major volcanic eruption may produce so much dust that the atmosphere is temporarily altered. In general, however, the natural sources of air pollution do not have a significant, long-term effect on air, which, like water, is able to cleanse itself.

Far more important than naturally occurring pollutants are the substances that people discharge into the air. These pollutants result primarily from burning fossil fuels (coal, gas, and oil) and other materials. Fossil fuels are burned in power plants that generate electricity, in many industrial plants, in home furnaces, and in cars, trucks, buses, and airplanes. Scientists estimate that about three-quarters of all air pollutants come from burning fossil fuels. The remaining pollutants largely result from industrial processes other than fuel burning, incinerating solid wastes, forest and agricultural fires, and the evaporation of solvents. Figure 5.10 depicts the major sources of air pollutants. Table 5.2 summarizes the major sources of six *primary* pollutants emitted in large quantities. Once they are in the atmosphere, these may react with other primary pollutants or with normal atmospheric constituents such as water vapor to form *secondary* pollutants.

Air pollution is a global problem. A recent study by the World Health Organization (WHO) concluded that more than 1.1 billion people live in urban areas with unhealthful air. Sulfur dioxide levels are considered unacceptable for some 625 million people in developing countries alone, and levels of smoke, dust, and other particulates are unacceptable for 1.25 billion people. Particularly at risk of breathing bad air are residents of such megacities as Mexico City, Cairo, Delhi, Seoul, Beijing, and Jakarta.

TABLE 5.2

MAJOR SOURCES OF PRIMARY AIR POLLUTANTS

TYPE OF POLLUTANT	SYMBOL	MAJOR SOURCES
Carbon dioxide	CO_2	Combustion of fossil fuels
Carbon monoxide	CO	Incomplete combustion of fossil fuels, mostly in vehicles
Hydrocarbons	HC	Combustion of fossil fuels; petroleum refineries
Nitrogen oxides	NO_x	Transportation vehicles; power plants
Particulates	—	Automobile exhaust; oil refining; coal-fired power plants
Sulfur oxides	SO_x	Combustion of sulfur-containing fuels, especially coal

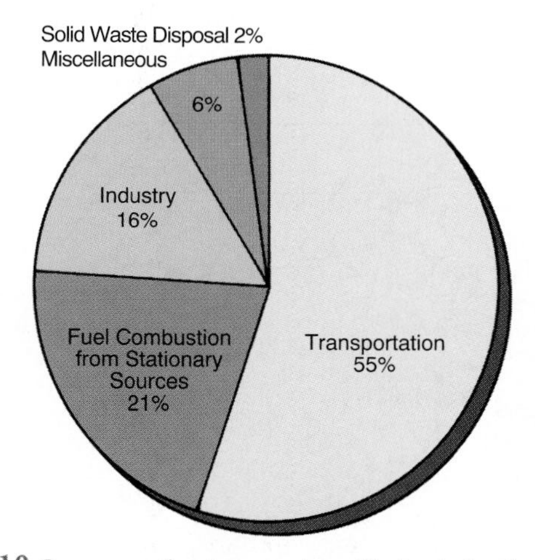

FIGURE 5.10 Sources of primary air pollutants in the United States. Transportation is the single largest source of human-caused air pollution. The second largest source is fuel combustion in stationary sources such as power plants and factories.

Redrawn from *Biosphere 2000*, Donald G. Kaufman and Cecilia M. Franz (NY: Harper Collins College Publishers, 1993), Figure 14.4, p. 257.

Like polluted water, dirty air kills people. According to the WHO, at least 2.7 million people die each year from illnesses caused by air pollution.

Factors Affecting Air Pollution

Many factors affect the type and the degree of air pollution found at a given place. Those over which people have relatively little control are climate, weather, wind patterns, and topography. These determine whether pollutants will be blown away or are likely to accumulate. Thus a city on a plain is less likely to experience a buildup than is a city in a valley.

Unusual weather can alter the normal patterns of pollutant dispersal. A *temperature inversion* magnifies the effects of air pollution. Under normal circumstances, air temperature decreases away from the earth's surface. A stationary layer of warm, dry air over a region, however, will prevent the normal rising and cooling of air from below. As described in Chapter 4, the air becomes stagnant during an inversion. Pollutants accumulate in the lowest layer instead of being blown away, so that the air becomes more and more contaminated. Normally, inversions last for only a few hours, although certain areas experience them much of the time. Temperature inversions occur often in Los Angeles in the fall and Denver in the winter (Figure 5.11). If an inversion lingers long enough, over several days, it can contribute to the accumulation of air pollutants to levels that seriously affect human health.

The air pollutants generated in one place may have their most serious effect in areas hundreds of kilometers away. Thus, the worst effects of the air pollution that originates in New York City are felt in Connecticut and parts of Massachusetts. The chemical reaction that produces smog takes a few hours, and by that time air currents have carried the pollutants away from New York. In a similar fashion, New York is the recipient of pollutants produced in other places. Much of the acid rain that affects New England and eastern Canada originates in the coal-fired power plants

FIGURE 5.11 An example of the effects of temperature inversion. A brown cloud hovers over Denver when temperature inversions keep air pollutants from dispersing. Denver frequently has the worst carbon monoxide level in the country from mid-November to mid-January, when cold winter air over the city is trapped by warm, still air at higher altitudes. The city has recently embarked on a campaign to reduce air pollution.

along the lower Great Lakes and in the Ohio Valley that use extremely high smokestacks to disperse sulfurous emissions. And the coal-based industries in Russia and Europe produce sulfate, carbon, and other pollutants that are transported by air currents to the land north of the Arctic Circle, where they result in a contamination known as Arctic haze.

Other factors that affect the type and the degree of air pollution at a given place are the levels of urbanization and industrialization. Population densities, traffic densities, the type and density of industries, and home-heating practices all help to determine the kinds of substances discharged into the air at a single point. In general, the more urbanized and industrialized a place is, the more responsible it is for pollution. The United States may contribute as much as one-third of the world's air pollution, a figure roughly equivalent to the proportion of the world's fossil fuel and mineral resources consumed in the country.

The sources of pollution are so many and varied that we cannot begin to discuss them all in this chapter. Instead, we will describe three types of air pollution and their associated effects.

Acid Rain

Although acid *precipitation* is a more precise description, **acid rain** is the term generally used for pollutants, chiefly oxides of sulfur and nitrogen, that are created by burning fossil fuels and that change chemically as they are transported through the atmosphere and fall back to earth as acidic rain, snow, fog, or dust. The main sources of these pollutants are vehicles, industries, power plants, and ore-smelting facilities. When sulfur dioxide is absorbed into water vapor in the atmosphere, it becomes sulfuric acid, which is highly corrosive. Sulfur dioxide contributes about two-thirds of the acids in the rain. About one-third comes from nitrogen oxides, transformed into nitric acid in the atmosphere.

Once the pollutants are airborne, winds can carry them hundreds of kilometers, depositing them far from their source. In North America, most of the prevailing winds are westerlies, which means that much of the acid rain that falls on the eastern seaboard and eastern Canada originates in ten states in the central and upper Midwest (Figure 5.12). Similarly, airborne pollutants from Great Britain, France, and Germany cause acidification problems in Scandinavia.

Acid rain has three kinds of effects: terrestrial, aquatic, and material. The acids change the *pH factor* (the measure of acidity/alkalinity on a scale of 1 to 14) of both soil and water, setting off a chain of chemical and biological reactions (Figure 5.13). It is important to note that the pH scale is logarithmic, which means every step on the scale represents a factor of 10. Thus, 4.0 is ten times more acidic than 5.0, and 100 times more acidic than 6.0. The average pH of

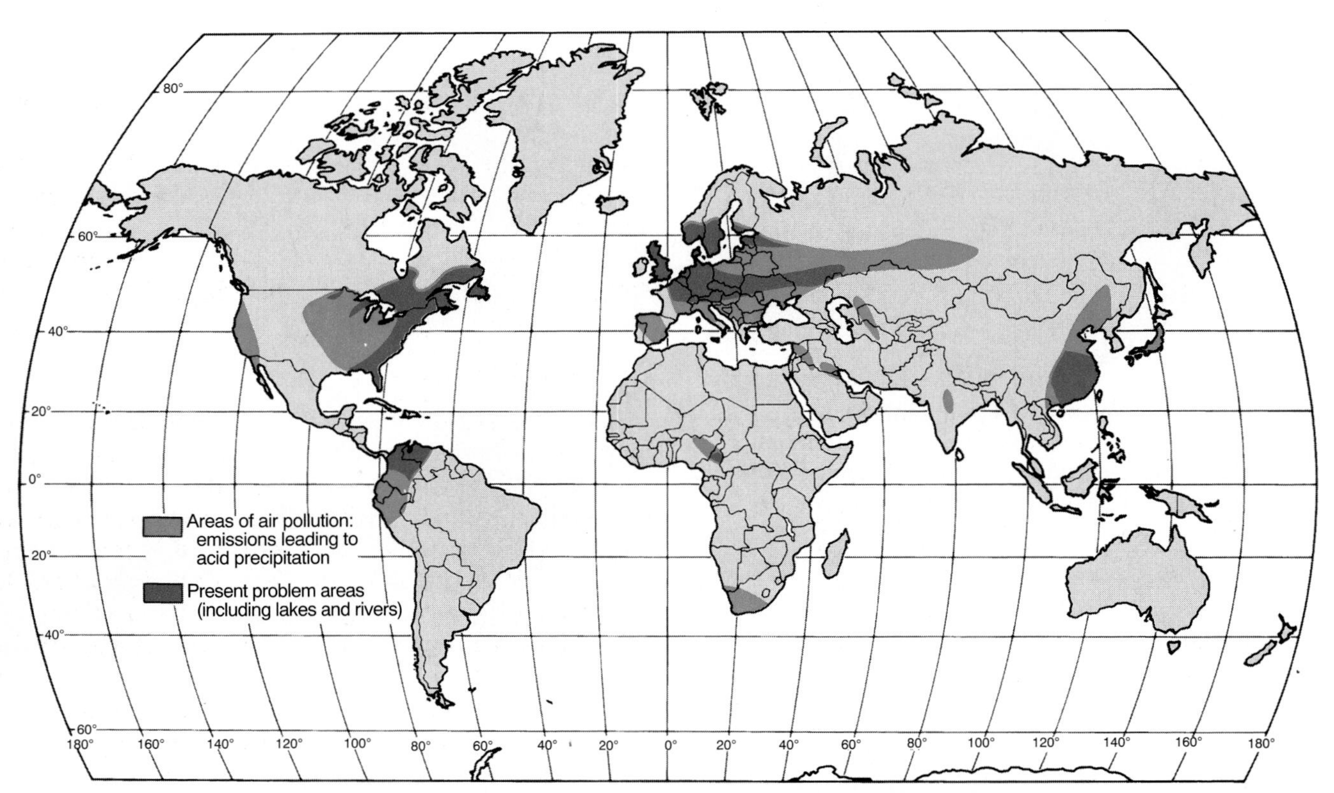

FIGURE 5.12 **Acid precipitation: points of origin and current problem areas.** Prevailing winds account for the fact that acid precipitation can be deposited far from its area of origination.

Redrawn from *Biosphere 2000*, Donald G. Kaufman and Cecilia M. Franz (NY: Harper Collins College Publishers, 1993), Figure 14.9, p. 263.

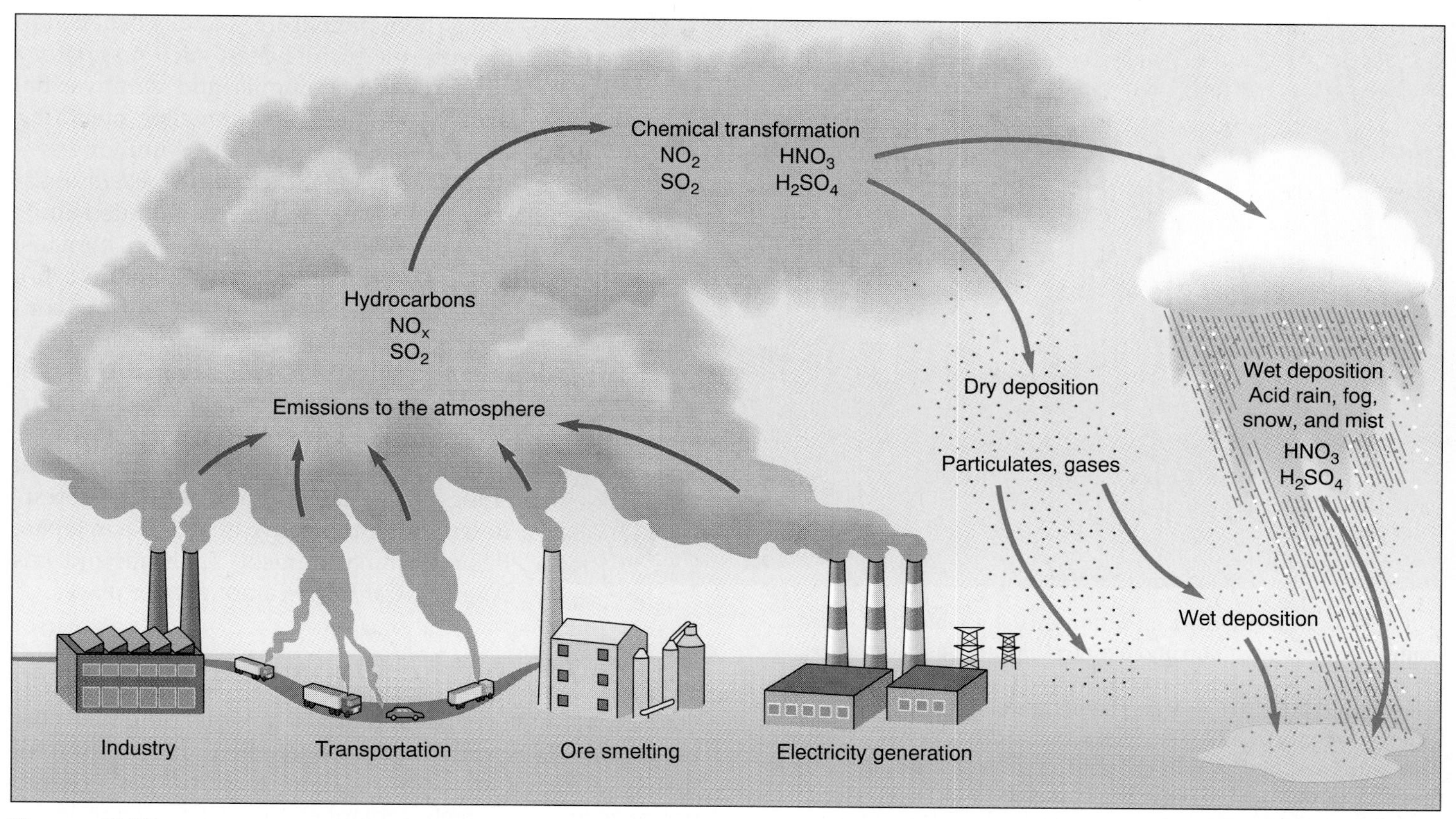

FIGURE 5.13 **The formation of acid precipitation.** Sulfur dioxide and nitrogen oxides produced by the combustion of fossil fuels are transformed into sulfate and nitrate particles; when the particles react with water vapor, they form sulfuric and nitric acids, which then fall to earth. The acids in the precipitation damage water, soil, vegetation, and buildings.
Redrawn from *Biosphere 2000,* Donald G. Kaufman and Cecilia M. Franz (NY: Harper Collins College Publishers, 1993), Figure 14.8, p. 262.

normal rainfall is 5.6, categorized as slightly acidic, but acid rainfalls with a pH of 1.5 (far more acidic than vinegar or lemon juice) have been recorded. Acid deposition harms soils and vegetation by acidifying soils and by coating the ground with particles of aluminum and toxic heavy metals such as cadmium and lead. It kills microorganisms in the soil that break down organic matter and recycle nutrients through the ecosystem. Significant forest damage has occurred in the eastern United States, northern and western Europe, Russia, and China (Figure 5.14).

The aquatic effects of acid rain are manifold. The acidity of a lake or stream need not increase much before it begins to interfere with the early reproductive stages of fish. Also, the food chain is disrupted as acidification kills the plants and insects upon which fish feed. Acid rains have been linked to the disappearance of fish in thousands of lakes and streams in New England, Canada, and Scandinavia, and to a decline of fish populations elsewhere.

While acid rain has been recognized for some years as a threat to freshwater organisms, recent evidence suggests that it may also damage marine life in coastal salt water. In this case, the damage stems not from acidity but from nitrogen in the rain, which causes eutrophication (discussed earlier in this chapter). Nitrogen is a nutrient that stimulates the excessive growth of algae. The algae, in turn, use up the supply of dissolved oxygen and reduce the amount of sunlight that penetrates the surface of the water.

The material effects of atmospheric acid are evident in damage to buildings and monuments. The acid etches and corrodes many building materials, including marble, limestone, steel, and bronze. Worldwide, tens of thousands of structures are slowly being dissolved by acid precipitation.

Photochemical Smog

While sulfur oxides are the chief cause of acid rain, oxides of nitrogen are responsible for the formation of **photochemical smog.** This type of air pollution is created when nitrogen oxides react with the oxygen present in water vapor in the air to form nitrogen dioxide. In the presence of sunlight, nitrogen dioxide reacts with hydrocarbons from automobile exhausts and industry to form new compounds, such as **ozone.** The primary component of photochemical smog, ozone is a molecule consisting of three oxygen atoms rather than the two of normal oxygen. Warm, dry weather and poor air circulation promote ozone formation. The hotter and sunnier the weather, the more ozone and smog are created. In general, therefore, more ozone is produced during the summer months than during the rest of the year.

FIGURE 5.14 **Dead trees on Mount Mitchell, North Carolina.** A combination of acid rain and ozone pollution has caused extensive damage to forests along the crest of the Appalachian Mountains from Maine to Georgia. The pollution weakens trees to the point where they cannot survive such natural stresses as temperature extremes, high winds, drought, or insects. In addition, the high levels of lead and other heavy metals in the soil make it difficult for the forests to regenerate. The impact of acid deposition on forests has been particularly severe in Europe. In Germany and Switzerland, for example, up to 50% and 30% respectively of the forests are dead or dying.

Because the primary sources of the nitrogen oxides and hydrocarbons are motor vehicles and industries, photochemical smog tends to be an urban problem. It occurs around the world, affecting cities such as Ankara, Turkey; New Delhi, India; Mexico City, Mexico; and Santiago, Chile. As many as 76 million Americans, or about one-third of the population, live in areas where ozone pollution exceeds the limit established by the federal Clean Air Act, 0.12 parts of ozone per million parts of air. The climate and topography of California are particularly conducive to ozone pollution. Its valleys are encircled by mountains that help hold air pollutants in the basins. When temperature inversions occur, the pollutants are effectively trapped, unable to escape to the stratosphere. Ozone levels in Los Angeles exceed the acceptable level nearly one-quarter of the days in the year, and sometimes reach three times the acceptable level. Other American metropolitan areas subject to ozone pollution are shown in Figure 5.15.

Photochemical smog damages both human health and vegetation. It has long been known to aggravate coughing and breathing problems for asthmatics, but recent studies indicate that exposure to ozone over a period of 6–8 hours also causes respiratory problems in healthy people. Chronic exposure to smog causes permanent damage to lungs, aging them prematurely, and is believed to increase the incidence of such respiratory ailments as pneumonia and emphysema. Because children have smaller breathing passages and less developed immune systems than adults, they are especially susceptible to damage from the polluted air.

In addition to its effects on humans, ozone harms vegetation (see Figure 5.14). Exposure over several days to ozone concentrations as low as .1 parts per million damages trees, plants, and crops. Although smog originates in urban industrial centers, it can affect areas downwind from them. Damage associated with photochemical smog has been documented in forests downwind from Tokyo and Osaka in Japan; Beijing, China; Karachi, Pakistan; and Los Angeles, California, among other places.

Depletion of the Ozone Layer

Ozone, the same chemical that is a noxious pollutant near the ground, is essential in the stratosphere. There, approximately 10–24 kilometers (6–15 mi) above the ground, ozone forms a protective blanket called the **ozone layer,** which shields all forms of life on earth from overexposure to lethal ultraviolet (UV) radiation from the sun. Mounting evidence indicates that emissions from a variety of chemicals are destroying the ozone layer. Most important are a family of synthetic chemicals developed in 1931 and known as **chlorofluorocarbons (CFCs).** CFCs are found in hundreds of products. They are used as coolants for refrigerators and air conditioners; as aerosol spray propellants;

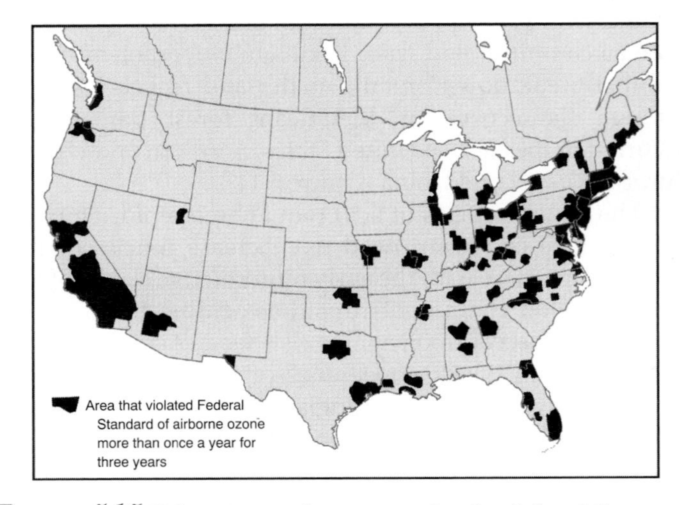

FIGURE 5.15 **Urban areas where ozone levels violated the federal standard more than once a year for three years.** By far the worst ozone pollution in the country occurs in the large Los Angeles basin, which often exceeds the federal ozone standard nearly one-fourth of the year.
Source: Environmental Protection Agency.

and as a component in foam packaging, home insulation, and upholstery. In liquified form, they are used to sterilize surgical equipment and to clean computer chips and other microelectronic equipment.

Also implicated in the depletion of the ozone layer are *halons,* used in fire extinguishers, and carbon tetrachloride and methyl chloroform, used as solvents and cleaning agents. CFCs, however, are by far the most important.

After the gases are released into the air, they rise through the lower atmosphere and, after a period of 7–15 years, reach the stratosphere (Figure 5.16). There, UV radiation breaks the molecules apart, producing free chlorine and bromine atoms. Over time, a single one of these atoms can destroy tens of thousands, if not a potentially infinite number, of ozone molecules.

Every Southern Hemisphere spring, beginning in late August, the atmosphere over the Antarctic loses more and more ozone. In 1985, researchers discovered what is popularly termed a "hole" as big as the continental United States in the ozone layer over Antarctica, extending northward as far as populated areas of South America (Figure 5.17). The hole disappears in late November, when the winds change and the ozone-deficient air mixes with the surrounding atmosphere. A less dramatic but still serious depletion of the ozone shield occurs over the North Pole.

A depleted ozone layer allows more UV radiation to reach the earth's surface. Although the exact consequences

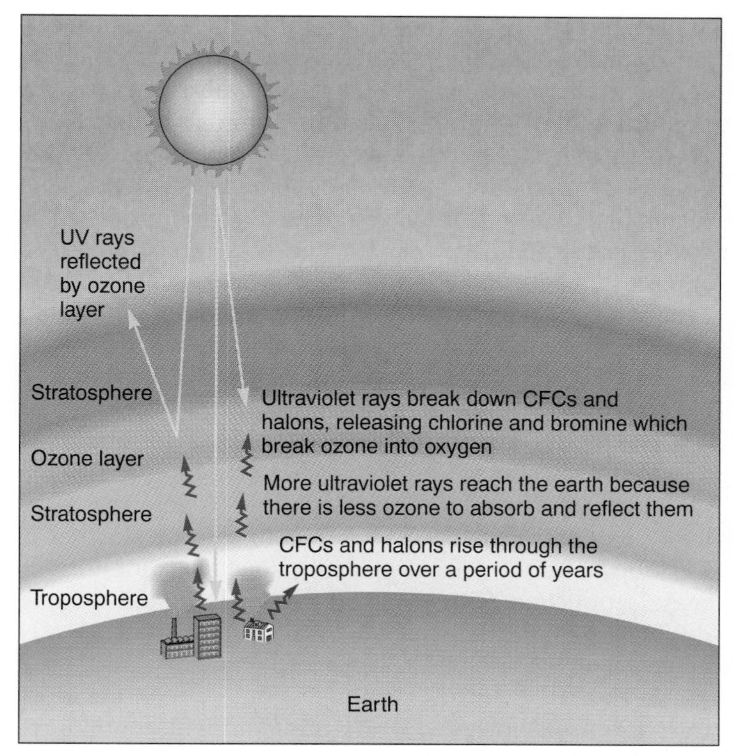

FIGURE 5.16 **How ozone is lost.** CFCs and halons released into the air rise through the troposphere without breaking down (as most pollutants do) and eventually enter the stratosphere. Once they reach the ozone layer, UV rays break them down, releasing chlorine (from CFCs) and bromine (from halons). These elements, in turn, disrupt ozone molecules, breaking them up into molecular oxygen, and thus, deplete the ozone layer.

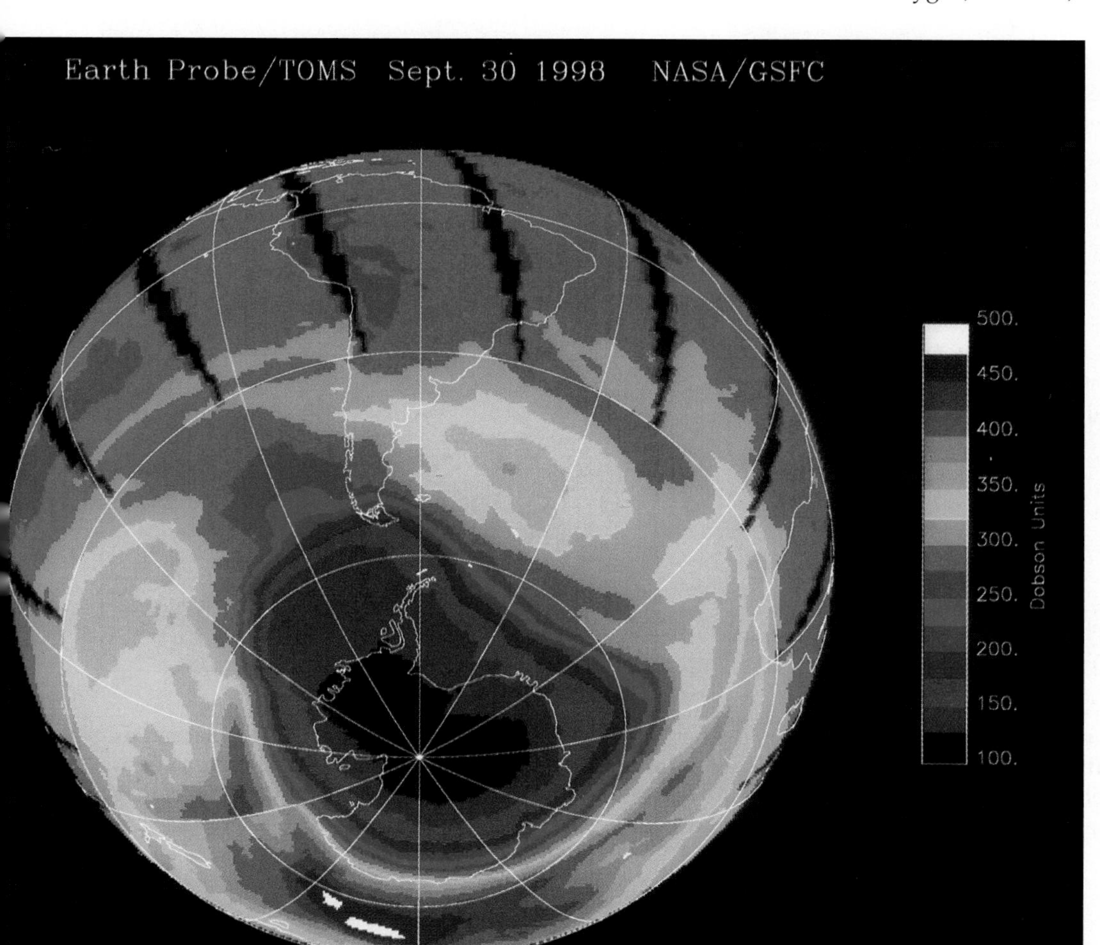

FIGURE 5.17 **Ozone depletion over the Southern Hemisphere on September 30, 1998.** The color scale to the right of the image shows the total ozone levels. The dark blues and purples indicate the areas of greatest ozone depletion. The black areas in the image are zones where no data were available from the spacecraft for that day. Each year, the area of significant depletion has become larger, lasted longer, and spread farther outward toward South America and Australia. Although depletion of the ozone layer has been particularly severe over Antarctica, a decline in stratospheric ozone has also been observed over other parts of the world.

of that increase will not be known for years, it is almost certain to cause a dramatic rise in the incidence of skin cancers and eye cataracts. Some fear it may also damage the immune system of humans and other animals by impairing the cells that fight viral infections and parasitic disease. Because UV radiation also causes cell and tissue damage in plants, it is likely to reduce agricultural production. The most serious damage may occur in oceans. Increased amounts of UV radiation affect the photosynthesis and metabolism of the microscopic plants called phytoplankton that flourish just below the surface of the Antarctic Ocean. Phytoplankton form the base of the oceanic food chain, and also play a central role in the earth's CO_2 cycle.

The production of CFCs is being phased out under the *Montreal Protocol on the Depletion of the Ozone Layer* of 1987, an international agreement endorsed by 146 countries. The treaty required developed countries to stop production of CFCs by January 1, 1996; developing states have until 2010 to cease production. The Montreal Protocol spurred a rapid decline in CFC output. By 1998, CFC production had already fallen about 90% from its peak a decade earlier.

Even if all countries eventually comply with the Montreal Protocol, however, past emissions will continue to cause ozone degradation for years to come. The two most widely used forms of CFCs stay in the stratosphere, breaking down ozone molecules, for up to 100 years. Thus, full recovery of the ozone layer cannot be expected until well into the next century.

In addition to their effect on the ozone layer, CFCs may have an impact on climate by contributing to the "greenhouse effect," the subject we explore next.

The Greenhouse Effect

One of the most hotly debated topics in recent years has been what is popularly termed the **greenhouse effect**. Put simply, the theory is that certain gases concentrate in the atmosphere, where they function as an insulating barrier, trapping infrared radiation that would otherwise be radiated back into the upper atmosphere and reradiating it earthward. In other words, like glass in a greenhouse, the gases admit incoming solar radiation but retard its reradiation back into space. You have experienced such a greenhouse effect if you have gotten into a car on a cold but sunny day; the car's interior is warmer than the outside air.

The earth has a natural greenhouse effect, provided mainly by water vapor that has evaporated from the ocean or evapotranspired from land (Figure 5.18). The water

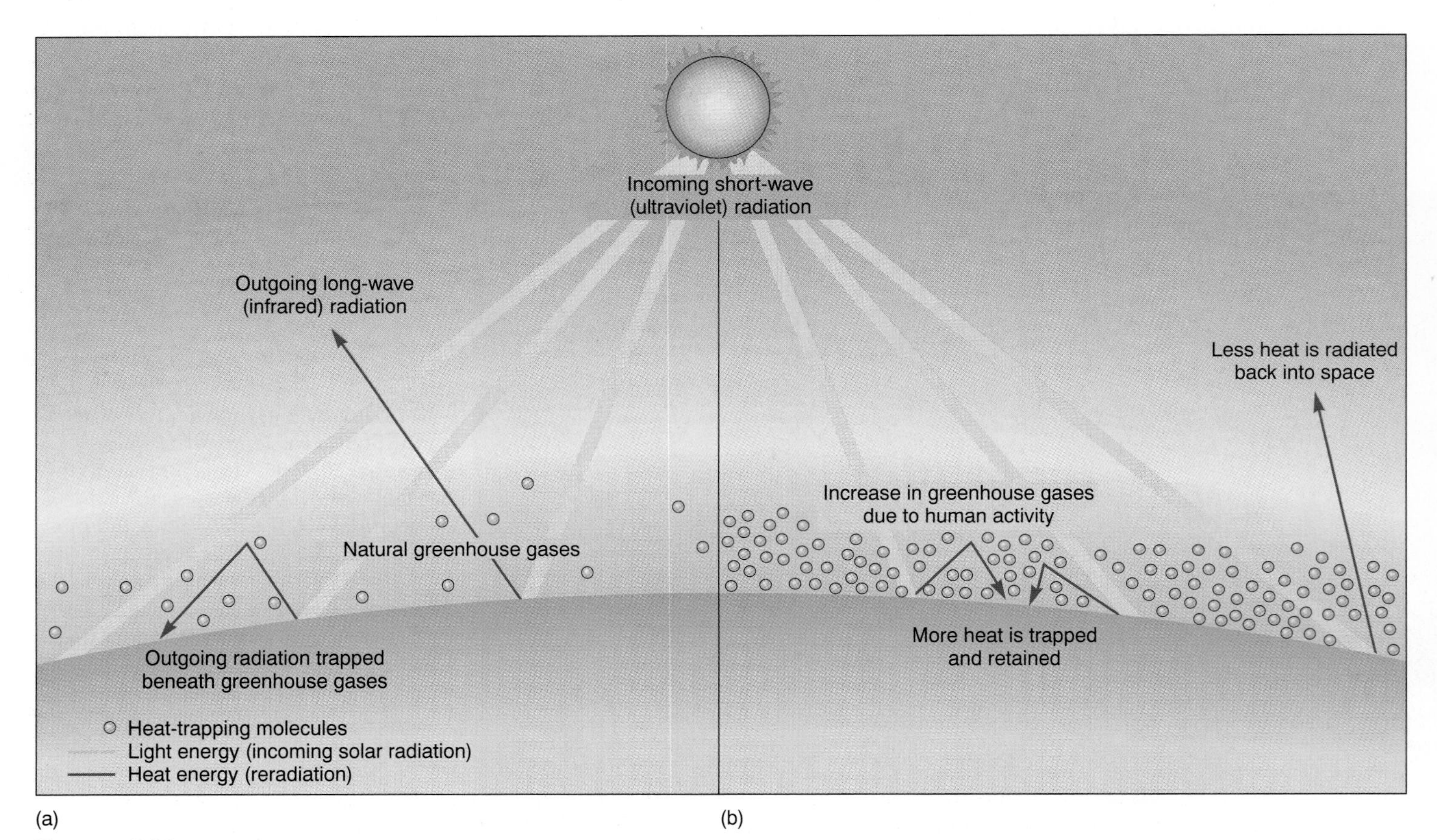

FIGURE 5.18 **How the greenhouse effect works.** When the level of CO_2 and other gases in the air is low, as in (a), incoming solar radiation strikes the earth's surface, and the earth radiates the energy back into space as infrared light (heat). The enhanced greenhouse effect, depicted in (b), is the result of the billions of tons of CO_2 and other gases that are released into the air each year. They form a blanket that deflects more of the energy downward, preventing it from escaping into the atmosphere.

vapor remains a constant, but during the last 150 years or so, human activities have increased the amount of other greenhouse gases in the atmosphere, augmenting its heat-trapping ability. Many scientists fear that an *enhanced* greenhouse effect could result in a gradual increase in the earth's average surface temperature, with significant impacts on the earth's ecosystems, a process called **global warming.**

Carbon dioxide (CO_2) is the primary greenhouse gas whose amount has been increased by human activities. Although it occurs naturally, excessive quantities of it are released by burning fossil fuels. Beginning with the Industrial Revolution in the mid-1700s, large amounts of coal, petroleum, and natural gas have been burned to power industry, heat and cool cities, and drive vehicles. Their combustion has turned fuels into carbon dioxide and water vapor. At the same time, much of the world's forests have been destroyed by logging and to clear land for agriculture. Deforestation adds to the greenhouse effect in two ways: it means there are fewer trees to capture carbon dioxide and produce oxygen, and burning the wood sends CO_2 back into the atmosphere at an accelerated rate.

Other greenhouse gases influenced by human activity are:

1. methane, from natural gas and coal mining, agriculture and livestock, swamps, and landfills;
2. nitrous oxides, from motor vehicles, industry, and chemical fertilizers;
3. chlorofluorocarbons and halons, widely used industrial chemicals.

Although these gases may be present in small amounts, some of them trap heat thousands of times more effectively than does CO_2. Fluorocarbon 12, for example, has 20,000 times the capacity of CO_2 to trap heat, and fluorocarbon 11 has 17,500 times the capacity of CO_2. Nitrous oxide is 300 times more potent than CO_2, and even methane is thirty times more potent than CO_2 in absorbing heat close to the earth.

As the Industrial Revolution gained momentum in Europe and North America during the 19th century, the concentration of CO_2 in the atmosphere rose from its preindustrial level of about 274 parts per million (ppm) to over 360 ppm in 1995 (Figure 5.19). (Just since 1958, concentrations of CO_2 have increased from 315 ppm.) The methane concentration in the lower atmosphere has *already* more than doubled from its preindustrial level and is currently increasing by just over 1% per year.

There is little dispute over the facts just presented, but there is less agreement on their import. Many, perhaps most, of the scientists who have studied the question believe that when CO_2 concentrations reach about 550 ppm (double pre-Industrial Revolution levels), average annual global temperatures will rise by 1.15°C to 3.9°C (2°–9°F). Predictions of when this doubling will occur vary. The year 2050 is commonly cited, but because gases other than CO_2 are contributing to the greenhouse effect, the warming may occur as early as 2030.

Proponents of the greenhouse effect theory believe that human activity has already put enough of the various gases to the atmosphere that there is no way to avoid a significant rise in temperature in the next century. That is, some warming is inevitable even if all emissions were to stop today, because the greenhouse gases are already in the atmosphere. Consequently, they say, serious environmental and economic damage are a foregone conclusion, although the worst problems are not expected to appear until the next century. These grim predictions were the background for a recent international conference and treaty proposal seeking to address and limit the dangers prophesied (see "The Kyoto Protocol").

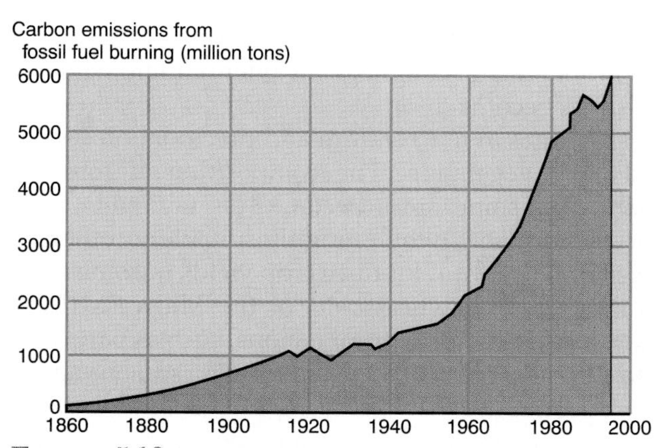

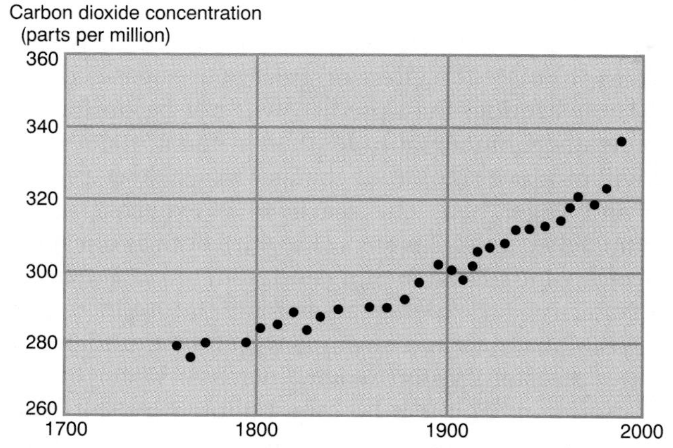

FIGURE 5.19 **Global emissions of carbon, 1860–1995, and atmospheric concentration of carbon dioxide, 1760–1995.** By 1998, carbon emissions had climbed to 6.3 billion tons, and atmospheric concentrations of CO_2 had reached 364 parts per million. The world's leading emitter of carbon is the United States, with 23% of the total in 1997. When the Industrial Revolution began, CO_2 concentration in the atmosphere was about 274 ppm. If present trends continue, the atmospheric CO_2 concentration will double from preindustrial levels by around the year 2050. That doubling, it is predicted, will lead to an increase of 2–5°C (4–9°F) in *average* world temperatures within 30 years; the increase would be uneven, rising most at the poles and least around the equator.

Redrawn from *Biosphere 2000,* Donald G. Kaufman and Cecilia M. Franz (NY: Harper Collins College Publishers, 1993), Figure 14.6, p. 260.

The Kyoto Protocol

Accumulating evidence of global warming, projections about its long-term effect, and growing public and political determination to address its causes led to the Kyoto Climate Change Summit of 1997.

Representatives from 171 countries met in Japan for ten days of intense and contentious negotiations. They came with different interests, agendas, and bargaining positions. The European Union, for example, proposed that industrial states reduce emissions of carbon dioxide and other heat-trapping gases by 15% below their 1990 levels within 12 years. The United States, in its turn, suggested that emissions be reduced no lower than 1990 levels and not until some time between 2008 and 2012. The developing countries demanded that industrialized countries collectively achieve a 35% emissions reduction by 2020 and rejected being subject to treaty provisions that would hamper their economic growth by limiting industrialization and the rapid expansion of fossil fuel use such growth implies.

The Kyoto Protocol, as the agreement reached at the climate summit is known, represents compromises among the various extreme positions originally held. It acknowledged the diversity of concerns among developed and developing economies. Thirty-eight Western industrial and former Eastern bloc countries are required collectively to reduce greenhouse emissions by an annual average of 5.2% below 1990 levels between the years 2008 and 2012. The actual targets differ among them, however. The European Union's goal is an 8% reduction, that of the United States is 7%, and Japan's is 6%. Some industrial states would have smaller reductions, and a few would not face any cuts immediately. New Zealand, for example, is permitted to keep its emissions at 1990 levels, while Australia was granted an increase of 8%.

Forested countries will have reduced quotas because their trees absorb carbon dioxide and thus help worldwide emissions controls. No specific goals were set for developing countries, including China and India, the two fastest growing sources of greenhouse gas emissions, though as a group they were asked to set voluntary reduction quotas.

Although methods of achieving their assigned goals are not specified, the Kyoto Protocol does permit "emissions trading" that allows countries that do not meet their own emission targets to purchase "saved" emissions from countries that do better than required. This provision would let countries such as the United States and Canada buy emission credits from Russia and former Eastern bloc countries, whose emissions are substantially below 1990 levels.

The accord approved by the Kyoto conference takes effect once it is ratified by at least 55 countries that represent at least 55% of 1990 carbon dioxide emissions. However, the Protocol assures that its terms are not binding on any individual country unless its government ratifies the accord. In the United States, responsible for nearly 23% of world carbon emissions, the fate of the Kyoto agreement was immediately placed in doubt by the forceful opposition expressed by many members of Congress, some of whom declared the Protocol "dead on arrival" because of its presumed negative impact on the country's economic health and its failure to include controls on the contributions of developing countries to greenhouse gas increases.

Researchers have developed various mathematical models to simulate the effect of greenhouses gases on the earth. The warming, they predict, will not be uniform. It will be greater at higher latitudes than in equatorial regions, and it will produce significant changes in sea level, precipitation, and vegetation. The sea level is expected to rise approximately 0.3 to 1.2 meters (1 to 4 ft), both as a result of some ice-cap and glacial melting and from *thermal expansion* of the water (water expands as its temperature increases). Most coastal marshes and swamps would be inundated by salt water; coastal erosion would increase. Water quality would decline as aquifers become polluted by salt. Such low-lying regions as the North American Gulf Coast, the Netherlands, the Nile Delta, Bangladesh, and much of Southeast Asia could lose substantial amounts of land. Many major ports might be flooded.

Warming of lakes and oceans would speed evaporation, causing more active convection currents in the atmosphere and thus creating fiercer storms. Important regional changes in precipitation would occur, with some areas receiving more precipitation and others less. Polar and equatorial regions might get heavier rainfall, while the midlatitudes become drier.

Changes in temperature and precipitation would affect soils and vegetation. The composition of forests would change, as some areas become less favorable for certain species of plants, while more hospitable to others. Hotter, drier weather would reduce crop yields in some areas, such as the corn and wheat belts of the Midwest. Conversely, more northerly agricultural regions, such as parts of Canada and Russia, might have longer growing seasons and become more productive.

Climate prediction is not an exact science, however, and not all scientists agree with the scenarios of change just presented. Temperature differences are the engine driving the global circulation of winds and ocean currents and help create conditions inducing or inhibiting winter and summer precipitation and daily weather conditions. Exactly how those vital climate details would express themselves locally and regionally is uncertain.

In addition, some scientists argue that global temperatures might stabilize or even decrease as the concentration of greenhouse gases increases. A hotter atmosphere, they say, would increase evaporation, sending up more water vapor that could condense into clouds. The increased cloud cover might reflect so much sunlight that it would slow the rate at which the earth would be heated. Others contend that the increased evaporation would produce more rainfall. As it fell, the rain would cool the land and subsequently cool the air over the land.

Finally, there are skeptics who dispute the role of humans in global warming. They believe that the geological record shows that large fluctuations in global temperature have always occurred independently of human activity, never as a result of it. These fluctuations are caused by such unpredictable events as variations in solar radiation, shifts in the earth's orbit and in ocean currents, meteoric activity, and volcanic eruptions. Doubters further note that every millennium since the end of the last Ice Age has had one or two centuries in which temperatures have risen by at least as much as they have in the last century. It is reasonable, they claim, to assume that recent atmospheric temperature increases are part of a natural warming cycle and have nothing to do with greenhouse gases.

Controlling Air Pollution

A number of developments in recent years have given rise to the hope that people can reverse the decline in air quality. The total amount of lead added to gasoline has dropped worldwide by 75% since 1970. Several countries, both industrialized and developing, eliminated leaded gasoline from their markets. Many others reduced the lead content and/or introduced unleaded gasoline. The development is significant because exposure to the microscopic particles of lead that are emitted into the atmosphere when leaded gasoline is burned contributes to mental retardation, high blood pressure, and an increased risk of heart attacks and strokes.

As discussed earlier, the Montreal Protocol of 1987 called for global efforts to eliminate the production of chlorofluorocarbons in order to protect the ozone layer. The treaty has made a significant difference. By 1995, total world production of CFCs had fallen 77% from its peak in 1988. The drop has been most pronounced in the European Union and the United States; consumption of CFCs has actually risen in China, India, and other developing economies during the last decade.

Another successful international accord is the 1979 Convention on Long-Range Transboundary Air Pollution, signed by 33 countries in Europe and North America and intended to reduce the emissions of nitrous oxides and sulfur dioxides. During the 1980s, air pollution in Europe was reduced as, for example, Austria, West Germany, Sweden, and Norway cut their SO_2 emissions by more than 50%. Emissions of nitrous oxides have proven more difficult to control.

A less successful attempt to improve air quality is represented by the Framework Convention on Climate Change, which was signed at the Earth Summit held in Rio de Janeiro in 1992 and subsequently ratified by more than 120 countries. It calls on industrialized states to try to cap emissions of greenhouse gases at 1990 levels by the year 2000. Recent assessments indicate that countries are not yet meeting that goal. By 1995, emissions of CO_2 were as much as 5% above the 1990 level in some developed countries, and up to 40% higher in some developing countries, which are not yet bound by any precise targets or timetables.

The United States has made significant progress in cleaning up its air in the last 30 years. A series of Clean Air Acts (1963, 1965, 1970, 1977) and amendments identified major pollutants and established national air quality standards. After many years of debate, Congress in 1990 passed a Clean Air Act that represented the most sweeping legislation to date. It set forth goals to protect public health and the environment by reducing the amount of air pollutants that can be released, and it established a timetable for reaching those goals. Major provisions call for:

- reducing urban smog by 15% by 1996, and 3% each year thereafter until federal air quality standards are met;
- stipulating that passenger cars must emit 60% less nitrogen oxide and 40% less hydrocarbons by the year 2003;
- using cleaner-burning fuels in the most polluted cities; and
- requiring utilities to reduce by half their emissions of NO_x and SO_2.

Reaching these goals will require reducing the type and volume of air pollutants from both stationary and nonstationary sources. A number of strategies can be employed to clean up stationary sources. Technological options include switching to cleaner-burning fuels; coal washing, which removes much of the sulfur in coal before it is burned; and removing pollutants from effluent gases in the smokestack by using scrubbers, precipitators, and filters. Another approach is to reduce energy consumption through the use of more efficient appliances, installing weatherstripping and insulation, and strengthening energy performance standards in the building codes for new buildings.

Reducing emissions from nonstationary sources—mainly motor vehicles of all types—can be accomplished in a number of ways. These include conforming to tighter tailpipe emission standards by retiring older automobiles, phasing out leaded gas, and implementing rigorous vehicle inspection programs. Catalytic converters have sharply reduced smog from cars (although recent evidence indicates they contribute to the greenhouse effect by rearranging the compounds of nitrogen and oxygen from car exhaust to form nitrous oxide). An increase in the price of gasoline would reduce gas consumption. Travel can be made more energy efficient if a community is committed to rewarding those who carpool or use alternative means of transport, such as bicycles or mass transit.

Impact on Land and Soils

People have affected the earth wherever they have lived. Whatever we do, or have done in the past, to satisfy our basic needs has had an impact on the landscape. To provide food, clothing, shelter, transportation, and defense, we have cleared the land and replanted it, rechanneled waterways, and built roads, fortresses, and cities. We have mined the earth's resources, logged entire forests, terraced mountainsides, even reclaimed land from the sea. The nature of the changes made in any single area depends on what was there to begin with and how people have used the land. Some of these changes are examined in the following sections.

Landforms Produced by Excavation

Although we tend to think of landforms as "givens," created by natural processes over millions of years, people have played and continue to play a significant role in shaping local physical landscapes. Some features are created deliberately, others unknowingly or indirectly. Pits, ponds, ridges and trenches, subsidence depressions, canals, and reservoirs are the chief landform features resulting from excavations. Some date back to neolithic times, when people dug into chalk pits to obtain flint for toolmaking. Excavation has had its greatest impact within the last two centuries, however, as earth-moving operations have been undertaken for mining; for building construction and agriculture; and for the construction of transport facilities such as railways, ship canals, and highways.

Surface mining, which involves the removal of vegetation, topsoil, and rocks from the earth's surface in order to get at the resources underneath, has perhaps had the greatest environmental impact. Open-pit mining and strip mining are the most commonly used methods of surface mining.

Open-pit mining is used primarily to obtain iron, copper, sand, gravel, and stone. As Figure 5.20 indicates, an enormous pit remains after the mining has been completed because most of the material has been removed for processing. *Strip* mining is increasingly being employed in the United States as a source of coal; more coal per year now comes from strip mines than from underground mines. Phosphate is also mined in this way. A trench is dug, the material is excavated, and another trench is dug, the soil and waste rock being deposited in the first trench, and so on. Unless reclamation is practiced, the result is a ridged landscape.

Landscapes marred by vast open pits or unevenly filled trenches are one of the most visible results of surface mining. Thousands of square miles of land have been affected, with the prospect of thousands more to come as the amount of surface mining increases. Damage to the aesthetic value of an area is not the only liability of surface mining. If the area is large, wildlife habitats are disrupted, and surface and subsurface drainage patterns are disturbed. In the United States in recent years, concern over the effect of strip mining has prompted federal and state legislation to increase regulation and to stop the worst abuses. Strip-mining companies are now expected to restore mined land to its original contours and to replant vegetation.

Landforms Produced by Dumping

Excavation in one area often leads to the creation, via dumping, of landforms nearby. Both surface and subsurface mining produce tons of waste and enormous spoil piles. In fact, in terms of tonnage, mining is the single greatest contributor to solid wastes, with about 2 billion tons per year left to be disposed of in the United States alone. The normal custom is to dump waste rocks and mill tailings in huge heaps near the mine sites. Unfortunately, this practice has secondary effects on the environment.

Carried by wind and water, dust from the wastes pollutes the air, and dissolved minerals pollute nearby water sources. Occasionally the wastes cause greater damage, as happened in Wales in 1966, when slag heaps from the coal mines slid onto the village of Aberfan, burying more than 140 schoolchildren. Such tragedies call attention to the need for less potentially destructive ways of disposing of mine wastes.

Another example of the combined effect of excavating and filling on the landscape is the agricultural terrace characteristic of parts of Asia. In order to retain water and increase the amount of arable land, terraces are cut into the slopes of hills and mountains. Low walls protect the patches of level land. The *tells* of the Middle East, pictured in Figure 5.21, are another type of landform produced by the accumulation of waste material.

Human impact on land has been particularly strong in areas where land and water meet. Dredging and filling operations undertaken for purposes of water control create landscape features like embankments and dikes. In many places, the actual shape of the shoreline has been altered, as builders in need of additional land have dumped solid wastes into landfills. In the Netherlands, millions of acres of land have been reclaimed from the sea by the building of dikes to enclose polders and canals to drain them. Farming practices in river valleys have had significant effects on deltas. For example, increased sedimentation has often extended the area of land into the sea.

Formation of Surface Depressions

The extraction of material from beneath the ground can lead to **subsidence,** the settling or sinking of a portion of the land surface. Many of the world's great cities are sinking because of the removal of *fluids* (groundwater, oil, and gas) from beneath them. Cities threatened by such subsidence are located on unconsolidated sediments (New

FIGURE 5.20 (a) An aerial view of the Bingham Canyon open-pit copper mine in Utah, said to be the largest human-made excavation in the world. Since mining began at the site, approximately six billion tons of material have been removed, creating a pit more than 800 meters (2600 ft) deep and 4 kilometers (2.5 mi) wide. The mine has yielded vast quantities of copper, gold, silver, and molybdenum. (b) About 390 square kilometers (150 sq mi) of land surface in the United States are lost each year to the strip mining of coal and other resources; far more is disrupted worldwide. Besides altering the topography, strip mining interrupts surface and subsurface drainage patterns, destroys vegetation, and places sterile and often highly acidic subsoil and rock on top of the new ground surface.

(a) Courtesy of Kennicott.

(b) © Stephen Trimble.

FIGURE 5.21 **Aerial view of Erbil, Iraq.** Here, and elsewhere in the Middle East, the debris of millennia of human settlement gradually raised the level of the land surface, producing *tells,* or occupation mounds. The city was constantly rebuilt at higher elevations directly on the accumulation of earlier occupants' refuse. In some cases, the striking landforms rise hundreds of feet above the surrounding plains.

©J. Baylor Roberts/National Geographic Image Collection.

Orleans and Bangkok), coastal marshes (Venice and Tokyo), or lake beds (Mexico City). When the fluids are removed, the sediments compact and the land surface sinks. Because many of the cities are on coasts or estuaries and are often only a few feet above sea level, subsidence makes them more vulnerable to flooding from the sea.

Groundwater abstraction has created serious subsidence in many places. Recent evidence indicates that the withdrawal of trillions of gallons of water from a large area of Arizona has resulted in widespread ground subsidence and the formation of more than a hundred earth fissures, jagged ground cracks that can be as much as 14.5 kilometers (9 mi) long and 120 meters (400 ft) deep.

The removal of *solids* (such as coal, salt, and gold) by underground mining may result in the collapse of land over the mine. *Sinkholes* or *pits* (circular, steep-walled depressions) and *sags* (larger and shallower depressions) are two types of landscape features produced by such collapse. If surface drainage patterns are disrupted, subsidence *lakes* may form in the depressions. Subsidence has become a more serious problem as towns and cities have expanded over mined-out areas.

As one might expect, subsidence damages structures built on the land, including buildings, roads, and sewage lines. A dramatic example occurred in Los Angeles in 1963, when subsidence caused the dam at the Baldwin Hills Reservoir to crack. In less than two hours, the water emptied into the city, resulting in millions of dollars worth of property damage. The withdrawal of groundwater from beneath Mexico City has led to severe though differential subsidence (Figure 5.22). One of the reasons the 1985 earthquake in that city was so damaging was that subsidence had weakened building structures.

Soils

By design or by accident, people have brought about many changes in the physical, chemical, and biochemical nature of the soil and altered its structure, fertility, and drainage characteristics. The exact nature of the changes in any area depends on past practices as well as on the original nature of the land.

Over much of the earth's surface, the thin layer of topsoil upon which life depends is only a few inches deep, usually less than 30 centimeters (1 ft). Below it, the lithosphere is a complex mixture of rock particles, inorganic mineral matter, organic material, living organisms, air, and water. Under natural conditions, soil is constantly being formed by the physical and chemical decomposition of rock material and by the decay of organic matter. It is simultaneously being eroded, for **soil erosion**—the removal of soil particles, usually by wind or running water—is as natural a process as soil formation, and occurs even when land is totally covered by forests or grass. Under most natural conditions, however, the rate of soil formation equals or exceeds the rate of soil erosion, so that soil depth and fertility tend to increase with time.

FIGURE 5.22 **Subsidence in Mexico City** is caused by the pumping out of the aquifer underneath the metropolitan area to supply water. The portion of Line 2 of the city Metro system shown here was horizontal when it was built in the 1960s. Parts of the city have sunk as much as 9.15 meters (30 ft). The subsidence has damaged or destroyed hundreds of buildings, particularly in the colonial-era city center, and continues to rupture water pipes, sewer lines, and subway tunnels.

©Gerardo Magallón/The New York Times.

When land is cleared and planted to crops, or when the vegetative cover is broken by overgrazing or other disturbances, the process of erosion accelerates. When its rate exceeds that of soil formation, the topsoil becomes thinner and eventually disappears, leaving behind only sterile subsoil or barren rock. At that point, the renewable soil resource has been converted through human impact into a nonrenewable and dissipated asset. Carried to the extreme of bare rock hillsides or wind-denuded plains, erosion spells the total end of agricultural use of the land.

Such massive destruction of the soil resource could endanger the survival of the civilization it has supported. For the most part, however, farmers devise ingenious ways to preserve and even improve the soil resource upon which their lives and livelihoods depend. Farming skills have not declined in recent years, but pressures upon farm lands have increased with population growth. Farming has been forced higher up onto steeper slopes, more forest land has been converted to cultivation, grazing and crops have been pushed farther and more intensively into semiarid areas, and existing fields have had to be worked more intensively and less carefully. Many traditional agricultural systems and areas that were ecologically stable and secure as recently as 1950, when world population stood at 2.5 billion people, are disintegrating under the pressures of more than 5 billion people.

The pressure of growing population numbers is having an especially destructive effect on tropical rain forests. Expanded demand for fuel and commercial wood, and a midlatitude market for beef that can be satisfied profitably by replacing tropical forest with cleared grazing land are responsible for some of the loss, but the major cause of *deforestation* is clearing the land for crops. Extending across parts of Asia, Africa, and Latin America, the tropical rain forests are the most biologically diverse places on earth, but some 100,000 square kilometers (40,000 sq mi) are being destroyed every year. About 45% of their original expanse has already been cleared or degraded. Deforestation is discussed in more detail in Chapter 11, but it is important to note here that accelerated soil erosion quickly removes tropical forest soils from deforested areas. Lands cleared for agriculture almost immediately become unsuitable for that use partially because of soil loss (Figure 5.23)

The tropical rain forests can succumb to deliberate massive human assaults and be irretrievably lost. With much less effort, and with no intent to destroy or alter the environment, humans are similarly affecting the arid and semiarid regions of the world. The process is called **desertification,** the spread of desertlike landscapes into arid and semiarid environments. Climatic change—unpredictable cycles of rainfall and drought—is usually a contributing cause, but desertification accelerates because of human activity, mainly overgrazing, deforestation for fuel wood, clearing of original vegetation for cultivation, and burning. Desertification implies a continuum of ecological alteration from slight to extreme (Figure 5.24).

Whatever its degree of development, when the process results from human rather than climatic change, it begins in the same fashion: the disruption or removal of the native cover of grasses and shrubs through farming or overgrazing (Figure 5.25). If the disruption is severe enough, the original vegetation cannot reestablish itself, and the exposed soil is made susceptible to erosion during the brief, heavy rains that dominate precipitation patterns in semiarid regions. Water runs off the land surface instead of seeping in, carrying soil particles with it. When the water is lost through surface flow rather than seepage downward, the water table is lowered. Eventually, even deep-rooted bushes are unable to reach groundwater, and all natural vegetation is lost. The process is accentuated when too many grazing animals pack the earth down with their hooves, blocking the passage of air and water through the soil. When both plant cover and soil moisture are lost, desertification has occurred.

FIGURE 5.23 **Wholesale destruction of the tropical rain forest** guarantees environmental degradation so severe that the forest can never naturally regenerate itself. Exposed soils quickly deteriorate in structure and fertility and are easily eroded.
©Philip Bailey/Stock, Boston.

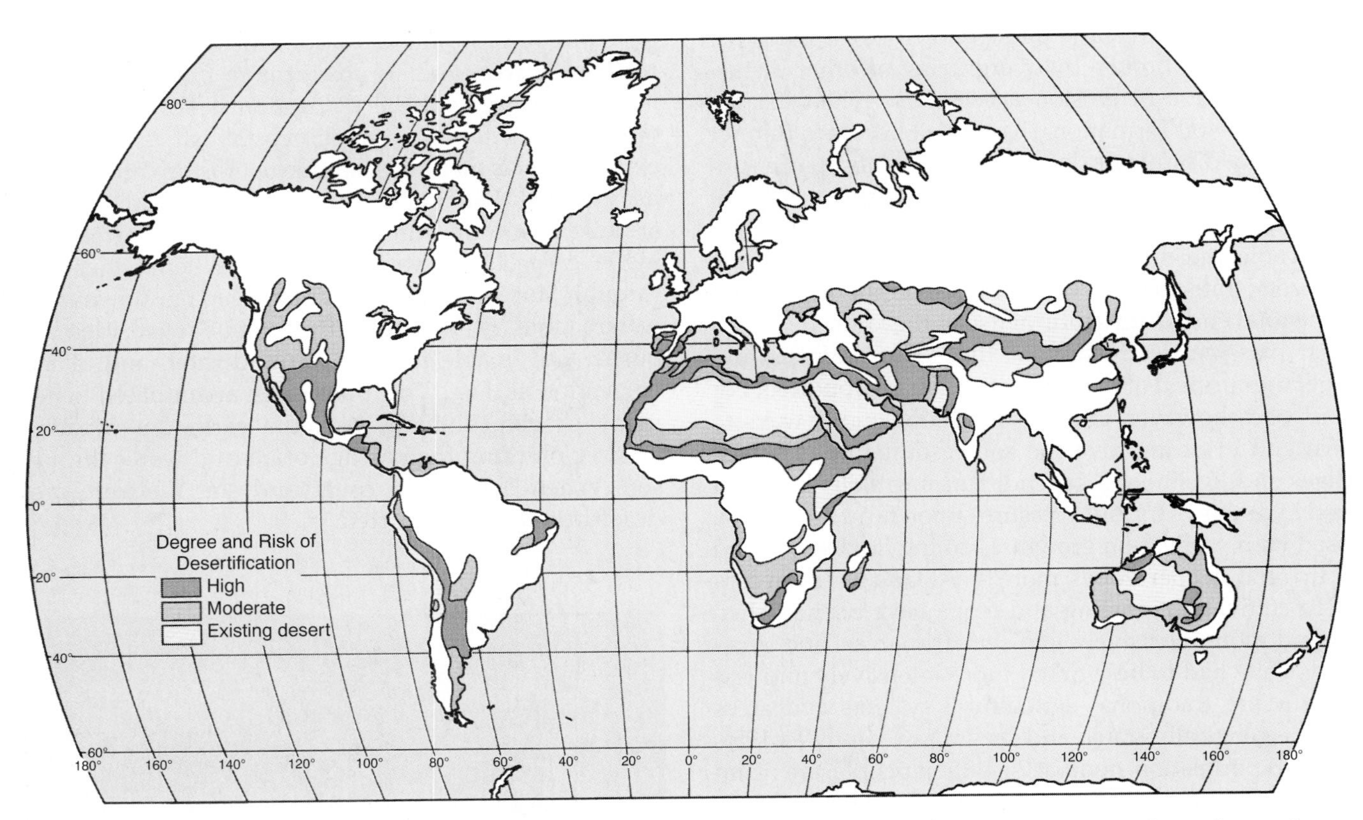

FIGURE 5.24 **Desertification** threatens about one-third of the world's land surface. Desertification is usually understood to imply the steady advance of the margins of the world's deserts into their bordering drylands, converting formerly usable pastures and croplands into barren landscapes. Satellite measurements, however, are forcing a reassessment of what is happening to arid and semiarid drylands along the perimeter of the Sahara. Imagery from 1980 through 1990 indicates that the Sahel drylands region on the southern border of the Sahara, for example, did not move steadily south as usually assumed. Rather, the vegetation line fluctuated back and forth in response to variable rainfall patterns. Many scientists now believe climate variation keeps the drylands in a continual state of disequilibrium, and some prefer to speak of an "ebb and flow" of the Sahara margins and of land degradation rather than of permanent conversion to true desert.

Sources: Based upon H. E. Dregne, *Desertification of Arid Lands,* Figure 1.2, copyright 1983 Harwood Academic Publishers; and *A World Map of Desertification,* UNESCO/FAO.

It happens with increasing frequency in many areas of the earth as pressures upon the land continue. Africa is most at risk; the United Nations has estimated that 40% of that continent's nondesert land is in danger of desertification. But nearly a third of Asia and a fifth of Latin America's land are similarly endangered. In countries where desertification is particularly extensive and severe (Algeria, Ethiopia, Iraq, Jordan, Lebanon, Mali, and Niger), per capita food production declined by some 40% between 1950 and the mid-1990s. The resulting threat of starvation spurs populations of the affected areas to increase their farming and livestock pressures on the denuded land, further contributing to their desertification.

Desertification is but one expression of land deterioration leading to accelerated soil erosion. The evidence of that deterioration is found in all parts of the world. In Guatemala, for example, some 40% of the productive capacity of the land has been lost through erosion, and several areas of the country have been abandoned because agriculture has become economically impracticable. The figure is 50% in El Salvador, and Haiti has no high-value soil left at all. In Turkey, about half of the land is severely or very severely eroded. A full one-quarter of India's total land area has been significantly eroded.

In recent years, soil erosion in the United States has been at an all-time high (Figure 5.26) (see "Maintaining Soil Productivity"). Wind and water are blowing and washing soil off pasturelands in the Great Plains, ranches in Texas, and farms in the Southeast. America's croplands lose almost 2 billion tons of soil per year to erosion, an average annual loss of over 4 tons per acre. In some areas, the average is 15–20 tons per acre. Of the roughly 167 million hectares (413 million acres) of land that are intensively cropped in the United States, more than one-third are losing topsoil faster than it can be replaced naturally (it can take up to a century to replace an inch of topsoil). In parts of Illinois and Iowa where the topsoil was once 30 centimeters (1 ft) deep, less than half of it remains. Every hour about 40,000 tons of topsoil wash into the Mississippi River.

Like most processes, soil erosion has secondary effects. As the soil quality and quantity decline, croplands become less productive and yields drop. Streams and reservoirs experience accelerated siltation. In countries where the topsoil is heavily laden with agricultural chemicals, erosion-borne silt pollutes water supplies.

FIGURE 5.25 Windblown dust is engulfing the scrub forest in this drought-stricken area of Mali, near Timbuktu. The district is part of the Sahel region of Africa where desertification has been accelerated by both climate and human pressures on the land. The cultivation of marginal land, overgrazing by livestock, and recurring droughts have led to the destruction of native vegetation, erosion, and land degradation.

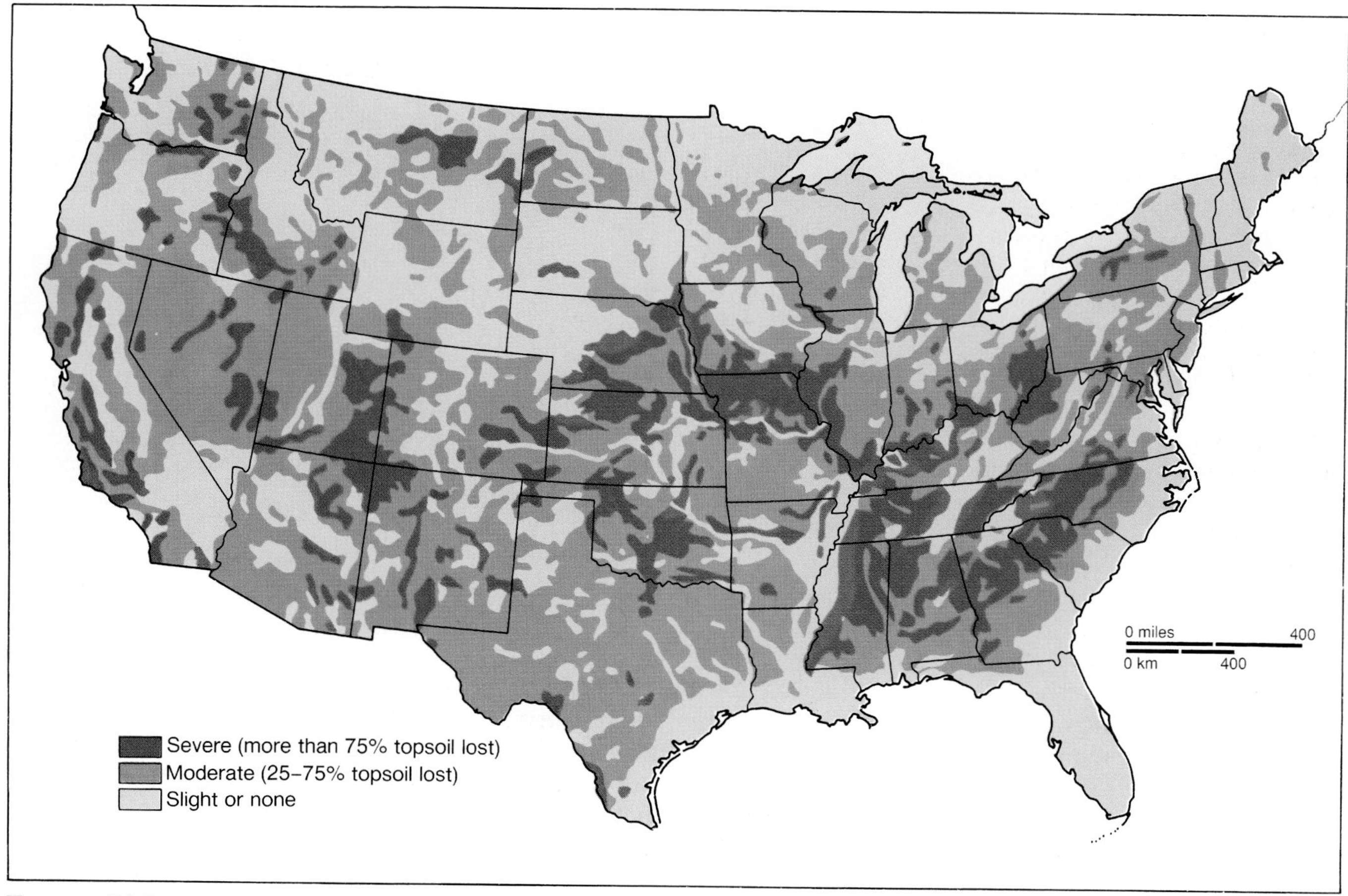

FIGURE 5.26 **Soil erosion in the United States.** Although many activities (mining, construction, and urbanization) contribute to erosion, agriculture and deforestation are particularly significant. In recent years, erosion has been most severe not in the southwestern dust bowl, but in the moist, rolling-hill regions of western Mississippi, western Tennessee, and Missouri. Each of these areas loses about ten tons of topsoil per acre of cropland annually. However, soil resource stress and depletion affect all parts of the country.

Sources: Based on data from 1934 Reconnaissance Erosion Survey of the United States and other soil conservation surveys by the U.S. Soil Conservation Service.

Maintaining Soil Productivity

In much of the world, increasing population numbers are largely responsible for accelerated soil erosion. In the United States in the 1970s and 1980s, economic conditions contributed to a rate of soil erosion equal to that registered during the Dust Bowl era of the 1930s. Federal tax laws and the high farmland values of the 1970s encouraged farmers to plow virgin grasslands and to tear down windbreaks to increase their cultivable land and yields. The secretary of agriculture exhorted farmers to plant all of their land, "from fencerow to fencerow," to produce more grain for export. Land was converted from cattle grazing to corn and soybean production as livestock prices declined.

When prices of both land and agricultural products declined in the 1980s, farmers felt impelled to produce as much as they could in order to meet their debts and make any profit at all. To maintain or increase their productivity, many neglected conservation practices, plowing under marginal lands and using fields for the same crops every year.

Conservation techniques, however, have not been forgotten. They are practiced by many and persistently advocated by farm organizations and soil conservation groups. Techniques to reduce erosion by holding the soils in place are well known. They include contour plowing, terracing, strip-cropping and crop rotation, erecting windbreaks and constructing water diversion channels, and practicing no-till farming (allowing crop residue such as cut corn stalks to remain on the soil surface throughout the winter). In addition, farmers can be paid to idle marginal, highly erodible land. Only by employing such practices can the country maintain the long-term productivity of soil, the resource base upon which all depend.

Strip-cropping.
©Robert Frerck/Odyssey Productions, Chicago.

No-till farming.
Gene Alexander/Soil Conservation Service, USDA.

The danger of floods increases as bottomlands fill with silt, and the costs of maintaining navigation channels grow.

Accelerated erosion is a primary cause of agricultural soil deterioration, but in arid and semiarid areas, salt accumulation can be a contributing factor. **Salinization** is the concentration of salts in the topsoil as a result of the evaporation of surface water. It occurs in poorly drained soils in dry climates, where evaporation exceeds precipitation. As water evaporates, some of the salts are left behind to form a white crust on the surface of the soil.

Like erosion, salinization is a natural process that has been accelerated by human activities. Poorly drained irrigation systems are the primary culprit, because irrigation water tends to move slowly and thus to evaporate rapidly. All irrigation water contains dissolved salts, which are left behind on the surface when water evaporates. Mild or moderate salinity makes soil less productive and lowers crop yields; extreme salinity ultimately can render the land unsuitable for agriculture. Thousands of once-fertile acres have been abandoned in Iran and Iraq; over 25% of the irrigated areas of India, Pakistan, Syria, and Egypt are affected by salinization. Approximately 1.6 million hectares (4 million acres) of cultivated soils in the Canadian provinces of Saskatchewan and Alberta are classified as overly saline. Areas of serious salinization also appear in the U.S. Southwest, particularly in the Colorado River drainage basin and in the Central Valley of California. Ironically, the irrigation water that transformed that arid, 430-kilometer-long (270-mi-long) valley into one of the country's most productive farming regions now threatens to make portions of it worthless again.

Impact on Plants and Animals

People have affected plant and animal life on the earth in several different ways. When human impact is severe enough, a species becomes *extinct,* that is, it no longer exists. Although fossil records show that extinction is a natural feature of life on earth, scientists estimate that in recent history the rate of extinction has increased a million times—an increase due to human activity. *Endangered* species are those likely to become extinct if the causes of endangerment continue, while *threatened* or *vulnerable* species are likely to become endangered within the foreseeable future.

The Species Survival Commission of the World Conservation Union in 1996 assessed the state of wildlife on every continent and reported that 24% of all known mammal species are considered threatened, as are 11% of the world's bird species. Among the different orders of mammals, listed as threatened are:

- 11 species of hoofed animals (rhinoceroses, zebras, etc.);
- 65 species of carnivores (wild cats and dogs, bears, etc.);
- 70 species of ungulates (hippopotamuses, deer, sheep, etc.);
- 96 species of primates (monkeys and apes).

The countries with the largest numbers of threatened mammal species are China, India, and Indonesia. Together they contain more than 40% of the world's human population, which is putting increasing pressure on critical habitats.

Although endangered mammals and birds have attracted public attention, many species of plants are also in jeopardy. According to a 1997 report by the World Conservation Union, a comprehensive global assessment revealed that of the world's 270,000 known plant species, at least one in eight is threatened by extinction. The United States ranks first among countries in the number of plants at risk—4669 of its 16,108 plant species (29%) are imperiled—but this may be due in part to the fact that plants have been better surveyed in the country than elsewhere.

In this section we examine some of the ways people have modified plant and animal life.

Habitat Disruption

One of the main causes of extinction has been the loss or alteration of habitats for wildlife. About two-thirds of all endangered and threatened species are affected by habitat degradation or loss. By clearing forest land, draining wetlands, extending farmland, and building cities, people modify or destroy the habitats in which plants and animals have lived.

Tidal marshes have been subjected to dredging and filling for residential and industrial development. The loss of such areas reduces the essential habitat of fish, crustaceans, and mollusks. The whooping crane has been virtually eliminated in the United States because the marshes where it nested were drained, and roads and canals brought intruders into its habitat. Its comeback, sought by breeding programs in the United States and Canada, is still uncertain.

Many people fear that as countries in Africa and South America become more industrialized and more urbanized and expand their areas under cultivation, there will be an increasingly negative impact on wildlife. It is already known that in Africa, wild animals are vanishing fast, in part the victims of habitat destruction. In Botswana, for example, 250,000 antelope and zebra died in a decade, disoriented by fences erected to protect cattle. As selected animal species decline in numbers, balances among species are upset and entire ecosystems are disrupted.

The destruction of the world's rain forests, the most biologically diverse places on earth, is by some estimates causing the extinction of about 1000 plant and animal species every year. For every plant that becomes extinct, approximately 20 animals that depend on the plant also vanish.

Hunting and Commercial Exploitation

Another way in which people have affected plants and animals is through their deliberate destruction. We have overhunted and overfished, for food, fur, hides, jewelry, and trophies. In the past, unregulated hunting harmed wildlife all over the world and was responsible for the destruction of many populations and species. Beavers, sea otters, alligators, and buffalo are among the species brought to the edge of extinction in the United States by thoughtless exploitation. Under protective legislation, their populations are now increasing, but hunting in developing countries still poses a threat to a number of species.

Three African animals whose existence is threatened by hunting, most of it illegal, are the elephant, black rhinoceros, and mountain gorilla. Prized for its ivory tusks, the African elephant has been ruthlessly slaughtered. Ten million of the elephants are estimated to have been alive in the 1930s. By 1979 the population had dropped to 1.5 million, and a decade later, to only one-half million (Figure 5.27). The black rhinoceros, killed for its horn, is now an endangered species. The population has declined from nearly a million in sub-Saharan Africa a century ago to about 2500 today. Only 400 mountain gorillas are believed to exist, most of them in the Virunga Mountains of Uganda and Rwanda.

For over a decade, commercial fishermen from Japan, South Korea, and Taiwan have used drift nets to capture

FIGURE 5.27 **Park rangers inspecting an elephant killed by poachers in Kenya.** Although habitat alteration has contributed to the decline of the elephant population, the greatest threat comes from illegal poaching. By their voracious consumption of vegetation, elephants help shape ecosystems in both savanna woodlands and rain forests. As elephants disappear, so will many species that share their habitat, including zebras, gazelles, and giraffes.
© Marc Chamberlain/Tony Stone Images.

squid, tuna, and salmon. The nets, which can stretch up to 65 kilometers (40 mi), have been called "curtains of death" because they essentially scoop up all life in their path, including millions of nontarget fish, birds, and such marine mammals as whales and dolphins. Devastatingly effective, the nets have also seriously depleted stocks of the target fish. The United Nations has condemned the drift nets and called for a moratorium on their use.

Introduction of New Species

The deliberate or inadvertent introduction of a plant or animal into an area where it did not previously exist can have damaging and unforeseen consequences. Introduced species have often left behind their natural enemies—predators and diseases—giving them an advantage over native species that are held in check by biological controls. The rabbit, for example, was purposely introduced into Australia in 1859. The original dozen pairs multiplied to a population in the thousands in only a few years and, despite programs of control, to an estimated one billion by 1950 (Figure 5.28). Inasmuch as five rabbits eat about as much as one sheep, a national problem had been created. Rabbits had denuded much of the grassland on which sheep could graze.

Invasions of alien species have multiplied as the speed and range of travel have increased. In the United States alone, at least 59 harmful invaders have been discovered since 1980—harmful because they consume or outcompete native species (see "On Becoming Mussel-Bound"). They include:

- the Asian gypsy moth, a voracious eater of trees and some crops, which arrived in Oregon, Washington, and British Columbia in 1991;
- the Asian tiger mosquito, which was discovered in Florida in 1986 and is a carrier of dengue fever;
- colonies of South American fire ants, which kill wildlife and pets and have settled throughout the South;
- Africanized honeybees, more aggressive and venomous than European bees, finally reached the southwestern United States in the 1990s after escaping from an experimental station in Brazil in 1957;
- the round goby, an aggressive fish that reached U.S. waters in ballast chambers of ships from the Black and Caspian Seas and threatens to disrupt the Great Lakes' ecosystem.

Plants and animals found on islands are particularly prone to extinction. Island plants and animals evolved in isolation, with few diseases or predators. Furthermore, because island species often occur on just one or a few islands, the loss of only a few individuals can be devastating to small populations. The Hawaiian islands are plagued by a number of alien species, including feral pigs, mongooses,

FIGURE 5.28 **Rabbits converging on a water hole in Australia during a drought.** Deliberately imported to Australia, the rabbit became an economic burden and environmental menace, competing with sheep for grazing land and stimulating soil erosion. In the 1950s, scientists released *myxomatosis,* a virus that is deadly to rabbits. Within two years, fewer than 100 million rabbits remained. As the survivors began building up immunity to the virus, however, the number of rabbits increased again. A new virus, *calicivirus,* was released into Australia in 1995; it has already killed more than 100 million rabbits, but some fear the virus could infect native animals or humans.
© Mark Boulton/Photo Researchers, Inc.

On Becoming Mussel-Bound

Zebra mussels appear to be harmless, even attractive creatures, with their tiny striped shells. But ecologists view their recent entry into North America as nothing but a disaster.

Native to the Caspian and Black Seas, zebra mussels (*Dreissena polymorpha*) were first detected in North America in 1988. They are believed to have been stowaways on an East European freighter that dumped ballast water into Lake St. Clair, near Detroit.

In just a few years, the mussels had spread to the rest of the Great Lakes. The floods of 1993 carried the mussel larvae into the Mississippi River system, extending their range as far south as New Orleans.

The rapid advance of the mussels is due both to a scarcity of natural predators, such as diving ducks and crawfish, and to the mussels' prodigious reproductive capacity. Adult females produce from 30,000 to 40,000 eggs a year, and males contribute a like amount of sperm.

The mussels have an unfortunate tendency to attach themselves to anything hard, including water intake pipes. Colonies of mussels now clog the underwater intake pipes of power plants, water treatment plants, and industrial plants in the Great Lakes. A square meter (about 10 sq ft) of wall at one utility plant was found to contain over 700,000 mussels, and a single intake pipe at an Ontario water plant was clogged by 30 tons of the shellfish.

The zebra mussels also endanger the Great Lakes fisheries, the largest freshwater fisheries in the world. Because they devour phytoplankton, the microscopic plants at the base of the aquatic food chain, the mussels compete with algae-eating fish for both food and oxygen. Accidentally introduced into the region, the zebra mussel might significantly alter the ecosystem of the entire Great Lakes.

Zebra mussels.
Courtesy of Detroit Edison.

eucalyptus, ginger, and gorse. Miconia, a large-leaved plant, was introduced as a tropical ornamental. Outside the confines of a flowerpot, it grows to 50 feet or more, and its mammoth leaves (1 meter or 3 feet across) cast dense shade, killing native vegetation beneath it, promoting water runoff and erosion. Small colonies of these plants are scattered throughout the islands, and there is a massive eradication program underway in an attempt to curb its spread. The fear is that miconia might do to Hawaii what it has done in Tahiti, where it has destroyed 70% of the rain forest.

As this discussion indicates, introduced plants, as well as animals, can alter vegetative patterns. Some 300 species of invasive plants now threaten native ecosystems in the mainland United States and Canada. At least half were deliberately imported, including purple loosestrife, the melaleuca tree, Norway maple, and water hyacinth. These and other imports have arrived without their natural enemies and spread uncontrolled, driving out native species.

The Asiatic chestnut blight has destroyed most of the native American chestnut trees in the United States, trees with significant commercial as well as aesthetic value. The cause was the importation of some chestnut trees from China to the United States. They carried a fungus that was fatal to the American chestnut tree but not to the Asiatic variety, which is largely immune to it.

An aquatic vine, hydrilla, imported into Florida from Sri Lanka for use in aquariums, was dumped into a canal in Tampa in 1951. Also known as water thyme, it has overgrown more than 40% of Florida's rivers and lakes and continues to spread rapidly. The state spends millions of dollars annually to fight the vine, which grows into dense mats, clogging boat propellers and preventing sunlight from reaching the water bottom. By monopolizing the dissolved oxygen that fish and aquatic plants require to thrive, hydrilla reduces native diversity.

In an attempt at biological control, Florida officials imported tilapia, a weed-eating fish, to solve the problems created by hydrilla. While the fish have done little to clear Florida waterways, they have driven out many types of native fish, especially large-mouth bass.

These are just a few of the many examples that illustrate an often ignored ecological truth: plant and animal life are so interrelated that when people introduce a new species to a region, whether by choice or by chance, there may be unforeseen and far-reaching consequences.

It is important to note, however, that not all introduced species are harmful. They are of little concern if they assimilate well and coexist with the native stock. Further, an import may become a problem in one area but not be highly invasive elsewhere.

Poisoning and Contamination

Humans have also affected plant and animal life by poisoning or contamination. In the last several years, we have become acutely conscious of the effect of insecticides, rodenticides, and herbicides, known collectively as *biocides.* The best known and most widely used has been DDT, although there are now thousands of compounds in use.

DDT was first used during World War II to kill insects that carried diseases such as malaria and yellow fever. In the years following the war, tons of DDT and other biocides were used, sometimes to combat disease, sometimes to increase agricultural yields. Insects, after all, can destroy a significant percentage of a given crop, either when it is in the field or after it has been harvested.

In the last twenty years, some of the side effects of these biocides have been well enough documented for us to question their indiscriminate use. Once used, a biocide settles into the soil, where it may remain or may be washed into a body of water. In either case, it is absorbed by organisms living in the soil or the mud. Through a process known as **biological magnification,** the biocide accumulates and is concentrated at progressively higher levels in the food chain (Figure 5.29). By remaining as residue in fatty tissues, a very small amount of a biocide produces unexpected effects, with predators accumulating larger amounts than their prey. The higher the level of an organism in the food chain, the greater the concentration of DDT will be, and that concentration may be lethal. Robins and other small birds die when they eat earthworms that have ingested DDT that has settled into the earth after being sprayed on trees.

DDT also causes a decrease in the thickness of the eggshells of some of the larger birds, causing a greater number of eggs to break than normally would. Peregrine falcons, bald eagles, and brown pelicans were among the birds nearly made extinct by this disruption of the reproductive process.

Although the use of DDT has declined as its effects have become apparent, other chlorinated hydrocarbon compounds have been developed and are in wide use. Indeed, the use of pesticides in the United States has more than doubled in the last 25 years. More than 900 million kilograms (2 billion lbs) containing more than 600 active ingredients are applied each year. The pesticides pollute water supplies, often contaminate the crops they are meant to protect, and sometimes sicken the farm workers who apply them. In addition, too often they are only temporarily effective.

Biocides may, in fact, exacerbate the problem their use is designed to eradicate. By altering the natural processes that determine which insects in a population will survive, biocides spur the development of resistant species. If all but 5% of the mosquito population in an area are killed by an insecticide, the ones that survive are the most resistant individuals, and they are the ones that will produce the succeeding generations.

There are now insects whose total resistance to certain pesticides has led some scientists to conclude that the entire process of insecticide development may be self-defeating. Despite the enormous growth in the use of pesticides, crop loss to insect and weed pests has actually grown. According to Department of Agriculture figures, 32% of crops were lost to pests in 1945; 45 years later, such losses had increased to 37%.

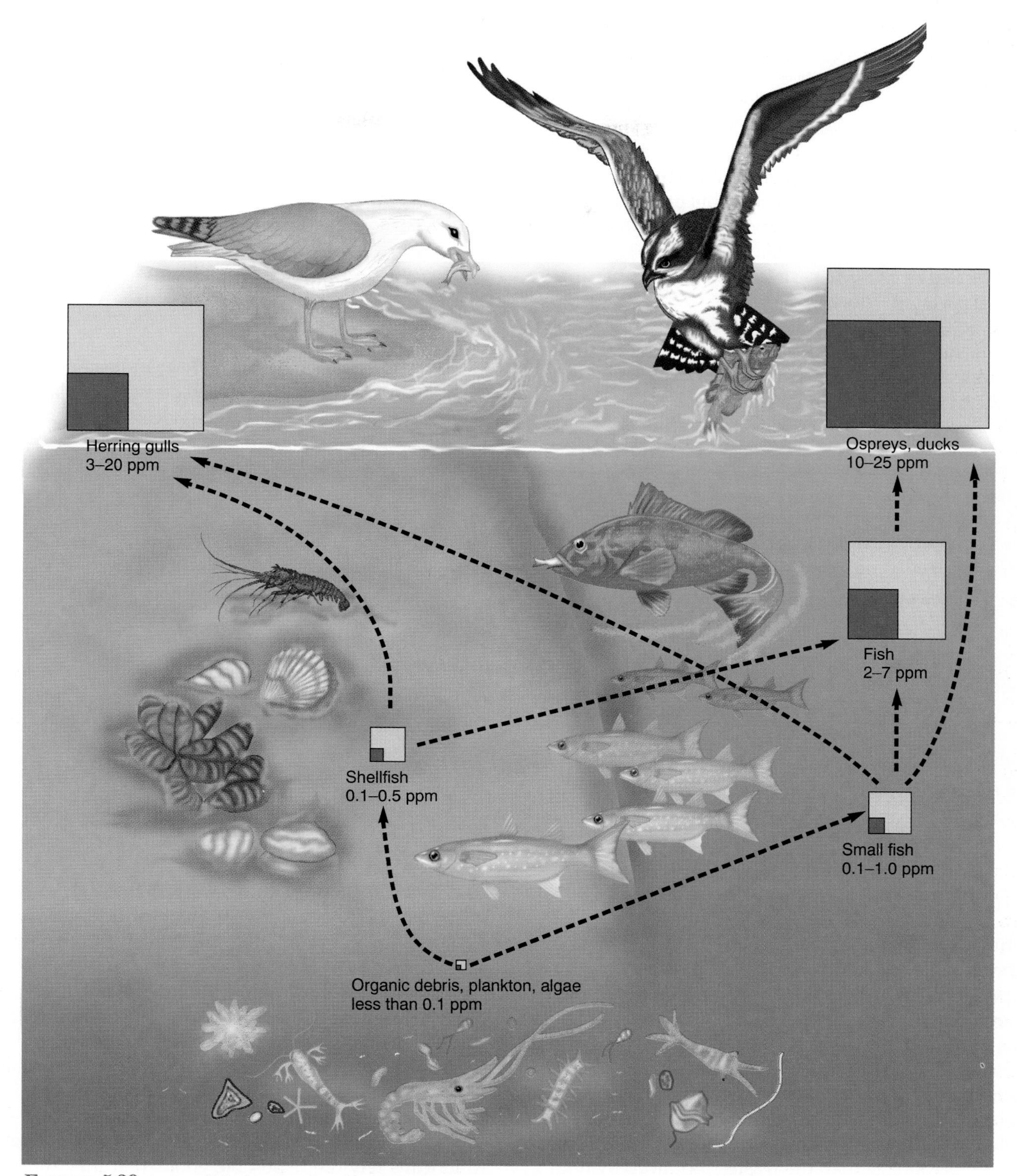

FIGURE 5.29 **A simplified example of biological magnification.** Although the level of DDT in the water and mud may be low, the impact on organisms at the top of the food chain can be significant. In this example, birds at the top of the chain have concentrations of residues as much as 250 times greater than the concentration in the water. PCBs and radioisotopes, such as strontium-90 and cesium-137, undergo magnification in the food chain just as insecticides do.

On-Line HUMAN IMPACT ON THE ENVIRONMENT

Internet resources contain a wealth of information about the ways that people impact and modify the physical environment. Many of the websites are maintained by governmental organizations, academic institutions, and nonprofit organizations concerned with various aspects of the environment. Some of the most useful sites are noted here, but many others also exist and can be located by following the Internet links included in the sites that are listed in this box.

EnviroLink advertises itself as the world's largest environmental information archive on-line and the clearinghouse for all on-line environmental information. Collections include the EnviroLink Library and the EnviroNews Service.

http://www.envirolink.org/

Environmental Organization WebDirectory contains a comprehensive listing of thousands of environmental sites on the World Wide Web, ranging alphabetically from agriculture to wildlife. Each subject category listed on its home page leads to an extensive set of links usually arranged by organizations, area, and subtopics.

http://www.webdirectory.com/

The *WWW Virtual Library:Environment* maintained by Earth Systems, Inc., is divided by subject, including atmosphere, biosphere, hydrosphere, and lithosphere. Use them for links to a wide variety of organizations, topics, and reports. The site also has an exhaustive "List of Environmental Resources," and links to other related Virtual Library subjects.

http://earthsystems.org/Environment.html

The *National Institute for the Environment* is a national, non-profit, nonregulatory organization concerned with disseminating environmental information, particularly through its developing "National Library for the Environment." The library features Congressional Research reports, texts of environmental laws, and population, environment, and biodiversity linkages.

http://www.cnie.org/

Federal agencies maintain many useful environmental websites, some of which are noted in this box.

The *U.S. Environmental Protection Agency* home page includes information about agency programs concerning land, air, and water systems. Resources for students and teachers are also noted.

http://www.epa.gov/

The *U.S. Fish and Wildlife Service,* through its many offices, supplies thousands of pages of Internet information, including material and linkages on resource management, migratory birds, waterfowl and wetlands, and endangered species. All topics are easily accessed through the search link from the Service's home page.

http://www.fws.gov/

The *U.S. Global Change Research Information Office* (*GCRIO*), a product of the Global Change Research Program and several governmental agencies and organizations, provides access to data and information on global change research and educa-

SOLID-WASTE DISPOSAL

Modern technologies and the societies that have developed them produce enormous amounts of solid wastes that must be disposed of. The rubbish heaps of past cultures suggest that humankind has always been faced with the problem of ridding itself of materials it no longer needs. The problem for advanced societies, with their ever-greater variety, amount, and durability of refuse, is even more serious: how to dispose of the solid wastes produced by residential, commercial, and industrial processes. Although these account for much less tonnage than the wastes produced by mining or agriculture, they are everywhere, a problem with which each individual and each municipality must deal.

Municipal Waste

The wastes that communities must somehow dispose of include newspapers and beer cans, toothpaste tubes and old television sets, broken refrigerators and rusted cars (Figure 5.30). American communities are now facing twin crises in disposing of these wastes: the sheer volume of trash and the toxic nature of much of it. Solid-waste disposal is a greater problem in the United States than in any other country, because Americans throw away more trash per person than any country in the world. Solid-waste disposal costs are now the second largest expenditure of most local governments. Americans generate more than twice as much waste per person as do Japanese and Europeans, four times as much as Pakistanis or Indonesians.

Our volume of trash is the result of three factors—affluence, packaging, and open space. Craving convenience, Americans rely on disposable goods that they throw away after very limited use. Thus, although readily available substitutes are more economical, Americans annually throw out 16 billion baby diapers, 2 billion razors, 1.6 billion pens, and a million tons of paper towels and napkins. People in less affluent countries repair and recycle a far greater proportion of domestic products. In addition, nearly all consumer goods are encased in some sort of wrapping, whether it be paper, cardboard, plastic, or foam. An astounding one-third of the yearly volume of trash consists of these packaging

On-Line *Continued*

tional resources. A changing selection of "Showcase Links" is a feature of the home page, and its "Global Change Resources" provides access to bibliographic databases and to selected documents and publications.

http://www.gcrio.org/

The Institute for the Environment is a joint public-private partnership between The George Washington University in Washington, D.C., and the U.S. Environmental Protection Agency. The "Environmental Information Resources" site provides information about the environment and has links to several hundred other sites around the world, with more being added daily.

http://www.gwu.edu/~greenu/

EE-Link provides environmental education facts, data, and activities for use in K–12 classrooms, from the National Consortium for Environmental Education and Training.

http://eelink.net/ee-linkintroduction.html

Canada has an extensive set of governmental and nongovernmental environmental agencies and websites. A good starting point is Environment Canada's *Green Lane (on the Information Highway).* This website reviews current issues and data sources, links to regional "Green Lane" sites, and the text of important official environmental publications.

http://www.doe.ca/

The Canada Centre for Inland Waters is a major source for water research, with particular emphasis on the Great Lakes through its "Great Lakes Information Management Resource." That and other information sources are accessible through the Centre's home page.

http://www.cciw.ca/

Many nonprofit and nongovernmental organizations, only a few of which are listed here, maintain home pages. You will find reference to most of them through the linked websites already mentioned.

Center for Marine Conservation: **http://www.cmc-ocean.org**

Greenpeace International: **http://www.greenpeace.org/**

National Wildlife Federation: **http://www.nwf.org**

Nature Conservancy's *Wired for Conservation:* **http://www.tnc.org/**

Resources for the Future: **http://www.rff.org**

Sierra Club home page: **http://www.sierraclub.org/**

World Resources Institute: **http://www.wri.org/**

materials. Finally, the United States has traditionally had ample space in which to dump unwanted materials. Countries that ran short of such space decades ago have made greater progress in reducing the volume of waste.

Although ordinary household trash does not meet the governmental designation of **hazardous waste**—defined as discarded material that may pose a substantial threat to human health or to the environment when improperly stored, transported, or disposed of—much of it is hazardous nonetheless. Products containing toxic chemicals include paint thinners and removers, furniture polishes, bleaches, oven and drain cleaners, used motor oil, and garden weed killers and pesticides.

Countries use various methods of disposing of solid wastes; each has its own impact on the environment. Loading wastes onto barges and dumping them in the sea, long a practice for coastal communities, inevitably pollutes oceans. Open dumps on land are a menace to public health, for they harbor disease-carrying rats and insects. Burning combustibles discharges chemicals and particulates into the air. In the United States, three methods of solid-waste disposal are employed: landfills, incineration, and recycling (Figure 5.31).

Landfills

An estimated 71% of U.S. municipal solid waste is deposited in *sanitary landfills,* where each day's waste is compacted and covered by a layer of soil (Figure 5.32). "Sanitary" is a deceptive word. Until recently, there were no federal standards to which local landfills had to adhere, and while some communities and states regulated the environmental impact of dumps, many did not. Even if no commercial or industrial waste has been dumped at the site, most landfills eventually produce *leachate,* liquids that contaminate the groundwater. The EPA estimates that only 15% of the country's landfills have liners to prevent contaminants from seeping into groundwater supplies. Indeed, more than two-thirds of the dumps lack any type of system to monitor groundwater quality. Leachate forms when precipitation entering the landfill interacts with the decomposing materials. Thus, heavy metals are leached from batteries and old electrical parts, while vinyl chlorides come from the plastic

FIGURE 5.30 Some of the 240 million tires Americans replace each year. Tire piles such as this one are breeding grounds for rats and mosquitoes and are susceptible to fire. Most states have enacted scrap-tire management programs, financed by a levy consumers pay when buying new tires. Most scrap tires are now either retreaded, shredded and burned to generate electricity, or exported abroad. Recently, entrepreneurs have begun to develop other ways to use old tires: grinding them into tire crumbs that are added to highway asphalt and the soils of athletic fields, for example, or heating them to produce oil.
© Don Kohlbauer/San Diego Union-Tribune.

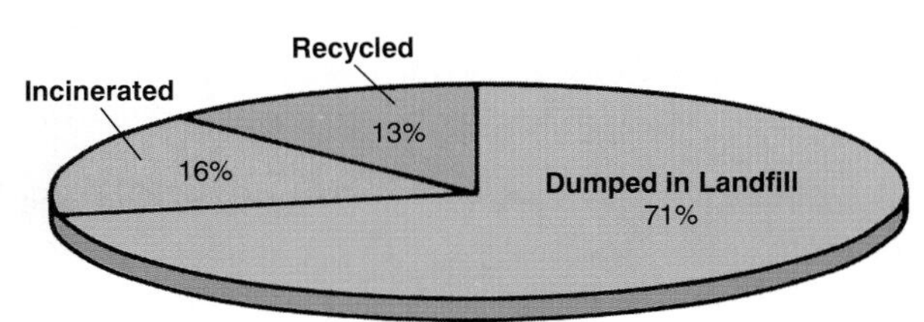

FIGURE 5.31 **Methods of solid-waste disposal in the United States.**

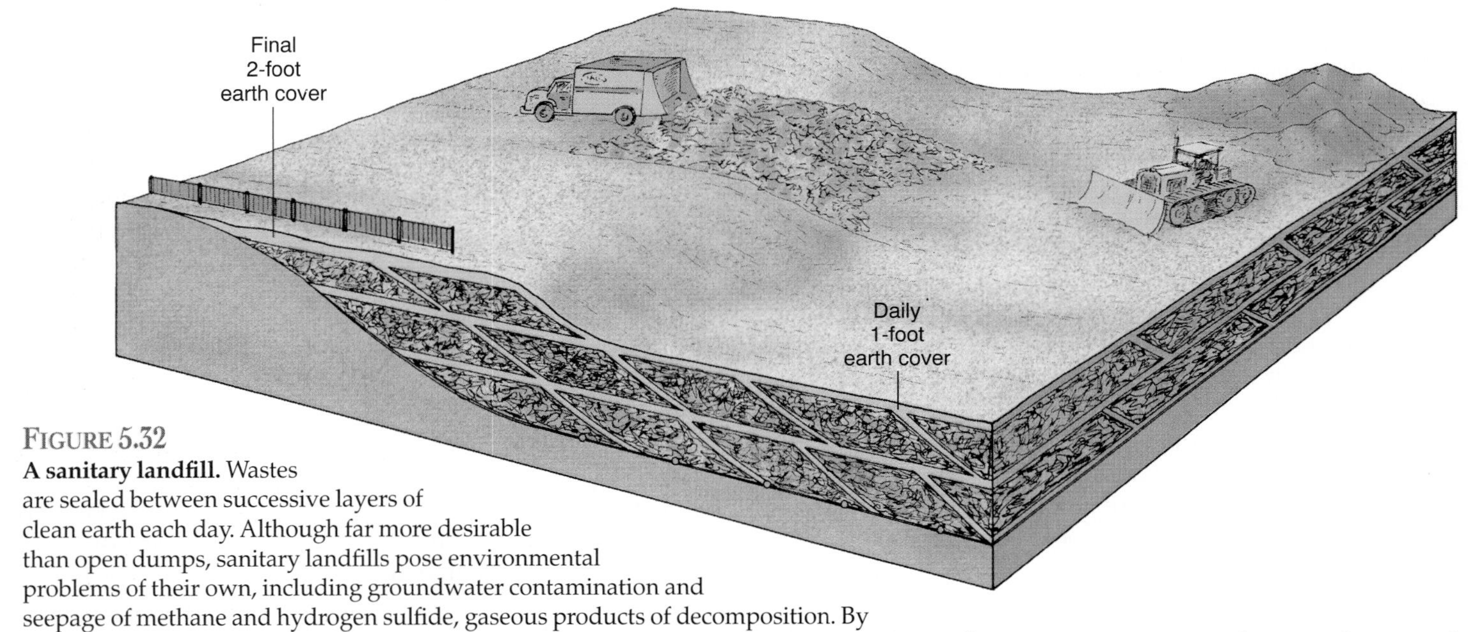

FIGURE 5.32
A sanitary landfill. Wastes are sealed between successive layers of clean earth each day. Although far more desirable than open dumps, sanitary landfills pose environmental problems of their own, including groundwater contamination and seepage of methane and hydrogen sulfide, gaseous products of decomposition. By federal law, modern landfills must be lined with clay and plastic, equipped with leachate collection systems to protect the groundwater, and monitored regularly for underground leaks—requirements that have increased significantly the cost of constructing and operating landfills.

in household products. Typical leachate contains more than 40 organic chemicals, many of them poisonous.

New York City's largest landfill, the Fresh Kills on Staten Island, illustrates the problem. Each day, approximately 13,000 tons of waste are trucked to the site. More than 50 years old, the dump was not constructed to hold its contents securely, and it does not comply with current state landfill regulations. It is located in an ecologically sensitive wetland area and is adjacent to residential communities. Every day, thousands of gallons of contaminated leachate seep into the groundwater beneath the landfill. The decomposition of organic material generates 140,000 cubic meters (5 million cu ft) of methane gas annually. The rotting garbage produces odors that include the fishy smells of amines, the goatlike aroma of organic acids, and the sickeningly sweet scents of aldehydes. When Fresh Kills is closed in the year 2002, it will cover almost 900 hectares (2200 acres) and will stand 170 meters (505 ft) high. Taller than the Statue of Liberty, taller even than the largest of the great pyramids of Egypt, Fresh Kills will be a veritable monument to trash.

Municipal landfill capacity is shrinking dramatically. The number of landfills in the United States fell from 18,000 in the late 1970s to 6000 in 1990, increasing pressure on those that remain open. One half of these will be closed within the next ten years, either because they are full or because their design and operation pose a threat to the environment. Old landfills often are not replaced by new ones for a variety of reasons. These include a lack of suitable (geologically sound) sites, rising economic costs of operating fills, and community opposition to the siting of new landfills. Increasingly, communities are being forced to look for alternative methods of waste disposal.

Incineration

The quickest way to reduce the volume of trash is to burn it, a practice that was common at open dumps in the United States until it was halted by the Clean Air Act of 1970. Concern over air pollution also forced the closure of old, inefficient *incinerators* (facilities designed to burn waste), providing an impetus to designing a new generation of incinerators. More than 150 municipal incinerators now operating in the United States, mostly in the northeast, burn about 16% of the national total of trash. Connecticut burns about 60% of its household trash; Massachusetts, New Jersey, and New York may soon do likewise. A Dade County facility burns 25% of the waste from 26 municipalities, including Miami. Most municipal incinerators are of the waste-to-energy type, which use extra-high (980°C, 1800°) temperatures to reduce trash to ash and simultaneously generate electricity or steam that is sold to help pay operating costs.

A decade ago, incinerators were hailed as the ideal solution to overflowing landfills, but it has become apparent that they pose environmental problems of their own by generating toxic pollutants in both air emissions and ash. Air emissions from incinerator stacks have been found to contain an alphabet soup of highly toxic elements, ranging from a (arsenic) to z (zinc), and including, among others, cadmium, dioxins, lead, and mercury, as well as significant amounts of such gases as carbon monoxide, sulfur dioxide, and nitrogen oxides. Emissions can be kept to acceptably low limits by installing electrostatic precipitators, filters, and scrubbers to capture pollutants before they are released into the outside air, although the devices add significantly to the cost of the plant.

A greater problem is created by the concentration of toxins, particularly lead and cadmium, in the ash residue of burning. Incinerators typically reduce trash by 90%; one-tenth remains as ash, which then must be buried in a landfill. In 1994, the U.S. Supreme Court ruled that the ash must be tested for toxicity and handled as hazardous waste if it exceeds federal safety standards. This means it must be disposed of in licensed hazardous waste landfills, which have double plastic linings, moisture collection systems, and tighter operating procedures than do ordinary municipal landfills.

Source Reduction and Recycling

Awareness of the problems associated with landfills and incinerators has spurred interest in two alternative waste-management strategies: source reduction and recycling. By *source reduction,* we mean producing less waste in the first place so as to shrink the volume of the waste stream. Manufacturers can reduce the amount of paper, plastic, glass, and metal they use to package food and consumer products. Over the last 30 years, for example, the weight of plastic soft-drink bottles and of aluminum beverage cans has been reduced by 20–30%. Some products, such as detergents and beverages, can be produced in concentrated form and packaged in smaller containers.

Recycling reduces the amount of waste needing disposal by making a portion of it available for reuse. Thousands of businesses and hundreds of communities in the United States have initiated recycling programs for a variety of materials. Most programs collect paper, aluminum beverage cans, and glass. Some also collect leaves and other yard waste. Although recycling of plastics is more difficult, plastic milk and soda bottles are easily recycled.

Compared to other methods of solid-waste disposal, recycling has a beneficent impact on the environment. It saves natural resources by making it possible to cut fewer trees, burn less oil, mine less ore. Because it takes less energy to make things out of recycled material than out of virgin materials, recycling saves energy. It reduces the pollution of air, water, and land that stems from the manufacture of new materials and from other methods of waste disposal, and saves increasingly scarce landfill space for materials that cannot be recycled. These include hazardous wastes.

Recycling plays a significant role in reducing the amount of solid waste that needs to be disposed of in many cities in developing countries. Wastepickers sort through trash looking for material they can sell to small-scale businesses and industries (Figure 5.33). It is estimated that in Indonesian cities, for example, wastepickers reduce total urban refuse by one-third. Unfortunately, the environmental conditions under which wastepickers work are not good, and they run the risk of illness from disease-carrying organisms and injuries from broken glass and metal.

Hazardous Waste

The EPA has classified more than 400 substances as hazardous, posing a threat to human health or the environment. Industrial wastes have grown steadily more toxic. Currently, about 10% of industrial waste materials are considered hazardous.

Every facility that either uses or produces radioactive materials generates *low-level waste,* material whose radioactivity will decay to safe levels in 100 years or less. Nuclear power plants produce about half the total low-level waste in the form of used resins, filter sludges, lubricating oils, and detergent wastes. Industries that manufacture radiopharmaceuticals, smoke alarms, radium watch dials, and other consumer goods produce low-level waste consisting of machinery parts, plastics, and organic solvents. Research establishments, universities, and hospitals also produce radioactive waste materials.

High-level waste can remain radioactive for 10,000 years and more; plutonium stays dangerously radioactive for 240,000 years. It consists primarily of spent fuel assemblies of nuclear power reactors—termed "civilian waste"—and waste generated as a by-product of the manufacture of nuclear weapons, or "military waste." The volume of civilian waste alone is not only great but increasing rapidly, because approximately one-third of a reactor's rods need to be disposed of every year.

By 1995, more than 70,000 spent-fuel assemblies were being stored in the containment pools of America's commercial nuclear power reactors, awaiting more permanent disposition. Some 6000 more are added annually. "Spent fuel" is a misleading term: the assemblies are removed from commercial reactors not because their radiation is spent, but because they have become too radioactive for further use. The assemblies will remain radioactively "hot" for thousands of years.

Unfortunately, no satisfactory method for disposing of any hazardous waste has yet been devised (see "Yucca Mountain"). Some wastes have been sealed in protective tanks and dumped at sea, a practice that recently has been banned worldwide. Cardboard boxes containing wastes contaminated with plutonium have been rototilled into the soil, on the assumption that the earth would dilute and absorb the radioactivity. Much low-level radioactive waste has been placed in tanks and buried in the earth at 13 sites operated by the U.S. Department of Energy and three sites run by private firms. Millions of cubic feet of high-level military waste are temporarily stored in underground tanks at four sites: Hanford, Washington; Savannah River, South Carolina; Idaho Falls, Idaho; and West Valley, New York (Figure 5.34). Several of these storage areas have experienced leakages, with seepage of waste into the surrounding soil and groundwater.

In the absence of an effective national program of radioactive waste storage, a 1995 decision by a federal Court of Appeals has made more than 70 communities near nuclear generating plants the long-term repositories for spent nuclear fuel by permitting utilities to store radioactive wastes indefinitely at their nuclear power plants—without holding public hearings or conducting any environmental assessment. By 1998, more than 40,000 tons of spent nuclear fuel had already accumulated at 71 civilian nuclear power plants in 34 states; the amount is expected to double in the next 20 years. Because the containment pools at those plants, originally intended to store wastes only temporarily, are becoming full, the government is now encouraging construction of aboveground sites where the high-level waste will be stored in steel barrels encased in concrete shells.

Another method of waste disposal, used for chemicals as well as for radioactive wastes, has been to inject them into deep steel- or concrete-lined wells. Because underground injection poses a threat of groundwater contamination and may contribute to earth tremors, the injection of wastes into or above strata that contain aquifers is being phased out.

Because low-level waste is generated by so many sources, its disposal is particularly difficult to control. Evidence indicates that much of it has been placed in landfills, often the local municipal dump, where the waste chemicals may leach through the soil and into the groundwater. The EPA estimates that about 85,000 known or suspected hazardous waste sites are scattered across the United States. More than 1300 of the worst are listed as Superfund sites, named afer the 1980 law created to clean up the country's worst toxic dumps. Sixty-five million Americans—one in four—live within 6.4 kilometers (4 mi) of a Superfund site, each a potential ecological disaster.

Solid waste will never cease to be a problem, but its impact on the environment can be lessened by reducing the volume of waste that is generated, eliminating or reducing the production of toxic residues, halting irresponsible dumping, and finding ways to reuse the resources that waste contains. Until then, current methods of waste disposal will continue to pollute soil, air, and water.

FIGURE 5.33 Wastepickers scavenge the Smokey Mountain dump in Manila, the Philippines, looking for items to resell. Metal, glass, plastic, paper, and cloth are all candidates. Wastepicking is an important source of income for many poor families, but the pickers, many of them women and children, usually work long hours in unhealthful conditions.
Nigel Dickinson/Tony Stone Worldwide.

Geography and Public Policy

YUCCA MOUNTAIN

If the U.S. federal government has its way, a long, low ridge in the basin-and-range region of Nevada will become America's first permanent repository for the deadly radioactive waste that nuclear power plants generate. If the opponents of the project have their way, however, Yucca Mountain will become a symbol of society's inability to solve a basic problem posed by nuclear power production: where to dispose of the used fuel. At present, no permanent disposal site for radioactive waste exists anywhere in the world.

In 1982, Congress ordered the Department of Energy (DOE) to construct by 1998 a permanent repository for the spent fuel of civilian nuclear power plants as well as vast quantities of waste from the production of nuclear weapons. Yucca Mountain in southern Nevada was selected as the site for this high-level waste facility, which is intended to safely store wastes for 10,000 years, until radioactive decay has rendered them less hazardous than they are today. Most of the waste would be in the form of radioactive fuel pellets sealed in metal rods; these would be encased in extremely strong glass and placed in steel canisters entombed in chambers 300 meters (1000 ft) below the Nevada desert. The steel containers would corrode in one or two centuries, after which the volcanic rock of the mountain would be responsible for containing the radioactivity.

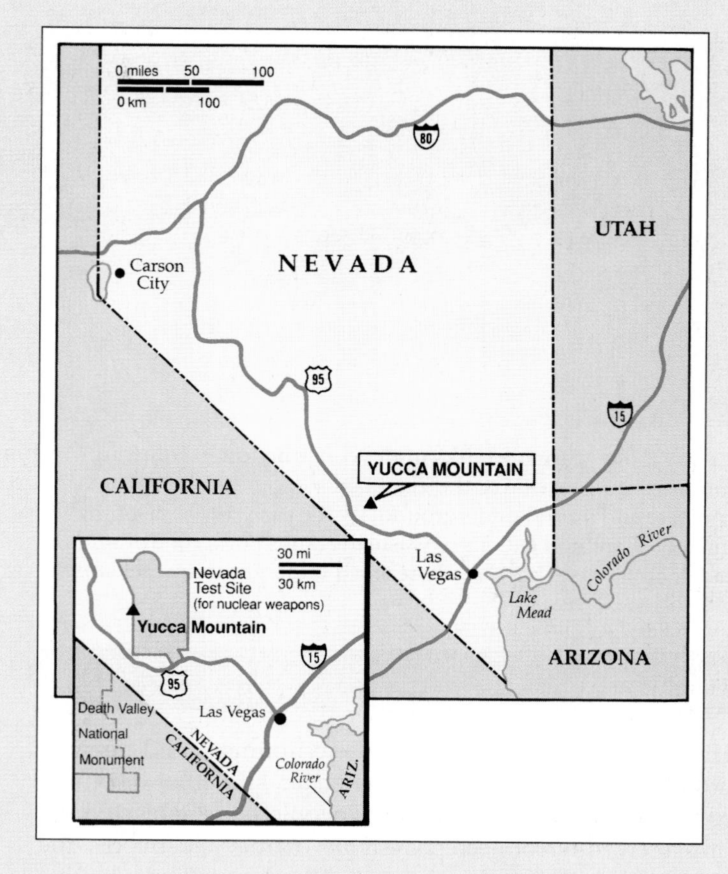

Plans now call for opening the repository in 2010, but many doubt that the Yucca Mountain facility will ever be completed and licensed. Three areas of concern have emerged. First, the area is vulnerable to both volcanic and earthquake activity, which could cause groundwater to well up suddenly and flood the repository. Yucca Mountain itself was formed from volcanic eruptions that occurred about 12–15 million years ago; some geologists are concerned that a new volcano could erupt within the mountain. Seven small cinder cones in the immediate area have erupted in recent times, the latest just 10,000 years ago. In addition, a number of seismic faults lie close to Yucca Mountain. One, the Ghost Dance Fault, runs right through the depository site. The epicenter of the 1992 earthquake at Little Skull Mountain was only 19 kilometers (12 mi) from the proposed dump site.

Second, rainwater percolating down through the mountain could penetrate the vaults holding the waste. Over centuries, water could dissolve the waste itself, and the resulting toxic brew could seep down into the water table.

Finally, the Yucca Mountain site lies between the Nevada Test Site, which the DOE uses as a nuclear bomb testing range, and the Nellis Air Force Base Bombing and Gunnery Range. Questions have been raised about the wisdom of locating a waste repository just a few kilometers from areas subject to underground explosions or aerial bombardment.

QUESTIONS TO CONSIDER

1. What are the advantages and disadvantages of countries operating nuclear power plants when no system for disposing of the hazardous wastes they generate exists?
2. Comment on the paradox that plutonium remains dangerously radioactive for 240,000 years, yet it might be stored in a repository that is intended to safely hold wastes for only 10,000 years.
3. Considering the uncertainties that would attend the irreversible underground entombment of high-level waste, do you think the government should pursue aboveground storage in a form that would allow for the continuous monitoring and retrieval of the wastes? Why or why not?
4. High-level waste is produced at relatively few sites in the United States. Low-level waste is much more ubiquitous; it is produced at more than 2000 sites in California alone. Which do you think represents the greater hazard, high- or low-level waste?

FIGURE 5.34 **Storage tanks under construction in Hanford, Washington.** Built to contain high-level radioactive wastes from the Hanford nuclear reservation, the giant tanks—some with a capacity of up to 1 million gallons—are shown here before they were encased in concrete and buried underground. By the early 1990s, 66 of the 177 underground tanks were already known to be leaking. Of the approximately 55 million gallons of waste the tanks hold, about 1 million gallons of liquids have seeped into the soil, raising the fear that the radioactive waste has already reached underground water supplies and is flowing toward the Columbia River.
Illustrator's Stock Photos.

SUMMARY

Humans are part of the natural environment and depend, literally, for their lives on the water, air, and other resources contained in the biosphere. But people have subjected the intricately interconnected systems of that biosphere—the troposphere, the hydrosphere, and the lithosphere—to profound and frequently unwittingly destructive alteration. All human activities have effects on the environment, effects that are complex and never isolated. An external action that impinges on any part of the web of nature inevitably triggers chain reactions, the ultimate consequences of which appear never to be fully anticipated. The simple desire to suppress summer dust in the town of Times Beach, Missouri, led in the end to the demise of the community.

Efforts to control the supply of water alter both the quantity and quality of water in streams, and structures such as dams and reservoirs often have unintended side effects. In many parts of the world, increased demand for fresh water has led to a lack of adequate supplies. Pollutants associated with agriculture, industry, and other activities have degraded the quality of freshwater supplies, although regulatory efforts have brought about major improvements in some areas in recent years.

Combustion of fossil fuels has contributed to serious problems of air pollution. Some manifestations of that pollution, such as acid rain and depletion of the ozone layer, are matters of global concern. If the theory of the greenhouse effect proves to be valid, air pollution will alter normal patterns of temperature and precipitation sometime during the next century.

Activities such as agriculture and mining have long helped to shape local landscapes, producing a variety of landforms, and have also contributed to degradation of air, water, and soil. In this century, pressure exerted by world population growth and economic expansion have accelerated deforestation, soil erosion, and desertification over much of the world.

People affect other living things—plants and animals—by importing them to areas where they did not previously exist, disrupting their habitats, hunting, and using biocides to eradicate them. The greatest human impact is occurring in the tropical rain forests, where agriculture, industrialization, and urbanization are contributing to the extinction of at least 1000 species a year.

Finally, all the common methods of disposing of the wastes that humans produce release contaminants into the surrounding environment, providing further evidence of the fact that people cannot manipulate, distort, pollute, or destroy any part of the ecosystem without diminishing its quality or disrupting its structure.

KEY WORDS

acid rain 160
biological magnification 178
biosphere 147
channelization 151
chlorofluorocarbons (CFCs) 162

desertification 171
ecology 147
ecosystem 148
environmental pollution 151
eutrophication 153
food chain 148
global warming 165
greenhouse effect 164
hazardous waste 181
hydrologic cycle 149
ozone 161
ozone layer 162
photochemical smog 161
polychlorinated biphenyls (PCBs) 156
salinization 174
soil erosion 170
subsidence 168
thermal pollution 156
troposphere 158

For Review and Consideration

1. Sketch and label a diagram of the *biosphere.* Briefly indicate the content of its component parts. Is that content permanent and unchanging? Explain.
2. How are the concepts of *ecosystem, niche,* and *food chain* related? How does each add to our understanding of the "web of nature"?
3. Draw a diagram of or briefly describe the *hydrologic cycle.* How do people have an impact on that cycle? What effect does urbanization have on it?
4. Is all environmental pollution the result of human action? When can we say that pollution of a part of the biosphere has occurred?
5. Describe the chief sources of water pollution. What steps have the United States and other countries taken to control water pollution?
6. What factors affect the type and degree of air pollution found at a place? What is *acid rain,* and where is it a problem? Describe the relationship of *ozone* to *photochemical smog.* Why has the ozone layer been depleted?
7. What causes the greenhouse effect? What impact might it have on the environment?
8. What kinds of landforms has excavation produced? Dumping? What are the chief causes and effects of subsidence?
9. How have people diminished the amount and productivity of soil? What types of areas are particularly subject to desertification? Why has soil erosion accelerated in recent decades?
10. Briefly describe the chief ways that humans affect plant and animal life. What is meant by *biological magnification?* Why may the use of biocides be self-defeating?
11. What methods do communities use to dispose of solid waste? What ecological problems does solid-waste disposal present? How does the government define *hazardous waste,* and how is it disposed of?

Selected References

Brown, Lester R., et al. *State of the World.* Washington, D.C.: Worldwatch Institute. Annual.

Calvin, William H. "The Great Climate Flip-flop." *Atlantic Monthly* 281, no. 1 (January 1998):47–64.

Cunningham, William P., and Barbara W. Saigo. *Environmental Science: A Global Concern.* 4th ed. Dubuque, Iowa: Wm. C. Brown Publishers, 1997.

Enger, Eldon D., and Bradley F. Smith. *Environmental Science: A Study of Interrelationships.* 5th ed. Dubuque, Iowa: Wm. C. Brown Publishers, 1995.

Falkenmark, Malin, and Carl Widstrand. "Population and Water Resources: A Delicate Balance." *Population Bulletin* 47, no. 1. Washington, D.C.: Population Reference Bureau, 1992.

Goudie, Andrew, and Heather Viles. *The Earth Transformed: An Introduction to the Human Impact on the Environment.* Cambridge, Mass.: Blackwell, 1997.

Hulme, Mick, and Mick Kelly. "Desertification and Climate Change." *Environment* 35, no. 6 (July/August 1993):4–11, 39–45.

Kaufman, Donald G., and Cecilia M. Franz. *Biosphere 2000: Protecting Our Global Environment.* New York: Harper Collins, 1993.

Kemp, David D. *Global Environmental Issues.* 2d ed. New York: Routledge, 1995.

Middleton, Nick. *The Global Casino: An Introduction to Environmental Issues.* New York: John Wiley & Sons, 1995.

Miller, G. Tyler, Jr. *Living in the Environment.* 6th ed. Belmont, Calif.: Wadsworth Publishing Co., 1990.

Park, Chris C. *The Environment.* London and New York: Routledge, 1997.

Pickering, Kevin T., and Lewis A. Owen. *An Introduction to Global Environmental Issues.* 2d ed. New York: Routledge, 1997.

Postel, Sandra. *Last Oasis: Facing Water Scarcity.* New York: W. W. Norton, 1992.

Roberts, Neil, ed. *The Changing Global Environment.* Cambridge, Mass.: Blackwell, 1994.

Scientific American 261, no. 3 (September 1989). Special issue: *Managing Planet Earth.*

Simmons, I. G. *Changing the Face of the Earth: Culture, Environment, History.* Cambridge, Mass.: Blackwell, 1996.

Thomas, William, ed. *Man's Role in Changing the Face of the Earth.* Chicago: University of Chicago Press, 1956.

Vogel, S. "Has Global Warming Begun?" *Earth* 4, no. 7 (December 1995): 24–34.

"Water: The Power, Promise, and Turmoil of North America's Fresh Water." Special edition of *National Geographic* 184, no. 5A (November 1993).

Wellburn, Alan. *Air Pollution and Acid Rain.* 2d ed. Reading, Mass.: Addison, Wesley, Longman, 1994.

Whyte, Ian D. *Climatic Change and Human Society.* New York: John Wiley & Sons, 1996.

World Bank. *Monitoring Environmental Progress: A Report on Work in Progress.* Washington, D.C.: The World Bank, 1995.

World Resources Institute/International Institute for Environment and Development. *World Resources.* Annual or biennial. New York: Oxford University Press.

Worldwatch Institute. *Worldwatch Papers* issued several times a year provide in-depth analysis of a variety of environmental issues. Washington, D.C.

Don't forget about Dushkin's *Annual Editions Online: Geography* at http://www.dushkin.com/aeonline/. See preface for details.

PART TWO

The Culture–Environment Tradition

The Crow country. The Great Spirit put it exactly in the right place; while you are in it, you fare well; whenever you get out of it, whichever way you travel, you fare worse. . . . The Crow country is in exactly the right place. It has snowy mountains and sunny plains; all kinds of climates and good things for every season. When the summer heats scorch the prairies, you can draw up under the mountains, where the air is sweet and cool. . . . In the autumn when your horses are fat and strong from the mountain pastures, you can go down on the plains and hunt the buffalo or trap beaver on the streams. And when winter comes on, you can take shelter in the woody bottoms along the rivers.

The Crow country is exactly in the right place. Everything good is found there. There is no country like the Crow country.

Such was the opinion of Arapoosh, Chief of the Crows, speaking of the Big Horn basin country of Wyoming in the early 19th century. In the 1860s, Captain Raynolds reported to the Secretary of War that the basin was "repelling in all its characteristics, surrounded on all sides by mountain ridges [and presenting] but few agricultural advantages."

In the three chapters of Part I, our primary concern was with the physical landscape. But while the physical environment may be described by process and data, it takes on human meaning only through the filter of culture. Arapoosh and Raynolds viewed the same landscape, but from the standpoints of their separate cultures and conditionings. Culture is like a piece of tinted glass, affecting and distorting our view of the earth. Culture conditions the way people think about the land, the way they use and alter the land, and the way they interact with one another upon the land.

Such conditioning is the focus of the culture–environment tradition of geography, a tradition still concerned with the landscape but not in the physical science sense of Part I of this book. In Part II, humans are the focus. Of course, the physical environment is always in our minds as we develop the notion of cultural difference and reality. Our landscapes, however, take on an added dimension and become human rather than purely physical.

The four chapters in this part of our study, therefore, concentrate upon the "people" portion of geography's environment–culture–people relationship. In Chapter 6, "Population Geography," we start with the basics of human populations in their numbers, compositions, distributions, growth trends, and the pressures they exert upon the resources of the lands they occupy. These quantitative and distributional aspects of peoples are important current concerns, as frequent popular reference to a "population explosion," public debate about legal and undocumented immigration, and speculation about population growth and food availability attest. More fundamentally, of course, numbers and locations of people are the essential background to all other understandings of human geography.

People are, however, more than numbers in a global counting parlor. They are, separately, individuals who think, react, and behave in response to the physical and social environments they occupy. In turn, those thoughts, actions, and responses are strongly conditioned by the standards and structures of the cultures to which the individuals belong. The world is a mosaic of culture groups and human landscapes that invite geographic study. Chapter 7, "Cultural Geography," introduces that study by examining the components and subsystems of culture, the ways in which culture changes, and the key variables in defining a culture and in producing variations in culture from place to place.

The world patterns of those variations are not permanent. They represent merely the current consequences of past movements of people, export and adaptation of ideas and technologies, and histories of conquest and settlement. They represent, that is, the results of past and present interaction of people and their cultures within and across earth space. Regularities of human spatial behavior and the factors that account for the manner in which people use earth area are the subject of Chapter 8, "Geography of Spatial

—*Continued on next page*

The annual trek downslope from summer pastures in the Karakoram Range, Pakistan.

Continued from previous page—

Behavior." The individual and group behaviors of humans create the cultural landscape in all its variations, including those of production, resource utilization, and urban settlement that will be the topics of Part III of this book.

A controlling setting for those later cultural considerations is explored in Chapter 9, "Political Geography." Political systems and processes strongly influence the form and distribution of many elements of culture. Economic and transportation systems coincide with national boundaries. Political regulations, whether detailed zoning codes or broad environmental protection laws, have a marked effect on the economic and urban landscape. In some countries, even such apparently nonpolitical matters as religion, literature, music, and the fine arts may be conditioned by governmental support for certain forms of expression and rejection and prohibition of others. And certainly governmental direction and political ideology are essential elements in the establishment and maintenance of the economic systems that determine resource exploitation and livelihood patterns of people across the world, systems subject to change as governments fall or as supranational political associations—the European Union, for example—impose new patterns of behavior on established national cultures.

Our theme in Part II, therefore, is people and the collective and personal cultural landscapes they create or envision. Let us begin with people themselves before we move on to consider their cultures and behaviors.

CHAPTER 6

POPULATION GEOGRAPHY

Shoppers throng the busy main street of the Ginza district, Tokyo, Japan.

Zero, possibly even negative [population] growth" was the 1972 slogan proposed by the prime minister of Singapore, an island country in Southeast Asia. His nation's population, which stood at 1 million at the end of World War II (1945), had doubled by the mid-1960s. To avoid the overpopulation he foresaw, the government decreed "Boy or girl, two is enough" and refused maternity leaves and access to health insurance for third or subsequent births. Abortion and sterilization were legalized, and children born fourth or later in a family were to be discriminated against in school admissions policy. In response, birth rates by the mid-1980s fell to below the level necessary to replace the population, and abortions were terminating more than one-third of all pregnancies.

"At least two. Better three. Four if you can afford it" was the national slogan proposed by that same prime minister in 1986, reflecting fears that the stringencies of the earlier campaign had gone too far. From concern that overpopulation would doom the country to perpetual Third World poverty, Prime Minister Lee Kuan Yew was moved to worry that population limitation would deprive it of the growth potential and national strength implicit in a youthful, educated workforce adequate to replace and support the present aging population. His 1990 national budget provided for sizable long-term tax rebates for second children born to mothers under 28. Not certain that financial inducements alone would suffice to increase population, the Singapore government annually renewed its offer to take 100,000 Hong Kong Chinese who might choose to leave when China took over that territory in 1997.

The policy reversal in Singapore reflects an inflexible population reality: The structure of the present determines the content of the future. The size, characteristics, growth trends, and migrations of today's populations help shape the well-being of peoples yet unborn but whose numbers and distributions are now being determined. The numbers, age, and sex distribution of people; patterns and trends in their fertility and mortality; their density of settlement and rate of growth all affect and are affected by the social, political, and economic organization of a society. Through them, we begin to understand how the people in a given area live, how they may interact with one another, how they use the land, what pressure on resources exists, and what the future may bring.

Population geography provides the background tools and understandings of those interests. It focuses on the number, composition, and distribution of human beings in relation to variations in the conditions of earth space. It differs from **demography**, the statistical study of human population, in its concern with *spatial* analysis—the relationship of numbers to area. Regional circumstances of resource base, type of economic development, level of living, food supply, and conditions of health and well-being are basic to geography's population concerns. They are, as well, fundamental expressions of the human–environmental relationships that are the substance of all human geographic inquiry.

Population Growth

Some time during the autumn of 1999, a human birth raised the earth's population to 6.0 billion people. In 1987 the 5 billion plateau was reached. That is, over the 12 years between those two dates, the world's population grew on average by about 84 million people annually, or some 230,000 per day. Although global fertility and growth rates have been declining in recent years, the United Nations still projects that the world will contain almost 9 billion inhabitants in 2050; in 1950 it had 2.5 billion. Demographers assume that world population will stabilize near 11 billion around the year 2200, with over 95% of the growth occurring in countries now considered "developing" (Figure 6.1). We will return to these projections, and the difficulties inherent in making them, later in this chapter.

Just what is implied by numbers in the millions and billions? With what can we equate the 1998 population of Gabon in Africa (about 1.25 million) or of China (about 1.25 billion)? Unless we have some grasp of their scale and meaning, our understanding of the data and data manipulations of the population geographer can at best be superficial. It is difficult to appreciate a number as vast as 1 million or 1 billion, and the great distinction between them. Some examples offered by the Population Reference Bureau may help in visualizing their immensity and implications.

- A 2.5 centimeter (1-inch) stack of U.S. paper currency contains 233 bills. If you had a *million* dollars in thousand-dollar bills, the stack would be 11 centimeters (4.3 inches) high. If you had a *billion* dollars in thousand-dollar bills, your pile of money would reach 109 meters (357 feet)—about the length of a football field.
- You had lived a *million* seconds when you were 11.6 days old. You won't be a *billion* seconds old until you are 31.7 years of age.
- The supersonic airplane, the Concorde, could theoretically circle the globe in only 18.5 hours at its cruising speed of 2150 kilometers (1340 mi) per hour. It would take 31 days for a passenger to journey a *million* miles on the Concorde, while a trip of a *billion* miles would last 85 years.

The implications of the present numbers and the potential increases in population are of vital current social, political, and ecological concern. Population numbers were much smaller some 11,000 years ago when continental glaciers began their retreat, people spread to formerly unoccu-

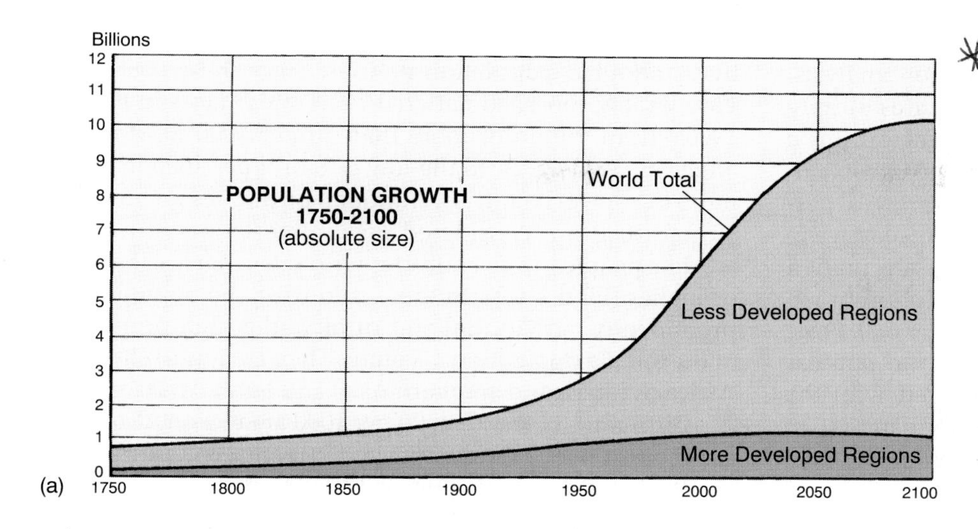

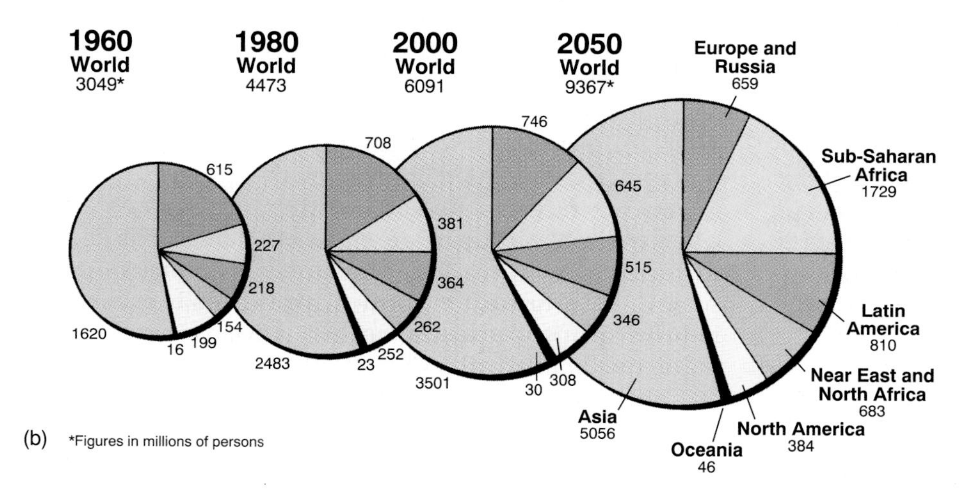

FIGURE 6.1 World population numbers and projections. (a) After two centuries of slow growth, world population began explosive expansion after World War II. United Nations demographers project a global population of nearly 9.4 billion in 2050. The total may rise to almost 11 billion by the end of the 21st century. (b) The greatest numerical growth during the 1990s occurred in Asia, and sub-Saharan Africa had the highest percentage increase. Differential growth rates will alter the relative A.D. 2000 rankings of regions by the year 2050, with sub-Saharan Africa replacing Europe/Russia in second position and Latin America moving from fourth to third. "Europe and Russia" includes the Eastern European and Caucasian states that were republics of the former Soviet Union and all of Russia, including Siberia.

Sources: (a) Estimates from Population Reference Bureau and United Nations Population Fund; (b) Based on United Nations, U.S. Bureau of the Census, and World Bank projections.

pied portions of the globe, and human experimentation with food sources initiated the Agricultural Revolution. The 5 or 10 million people who then constituted all of humanity obviously had considerable potential to expand their numbers. In retrospect, we see that the natural resource base of the earth had a population-supporting capacity far in excess of the pressures exerted on it by early hunting and gathering groups.

Some observers maintain that despite present numbers or even those we can reasonably anticipate for the future, the adaptive and exploitive ingenuity of humans is in no danger of being taxed. Others, however, compare the earth to a self-contained spaceship and declare with chilling conviction that a finite vessel cannot bear an ever-increasing number of passengers. They point to recurring problems of malnutrition and starvation (though these are realistically more a matter of failures of distribution than of inability to produce enough foodstuffs worldwide). They cite dangerous conditions of air and water pollution, the loss of forest and farmland, the apparent nearing exhaustion of many minerals and fossil fuels, and other evidences of strains on world resources as foretelling the discernible outer limits of population growth.

Why are we now at the start of the 21st century confronted with what seems to many an unmanageable problem—the apparently unending tendency of humankind to increase in numbers? On a worldwide basis, populations grow only one way: The number of births in a given period exceeds the number of deaths. Ignoring for the moment regional population changes resulting from migration, we can conclude that the observed and projected dramatic increases in population must result from the failure of natural controls to limit the number of births or to increase the number of deaths, or from the success of human ingenuity in circumventing such controls when they exist. The implications of these observations will become clearer after we define some terms important in the study of world population and explore their significance.

SOME POPULATION DEFINITIONS

Demographers employ a wide range of measures of population composition and trends, though all their calculations start with a count of events: of individuals in the population, of births, deaths, marriages, and so on. To those basic counts, demographers bring refinements that make the figures more meaningful and useful in population analysis. Among them are *rates* and *cohort measures.* **Rates** simply record the frequency of occurrence of an event during a given time frame for a designated population—for example, the marriage rate as the number of marriages performed per 1000 population in the United States last year. **Cohort** measures refer data to a population group unified by a specified common characteristic—the age cohort of 1–5 years, perhaps, or the college class of 2002 (Figure 6.2). Basic numbers and rates useful in the analysis of world population and population trends have been reprinted with the permission of the Population Reference Bureau as an appendix to this book. Examination of them will help illustrate the discussion that follows.

Birth Rates

The **crude birth rate (CBR)**, often referred to simply as the birth rate, is the annual number of live births per 1000 population. It is "crude" because it relates births to total population without regard to the age or sex composition of that population. A country with a population of 2 million and with 40,000 births a year would have a crude birth rate of 20 per 1000.

$$\frac{40{,}000}{2{,}000{,}000} = 20 \text{ per } 1000$$

The birth rate of a country is, of course, strongly influenced by the age and sex structure of its population, by the customs and family size expectations of its inhabitants, and by its adopted population policies. Since these conditions vary widely, recorded national birth rates vary—in the late 1990s, from a high of more than 50 per 1000 in Mali and Niger in West Africa to the low of 8 or 9 per 1000 in ten or more European countries. Although birth rates greater than 30 per 1000 are considered *high,* almost one-fifth of the world's people live in countries with rates that are that high or higher (Figure 6.3). In these countries, the population is prominently agricultural and rural, and a high proportion of the female population is young. They are found chiefly in Africa, western and southern Asia, and Latin America.

Birth rates of less than 20 per 1000 are reckoned *low* and are characteristic of industrialized, urbanized countries. All European countries including Russia, as well as Anglo America, Japan, Australia, and New Zealand have low rates as, importantly, do an increasing number of developing states such as China (see "China's Way—and Others") that have adopted stringent family planning programs. *Transitional* birth rates (between 20 and 30 per 1000) characterize some, mainly smaller, "developing" countries, though giant India entered that group in 1994.

As the recent population histories of Singapore and China indicate, birth rates are subject to change. The decline to current low birth rates of European countries and of some of the areas that they colonized is usually ascribed to industrialization, urbanization, and in recent years, maturing populations. While restrictive family planning policies in China rapidly reduced the birth rate from over 33 per 1000 in 1970 to 18 per 1000 in 1986, industrializing Japan experienced a comparable 15-point decline in the decade

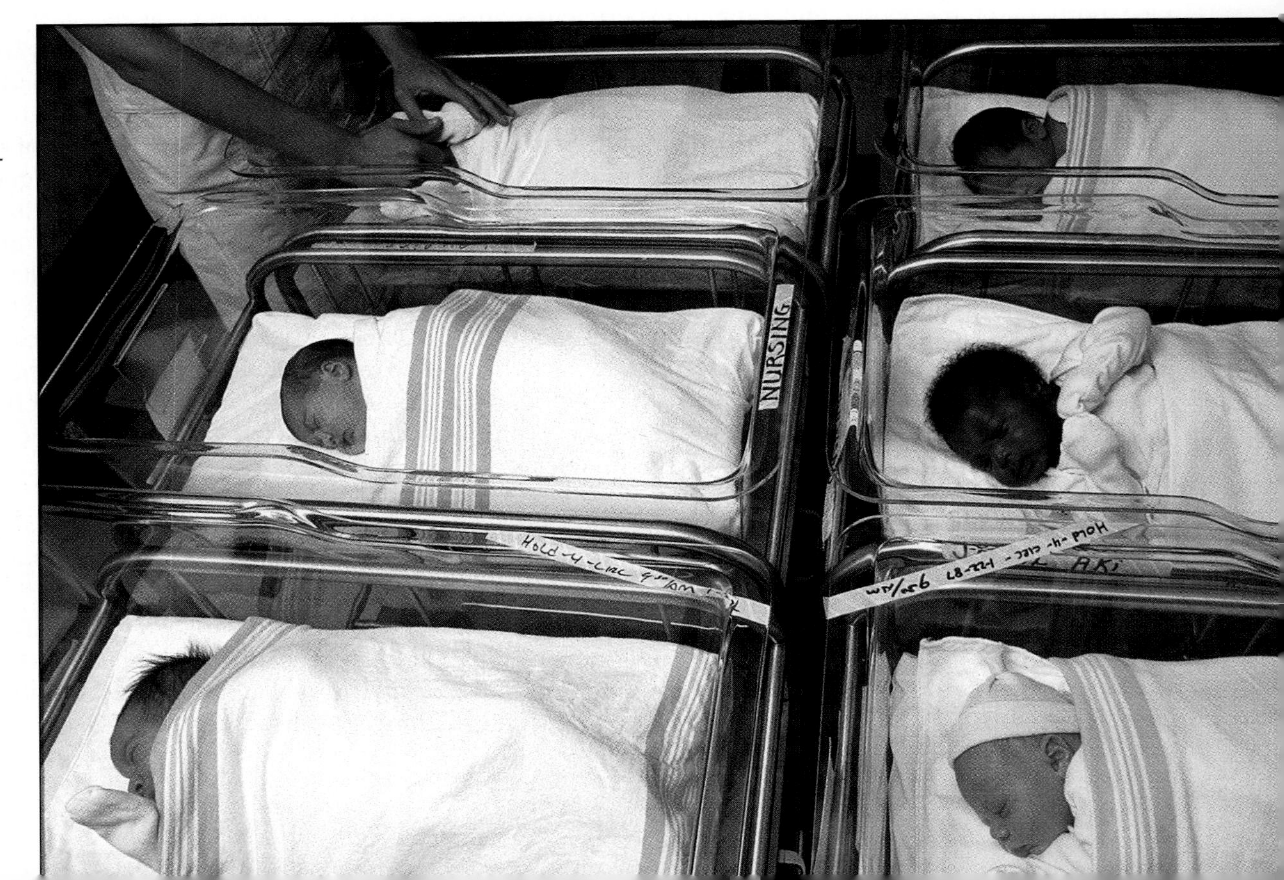

FIGURE 6.2 Whatever their differences may be by race, sex, or ethnicity, these babies will forever be clustered demographically into a single *birth cohort.*
©Herb Snitzer/Stock, Boston.

China's Way—and Others

An ever larger population is "a good thing," Chairman Mao announced in 1965 when China's birth rate was 37 per 1000 and population totaled 540 million. At Mao's death in 1976, numbers reached 852 million. During the 1970s, when it became evident that population growth was consuming more than half of the annual increase in the country's gross domestic product, China introduced a well-publicized campaign advocating the "two-child family" and providing services, including abortions, supporting that program. In response, China's growth rate dropped to 15.7 per 1000 by the late 1970s.

"One couple, one child" became the slogan of a new and more vigorous population control drive launched in 1979, backed by both incentives and penalties to assure its success in China's tightly controlled society. Late marriages were encouraged; free contraceptives, cash awards, abortions, and sterilizations were provided to families limited to a single child. Penalties, including steep fines, were levied for second births. At the campaign's height in 1983, the government ordered the sterilization of either husband or wife for couples with more than one child. Tragically, infanticide—particularly the exposure or murder of female babies—was a reported means both of conforming to a one child limit and of increasing the chances that the one child would be male. By 1986, China's officially-reported growth rate had fallen to 1%, far below the 2.4% then registered among the rest of the world's less developed countries. Unofficially, it is reported that the one-child policy was effectively dropped in 1984 in rural areas where 70% of Chinese population still resides. In 1996, President Jiang spoke of "reestablishing" the one-child restriction, and Chinese fertility rates now are most likely well above the officially cited figures of between 1.8 and 2.0.

Concerned with their own growing numbers, many developing countries have introduced their own less extreme programs of family planning, stressing access to contraception and sterilization. International agencies have encouraged these programs, buoyed by such presumed success as the 21% fall in fertility rates in Bangladesh from 1970 to 1990 as the proportion of married women of reproductive age using contraceptives rose from 3% to 40% under intensive family planning encouragement and frequent adviser visits. The costs per birth averted, however, were reckoned at an unsupportable $180 in 1987, about 120% of the country's per capita gross domestic product.

Research suggests that fertility falls because women decide they want smaller families, not because they have unmet needs for contraceptive advice and devices. Nineteenth-century northern Europeans without the aid of science, it is observed, had lower fertility rates than their counterparts today in middle income countries. With some convincing evidence, improved women's education has been proposed as a surer way to reduce fertility than either encouraged contraception or China's coercive efforts. Studies from individual countries indicate that one year of female schooling can reduce the fertility rate by between 5% and 10%. Yet the fertility rate of uneducated Thai women is only two-thirds that of Ugandan women with secondary education. Obviously, the demand for babies is not solely a function of ignorance.

Instead, that demand seems closely tied to the use value placed on children by poor families in some parts of the developing world. Where those families share in such communal resources as firewood, animal fodder, grazing land, fish, and the like, the more of those collective resources that can be converted to private family property and use, the better off is the family. Indeed, the more communal resources that are available for "capture," the greater are the incentives for a household to have more children to appropriate them. Some population economists conclude that only when population numbers increase to the point of total conversion of communal resources to private property—and children have to be supported and educated rather than employed—will poor families in developing countries want fewer children. If so, coercion, contraception, and education may be less effective as checks on fertility than the economic consequence of population increase itself.

Owen Franken/Corbis.

1948–1958 with little governmental intervention. Indeed, the stage of economic development appears closely related to variations in birth rates among countries, although rigorous testing of this relationship proves it to be imperfect (Figure 6.3). As a group, the more developed states of the world showed a crude birth rate of 11 per 1000 near the end of the 20th century; less developed countries (excluding China) registered nearly 30 per 1000.

Religious and political beliefs can also affect birth rates. The convictions of many Roman Catholics and Muslims that their religion forbids the use of artificial birth control techniques often lead to high birth rates among believers. However, dominantly Catholic Italy shows Europe's lowest birth rate, and Islam itself does not prohibit contraception. Similarly, some European governments—concerned about birth rates too low to sustain present population levels—subsidize births in an attempt to raise those rates. Regional variations in projected percentage contributions to world population growth are summarized in Figure 6.4.

Fertility Rates

Crude birth rates may display such regional variability because of differences in age and sex composition or disparities in births among the reproductive-age, rather than total, population. **Total fertility rate** (TFR) is a more accurate statement than the birth rate in showing the amount of reproduction in the population (Figure 6.5). The TFR tells us the average number of children that would be born to each woman if, during her childbearing years, she bore children at the current year's rate for women that age. The fertility rate minimizes the effects of fluctuation in the population structure and is thus a more reliable figure for regional comparative and predictive purposes than the crude birth rate.

A total fertility rate of 2.1 is necessary just to replace present population. On a worldwide basis, the TFR at the end of the 1990s was just below 3.0. The more developed countries recorded a 1.6 rate, while less developed states (excluding China) had a collective TFR of slightly less than 4.0, down from 5.0 in the mid-1980s. Indeed, the fertility rates for many less developed countries and regions have dropped dramatically since the early 1960s (Table 6.1). China's decrease from a TFR of 5.4 births per woman in the period 1960–1965 to (officially) below 2.0 in the early 1990s was most impressive and demographically very significant because of China's status as the world's foremost state in population.

Although fertility rates are still 5 or over per woman in much of sub-Saharan Africa and in many smaller Asian countries (see Appendix), fertility has been declining recently in all parts of the world, although not uniformly among all countries. Collectively, however, since 1980 Asia,

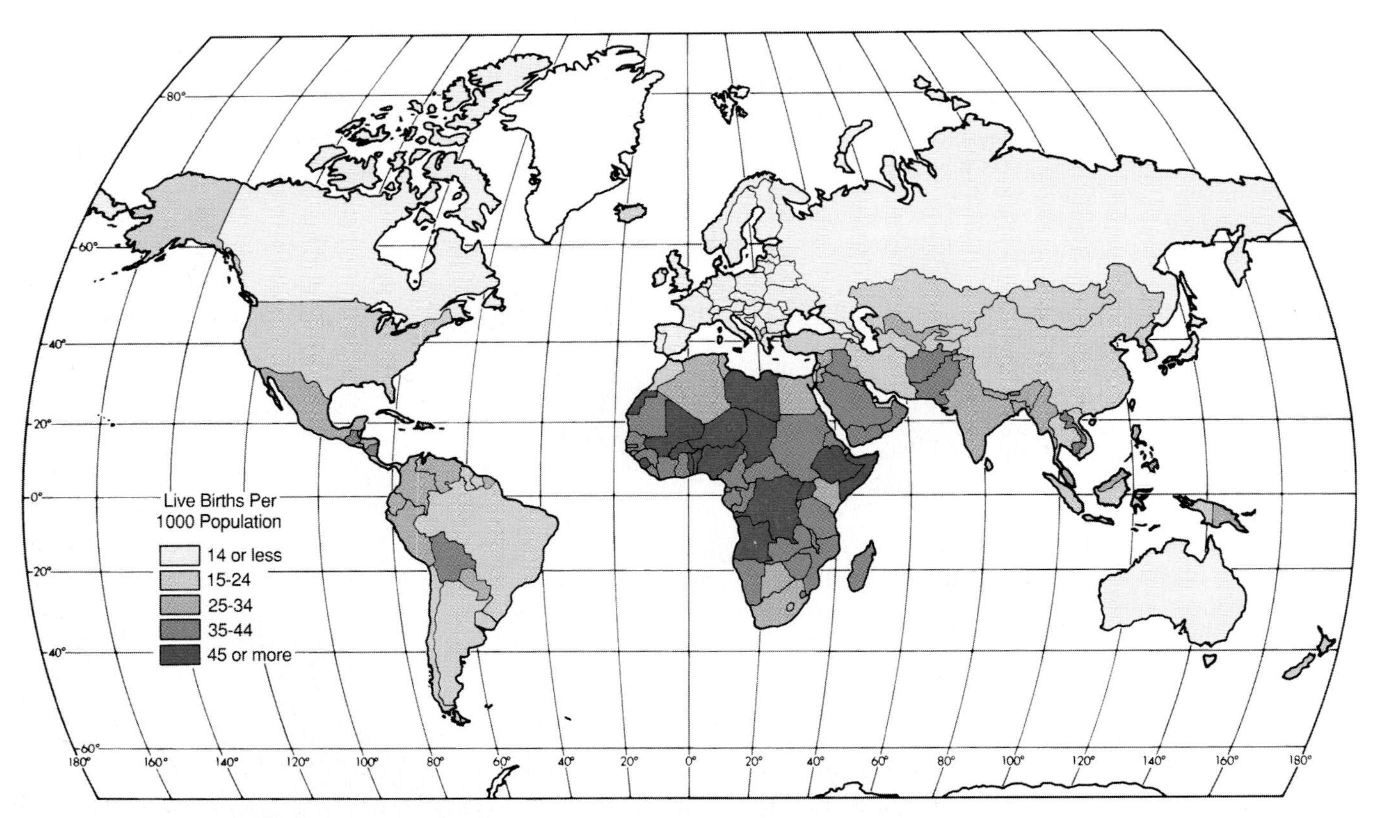

FIGURE 6.3 **Crude birth rates.** The map suggests a degree of precision that is misleading in the absence of reliable, universal registration of births. The pattern shown serves, however, as a generally useful summary of comparative reproduction patterns if class divisions are not taken too literally. Reported or estimated population data vary annually, so this and other population maps may not agree in all details with the figures recorded in the Appendix.

Source: Data from Population Reference Bureau.

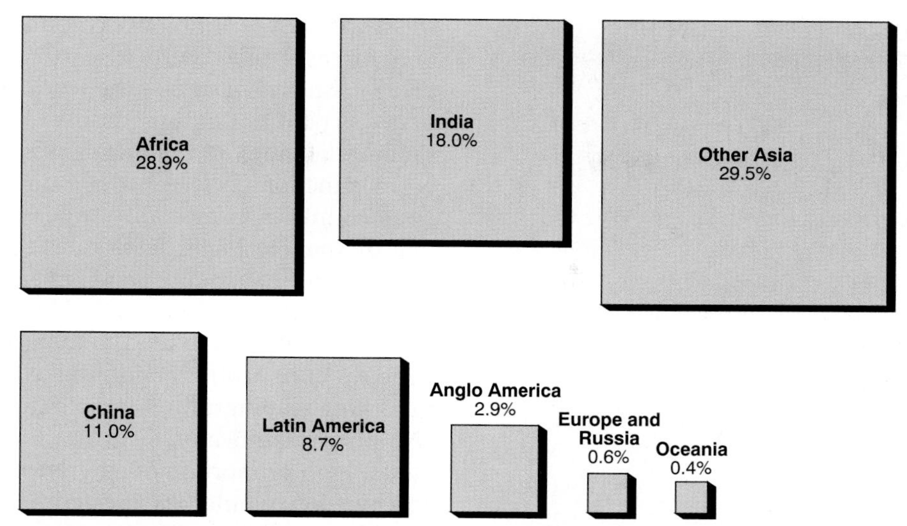

FIGURE 6.4 **Projected percentage contributions to world population growth, by region, 2000–2015.** Birth rate changes affecting differently sized regional populations are altering the world pattern of population increase. Africa, containing 14% of world population in 1998, will probably account for 29% of total world increase between 2000 and 2015. Between 1965 and 1975, China's contribution to world growth was 2.5 times that of Africa; between 2000 and 2015, Africa's numerical growth will be 2.6 times that of China. China added 65 million more people to world population than did India between 1970 and 1980. Between 2000 and 2015, India will add at least 80 million more people than China and will have overtaken China as the world's most populous country by A.D. 2050. Between 1990 and 2025, the UN projects, no less than 95% of global population growth will be in the developing countries of Africa, Asia, and Latin America.

Source: Projections based on World Bank and United Nations figures.

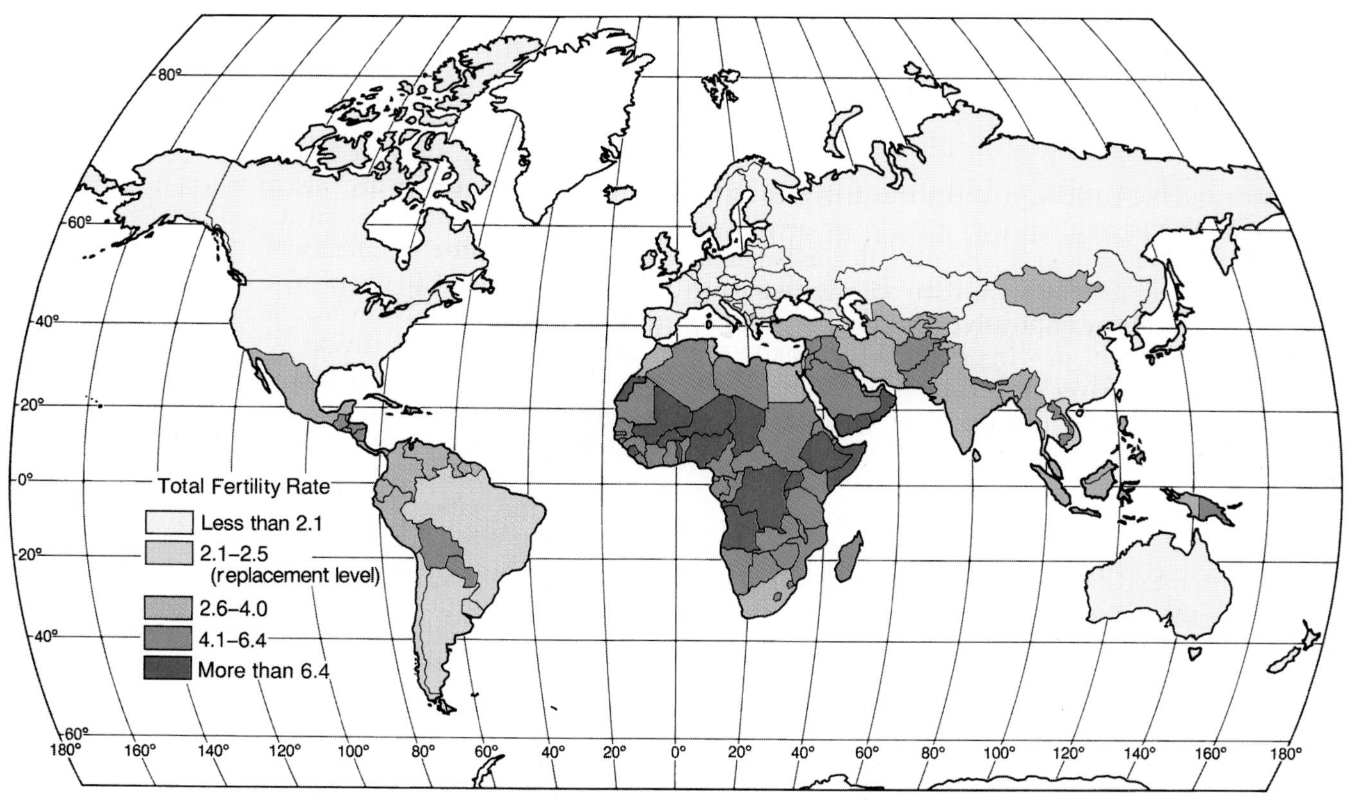

FIGURE 6.5 **Total fertility rate (TFR)** indicates the average number of children that would be born to each woman if, during her child-bearing years, she bore children at the same rate as women of those ages actually did in a given year. Since the TFR is age-adjusted, two countries with identical birth rates may have quite different fertility rates and therefore different prospects for growth. Depending on mortality conditions, a TFR of 2.1 to 2.5 children per woman is considered the "replacement level," at which a population will eventually stop growing.

Source: Data from Population Reference Bureau.

TABLE 6.1

Total Fertility Rates and Change in Selected Less Developed Countries and Regions

Region or Country	Early 1960s[a]	Late 1990s[b]	Percent Change
East Asia	5.3	1.8	–66.0
China	5.9	1.8	–69.5
South Asia	6.1	3.6	–41.0
Afghanistan	7.0	6.1	–12.9
Bangladesh	6.7	3.7	–50.7
India	5.8	3.4	–41.4
Nepal	5.9	4.6	–22.0
Thailand	6.4	2.0	–68.8
Africa	6.7	5.6	–16.4
Egypt	7.1	3.6	–49.3
Kenya	8.2	5.0	–39.0
Nigeria	6.9	6.5	–5.8
Latin America	5.9	3.0	–49.2
Brazil	6.2	2.5	–59.7
Guatemala	6.8	5.1	–25.0
Mexico	6.7	3.1	–53.7

Sources: [a]Thomas W. Merrick, with PRB staff, "World Population in Transition," Population Bulletin 41, no. 2 (Washington, D.C.: Population Reference Bureau, 1986).
[b]Population Reference Bureau and United Nations Population Fund.

Latin America and the Caribbean, and northern Africa have reduced the number of children born to the average woman from more than 5 to less than 3. And even in sub-Saharan Africa, many individual states are dramatically reducing fertility levels. Overall, the impressive recent fertility reductions have dropped world growth projections by nearly 5% from their early 1990s level and make it likely that total population will stabilize at just under 11 billion persons around 2200.

Death Rates

The **crude death rate** (CDR), also called the **mortality rate,** is calculated in the same way as the crude birth rate: the annual number of events per 1000 population. In the past, a valid generalization was that the death rate, like the birth rate, varied with national levels of development. Characteristically, highest rates (over 20 per 1000) were found in the less developed countries of Africa, Asia, and Latin America; lowest rates (less than 10) were associated with developed states of Europe and Anglo America. That correlation became decreasingly valid as dramatic reductions in death rates occurred in developing countries in the years following World War II. Infant mortality rates and life expectancies improved as antibiotics, vaccinations, and pesticides to treat diseases and control disease carriers were made available in almost all parts of the world and as increased attention was paid to funding improvements in urban and rural sanitary facilities and safe water supplies.

Distinctions between more developed and less developed countries in mortality, indeed, have been so reduced that by the late 1990s death rates for less developed countries as a group actually were lower than those for the more developed states (Figure 6.6). Notably and tragically, that reduction does not extend to maternal mortality rates (see "The Risks of Motherhood"). Like crude birth rates, death rates are meaningful for comparative purposes only when we study identically structured populations. Countries with a high proportion of elderly people, such as Denmark and Sweden, would be expected to have higher death rates than those with a high proportion of young people, such as Iceland, assuming equality in other national conditions affecting health and longevity. The pronounced youthfulness of populations in developing countries, as much as improvements in sanitary and health conditions, is an important factor in the recently reduced mortality rates of those areas.

To overcome that lack of comparability, death rates can be calculated for specific age groups. The *infant mortality rate,* for example, is the ratio of deaths of infants aged 1 year or under per 1000 live births:

$$\frac{\text{deaths age 1 year or less}}{\text{1000 live births}}$$

Infant mortality rates are significant because it is at these ages that the greatest declines in mortality have occurred, largely as a result of the increased availability of health services. The drop in infant mortality accounts for a large part of the decline in the general death rate in the last few decades, for mortality during the first year of life is usually greater than in any other year.

Two centuries ago, it was not uncommon for 200–300 infants per 1000 to die in their first year. Even today, despite significant declines in those rates over the last 60 years in many individual countries (Figure 6.7), striking world regional and national variations remain. For all of Africa, infant mortality rates exceed 90 per 1000, and individual African states (for example, Malawi, Guinea, and Sierra Leone) showed rates above 140 in the late 1990s. Nor are rates uniform within single countries. The former Soviet Union reported a national infant mortality rate of 23 (1991), but it registered above 110 in parts of its Central Asian region. In contrast, infant mortality rates in Anglo America and Western and Northern Europe are more uniformly in the 4–7 range.

Modern medicine and sanitation have increased life expectancy and altered age-old relationships between birth and death rates. In the early 1950s, only five countries, all in northern Europe, had life expectancies at birth of over 70 years. By the late 1990s, some 60 countries outside of

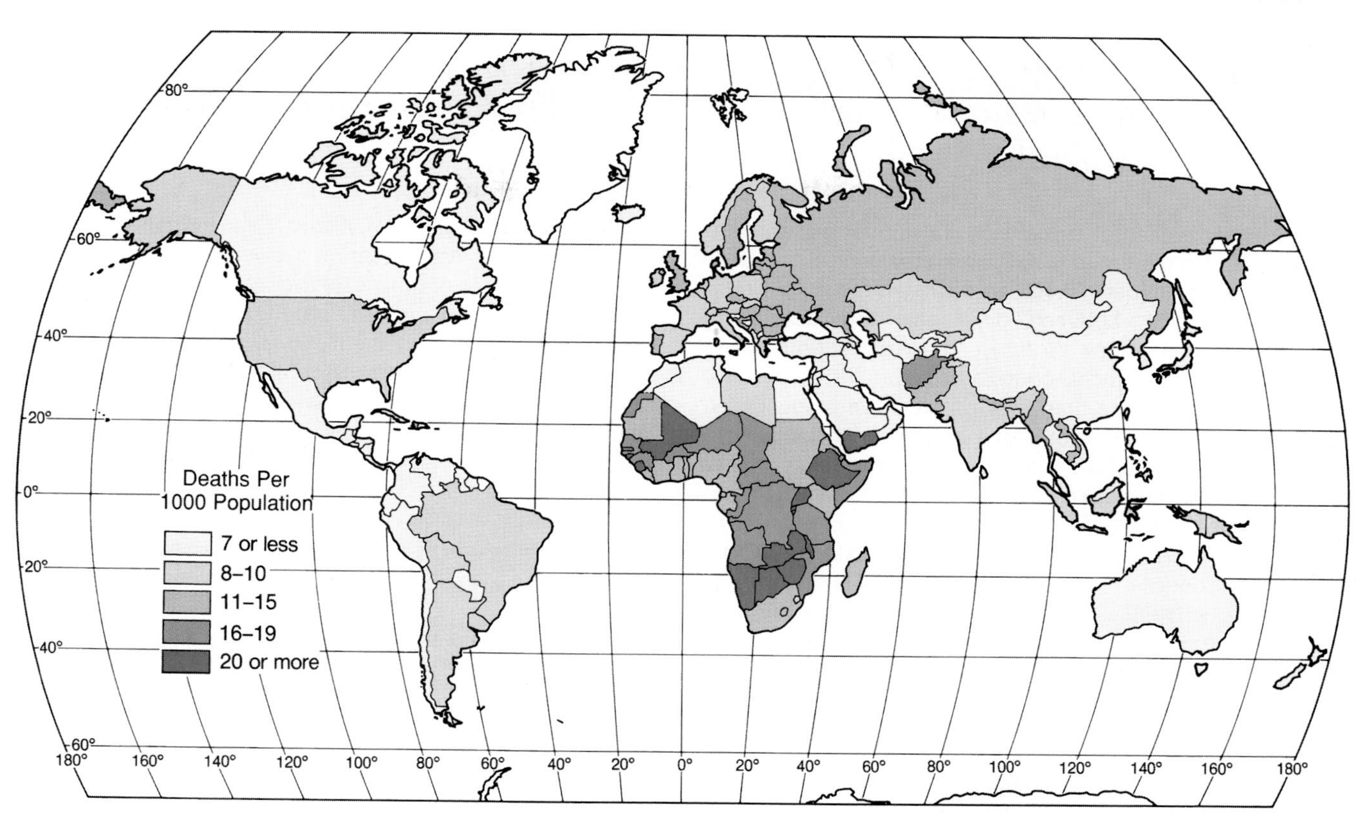

FIGURE 6.6 **Crude death rates** show less worldwide variability than do the birth rates displayed in Figure 6.3, the result of widespread availability of at least minimal health protection measures and a generally youthful population in the developing countries, where death rates are frequently lower than in "old age" Europe.
Source: Data from Population Reference Bureau.

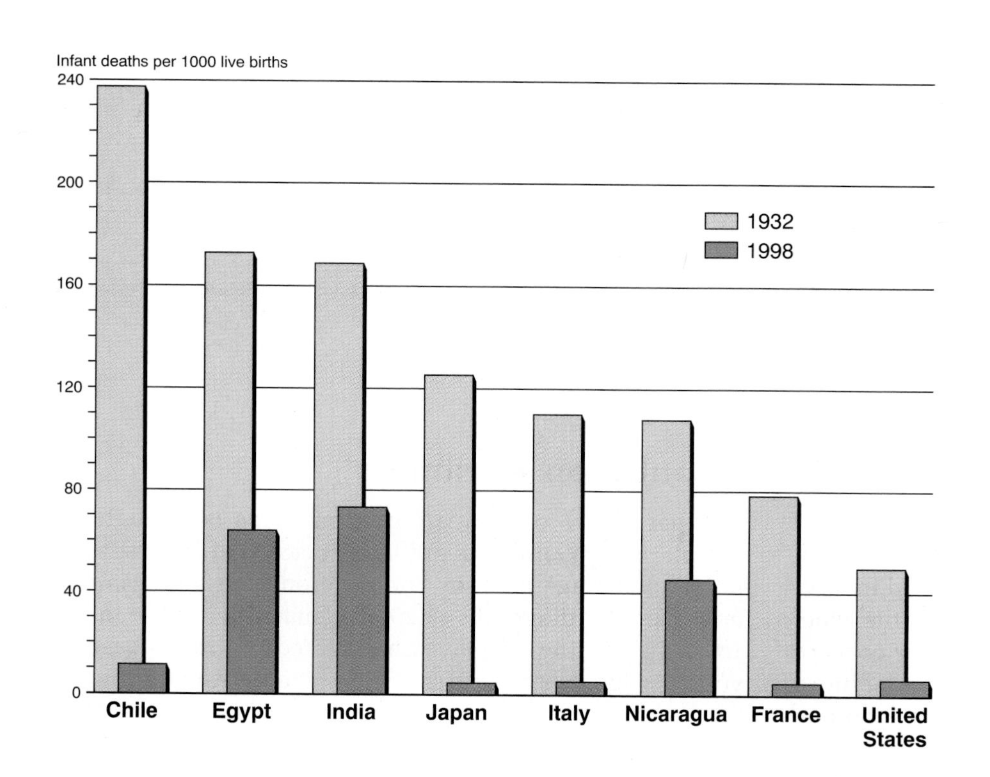

FIGURE 6.7 **Infant mortality rates for selected countries.** Dramatic declines in the rate have occurred in all countries, a result of international programs of health care delivery aimed at infants and children in developing states. Nevertheless, the decreases have been proportionately greatest in the urbanized, industrialized countries, where sanitation, safe water, and quality health care are widely available.
Source: Data from U.S. Bureau of the Census and Population Reference Bureau.

The Risks of Motherhood

The worldwide leveling of crude death rates does not apply to pregnancy-related deaths. In fact, the maternal mortality ratio—the number of deaths per 100,000 live births—is the single greatest health disparity between developed and developing countries. According to the World Health Organization, nearly 600,000 women die each year from causes related to pregnancy or its management; more than 99% of them live in less developed states where, as a group, the maternal mortality ratio (maternal deaths per 100,000 live births) is some 40 times greater than in the more developed countries. Complications of pregnancy, childbirth, and unsafe abortions are the leading slayers of women of reproductive age throughout the developing world, though the incidence of maternal mortality is by no means uniform, as the charts indicate. Country-level differences are even more striking: in Ethiopia, for example, 1 out of every 9 women dies from pregnancy-related complications compared to 1 in 8700 in Switzerland.

Excluding China, less developed countries as a group in the early 1990s had a maternal mortality ratio of 580. While 55% of all maternal deaths occurred in Asia (which accounts for about 61% of the world's births), sub-Saharan African women, burdened with 37% of world maternal mortality, were at greatest statistical risk. There, maternal death ratios reach above 1600 in Guinea and Somalia, and 1 in 13 women in sub-Saharan Africa dies of maternal causes. In contrast, the maternal mortality ratio in developed countries as a group is 10, and in some—Belgium and Ireland, for example—it is as low as 4 or fewer (it was 5 in Canada and 8 in the United States in the mid-1990s).

The vast majority of maternal deaths in the developing world are preventable. Most result from causes rooted in the social, cultural, and economic barriers confronting females in their home environment throughout their lifetimes: malnutrition, anemia, lack of access to timely basic maternal health care, physical immaturity due to stunted growth, and unavailability of adequate prenatal care or trained medical assistance at birth. Part of the problem is that women are considered expendable in societies where their status is low, although the correlation between women's status (Figure 7.41) and maternal mortality is not exact. In those cultures, little attention is given to women's health or their nutrition, and pregnancy, although a major cause of death, is simply considered a normal condition warranting no special consideration or management. To alter that perception and increase awareness of the affordable measures available to reduce maternal mortality worldwide, 1998 was designated "The Year of Safe Motherhood" by a United Nations interagency group.

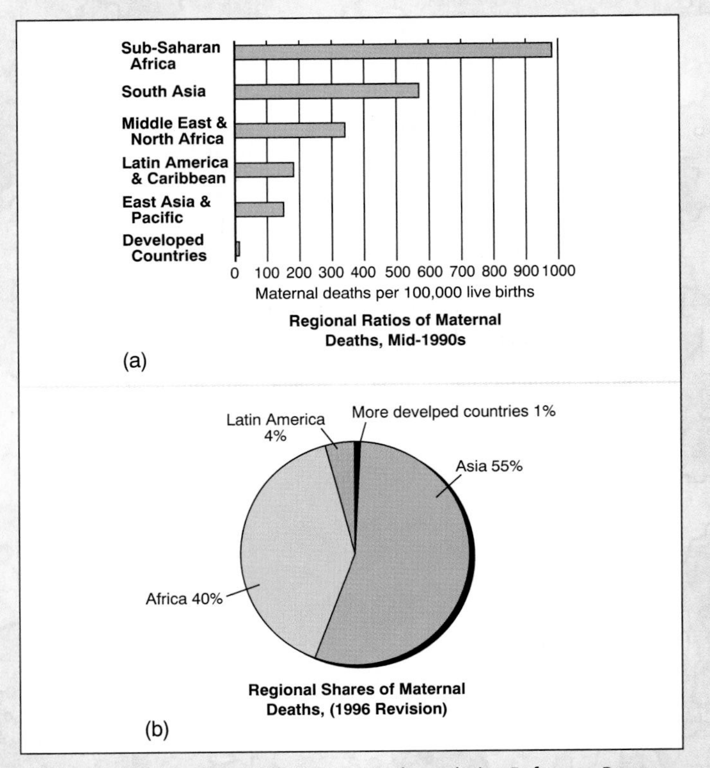

Sources: Graph data from UNICEF, WHO, and Population Reference Bureau.

Europe and North America—though none in sub-Saharan Africa—were on that list. The availability and employment of modern methods of health and sanitation have varied regionally, and the least developed countries have least benefited from them. In such underdeveloped and impoverished areas as much of sub-Saharan Africa, the chief causes of death are those no longer of immediate concern in more developed lands: diseases such as malaria, intestinal infections, typhoid, cholera, and especially among infants and children, malnutrition and dehydration from diarrhea.

Population Pyramids

Another means of comparing populations is through the **population pyramid**, a graphic device that represents a population's age and sex composition. The term *pyramid* describes the diagram's shape for many countries in the 1800s, when the display was created: a broad base of younger age groups and a progressive narrowing toward the apex as older populations were thinned by death. Now many different shapes are formed, each reflecting a different population history (Figure 6.8). By grouping several generations of people, the pyramids highlight the impact of

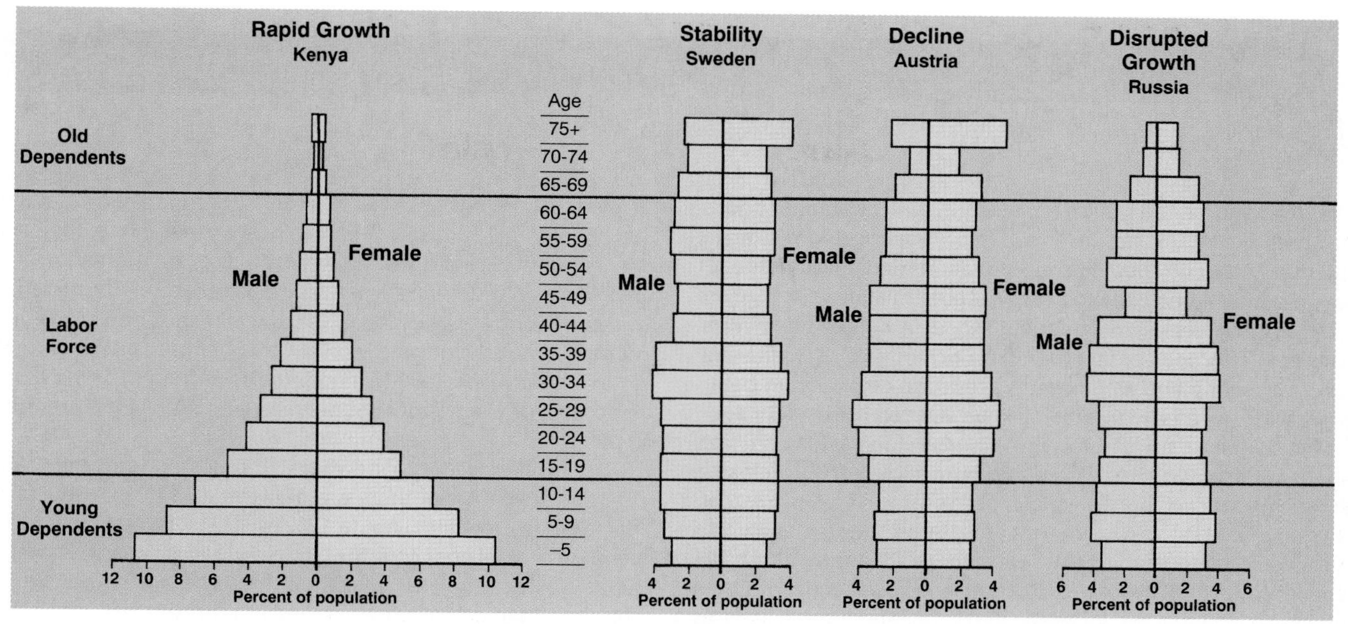

FIGURE 6.8 **Four patterns of population structure.** These diagrams show that population "pyramids" assume many shapes. The age distribution of national populations reflects the past, records the present, and foretells the future. In countries like Kenya, social costs related to the young are important and economic expansion is vital to provide employment for new entrants in the labor force. Austria's negative growth means a future with fewer workers to support a growing demand for social services for the elderly. The 1992 pyramid for Russia reports the sharp decline in births during World War II as a "pinching" of the 45–49 cohort, and shows in the large deficits of men above age 65 the heavy male mortality of both World Wars and the recent sharp reductions in Russian male longevity.

Sources: The World Bank; the United Nations; Population Reference Bureau; and Carl Haub, "Population Change in the Former Soviet Republics," *Population Bulletin* 49, no. 4 (1994).

"baby booms," population-reducing wars, birth-rate reductions, and external migrations.

A rapidly growing country such as Kenya has most people in the lowest age cohorts; the percentage in older age groups declines successively, yielding a pyramid with markedly sloping sides. Typically, female life expectancy is reduced in older cohorts of less developed countries, so that for Kenya the proportion of females in older age groups is lower than in, for example, Sweden. Female life expectancy and mortality rates may also be affected by cultural rather than economic developmental causes (see "100 Million Women Are Missing"). In Sweden, a wealthy country with a very slow rate of growth, the population is nearly equally divided among the age groups, giving a "pyramid" with almost vertical sides. Among older cohorts, as Austria shows, there may be an imbalance between men and women because of the greater life expectancy of the latter. The impacts of war, as Russia's pyramid vividly shows, are evident in that country's depleted age cohorts and male-female disparities. The sharp contrasts between the composite pyramids of sub-Saharan Africa and Western Europe summarize the differing population concerns of the developing and developed regions of the world (Figure 6.9).

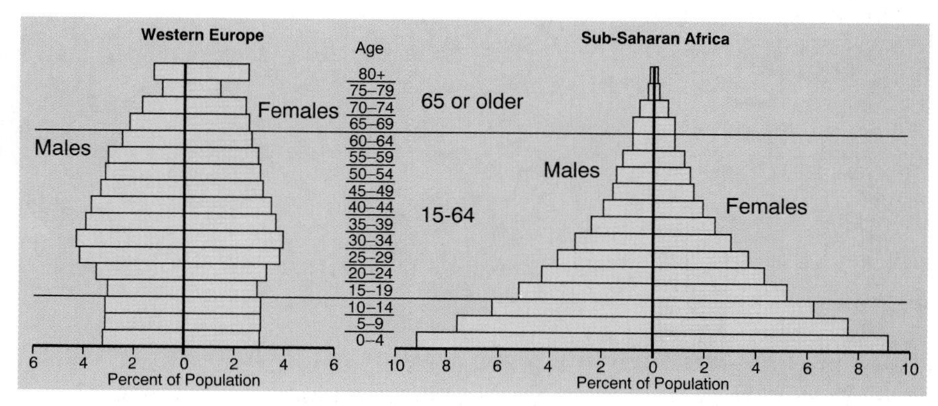

FIGURE 6.9 **Summary population pyramids.** The economically less developed countries of sub-Saharan Africa show a much younger age profile than do more developed Western European countries. In the late 1990s, nearly half of their population was below age 15; in many developed regions, less than one-fifth was in that youthful cohort. In contrast, the proportion of population above 65 in Western Europe is five times that of sub-Saharan countries.

Source: Lori S. Ashford, "New Perspectives on Population: Lessons from Cairo," *Population Bulletin* 50, no. 1 (1995), Figure 3.

100 Million Women are Missing

Worldwide, some 100 million females are missing, victims of nothing more than their sex. In China, India, Pakistan, New Guinea, and many other developing countries a traditional preference for boys has meant neglect and death for girls, millions of whom are killed at birth, deprived of adequate food, or denied the medical attention afforded to favored sons. Increasingly in China and India ultrasound and amniocentesis tests are employed to determine the sex of a fetus so that it can be aborted if it is a female.

The evidence for the missing women starts with one fact: About 105 males are conceived and born for every 100 females. Normally, girls are hardier and more resistant to disease than boys, and in populations where the sexes are treated equally in matters of nutrition and health care, there are about 105 to 106 females for every 100 males. However, the 1990 census of China found just 93.8 females for every 100 males, and the 1991 census of India found just 92.9 females for every 100 males. In both cases, the ratios were more unfavorable than they had been in censuses taken just a decade earlier. The disparity is greatest at the younger age cohorts; in China, nearly 10% of all girls of the 1995 birth cohort are "missing."

Ratio deviations are most striking for second and subsequent births. In China, South Korea, Taiwan, and Hong Kong, for example, the most recent figures for first-child sex ratios are near normal, but rise to 121 boys per 100 girls for a second Chinese child, and to 185 for a third Korean. On that evidence, the problem of missing females is getting worse. Conservative calculations suggest there are nearly 50 million females missing in China alone, about 4% of the national population and more than are unaccounted for in any other country.

The problem is seen elsewhere. In much of South and West Asia and North Africa there are only some 94 females for every 100 males, a shortfall of about 12% of normal (Western) expectations. But not all poor countries show the same disparities. In sub-Saharan Africa, where poverty and disease are perhaps more prevalent than on any other continent, there are 102 females for every 100 males and in Latin America and the Caribbean there are equal numbers of males and females. Cultural norms and practices, not poverty or underdevelopment, seem to determine the fate and swell the numbers of the world's 100 million missing women.

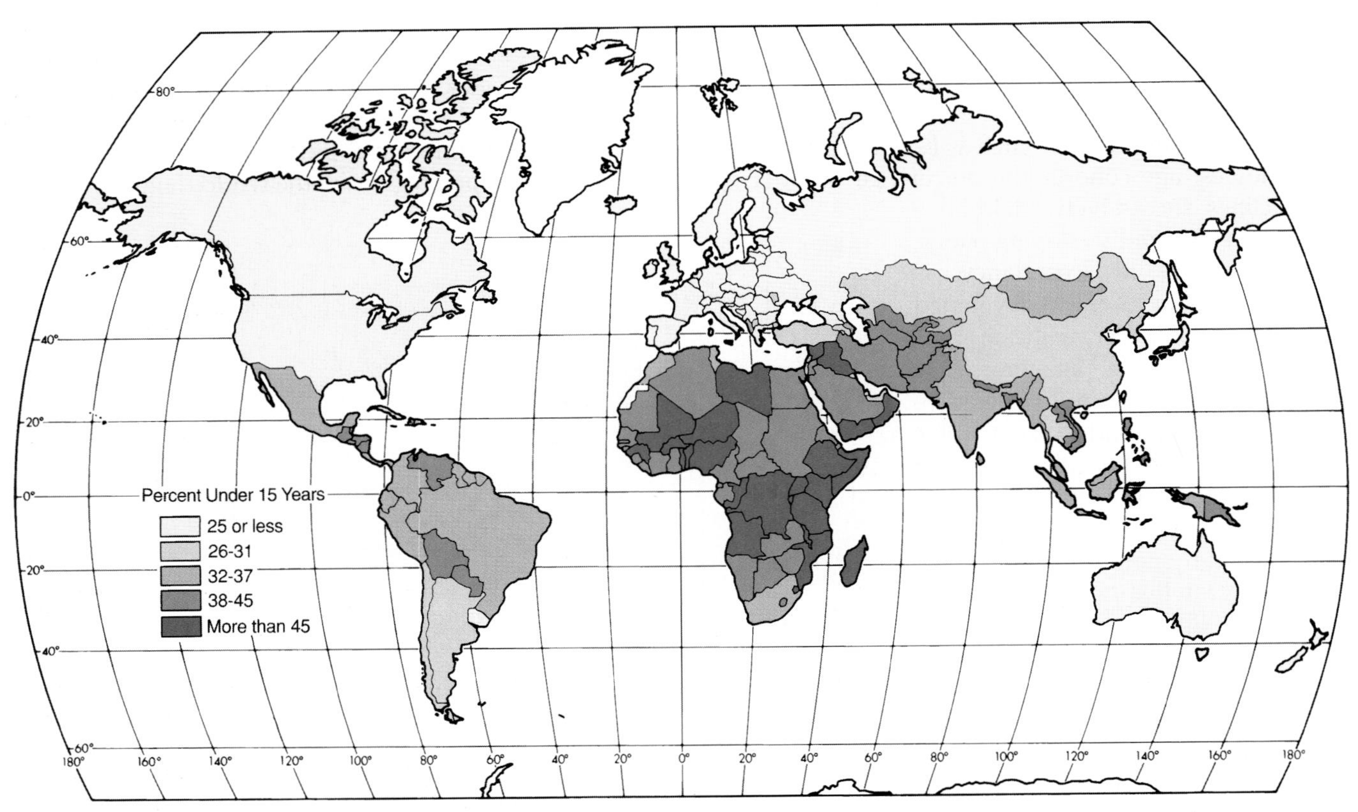

FIGURE 6.10 Percentage of population under 15 years of age. A high proportion of a country's population under 15 increases the dependency ratio of that state and promises future population growth as the youthful cohorts enter childbearing years.

Source: Data from Population Reference Bureau.

The population pyramid provides a quickly visualized demographic picture of immediate practical and predictive value. For example, the percentage of a country's population in each age group strongly influences demand for goods and services within that national economy. A country with a high proportion of young has a high demand for educational facilities and certain types of health delivery services. In addition, of course, a large portion of the population is too young to be employed (Figures 6.9 and 6.10). On the other hand, a population with a high percentage of elderly people also requires medical goods and services specific to that age group (Figure 6.11), and these people must be supported by a smaller proportion of workers. The **dependency ratio** is a simple measure of the number of dependents, old or young, that each 100 people in the productive years (usually, 15–64) must support. Population pyramids give quick visual evidence of that ratio.

They also foretell future problems resulting from present population policies or practices. The strict family-size rules and widespread preferences for sons in China, for example, skews the pyramid in favor of males. At current trends, about 1 million excess males a year will enter an imbalanced marriage market in China beginning about 2010. Millions of bachelors, unconnected to society by wives and children, may pose threats to social order and, perhaps, national stability not foreseen or planned when family control programs were put in place, but clearly suggested when made evident by population pyramid distortions.

Natural Increase

Knowledge of a country's sex and age distributions also enables demographers to forecast its future population levels, though the reliability of projections decreases with increasing length of forecast (Figure 6.12). Thus, a country with a high proportion of young people will experience a high rate of natural increase unless there is a very high mortality rate among infants and juveniles or fertility and birth rates change materially. The **rate of natural increase** of a population is derived by subtracting the crude death rate from the crude birth rate. *Natural* means that increases or

FIGURE 6.11 As these Dutch senior citizens exemplify, Europe is an aging continent with an ever-growing proportion of the elderly dependent on the financial support of a reduced working-age population. Rapidly growing developing countries, in contrast, face increasing costs for the needs of the young.
©Patrick Ward/Stock, Boston.

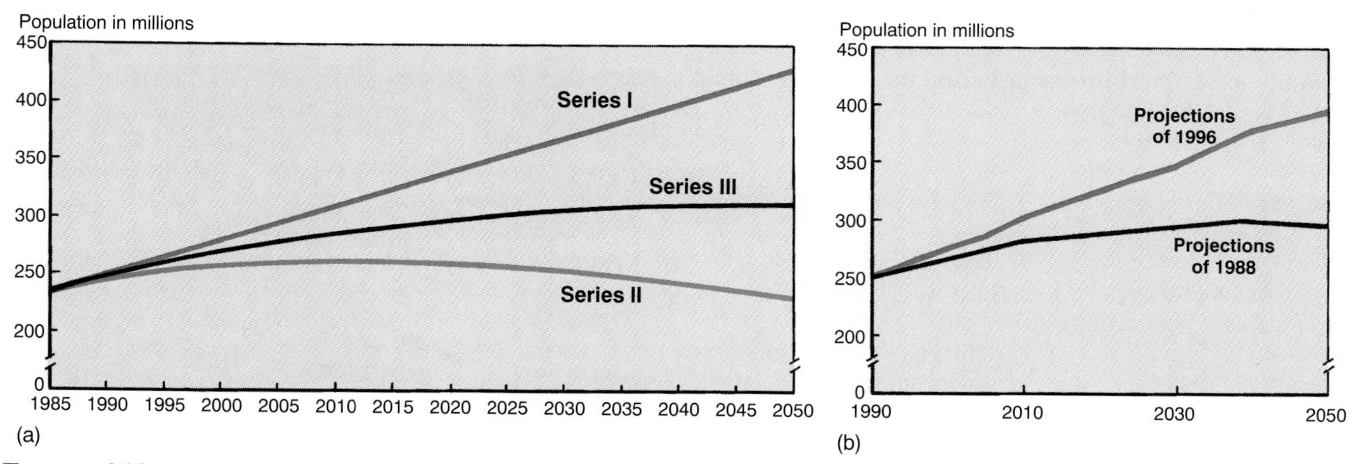

FIGURE 6.12 **Possible population futures for the United States.** As these population projections to 2050 illustrate, expected future numbers vary greatly because the birth and death rate and immigration flow assumptions they are based on are different. (a) Depending on the assumptions, 1985 Census Bureau projections of U.S. population in 2050 ranged from 231 million (low series) to 429 million (high series). (b) The Bureau's revised 1988 middle series projection was again adjusted in late 1996, reflecting actual population counts and new assumptions about fertility, immigration, and racial and ethnic differentials in births and deaths. Those counts and assumption revisions increased the earlier (1988) A.D. 2050 projection by 32%. While the Census Bureau's 1996 middle series projection called for 393.9 million Americans in 2050, its "highest series" guess was for 518.9 million.

Source: U.S. Bureau of the Census.

decreases due to migration are not included. If a country had a birth rate of 22 per 1000 and a death rate of 12 per 1000 for a given year, the rate of natural increase would be 10 per 1000. This rate is usually expressed as a percentage, that is, as a rate per 100 rather than per 1000. In the example given, the annual increase would be 1%.

Doubling Times

The rate of increase can be related to the time it takes for a population to double, that is, the **doubling time.** Table 6.2 shows that it would take 70 years for a population with a rate of increase of 1% (approximately the rate of growth of Taiwan or Argentina in the late 1990s) to double. A 2% rate of increase—recorded in 1998 by the developing world (excluding China)—means that the population will double in only 35 years. (Population doubling time can be closely determined by dividing the growth rate into the number 69. Thus, $69 \div 2 \approx 35$ years.) How could adding only 20 people per 1000 cause a population to grow so quickly? The principle is the same as that used to compound interest in a bank. Table 6.3 shows the number yielded by a 2% rate of increase at the end of successive five-year periods.

For the world as a whole, the rates of increase have risen over the span of human history. Therefore, the doubling time has decreased. Note in Table 6.4 how the population of the world has doubled in successively shorter periods of time. It will approach 9.4 billion by the middle of the 21st century if the present rate of growth continues (Figure 6.1). In countries with high rates of increase (Figure 6.13), the doubling time is less than the 50 years projected for the world as a whole (at end-of-century growth rates). Should world fertility rates decline (as they have in recent years), population doubling time will correspondingly increase, as it has since 1990 (Figure 6.14).

Here, then, lies the answer to the question posed earlier. Even small annual additions accumulate to large total increments because we are dealing with geometric or exponential (1, 2, 4, 8) rather than arithmetic (1, 2, 3, 4) growth.

The ever-increasing base population has reached such a size that each additional doubling results in an astronomical increase in the total. A simple mental exercise suggests the inevitable consequences of such doubling, or **J-curve,** growth. Take a very large sheet of the thinnest paper you can find and fold it in half. Fold it in half again. After seven or eight folds the sheet will have become as thick as a book—too thick for further folding by hand. If you could make 20 folds, the stack would be nearly as high as a football field is long. From then on, the results of further doubling are astounding. At 40 folds, the stack would be well on the way to the moon and at 70 it would reach twice as far as the distance to the nearest star. After 100 folds, our

TABLE 6.2

Doubling Time in Years at Different Rates of Increase

Annual Percentage Increase	Doubling Time (Years)
0.5	140
1.0	70
2.0	35
3.0	24
4.0	17
5.0	14
10.0	7

TABLE 6.3

Population Growth Yielded by a 2% Rate of Increase

Year	Population
0	1000
5	1104
10	1219
15	1345
20	1485
25	1640
30	1810
35	2000

TABLE 6.4

Population Growth and Approximate Doubling Times Since A.D. 1

Year	Estimated Population	Doubling Time (Years)
1	250 million	
1650	500 million	1650
1804	1 billion	154
1927	2 billion	123
1974	4 billion	47
World population may reach:		
2027	8 billion	53[a]

[a]The near-leveling of doubling time reflects assumptions of decreasing and stabilizing fertility rates. No current projections contemplate a further doubling to 16 billion people.

Source: United Nations.

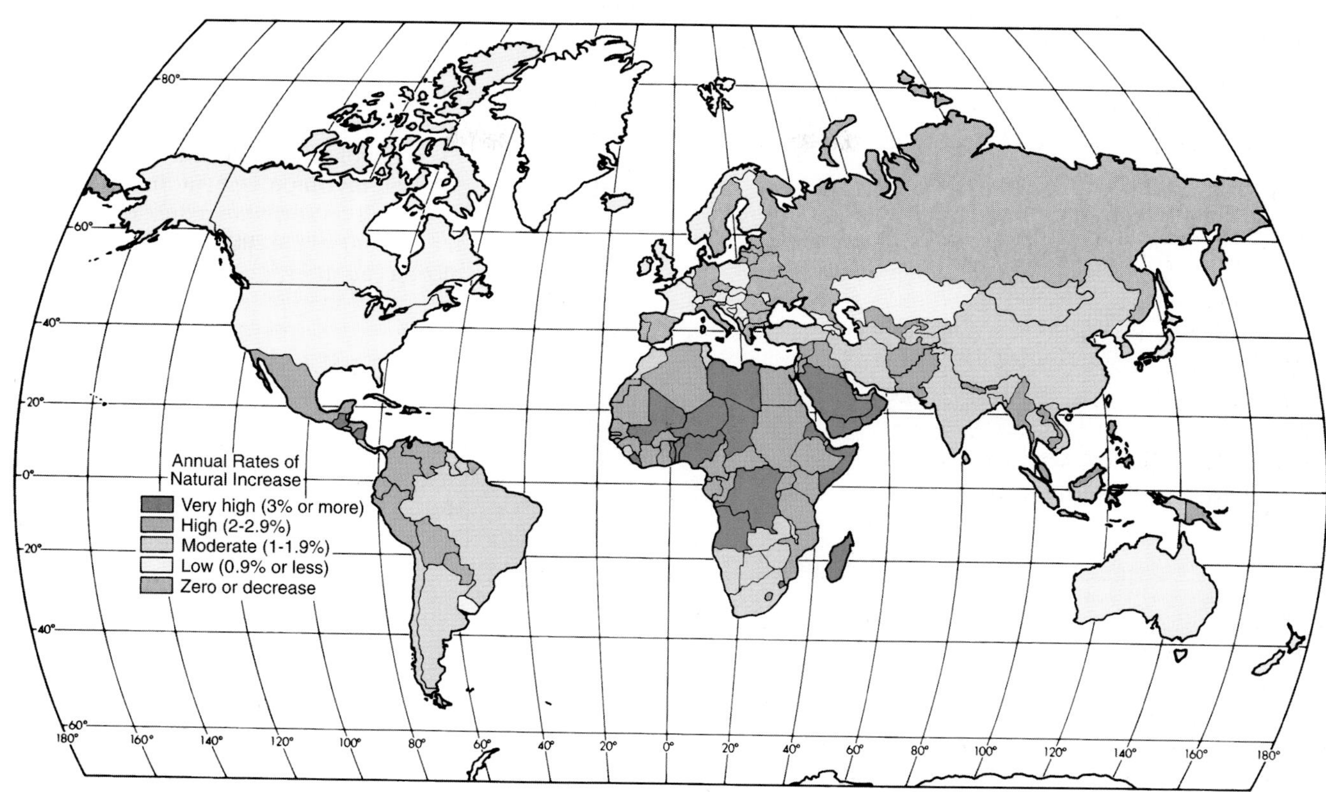

FIGURE 6.13 **Annual rates of natural increase.** The world's 1998 rate of natural increase (1.4%) would mean a doubling of population in 49 years. Since demographers now anticipate world population—currently 6 billion—will stabilize at less than 11 billion (in about A.D. 2200), the "doubling" implication and time frame of current rates of increase reflect mathematical, not realistic, projections. Many individual continents and countries, of course, deviate widely from the global average rate of growth and have vastly different doubling times. Africa as a whole has the highest rates of increase, followed by Central America and Western Asia. Europe and North America are prominent among the low-growth areas, with such countries as Italy actually experiencing negative growth or at best increases so small that doubling times must be measured in millennia. For regions and countries, rates of increase and doubling time projections have more valid implications than do those for the world as a whole.

Source: Data from Population Reference Bureau.

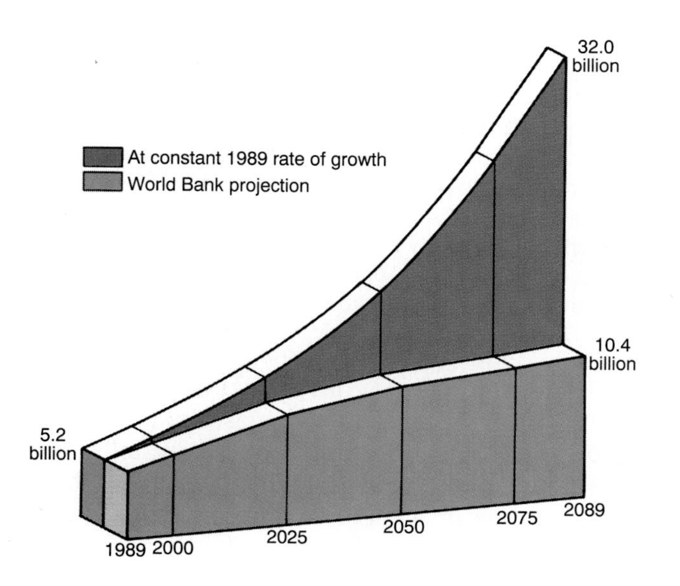

FIGURE 6.14 The "doubling time" calculation illustrates the long-range calculated effect of growth rates on populations. It should never be used to suggest a prediction of future population size, for population growth reflects not just birth rates, but death rates, age structure, collective family size decisions, and migration. Demographers generally assume that high present growth rates of developing countries will continue to be gradually reduced. Therefore, should their collective population double, it will take longer to do so than is suggested by a "doubling time" based on the current rate.

Source: Population Reference Bureau, *1989 World Population Data Sheet.*

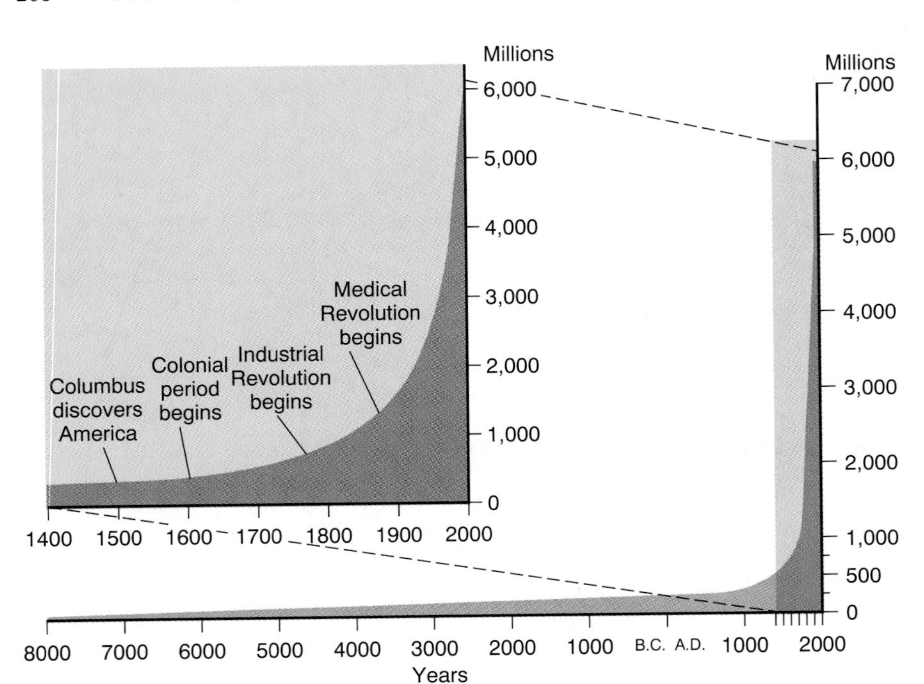

FIGURE 6.15 **World population growth 8000 B.C. to A.D. 2000.** Notice that the bend in the J-curve begins in about the mid-1700s when industrialization started to provide new means to support the population growth made possible by revolutionary changes in agriculture and food supply. Improvements in medical science and nutrition served to reduce death rates near the opening of the 20th century in the industrializing countries.

paper would be more than ten billion light years across and span the known universe. Rounding the bend on the J-curve, which world population has done (Figure 6.15), poses problems and has implications for human occupance of the earth of a vastly greater order of magnitude than ever faced before.

THE DEMOGRAPHIC TRANSITION

The theoretical consequence of exponential population growth cannot be realized. Some form of braking mechanism must necessarily operate to control totally unregulated population growth. If voluntary population limitation is not undertaken, involuntary controls of an unpleasant nature may be set in motion.

One attempt to summarize an historically observed voluntary limitation of population growth—and relating that control to economic development—is the **demographic transition** model. It traces the changing levels of human fertility and mortality presumably associated with industrialization and urbanization. Over time, the model assumes, high birth and death rates will gradually be replaced by low rates (Figure 6.16). The *first stage* of that replacement process—and of the demographic transition model—is characterized by high birth and high but fluctuating death rates.

As long as births only slightly exceed deaths, even when the rates of both are high, the population will grow only slowly. This was the case for most of human history until about A.D. 1750. Demographers think that it took from approximately A.D. 1 to A.D. 1650 for the population to increase from 250 million to 500 million, a doubling time of more than a millennium and a half. Growth was not steady, of course. There were periods of regional expansion that were usually offset by sometimes catastrophic decline. Wars, famine, and other disasters took heavy tolls. For example, the bubonic plague (the Black Death), which swept across Europe in the 14th century, is estimated to have killed over one-third of the population of that continent. The first stage of the demographic transition model is no longer found in any country. In the late 1990s, few countries—even in poorer regions of Africa and Asia—had death rates as high as 20 per 1000. However, in several

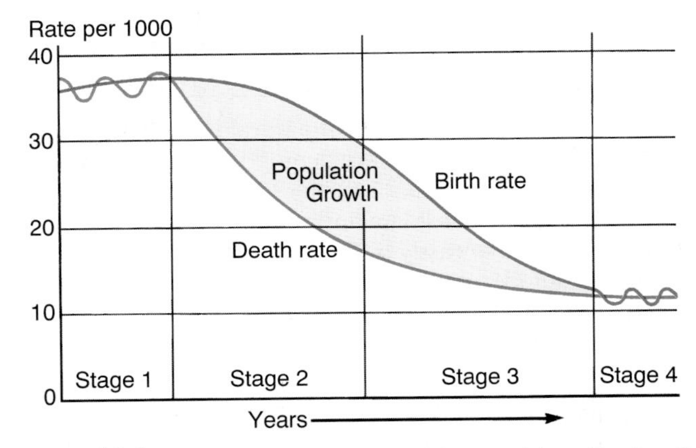

FIGURE 6.16 **Stages in the demographic transition.** During the first stage, birth and death rates are both high, and population grows slowly. When the death rate drops and the birth rate remains high, there is a rapid increase in numbers. During the third stage, birth rates decline and population growth is less rapid. The fourth stage is marked by low birth and death rates and, consequently, by a low rate of natural increase or even by decrease if death rates should exceed those of births. Indeed, the negative growth rates of many European countries have suggested to some that a fifth stage, one of population decline, is—at least regionally—a logical extension of the transition model.

states on those same continents birth rates approached or were above 50 per 1000.

The Western Experience

The demographic transition model was developed to explain the population history of Western Europe. That area entered a *second stage* with the industrialization that began about 1750. Its effects—declining death rates accompanied by continuing high birth rates—have been dispersed worldwide even without universal conversion to an industrial economy. Rapidly rising population during the second demographic stage results from dramatic increases in life expectancy. That, in turn, reflects falling death rates due to advances in medical and sanitation practices, improved foodstuff storage and distribution, a rising per capita income, and the urbanization that provides the environment in which sanitary, medical, and food distributional improvements are concentrated (Figure 6.17). Birth rates do not fall as soon as death rates; ingrained cultural patterns change more slowly than technologies. In many agrarian societies, large families are considered advantageous. Children contribute to the family by starting to work at an early age and by supporting their parents in old age.

Many countries in Latin America and southern and southwestern Asia display the characteristics of this second stage in the population model. Syria, with a birth rate of 33 and a death rate of 6, and Guatemala, with respective rates of 38 and 7 (1998 estimates), are typical. The annual rates of increase of such countries are near or above 30 per 1000, and their populations will double in about 20 to 25 years. Such rates, of course, do not mean that the full impact of the Industrial Revolution has been worldwide; they do mean that the underdeveloped societies have been beneficiaries of the life preservation techniques associated with it.

The *third stage* follows when birth rates decline as people begin to control family size. The advantages of having many children in an agrarian society are not so evident in urbanized, industrialized cultures. In fact, such cultures may view children as economic liabilities rather than assets. When the birth rate falls and the death rate remains low, the population size begins to level off. Chile, Sri Lanka, and Thailand are among the many countries now displaying the low death rates and transitional birth rates of the third stage.

The demographic transition model ends with a *fourth* and final stage. Essentially all European countries, Canada, Australia, and Japan are among the 50 or so states that have entered this phase. Because it is characterized by very low birth and death rates, it yields at best only very slight percentage increases in population. Population doubling times may be as long as a thousand years or more if those present low birth rates continue. In a few countries, indeed, death rates have begun to equal or exceed birth rates, and populations are declining, an extension of the fourth stage into a *fifth* of population decrease so far largely confined to the rich, industrialized, developed world (see "A Population Implosion?").

FIGURE 6.17 Paris, France, in the late 19th century. A modernizing Europe experienced improved living conditions and declining death rates during that century of progress.
©Roger Viollet/Gamma Liaison.

The demographic transition model describes the experience of northwest European countries as they went from rural-agrarian societies to urban-industrial ones. It may not fully reflect the prospects of contemporary developing countries. In Europe, church and municipal records, some dating from 16th century, show that people tended to marry late or not at all. In England before the Industrial Revolution, as many as half of all women in the 15–50 age cohort were unmarried. Infant mortality was high, life expectancy was low. With the coming of industrialization in the 18th and 19th centuries, immediate factory wages instead of long apprenticeship programs permitted earlier marriage and more children. Since improvements in sanitation and health came only slowly, death rates remained high. Around 1800, 25% of Swedish infancts died before their first birthday. Population growth rates ramined below 1% per year in France throughout the 19th century.

Beginning about 1860, first death rates and then birth rates began their significant, though gradual, decline. This "mortality revolution" came first, as an *epidemiologic transition* echoed the demographic transition with which it is associated. Many formerly fatal epidemic diseases became endemic, that is, essentially continual within a population. As people developed partial immunities, mortalities associated with them declined. Improvements in animal husbandry, crop rotation and other agricultural practices, and new foodstuffs (the potato was an early example) from overseas colonies raised the level of health of the European population in general.

At the same time, sewage systems and sanitary water supplies became common in larger cities, and general levels of hygiene improved everywhere (Figure 6.18). Deaths due to infectious, parasitic, and respiratory diseases and to malnutrition declined, while those related to chronic illnesses associated with a maturing and aging population increased. Western Europe passed from a first stage "Age of Pestilence and Famine" to an ultimate "Age of Degenerative and Human-Origin Diseases." However, recent increases in drug- and

A Population Implosion?

For much of the last half of the 20th century, demographers and economists have focused on a "population explosion" and its implied threat of a world with too many people and too few resources of food and minerals to sustain them. By the end of the century, those fears for some observers were being replaced by a new prediction of a world with too few rather than too many people.

That possibility is suggested by two related trends. The first became apparent by 1970 when it was noted that the total fertility rates (TFRs) of 19 countries, almost all of them in Europe, had fallen below the **replacement level**—the level of fertility at which populations replace themselves—of 2.1. Simultaneously, Europe's population pyramid began to become noticeably distorted, with a smaller proportion of young and a growing share of middle-aged and retired-age inhabitants. The decrease in native working-age cohorts had already, by 1970, encouraged the influx of non-European "guest workers" whose labor was needed to maintain economic growth and to sustain the generous security provisions guaranteed to what was becoming the oldest population of any continent.

Many countries of Western and Eastern Europe sought to reverse their birth rate declines by adopting pronatalist policies. The communist states of the East rewarded pregnancies and births with generous family allowances, free medical and hospital care, extended maternity leaves, and child care. France, Italy, the Scandinavian countries, and others gave similar bonuses or awards for first, second, and later births. Despite those inducements, however, reproduction rates continued to fall in Western states and, after the dissolution of communism in the East, to plummet there, too. Despite continuing large-scale in-migration of non-Europeans during the decade, the continent's total population remained essentially constant between 1990 and 1999. "In demographic terms," France's prime minister remarked, "Europe is vanishing."

Europe's experience soon was echoed in other societies of advanced economic development on all continents. By 1995, the United States, Canada, Australia, New Zealand, Japan, Taiwan, South Korea, Singapore, and other older and newly industrializing countries (NICs) registered fertility rates below the replacement level. As they have for Europe, simple projections foretold their aging and declining population. Japan's current slight increase, for example, will become a decrease in 2006 when its population will be older than Europe's; Taiwan forecasts negative growth by 2035.

The second trend indicating to many that world population numbers might stabilize and even decline during the lifetimes of today's college cohort is a simple extension of the first: TFRs are being reduced to or below the replacement levels in countries at all stages of economic development in all parts of the world. While only 18% of total world population in 1975 lived in countries with a fertility rate below replacement level, nearly one-half did so by the end of the century. By 2015, demographers estimate, nearly half the world's countries and over two-thirds of its population will show TFRs below 2.1 children per woman. Exceptions to the trend are and still will be found in Africa, especially sub-Saharan Africa, and in some areas of South, Central, and West Asia; but even in those regions fertility rates have been decreasing in recent years. "Powerful globalizing forces [are] at work pushing towards fertility reduction everywhere," was a 1997 observation from the French National Institute of Demographic Studies.

That conclusion is plausibly supported by assumptions of the United Nation's 1996 "low variant" world population projection. Noting that total fertility rates for more developed regions have already fallen to about 1.5 from 2.0 at start of the 1990s, the UN conjectured a further drop to about 1.4 by 2006. For the less developed regions, the rate dropped from 4.0 to 3.3 during the 1990s alone; the low variant model projects it will decline to about 2.0 in 2020 and 1.6 in 2050. Should those low variant assumptions prove valid, global depopulation could commence in less than 40 years. Between 2040 and 2050, the projection reports, world population would fall by about 85 million (roughly the amount it is increasing each year at the end of the 1990s), and shrink further by about 25% with each successive generation.

If the UN low variant scenario is realized in whole or in part—it is currently rejected by most demographers—a much different worldwide demographic and economic future is promised than that prophesied so recently by "population explosion" forecasts and by demographers' majority estimate of continuing growth to 11 billion in A.D. 2200. Declining rather than increasing pressure on world food and mineral resources would be in our future along with shrinking rather than expanding world, regional, and national economies. Even the achievement of **zero population growth,** a condition for individual countries when births plus immigration equal deaths plus emigration, has social and economic consequences not always perceived by its advocates. These inevitably include an increasing proportion of older citizens, fewer young people, a rise in the median age of the population, and a growing old-age dependency ratio with ever-increasing pension and social services costs borne by a shrinking labor force. Actual population decline, now the common European condition, would only exaggerate those consequences on a worldwide basis.

antibiotic-resistant diseases, pesticide resistance of disease-carrying insects, and such new scourges of both the less developed and more developed countries as AIDS (acquired immune deficiency syndrome) cast doubt on the finality of that "ultimate" stage (see "Our Delicate State of Health").

Even the resurgence of old and emergence of new scourges such as malaria, tuberculosis, and AIDS are unlikely to have serious foreseeable demographic consequences. The United Nations, for example, has estimated that in a hypothetical worst case—that is, if all of Africa

Our Delicate State of Health

Death rates have plummeted and the benefits of modern medicines, antibiotics, and sanitary practices have enhanced both the quality and expectancy of life in the developed and much of the developing world. Far from being won, however, the struggle against infectious and parasitic diseases is growing in intensity and is, perhaps, unwinnable. More than a half century after the discovery of antibiotics, the diseases they were to eradicate are on the rise, and both old and new disease-causing microorganisms are emerging and spreading all over the world. Infectious diseases kill between 16 and 20 million people each year; they officially account for one-third of global mortality and, because of poor diagnosis, certainly are responsible for far more. And their global incidence is rising.

The five leading infectious killers are acute respiratory infections such as pneumonia, diarrheal diseases, tuberculosis, malaria, and measles. In addition, AIDS is expected to kill 1.8 million persons yearly by 2000—more than measles and nearly as many as malaria. The incidence of infection, of course, is far greater than the occurrence of deaths. More than a third of the world's people—some 1.8 billion—for example, are infected with the bacterium that causes tuberculosis, but only 3.3 million are killed by the disease each year. Above 500 million people are infected with such tropical diseases as malaria, sleeping sickness, schistosomiasis, and river blindness, with perhaps 3 million annual deaths. Newer pathogens are constantly appearing, such as those causing Lassa fever, Rift Valley fever, Ebola, Hanta, and Hepatitis C, incapacitating and endangering far more than they kill. In fact, at least 30 new infectious diseases have appeared since the mid-1970s.

The spread and virulence of infectious diseases are linked to the dramatic changes so rapidly occurring in the earth's physical and social environments. Deforestation, water contamination, climatic change, wetland drainage, and other human-induced alterations to the physical environment disturb ecosystems and simultaneously disrupt the natural system of controls that keep infectious diseases in check. Rapid population growth and explosive urbanization, increasing global tourism, population-dislocating wars and migrations, and expanding world trade all increase interpersonal disease-transmitting contacts and the mobility and range of disease-causing microbes, including those brought from previously isolated areas by newly opened road systems and air routes. Add in poorly planned or executed public health programs, inadequate investment in sanitary infrastructures, and inefficient distribution of medical personnel and facilities, and the causative role of humans in many of the current disease epidemics is clearly visible.

In response, a worldwide Program for Monitoring Emerging Diseases (ProMED) was established in 1993 and developed a global on-line infectious disease network linking health workers and scientists in more than 100 countries to battle what has been called a growing "epidemic of epidemics." The most effective weapons in that battle are already known. They include improved health education; disease prevention and surveillance; research on disease vectors and incidence areas (including GIS and other mapping of habitats conducive to specific diseases); careful monitoring of drug therapy; mosquito control programs; provision of clean water supplies; and distribution of such simple and cheap remedies and preventatives as childhood immunizations, oral rehydration therapy, and vitamin A supplementation. All, however, require expanded investment and attention to those spreading infectious diseases—many with newly developed antibiotic-resistant strains—so recently thought to be no longer of concern.

were affected by AIDS on the same scale as its worst known affected areas—Africa's population growth rate would still be about 1.8% in 2000. On a global scale, reproduction rates seem certain to outpace disease mortality rates.

In Europe, the striking reduction in death rates was echoed by similar declines in birth rates as societies began to alter their traditional concepts of ideal family size. In cities, child labor laws and mandatory schooling meant that children became a burden, not a contribution, to family economies. As "poor-relief" legislation and other forms of public welfare substituted for family support structures, the insurance value of children declined. Family consumption

FIGURE 6.18 Pure piped water replacing individual or neighborhood wells, and sewers and waste treatment plants instead of privies, became increasingly common in urban Europe and North America during the 19th century. Their modern successors, such as the Windsor, Ontario, treatment plant shown here, helped complete the *epidemiologic transition* in developed countries.
©Steve McCutcheon / Visuals Unlimited.

patterns altered as the Industrial Revolution made more widely available goods that served consumption desires, not just basic living needs. Children hindered rather than aided the achievement of the age's promise of social mobility and lifestyle improvement. Perhaps most important, and by some measures preceding and independent of the implications of the Industrial Revolution, were changes in the status of women and in their spreading conviction that control over childbearing was within their power and to their benefit.

A World Divided

The demographic transition model described the presumed inevitable course of population events from the high birth and death rates of premodern (underdeveloped) societies to the low and stable rates of advanced (developed) countries. The model failed to anticipate, however, that by the 1990s many developing societies would seemingly be locked in the second stage of the model, unable to realize the economic gains and social changes necessary to progress to the third stage of falling birth rates. The population history of Europe was apparently not inevitably or fully applicable to all developing countries of the middle and late 20th century, though its demographic transition was replicated in Southeast and East Asian countries (Japan, Singapore, Taiwan, South Korea, and others) with very high rates of economic growth in the space of a single generation.

The introduction of Western technologies of medicine and public health, including antibiotics, insecticides, sanitation, immunization, infant and child health care, and eradication of smallpox, quickly and dramatically lowered the death rates in developing countries. Such imported technologies and treatments accomplished in a few years what it took Europe 50 or 100 years to experience. Sri Lanka, for example, sprayed extensively with DDT to combat malaria; life expectancy jumped from 44 years in 1946 to 60 only eight years later. With similar public health programs, India also experienced a steady reduction in its death rate after 1947. Simultaneously, with international sponsorship, food aid cut the death toll of developing states during drought and other disasters. The dramatic decline in mortality, which emerged only gradually throughout the European world but occurred so rapidly in contemporary developing countries, has been the most fundamental demographic change in human history.

Corresponding reductions in birth rates have been harder to achieve and depend less on supplied technology and assistance than they do on social acceptance of the idea of fewer children and smaller families (Figure 6.19). That acceptance has been growing broadly but unevenly worldwide. While only 18% of world population lived in countries with fertility rates at or below replacement levels (that is, countries that had achieved the demographic transition) in 1984, some 44% lived in such countries in 1997. Much of that increase is due to China's officially claimed birth rate reductions and, importantly, to comparable reductions in other Eastern and Southeastern Asian states of recent rapid economic growth.

The consequence is a world still polarized demographically. In those countries actively advancing economically—in Asia, Africa, and Latin America—significant birth rate reductions have been achieved largely through changing family attitudes. Even in some, particularly African, states not yet significantly industrializing, diminishing fertility rates are widespread and noticeable. In many countries, however, in all still developing parts of the world, birth rates remain high, averaging 1.5 to 2 times or more above the replacement level. In both instances, the established pattern tends to become self-reinforcing. Low growth permits the expansion of personal income and accumulation of capital that enhance the quality and security of life and make large families less attractive or essential.

When the population doubles each generation, as it must at the fertility rates of the highest-growth portion of the divided world, a different reinforcing mechanism operates. Population growth consumes in social services and assistance the investment capital that might promote economic expansion. Increasing populations place ever greater demands on limited soil, forest, water, grassland, and cropland resources. Those pressures may, through human-induced deforestation and desertification, for example, consume the environmental base itself. Productivity declines and population-supporting capacities are so diminished as to make difficult or impossible the economic progress on which the demographic transition depends, an apparent equation of increasing international concern (see "The Cairo Plan").

The Demographic Equation

Births and deaths among a region's population—natural increases or decreases—tell only part of the story of population change. Migration involves the long-distance movement of people from one residential location to another. When that relocation occurs across political boundaries, it affects the population structure of both the origin and destination jurisdictions. The **demographic equation** summarizes the contribution made to regional population change over time by the combination of *natural change* (difference between births and deaths) and *net migration* (difference between in-migration and out-migration). On a global scale, of course, all population change is accounted for by natural change. The impact of migration on the demographic equation increases as the population size of the areal unit studied decreases.

Population Relocation

In the past, emigration proved an important device for relieving the pressures of rapid population growth in at least some European countries (Figure 6.20). For example, in one 90-year span, 45% of the natural increase in the population of the British Isles emigrated, and between 1846 and 1935

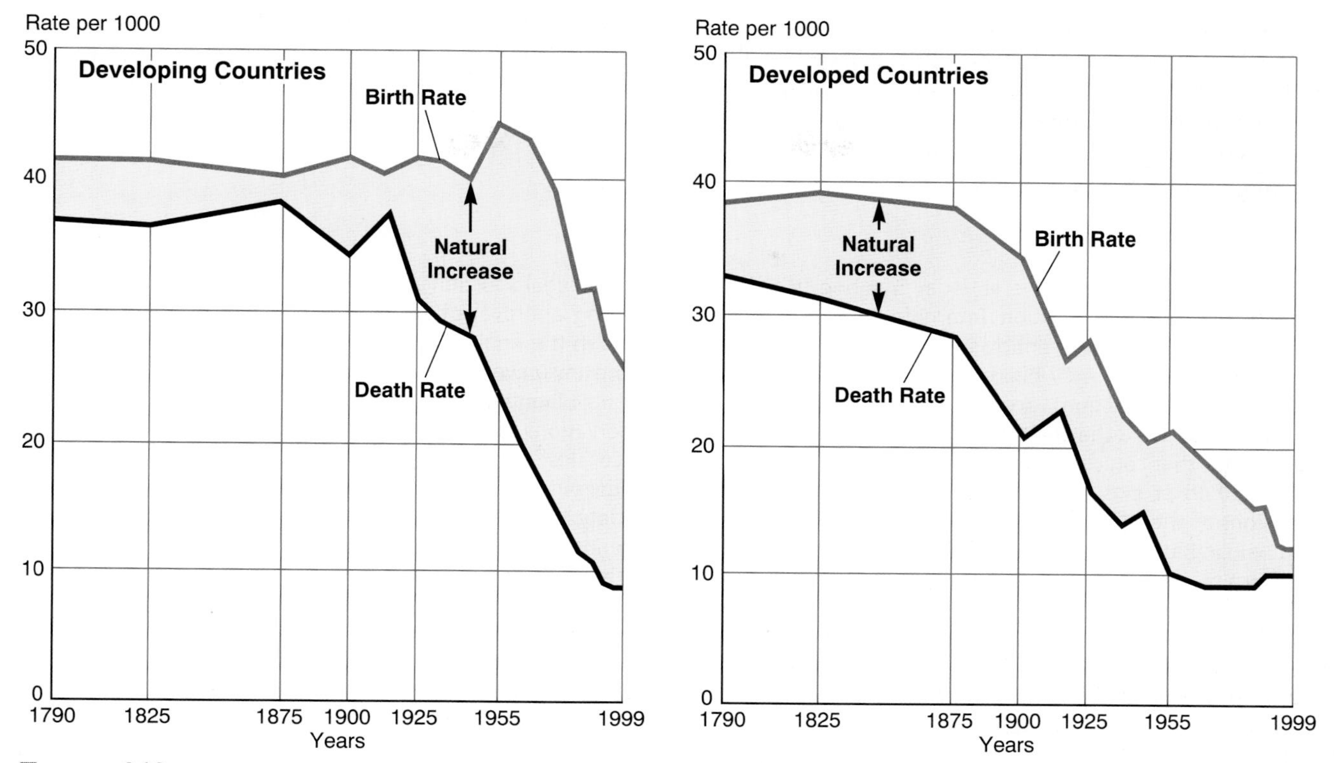

FIGURE 6.19 **World birth and death rates.** The "population explosion" after World War II (1939–1945) reflected the effects of drastically reduced death rates in developing countries without simultaneous and compensating reductions in births. By the end of the century, however, three interrelated trends had appeared in many developing world countries: (1) fertility had overall dropped further and faster than had been predicted 20 years earlier, (2) contraceptive acceptance and use had increased markedly, and (3) age at marriage was rising. In consequence, the demographic transition had been compressed from a century to a generation in some developing states. In others, fertility decline began to slacken in the mid-1970s, but continued to reflect the average number of children—four or more—still desired in many societies.

Source: Revised and redrawn from Elaine M. Murphy, *World Population: Toward the Next Century,* revised ed. (Washington, D.C.: Population Reference Bureau, 1989).

some 60 million Europeans of all nationalities left that continent. Despite recent massive movements of economic and political refugees across Asian, African, and Latin American boundaries, emigration today provides no comparable relief valve for developing countries (Table 6.5). Total population numbers are too great to be much affected by migrations of even millions of people. In only a few countries—Afghanistan, Cuba, El Salvador, and Haiti, for example—have as many as 10% of the population emigrated in recent decades. A more detailed treatment of the processes and patterns of international and intranational migrations as expressions of spatial interaction is presented in Chapter 8.

Immigration Impacts

Where cross-border movements are massive enough, migration may have a pronounced impact on the demographic equation and result in significant changes in the population structures of both the origin and destination regions. Past European and African migrations, for example, not only altered but substantially created the population structures of new, sparsely inhabited lands of colonization in the Western Hemisphere and Australasia. In some decades of the late

TABLE 6.5

Percentage of Natural Population Increase That Permanently Emigrated

Period	Europe	Asia[a]	Africa	Latin America[a]
1851–1880	11.7	0.4	b	0.3
1881–1910	19.5	0.3	b	0.9
1911–1940	14.4	0.1	b	1.8
1940–1960	2.7[c]	0.1	b	1.0
1960–1970	5.2	0.2	0.1	1.0
1970–1980	4.0	0.5	0.3	2.5

[a]The periods from 1851 to 1960 report emigration only to the United States.
[b]Less than 0.1 percent.
[c]Emigration only to the United States.

Source: World Bank, *World Development Report 1984,* p. 69. Note: Numbers are calculated from data on gross immigration in Australia, Canada, New Zealand, and the United States.

Geography *and* Public Policy

The Cairo Plan

After a sometimes rancorous nine-day meeting in Cairo in September 1994, the United Nations International Conference on Population and Development endorsed a strategy for stabilizing the world's population at 7.27 billion by no later than 2015. The 20-year "program of action" accepted by over 150 signatory countries seeks to avoid the environmental consequences of population growth that could put world totals at 11 to 12 billion or more by 2050. Its proposals were therefore linked to discussions and decisions of the UN Conference on Environment and Development held in Rio de Janeiro in June 1992.

The Cairo plan abandons several decades of policies that promoted "population control" (a phrase no longer employed) based on frequently coercive targets and quotas and, instead, embraces for the first time policies giving women greater control over their lives, greater economic equality and opportunity, and a greater voice in reproduction decisions. It recognizes that limiting population growth depends on programs that increase the educational and economic prospects and political rights of women, leading women to want fewer children and making them partners in economic development. In that recognition, the Conference builds on extensive evidence that increased educational access and economic opportunity for women are invariably linked to falling birth rates and smaller families. Earlier population conferences—1974 in Bucharest and 1984 in Mexico City—did not fully address these issues of equality, opportunity, education, and political rights; their adopted goals failed to achieve hoped-for changes in births in large part because women in many traditional societies had no power to enforce contraception and feared their other alternative, sterilization.

The earlier conferences carefully avoided or specifically excluded abortion as an acceptable family planning method. It was the more open discussion of abortion in Cairo that elicited much of the spirited debate that registered religious objections by the Vatican and many Muslim and Latin American states to the inclusion of legal abortion as part of health care, and to language suggesting approval of sexual relations outside of marriage. Even though the final text of the conference declaration did not promote any universal right to abortion and excluded it as a means of family planning, some 20 of the 150 delegations still registered reservations to its wording on both sex and abortion. At conference close, however, the Vatican endorsed the declaration's underlying principles, including the family as "the basic unit of society," the need to stimulate economic growth, and to promote "gender equality, equity, and the empowerment of women."

Islamic and Christian reservations on abortion were matched by even broader complaints that demands for action on population growth are apt to be tied to aid for poorer countries, selectively limiting their opportunities for economic development and progress. The argument runs that the burden of limiting population continues to fall on the poor while the richer industrialized states are, in reality, the greater danger to world environment and development because of their production of pollution and disproportionate demands on natural resources.

The Cairo plan assumes a three-fold increase in the world amount spent on population stabilization, from about $5 billion in 1994 to about $17 billion by 2000, and $22 billion by 2015 (all in 1994 dollars). Some $5.7 billion is to come from donor countries, while countries with existing successful programs combining family planning, education, and economic opportunities for women are to share their expertise. Both the World Bank and the U.S. Agency for International Development (A.I.D.) proposed launching integrated health, education, and loan programs for women, and the U.S. Congress approved $583 million for population aid in fiscal 1995, double the 1992 figure, but far short of the annual $1.9 billion U.S. contribution needed to meet Cairo's A.D. 2000 goal. U.S. funding in subsequent years has also fallen short of its hoped-for levels.

Questions to Consider

1. Do you think it is appropriate or useful for international bodies to promote policies affecting such purely personal or national concerns as reproduction and family planning? Why or why not?
2. Do you think that current international concerns over population growth, development, and the environment are sufficiently valid and pressing to risk the loss of long-enduring cultural norms and religious practices in many of the world's traditional societies? Why or why not?
3. Do you think the financial obligations implied for developed, donor countries by the Cairo plan are justified in light of the many other international needs and domestic concerns faced by their governments? Why or why not?
4. Many environmentalists see the world as a finite system unable to support ever-increasing populations; to exceed its limits would cause frightful environmental damage and global misery. Many economists counter that free markets will keep supplies of needed commodities in line with growing demand and that science will, as necessary, supply technological fixes in the form of substitutes or expansion of production. In light of such diametrically opposed views of population growth consequences, is it appropriate or wise to base international programs solely on one of them? Why or why not?

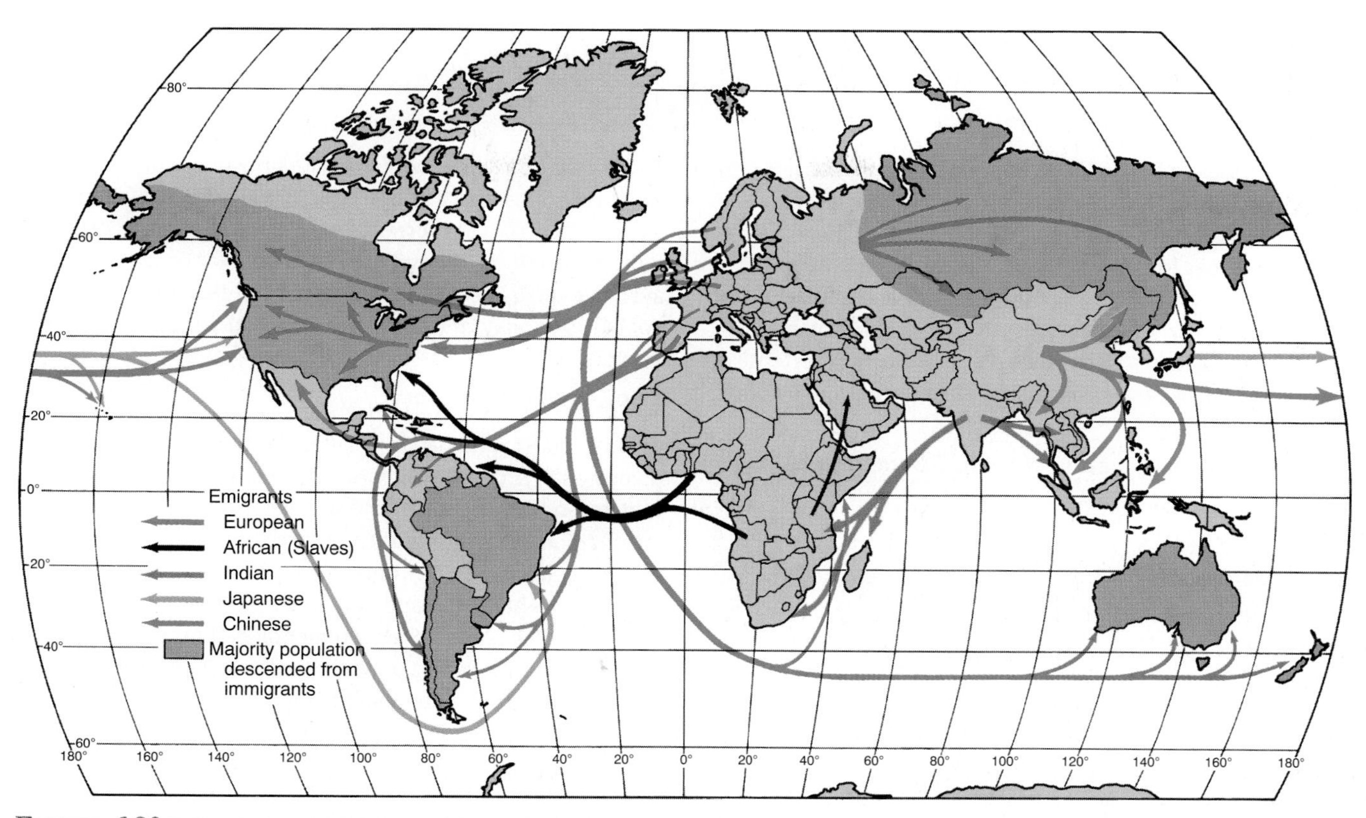

FIGURE 6.20 **Principal migrations of recent centuries.** The arrows suggest the major free and forced international population movements since about 1700. The shaded areas on the map are regions whose present population is more than 50% descended from the immigrants of recent centuries.

Source: Shaded zones after Daniel Noin, *Géographie de la Population*, p. 85, 1979, ©Masson, Paris.

18th and early 19th centuries 30% to more than 40% of population increase in the United States was accounted for by immigration. Similarly, eastward-moving Slavs colonized underpopulated Siberia and overwhelmed native peoples.

Migrants are rarely a representative cross section of the population group they leave, and they add an unbalanced age and sex component to the group they join. A recurrent research observation is that emigrant groups are heavily skewed in favor of young singles. Whether males or females dominate the outflow varies with circumstances. Although males traditionally far exceeded females in international flow, in recent years females have accounted for between 40% and 60% of all transborder migrants.

At the least, then, the receiving country will have its population structure altered by an outside increase in its younger age and, probably, unmarried cohorts. The results are both immediate in a modified population pyramid, and potential in future impact on reproduction rates and excess of births over deaths. The origin area will have lost a portion of its young, active members of childbearing years. It perhaps will have suffered distortion in its young adult sex ratios, and it certainly will have recorded a statistical aging of its population. The destination society will likely experience increases in births associated with the youthful newcomers and, in general, have its average age reduced.

World Population Distribution

The millions and billions of people of our discussion are not uniformly distributed over the earth. The most striking feature of the world population distribution map (Figure 6.21) is the very unevenness of the pattern. Some land areas are nearly uninhabited, others are sparsely settled, and still others contain dense agglomerations of people. More than half of the world's people are found—unevenly concentrated, to be sure—in rural areas. Some 45% are urbanites, however, and a constantly growing proportion are residents of very large cities of 1 million or more.

Earth regions of apparently very similar physical makeup show quite different population numbers and densities, perhaps the result of differently timed settlement or of settlement by different cultural groups. Had North America been settled by Chinese instead of Europeans, for example, it is likely that its western sections would be far more densely settled than they now are. Northern and Western Europe, inhabited thousands of years before North America, contain more people than the United States on 70% less land.

We can draw certain generalizing conclusions from the uneven but far from irrational distribution of population shown in Figure 6.21. First, almost 90% of all people live

On-Line Population Geography

Steadily increasing numbers of population-related websites, with constantly changing and expanding information content, are becoming available. We've listed here only a few of the more useful home pages from government and nongovernmental agencies, including universities and international organizations.

An efficient way of starting a search for population materials is to use a subject resource guide. Perhaps the most extensive is the *World Wide Web Virtual Library—Demography and Population Studies* catalog.

http://coombs.anu.edu.au/ResFacilities/DemographyPage.html.

Leading sources for U.S. population data include the following: *The Census Bureau Home Page* is a primary source for official social, economic, and demographic statistics of the U.S. population indexed by subject. It is as well a source of Census Bureau data maps and is linked to other population websites. Selected tables from the latest *Statistical Abstract* and *County and City Data Book* are included. Some lengthy reports need Adobe Acrobat Reader.

http://www.census.gov/

The Census Bureau's *State census data centers* home pages provide population estimates, employment reports, economic indicators, and other data at state, county, and city levels. Existing state data center websites may be accessed from the single source.

http://www.census.gov/sdc/www/

The *National Center for Health Statistics* website provides information on access to reports and statistics about births, deaths, marriages, fertility rates, etc.

http://www.cdc.gov/nchswww/

World population information is found at a number of sites. *United Nations Population Information Network* (*POPIN*) reports world, regional, and country-level demographic trends, and is a good source for historical world population growth, urbanization, child mortality estimates, AIDS impact, etc. Full-text regional reports and newsletters are also available, including *Country Health Profiles* of the Pan American Health Organization. The site is linked to many other population home pages and includes a worldwide directory of population organizations and institutions. It is well worth visiting.

http://www.undp.org/popin/

The United Nations Population Fund assists developing countries in their reproductive health and family planning services. Its website provides on-line access to its current "State of World Population" annual report, to various technical reports and general interest publications, and links to related UN and nongovernmental organization home pages.

http://www.unfpa.org/

The *Population Reference Bureau,* a principal source of demographic data used in this book and in many newspaper and journal reports, gives current-year demographic statistics for more than 190 countries in its *World Population Data Sheet* available on its website as well as the full-text PRB newsletter, *Population Today,* and a website "hot list" for population matters.

http://www.prb.org/prb/

PopNet, also maintained by the Population Reference Bureau, is dedicated to providing comprehensive data on global population issues. Dubbing itself "the source for global population information," it presents data on such topics as demographic statistics, education, environment, economics, gender, and reproductive health; in addition, it has multiple links to websites of governmental and nongovernmental domestic and international organizations and university centers. PopNet can be reached through the Population Reference Bureau website (above) or directly at its own address.

http://www.popnet.org

Demographic and Health Surveys is a primary information source on matters of fertility, maternal and child health, and household living conditions in developing countries.

http://www.macroint.com/dhs/

The *World Health Organization*'s WHOSIS website describes and provides access to statistical data and information available from the WHO and elsewhere in electronic and other forms.

http://www.who.int/whosis/

The *International Programs Center* (*IPC*) of the U.S. Bureau of the Census provides a wealth of comparative statistics for all world countries. Included are population, life tables, migration, ethnicity, language, religion, vital statistics, labor force and economic data, and more. Summary tables and maps are available, as are software and applications of interest to demographers.

http://www.census.gov/ftp/pub/ipc/www

Valuable text and statistical supplements to the population appendix in this book are to be found in the *CIA World Factbook* of the United States Central Intelligence Agency. The site contains demographic, economic, and social information for more than 260 countries, including data on population, vital statistics, ethnic composition, religions, languages, net migration, and more.

http://www.odci.gov/cia/publications/factbook/index.html

Population associations and information source guides may help you gain access to other useful databases, bibliographies, and agencies. Following are a few of potential interest.

The *Population Association of America* reports its activities in its full-text newsletter.

http://www.pop.psu.edu/general/pubs/PAA_Affairs

Internet Resources for Demographers is a collection of demographic Internet sites categorized under "North American Demography," "International Demography," "General Demography," etc.[a]

http://members.tripod.com/~tgryn/demog.html

The *Office of Population Research* of Princeton University hosts the *Population Index on the Web,* presenting on-line the most important bibliographic record to the world's population literature taken from the journal *Population Index.* The database from 1986 onward is searchable by author, subject matter, geographical region, and date.

http://popindex.princeton.edu/

Also useful are: the *Social Science Information Gateway (SOSIG)—Demography*

http://www.sosig.ac.uk/roads/subject-listing/World/demog.html

and the Johns Hopkins University Population Information Program *Popline,* a searchable bibliographic database of over 250,000 records covering worldwide literature on population, family planning, and health issues.

http://www.jhuccp.org/popline/index.stm

Finally, you might wish to download the *Intlpop* program developed at Virginia Tech, software that allows users to manipulate data on fertility, mortality, and migration levels of country populations and track the consequences of those changes on population size, growth, and age structure.

http://geosim.cs.vt.edu/index.html

[a]The site is an extension of an exhaustive printed document, "Internet Resources for Demographers," by Thomas A. Gryn that appears in *Population Index* 20, no. 2 (Summer, 1997):189–204.

The guidance of the Population Reference Bureau in the preparation of this listing is gratefully acknowledged.

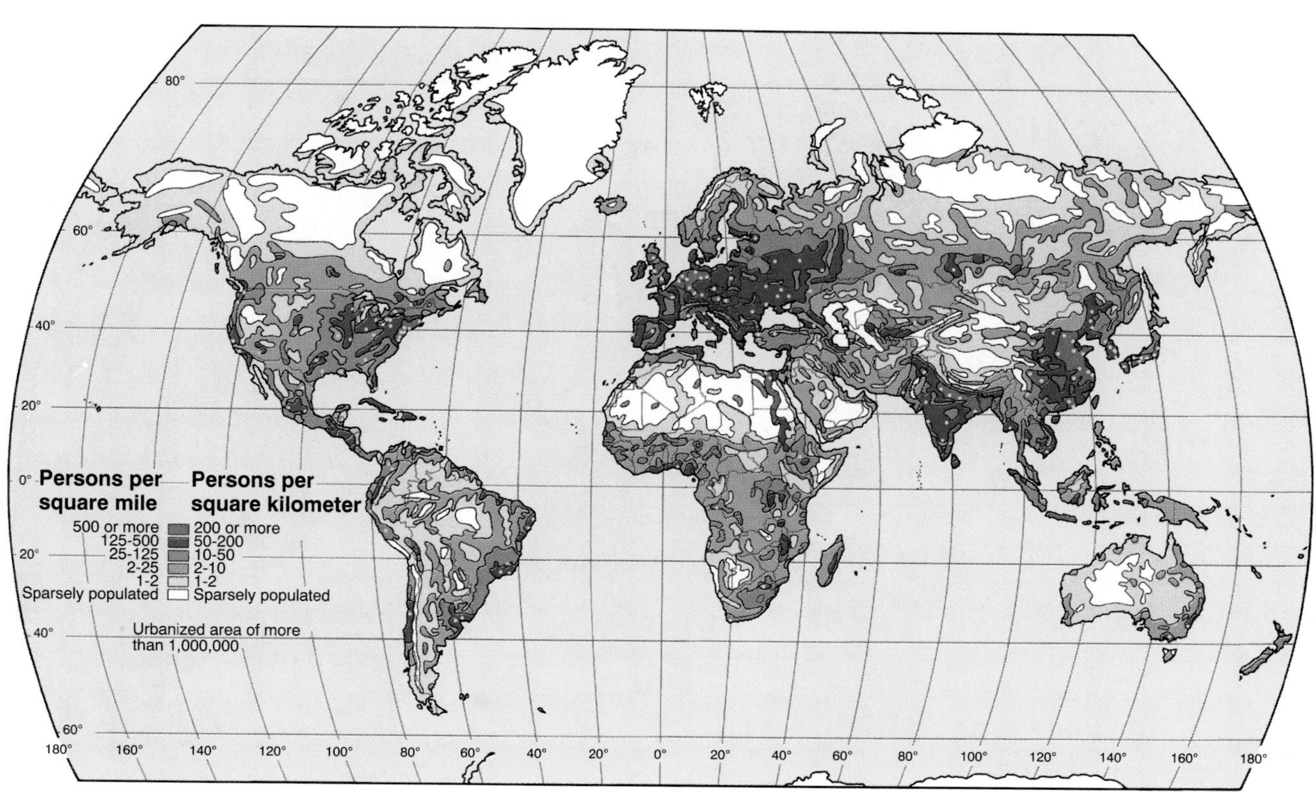

FIGURE 6.21 **World population density.**

north of the equator and two-thirds of the total dwell in the midlatitudes between 20° and 60° North (Figure 6.22). Second, a large majority of the world's inhabitants occupy only a small part of its land surface. More than half the people live on about 5% of the land, two-thirds on 10%, and almost nine-tenths on less than 20%. Third, people congregate in lowland areas; their numbers decrease sharply with increases in elevation. Temperature, length of growing season, slope and erosion problems, even oxygen reductions at very high altitudes, all appear to limit the habitability of higher elevations. One estimate is that between 50% and 60% of all people live below 200 meters (650 ft), a zone containing less than 30% of total land area. Nearly 80% reside below 500 meters (1650 ft).

Fourth, although low-lying areas are preferred settlement locations, not all such areas are equally favored. Continental margins have attracted densest settlement. About two-thirds of world population is concentrated within 500 kilometers (300 mi) of the ocean, much of it on alluvial lowlands and river valleys. Latitude, aridity, and elevation, however, limit the attractiveness of many seafront locations. Low temperatures and infertile soils of the extensive Arctic coastal lowlands of the Northern Hemisphere have restricted settlement there. Mountainous or desert coasts are sparsely occupied at any latitude, and some tropical lowlands and river valleys that are marshy, forested, and disease-infested are unevenly settled.

Within the sections of the world generally conducive to settlement, four areas contain great clusters of population: East Asia, South Asia, Europe, and northeastern United States/southeastern Canada. The *East Asia* zone, which includes Japan, China, Taiwan, and South Korea, is the largest cluster in both area and numbers. The four countries forming it contain 25% of all people on earth; China alone accounts for one in five of the world's inhabitants. The *South Asia* cluster is composed primarily of countries associated with the Indian subcontinent—Bangladesh, India, Pakistan, and the island state of Sri Lanka—though some might add to it the Southeast Asian countries of Cambodia, Myanmar, and Thailand. The four core countries alone account for another one-fifth, 21%, of the world's inhabitants. The South and the East Asian concentrations are thus home to nearly one-half of the world's people.

Europe—southern, western, and eastern through Ukraine and much of European Russia—is the third extensive world population concentration, with another 13% of its inhabitants. Much smaller in extent and total numbers is the cluster in *northeastern United States/southeastern Canada*. Other smaller but pronounced concentrations are found around the globe: on the island of Jawa (Java) in Indonesia, along the Nile River in Egypt, and in discontinuous pockets in Africa and Latin America.

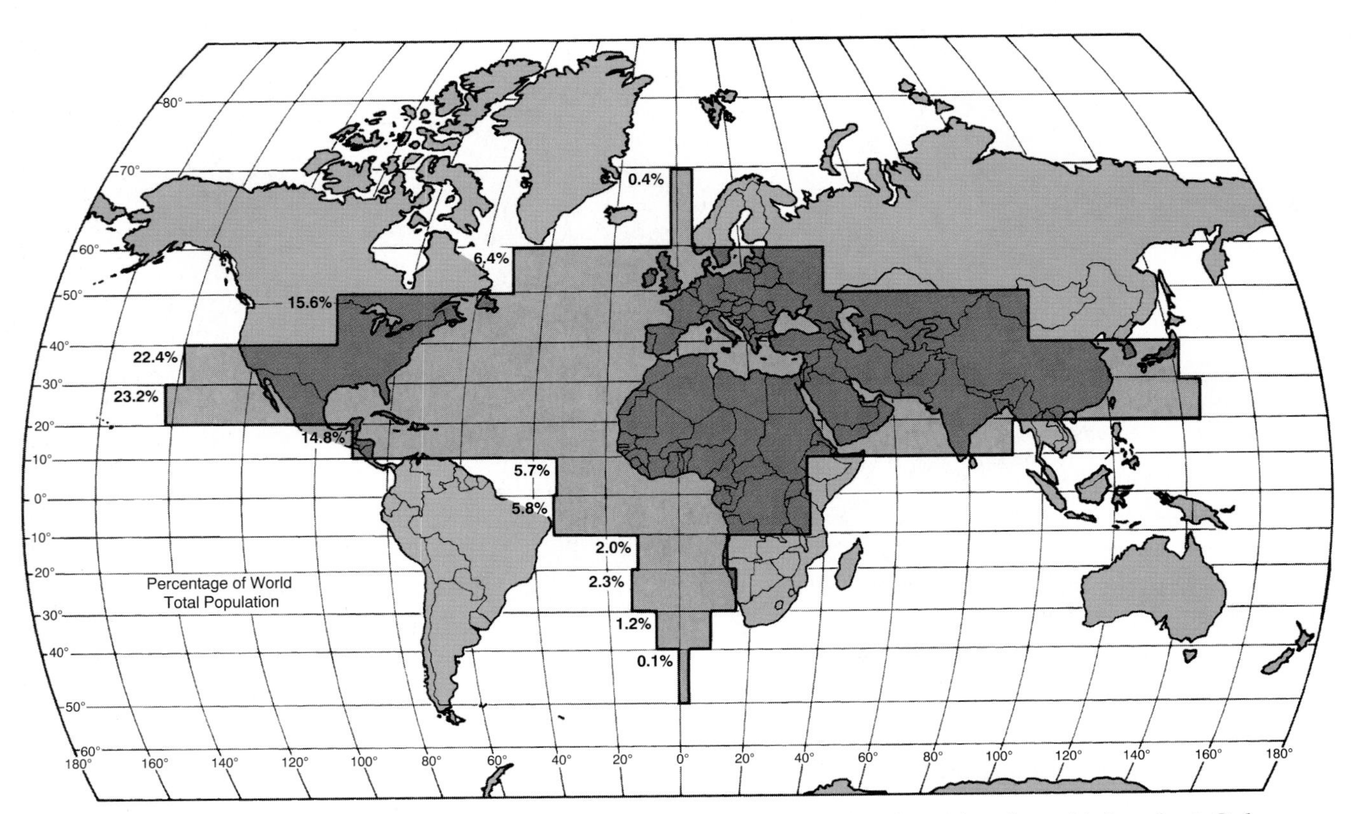

FIGURE 6.22 **The population dominance of the Northern Hemisphere** is strikingly evident from this bar chart. Only one out of nine people lives south of the equator—not because the Southern Hemisphere is underpopulated, but because it is mainly water.

The term **ecumene** is applied to permanently inhabited areas of the earth's surface. The ancient Greeks used the word, derived from their verb "to inhabit," to describe their known world between what they believed to be the unpopulated, searing southern equatorial lands and the permanently frozen northern polar reaches of the earth. Clearly, natural conditions are less restrictive than Greek geographers believed. Both ancient and modern technologies have rendered habitable areas that natural conditions make forbidding. Irrigation, terracing, diking, and draining are among the methods devised to extend the ecumene locally (Figure 6.23).

At the world scale, the ancient observation of habitability appears remarkably astute. The **nonecumene,** or *anecumene,* the uninhabited or very sparsely occupied zone, does include the permanent ice caps of the Far North and Antarctica and large segments of the tundra and coniferous forest of northern Asia and North America. But the nonecumene is not continuous, as the ancients supposed. It is discontinuously encountered in all portions of the globe and includes parts of the tropical rain forests of equatorial zones, midlatitude deserts of both the Northern and Southern Hemispheres, and high mountain areas.

Even parts of these unoccupied or sparsely occupied districts have localized dense settlement nodes or zones based on irrigation agriculture, mining and industrial activities, and the like. Perhaps the most anomalous case of settlement in the nonecumene world is that of the dense population in the Andes Mountains of South America and the plateau of Mexico. Here Native Americans found temperate conditions away from the dry coast regions and the hot, wet Amazon basin. The fertile high basins have served a large population for more than a thousand years.

FIGURE 6.23 Terracing of hillsides is one device to extend a naturally limited productive area. The technique is effectively used here at the Malegcong rice terraces on densely settled Luzon Island of the Philippines.

Even with these locally important exceptions, the nonecumene portion of the earth is extensive. Some 35–40% of all the world's land surface is inhospitable and without significant settlement. This is, admittedly, a smaller proportion of the earth than would have qualified as uninhabitable in ancient times or even during the 19th century. Since the end of the Ice Age some 11,000 years ago, humans have steadily expanded their areas of settlement.

Population Density

Margins of habitation could only be extended, of course, as humans learned to support themselves from the resources of new settlement areas. The numbers that could be sustained in old or new habitation zones were and are related to the resource potential of those areas and the cultural levels and technologies possessed by the occupying populations. The term **population density** expresses the relationship between number of inhabitants and the area they occupy.

Density figures are useful, if sometimes misleading, representations of regional variations of human distribution. The **crude density** or **arithmetic density** of population is the most common and least satisfying expression of that variation. It is the calculation of the number of people per unit area of land, usually within the boundaries of a political entity. It is an easily reckoned figure. All that is required is information on total population and total area, both commonly available for national or other political units. The figure can, however, be misleading and may obscure more of reality than it reveals. The calculation is an average that blankets a country's largely undevelopable or sparsely populated regions along with its intensively settled and developed districts. A national average density figure reveals nothing about either class of territory. In general, the larger the political unit for which crude or arithmetic population density is calculated, the less useful is the figure.

Various modifications may be made to refine density as a meaningful abstraction of distribution. Its descriptive precision is improved if the area in question can be subdivided into comparable regions or units. Thus it is more revealing to know that in the late 1990s New Jersey had a density of 419 and Wyoming of 2 persons per square kilometer (1085 and 5 per sq mi) of land area than to know only that the figure for the conterminous United States (48 states) was 35 per square kilometer (90 per sq mi). If Hawaii and large, sparsely populated Alaska are added, the U.S. density figure drops below 30 per square kilometer (76 per sq mi). The calculation may also be modified to provide density distinctions between classes of population—rural versus urban, for example. Rural densities in the United States rarely exceed 115 per square kilometer (300 per sq mi), while portions of major cities can have tens of thousands of people in equivalent space.

Another revealing refinement of crude density relates population not simply to total national territory but to that area of a country that is or may be cultivated, that is, to *arable* land. When total population is divided by arable land area alone, the resulting figure is the **physiological density** which is, in a sense, an expression of population pressure exerted on agricultural land. Table 6.6 makes evident that countries differ in physiological density and that the contrasts between crude and physiological densities of countries point up actual settlement pressures that are not revealed by arithmetic densities alone. The calculation of physiological density, however, depends on uncertain definitions of arable and cultivated land, assumes that all arable land is equally productive and comparably used, and includes only one part of a country's resource base.

Overpopulation

It is an easy and common step from concepts of population density to assumptions about overpopulation or overcrowding. It is wise to remember that **overpopulation** is a value judgment reflecting an observation or conviction that an environment or territory is unable to support its present population. (A related but opposite concept of *underpopulation* refers to the circumstance of too few people to sufficiently develop the resources of a country or region to improve the level of living of its inhabitants.)

Overpopulation is not the necessary and inevitable consequence of high density of population. Tiny Monaco, a principality in southern Europe about half the size of New

TABLE 6.6

Comparative Densities for Selected Countries

Country	Crude Density		Physiological Density[a]	
	Sq Mi	Km^2	Sq Mi	Km^2
Argentina	34	13	342	132
Australia	6	2	106	41
Bangladesh	2454	947	3670	1417
Canada	9	3	171	66
China	345	133	3450	1332
Egypt	171	66	5682	2194
India	861	332	1510	583
Iran	102	39	930	359
Japan	869	336	7236	2794
Nigeria	346	134	961	371
United Kingdom	634	245	2530	977
United States	76	29	363	140

[a]Includes arable land and land in permanent crops.

Sources: UN Food and Agriculture Organization (FAO), *Production Yearbook;* World Bank, *World Development Indicators;* and Population Reference Bureau, *World Population Data Sheet.*

York's Central Park, has a crude density of nearly 20,000 people per square kilometer (50,000 people per sq mi). Mongolia, a sizable state of 1,565,000 square kilometers (604,000 sq mi) between China and Siberian Russia, has 1.6 persons per square kilometer (4.1 per sq mi); Iran, only slightly larger, has 39 per square kilometer (102 per sq mi). Macao, an island possession of Portugal off the coast of China, has more than 32,000 persons per square kilometer (83,000 per sq mi); the Falkland Islands off the Atlantic coast of Argentina count at most 1 person for every 5 square kilometers (2 sq mi) of territory. No conclusions about conditions of life, levels of income, adequacy of food, or prospects for prosperity can be drawn from these density comparisons.

Overcrowding is a reflection not of numbers per unit area but of the **carrying capacity** of land—the number of people an area can support on a sustained basis given the prevailing technology. A region devoted to efficient, energy-intensive commercial agriculture that makes heavy use of irrigation, fertilizers, and biocides can support more people at a higher level of living than one engaged in the slash-and-burn agriculture described in Chapter 10. An industrial society that takes advantage of resources such as coal and iron ore and has access to imported food will not feel population pressure at the same density levels as a country with rudimentary technology.

Since carrying capacity is related to the level of economic development, maps such as Figure 6.21, displaying present patterns of population distribution and density, do not suggest a correlation with conditions of life. Many industrialized, urbanized countries have lower densities and higher levels of living than do less-developed ones. Densities in the United States, where there is a great deal of unused and unsettled land, are considerably lower than those in Bangladesh, where essentially all land is arable and which, with nearly 950 people per square kilometer (over 2450 per sq mi), is the most densely populated nonisland state in the world. At the same time, many African countries have low population densities and low levels of living, whereas Japan combines both high densities and wealth.

Overpopulation can be equated with levels of living or conditions of life that reflect a continuing imbalance between numbers of people and carrying capacity of the land. One measure of that imbalance might be the unavailability of food supplies sufficient in caloric content to meet individual daily energy requirements or so balanced as to satisfy normal nutritional needs. Unfortunately, dietary insufficiencies—with long-term adverse implications for life expectancy, physical vigor, and mental development—are most likely to be encountered in the developing countries, where much of the population is in the younger age cohorts (Figure 6.10).

If those developing countries simultaneously have rapidly increasing population numbers dependent on domestically produced foodstuffs, the prospects must be for continuing undernourishment and overpopulation. Much of sub-Saharan Africa finds itself in this circumstance. Its per capita food production decreased 12% between 1980 and 1995, with continuing decline predicted over the following quarter century as the population—food gap widens (Figure 6.24). The countries of North Africa are similarly strained. Egypt already must import well over half the food it consumes. Africa is not alone. The international Food and Agriculture Organization (FAO) estimates that in 2000, at least 65 separate countries with over 30% of the population of the developing world will be unable to feed their inhabitants from their own national territories at the low level of agricultural technology and inputs apt to be employed. Even rapidly industrializing China, an exporter of grain until 1994, now in most years is a net grain importer; if its massive and growing population continues its new dependence on imported basic foodstuffs, world grain supplies and prices and international food aid flows will be seriously affected.

In the contemporary world, insufficiency of domestic agricultural production to meet national caloric requirements cannot be considered a measure of overcrowding or poverty. Only a few countries are agriculturally self-sufficient. Japan, a leader among the advanced states, is the world's biggest food importer and supplies from its own production only 40% of the calories its population consumes. Its physiological density is high, as Table 6.6 indicates, but it

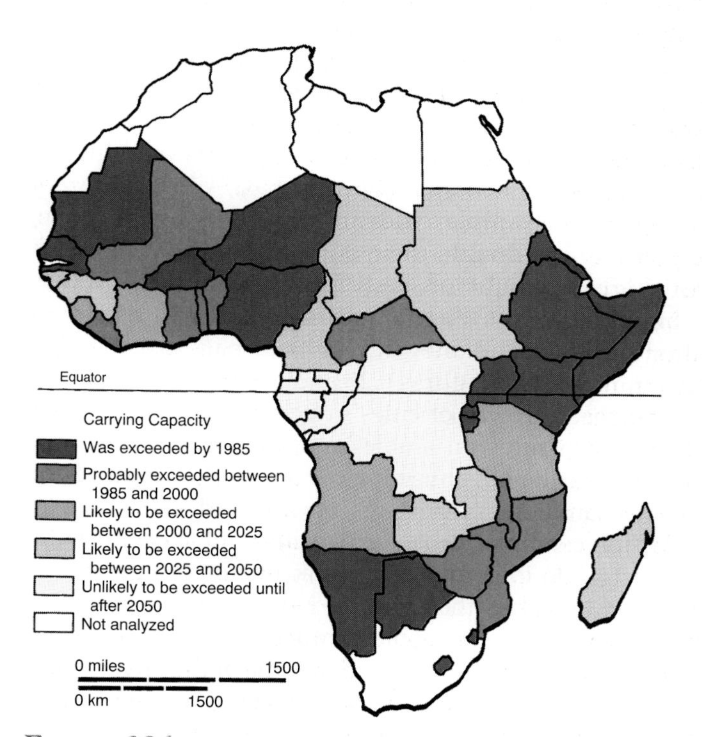

FIGURE 6.24 **Carrying capacity and potentials in sub-Saharan Africa.** The map assumes that (1) all cultivated land is used for growing food; (2) food imports are insignificant; (3) agriculture is conducted by low-technology methods.

Sources: World Bank; United Nations Development Programme; Food and Agriculture Organization (FAO); and Bread for the World Institute.

obviously does not rely on an arable land resource for its present development. Largely lacking in either agricultural or industrial resources, it nonetheless ranks well on all indicators of national well-being and prosperity. For countries such as Japan, a sudden cessation of the international trade that permits the exchange of industrial products for imported food and raw materials would be disastrous. Domestic food production could not maintain the dietary levels now enjoyed by their populations and they, more starkly than many underdeveloped countries, would be "overpopulated."

Urbanization

Pressures on the land resource of countries are increased not just by their growing populations but by the reduction of arable land caused by such growth. More and more of world population increase must be accommodated not in rural areas, but in cities that hold the promise of jobs and access to health, welfare, and other public services. As a result, the *urbanization* (transformation from rural to urban status) of population in developing countries is increasing dramatically. Since the 1950s, cities have grown faster than rural areas in nearly all developing states. Although Latin America, for example, has experienced substantial overall population increase, the size of its rural population is actually declining. Indeed, on UN projections, some 97% of all world population increase between 2000 and 2030 will be in urban areas and almost entirely within the developing regions and countries, continuing a pattern established by 1950 (Figure 6.25). In those areas collectively, cities are growing on average by over 3% a year, and the poorest regions are experiencing the fastest growth. In East, West, and Central Africa, for example, cities are expanding by 5% a year, a pace that can double their population every 14 years. Global urban population, just 750 million in 1950, grew to 2.7 billion by century's end and is projected to rise to 5.1 billion by 2030. The uneven results of past urbanization are summarized in Figures 6.26 and 12.2.

The sheer growth of cities in people and territory has increased pressures on arable land and adjusted upward both arithmetic and physiological densities. Urbanization consumes millions of hectares of cropland each year. In Egypt, for example, urban expansion and new development between 1965 and 1985 took out of production as much fertile soil as the massive Aswan dam on the Nile River made newly available through irrigation with the water it impounds. By themselves, some of the developing-world cities, often surrounded by concentrations of people living in uncontrolled settlements, slums, and shantytowns (Figure 6.27), are among the most densely populated areas in the world. They face massive problems in trying to provide housing, jobs, education, and adequate health and social services for their residents. These and other matters of urban geography are the topics of Chapter 12.

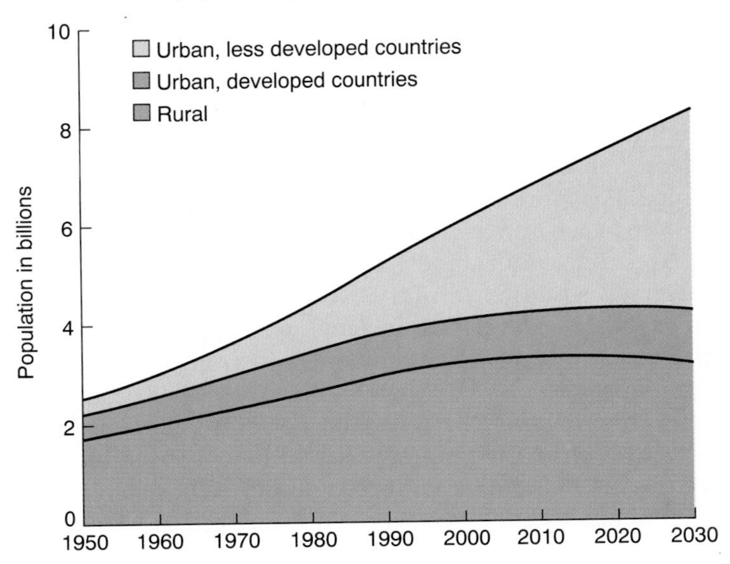

FIGURE 6.25 Past and projected urban and rural population growth. According to UN projections, some 65% of the world's total population will be urbanized by 2030.

Redrawn from *Population Bulletin* vol. 53, no. 1, Figure 3, page 12 (Population Reference Bureau, 1998).

Population Data and Projections

Population geographers, demographers, planners, governmental officials, and a host of others rely on detailed population data to make their assessments of present national and world population patterns and to estimate future conditions. Birth rates and death rates, rates of fertility and of natural increase, age and sex composition of the population, and other items are all necessary ingredients for their work.

Population Data

The data that students of population employ come primarily from the United Nations Statistical Office, the World Bank, the Population Reference Bureau, and ultimately, from national censuses and sample surveys. Unfortunately, the data as reported may on occasion be more misleading than informative. For much of the developing world, a national census is a massive undertaking. Isolation and poor transportation, insufficiency of funds and trained census personnel, high rates of illiteracy limiting the type of questions that can be asked, and populations suspicious of all things governmental serve to restrict the frequency, coverage, and accuracy of population reports.

However derived, detailed data are published by the major reporting agencies for all national units even when those figures are poorly based on fact or are essentially fictitious. For years, data on the total population, birth and death rates, and other vital statistics for Somalia were regularly reported and annually revised. The fact was, however, that Somalia had never had a census and had no system

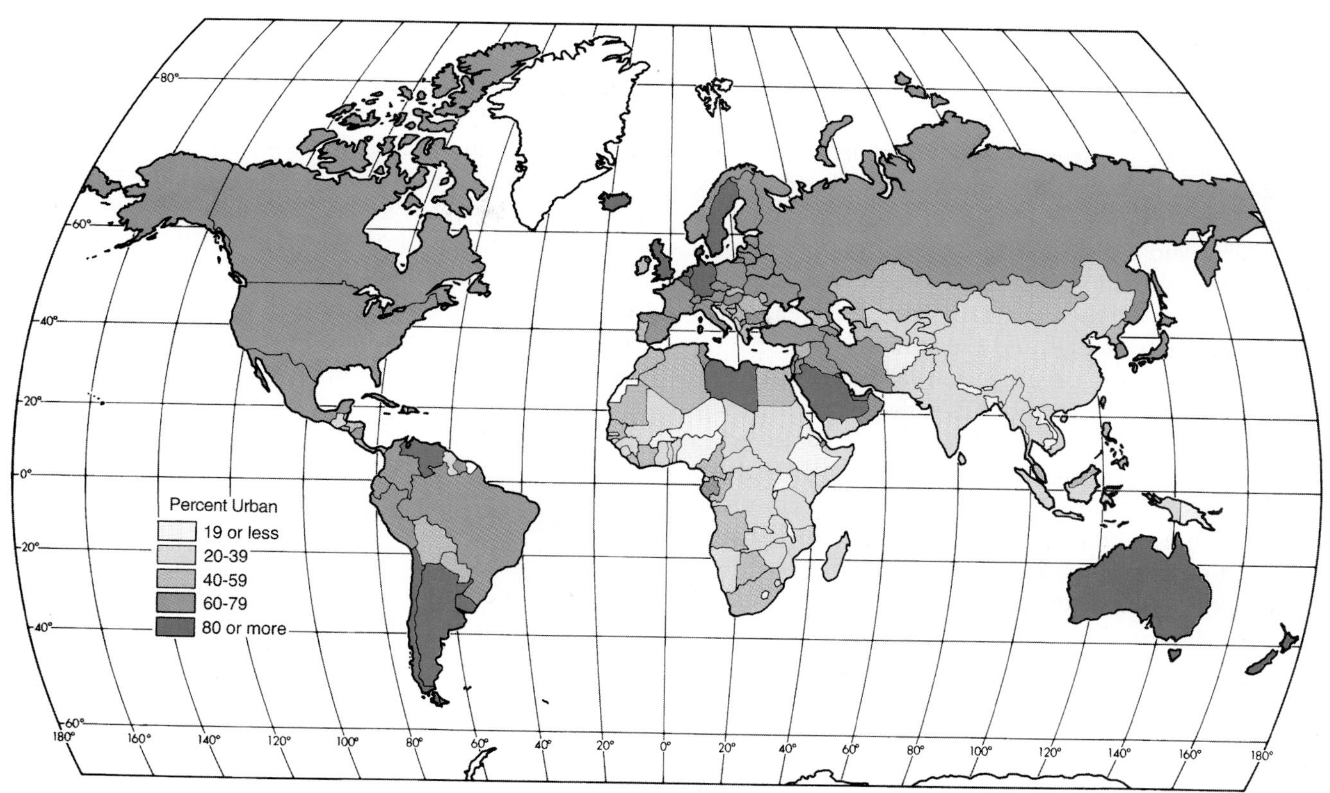

FIGURE 6.26 **Percentage of national population that is classified as urban.** Urbanization has been particularly rapid in the developing continents. In 1950, only 17% of Asians and 15% of Africans were urban; by the end of the 1990s some one-third of both Asians and Africans were city dwellers and collectively the less-developed areas contained two-thirds of the world's city population.
Source: Data from Population Reference Bureau.

FIGURE 6.27 Millions of people of the developing world live in shantytowns on the fringes of large cities, without benefit of running water, electricity, sewage systems, or other public services. The UN reports that up to 40% of all urban dwellers worldwide live in such squatter settlements and slums. The hillside slum pictured here is one of the many *favelas* that are home for nearly half of Rio de Janeiro's more than 11 million residents.

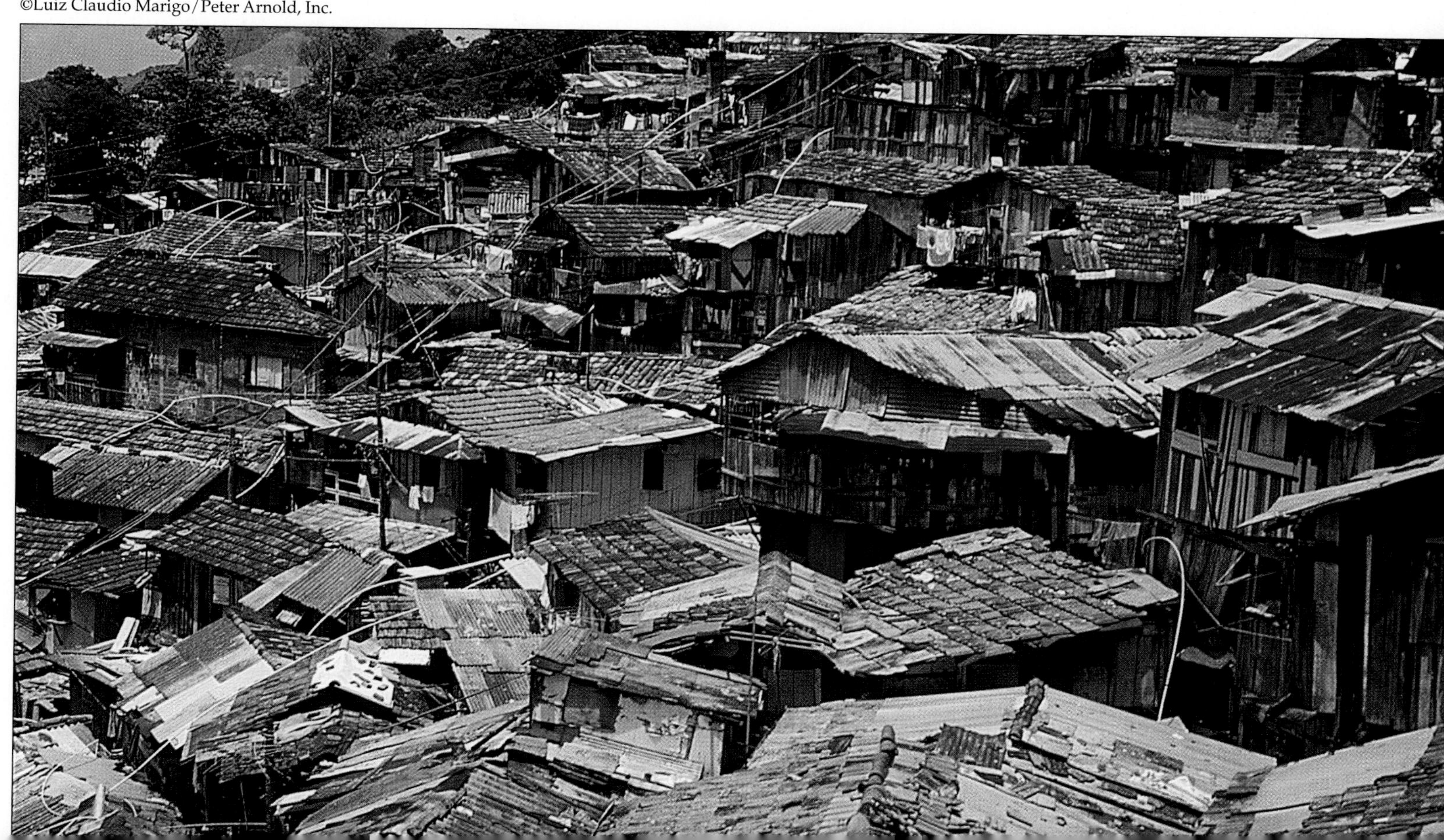

whatsoever for recording births. Seemingly precise data were regularly reported as well for Ethiopia. When that country had its first-ever census in 1985, at least one data source had to drop its estimate of the country's birth rate by 15% and increase its figure for Ethiopia's total population by more than 20%. And a disputed 1992 census of Nigeria officially reported a population of 88.5 million, still the largest in Africa but far below the then generally accepted and widely cited estimates of between 110 and 120 million Nigerians.

Fortunately, census coverage on a world basis is improving. Almost every county has now had at least one census of its population, and most have been subjected to periodic sample surveys (Figure 6.28). However, only about 10% of the developing world's population live in countries with anything approaching complete systems for registering births and deaths. Estimates are that 40% or less of live births in Indonesia, Pakistan, India, or the Philippines are officially recorded. Apparently, deaths are even less completely reported than births throughout Asia. And whatever the deficiencies of Asian states, African statistics are still less complete and reliable. It is, of course, on just these basic birth and death data that projections about population growth and composition are founded.

Even the age structure reported for national populations, so essential to many areas of population analysis, must be viewed with suspicion. In many societies, birthdays are not noted nor are years recorded by the Western calendar. Non-Western ways of counting age also confuse the record. The Chinese, for example, consider a person to be 1 year old at birth and increase that age by 1 year each (Chinese) New Year's Day. Bias and error arise from the common tendency of people after middle age to report their ages in round numbers ending in *0*. Also evident is a bias toward claiming an age ending in the number *5* or as an even number of years. Inaccuracy and noncomparability of reckoning added to incompleteness of survey and response conspire to cloud national comparisons in which age or the implications of age are important ingredients.

Population Projections

For all their inadequacies and imprecisions, current data reported for country units form the basis of **population projections,** estimates of future population size, age, and sex composition based on current data. Projections are not forecasts, and demographers are not the social science equivalent of meteorologists. Weather forecasters work with a myriad of accurate observations applied against a known, tested model of the atmosphere. The demographer, in contrast, works with sparse, imprecise, and missing data applied to human actions that will be unpredictably responsive to stimuli not yet evident.

FIGURE 6.28 Taking the census in rural China in 1982. The sign identifies the "Third Population Census. Mobile Registration Station." A new Fourth Population Census, requiring 7 million census workers to conduct, was undertaken on July 1, 1990.
United Nations Photo Library.

Population projections, therefore, are based on assumptions for the future applied to current data that are, themselves, frequently suspect. Since projections are not predictions, they can never be wrong. They are simply the inevitable result of calculations about fertility, mortality, and migration applied to each age cohort of a population now living, and the making of birth rate, survival, and migration assumptions about cohorts yet unborn. Of course, the perfectly valid *projections* of future population size and structure resulting from those calculations may be dead wrong as *predictions*.

Since those projections are invariably treated as scientific expectations by a public that ignores their underlying qualifying assumptions, agencies such as the UN that estimate the population of, say, Africa in the year 2025, do so by not one but by three or more projections: high, medium, and low, for example (see "World Population Projections"). For areas as large as Africa, a medium projection is assumed to benefit from compensating errors and statistically predictable behaviors of very large populations. For individual African countries and smaller populations, the medium projection may be much less satisfying. The usual tendency in projections is to assume that something like current conditions will be applicable in the future. Obviously, the more distant the future, the less likely is that assumption to remain true. The resulting observation should be that the further into the future the population structure of small areas is projected, the greater is the implicit and inevitable error (see Figure 6.12).

Population Controls

All population projections include an assumption that at some point in time population growth will cease and plateau at the replacement level. Without that assumption, future numbers become unthinkably large. For the world at unchecked present growth rates, there would be 1 trillion people three centuries from now, 4 trillion four centuries in the future, and so on. Although there is reasonable debate about whether the world is now overpopulated and about what either its optimum or maximum sustainable population should be, totals in the trillions are beyond any reasonable expectation.

Population pressures do not come from the amount of space humans occupy. It has been calculated, for example, that the entire human race could easily be accommodated within the boundaries of the state of Delaware. The problems stem from the food, energy, and other resources necessary to support the population and from the impact on the environment of the increasing demands and the technologies required to meet them. Rates of growth currently prevailing

World Population Projections

While the need for population projections is obvious, demographers face difficult decisions regarding the assumptions they use in preparing them. Assumptions must be made about the future course of birth and death rates and, in some cases, about migration.

Demographers must consider many factors when projecting a country's population. What is the present level of the birth rate, of literacy, and of education? Does the government have a policy to influence population growth? What is the status of women?

Along with these questions must be weighed the likelihood of socioeconomic change, for it is generally assumed that as a country "develops," a preference for smaller families will cause fertility to fall to the replacement level of about two children per woman. But when can one expect this to happen in less developed countries? And for the majority of more developed countries with fertility currently below replacement level, can one assume that fertility will rise to avert eventual disappearance of the population and, if so, when?

Predicting the pace of fertility decline is most important, as illustrated by one set of United Nations long-range projections for Africa. As with many projections, these were issued in a "series" to show the effects of different assumptions. The "low" projection for Africa assumed that replacement level fertility will be reached in 2030, which would put the continent's population at 1.4 billion in 2100. If attainment of replacement level fertility is delayed to 2065, the population would reach 4.4 billion in 2100. That difference of 3 billion should serve as a warning that using population projections requires caution and consideration of *all* the possibilities.

Unfortunately, demographers usually cast their projections in an environmental vacuum, ignoring the realities of soils, vegetation, water supplies, and climate that ultimately determine feasible or possible levels of population support. Inevitably, different analysts present different assessments of the absolute carrying capacity of the earth. At an unrealistically low level, the World Hunger Project calculated that the world's ecosystem can, with present agricultural technologies and with equal distribution of food supplies, support on a sustained basis no more than 5.5 billion people, a number already exceeded. Many agricultural economists, in contrast—citing present trends and prospective increases in crop yields, fertilizer efficiencies, and intensification of production methods—are confident that the earth can readily feed 10 billion or more on a sustained basis. Nearly all observers, however, agree that physical environmental realities make unrealistic purely demographically-based projections of a world population three or four times its present size.

in many countries make it nearly impossible for them to achieve the kind of social and economic development they would like.

Clearly, at some point population will have to stop increasing as fast as it has been. That is, either the self-induced limitations on expansion implicit in the demographic transition will be adopted or an equilibrium between population and resources will be established in more dramatic fashion. Recognition of this eventuality is not new. "[P]estilence, and famine, and wars, and earthquakes have to be regarded as a remedy for nations, as the means of pruning the luxuriance of the human race," was the opinion of the theologian Tertullian during the 2nd century A.D.

Thomas Robert **Malthus** (1766–1834), an English economist and demographer, put the problem succinctly in a treatise published in 1798: All biological populations have a potential for increase that exceeds the actual rate of increase, and the resources for the support of increase are limited. In later publications, Malthus amplified his thesis by noting the following:

1. Population is inevitably limited by the means of subsistence.
2. Populations invariably increase with increase in the means of subsistence unless prevented by powerful checks.
3. The checks that inhibit the reproductive capacity of populations and keep it in balance with means of subsistence are either "private" (moral restraint, celibacy, and chastity) or "destructive" (war, poverty, pestilence, and famine).

The deadly consequences of Malthus's dictum that unchecked population increases geometrically while food production can increase only arithmetically have been reported throughout human history, as they are today. Starvation, the ultimate expression of resource depletion, is no stranger to the past or present. By conservative estimate, 100 people worldwide will die of hunger and malnutrition during the two minutes it takes you to read this page; half will be children under five. They will, of course, be more than replaced numerically by new births during the same 2 minutes. Losses are always recouped. All battlefield casualties, perhaps 50 million, in all of humankind's wars over the last 300 years equal less than an eight-month replacement period at present rates of natural increase.

Yet, inevitably—following the logic of Malthus, the apparent evidence of history, and our observations of animal populations—equilibrium must be achieved between numbers and support resources. When overpopulation of any species occurs, a population dieback is inevitable. The madly ascending leg of the J-curve is bent to the horizontal, and the J-curve is converted to an S-curve. It has happened before in human history, as Figure 6.29 summarizes. The top of the **S-curve** represents a population size consistent with and supportable by the exploitable resource base. When the population is equivalent to the carrying capacity of the occupied area, it is said to have reached a **homeostatic plateau.**

In animals, overcrowding and environmental stress apparently release an automatic physiological suppressant of fertility. Although famine and chronic malnutrition may reduce fertility in humans, population limitation usually must be either forced or self-imposed. The demographic transition to low birth rates matching reduced death rates is cited as evidence that Malthus's first assumption was wrong: human populations do not inevitably grow geometrically. Fertility behavior, it was observed, is conditioned by social determinants, not solely by biological or resource imperatives.

Although Malthus's ideas were discarded as deficient by the end of the 19th century in light of the European population experience, the concerns he expressed were revived during the 1950s. Observations of population growth in underdeveloped countries and the strain that growth placed on their resources inspired the viewpoint that improvements in living standards could be achieved only by raising investment per worker. Rapid population growth was seen as a serious diversion of scarce resources away from capital investment and into unending social welfare programs. In order to lift living standards, the existing national efforts to lower mortality rates had to be balanced by governmental programs to reduce birth rates. **Neo-Malthusianism,** as this viewpoint became known, has been the underpinning of national and international programs of population limitation primarily through birth control and family planning (Figure 6.30).

Neo-Malthusianism has had a mixed reception. Asian countries, led by China and India, have in general—though with differing successes—adopted family planning programs and policies. In some instances, success has been declared complete. Singapore established its Population and Family Planning Board in 1965, when its fertility rate was 4.9 lifetime births per woman. By 1986, that rate had declined to 1.7, well below the 2.1 replacement level for developed countries, and the board was abolished as no

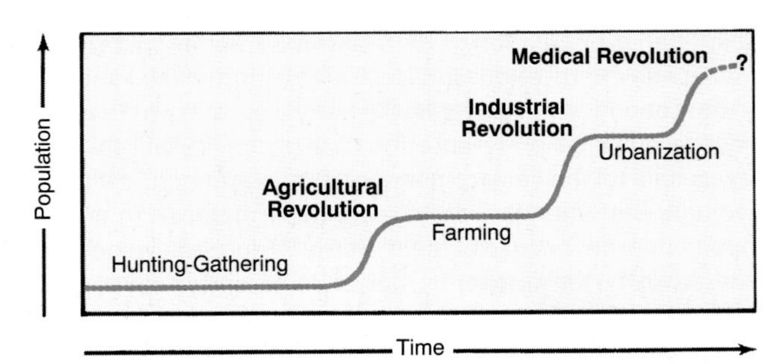

FIGURE 6.29 The steadily higher *homeostatic plateaus* (states of equilibrium) achieved by humans are evidence of their ability to increase the carrying capacity of the land through technological advance. Each new plateau represents the conversion of the J-curve into an S-curve.

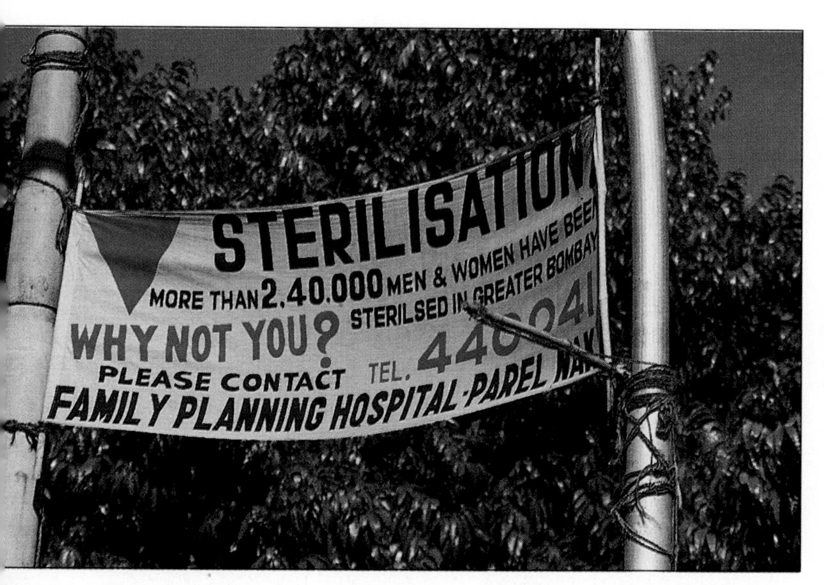

FIGURE 6.30 A Bombay, India, sign promoting the government's continuing program to reduce the country's high fertility rate. Sterilization is a major contraception practice in India, reportedly accounting for some 70% of all contraception there.
©Carl Purcell.

longer necessary. Caribbean and South American countries, except the poorest and most agrarian, have also experienced declining fertility rates, though often these reductions have been achieved despite pronatalist views of governments influenced by the Roman Catholic Church. Africa and the Middle East have generally been less responsive to the neo-Malthusian arguments because of ingrained cultural convictions among people, if not in all governmental circles, that large families—6 or 7 children—are desirable. Although total fertility rates have begun to decline in several sub-Saharan African states, they still remain nearly everywhere far above replacement levels. Islamic fundamentalism opposed to birth restrictions also is a cultural factor in the Near East and North Africa. However, the Muslim theocracy of Iran has recently endorsed a range of contraceptive procedures and developed one of the world's more aggressive programs to encourage small families.

Other barriers to fertility control exist. When first proposed by Western states, neo-Malthusian arguments that family planning was necessary for development were rejected by many less-developed countries. Reflecting both nationalistic and Marxist concepts, they maintained that remnant colonial-era social, economic, and class structures rather than population increase hindered development. Some government leaders think there is a correlation between population size and power and pursue pronatalist policies, as did Mao's China during the 1950s and early 1960s. And a number of American economists called *cornucopians* expressed the view, beginning in the 1980s, that population growth is a stimulus, not a deterrent, to development and that human minds and skills are the world's ultimate resource base. Since the time of Malthus, they observe, world population has grown from 900 million to 6 billion without the predicted dire consequences—proof that Malthus failed to recognize the importance of technology in raising the carrying capacity of the earth. Still higher population numbers, they suggest, are sustainable, perhaps even with improved standards of living for all.

A third view, modifying cornucopian optimism, admits that products of human ingenuity such as the Green Revolution (see page 357) increases in food production have managed to keep pace with rapid population growth since 1970. But its advocates argue that scientific and technical ingenuity to enhance food production does not automatically appear; both complacency and inadequate research support have hindered continuing progress in recent years. And even if further advances are made, they observe, not all countries or regions have the social and political will or capacity to take advantage of them. Those that do not, third-view advocates warn, will fail to keep pace with the needs of their populace and will sink into varying degrees of poverty and environmental decay, creating national and regional—though not necessarily global—crises.

Yet global crisis is exactly what is being predicted by some as the logical outcome of China's combination of expanding population and booming prosperity. Projecting from recent Chinese population trends, cropland and water scarcity, and increasing grain, dairy, and meat consumption, some worry that within 35 years China's demand for grain will so far exceed its own production capacity and place such massive demands on world grain supplies that global shortages and rocketing food costs will result. Ominously growing food scarcity, they fear, not military aggression, may be the real threat to future world economic and political stability.

POPULATION PROSPECTS

Regardless of population philosophies, theories, or cultural norms, the fact remains that in many parts of the world developing countries are showing significantly declining population growth rates. Global fertility and birth rates appear to be falling to an extent not anticipated by pessimistic Malthusians and at a pace that suggests stabilization of world population numbers sooner—and at lower totals—than previously projected (see "A Population Implosion?", p. 208). In all world regions, steady and continuous fertility declines have been recorded over the past quarter century, reducing fertility from global 5.0-children-per-woman levels in the early 1950s to less than 3 per woman by the end of the century.

But reducing fertility levels even to the replacement level of 2.1 births per woman does not mean an immediate end to population growth. Because of the age composition of many societies, numbers of births will continue to grow even as fertility rates per woman decline. The reason is to be found in **demographic** (or **population**) **momentum,** and the key to that is the age structure of a country's population.

When a high proportion of the population is young, the product of past high fertility rates, larger and larger numbers enter the childbearing age each year; that is the case for major parts of the world even at the end of the 20th century. The populations of developing countries are far younger than those of the established industrially developed regions (Figure 6.10), with about one-third (in Asia and Latin America) to almost one-half (in Africa) below the age of 15. The consequences of the fertility of these young people are yet to be realized. They will continue to be felt until the now youthful groups mature and work their way through the population pyramid.

Inevitably, while this is happening, even the most stringent national policies limiting growth cannot stop it entirely. A country with a large present population base will experience large numerical increases despite declining birth rates. Indeed, the higher fertility was to begin with and the sharper its drop to low levels, the greater will be the role of momentum even after rates drop below replacement. A simple comparison of South Korea and the United Kingdom may serve to demonstrate the point. The two countries had (in 1998) the same level of fertility, with women averaging about 1.7 children each. Between that year and 2025, the population of the U.K. is projected to decline by 2 million persons while much more youthful South Korea will continue growing, adding 5 million people.

Eventually, of course, young populations grow older, and even the youthful developing countries are beginning to face the consequences of that reality. The problems of a rapidly aging population that already confront the industrialized economies are now being realized in the developing world as well. According to UN projections, one in 10 persons worldwide will be age 65 or older in 2025; in 1999, the ratio was one in 15. However, the growth rate of people aged 55 and over is three times as high in developing countries as in the developed ones; in most, the rate is highest for those 75 and over. More than 1.2 million people worldwide reach the age of 55 each month; of that number, 80% live in developing countries that generally lack health, income, and social service support systems adequate to the needs of their older citizens. To the social and economic implications of their present population momentum, therefore, developing countries must add the aging consequences of past patterns and rates of growth (Figure 6.31).

FIGURE 6.31 These senior citizens practicing tai chi exercises in Beijing, China, are part of the rapidly aging population of many developing countries. Worldwide, the over-60 cohort will number nearly 2 billion by 2050, some 22% of total population and greater than the number of children less than 15 years of age. But by 2020, a third of Singapore citizens will be 55 or older and China will have as large a share of its population over 60—about one in four—as will Europe. Already, the numbers of old people in the world's poorer countries are beginning to dwarf those in the rich world. By 1999 there were 400 million persons over 60 in developing states, nearly twice as many as those in the advanced ones, but most are without the old-age assistance and welfare programs developed countries have put in place.

SUMMARY

Birth, death, fertility, and growth rates are important in understanding the numbers, composition, distribution, and spatial trends of population. Recent "explosive" increases in human numbers and the prospects of continuing population expansion may be traced to sharp reductions in death rates, increases in longevity, and the impact of demographic momentum on a youthful population largely concentrated in the developing world. Control of population numbers historically was accomplished through a demographic transition first experienced in European societies that adjusted their fertility rates downward as death rates fell and life expectancies increased. The introduction of advanced technologies of preventive and curative medicine, pesticides, and famine relief have reduced mortality rates in developing countries without, until recently, always a compensating reduction in birth rates. Recent fertility declines in many developing regions suggest the demographic transition is no longer limited to the advanced industrial countries and promise world population stability earlier and at lower numbers than envisioned just a few years ago.

Even with the advent of more widespread fertility declines, the 6 billion human beings present at the end of the 1990s will still likely grow to some 9.4 billion by the middle of the 21st century. That growth will largely reflect increases unavoidable because of the size and youth of populations in developing countries. Eventually, a new balance between population numbers and carrying capacity of the world will be reached, as it has always been following past periods of rapid population increase.

People are unevenly distributed over the earth. The ecumene, or permanently inhabited portion of the globe, is discontinuous and marked by pronounced differences in population concentrations and numbers. East Asia, South Asia, Europe, and northeastern United States/southeastern Canada represent the world's greatest population clusters, though smaller areas of great density are found in other regions and continents. Since growth rates are highest and population doubling times generally shorter in world regions outside these four present main concentrations, new patterns of population localization and dominance are taking form.

A respected geographer once commented that "population is the point of reference from which all other elements [of geography] are observed." Certainly, population geography is the essential starting point of the human component of the human–environment concerns of geography. But human populations are not merely collections of numerical units; nor are they to be understood solely through statistical analysis. Societies are distinguished not just by the abstract data of their numbers, rates, and trends, but by experiences, beliefs, understandings, and aspirations which collectively constitute that human spatial and behavioral variable called *culture.* It is to that fundamental human diversity that we next turn our attention.

KEY WORDS

arithmetic density 218
carrying capacity 219
cohort 194
crude birth rate (CBR) 194
crude death rate (CDR) 198
crude density 218
demographic equation 210
demographic (population) momentum 225
demographic transition 206
demography 192
dependency ratio 203
doubling time 204
ecumene 217
homeostatic plateau 224
J-curve 204
Malthus 224
mortality rate 198
natural increase 203
neo-Malthusianism 224
nonecumene 217
overpopulation 218
physiological density 218
population density 218
population geography 192
population projection 222
population pyramid 200
rate of natural increase 203
rates 194
replacement level 208
S-curve 224
total fertility rate (TFR) 196
zero population growth (ZPG) 208

FOR REVIEW AND CONSIDERATION

1. How do the *crude birth rate* and the *fertility rate* differ? Which measure is the more accurate statement of the amount of reproduction occurring in a population?
2. How is the *crude death rate* calculated? What factors account for the worldwide decline in death rates since 1945?
3. How is a *population pyramid* constructed? What shape of "pyramid" reflects the structure of a rapidly growing country? Of a population with a slow rate of growth? What can we tell about future population numbers from those shapes?
4. What variations do we discern in the spatial pattern of the *rate of natural increase* and, consequently, of population growth? What rate of natural increase would double population in 35 years?
5. How are population numbers projected from present conditions? Are projections the same as predictions? If not, in what ways do they differ?
6. Describe the stages in the demographic transition. Where has the final stage of the transition been achieved? What appears to be the applicability of the demographic transition to other parts of the world?
7. Contrast *crude population density* and *physiological density.* For what differing purposes might each be useful? How is *carrying capacity* related to the concept of density?
8. What was Malthus's underlying assumption concerning the relationship between population growth and food supply? In what ways do the arguments of *neo-Malthusians* differ from the original doctrine? What governmental policies are implicit in neo-Malthusianism?
9. Why is *demographic momentum* a matter of interest in population projections? In which world areas are the implications of demographic momentum most serious in calculating population growth, stability, or decline?

SELECTED REFERENCES

Ashford, Lori S. "New Perspectives on Population: Lessons from Cairo." *Population Bulletin* 50, no. 1. Washington, D.C.: Population Reference Bureau, 1995.

Bongaarts, John. "Population Pressure and the Food Supply System in the Developing World." *Population and Development Review* 22, no. 3 (1996):483–503.

Caldwell, John C., I. O. Orbulove, and Pat Caldwell. "Fertility Decline in Africa: A New Type of Transition?" *Population and Development Review* 19, no. 2 (1992):211–242.

Castles, Stephen, and Mark J. Miller. 2d ed. *The Age of Migration: International Population Movements in the Modern World.* New York: Guilford Press, 1998.

Cohen, Joel E. *How Many People Can the Earth Support?* New York: W. W. Norton, 1995.

Daugherty, Helen Ginn, and Kenneth C. Kammeyer. *An Introduction to Population.* 2d ed. New York: Guilford Publications, 1995.

Haub, Carl. "Understanding Population Projections." *Population Bulletin* 42, no. 4. Washington, D.C.: Population Reference Bureau, 1987.

Haupt, Arthur, and Thomas Kane. *Population Handbook.* 4th ed. Washington, D.C.: Population Reference Bureau, 1997.

Hirschman, Charles, and Philip Guest. "The Emerging Demographic Transitions of Southeast Asia." *Population and Development Review* 16, no. 1 (March 1990):121–152.

Hornby, William F., and Melvyn Jones. *An Introduction to Population Geography.* Cambridge, England: Cambridge University Press, 1993.

Keyfitz, Nathan, and Wilhelm Flinger. *World Population Growth and Aging: Demographic Trends in the Late 20th Century.* Chicago: University of Chicago Press, 1991.

Martin, Philip, and Jonas Widgren. "International Migration: A Global Challenge." *Population Bulletin* 51, no. 1. Washington, D.C.: Population Reference Bureau, 1996.

McFalls, Joseph A., Jr. "Population: A Lively Introduction." *Population Bulletin* 53, no. 3. 3d. ed. Washington, D.C.: Population Reference Bureau, 1998.

McIntosh, C. Alison, and Jason L. Finkle. "The Cairo Conference on Population and Development: A New Paradigm?" *Population and Development Review* 21, no. 2 (1995):223–260.

Olshansky, S. Jay, Bruce Carnes, Richard G. Rogers, and Len Smith. "Infectious Diseases—New and Ancient Threats to World Health." *Population Bulletin* 52, no. 2. Washington, D.C.: Population Reference Bureau, 1997.

"Population." *National Geographic.* October, 1998.

Pritchett, Lant H. "Desired Fertility and the Impact of Population Policies." *Population and Development Review* 20, no. 1 (March 1994):1–55.

Riley, Nancy E. "Gender, Power, and Population Change." *Population Bulletin* 52, no. 1. Washington, D.C.: Population Reference Bureau, 1997.

Robey, Bryant, Shea O. Rutstein, and Leo Morris. "The Fertility Decline in Developing Countries." *Scientific American* 269 (Dec. 1993):30–37.

United Nations. *Population and Women.* New York: United Nations, 1996.

United Nations Population Fund. *The State of World Population.* New York: United Nations. Annual.

World Resources Institute. "Population and Human Development." Chapter 8, pp. 173–199 in *World Resources 1996–97.* New York: Oxford University Press, 1996.

Don't forget about Dushkin's *Annual Editions Online: Geography* at http://www.dushkin.com/aeonline/. See preface for details.

CULTURAL GEOGRAPHY

CHAPTER

7

Noon prayers at Al Akhsa mosque on Temple Mount, Jerusalem.
©Erica Lansner/Tony Stone Images.

The Gauda's[1] son is eighteen months old. Every morning, a boy employed by the Gauda carries the Gauda's son through the streets of Gopalpur. The Gauda's son is clean; his clothing is elegant. When he is carried along the street, the old women stop their ceaseless grinding and pounding of grain and gather around. If the child wants something to play with, he is given it. If he cries, there is consternation. If he plays with another boy, watchful adults make sure that the other boy does nothing to annoy the Gauda's son.

Shielded by servants, protected and comforted by virtually everyone in the village, the Gauda's son soon learns that tears and rage will produce anything he wants. At the same time, he begins to learn that the same superiority which gives him license to direct others and to demand their services places him in a state of danger. The green mangoes eaten by all of the other children in the village will give him a fever; coarse and chewy substances are likely to give him a stomachache. While other children clothe themselves in mud and dirt, he finds himself constantly being washed. As a Brahmin [a religious leader], he is taught to avoid all forms of pollution and to carry out complicated daily rituals of bathing, eating, sleeping, and all other normal processes of life.

In time, the Gauda's son will enter school. He will sit motionless for hours, memorizing long passages from Sanskrit holy books and long poems in English and Urdu. He will learn to perform the rituals that are the duty of every Brahmin. He will bathe daily in the cold water of the private family well, reciting prayers and following a strict procedure. The gods in his house are major deities who must be worshipped every day, at length and with great care.[2]

The Gauda and his family are not Americans, as references and allusions in the preceding paragraphs make plain. The careful reader would infer, correctly, that they are Indians—and if one were to read more of the book from which this excerpt is taken, it would become evident that the Gauda lives in a village in southern India. The class structure, the religion, the language, the food, and other strands of the fabric of life mentioned in the passage place the Gauda, his family, and his village in a specific time and place. They bind the people of the region together as sharers of a common culture and set them off from those of other areas with different cultural heritages. The 6 billion people who were the subject of Chapter 6 are of a single human family, but it is a family differentiated into many branches, each characterized by a distinctive *culture.*

[1]Gauda = village headman

[2]Excerpt from *Gopalpur: A South Indian Village,* by Alan R. Beals, copyright © 1962 by Holt, Rinehart and Winston, Inc., and renewed 1990 by Alan R. Beals, reprinted by permission of the publisher.

To some writers in newspapers and the popular press, "culture" means the arts (literature, painting, music, etc.). To a social scientist, **culture** is the specialized behavioral patterns, understandings, and adaptations that summarize the way of life of a group of people. In this broader sense, culture is as much a part of the regional differentiation of the earth as are topography, climate, and other aspects of the physical environment. The visible and invisible evidences of culture—buildings and farming patterns, language and political organization—are elements in the spatial diversity that invites and is subject to geographic inquiry. Cultural differences in area result in human landscapes with variations as subtle as the differing "feel" of urban Paris, Moscow, and New York or as obvious as the sharp contrasts of rural Zimbabwe and the Prairie Provinces of Canada (Figure 7.1).

Since such differences exist, cultural geography exists, and one branch of cultural geography addresses a whole range of "why?" and "what?" and "how?" questions. Why, since humankind constitutes a single species, are cultures so varied? What are the most pronounced ways in which cultures and culture regions are distinguished? What were the origins of the different culture regions we now observe? How, from whatever limited areas in which single culture traits and amalgams developed, were they diffused over a wider portion of the globe? Why do cultural contrasts between recognizably distinct groups persist even in such presumed "melting pot" societies as that of the United States or in the outwardly homogeneous, long-established countries of Europe? These and similar questions are the concerns of the present chapter and, in part, of Chapters 8 and 9.

Components of Culture

Culture is transmitted within a society to succeeding generations by imitation, instruction, and example. It is learned, not biological, and has nothing to do with instinct or with genes. As members of a social group, individuals acquire integrated sets of behavioral patterns, environmental and social perceptions, and knowledge of existing technologies. Of necessity we learn the culture in which we are born and reared. But we need not—indeed, cannot—learn its totality. Age, sex, status, or occupation dictate the aspects of the cultural whole in which we become fully indoctrinated.

A culture, that is, despite overall generalized and identifying characteristics and even an outward appearance of uniformity, displays a social structure—a framework of roles and interrelationships of individuals and established groups. Each individual learns and adheres to the rules and conventions not only of the culture as a whole but, importantly, of those specific to the subgroup to which he or she belongs. And that subgroup may have its own recognized social structure.

Culture is a complexly interlocked web of behaviors and attitudes. Realistically, its full and diverse content cannot be

(a)

(b)

FIGURE 7.1 Cultural contrasts are clearly evident between (a) a subsistence maize plot in Zimbabwe and (b) the immense fields and mechanized farming of the Canadian prairies.
(a) ©Ian Murphy/Tony Stone Images.
(b) ©George Hunter/Tony Stone Images.

appreciated, and in fact may be wholly misunderstood, if we concentrate our attention only on limited, obvious traits. Distinctive eating utensils, the use of gestures, or the ritual of religious ceremony may summarize and characterize a culture for the casual observer. These are, however, individually insignificant parts of a much more complex structure that can be appreciated only when the whole is experienced.

Out of the richness and intricacy of human life we seek to isolate for special study those more fundamental cultural variables that give structure and spatial order to societies. We begin with culture traits, the smallest distinctive items of culture. **Culture traits** are units of learned behavior ranging from the language spoken to the tools used or to the games played. A trait may be an object (a fishhook, for example), a technique (weaving and knotting of a fish net), a belief (in the spirits resident in water bodies), or an attitude (a conviction that fish is superior to other animal flesh). Such traits are the most elementary expressions of culture, the building blocks of the complex behavioral patterns of distinctive groups of peoples.

Individual culture traits that are functionally interrelated comprise a **culture complex.** The existence of such complexes is universal. Keeping cattle was a *culture trait* of the Maasai of Kenya and Tanzania. Related traits included the measurement of personal wealth by the number of cattle owned; a diet containing the milk and blood of cattle; and disdain for labor unrelated to herding. The assemblage of these and other related traits yielded a *culture complex* descriptive of one aspect of Maasai society (Figure 7.2). In exactly the same way, religious complexes, business behavior

FIGURE 7.2 The formerly migratory Maasai of eastern Africa are now largely sedentary, partially urbanized, and frequently owners of fenced farms. Cattle formed the traditional basis of Maasai culture and were the evidence of wealth and social status. They provided as well the milk and blood important in the Maasai diet. Here, a herdsman catches blood released from a small neck incision he has just made.
©Kennan Ward/Corbis Media.

complexes, sports complexes, and others can easily be recognized in American or any other society.

Culture traits and complexes have spatial extent. When they are plotted on maps, the regional character of the components of culture is revealed. Geographers are interested in the spatial distribution of these individual elements, but their usual concern is with the **culture region,** a portion of the earth's surface occupied by people sharing recognizable and distinctive cultural characteristics that summarize their collective attributes or activities. Examples include the political organizations societies devise, their religions, their form of economy, and even their clothing, eating utensils, or housing. There are as many such culture regions as there are separate culture traits and complexes of population groups.

Finally, a set of culture regions showing related culture complexes and landscapes may be grouped to form a **culture realm.** The term recognizes a large segment of the earth's surface having fundamental uniformity in its cultural characteristics and showing a significant difference in them from adjacent realms. Culture realms are, in a sense, culture regions at the broadest scale of generalization. In fact, the scale is so broad and the diversity within the recognized realms so great that the very concept of realm may mislead more than it informs. One of the many possible divisions of human cultures into realms is offered in Figure 7.3. We will return to the idea and scope of culture realms near the end of this chapter.

Interaction of People and Environment

Culture develops in a physical environment that, in its way, contributes to differences among people. In primitive societies, the acquisition of food, shelter, and clothing—all parts of culture—depends on the utilization of the natural resources at hand. The interrelations of people with the environment of a given area, their perceptions and utilization of it, and their impact on it are consistent and interwoven themes of geography. They are the special concerns of those geographers exploring *cultural ecology,* the study of the relationship between a culture group and the natural environment it occupies.

Environments as Controls

Geographers have long dismissed as invalid and intellectually limiting the ideas of **environmental determinism**—the belief that the physical environment by itself shapes humans, their actions, and their thoughts. Environmental conditions alone cannot account for the cultural variations that occur around the world. Levels of technology, systems of social organization, and ideas about what is true and right have no obvious relationship to environmental circumstances.

The environment does place certain limitations on the human use of territory. However, such limitations must be

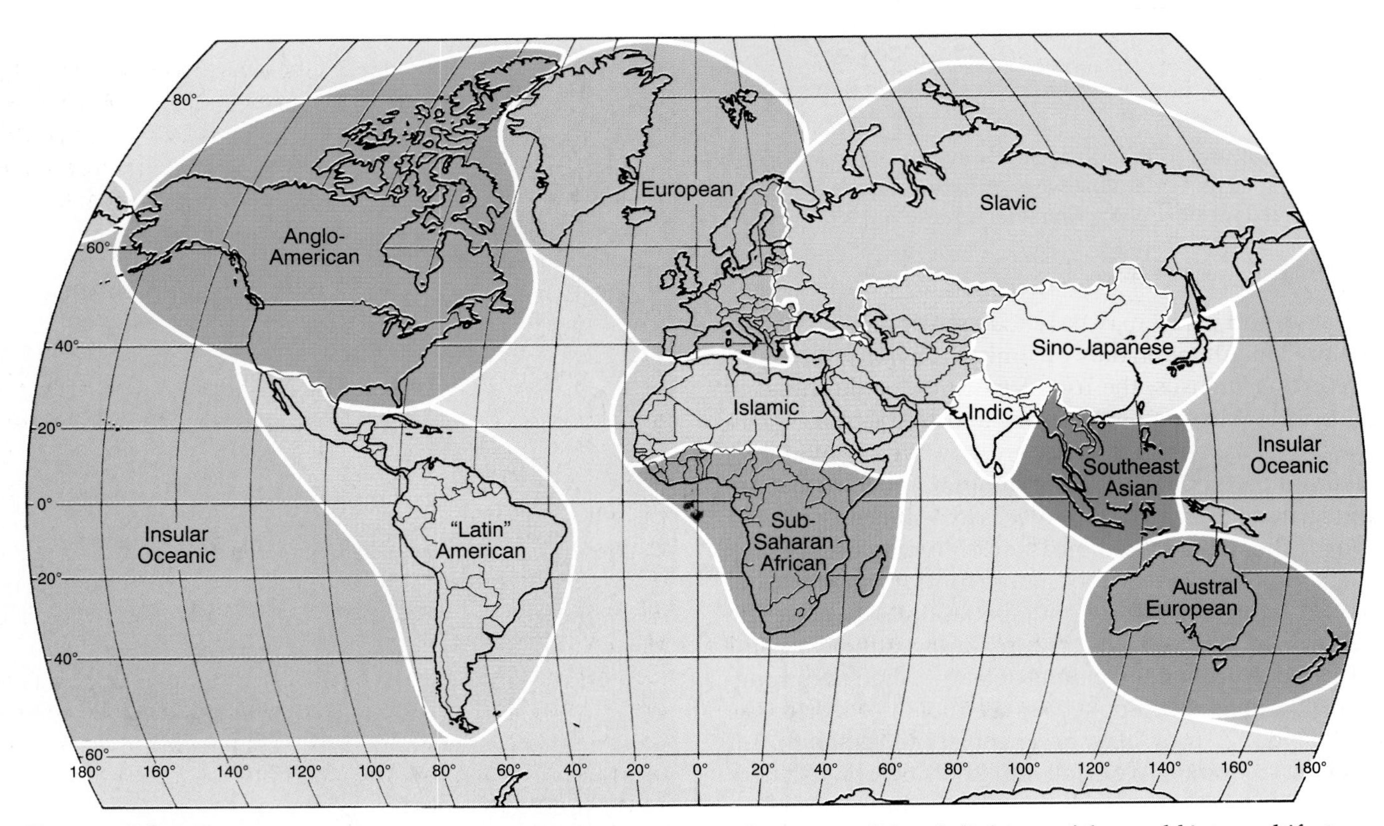

FIGURE 7.3 **Culture realms of the modern world.** This is just one of many possible subdivisions of the world into multifactor cultural regions.

seen not as absolute, enduring restrictions, but as relative to technologies, cost considerations, national aspirations, and linkages with the larger world. Human choices in the use of landscapes are affected by group perception of the possibility and desirability of their settlement and exploitation. These are not circumstances inherent in the land.

Possibilism is the viewpoint that people, not environments, are the dynamic forces of cultural development. The needs, traditions, and technological level of a culture affect how that culture both assesses the possibilities of an area and shapes the choices that it makes regarding them. Each society uses natural resources in accordance with its culture. Changes in a group's technical abilities or objectives bring about changes in its perceptions of the usefulness of the land. Of course, there are some environmental limitations on use of area. For example, if resources for feeding, clothing, or housing ourselves within an area are lacking, or if we do not recognize them there, there is no inducement for people to occupy that territory. Environments that do contain such resources provide the framework within which a culture operates.

FIGURE 7.4 The physical and cultural landscapes in juxtaposition. Advanced societies are capable of so altering the circumstances of nature that the cultural landscapes they create become controlling environments. The city of Cape Town, South Africa, is a "built environment" largely unrelated to its physical surroundings.
©Paul Almasy/Corbis Media.

Human Impacts

People are also able to modify their environment, and this is the other half of the human–environment relationship of geographic concern. Geography, including cultural geography, examines both the reactions of people to the physical environment and their impact on that environment. By using it we modify our environment—in part, through the material objects we place on the landscape: cities, farms, roads, and so on (Figure 7.4). The form these take is the product of the kind of culture group in which we live. The **cultural landscape,** the earth's surface as modified by human action, is the tangible, physical record of a given culture. House types, transportation networks, parks and cemeteries, and the size and distribution of settlements are among the indicators of the use that humans have made of the land.

As a rule, the more technologically advanced and complex the culture, the greater its impact on the environment, although preindustrial societies can and frequently do exert destructive pressures on the lands they occupy (see "Chaco Canyon Desolation"). In sprawling urban industrial societies, the cultural landscape has come to outweigh the natural physical environment in its impact on people's daily lives. It interposes itself between "nature" and humans. Residents of the cities of such societies—living and working in climate-controlled buildings, driving to enclosed shopping malls—can go through life with very little contact with or concern about the physical environment.

Subsystems of Culture

Understanding a culture fully is, perhaps, impossible for one who is not part of it. For analytical purposes, however, the traits and complexes of culture—its building blocks and expressions—may be grouped and examined as subsets of the whole. The anthropologist Leslie White suggested that a culture could be viewed as a three-part structure composed of subsystems that he termed *technological, sociological,* and *ideological.* In a similar classification, the biologist Julian Huxley identified three components of culture: *artifacts, sociofacts,* and *mentifacts.* Together, according to these interpretations, the subsystems—recognized by their separate components—comprise the structure of culture as a whole. But they are integrated; each reacts on the others and is affected by them in turn.

The **technological subsystem** is composed of the material objects and the techniques of their use by means of

Chaco Canyon Desolation

It is not certain when they first came, but by A.D. 1000 the Anasazi people were building a flourishing civilization in what are now the states of Arizona and New Mexico. In the Chaco Canyon alone they erected as many as 75 towns, all centered around pueblos, huge stone-and-adobe apartment buildings as tall as five stories and with as many as 800 rooms. These were the largest and tallest buildings of North America prior to the construction of iron-framed "cloud scrapers" in major cities at the end of the 19th century. An elaborate network of roads and irrigation canals connected and supported the pueblos. About A.D. 1200, the settlements were abruptly abandoned. The Anasazi, advanced in their skills of agriculture and communal dwelling, were—according to some scholars—forced to move on by the ecological disaster their pressures had brought to a fragile environment.

They needed forests for fuel and for the hundreds of thousands of logs used as beams and bulwarks in their dwellings. The pinyon-juniper woodland of the canyon was quickly depleted. For larger timbers needed for construction, they first harvested stands of ponderosa pine found some 40 kilometers (25 mi) away. As early as A.D. 1030 these, too, were exhausted, and the community switched to spruce and Douglas fir from mountaintops surrounding the canyon. When they were gone by 1200, the Anasazi fate was sealed—not only by the loss of forest, but by the irreversible ecological changes deforestation and agriculture had occasioned. With forest loss came erosion that destroyed the topsoil. The surface water channels that had been built for irrigation were deepened by accelerated erosion, converting them into enlarging arroyos, useless for agriculture.

The material roots of their culture destroyed, the Anasazi turned upon themselves and warfare convulsed the region. Smaller groups sought refuge elsewhere, recreating on reduced scale their pueblo way of life, but now in nearly inaccessible, highly defensible mesa and cliff locations. The destruction they had wrought destroyed the Anasazi in turn.

©Don Johnston/Photo/Nats.

which people are able to live. Such objects are the tools and other instruments that enable us to feed, clothe, house, defend, transport, and amuse ourselves. The **sociological subsystem** of a culture is the sum of those expected and accepted patterns of interpersonal relations that find their outlet in economic, political, military, religious, kinship, and other associations. The **ideological subsystem** consists of the ideas, beliefs, and knowledge of a culture and of the ways in which they are expressed in speech or other forms of communication.

The Technological Subsystem

Examination of variations in culture and in the manner of human existence from place to place centers on a series of commonplace questions: How do the people in an area make a living? What resources and what tools do they use to feed, clothe, and house themselves? Is a larger percentage of the population engaged in agriculture than in manufacturing? Do people travel to work in cars, on bicycles, or on foot? Do they shop for food or grow their own?

These questions concern the adaptive strategies used by different cultures in "making a living." In a broad sense, they address the technological subsystems at the disposal of those cultures—the instruments and tools people use in the daily cycle of existence. Huxley termed the material objects we employ to carry on our activities *artifacts.* For most of human history, people lived by hunting and gathering, taking the bounty of nature with only minimal dependence on weaponry, implements, and the controlled use of fire. Their adaptive skills were great, but their technological level was low. They had few specialized tools, could exploit only a limited range of potential resources, and had little or no control of nonhuman sources of energy. Their impact on the environment was small, but at the same time the "carrying capacity" of the land discussed in Chapter 6 was low everywhere, for technologies and artifacts were essentially the same among all groups.

The retreat of the last glaciers about 11,000 years ago marked the start of a period of unprecedented cultural development. It led from primitive hunting and gathering economies at the outset through the evolution of agriculture and animal husbandry to, ultimately, urbanization, industrialization, and the intricate complexity of the modern technological subsystem. Since not all cultures passed through all stages at the same time, or even at all, *cultural divergence* between human groups became evident.

Cultural diversity among ancient societies reflected the proliferation of technologies that followed a more assured food supply and made possible a more intensive and extensive utilization of resources. Different groups in separate environmental circumstances developed specialized tools and behaviors to exploit resources they recognized. Beginning with the Industrial Revolution of the 18th century, however, a reverse trend—toward commonality of technology—began.

Today, advanced societies are nearly indistinguishable in the tools and techniques at their command. They have experienced *cultural convergence*—the sharing of technologies, organizational structures, and even cultural traits and artifacts that is so evident among widely separated societies in a modern world united by instantaneous communication and efficient transportation. Those differences in technological traditions that still exist between developed and underdeveloped societies reflect, in part, national and personal wealth, stage of economic advancement and complexity, and, importantly, the level and type of energy used (Figure 7.5).

(a)

(b)

FIGURE 7.5 (a) This Balinese farmer, working with draft animals, employs tools typical of the low technological levels of subsistence economies. (b) Cultures with advanced technological subsystems use complex machinery to harness inanimate energy for productive use.

(a) ©Dave G. Houser/Hillstrom Stock Photos.
(b) ©R. Aguirre and G. Switkes/Amazonia.

In technologically advanced countries, many people are employed in manufacturing or allied service trades. Per capita incomes tend to be high, as do levels of education and nutrition, life expectancies, and medical services. These countries wield great economic and political power. In contrast, technologically less advanced countries have a high percentage of people engaged in farming, with much of the agriculture at a subsistence level (Figure 7.6). The gross national products or GNP (which measures the total domestic and foreign value added of all goods and services claimed by residents of a country during a year) of these countries are much lower than those of industrialized states. Per capita incomes (Figure 7.7), life expectancies, and literacy rates tend to be low.

Labels such as *advanced–less advanced, developed–underdeveloped,* or *industrial–nonindustrial* can mislead us into thinking in terms of "either–or." They may also be misinterpreted to mean, in general cultural terms, "superior–inferior." This belief is totally improper since the terms relate solely to economic and technological circumstances and bear no qualitative relationship to such vital aspects of culture as music, art, or religion.

Properly understood, however, terms and measures of economic development can reveal important national and world regional contrasts in the implied technological subsystems of different cultures and societies. Figure 7.8 suggests that technological status is relatively high in nearly all European countries and Japan, the United States, and Canada—the "North." Most of the less-developed countries are in Latin America, Africa, and southern Asia—the "South."[3]

It is important, however, to recognize that these implied national averages conceal internal contrasts. All countries include areas that are at different levels of development. We must also remember that technological development is a dynamic concept. It is most useful and accurate to think of the countries of the world as arrayed along an ever-changing continuum of technological levels and subsystems.

[3]The terms "North" and "South" to describe relative developmental levels were introduced in *North–South: A Programme for Survival* (sometimes called the *Brandt Report*), published in 1980 by the Independent Commission on International Development Issues. The former Soviet Union was at that time included within the North, and its successor states retain that association.

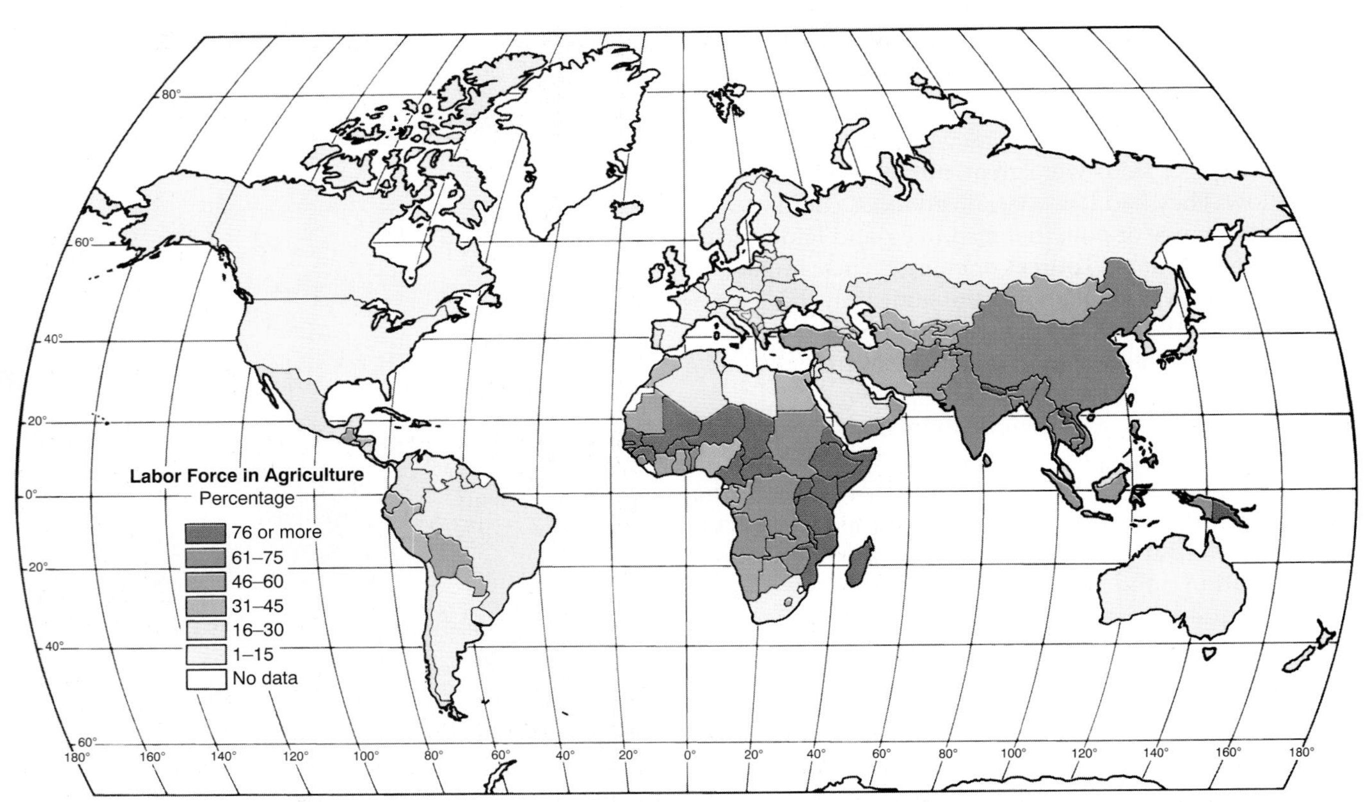

FIGURE 7.6 **Percentage of labor force engaged in agriculture.** Highly developed economies usually have relatively low proportions of their labor forces in the agricultural sector, but the contrast between advanced and underdeveloped countries in this labor force measure is diminishing. Rapid developing-world population growth has resulted in increased rural landlessness and poverty from which escape is sought by migration to cities. The resulting reduction in the agricultural labor force percentage is an expression of relocation of poverty and unemployment, not of economic advancement.

Source: Data from World Bank, and United Nations Development Programme.

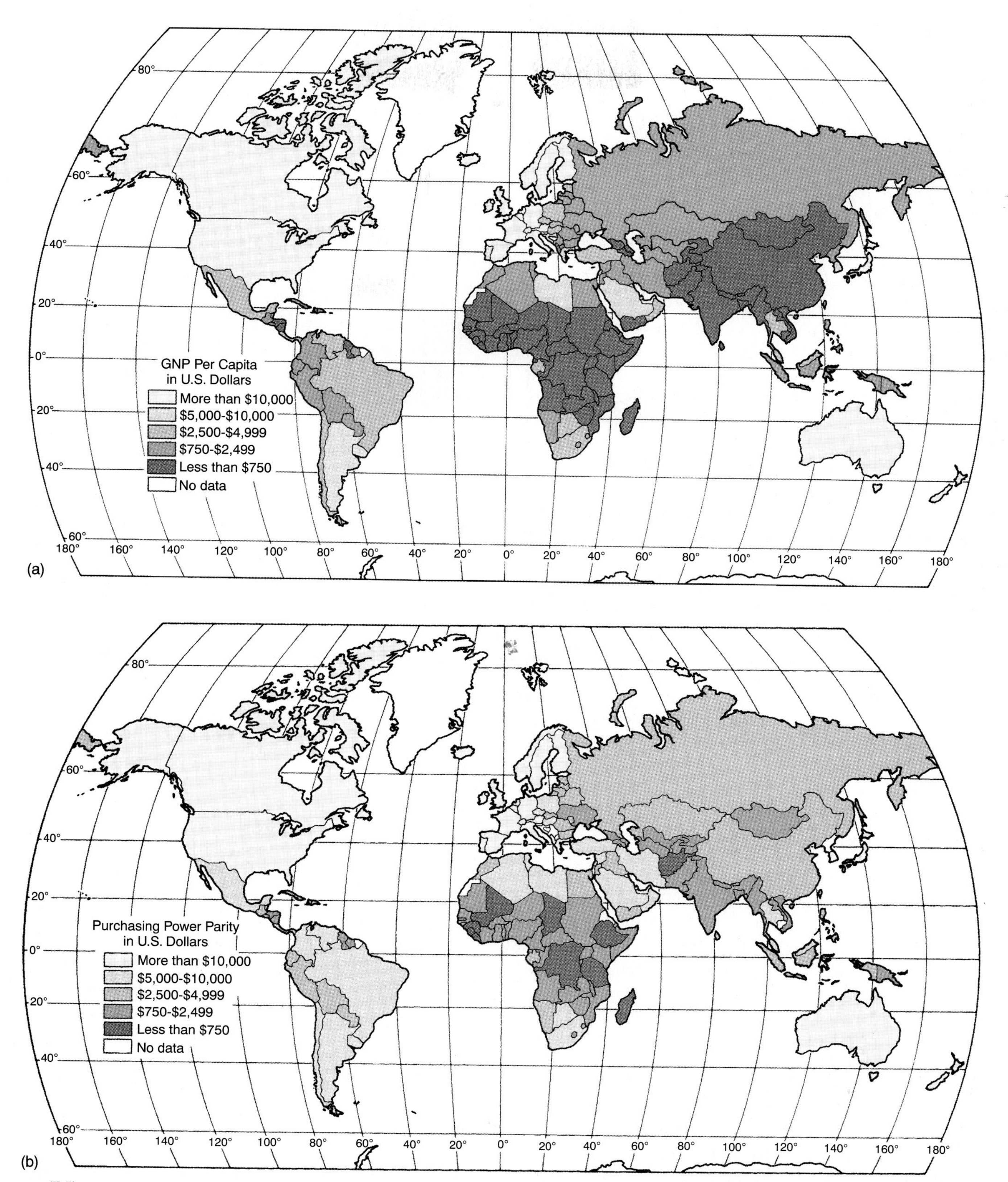

FIGURE 7.7 Two contrasting views of income. Both assume that total national income is evenly divided among all citizens, but each takes a different view of the meaning of the per capita income that division implies. (a) **Gross national product per capita.** GNP per capita expresses the individual's presumed income converted into U.S. dollars at official market exchange rates. Showing great contrasts between more- and less-advanced economies, it is a frequently employed summary of degree of technological development, though high incomes in sparsely populated, oil-rich countries may not have the same meaning in subsystem terms as do comparable per capita values in industrialized states. (b) **Purchasing Power Parity (PPP)** attempts to measure the relative domestic purchasing power of local currencies. By this more realistic measure, the abject poverty suggested by per capita GNP is seen to be much reduced in many developing countries.

(a) Sources: Data from the World Bank and Population Reference Bureau; (b) Sources: Data from World Bank and United Nations.

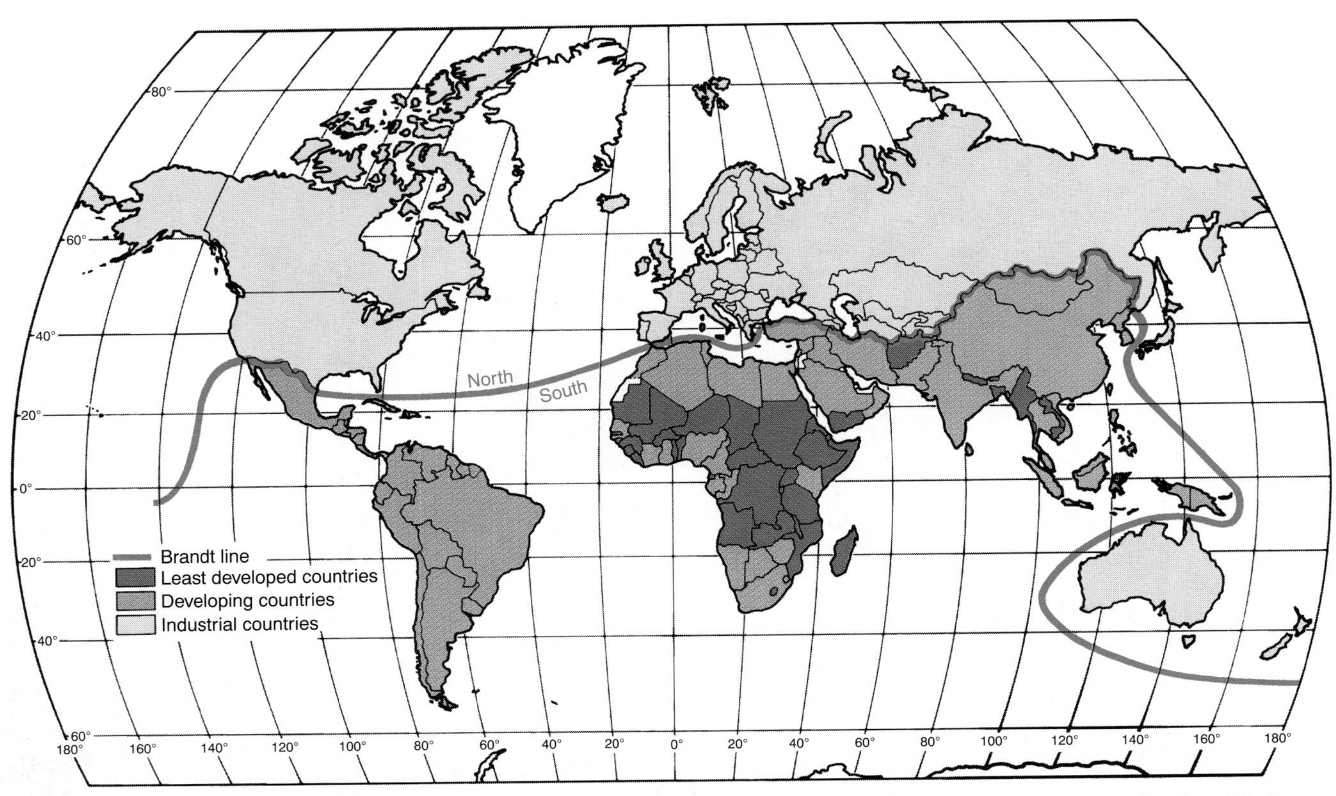

FIGURE 7.8 **Comparative development levels.** The "North–South" line of the 1980 *Brandt Report* suggested a simplified world contrast of development and underdevelopment based largely on degree of industrialization and per capita wealth. More recently, the United Nations General Assembly recognized 45 "least developed countries." That recognition reflects low ratings in three indicators: gross domestic product, share of manufacturing in the gross domestic product, and literacy rate. The "industrial countries" are those identified as most developed by the United Nations Development Program, a designation that ignores recent significant economic and social gains in several Asian and Latin American states raising them now to "developed" status.

The Sociological Subsystem

Continuum and change also characterize the religious, political, formal and informal educational, and other institutions that comprise the sociological subsystem of culture. Together, these *sociofacts* define the social organizations of a culture. They regulate how the individual functions relative to the group, whether it be family, church, or state.

There are no "givens" as far as patterns of interaction in any of these associations are concerned, except that most cultures possess a variety of formal and informal ways of structuring behavior. The importance to the society of the differing behavior sets varies among, and constitutes obvious differences between, cultures. Differing patterns of behavior are learned expressions of culture and are transmitted from one generation to the next by formal instruction or by example and expectation (Figure 7.9). The story of the Gauda's son that opened this chapter illustrates the point.

Social institutions are closely related to the technological system of a culture group. Thus, hunter-gatherers have one set of institutions and industrial societies quite different ones. Preagricultural societies tended to be composed of small bands based on kinship ties, with little social differentiation or specialization of function in the band; the San (Bushmen) of arid southern Africa and isolated rain forest groups in Amazonia might serve as modern examples (Figure 7.10).

The revolution in food production occasioned by plant and animal domestication beginning around 10,000 years ago touched off a social transformation that included increases in population, urbanization, work specialization, and structural differentiation within the society. Politically, the rules and institutions by which people were governed changed with the formation of sedentary, agricultural societies. Loyalty was transferred from the kinship group to the state; resources became possessions rather than the common property of all. Equally far-reaching changes occurred after the 18th-century Industrial Revolution, leading to the complex of human social organizations that we experience and are controlled by today in "developed" states and that increasingly affect all cultures everywhere.

Culture is a complexly intertwined whole. Each organizational form or institution affects, and is affected by, related culture traits and complexes in intricate and variable ways. Systems of land and property ownership and

(a)

(b)

(c)

(d)

FIGURE 7.9 All societies prepare their children for membership in the culture group. In each of these settings, certain values, beliefs, skills, and proper ways of acting are being transmitted to young people.

(a) ©Cary Wolinsky/Stock, Boston; (b) ©John Eastcott/Momatiuk/The Image Works; (c) ©Yigol Pardo/Hillstrom Stock Photo; (d) ©Paul Conklin/PhotoEdit.

control, for example, are institutional expressions of the sociological subsystem. They are, simultaneously, explicitly central to the classification of economies and to the understanding of spatial and structural patterns of economic development, as Chapter 10 will examine. Again, for each country the adopted system of laws and justice is a cultural variable identified with the sociological subsystem but extending its influence to all aspects of economic and social organization, including the political geographic systems discussed in Chapter 9.

The Ideological Subsystem

The third class of elements defining and identifying a culture comprises the ideological subsystem. This subsystem consists of ideas, beliefs, knowledge, and ways we express these things in our speech or other forms of communication. Mythologies, theologies, legend, literature, philosophy, folk wisdom, and commonsense knowledge make up this category. Passed on from generation to generation, these abstract belief systems, or *mentifacts*, tell us what we ought to believe, what we should value, and how we ought to act. Beliefs form the basis of the socialization process.

Often we know—or think we know—what the beliefs of a group are from written sources. Sometimes, however, we must depend on the actions or objectives of a group to tell us what its true ideas and values are. "Actions speak louder than words" and "Do as I say, not as I do" are commonplace recognitions of the fact that actions and words do not always coincide. The values of a group cannot be deduced from the written record alone.

Nothing in a culture stands totally alone. Changes in the ideas that a society holds may affect the sociological and technological systems just as, for example, changes in technology force changes in the social system. The abrupt alteration after World War I (1914–1918) of the ideological structure of Russia from a monarchical, agrarian, capitalistic system to an industrialized, communistic society involved sudden, interrelated alteration in all facets of that country's culture system. The equally abrupt disintegration of Russian

FIGURE 7.10 Hunter-gatherers practiced the most enduring lifestyle in human history, trading it for the more arduous life of farmers under the necessity to provide larger quantities of less diversified foodstuffs for a growing population. Unlike their settled farmer rivals and successors, among hunter-gatherers, age and sex differences not caste or economic status, were and are the primary basis for interpersonal relations and the division of labor. Here, a San (Bushman) hunter of Botswana, Africa, stalks his prey. Men also help contribute to the gathered food that constitutes 80% of the San diet.

communism in the early 1990s was similarly disruptive of all its established economic, social, and administrative organizations. The interlocking nature of all aspects of a culture is termed **cultural integration.**

The recognition of three distinctive subsystems of culture, while helping us to appreciate its structure and complexity, can at the same time obscure the many-sided nature of individual elements of culture. Cultural integration means that any cultural object or act may have a number of meanings. A dwelling, for example, is an artifact providing shelter for its occupants. It is, simultaneously, a sociofact reflecting the nature of the family or kinship group it is designed to house, and a mentifact summarizing a culture group's convictions about appropriate design, orientation, and building materials of dwelling units. In the same vein, clothing serves as an artifact of bodily protection appropriate to climatic conditions, available materials and techniques, or the activity in which the wearer is engaged. But garments also may be sociofacts, identifying an individual's role in the social structure of the community or culture, and mentifacts, evoking larger community value systems (Figure 7.11).

CULTURE CHANGE

The recurring theme of cultural geography is change. No culture is, or has been, characterized by a permanently fixed set of material objects, systems of organization, or even ideologies, although all of these may be long-enduring within a stable, isolated society. Such isolation and stability have always been rare. On the whole, while cultures are essentially conservative, they are always in a state of flux.

Many individual changes, of course, are so slight that they initially will be almost unnoticed, though collectively they may substantially alter the affected culture. Think of how the culture of the United States differs today from what it was in 1940—not in essentials, perhaps, but in the innumerable electric, electronic, and transportational devices that have been introduced and in the recreational, social, and behavioral adjustments they and other technological changes have wrought. Among these latter have been shifts in employment patterns to include greater participation by women in the waged workforce and associated changes in attitudes toward the role of women in the society at large. Such cumulative changes occur because the cultural traits of any group are not independent; they are clustered in a coherent and integrated pattern. Change on a small scale will have wide repercussions as associated traits also change to accommodate the adopted adjustment. Change, both major and minor, within cultures is induced by *innovation, spatial diffusion,* and *acculturation.*

Innovation

Innovation implies changes to a culture that result from ideas created within the social group itself and adopted by the culture. The novelty may be an invented improvement in material technology, like the bow and arrow or the jet engine. It may involve the development of nonmaterial forms of social structure and interaction: feudalism, for example, or Christianity.

(a)

(b)

(c)

FIGURE 7.11 (a) When clothing serves primarily to cover, protect, or assist in activities, it is an *artifact.* (b) Some garments are *sociofacts,* identifying a role or position within the social structure: the distinctive "uniforms" of a soldier, a cleric, or a beribboned ambassador immediately proclaim their respective roles in a culture's social organizations. (c) The mandatory chadors of Iranian women are *mentifacts,* indicative not specifically of the role of the wearer but of the values of the culture the wearer represents.
(a) ©Mark Antman/The Image Works.
(b) ©W. Marc Bernsau/The Image Works.
(c) ©Owen Franken/Stock, Boston.

Primitive and traditional societies characteristically are not innovative. In societies at equilibrium with their environment and with no unmet needs, change has no adaptive value and has no reason to occur. Indeed, all societies have an innate resistance to change. Complaints about youthful fads or the glorification of times past are familiar cases in point. However, when a social group is inappropriately unresponsive—mentally, psychologically, or economically—to changing circumstances and innovation, it is said to exhibit *cultural lag.*

Innovation—invention—frequently under stress, has marked the history of humankind. An expanded food base accompanied the pressures of growing populations at the end of the Ice Age. Domestication of plants and animals appears to have occurred independently in several recognizable areas of "invention" of agriculture, shown in Figure 7.12. From them, presumably, came a rapid diffusion of food types, production techniques, and new modes of economic and social organization as the majority of humankind changed from hunting-gathering to sedentary farming by no later than 2000

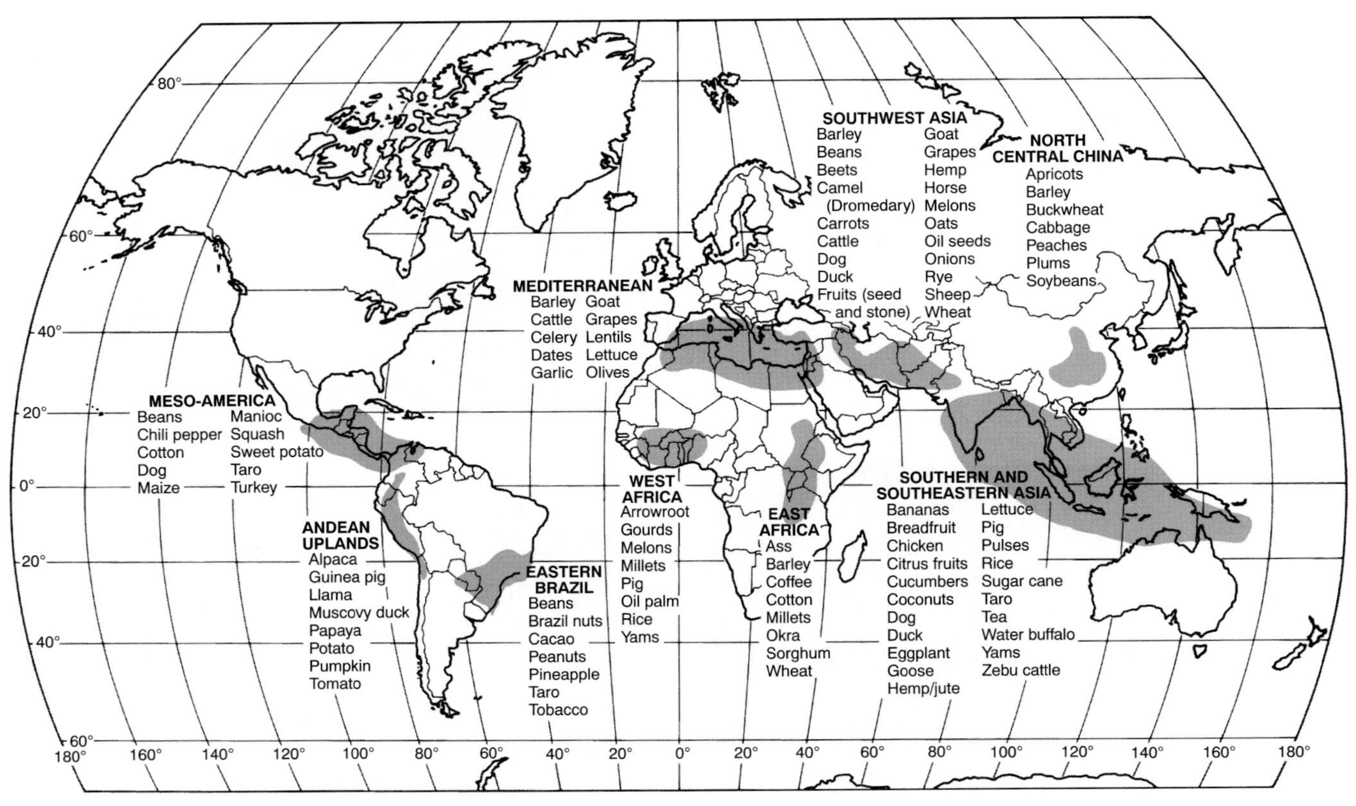

FIGURE 7.12 **Chief centers of plant and animal domestication.** The Southern and Southeastern Asian center was characterized by the domestication of plants such as taro, that are propagated by the division and replanting of existing plants (vegetative reproduction). Reproduction by the planting of seeds (e.g., maize and wheat) was more characteristic of Meso-America and Southwest Asia. The African and Andean areas developed crops reproduced by both methods. The lists of crops and livestock associated with the separate origin areas are selective, not exhaustive.

years ago. All innovation has a radiating impact upon the web of culture; the more basic the innovation, the more pervasive its consequences.

Few innovations in human history have been more basic than the Agricultural Revolution. It affected every aspect of society. Culture altered at an accelerating pace, and change itself became a way of life. Humans learned the arts of spinning and weaving plant and animal fibers. They learned to use the potter's wheel, to fire clay, and make utensils. They developed techniques of brick making, mortaring, and building construction. They discovered the skills of mining, smelting, and casting metals. Special local advantages in resources or products promoted the development of long-distance trading connections. On the foundation of such technical advancements a more complex exploitative culture appeared, including a stratified society to replace the rough equality of hunting and gathering economies.

The source regions of such social and technical revolutions were initially spatially confined. The term **culture hearth** is used to described those restricted areas of innovation from which key culture elements diffused to exert an influence on surrounding regions. The hearth may be viewed as the "cradle" of a culture group whose developed systems of livelihood and life created distinctive cultural landscapes. All hearth areas produced the trappings of *civilization,* which are usually assumed to include writing (or other forms of record keeping), metallurgy, long-distance trade connections, astronomy and mathematics, social stratification and labor specialization, formalized governmental systems, and a structured urban society. Several major culture hearths emerged, some as early as 7000–8000 years ago, following the initial revolution in food production. Prominent centers of early creativity were located in Egypt, Mesopotamia, the Indus Valley of the Indian subcontinent, northern China, southeastern Asia, and in several locations in Africa, the Americas, and elsewhere (Figure 7.13).

In most modern societies, innovative change has become common, expected, and inevitable, though it may be rejected by some of their separate culture groups (see "Folk and Popular Culture"). The rate of invention, at least as measured by the number of patents granted, has steadily increased, and the period between idea conception and product availability has been decreasing. A general axiom is that the more ideas available and the more minds able to exploit and combine them, the greater the rate of innovation. The spatial implication is that larger urban centers of

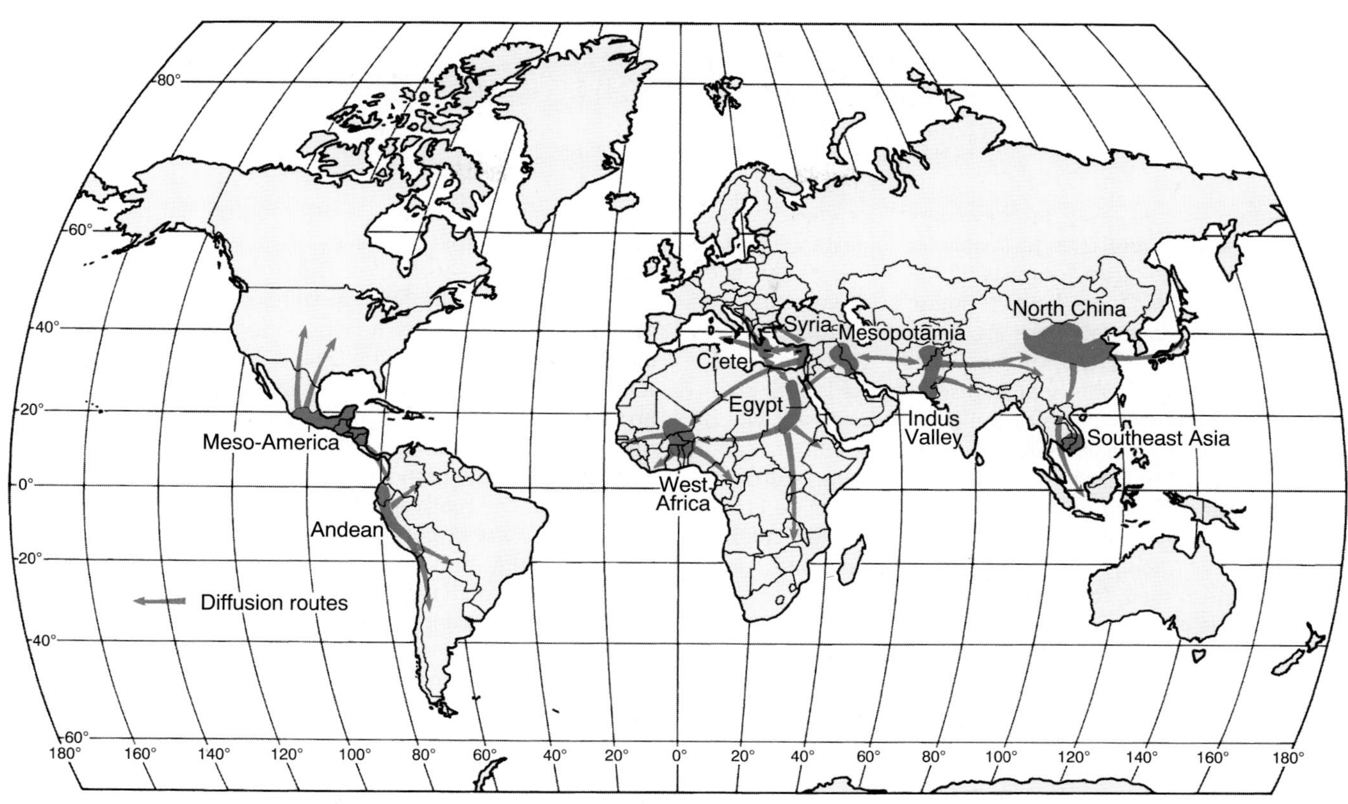

FIGURE 7.13 **Early culture hearths of the Old World and the Americas.**

advanced economies tend to be centers of innovation, not just because of their size but because of the number of ideas interchanged. Indeed, ideas not only stimulate new ideas, but also create circumstances in which new solutions must be developed to maintain the forward momentum of the society (Figure 7.14).

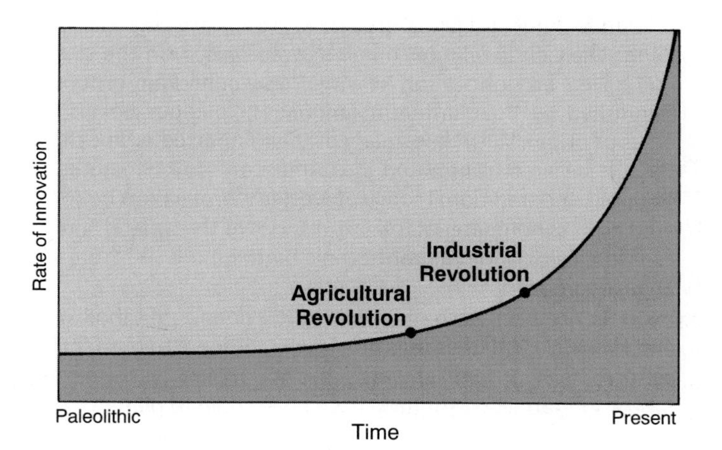

FIGURE 7.14 **The rate of innovation through human history.** Hunter-gatherers, living in easy equilibrium with their environment and their resource base, had little need for innovation and no necessity for cultural change. Increased population pressures were reduced by the Agricultural Revolution and the diffusion of the ideas and techniques of domestication, urbanization, and trade. With the Industrial Revolution, dramatic increases in innovation began to alter cultures throughout the world.

Diffusion

Spatial diffusion is the process by which a concept, practice, innovation, or substance spreads from its point of origin to new territories. Diffusion may assume a variety of forms, but basically two processes are involved. Either people move to a new area and take their culture with them (as the immigrants to the American colonies did), or information about an innovation (like barbed wire or hybrid corn) may spread throughout a culture. In either case, new ideas are transferred from their source region to new areas and to different culture groups. Spatial diffusion will be discussed in more detail in the following chapter.

It is not always possible to determine whether the existence of a culture trait in two different areas is the result of diffusion or of independent (or *parallel*) innovation. Cultural similarities do not necessarily prove that spatial diffusion has occurred. The pyramids of Egypt and of Central America most likely were separately conceived and are not evidence, as some have proposed, of pre-Columbian voyages from the Mediterranean to the Americas. A monument-building culture, after all, has only a limited number of shapes from which to choose.

Historical examples of independent, parallel invention are numerous: logarithms by Napier (1614) and Burgi (1620), the calculus by Newton (1672) and Leibnitz (1675),

Folk and Popular Culture

Not all culture groups, even in "developed" societies, readily adopt or adjust to cultural change. In general understanding, *culture* is understood to mean "our way of life"—how we act (and why), what we eat and wear, how we amuse ourselves, what we believe, whom we admire. There are distinctions to be made, however, on the universality of the "way of life" that is accepted.

Folk groups exist in either spatial or self-imposed social isolation from the common culture of the larger societies of which they are presumably a part. **Folk** connotes the traditional and nonfaddish way of life characteristic of a homogeneous, cohesive, largely self-sufficient group that is essentially isolated from or resistant to outside influences. Tradition controls folk culture, and resistance to change is strong. The homemade and handmade dominate in tools, food, music, story, and ritual. *Folk life* is a cultural whole composed of both tangible and intangible elements. **Material culture** is the tangible part, made up of physical, visible things: everything from musical instruments to furniture, tools, and buildings. In folk societies, these are products of the household or community itself, not of commercial mass-production. Their intangible **nonmaterial culture** comprises the mentifacts and sociofacts expressed in oral tradition, folk song and folk story, and customary behavior; ways of speech, patterns of worship, outlooks and philosophies are passed to following generations by teachings and examples.

Within Anglo America, true folk groups are few and dwindling. Their earlier, larger numbers were based on the customs and beliefs brought to the New World by immigrant groups distinguished by their different languages, religious beliefs, and areas of origin. With time, many of their imported *ethnic* characteristics became transmuted into American "folk" features. For example, the traditional songs of western Virginia can be considered both as nonmaterial folk expressions of the Upland South or as evidence of an immigrant ethnic heritage derived from rural English forebears.

In that respect, each of us bears the evidence of ethnic origin and folk life. Each of us uses proverbs traditional to our family or culture; each is familiar with childhood nursery rhymes and fables. We rap wood for luck, have heard how to plant a garden by phases of the moon, and know what is the "right" way to celebrate a holiday or prepare a favorite dish. For most, however, such evidences of folk culture are minor elements in our life, and only a few groups—such as the Old Order Amish with their rejection of electricity, the internal combustion engine, and other "worldly" accouterments in favor of buggy, hand tools, and traditional dress—remain in the United States as reminders of the folk cultural distinctions formerly widely recognizable. Canada, on the other hand, has retained a greater number of clearly recognizable ethnically unique folk and decorative arts traditions.

By general understanding, *popular culture* stands in opposition—and as the replacement for—folk culture. **Popular** implies the general mass of people, not the small-group distinctiveness and individuality of folk groups. It suggests a process of constantly adopting, conforming to, and quickly abandoning ever-changing fads and common modes of behavior. In that process, locally distinctive lifestyles and material and nonmaterial folk culture traits are largely replaced and lost; uniformity is substituted for differentiation, and small-group identity is eroded. For most of us, it is a sought-after uniformity. In the 1750s, George Washington wrote to his British agent to request . . . two pair of Work'd Ruffles . . . ; if work'd Ruffles shou'd be out of fashion send such as are not . . ." and "whatever goods you may send me . . . you will let them be fashionable." His desire, echoed today, was to fit in with the peer group and larger social milieu of which he was a part.

Popular culture may be seen as both a leveling and a liberating force. On the one hand, it serves to obliterate those locally distinctive folk culture lifestyles that emerge when groups remain isolated and self-sufficient. At the same time, however, individuals are exposed to a broader range of available opportunities—in clothing, foods, tools, recreations, and lifestyles—than ever are available to them in a cultural environment controlled by the restrictive and limited choices imposed by custom and isolation. Broad areal uniformity—in the form of the repetitive content of national discount stores, duplicate retailers in identical shopping malls, or the familiarity of ubiquitous fast-food chains—may displace folk cultural localisms. But it is a cultural uniformity vastly richer in content, diversity, and possibilities than any it replaces—though not necessarily an improvement in the social or religious values it contains or promotes.

the telephone by Elisha Gray and Alexander Graham Bell (1876) are commonly cited examples. It appears beyond doubt that agriculture was independently invented not only both in the New World and in the Old, but also in more than one culture hearth in each of the hemispheres.

All cultures are amalgams of innumerable innovations spread spatially from their points of origin and integrated into the structure of the receiving societies. It has been estimated that no more than 10% of the cultural items of any society are traceable to innovations created by its members, and that the other 90% come to the society through diffusion (see "A Homemade Culture").

Barriers to diffusion do exist, of course, as Chapter 8 explains. Generally, the closer and the more similar two cultural areas are to one another, the lower those barriers are and the greater is the likelihood of the adoption of an innovation, for diffusion is a selective process. Of course, the receiver culture may selectively adopt some goods or ideas from the donor society and reject others. The decision to adopt is governed by the receiving group's own culture.

A Homemade Culture

Reflecting on an average morning in the life of a "100% American," Ralph Linton noted:

> Our solid American citizen awakens in a bed built on a pattern which originated in the Near East but which was modified in Northern Europe before it was transmitted to America. He throws back covers made from cotton, domesticated in India, or linen, domesticated in the Near East, or wool from sheep, also domesticated in the Near East, or silk, the use of which was discovered in China. All of these materials have been spun and woven by processes invented in the Near East. . . . He takes off his pajamas, a garment invented in India, and washes with soap invented by the ancient Gauls. . . .
>
> Returning to the bedroom, . . . he puts on garments whose form originally derived from the skin clothing of the nomads of the Asiatic steppes [and] puts on shoes made from skins tanned by a process invented in ancient Egypt and cut to a pattern derived from the classical civilizations of the Mediterranean. . . . Before going out for breakfast he glances through the window, made of glass invented in Egypt, and if it is raining puts on overshoes made of rubber discovered by the Central American Indians and takes an umbrella invented in southeastern Asia. . . .
>
> [At breakfast] a whole new series of borrowed elements confronts him. His plate is made of a form of pottery invented in China. His knife is of steel, an alloy first made in southern India, his fork a medieval Italian invention, and his spoon a derivative of a Roman original. He begins breakfast with an orange, from the eastern Mediterranean, a cantaloupe from Persia, or perhaps a piece of African watermelon. With this he has coffee, an Abyssinian plant. . . . [H]e may have the egg of a species of bird domesticated in Indo-China, or thin strips of flesh of an animal domesticated in Eastern Asia which have been salted and smoked by a process developed in northern Europe.
>
> When our friend has finished eating . . . he reads the news of the day, imprinted in characters invented by the ancient Semites upon a material invented in China by a process invented in Germany. As he absorbs the accounts of foreign troubles he will, if he is a good conservative citizen, thank a Hebrew deity in an Indo-European language that he is 100 percent American.

Political restrictions, religious taboos, and other social customs are cultural barriers to diffusion. The French Canadians, although geographically close to many centers of diffusion, such as Toronto, New York, and Boston, are only minimally influenced by them. Both their language and culture complex govern French Canadian selective acceptance of Anglo influences. Traditional groups, perhaps controlled by firm religious conviction, may very largely reject culture traits and technologies of the larger society in whose midst they live (see "Folk and Popular Culture" and Figure 7.15).

Adopting cultures do not usually accept intact items originating from the outside. Diffused ideas and artifacts commonly undergo some alteration of meaning or form that makes them acceptable to a borrowing group. The process of the fusion of the old and new, called **syncretism,** is a major feature of culture change. It can be seen in alterations to religious ritual and dogma made by convert societies seeking acceptable conformity between old and new beliefs; the mixture of Catholic rites and voodooism in Haiti is an example. On a more familiar level, syncretism is reflected in subtle or blatant alterations of imported cuisines to make them conform to the demands of America's fast-food franchises.

Acculturation

Acculturation is the process by which one culture group undergoes a major modification by adopting many of the characteristics of another, usually dominant, culture group. In practice, acculturation may involve changes in the original cultural patterns of either or both of two groups involved in prolonged firsthand contact. Such contact and subsequent cultural alteration may occur in a conquered or colonized region. Very often the subordinate or subject population is forced to acculturate or does so voluntarily, overwhelmed by the superiority in numbers or the technical level of the conqueror.

The tribal Europeans in areas of Roman conquest, native populations in the wake of Slavic occupation of Siberia, and Native Americans following European settlement of North America experienced this kind of acculturation. In a different fashion, it is evident in the changes in Japanese political organization and philosophy imposed by occupying Americans after World War II or in the Japanese adoption of some more frivolous aspects of American life (Figure 7.16). In turn, American life was enriched by awareness of Japanese food, architecture, and philosophy, demonstrating the two-way nature of acculturation.

FIGURE 7.15 Motivated by religious conviction that the "good life" must be reduced to its simplest forms, the Amish community of east central Illinois shuns all modern luxuries of the majority, secular society around them. Children use horse and buggy, not school bus or automobile, on the daily trip to their rural school.
Courtesy of Jean Fellman.

FIGURE 7.16 Baseball, an import from America, is one of the most popular sports in Japan, attracting millions of spectators annually.
©Geoffrey Hiller/Leo de Wys, Inc.

On occasion, the invading group is assimilated into the conquered society as, for example, the older, richer Chinese culture prevailed over that of the conquering tribes of invading Mongols during the 13th and 14th centuries. The relationship of a mother country to its colony may also result in permanent changes in the culture of the colonizer even though little direct population contact is involved. The early European spread of tobacco addiction (see "Documenting Diffusion," in Chapter 8) may serve as an example, as can the impact on Old World diets and agriculture of potatoes, maize, and turkeys introduced from America.

On-Line Cultural Geography

The subject matter of cultural geography encompasses nearly all the topics and concerns of human life. Most of those topics have websites reflecting the myriad interests of private citizens and groups, academic and nongovernmental organizations, domestic and international agencies, and the like. Just a few of the ever-changing and multiplying home page resource lists providing links to those special agencies and groups—and that reflect the principal topics of this chapter—are indicated here.

Language *Foreign Language Resources on the Web* offers links to web pages with foreign language/culture-specific interests. Arranged by language, each citation contains a brief summary of the website contents and emphasis.

http://www.itp.berkeley.edu/~thorne/HumanResources.html

Language Links is a listing of multilanguage sites oriented towards language instructors but with significant cultural content.

http://polyglot.Lss.wisc.edu/lss/lang/langlink.html

Ethnologue is a near-complete review of all the world's languages; it lists 6703 of them. They can be viewed by area and by country. A small but increasing number of maps shows language distribution within individual countries, while a language-family index shows how the various languages are related—one guide to past human migration patterns.

http://www.sil.org/ethnologue/ethnologue.html

Religion Many universities and seminaries host home pages giving access to databases and websites concerned with religions and religious matters. Like many others, the University of Chicago Library *Religion* page provides links to other "Internet Sites of Interest."

http://www.lib.uchicago.edu/LibInfo/SourcesBySubject/Religion

World Internet Directory: Religion is an exhaustive listing of websites representing a massive array of religious, atheist, and agnostic interests.

http://www.tradenet.it:80/links/arsocu/religion.html

Ethnicity Several U.S. universities or their members maintain websites concerned with various aspects of ethnic studies. The University of California-Santa Barbara serves as host to Alan Liu's *Voice of the Shuttle: Minority Studies Page,* with links to general resource sites and ethnic-specific home pages (African American; Asian American; Chicano, Latino, Hispanic, etc.).

http://humanitas.ucsb.edu/shuttle/minority.html

Kansas State's *National Association for Ethnic Studies* provides linked references to other university ethnic studies programs and to a number of ethnic websites and other ethnic-related resources. It may be found at

http://www.ksu.edu/ameth/naes/ethnic.htm

Gender The *Feminist Internet Gateway* is a compilation of global feminism sites maintained by organizations. The sites are resource centers providing information on publications, position papers, and international meetings and programs.

http://www.feminist.org/gateway/gl_exec.html

Feminism and Women's Resources is a listing of women's studies and women-related sources on the Internet; wide-ranging, but with a Canadian emphasis.

http://www.ibd.nrc.ca/~mansfield/feminism/

Women Watch—The UN Internet Gateway on the Advancement and Empowerment of Women provides information on national and international efforts for women's interests and advancement. The site features information of UN global conferences on women, presents summaries of regional plans of action furthering female empowerment, and includes a collection of annotated related links.

http://www.un.org/womenwatch/

Popular culture *Sarah Zupko's Cultural Studies Center* is a collection of annotated links to popular culture and cultural studies, including journals, articles, academic programs, bibliographic references, film, television, etc.

http://www.popcultures.com

In the modern period the population relocations and immigration impacts discussed in Chapter 6 (p. 210) have resulted in unprecedented cultural mixings throughout the world. The traditional "melting pot"—more formally, **amalgamation theory**—view of immigrant integration into, for example, United States or Canadian society suggests that the receiving society and the varied arriving newcomer groups eventually merge into a composite mainstream culture incorporating the many traits of its collective components. More realistically, in order to be accepted, newcomer groups must learn the accustomed patterns of behavior and response and the dominating language of the workplace and government of the culture they have entered. Acculturation for them involves the adoption of the values, attitudes, ways of behavior, and speech of the receiving society. In that process, the immigrant group loses its separate cultural identity to the extent that it accepts over time the culture of the larger host community.

When that integration process is completed, **assimilation** has occurred. But assimilation does not necessarily mean that consciousness of original cultural identity is reduced or lost. *Competition theory,* in fact, suggests that as cultural minorities begin to achieve success and enter into mainstream social and economic life, awareness of cultural differences may be heightened, transforming the strengthening immigrant group into a self-assertive minority pursuing goals and interests that defend and protect its position within the larger society. Carried to extremes, militant minority self-assertion may result in the loss of the larger social and cultural integration the acculturation process seeks to assure.

Cultural Diversity

We began our discussion of culture with its subsystems of technological, sociological, and ideological content. We have learned that the distinctive makeup of those subsystems—the combinations and interactions of traits and complexes characteristic of particular cultures—is subjected to, and the product of, change through innovation, spatial diffusion, adoption, and acculturation. Those processes of cultural development and alteration have not, however, led to a homogenized world culture even after thousands of years of cultural contact and exchange since the origins of agriculture.

It is true, as we earlier observed, that in an increasingly integrated world, access to the material trappings and technologies of modern life and economy is widely available to all peoples and societies. As a result, important cultural commonalities have developed. Nevertheless, all of our experience and observation indicate a world still divided, not unified, in culture. Our concern as geographers is to identify those traits of a culture that have spatial expression and also to indicate how that culture is significantly different from other culture complexes.

We may reject as superficial and meaningless generalizations derived from trivialities: the foods people eat for breakfast, for example, or the kinds of eating implements they use. This rejection is a reflection of the kinds of understanding and the level of generalization we seek. There is no single most appropriate way to designate or recognize a culture or to delimit a culture region. As geographers concerned with world systems, we are interested in those aspects of culture that vary over extensive regions of the world and differentiate societies in a broad, summary fashion.

Language, religion, ethnicity, and gender meet our criteria and are among the most prominent of the differentiating cultural traits of societies and regions. Language and religion are basic components of culture, helping to identify who and what we are as individuals and clearly placing us within larger communities of persons with similar characteristics. In our earlier terminology, they are mentifacts, components of the ideological subsystem of culture that help shape the belief system of a society and transmit it to succeeding generations.

Ethnicity is a cultural summary rather than a single trait. It is based on the firm understanding by members of a group that they are in some fundamental ways different from others who do not share their distinguishing composite characteristics, which may include language, religion, national origin, unique customs, or other identifiers. Like language and religion, ethnicity has spatial identification. Like them, too, it may serve as an element of diversity and division within culturally complex societies and states.

Language, religion, and ethnicity treat all members (or all adult members) of a society uniformly. Yet among the most prominent strands in the fabric of culture are the social structures (sociofacts) and relationships that establish distinctions between males and females in the duties assigned and the rewards afforded to each. Gender is the reference term recognizing those socially created distinctions. It conditions the way people use space and assess economic and cultural possibilities of area, and assures that the status of women is a cultural spatial variable of fundamental significance.

Language

Forever changing and evolving, language in spoken or written form makes possible the cooperative efforts, the group understandings, and shared behavior patterns that distinguish culture groups. *Language,* defined simply as an organized system of speech by which people communicate with each other with mutual comprehension, is the most important medium by which culture is transmitted. Language enables parents to teach their children what the world they live in is like and what they must do to become functioning members of society. Some argue that the language of a society structures the perceptions of its speakers. By the words it contains and the concepts it can formulate, language is said to determine the attitudes, the understandings, and the responses of the society. Language therefore may be both a cause and a symbol of cultural differentiation (Figure 7.17).

If that conclusion is true, one aspect of cultural heterogeneity may be easily understood. The some 6 billion people on earth speak many thousands of different languages. Knowing that as many as 1500 languages and dialects are spoken in sub-Saharan Africa gives us a clearer appreciation of the political and social divisions in that continent. Europe alone has more than 100 languages and dialects. Language is a hallmark of cultural diversity, and the present world distribution of major languages (Figure 7.18) records not only the migrations and conquests of our linguistic ancestors but also the continuing dynamic pattern of human movements, settlements, and colonizations of more recent centuries.

Languages differ greatly in their relative importance, if "importance" can be taken to mean the number of people using them. More than half of the world's inhabitants are

FIGURE 7.17 In their mountainous homeland, the Basques have maintained a linguistic uniqueness despite more than 2000 years of encirclement by dominant lowland speakers of Latin or Romance languages. This sign of friendly farewell gives its message in both Spanish and the Basque language, Euskara.
©Mark Antman/The Image Works.

speakers of just eight of its thousands of tongues. Table 7.1 lists those languages spoken by more than 40 million people, a list that includes four-fifths of the world's population. At the other end of the scale are a number of rapidly declining languages whose speakers number in the hundreds or, at most, the few thousands.

The diversity of languages is simplified when we recognize among them related families. A **language** (or *linguistic*) **family** is a group of languages thought to have a common origin in a single, earlier tongue. The Indo-European family of languages is among the most prominent of such groupings, embracing most of the languages of Europe and a large part of Asia, and the introduced—not the native—languages of the Americas (Figure 7.19). All told, languages in the Indo-European family are spoken by about half the world's peoples.

By recognizing similar words in most Indo-European languages, linguists deduce that these languages derived from a common ancestor tongue called *proto-Indo-European,* which was spoken by people living somewhere in eastern Europe about 5000 years ago (though some conclude that central Turkey was the more likely site of origin). About 2500 B.C. their society apparently fragmented; the homeland was left and segments of the parent culture migrated in different directions. Wherever this remarkable people settled, they appear to have dominated local populations and imposed their language upon them.

Within a language family we can distinguish *subfamilies.* The Romance languages (including French, Spanish, and Italian)—offsprings of Latin—and the Germanic languages (such as English, German, and Dutch) are subfamilies of Indo-European. The languages in a subfamily often show similarities in sounds, grammatical structure, and vocabulary even though they are mutually unintelligible. English *daughter,* German *Tochter,* and Swedish *dotter* are Germanic examples.

Language Spread and Change

Language spread as a geographical event represents the increase or relocation through time in the area over which a language is spoken. The more than 300 Bantu languages found south of the "Bantu line" in sub-Saharan Africa, for example, are variants of a proto-Bantu carried by an expanding, culturally advanced people who displaced linguistically different preexisting populations (Figure 7.20). More recently, the languages of European colonists similarly replaced native tongues in their areas of settlement in North and South America, Australasia, and Siberia. That is, languages may spread because their speakers occupy new territory.

Latin, however, replaced earlier Celtic languages in western Europe not by force of numbers—Roman legionnaires, administrators, and settlers never represented a majority population—but by the gradual abandonment of their former tongues by native populations brought under the influence and control of the Roman Empire. Adoption rather than eviction of language appears the rule followed in the majority of historical and contemporary instances of language spread. That is, languages may spread because they acquire new speakers.

Either form of language spread—dispersion of speakers or acquisition of speakers—may, through segregation and isolation, give rise to separate, mutually incomprehensible tongues because the society speaking the parent protolanguage no longer remains unitary. Comparable changes occur normally and naturally within a single language in word meaning, pronunciation, vocabulary, and *syntax* (the way words are put together in phrases and sentences). Because they are gradual, such changes tend to go unremarked. Yet, cumulatively, they can result in language change so great that in the course of centuries an essentially new language has been created. The English of 17th-century Shakespearean writings or the King James Bible (1611) sounds stilted to our ears. Few of us can easily read Chaucer's 14th-century *Canterbury Tales* in its original Middle English, and 8th-century *Beowulf* is practically unintelligible.

Language evolution may be gradual and cumulative, with each generation deviating in small degree from the speech patterns and vocabulary of its parents, or it may be massive and abrupt—reflecting conquests, migrations, new trade contacts, and other disruptions of cultural isolation. English owes its form to the Celts, the original inhabitants of the British Isles, and to successive waves of invaders, including the Latin-speaking Romans and the Germanic Angles, Saxons, and Danes. The French-speaking Norman

FIGURE 7.18 **World language families.** Language families are groups of individual tongues that have a common but remote ancestor. By suggesting that the area assigned to a language or language family uses that language exclusively, the map pattern conceals important linguistic detail. Many countries and regions have local languages spoken in territories too small to be recorded at this scale. The map also fails to report that the population in many regions is fluent in more than one language, or that a second language serves as the necessary vehicle of commerce, education, or government. Nor is important information given about the number of speakers of different languages; the fact that there are more speakers of English in India or Africa than in Australia is not even hinted at by a map at this scale.

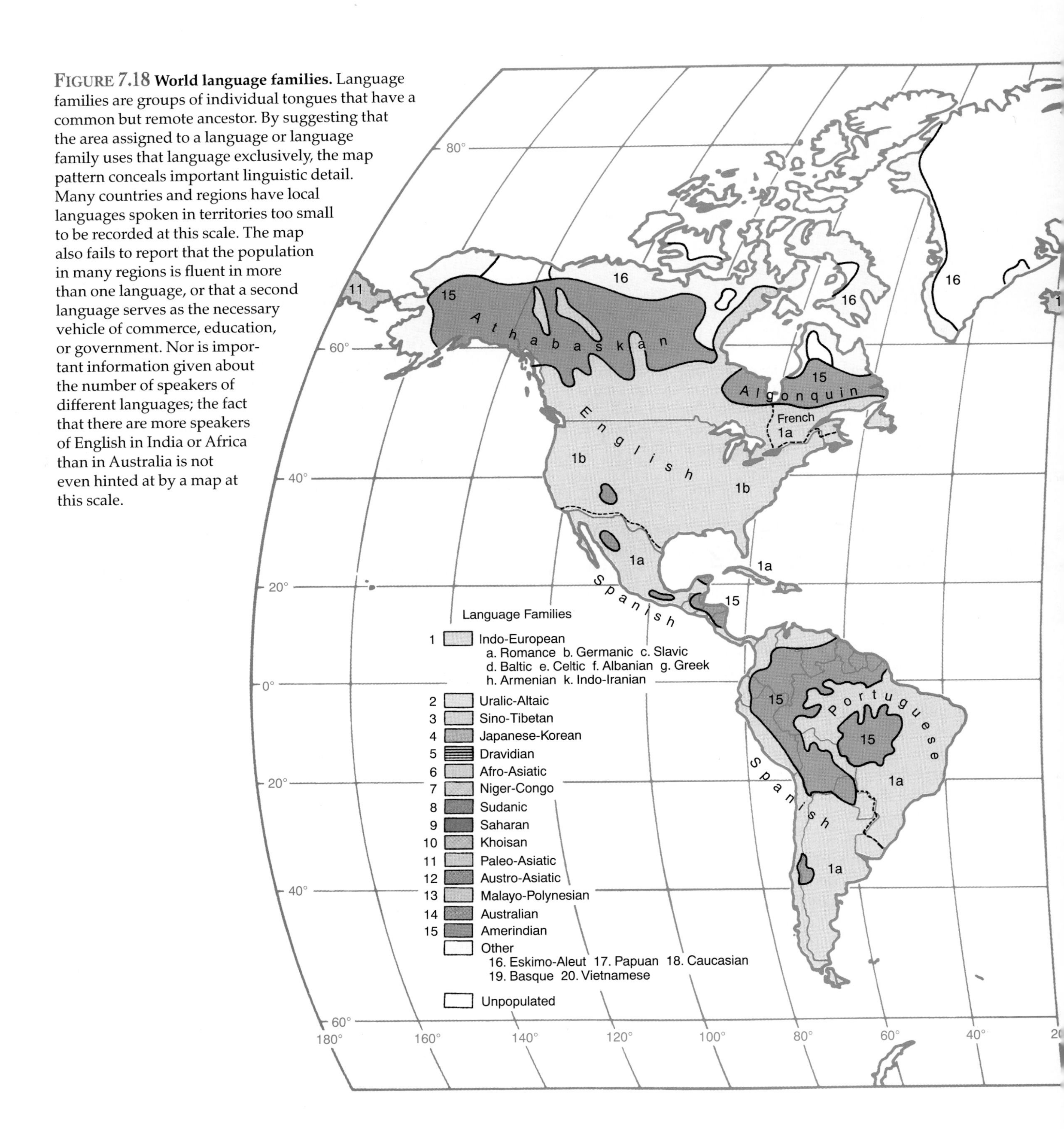

Norwegian
Swedish
Finnish
Yakut
Tungus
Manchu
Chukchi
Koryak
Byelo-Russian
Polish
Ukrainian
German
French
Italian
Spanish
Portuguese
Turkish
Kurdish
Kazakh
Uzbek
Turkmen
Kirgiz
Uigur
Mongol
Chinese
Korean
Japanese
Farsi
Pashto
Hindi
Tibetan
Telugu
Tamil
Arabic
Berber
Wolof
Fulani
Bambara
Akan
Yoruba
Amharic
Cushitic
Bantu
Congo
Ganda
Luba
Swahili
Mbundu
Bemba
Bushman
Shona
Hottentot
Zulu
Afrikaans
Malagasy
English
0°
20°
40°
60°
80°
100°
120°
140°
160°
180°

TABLE 7.1

Languages Spoken by More than 40 Million People, 1998

Language	Millions of Speakers	Language	Millions of Speakers
Mandarin (China)	1019	Marathi (India)	72
English	510	Cantonese (China)	72
Hindi[a] (India, Pakistan)	492	Wu (China)	70
Spanish	414	Vietnamese	69
Russian	268	Javanese	64
Arabic	239	Italian	63
Bengali (Bangladesh, India)	209	Turkish	62
Portuguese	188	Tagalog (Philippines)	57
Malay-Indonesian	171	Thai	53
French	126	Min (China)	51
Japanese	125	Swahili (East Africa)	50
German	124	Ukrainian	47
Urdu[a] (Pakistan, India)	108	Kannada (India)	46
Punjabi (India, Pakistan)	96	Gujarati (India, Pakistan)	45
Telugu (India)	76	Polish	41
Tamil (India, Sri Lanka)	75	Hausa (West Africa)	40
Korean (Korea, China, Japan)	74		

[a]Hindi and Urdu are basically the same language: Hindustani. Written in the Devangari script, it is called *Hindi,* the official language of India; in the Arabic script it is called *Urdu,* the official language of Pakistan.

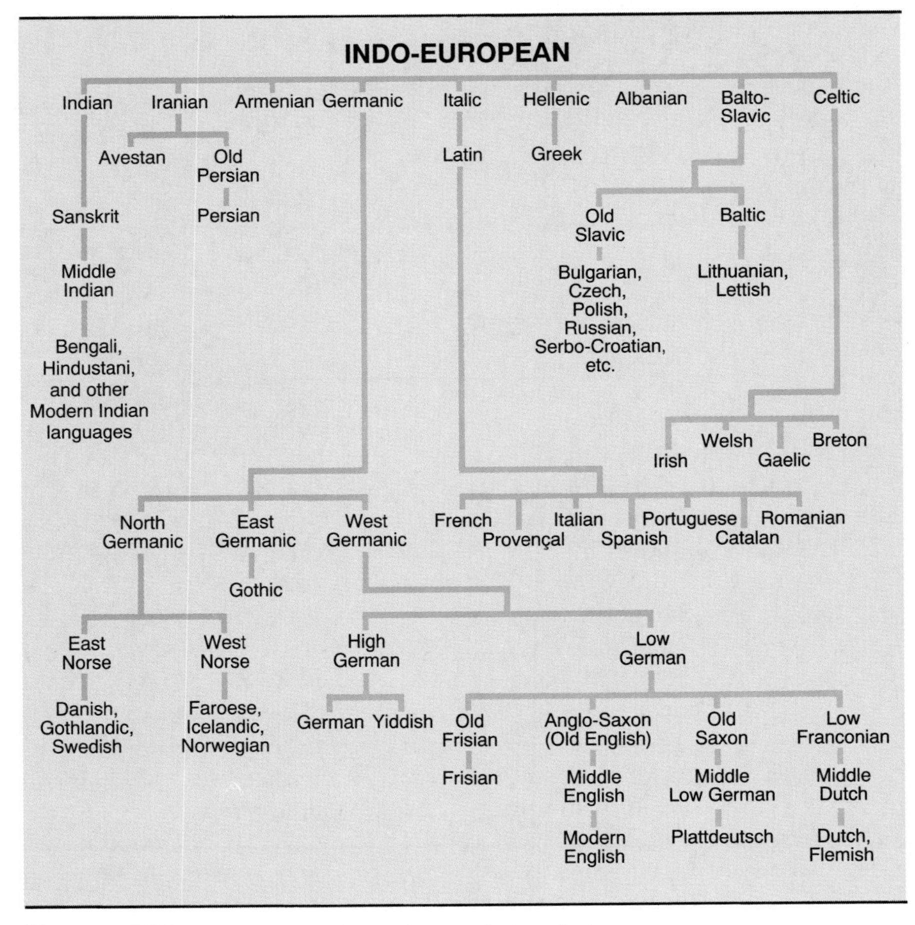

Figure 7.19 The Indo-European linguistic family tree.

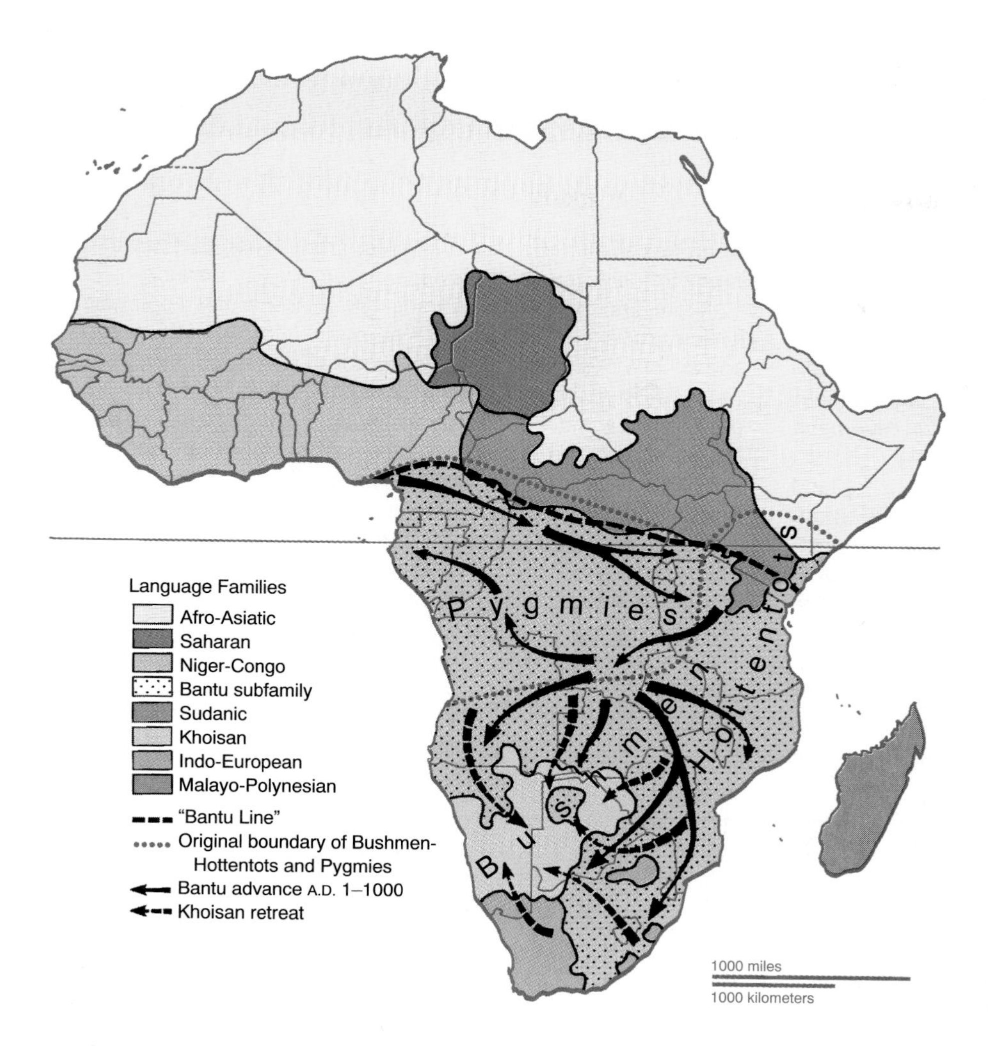

FIGURE 7.20 **Bantu advance, Khoisan retreat in Africa.** Linguistic evidence suggests that proto-*Bantu* speakers originated in the region of the Cameroon-Nigeria border, spread eastward across the southern Sudan, then turned southward to Central Africa. From there they dispersed slowly eastward, westward and, against slight resistance, southward. The earlier *Khoisan*-speaking occupants of sub-Saharan Africa were no match against the advancing metal-using Bantu agriculturalists. Pygmies, adopting a Bantu tongue, retreated deep into the forests. Bushmen and Hottentots retained their distinctive Khoisan "click" language but were forced out of forests and grasslands into the dry steppes and deserts of the southwest.

conquerors of the 11th century added about 10,000 new words to the evolving English tongue.

Discovery and colonization of new lands and continents in the 16th and 17th centuries greatly and necessarily expanded English as new foods, vegetation, animals, and artifacts were encountered and adopted along with their existing aboriginal American, Australian, Indian, or African names. The Indian languages of the Americas alone brought more than 200 relatively common daily words to English, 80 or more from the North American native tongues and the rest from Caribbean, Central, and South American languages. More than two thousand more specialized or localized words were also added. *Moose, raccoon, skunk, maize, squash, succotash, igloo, toboggan, hurricane, blizzard, hickory, pecan,* and a host of other names were taken directly into English; others were adopted secondhand from Spanish variants of South American native words: *cigar, potato, tobacco, hammock.*

A worldwide diffusion of the language resulted as English colonists carried it to the Western Hemisphere and Australasia and as trade, conquest, and territorial claim took it to Africa and Asia. In that areal spread, English was further enriched by its contacts with other languages. By becoming the accepted language of commerce and science, it contributed in turn to the common vocabularies of other tongues (see "Language Exchange"). Within some 400 years, English has developed from a localized language of 7 million islanders off the European coast to a truly international language with some 400 million native speakers, perhaps the same number who use it as a second language, and another 300–400 million who have reasonable competence in English as a foreign language. English serves as an official language of more than 60 countries (Figure 7.21), far exceeding in that role French (27), Arabic (21), or Spanish (20), the other leading current international languages. No other language in history has assumed so important a role on the world scene.

Standard and Variant Languages

People who speak a common language such as English are members of a *speech community,* but membership does not necessarily imply linguistic uniformity. A speech community usually possesses both a **standard language** comprising the accepted community norms of syntax, vocabulary, and pronunciation and a number of more or less distinctive **dialects,** the ordinary speech of areal, social, professional, or other subdivisions of the general population.

An official or unofficial standard language is the form carrying governmental, educational, or societal sanction. In Arab countries, for example, classical Arabic is the language of the mosque, of education, and of the newspapers and is standardized throughout the Arabic-speaking world. Colloquial Arabic is used at home, in the street, and at the market—and in its regional variants may be as widely different as are, for example, Portuguese and Italian. On the other hand, the United States, English-speaking Canada,

Language Exchange

English has a happily eclectic vocabulary. Its foundations are Anglo-Saxon (*was, that, eat, cow*) reinforced by Norse (*sky, get, bath, husband, skill*); its superstructure is Norman-French (*soldier, Parliament, prayer, beef*). The Norman aristocracy used their words for the food, but the Saxon serfs kept theirs for the animals. Its decor comes from Renaissance and Enlightenment Europe: 16th-century France yielded *etiquette, naive, reprimand* and *police*.

Italy provided *umbrella, duet, bandit* and *dilettante;* Holland gave *cruise, yacht, trigger, landscape,* and *decoy.* Its elaborations come from Latin and Greek: *misanthrope, meditate,* and *parenthesis* all first appeared during the 1560s. In this century, English adopted *penicillin* from Latin, *polystyrene* from Greek, and *sociology* and *television* from both. And English's ornaments come from all round the world: *slogan* and *spree* from Gaelic, *hammock* and *hurricane* from Caribbean languages, *caviar* and *kiosk* from Turkish, *dinghy* and *dungarees* from Hindi, *caravan* and *candy* from Persian, *mattress* and *masquerade* from Arabic.

Redressing the balance of trade, English is sharply stepping up its linguistic exports. Not just the necessary *imotokali* (motor car) and *izingilazi* (glasses) to Zulu; or *motokaa* and *shillingi* (shilling) to Swahili; but also *der Bestseller, der Kommunikations Manager, das Teeshirt* and *der Babysitter* to German; and, to Italian, *la pop art, il popcorn* and *la spray.* In some Spanish-speaking countries you might wear *un sueter* to *el beisbol,* or witness *un nocaut* at *el boxeo.* And in Russia, *biznesmen* prepare a *press rilis* on the *leptop kompyuter* and print it by *lazerny printer.* Indeed, a sort of global English wordlist can be drawn up: *airport, passport, hotel, telephone; bar, soda, cigarette; sport, golf, tennis; stop, OK,* and increasingly, *weekend, jeans, know-how, sex-appeal* and *no problem.*

Excerpted by permission from *The Economist,* London, December 20, 1986, p. 131.

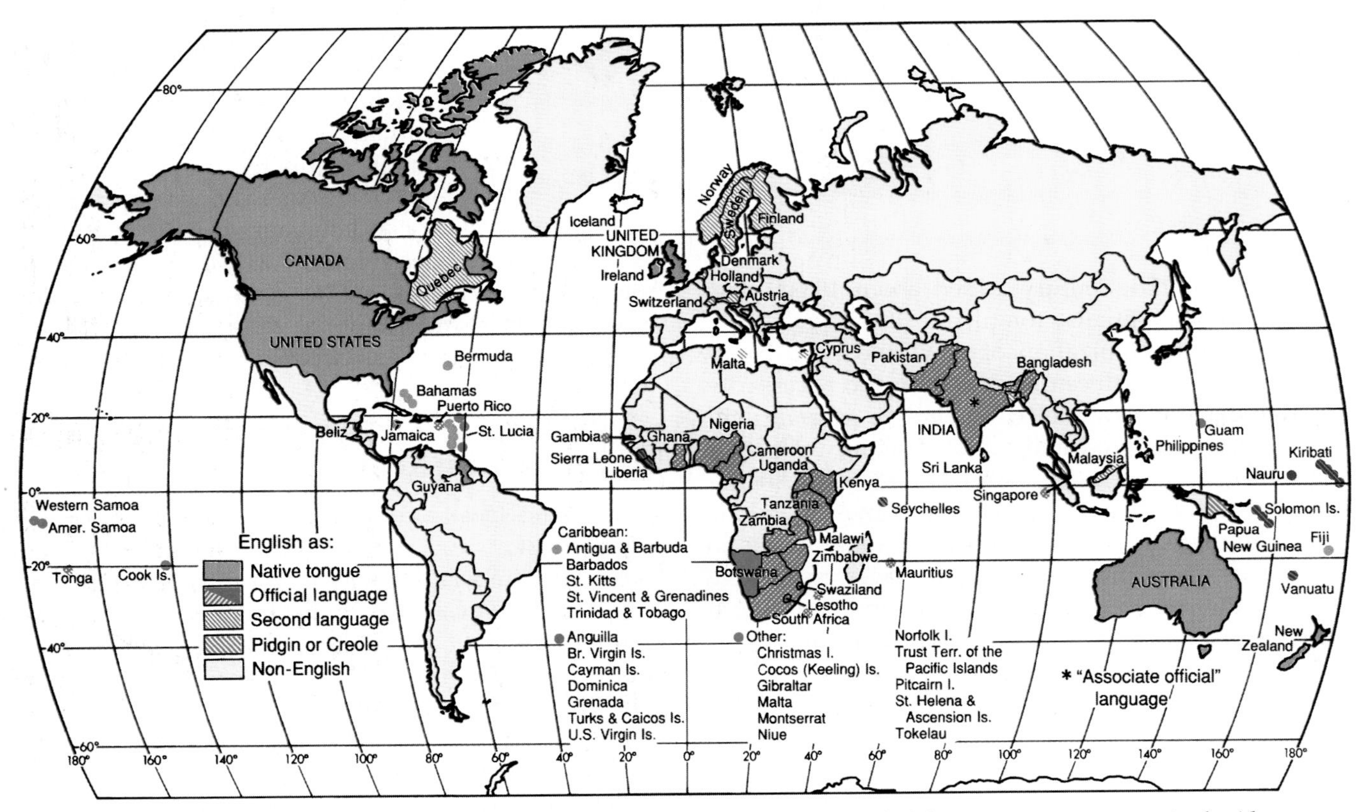

FIGURE 7.21 International English. In worldwide diffusion and acceptance, English has no past or present rivals. Along with French, it is one of the two working languages of the United Nations, and some two-thirds of all scientific papers are published in it. English is the sole or joint official language of more nations and territories, some too small to be shown here, than any other tongue. It also serves as the effective unofficial language of administration in other multilingual countries with different formal official languages. "English as a second language" is indicated for countries with near-universal or mandatory English instruction in public schools. The full extent of English penetration of Continental Europe, where over 80% of secondary school students study it as a second language and more than one-third of European Union residents can easily converse in it, is not evident on this map.

Australia, and the United Kingdom all have only slightly different forms of standard English.

Just as no two individuals talk exactly the same, all but the smallest and most closely knit speech communities display recognizable speech variants called *dialects.* Vocabulary, pronunciation, rhythm, and the speed at which the language is spoken may clearly set groups of speakers apart from one another. Dialects may coexist in space. Cockney and cultured English share the streets of London; black English ("Ebonics") and Standard American are heard in the same schoolyards throughout the United States. In many societies, *social dialects* denote social class and educational levels, with speakers of higher socioeconomic status or educational achievement most likely to follow the norms of their standard language. Less-educated or lower-status people are more apt to use the *vernacular*—nonstandard language or dialect adopted by the social group.

More commonly, we think of dialects in spatial terms. Speech is a geographic variable. Each locale is likely to have its own, perhaps slight, language differences from neighboring places. Such differences in pronunciation, vocabulary, word meanings, and other language characteristics help define the *linguistic geography*—the study of the character and spatial pattern of *geographic* or *regional dialects*—of a generalized speech community. Figure 7.22 records the variation in usage associated with just one phrase. In the United States, Southern English and New England speech are among the regional dialects that are most easily recognized by their distinctive accents. In some instances there may be so much variation among geographic dialects that some are almost foreign tongues to other speakers of the same language. Effort is required for Americans to understand Australian English or that spoken in Liverpool, England, or in Glasgow, Scotland. An interesting United States example is discussed in the regional study, "Gullah as Language," in Chapter 13.

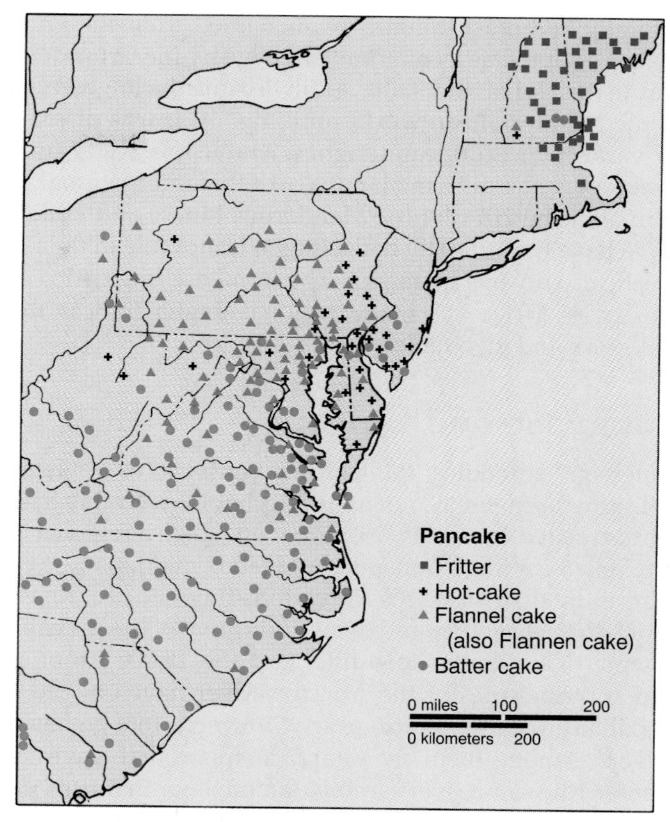

FIGURE 7.22 Maps such as this are used to record variations over space and among social classes in word usage, accent, and pronunciation. The differences are due not only to initial settlement patterns, but also to more recent large-scale movements of people—for example, from rural to urban areas and from the South to the North. Despite the presumed influence of national radio and television programs in promoting a "general" or "standard" American accent and usage, regional and ethnic language variations persist.

Redrawn with permission from *A Word Geography of the Eastern United States* by Hans Kurath. Copyright © 1949 by the University of Michigan Press.

Language is rarely a total barrier in communication between peoples. Bilingualism or multilingualism may permit skilled linguists to communicate in a jointly understood third language, but long-term contact between less able populations may require the creation of a new language—a pidgin—learned by both parties. A **pidgin** is an amalgam of languages, usually a simplified form of one of them, such as English or French, with borrowings from another, perhaps non-European local language. In its original form, a pidgin is not the mother tongue of any of its speakers; it is a second language for everyone who uses it, one generally restricted to such specific functions as commerce, administration, or work supervision.

Pidgins are characterized by a highly simplified grammatical structure and a sharply reduced vocabulary adequate to express basic ideas but not complex concepts. If a pidgin becomes the first language of a group of speakers—who may have lost their former native tongue through disuse—a **creole** has evolved. Creoles invariably acquire a more complex grammatical structure and enhanced vocabulary.

Creole languages have proved useful integrative tools in linguistically diverse areas; several have become symbols of nationhood. Swahili, a pidgin formed from a number of Bantu dialects, originated in the coastal areas of East Africa and spread by trade during the period of English and German colonial rules. When Kenya and Tanzania gained independence, they made Swahili the national language of administration and education. Other examples of creolization are Afrikaans (a pidginized form of 17th-century Dutch used in the Republic of South Africa); Haitian Creole (the language of Haiti, derived from the pidginized French used in the slave trade); and Bazaar Malay (a pidginized form of the Malay language, a version of which is the official national language of Indonesia).

A **lingua franca** is an established language used habitually for communication by people whose native tongues are mutually incomprehensible. For them it is a *second language,* one learned in addition to the native tongue. Lingua franca (literally "Frankish tongue") was named from the French dialect adopted as a common language by the Crusaders at war in the Holy Land. Later, Latin became the lingua franca of the Mediterranean world until, finally, it was displaced by vernacular European tongues. Arabic followed Muslim conquest as the unifying language of that international religion after the 7th century. Mandarin Chinese and Hindi in India have traditionally had a lingua franca role in their linguistically diverse countries. The immense linguistic complexity of Africa has made regional lingua francas there necessary and inevitable.

Language and Culture

Language embodies the culture complex of a people, reflecting both environment and technology. Arabic has 80 words related to camels, an animal on which a regional culture relied for food, transport, and labor, and Japanese contains more than 20 words for various types of rice. Russian is rich in terms for ice and snow, indicative of the prevailing climate of its linguistic cradle; and the 15,000 tributaries and subtributaries of the Amazon River have obliged the Brazilians to enrich Portuguese with words that go beyond "river." Among them are *paraná* (a stream that leaves and reenters the same river), *igarapé* (an offshoot that runs until it dries up), and *furo* (a waterway that connects two rivers).

Most—perhaps all—cultures display subtle or pronounced differences in ways males and females use language. Most have to do with vocabulary and with grammatical forms peculiar to individual cultures. For example, among the Caribs of the Caribbean, the Zulu of Africa, and elsewhere, men have words that women through custom or taboo are not permitted to use, and "the women have words and phrases which the men never use, or they would be laughed to scorn," an informant reports. Evidence from English and many other unrelated tongues indicates that as a rule female speakers use forms considered to be "better" or "more correct" than males of the same social class. The greater and more inflexible the difference in the social roles of men and women in a particular culture, the greater and more rigid are the observed linguistic differences between the sexes.

A common language fosters unity among people. It promotes a feeling for a region; if it is spoken throughout a country, it fosters nationalism. For this reason, languages often gain political significance and serve as a focus of opposition to what is perceived as foreign domination. Although nearly all people in Wales speak English, many also want to preserve Welsh because they consider it an important aspect of their culture. They think that if the language is forgotten, their entire culture may also be threatened. French Canadians received government recognition of their language and established it as the official language of Quebec Province; Canada itself is officially bilingual. In India, with 15 constitutional languages and 1652 other tongues, serious riots have occurred by people expressing opposition to the imposition of Hindi as the single official national language.

Bilingualism or multilingualism complicates national linguistic structure. Areas are considered bilingual if more than one language is spoken by a significant proportion of the population. In some countries—Belgium, for example, or Switzerland—there is more than one official language. In many others, such as the United States, only one language may have implicit or official government sanction, although several others are spoken (see "An Official U.S. Language?"). Speakers of one of these may be concentrated in restricted areas (e.g., most speakers of French in Canada live in Quebec Province). Less often, they may be distributed fairly evenly throughout the country. In some countries, the language in which instruction, commercial transactions, and government business take place is not a domestic language at all. In linguistically complex sub-Saharan Africa, nearly all countries have selected a European tongue—usually that of their former colonial governors—as an official language (Figure 7.23).

Toponyms—place-names—are language on the land, the record of past and present cultures whose namings endure as reminders of their passing and their existence. **Toponymy,** the study of place-names, therefore is a revealing tool of historical cultural geography, for place-names

FIGURE 7.23 **Europe in Africa through official languages.** Both the linguistic complexity of sub-Saharan Africa and the colonial histories of its present political units are implicit in the designation of a European language as the sole or joint "official" language of the different countries.

Geography and Public Policy

An Official U.S. Language?

In Lowell, Massachusetts, public school courses are offered in Spanish, Khmer, Lao, Portuguese, and Vietnamese, and all messages from schools to parents are translated into five languages. Polyglot New York City gives bilingual programs in Spanish, Chinese, Haitian Creole, Russian, Korean, Vietnamese, French, Greek, Arabic, and Bengali. In most states, it is possible to get a high school equivalency diploma without knowing English because tests are offered in French and Spanish. In 39 states, driving tests are available in foreign languages; California provides 30 varieties, New York 23, and Michigan 20, including Arabic and Finnish. And the 1965 federal Voting Rights Act requires multilingual ballots in 375 electoral jurisdictions.

These, and innumerable other evidences of governmentally sanctioned linguistic diversity, may come as a surprise to those many Americans who assume that English is the official language of the United States. It isn't; nowhere does the Constitution provide for an official language, and no federal law specifies one. The country was built by a great diversity of cultural and linguistic immigrants who nonetheless shared an eagerness to enter mainstream American life. In the 1990s, more than 15% of all American residents speak a language other than English in the home. In California public schools, 1 out of 3 students uses a non-English tongue within the family. In Washington, D.C. schools, students speak 127 languages and dialects, a linguistic diversity duplicated in other major city school systems.

Nationwide bilingual teaching began as an offshoot of the civil rights movement in the 1960s, was encouraged by a Supreme Court opinion authored by Justice William O. Douglas, and has been actively promoted by the U.S. Department of Education under the Bilingual Education Act of 1974 as an obligation of local school boards. Its purpose has been to teach subject matter to minority-language children in the language in which they think while introducing them to English, with the hope of achieving English proficiency in two or three years. Disappointment with the results achieved led to an anti-bilingual education initiative, Proposition 227, to abolish the program presented to California voters in 1998.

Opponents of the implications of governmentally encouraged multilingual education, bilingual ballots, and ethnic separatism argue that a common language is the unifying glue of the United States and all countries; without that glue, they fear, the process of "Americanization" and *acculturation*—the adoption by immigrants of the values, attitudes, ways of behavior, and speech of the receiving society—will be undermined. Convinced that early immersion and quick proficiency in English is the only sure way for minority newcomers to gain necessary access to jobs, higher education, and full integration into the economic and social life of the country, proponents of "English only" use in public education, voting, and state and local governmental agencies, successfully passed Official English laws and constitutional amendments in 23 states during the late 1980s and early 1990s.

©Bob Coyle and courtesy of Iowa Department of Transportation.

Although the amendments were supported by sizable majorities of the voting population, resistance to them—and to their political and cultural implications—was in every instance strong and persistent. Ethnic groups, particularly Hispanics, who are the largest of the linguistic groups affected, charged that they were evidence of blatant Anglo-centric racism, discriminatory and repressive in all regards. Some educators argued persuasively that all evidence proved that while immigrant children eventually acquire English proficiency in any event, they do so with less harm to their self-esteem and subject-matter acquisition when initially taught in their own language. Business people with strong minority labor and customer ties and political leaders—often themselves members of ethnic communities or with sizable minority constituencies—argued against "discriminatory" language restrictions.

And historians noted that it had all been unsuccessfully tried before. The anti-Chinese Workingmen's Party in 1870s California led the fight for English-only laws in that state. The influx of immigrants from central and southeastern Europe at the turn of the century led Congress to make oral English a requirement for naturalization, and anti-German sentiment during and after World War I led some states to ban any use of German. The Supreme Court struck down those laws in 1923, ruling that the "protection of the Constitution extends to all, to those who speak other languages as well as to those born with English on their tongue." Following suit, some of the recent state language amendments have also been voided by state courts. In ruling its state's English-only law unconstitutional, Arizona's Supreme Court in 1998 noted it "chills First Amendment rights."

To counter those judicial restraints and the possibility of an eventual multilingual, multicultural United States in which English

—Continued on page 258

Continued from page 257

and, likely, Spanish would have coequal status and recognition, U.S. English—an organization dedicated to the belief that "English is, and ever must remain, the only official language of the people of the United States"—actively supports the proposed U.S. Constitutional amendment first introduced in Congress by former Senator S. I. Hayakawa in 1981, and resubmitted by him and others in subsequent years. The proposed amendment would simply establish English as the official national language but would impose no duty on people to learn English and would not infringe on any right to use other languages. Whether or not these modern attempts to designate an official U.S. language eventually succeed, they represent a divisive subject of public debate affecting all sectors of American society.

Questions to Consider

1. Do you think that multiple languages and ethnic separatism represent a threat to America's cultural unity that can be avoided only by viewing English as a necessary unifying force? Or do you think making English the official language might divide its citizens and damage its legacy of tolerance and diversity?
2. Do you feel that immigrant children would learn English faster if bilingual classes were reduced and immersion in English was more complete? Or do you think that a slower pace of English acquisition is acceptable if subject-matter comprehension and cultural self-esteem is enhanced?
3. Do you think Official English laws serve to inflame prejudice against immigrants or to provide all newcomers with a common standard of admission to the country's political and cultural mainstream?

become a part of the cultural landscape that remains long after the name givers have passed from the scene.

In England, for example, place-names ending in *chester* (as in Winchester and Manchester) evolved from the Latin *castra,* meaning "camp." Common Anglo-Saxon suffixes for tribal and family settlements were *ing* (people or family) and *ham* (hamlet or, perhaps, meadow) as in Birmingham or Gillingham. Norse and Danish settlers contributed place-names ending in *thwaite* (meadow) and others denoting such landscape features as *fell* (an uncultivated hill) and *beck* (a small brook). The Arabs, sweeping out from Arabia across North Africa and into Iberia, left their imprint in place-names to mark their conquest and control. *Cairo* means "victorious," *Sudan* is "the land of the blacks," and *Sahara* is "wasteland" or "wilderness." In Spain, a corrupted version of the Arabic *wadi,* "watercourse," is found in *Guadalajara* and *Guadalquivir.*

In the New World, not one people but many placed names on landscape features and new settlements. In doing so they remembered their homes and homelands, honored their monarchs and heroes, borrowed and mispronounced from rivals, adopted and distorted Amerindian names, followed fads, and recalled the Bible. Homelands were honored in New England, New France, or New Holland; settlers' hometown memories brought Boston, New Bern, and New Rochelle from England, Switzerland, and France. Monarchs were remembered in Virginia for the Virgin Queen Elizabeth, Carolina for one English King, Georgia for another, and Louisiana for a King of France. Washington, D.C.; Jackson, Mississippi and Michigan; Austin, Texas; and Lincoln, Illinois, memorialized heroes and leaders.

Names given by the Dutch in New York were often distorted by the English; Breukelyn, Vlissingen, and Haarlem became Brooklyn, Flushing, and Harlem. French names underwent similar twisting or translation, and Spanish names were adopted, altered, or, later, put into such bilingual combinations as Hermosa Beach. Amerindian tribal names—the Yenrish, Maha, Kansa—were modified, first by French and later by English speakers—to Erie, Omaha, and Kansas. A faddish classical revival after the American Revolution gave us Troy, Athens, Rome, Sparta, and other ancient town names and later spread them across the country. Bethlehem, Ephrata, Nazareth, and Salem came from the Bible.

Religion

Enduring place-names are only one measure of the importance of language as a powerful unifying thread in the culture complex of people and as a fiercely defended symbol of the history and individuality of a distinctive social group. But language is not alone in that role. In some ways it yields to religion as a cultural rallying point. French Catholics and French Huguenots (Protestants) freely slaughtered each other in the name of religion in the 16th century. English Roman Catholics were hounded from the country after the establishment of the Anglican Church. Religious enmity between Muslims and Hindus forced the partition of the Indian subcontinent after the departure of the British in

1947. And the 1990s have witnessed continuing religious confrontations between, among many others, Catholic and Protestant Christian groups in Northern Ireland; Muslim sects in Lebanon, Iran, and Iraq; Muslims and Jews in Palestine; Christians and Muslims in the Philippines and Lebanon; and Buddhists and Hindus in Sri Lanka.

However, unlike language, which is an attribute of all people, religion varies in its cultural role—dominating among some societies, unimportant or rejected in others. All societies have value systems—common beliefs, understandings, expectations, and controls—that unite their members and set them off from other, different culture groups. Such a value system is termed a *religion* when it involves systems of formal or informal worship and faith in the sacred and divine.

Religion may intimately affect all facets of a culture. Religious belief is by definition an element of the ideological subsystem; formalized and organized religion is an institutional expression of the sociological subsystem. And religious beliefs strongly influence attitudes toward the tools and rewards of the technological subsystem.

Nonreligious value systems can exist—humanism or Marxism, for example—that are just as binding on the societies that espouse them as are more traditional religious beliefs. Even societies that largely reject religion, however, are strongly influenced by traditional values and customs set by predecessor religions—in days of work and rest or in legal principles, for example.

Since religions are formalized views about the relation of the individual to this world and to the hereafter, each carries a distinct conception of the meaning and value of this life, and most contain strictures about what must be done to achieve salvation (Figure 7.24). These rules become interwoven with the traditions of a culture. One cannot understand India without a knowledge of Hinduism, or Israel without an appreciation of Judaism.

Economic patterns may be intertwined with past or present religious beliefs. Traditional restrictions on food and drink may affect the kinds of animals that are raised or avoided, the crops that are grown, and the importance of those crops in the daily diet. Occupational assignment in the Hindu caste system is in part religiously supported. In

FIGURE 7.24 Worshipers gathered during *hajj*, the pilgrimage to Mecca. The black structure is the Ka'bab, the symbol of Allah's (God's) oneness and the unity of God and humans. Many rules concerning daily life are given in the Koran, the holy book of the Muslims. All Muslims are expected to observe the five pillars of the faith: (1) repeated saying of the basic creed; (2) prayers five times daily, facing Mecca; (3) a month of daytime fasting (Ramadan); (4) almsgiving; and (5) if possible, a pilgrimage to Mecca.

many countries, there is a state religion; that is, religious and political structures are intertwined. Buddhism, for example, has been the state religion in Myanmar, Laos, and Thailand. By their official names, the Islamic Republic of Pakistan and the Islamic Republic of Iran proclaim their identity of church and government. Despite the country's overwhelming Muslim majority, Indonesia seeks domestic harmony by recognizing five official religions and a state ideology—*pancasila*—whose first tenet is belief in one god.

Classification and Distribution of Religions

Religions are cultural innovations. They may be unique to a single culture group, closely related to the faiths professed in nearby areas, or derived from or identical to belief systems spatially far removed. Although interconnections and derivations among religions can frequently be discerned—as Christianity and Islam can trace descent from Judaism—family groupings are not as useful to us in classifying religions as they were in studying languages. A distinction between *monotheism,* belief in a single deity, and *polytheism,* belief in many gods, is frequent, but not particularly spatially relevant. It is more useful for the spatial interests of geographers to categorize religions as *universalizing, ethnic,* or *tribal* (*traditional*).

Christianity, Islam, and Buddhism are the major world **universalizing religions,** faiths that claim applicability to all humans and that seek to transmit their beliefs to all lands through missionary work and conversion. Membership in universalizing religions is open to anyone who chooses to make some sort of symbolic commitment, such as baptism in Christianity. No one is excluded because of nationality, ethnicity, or previous religious belief.

Ethnic religions have strong territorial and cultural group identification. One becomes a member of an ethnic religion by birth or by adoption of a complex lifestyle and cultural identity, not by a simple declaration of faith. These religions do not usually proselytize (convert nonbelievers), and their members form distinctive closed communities identified with a particular ethnic group, region, or political unit. An ethnic religion—for example, Judaism, Indian Hinduism, or Japanese Shinto—is an integral element of a specific culture. To be part of the religion is to be immersed in the totality of the culture.

Tribal (or *traditional*) **religions** are special forms of ethnic religions distinguished by their small size, their unique identity with localized culture groups not yet fully absorbed into modern society, and their close ties to nature. *Animism* is the name given to their belief that life exists in all objects, from rocks and trees to lakes and mountains, or that such objects are the abode of the dead, of spirits, and of gods. *Shamanism* is a form of tribal religion that involves community acceptance of a *shaman* who, through special powers, can intercede with and interpret the spirit world.

The nature of the different classes of religions is reflected in their distributions over the world (Figure 7.25)

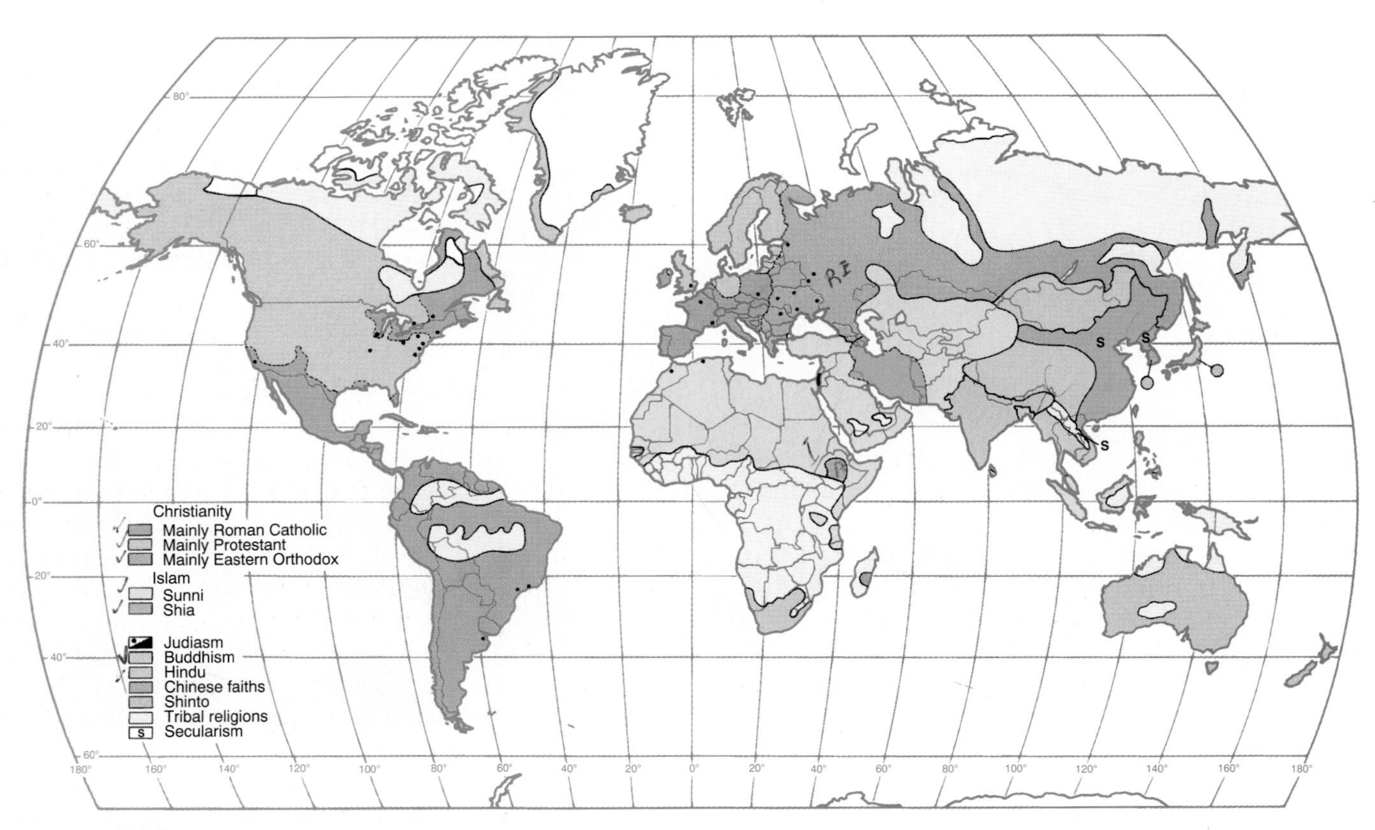

FIGURE 7.25 **The pattern of principal world religions.**

and in their number of adherents. Universalizing religions tend to be expansionary, carrying their message to new peoples and areas. Ethnic religions, unless their adherents are dispersed, tend to be regionally confined or to expand only slowly and over long periods. Tribal religions tend to contract spatially as their adherents are incorporated increasingly into modern society and converted by proselytizing faiths.

As we expect in cultural geography, the map records only the latest stage of a constantly changing reality. While established religious institutions tend to be conservative and resistant to change, religion as a culture trait is dynamic. Personal and collective beliefs may alter in response to developing individual and societal needs and challenges. Religions may be imposed by conquest, adopted by conversion, or be defended and preserved in the face of surrounding hostility or indifference.

Nor does the map present a full picture even of current religious regionalization or affiliation. Few societies are homogeneous, and most modern ones contain a variety of different faiths or, at least, variants of the dominant professed religion. Frequently, members of a particular religion show areal concentration within a country. Thus, in urban Northern Ireland, Protestants and Catholics reside in separate areas whose boundaries are clearly understood and respected. The "Green Line" in Beirut, Lebanon, marked a guarded border between the Christian East and the Muslim West sides of the city, while within the country as a whole regional concentrations of adherents of different faiths and sects are clearly recognized (Figure 7.26). Religious diversity within countries may reflect the degree of toleration a majority culture affords minority religions. In dominantly (90%) Muslim Indonesia, Christian Bataks, Hindu Balinese, and Muslim Javanese have lived in easy coexistence. By contrast, the fundamentalist Islamic regime in Iran has persecuted and executed those of the Baha'i faith.

One cannot assume that all people within a mapped religious region are adherents of the designated faith, nor can it be assumed that membership in a religious community means active participation in its belief system. *Secularism,* an indifference to or rejection of religion and religious belief, is an increasing part of many modern societies, particularly of the industrialized countries and those now or recently under communist regimes. The incidence of secularism in a few Asian societies is suggested on the map (Figure 7.25) by letter symbol; its widespread occurrence in other, largely Christian, countries should be understood though it is not mapped. In England, for example, the state Church of England claims only 20% of the British as communicants, and less than 2% of the population attends its Sunday services. Even in devoutly Roman Catholic South American states, low church attendance attests to the rise of at least informal secularism. In Colombia, only 18% of the people attend Sunday services; in Chile, the figure is 12%, in Mexico 11%, and Bolivia 5%.

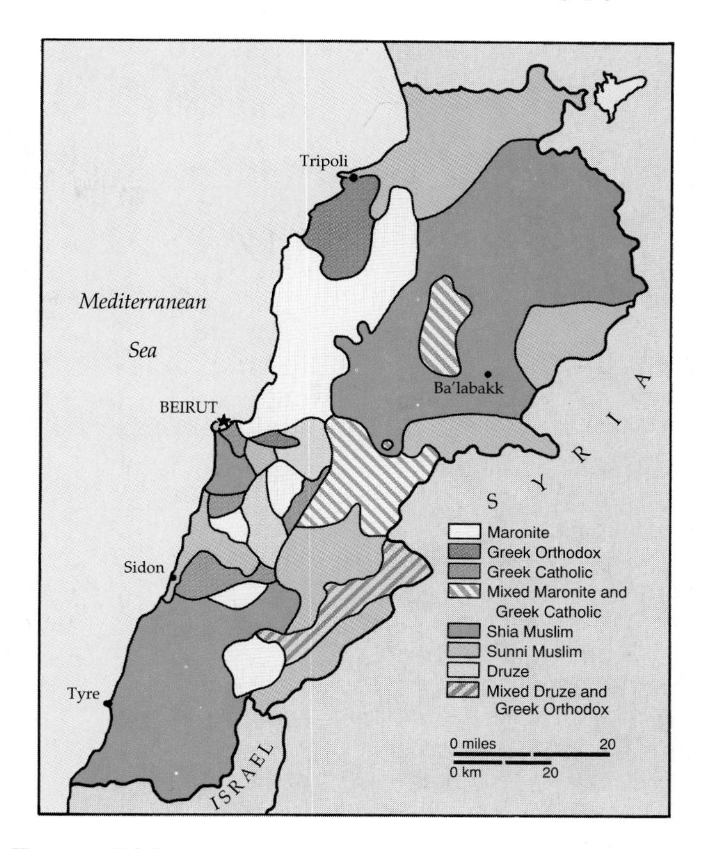

FIGURE 7.26 **Religious regions of Lebanon.** Religious territoriality and rivalry contributed to a prolonged period of conflict and animosity in this eastern Mediterranean country.

The Principal Religions

Each of the major religions has its own unique mix of cultural values and expressions, each has had its own pattern of innovation and spatial diffusion (Figure 7.27), and each has had its own impact on the cultural landscape. Together they contribute importantly to the worldwide pattern of human diversity.

Judaism

We may begin our review of world faiths with *Judaism,* whose belief in a single God laid the foundation for both Christianity and Islam. Unlike its universalizing offspring, Judaism is closely identified with a single ethnic group and with a complex and restrictive set of beliefs and laws. It emerged some 3000 to 3500 years ago in the Near East, one of the ancient cultural hearth regions (Figure 7.13).

Judaism is a distinctively *ethnic* religion, the determining factors of which are descent from Israel (the patriarch Jacob), the Torah (law and scripture), and the traditions of the culture and the faith. Early military success gave the Jews a sense of territorial and political identity to supplement their religious self-awareness. Later conquest by nonbelievers led to their dispersion (*diaspora*) to much of the Mediterranean world and farther east into Asia by A.D. 500 (Figure 7.28).

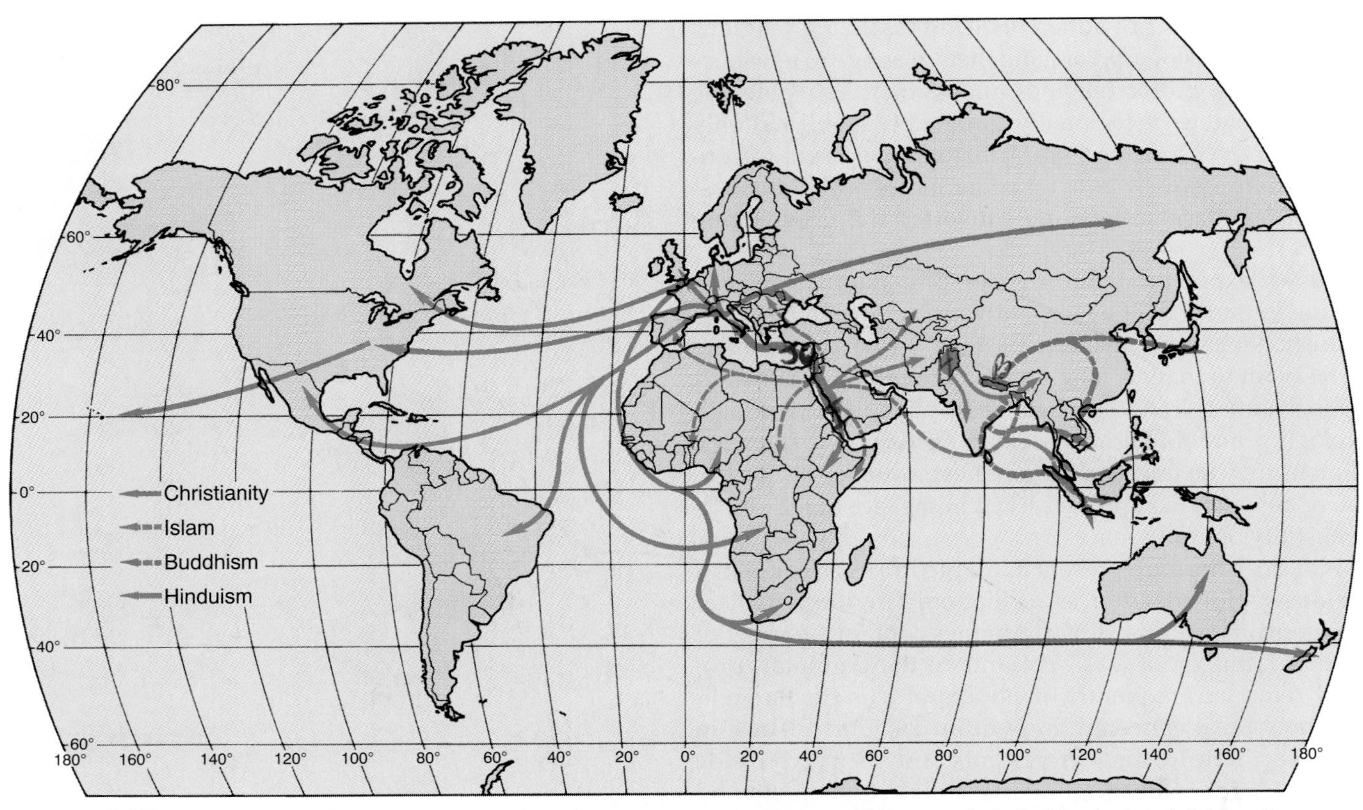

FIGURE 7.27 **Innovation areas and diffusion routes of major world religions.** The monotheistic (single deity) faiths of Judaism, Christianity, and Islam arose in southwestern Asia, the first two in Palestine in the eastern Mediterranean region, and the latter in western Arabia near the Red Sea. Hinduism and Buddhism originated within a confined hearth region in the northern part of the Indian subcontinent. Their rates, extent, and directions of spread are suggested here and detailed on later maps.

Between the 13th and 14th centuries, many Jews sought refuge in Poland and Russia from persecution in western and central Europe; during the later 19th and early 20th centuries, Jews were important elements of the European immigrant stream to the Western Hemisphere. The establishment of the state of Israel in 1948 was a fulfillment of the goal of *Zionism,* the belief in the need to create an autonomous Jewish state in Palestine. It demonstrated a determination that Jews not lose their identity by absorption into alien cultures and societies.

Judaism's imprint upon the cultural landscape has been subtle and unobtrusive. The Jewish community reserves space for the practice of communal burial; the spread of the cultivated citron in the Mediterranean area during Roman times has been traced to Jewish ritual needs; and the religious use of grape wine assured the cultivation of the vine in their areas of settlement. The synagogue as place of worship has tended to be less elaborate than its Christian counterpart. The essential for religious service is a community of at least ten adult males, not a specific structure.

Christianity

Christianity had its origin in the life and teachings of Jesus, a Jewish preacher of the 1st century of the modern era, whom his followers believed was the messiah promised by God. The new covenant he preached was not a rejection of traditional Judaism, but a promise of salvation to all humankind rather than to just a chosen people.

Christianity's mission was conversion. As a universal religion of salvation and hope it spread quickly among the underclasses of both the eastern and western parts of the Roman Empire, carried to major cities and ports along the excellent system of Roman roads and sea lanes (Figure 7.29). In A.D. 313, the Emperor Constantine proclaimed Christianity the state religion. Much later, of course, the faith was brought to the New World with European settlement (Figure 7.27).

The dissolution of the Roman Empire into a western and eastern half after the fall of Rome also divided Christianity. The Western Church, based in Rome, was one of the very few stabilizing and civilizing forces uniting western Europe during the Dark Ages. Its bishops became the civil as well as ecclesiastical authorities over vast areas devoid of other effective government. Parish churches were the focus of rural and urban life, and the cathedrals replaced Roman monuments and temples as the symbols of the social order.

Secular imperial control endured in the eastern empire, whose capital was Constantinople. Thriving under its protection, the Eastern Church expanded into the Balkans,

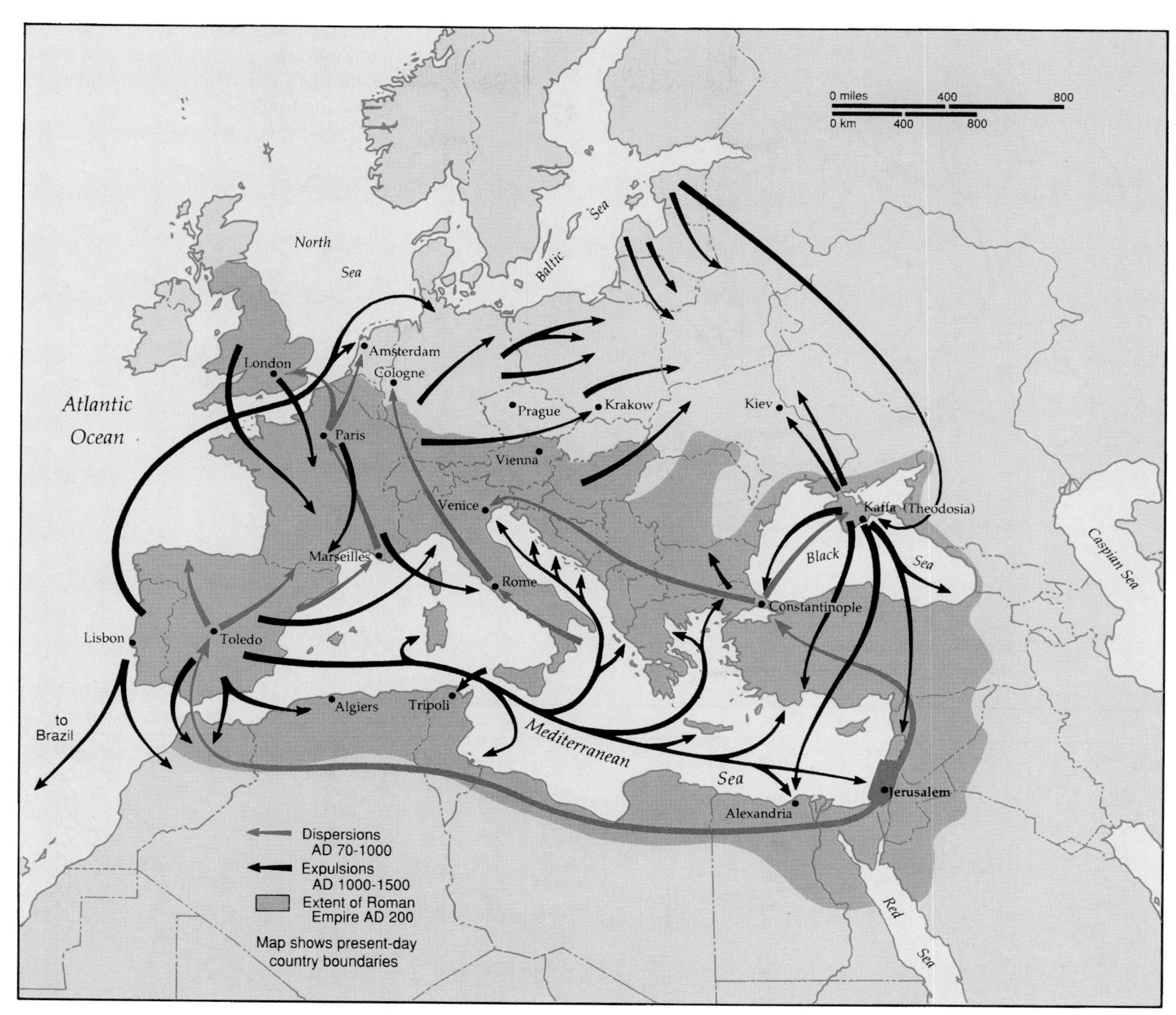

FIGURE 7.28 **Jewish dispersions, A.D. 70–1500.** A revolt against Roman rule in A.D. 66 was followed by the destruction of the Jewish Temple four years later and an Imperial decision to Romanize the city of Jerusalem. Judaism spread from the hearth region, carried by its adherents dispersing from their homeland to Europe, Africa, and eventually in great numbers to the Western Hemisphere. Although Jews established themselves and their religion in new lands, they did not lose their sense of cultural identity nor did they seek to attract converts to their faith.

eastern Europe, Russia, and the Near East. The fall of the eastern empire to the Turks in the 15th century opened eastern Europe temporarily to Islam, though the Eastern Orthodox Church (the direct descendant of the Byzantine State Church) remains, in its various ethnic branches, a major component of Christianity.

The Protestant Reformation of the 15th and 16th centuries split the church in the west, leaving Roman Catholicism supreme in southern Europe but installing a variety of Protestant denominations and national churches in western and northern Europe. The split was reflected in the subsequent worldwide dispersion of Christianity. Catholic Spain and Portugal colonized Latin America, bringing both their languages and the Roman church to that area (Figure 7.27), as they did to colonial outposts in the Philippines, India, and Africa. Catholic France colonized Quebec in North America. Protestants, many of them fleeing Catholic or repressive Protestant state churches, were primary early settlers of Anglo America, Australia, New Zealand, Oceania, and South Africa.

Although religious intermingling rather than rigid territorial division is characteristic of the contemporary American scene, the beliefs and practices of various immigrant groups and the innovations of domestic congregations have created a particularly varied spatial patterning of "religious regions" in the United States (Figure 7.30).

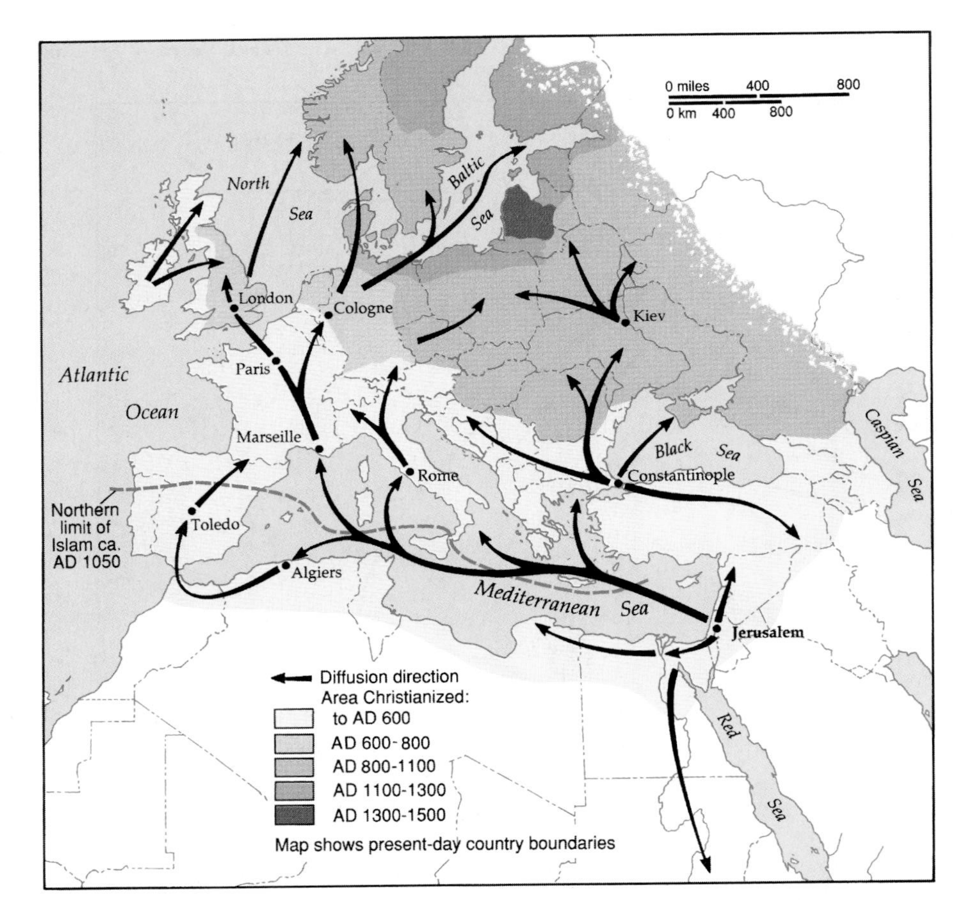

FIGURE 7.29 Diffusion paths of Christianity, A.D. 100–1500. Routes and dates are for Christianity as a composite faith. No distinction is made between the Western Church and the various subdivisions of the Eastern Orthodox denominations.

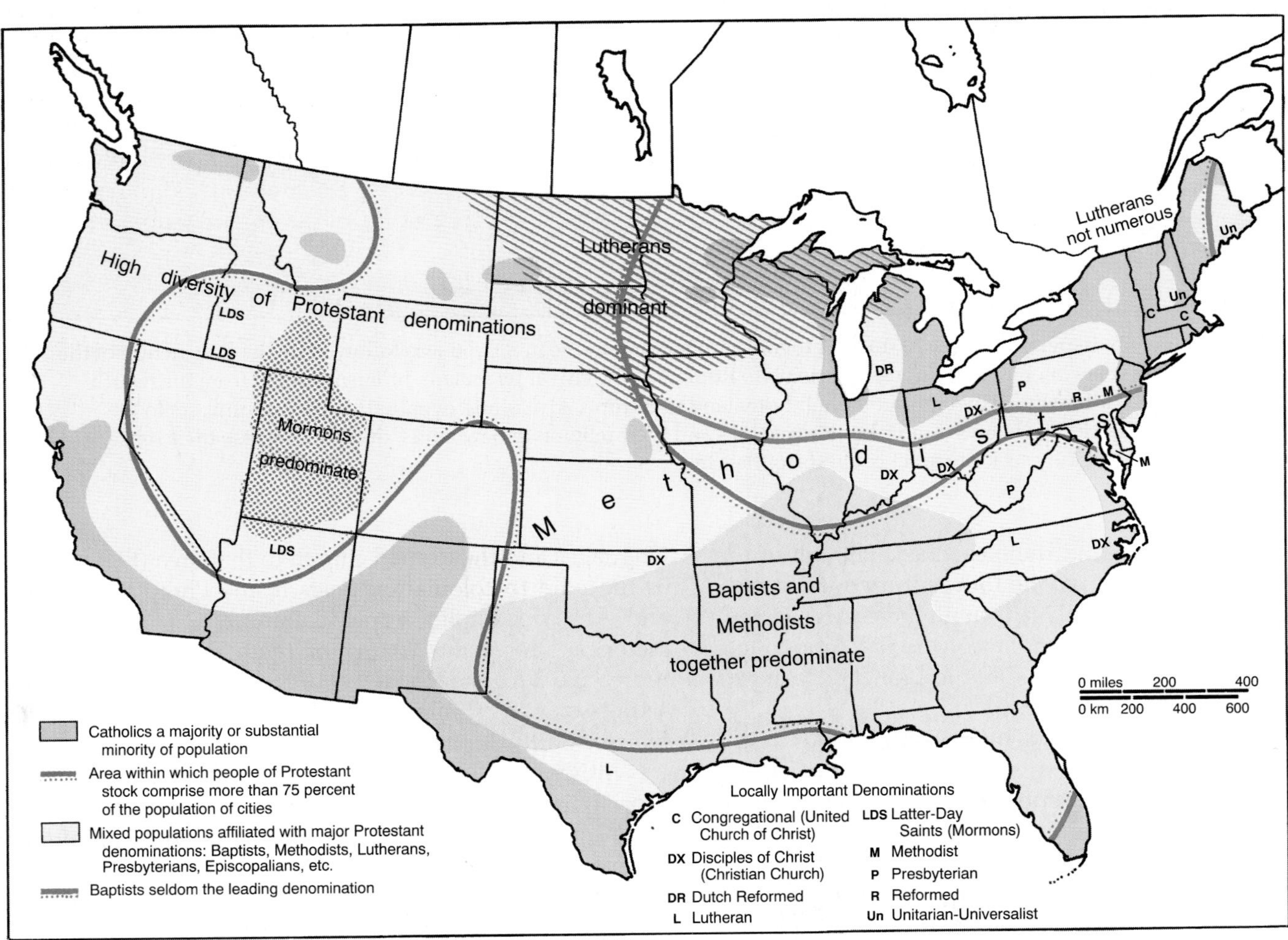

FIGURE 7.30 Religious affiliation in the conterminous United States.
Source: Redrawn from *Historical Atlas of the Religions of the World*, ed. by Isma'il al-Faruqi and David E. Sopher, Macmillan, 1974, p. 235.

The mark of Christianity on the cultural landscape has been conspicuous and enduring. In pre-Reformation Catholic Europe, the parish church formed the center of life for small neighborhoods of every town, the village church was the centerpiece of every rural community, and in larger cities the central cathedral served simultaneously as a glorification of God, a symbol of piety, and the focus of religious and secular life (Figure 7.31a).

Protestantism placed less importance on the church as a monument and symbol, although in many communities—colonial New England, for example—the churches of the principal denominations were at the village center (Figure 7.31b). Frequently they were adjoined by a cemetery, for Christians—in common with Muslims and Jews—practice burial in areas reserved for the dead. In Christian countries, particularly, the cemetery—whether connected to the church, separate from it, or unrelated to a specific denomination—has traditionally been a significant land use within urban areas.

Islam

Islam springs from the same Judaic roots as Christianity and embodies many of the same beliefs: there is only one God, who may be revealed to humans through prophets; Adam was the first human; Abraham was one of his descendants. Mohammed is revered as the prophet of *Allah* (God), succeeding and completing the work of earlier prophets of Judaism and Christianity, including Moses, David, and Jesus. The Koran, the word of Allah revealed to Mohammed, contains not only rules of worship and details of doctrine but also instructions on the conduct of human affairs. For fundamentalists, it thus becomes the unquestioned guide to matters both religious and secular. Observance of the "five pillars" (Figure 7.24) and surrender to the will of Allah unites the faithful into a brotherhood that has no concern with race, color, or caste.

It was that law of brotherhood that served to unify an Arab world sorely divided by tribes, social ranks, and multiple local deities. Mohammed was a resident of Mecca, but fled in A.D. 622 to Medina, where the Prophet proclaimed a constitution and announced the universal mission of the Islamic community. That flight—*Hegira*—marks the starting point of the Islamic (lunar) calendar. By the time of Mohammed's death in A.H. 11 (*Anno*—in the year of—*Hegira,* or A.D. 632), all of Arabia had joined Islam. The new religion swept quickly outward from that source region over most of Central Asia and, at the expense of Hinduism, into northern India (Figure 7.32). Its advance westward was particularly rapid and inclusive in North Africa. Later, Islam dispersed into Indonesia, southern Africa, and the Western Hemisphere. It continues its spatial spread as the fastest growing major religion at the present time.

The mosque—place of worship, community clubhouse, meeting hall, school—is the focal point of Islamic communal life and the primary imprint of the religion on the cultural landscape. Its principal purpose is to accommodate the Friday communal service, mandatory for all male Muslims. It is the congregation rather than the structure that is important; small or poor communities are as well served by a bare whitewashed room as are larger cities by architecturally splendid mosques. With its perfectly proportioned, frequently gilded or tiled domes, its graceful, soaring towers and minarets (from which the faithful are called to prayer), and its delicately wrought parapets and cupolas, the carefully tended mosque is frequently the most elaborate and imposing structure of the town (Figure 7.33).

FIGURE 7.31 In Christian societies, the church assumes a prominent central position in the cultural landscape. (a) The building of Notre Dame Cathedral in Paris, France, begun in 1163, took more than 100 years to complete. Between 1170 and 1270, some 80 cathedrals were constructed in France alone. The cathedrals in all of Catholic Europe were located in the center of major cities. Their plazas were the sites of markets, public meetings, and religious ceremonies. (b) Individually less imposing than the central cathedral of Catholic areas, the several Protestant churches common in small and large American towns collectively constitute an important land use frequently sited in the center of the community. The church shown here is in Waitsfield, Vermont.

(a) ©David A. Burney.

(b) ©David Brownell.

(a)

(b)

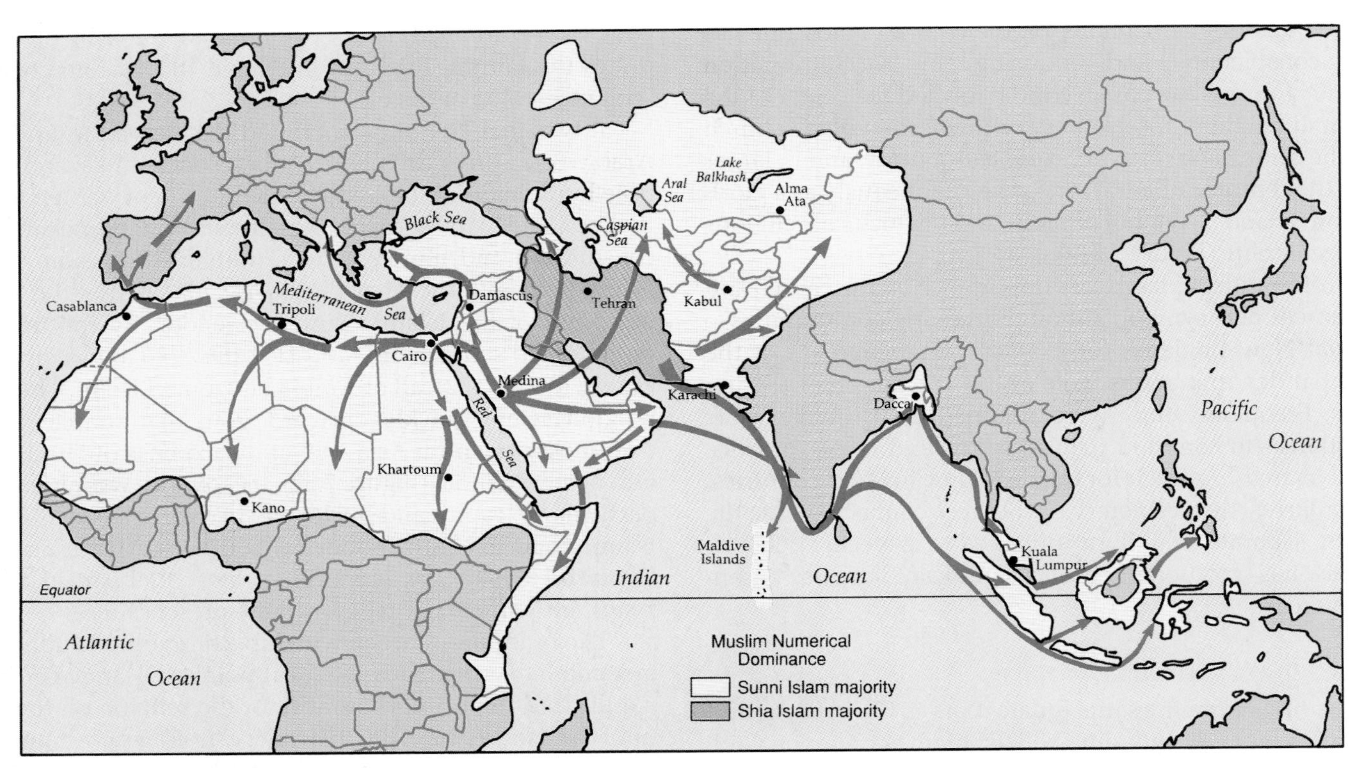

FIGURE 7.32 **Spread and extent of Islam.** Islam predominates in over 35 countries along a band across northern Africa to Central Asia and the northern part of the Indian subcontinent. Still farther east, Indonesia has the largest Muslim population of any country. Islam's greatest development is in Asia, where it is second only to Hinduism, and in Africa, where, some observers suggest, it may be the leading faith. Current Islamic expansion is particularly rapid in the Southern Hemisphere.

FIGURE 7.33 The common architectural features of the mosque make it an unmistakable landscape evidence of the presence of Islam in any local culture. The Badashi Mosque in Lahore, Pakistan, would not be out of place architecturally in Muslim Malaysia or Indonesia.

Hinduism

Hinduism is the world's oldest major religion. Though it has no datable founding event or initial prophet, some evidence traces its origin back 5000 or more years. Hinduism is an ethnic religion, an intricate web of religious, philosophical, social, economic, and artistic elements comprising a distinctive Indian civilization. Its estimated one billion adherents are primarily Asian and largely confined to India, where it claims 80% of the population.

From its cradle area in the valley of the Indus River, Hinduism spread eastward down the Ganges River and southward throughout the subcontinent and adjacent regions by amalgamating, absorbing, and eventually supplanting earlier native religions and customs. Its practice eventually spread throughout Southeast Asia, into Indonesia, Malaysia, Cambodia, Thailand, Laos, and Vietnam, as well as into neighboring Myanmar and Sri Lanka. The largest Hindu temple complex is in Cambodia, not India, and Bali remains a Hindu pocket in dominantly Islamic Indonesia.

There is no common creed, single doctrine, or central ecclesiastical organization defining the Hindu. A Hindu is one born into a caste, a member of a complex social and economic—as well as religious—community. Hinduism accepts and incorporates all forms of belief; adherents may believe in one god or many or none. The *caste* (meaning "birth") structure of society is an expression of the eternal transmigration of souls. For the Hindu, the primary aim of this life is to conform to prescribed social and ritual duties and to the rules of conduct for the assigned caste and profession. Those requirements comprise that individual's *dharma*—law and duties. Traditionally, each craft or profession is the property of a particular caste.

The practice of Hinduism is rich with rites and ceremonies, festivals and feasts, processions and ritual gatherings of literally millions of celebrants. It involves careful observance of food and marriage rules and the performance of duties within the framework of the caste system. Worship in the temples and shrines that are found in every village (Figure 7.34) and the leaving of offerings to secure

FIGURE 7.34 The Hindu temple complex at Khajraho in central India. The creation of temples and the images they house has been a principal outlet of Indian artistry for more than 3000 years. At the village level, the structure may be simple, containing only the windowless central cell housing the divine image, a surmounting spire, and the temple porch or stoop to protect the doorway of the cell. The great temples, of immense size, are ornate extensions of the same basic design.

merit from the gods are required. The temples, shrines, daily rituals and worship, numerous specially garbed or marked holy men and ascetics, and the ever-present sacred animals mark the cultural landscape of Hindu societies, a landscape infused with religious symbols and sights that are part of a total cultural experience.

Buddhism

Numerous reform movements have derived from Hinduism over the centuries, some of which have endured to the present day as major religions on a regional or world scale. For example, *Sikhism* developed in the Punjab area of northwestern India in the late 15th century A.D., rejecting the formalism of both Hinduism and Islam and proclaiming a gospel of universal toleration. The great majority of some 20 million Sikhs still live in India, mostly in the Punjab, though others have settled in Malaysia, Singapore, East Africa, the United Kingdom, and North America.

The largest and most influential of the dissident movements has been *Buddhism,* a universalizing faith founded in the 6th century B.C. in northern India by Siddhartha Gautama, the Buddha ("Enlightened One"). The Buddha's teachings were more a moral philosophy that offered an explanation for evil and human suffering than a formal religion. He viewed the road to enlightenment and salvation to lie in understanding the "four noble truths": existence involves suffering; suffering is the result of desire; pain ceases when desire is destroyed; the destruction of desire comes through knowledge of correct behavior and correct thoughts. The Buddha instructed his followers to carry his message as missionaries of a doctrine open to all castes, for no distinction among people was recognized. In that message all could aspire to ultimate enlightenment, a promise of salvation that raised the Buddha in popular imagination from teacher to savior and Buddhism from philosophy to universalizing religion.

The belief system spread throughout India, where it was made the state religion in the 3rd century B.C. It was carried elsewhere into Asia by missionaries, monks, and merchants. While expanding abroad, Buddhism began to decline at home as early as the 4th century A.D., slowly but irreversibly reabsorbed into a revived Hinduism. By the 8th century its dominance in northern India was broken by conversions to Islam, and by the 15th century it had essentially disappeared from all of the subcontinent.

Present-day spatial patterns of Buddhist adherence reflect the schools of thought, or *vehicles,* that were dominant during different periods of dispersion of the basic belief system (Figure 7.35). In all of its many variants, Buddhism imprints its presence vividly on the cultural landscape. Buddha images in stylized human form began to appear in the 1st century A.D. and are common in painting and sculpture throughout the Buddhist world. Equally widespread are the three main types of buildings and monuments: the *stupa,* a commemorative shrine; the temple or pagoda enshrining an image or relic of the Buddha (Figure 7.36); and the monastery, some of them the size of small cities.

East Asian Ethnic Religions

When Buddhism reached China from the south some 1500 to 2000 years ago and was carried to Japan from Korea in the 6th century, it encountered and later amalgamated with already well established ethical belief systems. The Far Eastern ethnic religions are *syncretisms,* combinations of different forms of belief and practice. In China the union was with Confucianism and Taoism, themselves becoming intermingled by the time of Buddhism's arrival, and in Japan it was with Shinto, a polytheistic animism and shamanism.

Chinese belief systems address not so much the hereafter as the achievement of the best possible way of life in

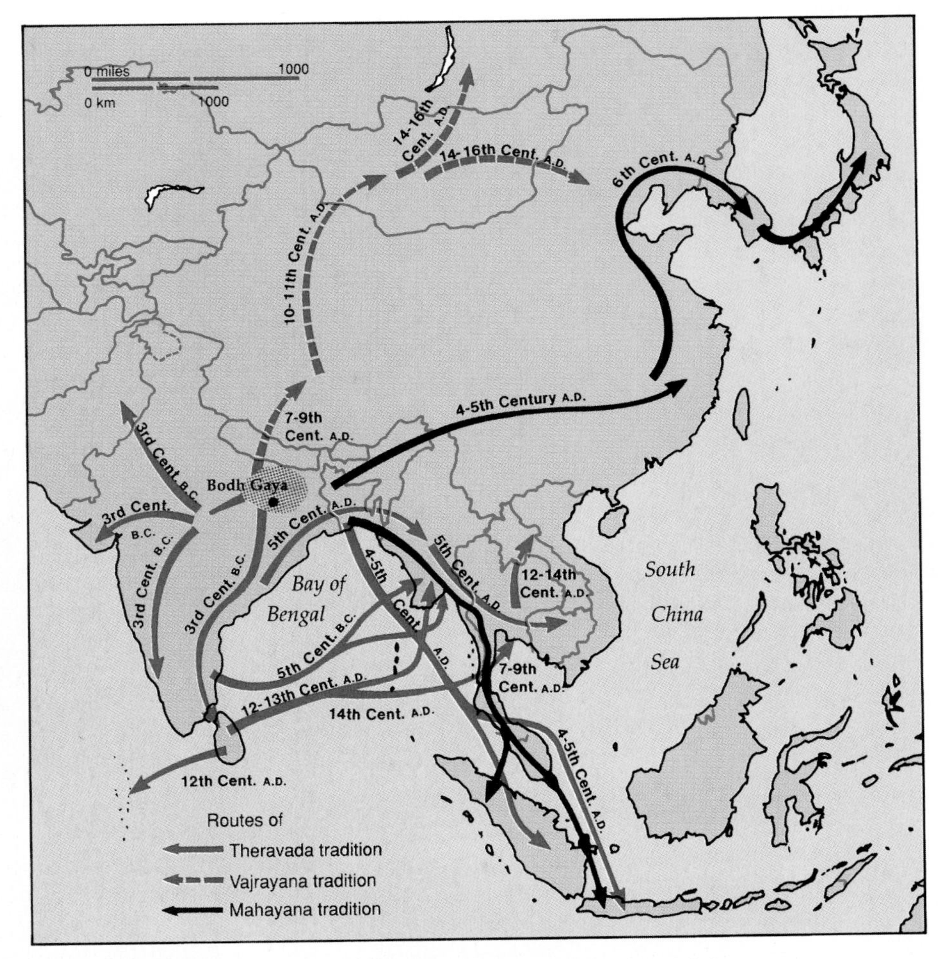

FIGURE 7.35 **Diffusion paths, times, and "vehicles" of Buddhism.**

FIGURE 7.36 The golden stupas of the Swedagon pagoda, Yangon, Myanmar (Rangoon, Burma).

the present existence. They are more ethical or philosophical than religious in the pure sense. Confucius (K'ung Fu-tzu), a compiler of traditional wisdom who lived about the same time as Gautama Buddha, emphasized the importance of proper conduct between ruler and subjects and between family members. The family was extolled as the nucleus of the state, and filial piety was the loftiest of virtues. There are no churches or clergy in *Confucianism,* though its founder believed in a heaven seen in naturalistic terms, and the Chinese custom of ancestor worship as a mark of gratitude and respect was encouraged.

Confucianism was joined by, or blended with, *Taoism,* an ideology that according to legend was first taught by Lao Tsu in the 6th century B.C. Its central theme is *Tao* (the Way), a philosophy teaching that eternal happiness lies in total identification with nature and deploring passion, unnecessary invention, unneeded knowledge, and government interference in the simple life of individuals. Buddhism, stripped by Chinese pragmatism of much of its Indian otherworldliness and defining a *nirvana* achievable in this life, was easily accepted as a companion to these traditional Chinese belief systems. Along with Confucianism and Taoism, Buddhism became one of the honored Three Teachings, and to the average person there was no distinction in meaning or importance between a Confucian temple, Taoist shrine, or Buddhist stupa.

Buddhism also joined and influenced Japanese Shinto, the traditional religion of Japan that developed out of nature and ancestor worship. *Shinto*—The Way of the Gods—is basically a structure of customs and rituals rather than an ethical or moral system. It observes a complex set of deities, including deified emperors, family spirits, and the divinities residing in rivers, trees, certain animals, mountains and, particularly, the sun and moon. At first resisted, Buddhism was later amalgamated with traditional Shinto. Buddhist deities were seen as Japanese gods in a different form, and Buddhist priests formerly but no longer assumed control of most of the numerous Shinto shrines in which the gods are believed to dwell and which are approached through ceremonial *torii,* or gateway arches (Figure 7.37).

ETHNICITY

Any discussion of cultural diversity would be incomplete without the mention of **ethnicity.** Based on the root word *ethnos,* meaning "people" or "nation," the term is usually used to refer to the ancestry of a particular people who have in common distinguishing characteristics associated with their heritage. No single trait denotes ethnicity. Recognition of ethnic communities may be based on language, religion, national origin, unique customs, or an ill-defined concept of

FIGURE 7.37 A Shinto shrine in Nikko, Honshu Island, Japan.

"race" (see "The Matter of Race"). Whatever the unifying thread, ethnic groups may strive to preserve their special shared ancestry and cultural heritage through the collective retention of language, religion, festivals, cuisines, traditions, and in-group work relationships, friendships, and marriages. Those preserved associations are fostered by and support *ethnocentrism,* the feeling that one's own ethnic group is superior.

Normally, reference to ethnic communities is recognition of their minority status within a country or region dominated by a different, majority culture group. We do not identify Koreans living in Korea as an ethnic group because theirs is the dominant culture in their own land. Koreans living in Japan, however, constitute a discerned and segregated group in that foreign country. Ethnicity, therefore, is an evidence of areal cultural diversity and a reminder that culture regions are rarely homogeneous in the characteristics displayed by all of their occupants.

Territorial segregation is a strong and sustaining trait of ethnic identity, and one that assists groups to retain their distinction. On the world scene, indigenous ethnic groups have developed over time in specific locations and have established themselves in their own and others' eyes as distinctive peoples with defined homeland areas. The boundaries

The Matter of Race

Human populations may be differentiated from one another on any number of bases: gender, nationality, stage of economic development, and so on. One common form of differentiation is based on recognizable inherent physical characteristics, or *race.*

A **race** is usually understood to be a population subset whose members have in common some hereditary biological characteristics that set them apart physically from other human groups. The spread of human beings over the earth and their occupation of different environments were accompanied by the development of physical variations in skin pigmentation, hair texture, facial characteristics, blood composition, and other traits largely related to soft tissue. Some subtle skeletal differences among peoples also exist. Such differences have formed the basis for the segregation, by some anthropologists, of humanity into different racial groups. Caucasoid, Negroid, Mongoloid, Amerindian, Australoid and other races have been recognized in a process of classification invention, modification, and refinement that began at least two centuries ago. Racial differentiation, as commonly understood, is old and can reasonably be dated at least to the Paleolithic (100,000 to about 11,000 years ago) spread and isolation of population groups.

Although racial classificaitons vary by author, most are based on recognized geographical variations of populations. Thus, Mongoloids are associated with northern and eastern Asia; Australoids are the aboriginal people of Australasia; Amerindians developed in the Americas, and so on. If all of humankind belongs to a single species that can freely interbreed and produce fertile offspring, how did this areal differentiation by race occur? Why is it that despite millennia of mixing and migration, people with distinct combinations of physical traits appear to be clustered in particular areas of the world?

Two causative forces appear to be most important. First, through evolutionary **natural selection** or **adaptation,** characteristics are transmitted that enable people to adapt to particular environment conditions, such as climate. Studies have suggested some plausible relationship between, for example, solar radiation and skin color, and between temperature and body size. In its carrier state, sickle-cell anemia, afflicting mainly people of African descent, protects against malaria. The second force, **genetic drift,** refers to a heritable trait (such as flatness of face) that appears by chance in one group and becomes accentuated through inbreeding. If two populations are too separated spatially for much interaction to occur (*isolation*), a trait may develop in one but not in the other. Unlike natural selection, genetic drift differentiates populations in nonadaptive ways.

Natural selection and genetic drift promote racial differentiation. Countering them is **gene flow** via interbreeding (also called *admixture*), which acts to homogenize neighboring populations. Genetically, it has been observed, there is no such thing as "pure" race since people breed freely outside their local group. Opportunities for interbreeding, always part of the spread and intermingling of human populations, have accelerated with the growing mobility and migrations of people in the past few centuries. While we may have an urge to group humans "racially," we cannot use biology to justify it, and anthropologists have largely abandoned the idea of race as a scientific concept.

Nor does race have meaningful application to any human characteristics that are culturally acquired. That is, race is *not* equivalent to ethnicity or nationality and has no bearing on differences in religion or language. There is no "Irish" or "Hispanic" race, for example. Such groupings are based on culture, not genes. Culture summarizes the way of life of a group of people, and members of the group may adopt it irrespective of their individual genetic heritage, or race.

of most countries of the world encompass a number of racial or ethnic minorities (Figure 7.38). Their demands for special territorial recognition have sometimes increased with advances in economic development and self-awareness, as Chapter 9, "Political Geography," points out. Where clear territorial separation does not exist but ethnic identities are distinct and animosities bitter, tragic conflict within single political units can erupt. Recent histories of deadly warfare between Tutsi and Hutu in Rwanda, or Serb and Croat in Bosnia, make vivid the often continuing reality of ethnic discord and separatism.

Increasingly in a world of movement, ethnicity is less a matter of indigenous populations and more one of outsiders in an alien culture. Immigrants, legal and illegal, and refugees from war, famine, or persecution are a growing presence in countries throughout the world. Immigrants to a country typically have one of two choices. They may hope for *assimilation* by giving up many of their past cultural traits, losing their distinguishing characteristics and merging into the mainstream of the dominant culture, as reviewed earlier on p. 247. Or, they may try to retain their distinctive cultural heritage. In either case, they usually settle initially in an area where other members of their ethnic group live, as a place of refuge and learning (Figures 7.39 and 12.27). With the passage of time, they may leave their protected community and move out among the general population.

The Chinatowns and Little Italys of Anglo American cities have provided the support systems essential to new immigrants in an alien culture region. Japanese, Italians, Germans, and other ethnics have formed agricultural colonies in Brazil in much the same spirit. Such ethnic enclaves may provide an entry station, allowing both individuals and the groups to which they belong to undergo

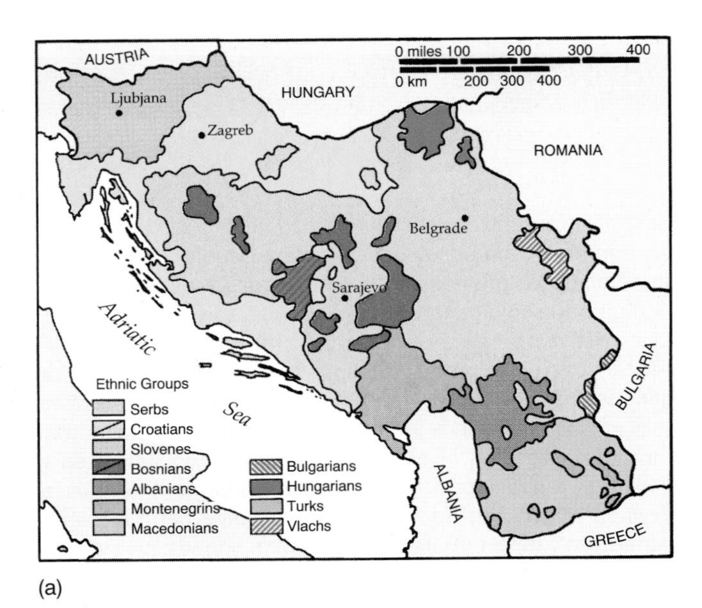

(a)

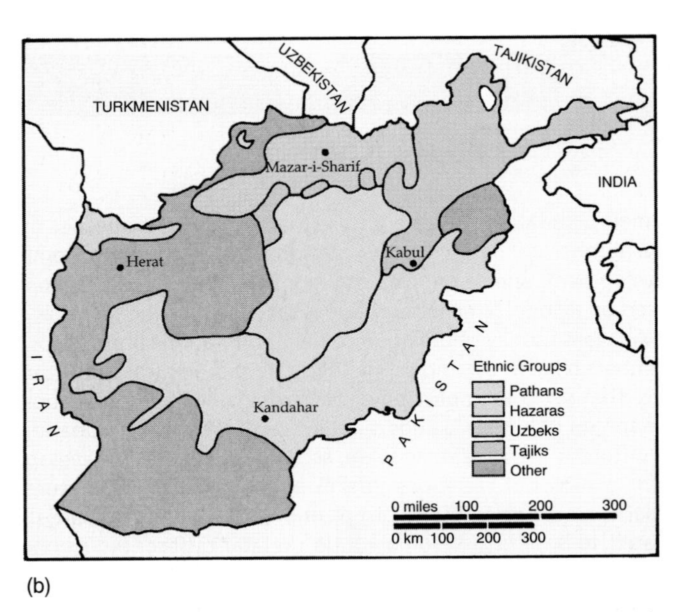

(b)

FIGURE 7.38 (a) **Ethnicity in former Yugoslavia.** Yugoslavia was formed after World War I (1914–1918) from a patchwork of Balkan states and territories, including the former kingdoms of Serbia and Montenegro, Bosnia and Herzegovina, Croatia-Slavonia, and Dalmatia. The authoritarian central government created in 1945 began to disintegrate in 1991 as non-Serb minorities voted for regional independence and national separation. Religious differences between Eastern Orthodox, Roman Catholic, and Muslim adherents compounded resulting conflicts rooted in nationality and rival claims to ethnic homelands. (b) **Afghanistan** houses Pathan, Tajik, Uzbek, and Hazara ethnic groups (among others) speaking Pashto, Dari Persian, Uzbek, and several minor languages, and split between majority Sunni and minority Shia Muslim believers.

FIGURE 7.39 Cuban and other Spanish-speaking immigrants have placed their distinctive ethnic marks on the south Florida urban landscape.
©Barry Barker.

cultural and social modifications sufficient to enable them to operate effectively in the new, majority society. Sometimes, of course, settlers have no desire to assimilate or are not allowed to assimilate, so that they and their descendants form a more or less permanent subculture in the larger society. The Chinese in Malaysia belong to this category. Ethnicity in the context of nationality is discussed more fully in Chapter 9.

GENDER AND CULTURE

Gender refers to socially created—not biologically based—distinctions between femininity and masculinity. Because gender relationships and role assignments differ among societies, the status of women is a cultural spatial variable and becomes, therefore, a topic of geographic interest and inquiry. Gender distinctions are complex, and the role and reward assignments of males and females differ from society to society. In many fundamental ways those assignments are conditioned by areally different levels of economic development. Therefore, we might well assume a close similarity between the economic roles and production assignments of males and females—and the status of women—in different cultures that are at the same level of technological advancement. Indeed, it has been observed that modern African or Asian subsistence agricultural groups and those of 18th-century frontier America show more similarities in gender relationships than do pioneer rural and contemporary postindustrial American society.

It may further be logical to believe that advancement in the technological sense would be reflected in an enhancement of the status and rewards of both men and women in all developing societies. The pattern that we actually observe is not quite that simple or straightforward, however. In addition to a culture's economic stage, religion and custom play their important roles in determining gender relationships and female prestige. Further, it appears that at least in the earlier phases of technological change and development, women generally lose rather than gain in standing and rewards. Only recently and only in the most developed countries have gender contrasts been reduced within and between societies.

Hunting and gathering cultures observed a general egalitarianism; each sex had a respected, productive, coequal role in the kinship group (see Figure 7.9). Gender is more involved and changeable in agricultural societies (see "Women and the Green Revolution," page 360). The Agricultural Revolution—a major change in the technological subsystem—altered the earlier structure of gender-related responsibilities. In the hoe agriculture that was the first advance over hunting and gathering and is today found in much of sub-Saharan Africa and in South and Southeast Asia, women became responsible for most of the actual field work, while still retaining their traditional duties in child rearing, food preparation, and the like; their economic role and status remained equivalent to males. Plow agriculture, on the other hand, tended to subordinate the role of women and diminish their level of equality. Women might have hoed, but men plowed, and female participation in farm work was drastically reduced. This is the case today in Latin America and, increasingly, in sub-Saharan Africa where women are often more visibly productive in the market than in the field (Figure 7.40). As women's agricultural productive role declined, they were afforded less domestic authority, less control over their own lives, and few if any property rights independent of male family members.

Western industrial—"developed"—society emerged directly from the agricultural tradition of the subordinate female who was not considered an important element in the economically active population, no matter how arduous or essential the domestic tasks assigned. With the growth of cities and industry in 19th-century America, for example, as women began to enter the workforce in increasing numbers, a "cult of true womanhood" developed as a reaction to the competitive pressures of the marketplace and factory floor. It held that women were morally superior to men; their role was private, not public. A woman's job was to rear children, attend church, and above all to keep a sober, virtuous, and cultured home, a place that offered the male breadwinner refuge, security, and privacy. This Victorian ideal fostered in America and much of Western Europe both social and economic discrimination against working women.

Only within the later 20th century, and then only in the more developed countries, did that subordinate role pattern change. Women have become increasingly economically active and have placed themselves as never before in direct competition with men for similar occupations and wages. The feminist movement in modern industrialized societies has been the direct response to the barriers that formerly restricted favored economic and legal positions to men. Even though women made up over 46% of the total employed labor force in the United States in the late 1990s,

FIGURE 7.40 Women dominate the once-a-week *periodic* markets in nearly all developing countries. Here they sell produce from their gardens or the family farm and frequently offer for sale processed goods to which their labor has added value: oil pressed from seeds or—in Niger, for example—from peanuts grown on their own fields; cooked, dried, or preserved foods; simple pottery, baskets, or decorated gourds. The market shown here is in Freetown, Sierra Leone. More than half the economically active women in sub-Saharan Africa and southern Asia and about one-third in northern Africa and the rest of Asia are self-employed, working primarily in the informal sector. In the developed world, only about 14% of active women are self-employed.
©Topham/The Image Works.

their representation in higher paying, more prestigious positions has been much lower despite increasing acceptance of the standard of equal pay for equal work. The elimination of remaining legal, social, and economic discriminations has been a primary objective of the North American "feminist revolution."

Such a revolution is much less likely or possible in strongly conservative and traditional economies and societies. The present world pattern of gender-related institutional and economic role assignments is influenced not only by a country's level of economic development, but also by the persistence of religious and customary restrictions its culture imposes on women, and by the specific nature of its economic—particularly agricultural—base. The first control is reflected in contrasts between the developed and developing world; the second and third are evidenced in variations within the developing world itself.

A distinct gender-specific regionalization has emerged. Among the Arab or Arab-influenced Muslim areas of western Asia and North Africa, the proportion of the female population that is economically active is low; religious tradition restricts women's acceptance in economic activities outside the home. The same cultural limitations do not apply under the different rural economic conditions of Muslims in southern and southeastern Asia, where labor force participation by women in Indonesia and Bangladesh, for example, is much higher than it is among the western Muslims. In Latin America, well-known for a patriarchal social structure, women have been overcoming cultural restrictions on their growing employment outside the home.

Sub-Saharan Africa, highly diverse culturally and economically, in general is highly dependent on female farm labor and market income. The traditional role of strongly independent, property-owning females formerly encountered under traditional agricultural and village systems, however, has increasingly been replaced by subordination of women with modernization of agricultural techniques and introduction of formal, male-dominated financial and administrative agricultural institutions. For the developing world in general, a series of indicators has been combined to establish a "gender empowerment measure" and ranking (Figure 7.41); it clearly displays regional differentials in the position of women in different cultures and world areas.

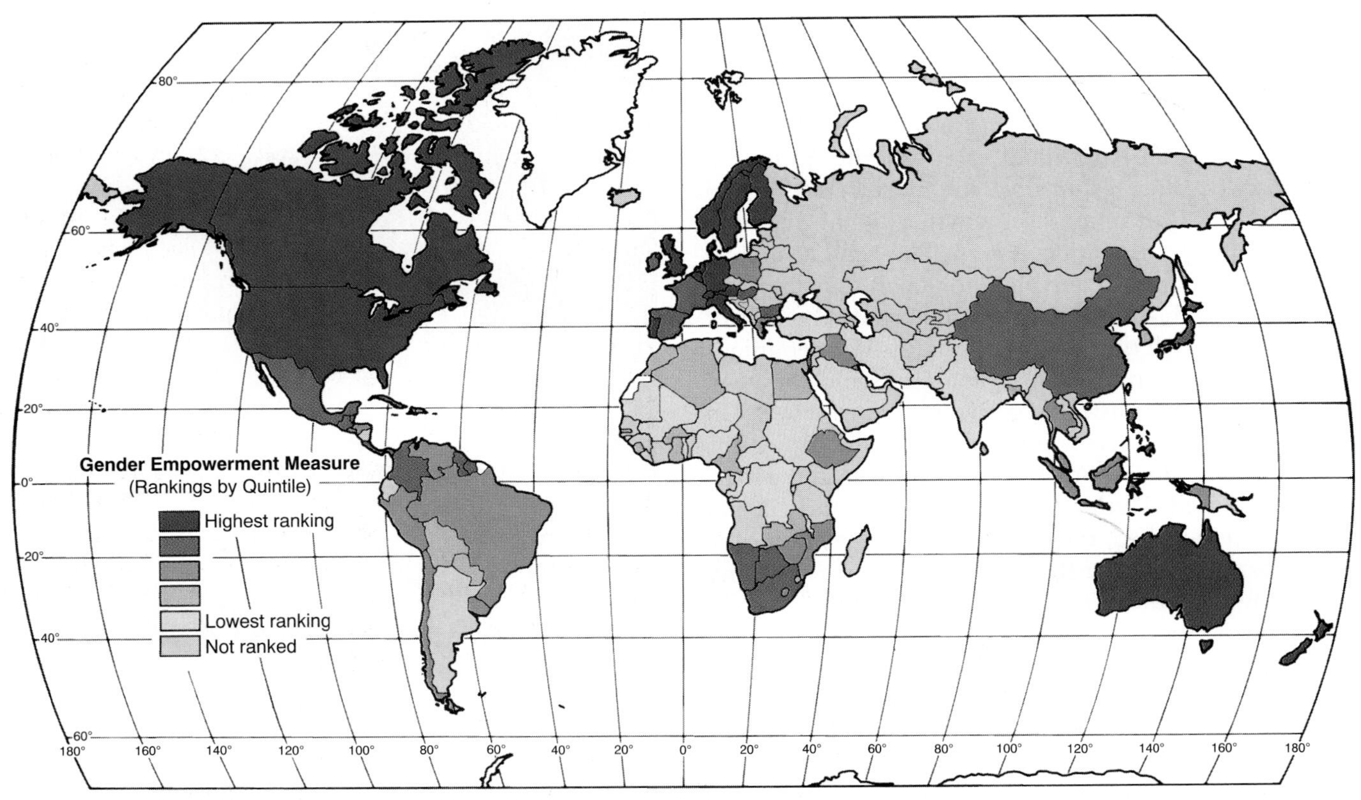

FIGURE 7.41 **The gender empowerment measure (GEM)** summarizes women's access to political and economic power based on three variables: female share of parliamentary seats; share of professional, technical, administrative, and managerial jobs; and earning power, determined by access to jobs and wages. The GEM rankings show that gender equality in political, economic, and professional activities is not necessarily related to level of national wealth or development. Some developing countries according to this measure—China, for example, where women are afforded a large share of legislative seats and political administrative positions—outperform industrialized France and Spain. Only 116 countries are ranked; in most, women are in a distinct minority in the exercise of economic power and decision-making.

Source: Rankings from United Nations Development Programme.

Regional and cultural contrasts in gender relationships are also encountered in the advanced economies and the industrial components of developing countries. In modernizing Eastern Asian states, for example, women have yet to achieve the status they enjoy in most Western economies. In China, although it ranks high in "gender empowerment" (Figure 7.41), women are generally not effectively competitive with men and are largely absent from highest managerial and administrative levels; in Japan, males nearly exclusively run the huge industrial and political machinery of the country. In contrast, economic and social gender equality is more advanced in the Scandinavian countries than perhaps in any other portion of the industrialized world.

Other Aspects of Diversity

Culture is the sum total of the way of life of a society. It is misleading to isolate, as we have done, only a few elements of the technological, sociological, and ideological subsystems and imply by that isolation that they are identifying characteristics differentiating culture groups. Economic developmental levels, language, religion, ethnicity, and gender all are important and common distinguishing cultural traits, but they tell only a partial story. Other suggestive, though perhaps less pervasive, basic elements exist.

Architectural styles in public and private buildings are evocative of region of origin even when they are indiscriminately mixed together in American cities. The Gothic and New England churches, the neoclassical bank, and the skyscraper office building suggest not only the functions they house but the culturally and regionally variant design solutions that gave them form. The Spanish, Tudor, French Provincial, or ranch-style residence may not reveal the ethnic background of its American occupant, but it does constitute a culture statement of the area and the society from which it diffused.

Music, food, games, and other evidences of the joys of life, too, are cultural indicators associated with particular world or national areas. Music is an emotional form of communication found in all societies, but, being culturally patterned, it varies among them. Instruments, scales, and types of compositions are technical forms of variants; the emotions aroused and the responses evoked among peoples to musical cues are learned behaviors. The Christian hymn means nothing emotionally to a pagan New Guinea clan. The music of a Chinese opera may be simply noise to the European ear. Where there is sufficient similarity between musical styles and instrumentation, blending (syncretism) and transferal may occur. American jazz represents a blend; calypso and flamenco music have been transferred to the Anglo American scene. Foods identified with other culture regions have similarly been transferred to become part of the culinary environment of the American "melting pot."

These are but a few additional minor statements of the variety and the intricate interrelationships of that human mosaic called culture. Individually and collectively they are, in their areal expressions and variations, the subject matter of the cultural geographer.

Culture Realms

Our discussion of culture has had one consistent and recurrent message: culture has spatial expression in all of its details and composites. The individual culture traits we examined and mapped show the subdivision of earth space into special-purpose regions. Of course, the same trait—the Christian religion, perhaps, or the Spanish language—may be part of more than one culture, but each separate culture will be marked by a distinctive complex of such individual traits, setting it off spatially from adjacent cultures with their own identifying composites of traits.

If two or more culture complexes have a number of such traits in common, a **culture system** may be recognized as a larger spatial reality and generalization. Multiethnic societies, perhaps further subdivided by linguistic differences, varied food preferences, and a host of other internal differentiations, may nonetheless share enough common characteristics of the subsystems of culture to be recognizably distinctive cultural entities to themselves and others. Certainly, citizens of "melting-pot" United States would identify themselves as *Americans,* together comprising a unique culture system on the world scene.

Culture regions and complexes are elements in the spatial hierarchy of cultural geography. They may, at a still higher level of generalization, be combined into composite world regions, into **culture realms** that are, simply, regionally discrete areas that are more alike internally than they are like other realms. At the level of generalization suggested in Figure 7.3 and used as the background for the separate topics discussed in this chapter, cultural specifics become obscured and perceptions of world regional differentiation come into play. Culture realms attempt to document those perceptions of world-scale cultural contrasts by identifying groups of culture systems with enough distinctive characteristics in common to set them apart from other realms with differing sets of identifying generalizations.

Clearly, our present database is inadequate to the task of definitive world cultural regionalization. Political structure, economic orientation, patterns of behavior, levels of urbanization—all aspects of contemporary culture—are yet to be considered. Figure 7.3, therefore, can be considered only a preliminary recognition of composite culture realms, one limited by dependence only on the fundamental differentiating characteristics of development level, language, religion, ethnicity, and gender already discussed. We can separately extend and refine our views of world culture realms as we consider the topics yet to be discussed in the following chapters.

Summary

Culture is the learned behaviors and beliefs of distinctive groups of people. Culture traits, the smallest differentiating items of culture, are the building blocks of integrated culture complexes. Together, traits and complexes in their spatial patterns create human—"cultural"—landscapes, define culture regions, and distinguish culture groups. Those landscapes, regions, and group characteristics change through time as human societies interact with their environment, develop new solutions to collective needs, or are altered through innovations adopted from outside the group.

The detailed complexities of culture can be simplified by recognition of its component subsystems. The technological subsystem is composed of the material objects (artifacts) and techniques of livelihood. The sociological subsystem comprises the formal and informal institutions (sociofacts) that control the social organization of a culture group. The ideological subsystem consists of the ideas and beliefs (mentifacts) a culture expresses in speech and through belief systems.

The presumed cultural uniformity of a preagricultural world was lost as domestication of plants and animals in many world areas led to the emergence of culture hearths of wide-ranging innovation and to a cultural divergence between different groups. While modern-day shared technologies contribute to cultural convergence throughout the world, many elements of cultural distinction remain to identify and separate social groups. Among the most prominent of the differentiating cultural traits are language, religion, ethnicity, and gender.

Language and religion are both transmitters of culture and identifying traits of culture groups. Both have distinctive spatial patterns, reflecting past and present processes of interaction and change. Although languages may be grouped by origins and historical development, their world distributions depend as much on the movements of peoples and histories of conquest and colonization as they do on linguistic evolution. Toponymy, the study of place-names, helps document that history of movement. Linguistic geography studies spatial variations in languages, variations that may be minimized by encouragement of standard languages or overcome by pidgins, creoles, and lingua francas.

Religion is a less pronounced identifier or transmitter of culture than is language, but even in secular societies religion may influence economic activities, legal systems, holiday observances, and the like. Although religions do not lend themselves to easy classification, their spatial patterns are distinct and reveal past and present histories of migration, conquest, and diffusion. Those patterns are also important components in the spatially distinctive cultural landscapes created in response to different religious belief systems.

Ethnicity, affiliation in a group sharing common identifying cultural traits, is fostered by territorial separation or isolation and preserved in ethnically complex societies by a feeling that one's own ethnic group is superior to others. Ethnic diversity is a reality in most countries of the world and is increasing in many of them. Many ethnic minority groups may seek absorption into their surrounding majority culture through acculturation and assimilation, but other groups choose to preserve their identifying distinctions through spatial separation or overt rejection of the majority cultural traits.

Gender, the culturally based social distinctions between men and women, reflects religion, custom, and, importantly, the stage of economic development of a society and the productive role assigned to women within that economy. Gender roles change as the economic structure changes, though their modification is often resisted by conservative forces within a culture. Regional variation in the status of women is both distinctive and an important contributor to the pattern of world culture realms.

The culture realms themselves, as we have seen, are ever-changing. They are but temporary reflections of the migrations of ethnic and cultural groups, the diffusion or adoption of languages and religions, the spread and acceptance of new technologies, and the alteration of gender relationships as economies modernize and cultural traditions respond. Such movements, diffusions, adoptions, and responses are themselves expressions of broader concepts and patterns of the geography of spatial behavior, an essential component of the culture–environment tradition to which we next turn our attention.

Key Words

FOR REVIEW AND CONSIDERATION

1. What is included in the concept of *culture?* How is culture transmitted? What personal characteristics affect the aspects of culture that any single individual acquires or fully masters?
2. What is a *culture hearth?* What new traits of culture characterized the early hearths? In the cultural geographic sense, what is meant by *innovation?*
3. Differentiate between *culture traits* and *culture complexes.* Between *environmental determinism* and *possibilism.*
4. What are the components or subsystems of the three-part system of culture? What characteristics—aspects of culture—are included in each of the subsystems?
5. Why might one consider language the dominant differentiating element of culture separating societies?
6. In what way may religion affect other cultural traits of a society?
7. How does the classification of religions as *universalizing, ethnic,* or *tribal* help us understand their patterns of distribution and spread?
8. How does *acculturation* occur? Is *ethnocentrism* likely to be an obstacle in the acculturation process? How do acculturation and *assimilation* differ?
9. How are the concepts of *ethnicity, race,* and *culture* related?

SELECTED REFERENCES

Allen, James P., and Eugene Turner. *We the People: An Atlas of America's Ethnic Diversity.* New York: Macmillan, 1988.

al-Faruqi, Isma'il R., and David E. Sopher, eds. *Historical Atlas of the Religions of the World.* New York: Macmillan Publishing Co., 1974.

Fellmann, Jerome D., Arthur Getis, and Judith Getis. *Human Geography.* 6th ed. Dubuque, Iowa: WCB/McGraw-Hill, 1999.

Gaustad, Edwin S. *Historical Atlas of Religion in America.* New York: Harper & Row, 1976.

Hanson, Susan, and Geraldine Pratt. *Gender, Work and Space.* New York: Routledge, 1995.

Journal of Cultural Geography (Popular Culture Association and the American Cultural Association) 7, no. 1 (Fall/Winter, 1986). Special issue devoted to geography and religion.

Lieberson, Stanley, and Mary C. Waters. *From Many Strands: Ethnic and Racial Groups in Contemporary America.* Census Monograph Series. New York: Russell Sage Foundation, 1988.

McCrum, Robert, William Cran, and Robert MacNeil. *The Story of English.* New York: Elizabeth Sifton Books/Viking, 1986.

Momsen, Janet H. *Women and Development in the Third World.* New York: Routledge, 1991.

Moseley, Christopher, and R. E. Asher, eds. *Atlas of the World's Languages.* London, England and New York: Routledge, 1994.

Noble, Allen G., ed. *To Build in a New Land: Ethnic Landscapes in North America.* Baltimore, Md.: Johns Hopkins University Press, 1992.

Park, Chris C. *Sacred Worlds: An Introduction to Geography and Religion.* London and New York: Routledge, 1994.

Rogers, Alisdair, ed. *Peoples and Cultures.* New York: Oxford University Press, 1992. The Illustrated Encyclopedia of World Geography.

Sauer, Carl. *Agricultural Origins and Dispersals.* New York: American Geographical Society, 1952.

Sopher, David E. *The Geography of Religions.* Englewood Cliffs, N.J.: Prentice-Hall, 1967.

Stewart, George R. *Names on the Globe.* New York: Oxford University Press, 1975.

Stewart, George R. *Names on the Land.* 4th ed. San Francisco: Lexikos, 1982.

Thomas, William L., Jr., ed. *Man's Role in Changing the Face of the Earth.* Chicago: University of Chicago Press, 1956.

United Nations. *The World's Women 1995: Trends and Statistics.* Social Statistics and Indicators, Series K, no. 12. New York: United Nations, 1995.

Zelinsky, Wilbur. *The Cultural Geography of the United States.* Rev. ed. Englewood Cliffs, N.J.: Prentice-Hall, 1992.

Don't forget about Dushkin's *Annual Editions Online: Geography* at http://dushkin.com/aeonline/. See preface for details.

CHAPTER

8

GEOGRAPHY OF SPATIAL BEHAVIOR

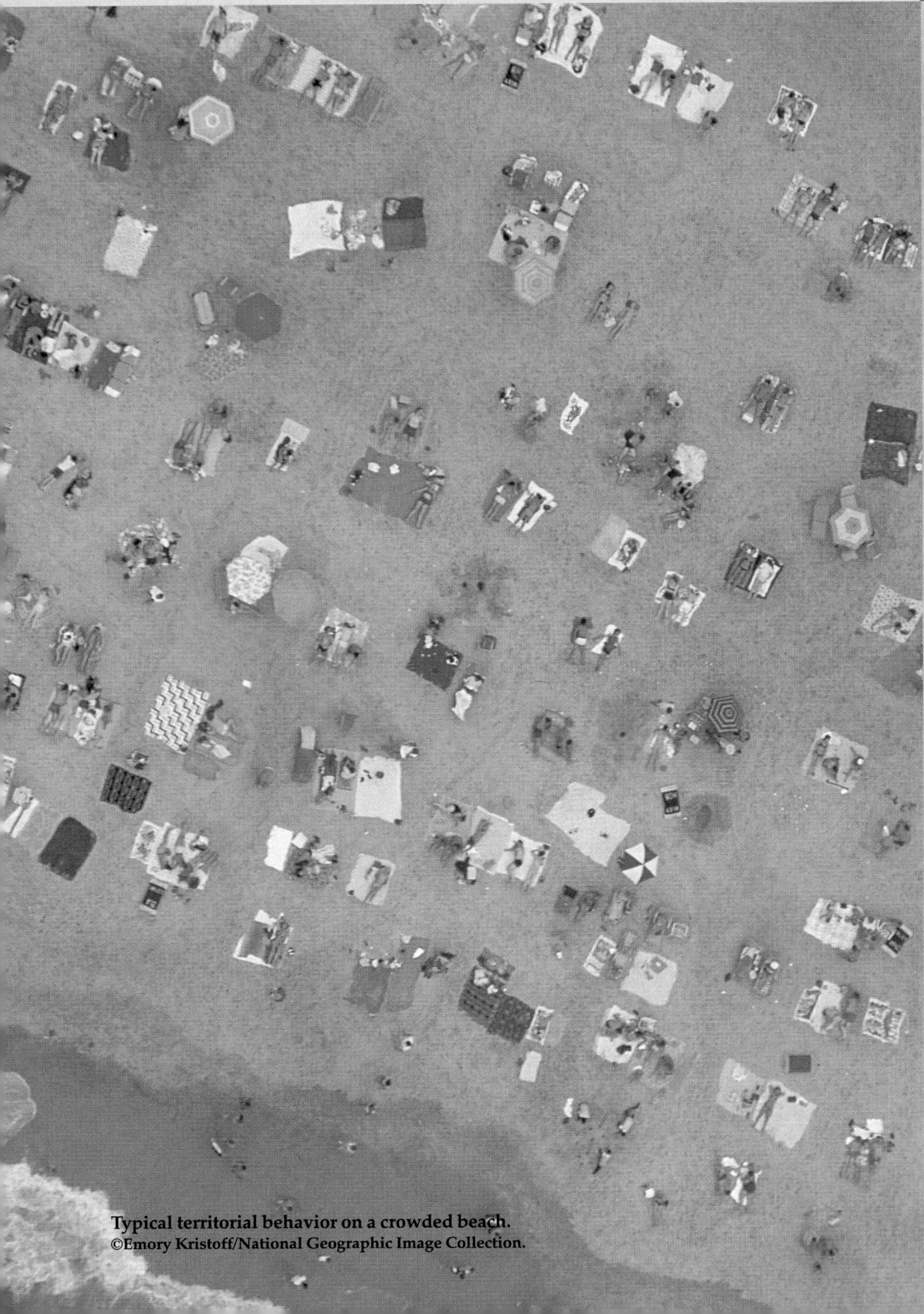

Typical territorial behavior on a crowded beach.
©Emory Kristoff/National Geographic Image Collection.

*Early in January of 1849 we first thought of migrating to California. It was a period of National hard times . . . and we longed to go to the new El Dorado and "pick up" gold enough with which to return and pay off our debts. Our discontent and restlessness were enhanced by the fact that my health was not good. . . . The physician advised an entire change of climate thus to avoid the intense cold of Iowa, and recommended a sea voyage, but finally approved of our contemplated trip across the plains in a "prairie schooner." Full of the energy and enthusiasm of youth, the prospects of so hazardous an undertaking had no terror for us, indeed, as we had been married but a few months, it appealed to us as a romantic wedding tour.**

So begins Catherine Haun's account of her journey from Clinton, Iowa, to California in 1849, a trip that was to last eight months and cover 3900 kilometers (2400 mi). The Hauns were just two of the 250,000 people who traveled across the continent on the Overland Trail in one of the world's great migrations. The dangers inherent in such a trip were numerous; thousands were to die en route, and many others stopped or turned back short of their goal. The migrants faced at least six months, and often more, of grueling travel over badly marked routes that crossed swollen rivers, deserts, and mountains. The weather was often foul, with hailstorms, drenching rains, and summer temperatures that could exceed 43°C (110°F) inside the covered wagons. Wagon breakdowns were frequent. Graves along the route were a silent testimony to the lives claimed by buffalo stampedes, Indian skirmishes, cholera epidemics, and other disasters.

What inducements were so great as to make emigrants leave behind all that was familiar and risk their lives on an uncertain venture? Catherine Haun's account is unusual in that it gives the reasons for their trip. She alludes to economic hard times; the depression that swept the United States in 1837 inaugurated a prolonged period of bank closures, depressed prices for agricultural goods, and high rates of unemployment. The Hauns hoped to strike it rich by mining gold. Other migrants were attracted by reports of free land and rich soil in the Oregon and California territories, productive fishing grounds, and ample furs for trapping. Like other migrants, the Hauns were also attracted by the climate in the West, which was said to be always sunny and free of disease. Finally, like most who undertook the trip along the Overland Trail, the Hauns were young, moved by restlessness and a sense of adventure.

*From Catherine Haun, "A Woman's Trip Across the Plains in 1849," as quoted in Lillian Schlissel, *Women's Diaries of the Westward Journey,* New York: Schocken Books, Inc., 1982. By permission of the Huntington Library, San Marino, CA.

Catherine Haun's story is unique only in its particulars. As did her predecessors back to the beginnings of humankind, she and her family acted in space and over space on the basis of acquired information and awareness of opportunity (Figure 8.1). Her story summarizes the content of this chapter, a survey of how individuals make spatial behavioral decisions and how those separate decisions may be summarized by models and generalizations to explain collective actions.

The geography of spatial behavior is a relatively new field of the discipline of geography. In the last 30 years, geographers have become more interested in studying culture from the viewpoint of the individual. In Chapter 7, we outlined what might be called the general patterns of cultural geography. In this chapter, we pose a question basic to the themes that we have been exploring in preceding chapters: What considerations influence how individual human beings use space and act within it? Implicit in this question are subsidiary analytical concerns: How do individuals view (perceive) their environment? How is information transmitted through space and acted upon? How might all the separate decisions of many individuals be summarized so that we may understand the order that underlies the seeming randomness of individual action?

We will begin with a discussion of human environmental perception. This topic leads to a consideration of how people define the space within which they carry out their activities. Then we turn to the way in which ideas are transmitted over space. Finally, we bring together all of these concepts in a section devoted to the migration process. Before the discussion begins, however, we must emphasize that topics related to cultural diversity, such as language, religion, ethnicity, and gender, while important elements of spatial behavior, can only be touched upon.

FIGURE 8.1 Cross-country movement was slow, arduous, and dangerous early in the 19th century, and the price of long-distance spatial interaction was far higher in time and risks than a comparable journey today.
©Melinda Berge/Photographers Aspen, Inc.

Perception of the Environment

The term **environmental perception** refers to our awareness, as individuals, of home and distant places and the beliefs we have about them. It involves our feelings, reasoned or irrational, about the complex of natural and cultural characteristics of an area. Whether our view accords with that of others or truly reflects the "real" world seen in abstract descriptive terms is not the major concern. Our perceptions are our reality. The decisions people make about the use of their lives are based not necessarily upon reality but on their perceptions of reality.

Geographers interested in determining how we arrive at our environmental understanding employ the concept, taken from psychology, termed **cognition.** Cognition refers to the way information, once received by an individual, is given mental meaning. Perceptions of the environment are sent to the brain, where they are stored together with our previously accumulated environmental knowledge. The mental or cognitive structures influence the way we process and recall our perceptions. One might perceive a college classroom by physically being there, but later recalling our perception of the classroom depends on the way we have mentally organized our perceptions.

The fact that people have different experiences has much to do with the way they perceive their environment. As people order new information, they store it within their own developing mental structures. Thus, it is not unusual for people to interpret similar phenomena in different ways. The more uniform the life experiences of a group of people, the greater the chance they will have similar perceptions.

In technologically advanced societies, television and radio, magazines and newspapers, books and lectures, the Internet, travel brochures, and hearsay all combine to help develop a mental picture of unfamiliar places. The most effectively transmitted information seems to come from word-of-mouth reports. These may be in the form of letters, e-mail, or conversations with relatives, friends, and associates. Probably the strongest lines of attachment to relatively unknown regions, whether nearby or far away, develop through the information supplied by family members and friends.

Of course, our knowledge of close places is greater than our knowledge of far places. But barriers to information flow give rise to *directional biases.* Not having friends or relatives in one part of a country may represent a barrier to individuals, so that interest in and knowledge of the area beyond the "unknown" region are sketchy. In the United States, both northerners and southerners tend to be less well informed about each other's areas than about the western part of the country. Traditional communication lines in the United States follow an east-west rather than a north-south direction, which is the result of early migration patterns, business connections, and the pattern of the development of principal cities.

In wayfinding experiments, behavioral geographers have found that certain primary *anchor points* such as the home, workplace, and shopping places, are the nodes on which understanding of the environment is based. Secondary anchor points may include commonly recognized and often-used places such as schools and well-known landmarks. We learn about our neighborhood, community, region, and so on, first from our knowledge of anchor points, second, from the paths connecting them, and third, from the areas surrounding the anchor points and the paths.

Mental Maps

When information about a place is sketchy, blurred pictures develop. These influence the impression we have of places and cannot be discounted. We may say that each individual has a **mental map** of the world. No single person, of course, has a true and complete image of the world; therefore having a completely accurate mental map is impossible. In fact, the best mental map that most individuals have is that of their own residential neighborhood, the place where they spend the most time.

No one can reproduce on paper an exact replica of the mental image that he or she has of an area. The study of mental maps must, by necessity, be indirect. If we want to know how particular people envisage their town, we must either ask them questions about the town or ask them to draw sketch maps. Although the result will not be a completely accurate picture of what they have in mind, it will be suggestive of the mental map.

Whenever individuals think about a place or how to get to a place, they produce a mental map. What are believed to be unnecessary details are left out, and only the important elements are incorporated (Figure 8.2). Those elements usually include awareness that the object or the destination does indeed exist, some conception of the distances separating the starting point and the named object(s), and a feeling for the directional relationships between points. A mental route map may also include reference points to be encountered on the chosen path of connection or on alternate lines of travel. Although mental maps are highly personalized, people with similar experiences tend to give similar answers to questions about the environment and to produce roughly comparable sketch maps.

Awareness of places is usually accompanied by opinions about them, but there is no necessary relationship between the depth of knowledge and the perceptions held. In general, the more familiar we are with a locale, the more sound will be the factual basis of our mental image of it. But individuals form firm impressions of places totally unknown by them personally, and these may affect travel or migration decisions.

One way to ascertain how individuals envisage the environment is to ask them what they think of various places. For instance, they might be asked to rate places according to desirability—perhaps residential desirability—or to make a list of the best and worst places in a region such as the United States. Certain regularities appear in such studies. Figure 8.3 presents some residential desirability data as elicited from college students in three provinces of Canada. These and comparable mental maps suggest that near places are preferred to far places unless much information is available about the far places. Places with similar cultural forms are preferred, as are places with high standards of living. Individuals tend to be indifferent to unfamiliar places, and to dislike unfamiliar areas that have competing cultural interests (such as disliked political and military activities) or a physical environment known to be unpleasant.

People mentally tend to increase the size of the familiar and to decrease the size of all else. They tend to place their own location in a central position and to increase the size of things nearby. The regional study, "The Yurok World View" in Chapter 13, demonstrates these conclusions in the collective mental map of a primitive society. Figure 8.4 gives four examples of the ways people are affected

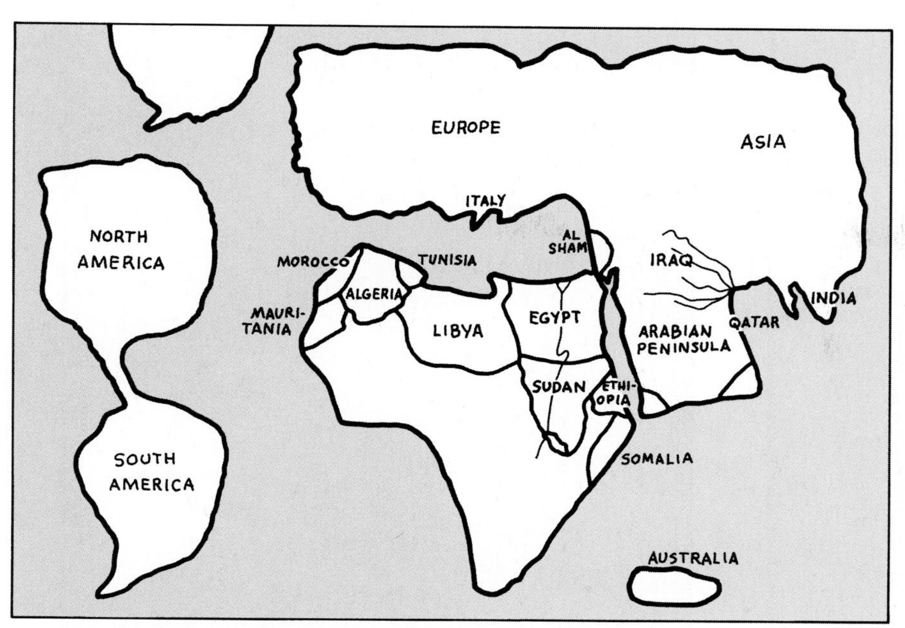

FIGURE 8.2 **A mental map of the world** as drawn by a Palestinian high school student from Gaza. The map reflects the secondary school education the author is receiving, which conforms to the Egyptian national school curriculum and thus is influenced by the importance of the Nile River and pan-Arabism. Al Sham is the old, but still used, name for the area including Syria, Lebanon, and Palestine. The map might be quite different if the Gaza school curriculum were designed by Palestinians or if an Israeli drew it.

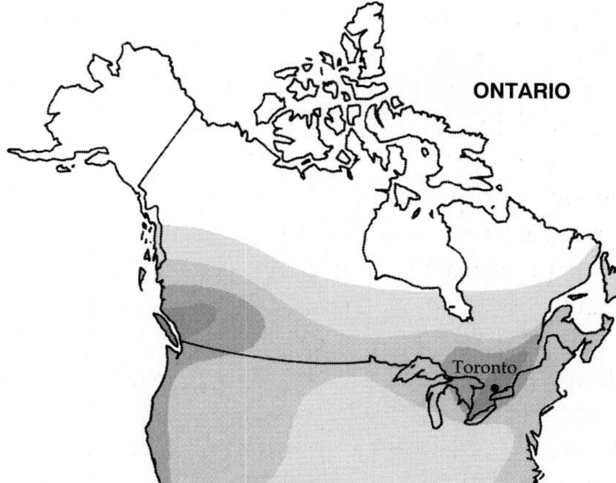

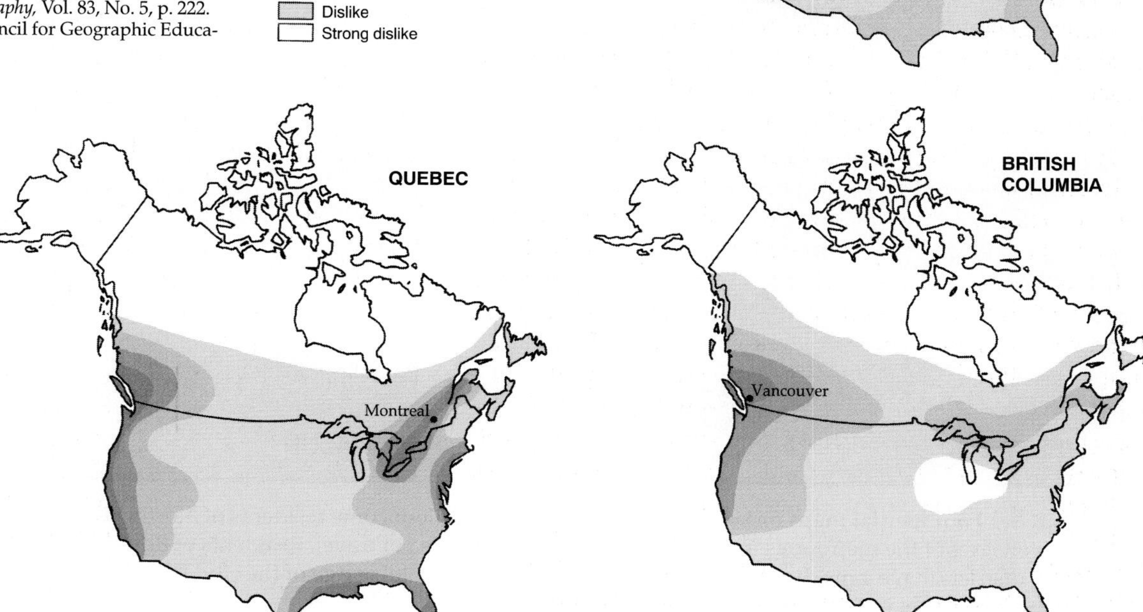

FIGURE 8.3 Each of these maps shows the residential preference of a sampled group of Canadians from the provinces of Ontario, Quebec, and British Columbia, respectively. Note that each group of respondents prefers its own area but that all like the Canadian and U.S. west coasts.

Redrawn with permission from Herbert A. Whitney, "Preferred Locations in North America: Canadians, Clues, and Conjectures," in *Journal of Geography*, Vol. 83, No. 5, p. 222. Copyright © 1984 National Council for Geographic Education, Indiana, PA.

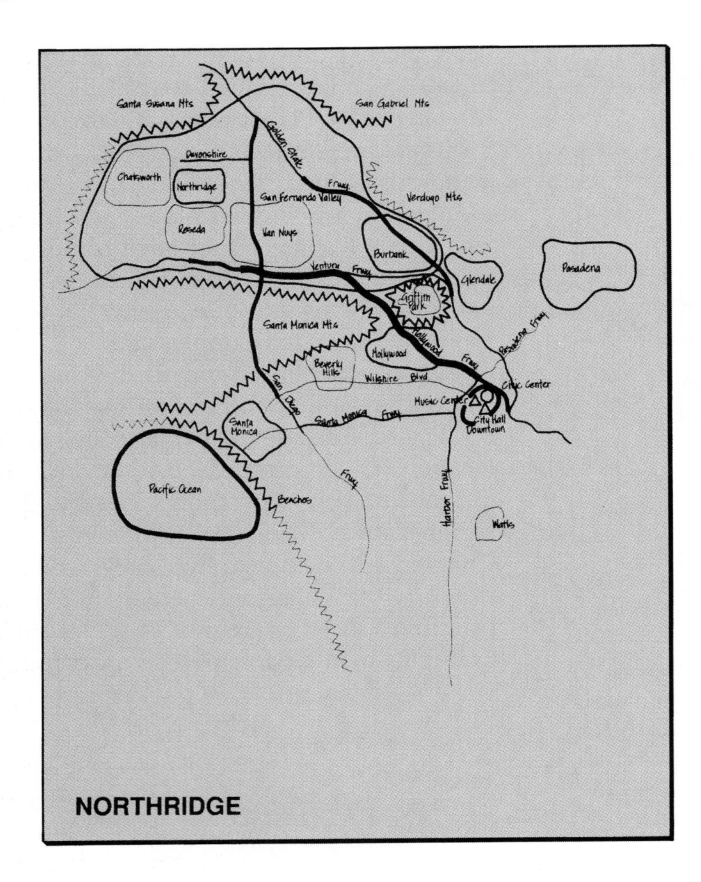

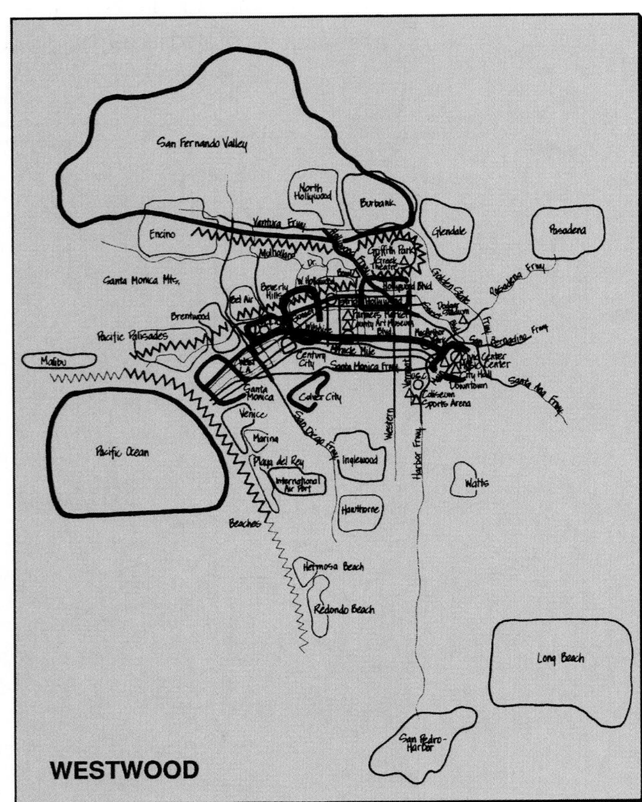

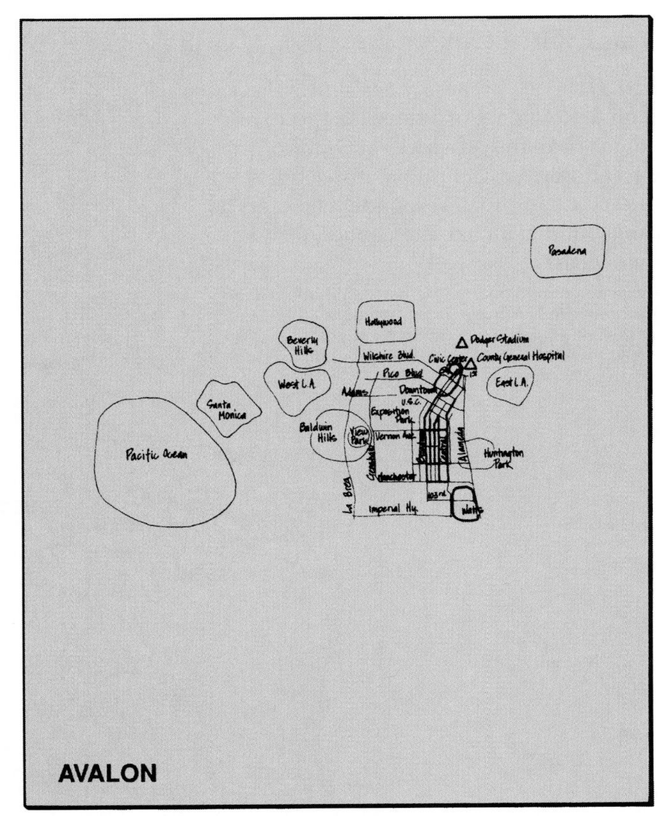

FIGURE 8.4 **Four mental maps of Los Angeles.** The upper-middle-income residents of Northridge and Westwood have expansive views of the metropolis, reflecting their mobility and area of travel. Residents of Boyle Heights and Avalon, both minority districts, have a much more restricted and incomplete mental image of the city. Their limited mental maps reflect and reinforce their spatial isolation within the metropolitan area.

From the Department of City Planning, City of Los Angeles, *The Visual Environment of Los Angeles*, 1971.

by the familiar. Also, as we grow older our perspectives change, as shown in Figure 8.5.

Perception of Natural Hazards

An intriguing area of research to which geographers have addressed themselves deals with how people perceive *natural hazards,* defined as processes or events in the physical environment that are not caused by humans but that have consequences harmful to them. Most climatic (e.g., hurricanes, tornadoes, and blizzards) and geological (earthquakes and volcanic eruptions) hazards cannot be prevented; their consequences may be disastrous. The hurricanes that struck the delta area of Bangladesh in 1970 and again in 1985 and 1991 resulted in at least 500,000 deaths; the 1976 earthquake in the Tangshan area of China devastated a major urban-industrial complex with deaths estimated at 242,000. These were major and exceptional natural hazards, but more common occurrences are experienced and apparently discounted by those in their areas of effect. For example, Johnstown, Pennsylvania, has suffered recurrent floods, and yet its residents rebuild; and violent storms strike the Gulf and East Coasts of the United States, and people remain or return (Figure 8.6).

Why do people choose to settle in high-hazard areas, despite the potential threat to their lives and property? Why do hundreds of thousands of people live along the San Andreas Fault in California, build houses in Pacific coastal areas known to experience severe erosion during storms, or farm in flood-prone areas adjacent to the Mississippi River? What is it that makes the risks worth taking?

There are many reasons that hazardous areas are perceived differently by people, and thus that some people choose to settle in them. Of major importance is the fact that specific hazards are relatively rare occurrences. Many people think that the likelihood of an earthquake, flood, or some other natural calamity is sufficiently remote so that it is not economically feasible to protect themselves against it. They are also influenced by the fact that the scientists who study such hazards may themselves differ on the probability of an event or on the damage that it may inflict. And, in fact, the prediction of hazards is not an exact science, being based on calculations of the probability of occurrences of uncommon events.

People are also influenced by their past experiences in high-hazard areas. If they have not suffered much damage in the past, they may be optimistic about the future. If, on the other hand, past damage has been great, they may think the probability of similar occurrences in the future is low (Table 8.1). People's memories can be short. In the years following an earthquake, for example, a sense of security grows, building codes or their interpretation are relaxed, and population in the area increases.

High-hazard areas are often sought out not because they pose risks, but because they possess desirable topography or scenic views; for instance, the Atlantic and Pacific

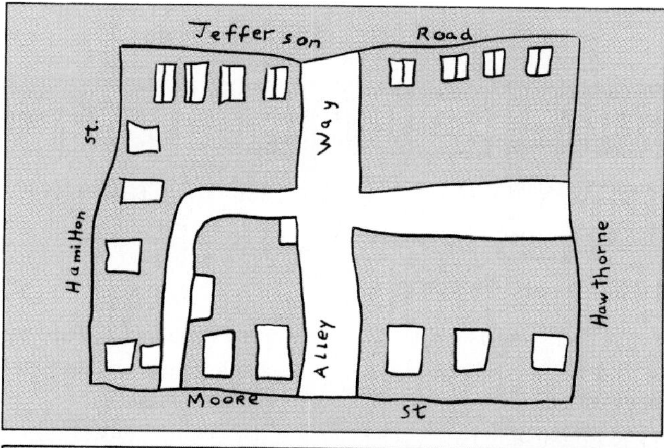

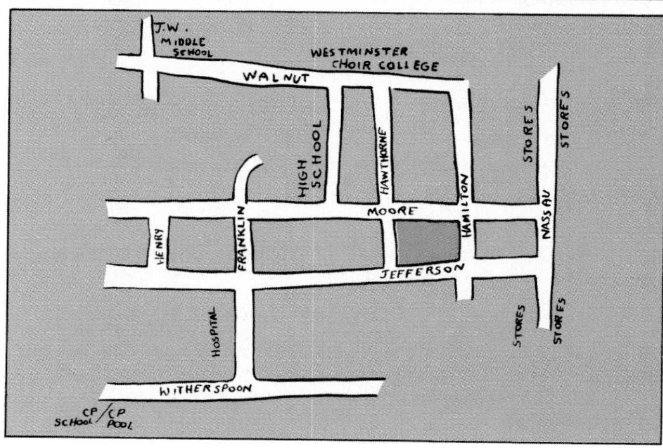

FIGURE 8.5 Three children, ages 6, 10, and 13, who lived in the same house, were asked to draw maps of their neighborhood. No further instructions were given. Notice how perspectives broaden and neighborhoods expand with age. For the 6-year-old, the neighborhood consisted of the houses on either side of her own. The square block on which she lived was the neighborhood for the 10-year-old. The wider activity space of the 13-year-old is also evident. The square block that the 10-year-old drew is shaded in the 13-year-old's sketch.

FIGURE 8.6 **The aftermath of Hurricane Fran, which struck the coast of North Carolina in September 1996.** With gusts of wind that reached nearly 200 kilometers per hour (120 mph), Fran cut a swath of destruction across North and South Carolina, Virginia and West Virginia, and Pennsylvania. At least 22 people died, and severe flooding damaged or destroyed thousands of houses. The total cost of the hurricane was estimated to be more than $1 billion.

coasts entice people. Once people have purchased property in a known hazard area, they may be unable to sell it for a reasonable price even if they so desire. They think they have no choice but to remain and protect their investment. The cultural hazard—loss of livelihood and investment—appears to be more serious than whatever natural hazards may exist.

TABLE 8.1

COMMON RESPONSES TO THE UNCERTAINTY OF NATURAL HAZARDS

ELIMINATE THE HAZARD	
Deny or Denigrate Its Existence	*Deny or Denigrate Its Recurrence*
We have no floods here, only high water."	"Lightning never strikes twice in the same place."
"It can't happen here."	"It's a freak of nature."
ELIMINATE THE UNCERTAINTY	
Make It Determinate and Knowable	*Transfer Uncertainty to a Higher Power*
"Seven years of great plenty. . . . After them seven years of famine."	"It's in the hands of God."
"Floods come every five years."	"The government is taking care of it."

Ian Burton and Robert Kates, "The Perception of Natural Hazards in Resource Management." Reprinted with permission from 3 *Natural Resources Journal* 435 (1964), published by the University of New Mexico School of Law, Albuquerque, N.M.

INDIVIDUAL ACTIVITY SPACE

We will see in Chapter 9 that groups and countries draw boundaries around themselves to divide space into territories that are, if necessary, defended. The concept of **territoriality**—the emotional attachment to, and the defense of, home ground—has been seen by some as a root explanation of many human actions and responses. It is true that some collective activity appears to be governed by territorial defense responses: the conflict between street groups in claiming and protecting their "turf" (and the fear for their lives when venturing beyond it) and the sometimes violent rejection by ethnic urban neighborhoods of an encroaching black, Hispanic, or other population group.

But for most, our personal sense of territoriality is a tempered one. Homes and property are regarded as defensible private domains but are opened to innocent visitors, known or unknown, or to those on private or official business. Nor do we confine our activities so exclusively within controlled home territories as street-gang members do within theirs. Rather, we have a more-or-less extended home range, an **activity space** within which we move freely on our rounds of regular activity, sharing that space with others who are also about their daily affairs.

Figure 8.7 suggests a probable activity space for a suburban family of five for one day. Note that the activity space

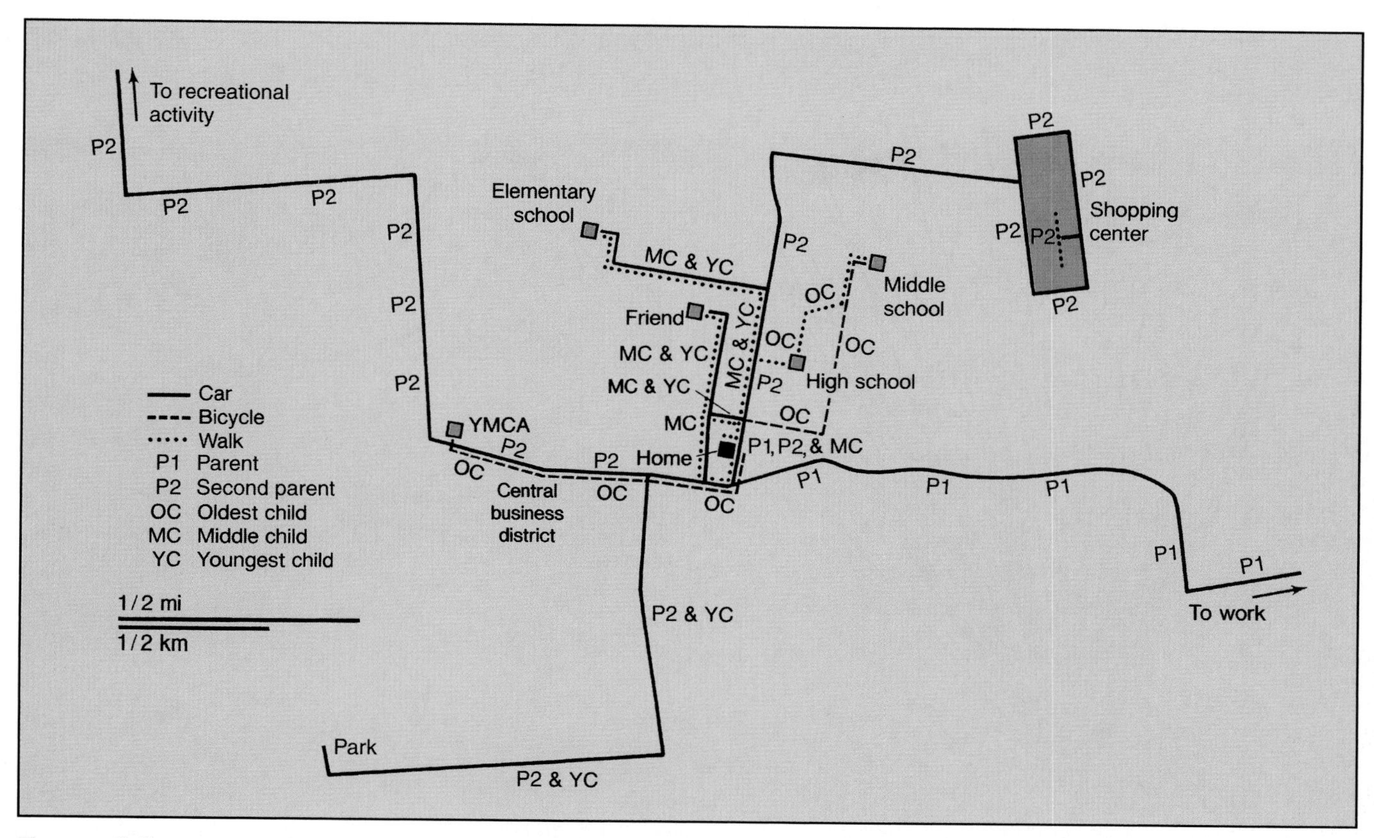

FIGURE 8.7 **Activity space for each member of a family of five for a typical weekday.** One parent (P1) commutes to work, while the other parent (P2) works at home. Routes of regular movement and areas recurrently visited help to foster a sense of territoriality and to affect one's perceptions of space.

for each individual for one day is rather limited, even though two members of the family use automobiles. If one week's activity were shown, more paths would have to be added to the map, and in a year's time, several long trips would probably have to be noted. Because long trips are taken irregularly, we will confine our idea of activity space to often-visited places.

The kind of activities individuals engage in can be classified according to type of trip: journeys to work, school, or shops, or for recreation, and so on. People in nearly all parts of the world make the same types of journeys, though the spatially variable requirements of culture and economy dictate their frequency, duration, and significance in the time budget of an individual. Figures 8.8 and 8.9 illustrate this point.

Figure 8.8 depicts variations in travel patterns in two different culture groups in rural midwestern Canada. It suggests that "modern" rural Canadians, who want to take advantage of the variety of goods offered in the regional capital, are willing to travel longer distances than are people of a traditional culture who have different tastes in clothing and consumer goods and whose demands are satisfied in local settlements. In addition, the traditionalists in this case do not own cars, thus limiting their spatial range.

Figure 8.9 suggests the importance of the journey to work in an urban population. It had long been argued that the journey to work, as measured from homeplace to workplace, constitutes the fundamental building block for developing idealized or planned city structures. In more recent years, however, it has become evident that for many, the journey to work is really a multipurpose trip which may include sidetrips to day-care centers, cleaners, schools, and shops of various kinds.

The types of trips individuals make, and thus the extent of their activity space, depend on at least three variables: their stage in life course; the means of mobility at their command; and the demands or opportunities implicit in their daily activities. The first, *stage in life course,* refers to membership in specific age groups. Stages include preschool-age, school-age, young adult, adult, and elderly. Preschoolers stay close to home unless they accompany their parents. School-age children usually travel short distances to lower schools and longer distances to upper-level schools. After-school activities tend to be limited to walking or bicycle trips to nearby locations. High-school students, or young adults, are usually more mobile and take part in more activities than do younger children. Adults responsible for household duties make shopping trips and trips related to child care, as well as journeys away from home for social, cultural, or recreational purposes. Wage-earning adults usually travel farther from home than other family members. Elderly people normally do not find it feasible or desirable to have extended activity spaces (Figure 8.10).

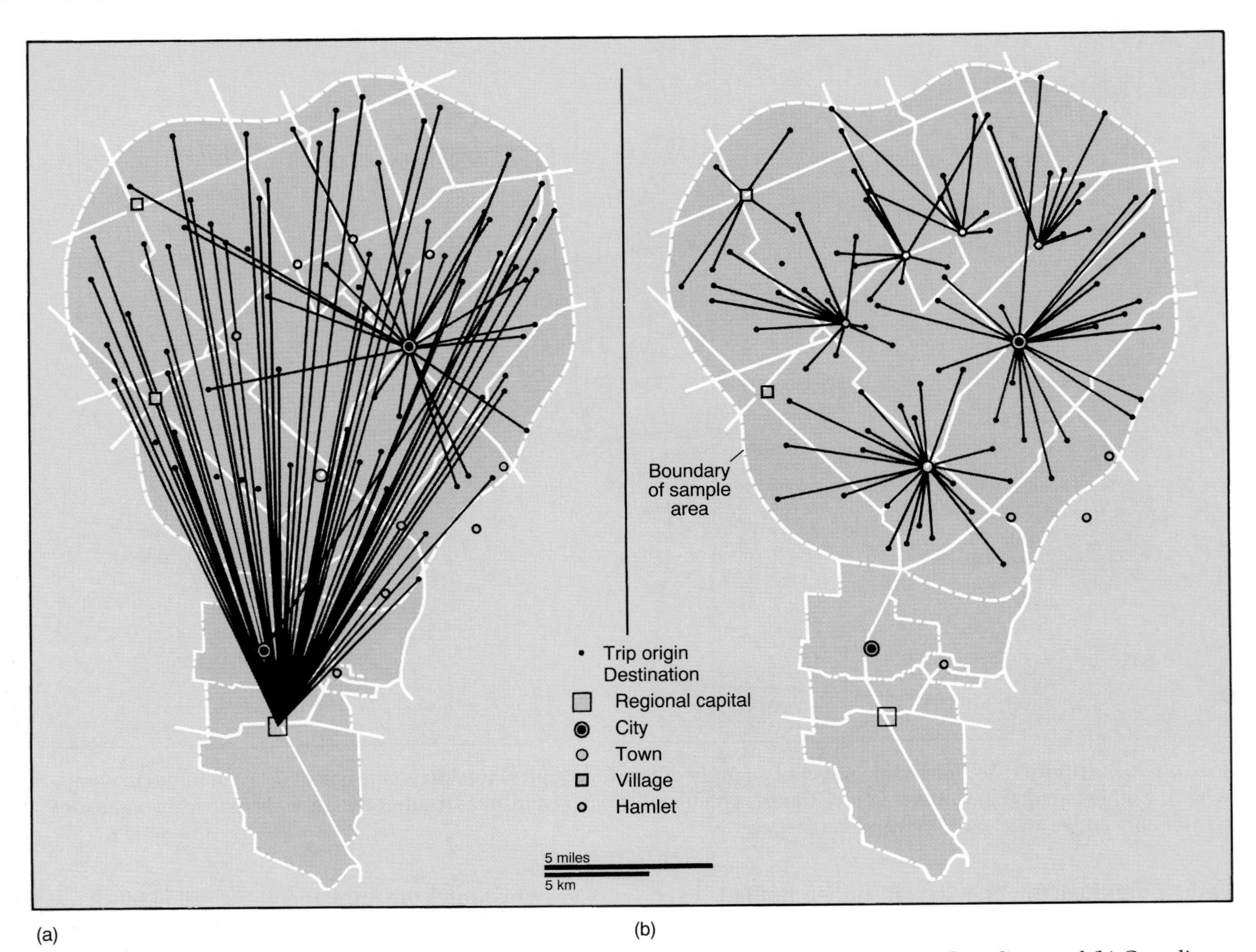

FIGURE 8.8 **Travel patterns for purchase of clothing and yard goods** of (a) rural cash-economy Canadians and (b) Canadians of the old-order Mennonite sect. These strikingly different travel behaviors demonstrate the great differences that may exist in the action spaces of different culture groups occupying the same territory.

Redrawn with permission from Robert A. Murdie, "Cultural Differences in Consumer Travel" in *Economic Geography,* Vol. 41, No. 3, p. 221. Copyright © 1965 Clark University, Worcester, MA.

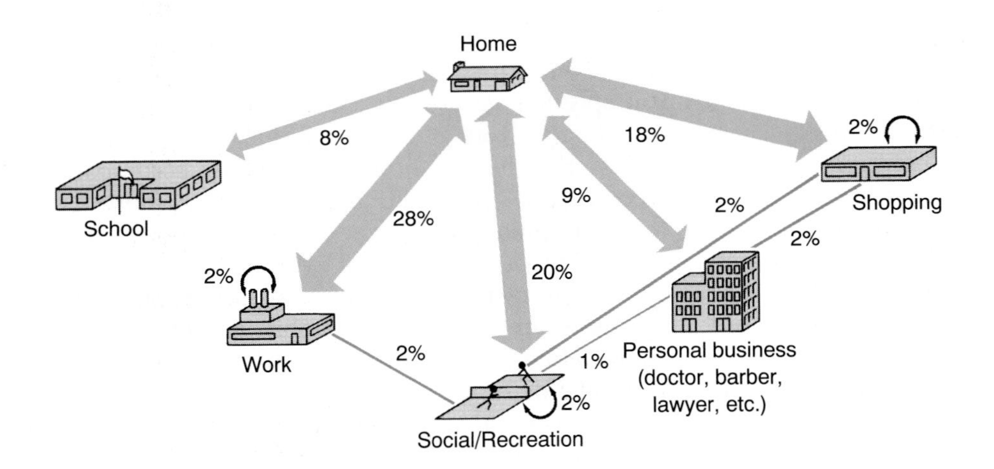

FIGURE 8.9 **Chicago travel patterns.** The numbers are the percentages of all urban trips taken in Chicago. The greatest single movement is the journey to and from work. More than 96% of all trips are represented on the diagram.

Source: Data from Chicago Area Transportation Study, *1970 Travel Characteristics.*

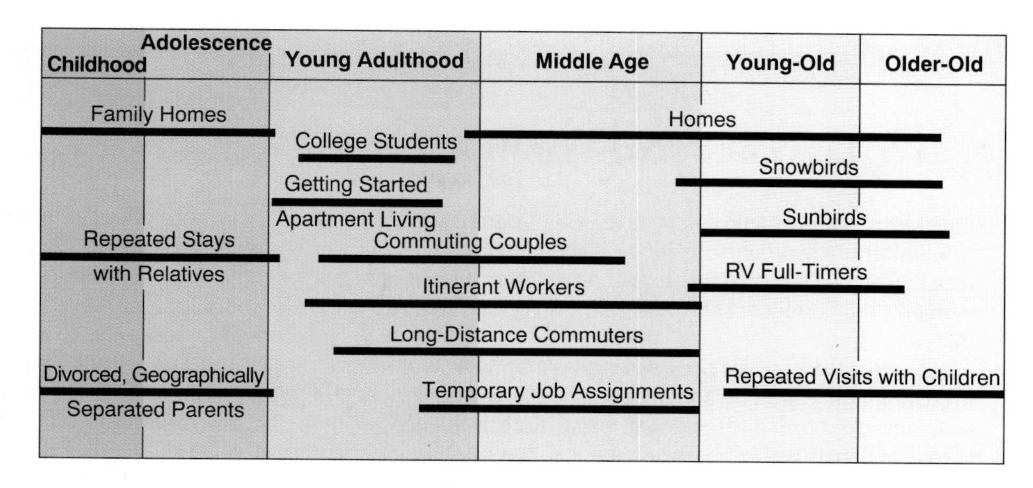

FIGURE 8.10 Examples of multiple residence by life course phase. Each horizontal line represents a period of time in a possible new residence.

From K. McHugh, T. Hogan, S. Happel, "Multiple Residence and Cyclical Migration," *The Professional Geographer,* vol. 47, no. 3 (August 1995), Fig. 1, p. 253. Association of American Geographers, 1995.

The second variable that affects the extent of activity space is *mobility,* or the ability to travel. An informal consideration of the cost and effort required to overcome the friction of distance is implicit. Where incomes are high, automobiles are available, and the cost of fuel is a minor item in the family budget, mobility may be great and individual action space can be large. In societies where cars are not a standard means of personal conveyance, the daily activity space may be limited to the shorter range afforded by bicycles or walking. Obviously, both intensity of purpose and the condition of the roadway affect the execution of movement decisions.

The mobility of individuals in countries or in sections of countries with high incomes is relatively great; people's activity space horizons are broad. These horizons, however, are not limitless. There are a fixed number of hours in a day, most of them consumed in performing work, preparing and eating food, and sleeping. In addition, there are a fixed number of road, rail, and air routes, so even the most mobile individuals are constrained in the amount of activity space they can use. No one can easily claim the world as his or her activity space. An example of this limitation is that of women living in suburban communities who must balance family obligations, such as preparing meals and caring for children, with their workforce activities. In this case, women's mobility is restricted, and, as a result, their occupational opportunities are limited (see "Space, Time, and Women").

The third factor limiting activity space is the individual assessment of the availability of possible activities or *opportunities.* In the subsistence economies discussed in Chapter 10, the needs of daily life are satisfied at home; the impetus for journeys away from the residence is minimal. If there are no stores, schools, factories, or roads, one's expectations and opportunities are limited, and activity space is therefore reduced. In impoverished countries or neighborhoods, low incomes limit the inducements, opportunities, destinations, and necessity of travel.

DISTANCE AND SPATIAL INTERACTION

Because people make many more short-distance trips than long ones, there is greater human interaction over short distances than long distances. If we drew a boundary line around our activity space, it would be evident that trips to the boundary are taken much less often than short-distance trips near the home. Think of activity space as more intensively used (greater spatial interaction) near one's home place or base, and as declining in use as distance from the base increases. This is the principle of **distance decay** (introduced briefly in Chapter 1), the exponential decline of an activity, function, or amount of interaction with increasing distance from the point of origin. The tendency is for the frequency of trips to decrease very rapidly beyond an individual's **critical distance.** Figure 8.11 illustrates this principle with regard to journeys from the homesite.

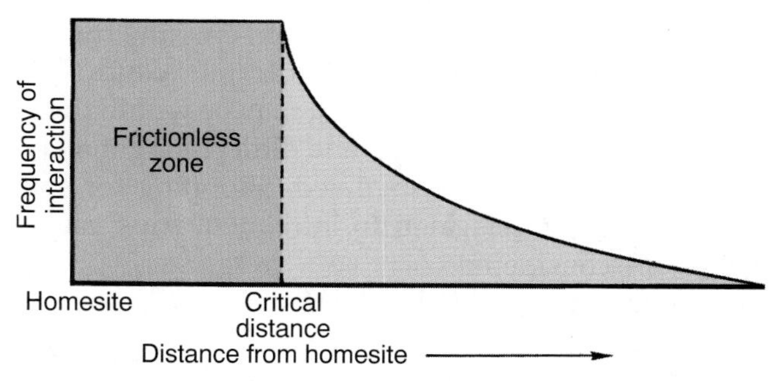

FIGURE 8.11 This general diagram indicates how distance is observed by most people. For each activity, there is a distance beyond which the intensity of contact declines. This is called the *critical distance,* if distance alone is being considered, or the *critical isochrone,* if time is the measuring rod. For the distance up to the critical distance, a frictionless zone is identified in which time or distance considerations do not effectively figure into the trip decision.

SPACE, TIME, AND WOMEN

From a time-geographic perspective, it is apparent that many of the limitations women face in their choices of employment or other activities outside the home reflect the restrictions that women's time budgets and travel paths place on their individual daily activities.

Consider the case* of the working woman with one or more preschool-age children. The location and operating hours of available child-care facilities may have more influence on her choice of job than do her labor skills or the relative merits of alternative employment opportunities. For example, the woman may not be able to leave her home base before a given hour because the only available full-day child-care service is not open earlier. She must return at the specified child pick-up time and arrive home to prepare food at a reasonable (for the child) dinner time. Her travel mode and speed determine the outer limits of her possible daily activity space.

Suppose that two solid job offers both have the same working hours and fall within her possible activity space. She cannot accept the preferred, better-paying job because drop-off time at the child-care center would make her late for work, and work hours would make her miss the center's closing time. On the other hand, although the other job is acceptable from a child-care standpoint, it leaves no time (or store options) for shopping or errands except during the lunch hour. Job choice and shopping opportunities are thus determined not by the woman's labor skills or awareness of store price comparisons, but by her time-geographic constraints. Other women in other job-skill, parenthood, locational, or mobility circumstances experience different but comparable activity space restrictions.

Mobility is a key to activity mix, time-budget, and activity space configurations. Again, research indicates that women are frequently disadvantaged. Because of their multiple work, child-care, and home maintenance tasks, women on average make more—though shorter—trips than men, leaving less time for alternate activities.

Although an automobile reduces time constraints, women have less access to cars than do males, in part because, in many cities, they are less likely to have a driver's license and also because women typically cede use of a single family car to husbands. The lower income level of many single women, with or without children, limits their ability to own cars, which leads them to use public transit disproportionately to their numbers—to the detriment of both their money and time-space budgets. Women are, it has been observed, "transportation deprived and transit dependent."

*Suggested by Risa Palm and Allan Pred, *A Time-Geographic Perspective on Problems of Inequality for Women.* Institute of Urban and Regional Development, Working Paper No. 236. University of California, Berkeley, 1974.

The critical distance is the distance beyond which cost, effort, and perception play an overriding role in our willingness to travel. A small child, for example, will make many trips up and down the block, but he or she will be inhibited by parental admonitions from crossing the street. Different but equally effective constraints control adult behavior. Daily or weekly shopping may be within the critical distance of an individual, and little thought may be given to the cost or effort involved. Shopping for special goods, however, is relegated to infrequent trips, and cost and effort are considered.

Effort may be measured in terms of *time-distance,* that is, the time required to complete the trip. For the journey-to-work, time rather than cost often plays the major role in determining the critical distance. When significant differences between our cognition of distance and real distance are evident, we use the term *psychological distance* to describe our perception of distance. A number of studies show that people tend to psychologically consider known places as nearer than they really are, and little-known places as farther than true distance.

SPATIAL INTERACTION AND THE ACCUMULATION OF INFORMATION

The critical distance is different for each person. The variables of life-course stage, mobility, and opportunity, together with an individual's interests and demands, help to define how much and how far a person will travel. On the basis of these variables, we can make inferences about the likelihood that a person will gain more or less information about his or her activity space and the space beyond.

We gain information about the world from many sources. Although information obtained from radio, television, and newspapers is important to us, face-to-face contact is assumed to be the most effective means of communication. If we combine the ideas of activity space and distance decay, we see that as the distance away from the home place increases, the number of possible face-to-face contacts usually decreases. We expect more human interactions at short distances than at long distances.

towns, regional centers, metropolises). By **hierarchical diffusion** we mean the spread of innovation up or down a hierarchy of places.

As an example, let us suppose that a new way of processing passengers is adopted at an airport in a major city. Information on the innovation is spread, but only officials at comparably-sized airports are in a position to accept the idea at first. It may be that the quality of information diffused to larger airports is better, or that larger airports are more financially able to adopt the idea than smaller airports. Eventually, the innovation is adopted at airports of somewhat smaller cities, and so on down the hierarchy as it becomes better known or more financially feasible. A hypothetical scheme showing how a four-level hierarchy may be connected in the flow of information is presented in Figure 8.16. Note that the lowest-level centers are connected to higher-level centers but not to each other. Observe, too, that connections may bypass intermediate levels and link only with the highest-level center.

Many times, hierarchical diffusion takes place simultaneously with contagious diffusion. One might expect variations when the density of high-level centers is great and when distances between centers are short. A quick and inexpensive way to spread an idea is to communicate information about it at high-order hierarchical levels. Then the three types of diffusion processes may be used most effectively; even while an idea is diffusing through a high level in the hierarchy, it is also spreading outward from high-level centers. Consequently, low-level centers that are a short distance from high-level centers may be apprised of the innovation before more distant medium-level centers. People living in suburbs and small towns near a large city are privy to much that is new in the large city, as are individuals in other large cities half a continent away. Figures 8.17 and 8.18 show these patterns for a case taken from Japan.

These forms of diffusion operate in the spread of culture. The consequences are the spatial interaction and innovation discussed earlier in this chapter. We should also recall from Chapter 7 that migration, invasions, selective cultural adoptions, and cultural transference aid the diffusion of innovation. These broader movements and exchanges represent interactions of people beyond their usual activity spaces (see "Documenting Diffusion").

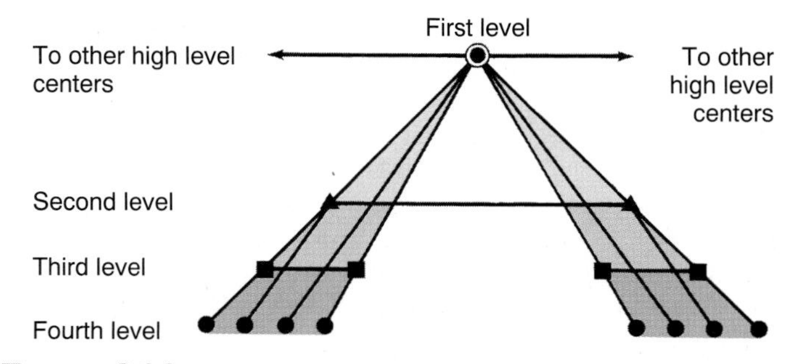

FIGURE 8.16 **A four-level communication hierarchy.**

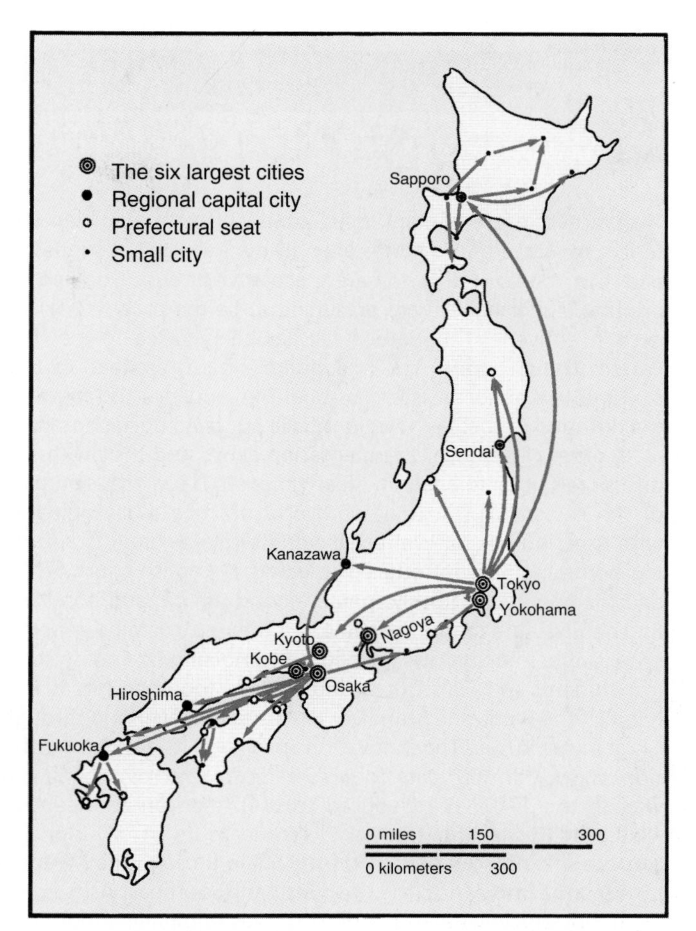

FIGURE 8.17 Rotary Clubs, members of the international service association, were established in the large cities of Japan during the 1920s. New clubs were established under the sponsorship of the original clubs. This map shows the pattern of diffusion. Note the regional and spatial effects of the diffusion process.

Redrawn with permission from Yoshio Sugiura, "Diffusion of Rotary Clubs in Japan, 1920–1940: A Case of Non-Profit Motivated Innovation Diffusion Under a Decentralized Decision-Making Structure," in *Economic Geography,* Vol. 62, No. 2, p. 128. Copyright © 1986 Clark University, Worcester, MA.

MIGRATION

An important aspect of human history has been the migration of peoples, the evolution of their separate cultures, and the relocation diffusion of those cultures. Portions of that history have been mentioned in Chapter 6. The settlement of North America, Australia, and New Zealand involved great long-distance movements of peoples. The flight of refugees from past and recent wars, the settlement of Jews in Israel, the movement south in Africa from the drought-stricken Sahel, the current migration of workers to the United States from overpopulated Mexico, and innumerable other examples of mass movement come quickly to mind. In all cases, societies transplanted their cultures to the new areas, their cultures therefore diffused and intermixed, and history was altered. The vast numbers who suffered the hardships of movement to the American West, like Catherine Haun, were the predecessors of the regionally relocating job seekers and retirees of today.

DOCUMENTING DIFFUSION

The places of origin of many ideas, items, and technologies important in contemporary cultures are only dimly known or supposed, and their routes of diffusion are speculative at best. Gunpowder, printing, and spaghetti are presumed to be the products of Chinese inventiveness; the lateen sail has been traced to the Near Eastern culture world. The moldboard plow is ascribed to 6th-century Slavs of northeastern Europe. The sequence and routes of the diffusion of these innovations have not been documented.

In other cases, such documentation exists, and the process of diffusion is open to analysis. Clearly marked is the diffusion path of the custom of smoking tobacco, a practice that originated with Amerindians. Sir Walter Raleigh's Virginia colonists, returning home in 1586, introduced smoking in English court circles, and the habit very quickly spread among the general populace. England became the source region of the new custom for northern Europe; smoking was introduced to Holland by English medical students in 1590. Dutch and English together spread the habit to the Baltic and Scandinavian areas and overland through Germany to Russia. The innovation continued its eastward diffusion, and within 100 years, tobacco had spread across Siberia and was, in the 1740s, reintroduced to the American continent at Alaska by Russian fur traders. A second route of diffusion for tobacco smoking can be traced from Spain through the Mediterranean area into Africa, the Near East, and Southeast Asia.

In more recent times, hybrid corn was originally adopted by imaginative farmers of northern Illinois and eastern Iowa in the mid-1930s. By the late 1930s and early 1940s, the new seeds were being planted as far east as Ohio and north to Minnesota, Wisconsin, and northern Michigan. By the late 1940s, all commercial corn-growing districts of the United States and southern Canada were cultivating hybrid varieties.

Another study traced the spread of Wal-Mart stores, a new type of discount retail store, across the United States from its origins in Arkansas in 1962.

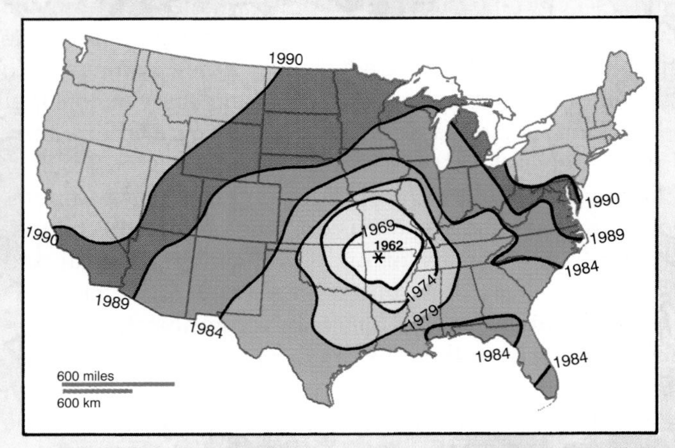

Redrawn from data in Thomas O. Graff and Dub Ashton, "Spatial Diffusion of Wal-Mart," *The Professional Geographer*, vol. 46, no. 1, pp. 19–29. Association of American Geographers, 1994.

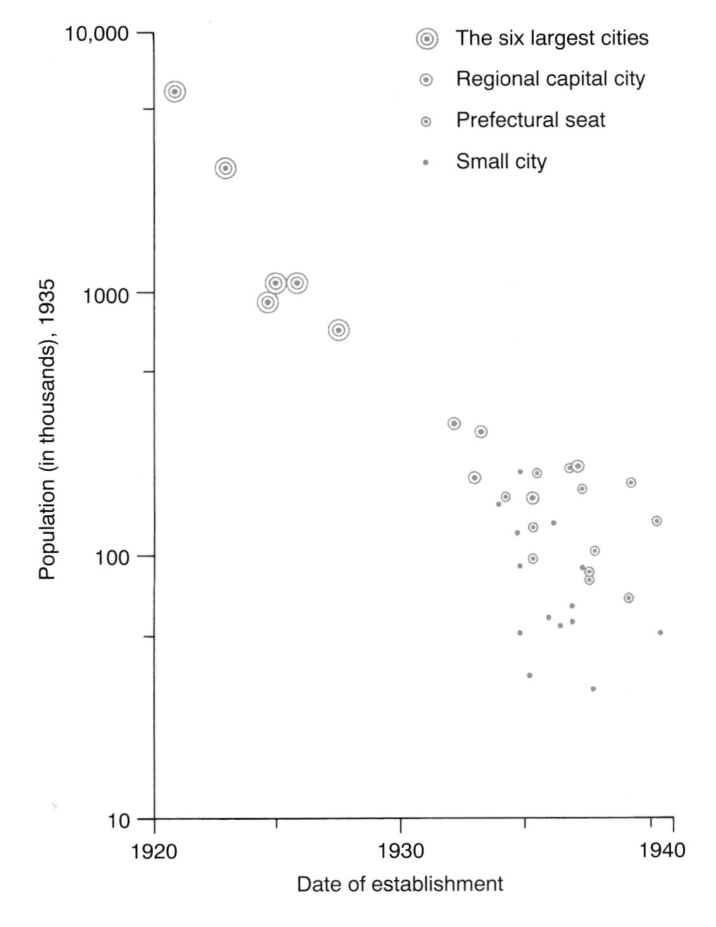

FIGURE 8.18 This diagram shows the hierarchical diffusion of Rotary Clubs in Japan. The largest cities were the first centers of Rotary Club activity, followed by cities at lower and lower levels of urban population and city function.

Redrawn with permission from Yoshio Sugiura, "Diffusion of Rotary Clubs in Japan, 1920–1940: A Case of Non-Profit Motivated Innovation Diffusion Under a Decentralized Decision-Making Structure," in *Economic Geography,* Vol. 62, No. 2, p. 128. Copyright © 1986 Clark University, Worcester, MA.

The *planned two-way trip* is a long-distance movement that is not considered to be a migration. It includes the exciting, perhaps historically decisive, daring journey of exploration that we read about in history, or the more common contemporary vacation or business or social trip in which modern societies indulge so freely. The latter sort of trip, of course, enhances the mental maps and enlarges the awareness space of the participants and may be the prelude to migration. It may also contribute to the diffusion of information through cultural contact, but its individual impact is small because it is so transitory. Having much more impact is the larger scale back and forth movement of workers to industrial countries from developing countries.

Much more important than the planned two-way trip is **migration:** a relocation of both residence and activity space. Naturally, the length of the move and its degree of disruption of normal household activities raise distinctions important in the study of migration. A change of residence from the central city to the suburbs certainly changes the activity space of schoolchildren and of adults in many of their nonworking activities, but the workers may still retain the city—indeed, the same place of employment there—as an action space. On the other hand, immigration from Europe to the United States and the massive farm-to-city movements of rural Americans late in the last and early in the present century meant a total change of all aspects of behavioral patterns.

The Decision to Migrate

The decision to move is a cultural and temporal variable. Nomads fleeing famine and spreading deserts in the Sahel obviously are motivated by different considerations than those of the executive receiving a job transfer to Chicago, the resident of Appalachia seeking factory employment in the city, or the retired couple searching for sun and sand. Mexican immigrants heading north to the United States have given the decision to migrate a great deal of thought. In general, people who decide to migrate are seeking better economic, political, or cultural conditions or certain amenities. Of course, as in the case of the Hauns, the reasons for migration are frequently a combination of several of these categories.

Economic causes have impelled more migrations than any other single category. If migrants face unsatisfactory conditions at home (e.g., unemployment or famine) and believe that the economic opportunities are better elsewhere, they will be attracted to the thought of moving. To a nomadic herdsman, better opportunities might be abundant grass and water; to a Western emigrant, rich soil; to a modern worker, the promise of a high-paying job in a Seattle suburb. Attractions such as these are called **pull factors,** and those that contribute to dissatisfaction at home are called **push factors.** Very often, migration is a result of both push and pull factors (Figure 8.19).

The desire to escape war and persecution at home and to pursue the promise of freedom in a new location is a political incentive for migration (Figure 8.20). Americans are familiar with the history of settlers who emigrated to North America seeking religious and political freedom. In more recent times, the United States has received hundreds of thousands of refugees from countries such as Hungary, following the uprising of 1956; Cuba, after its takeover by Fidel Castro; and Vietnam, after the fall of South Vietnam. The massive movements of Hindus and Muslims across the Indian subcontinent in 1947, when Pakistan and India were established as governing entities, and the exodus of Jews fleeing persecution in Nazi Germany in the 1930s are other examples of politically inspired moves. In more recent times, ethnic Muslims living in Bosnia (part of the former Yugoslavia) have been pushed out of their ancestral homes by Serbs; about one million Hutus fled into neighboring African countries after ethnic Tutsis took over Rwanda's government; and many Haitians, under severe economic privation during a political crisis, have left for the United States. Figure 8.21 identifies recent major forced migrations in Africa.

Migration normally but not always involves a hierarchy of decisions. Once people have decided to move and have selected a general destination (e.g., America or the West), they must still choose a particular site in which to settle. At this scale, cultural variables can be important pull factors. Migrants tend to be attracted to areas where the language, religion, and racial or ethnic background of the inhabitants is similar to their own. This similarity can help migrants feel at home when they arrive at their destination, and it may make it easier to find a job and to become assimilated into the new culture. The Chinatowns and little Italies of large cities attest to the drawing power of cultural factors, as discussed in Chapter 7.

Another set of inducements is grouped under the heading *amenities,* the particularly attractive or agreeable features that are characteristic of a place. Amenities may be natural (mountains, oceans, climate, and the like) or cultural (e.g., the arts and music opportunities available in

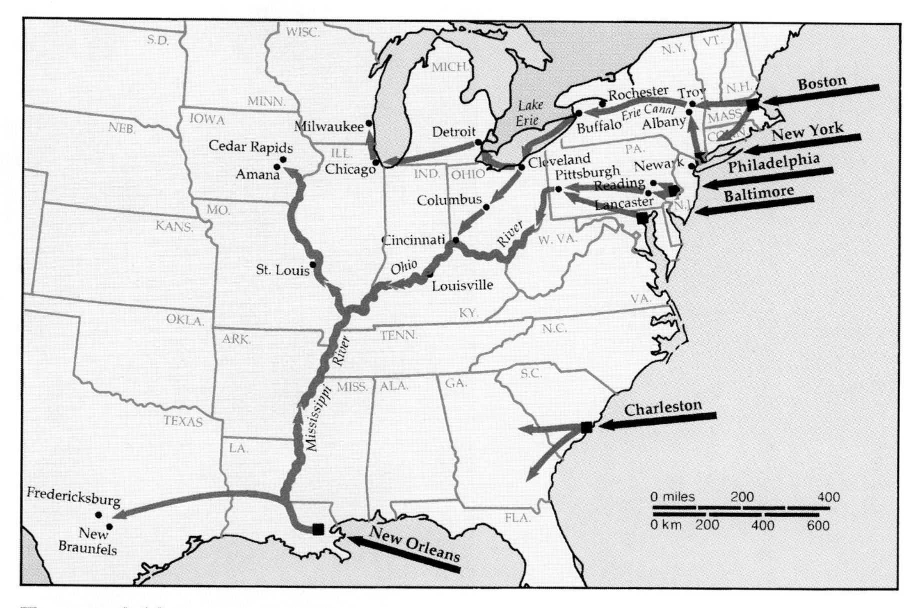

FIGURE 8.19 **The major paths of the early migration of Germans to America** constitute a relocation diffusion process. Most emigrants left Germany because of religious and political persecution. They chose the United States not only because immigrants were made welcome, but also because labor was in demand and farmland was available. The first immigrants landed, and many settled, in Boston, New York, Philadelphia, Baltimore, Charleston, and New Orleans. As is characteristic of a relocation diffusion process, the migrants carried with them such aspects of their culture as religion, language, and food preferences.

FIGURE 8.20 Rwandan refugees near the border of Rwanda and Tanzania. Approximately 1 million Rwandans fled into Zaire, Tanzania, Uganda, and Burundi in 1994 to escape the civil war in Rwanda. The Office of the United Nations High Commissioner for Refugees in 1994 estimated that, worldwide, there were about 23 million refugees, people who had left their country because of war, famine, or the well-founded fear of being persecuted for their race, religion, nationality, political views, or membership in certain social groups. An additional 26 million people are "internally displaced." In their search for security or sustenance, they have left their home areas but not crossed an international boundary.
Newsphotos/Bettmann.

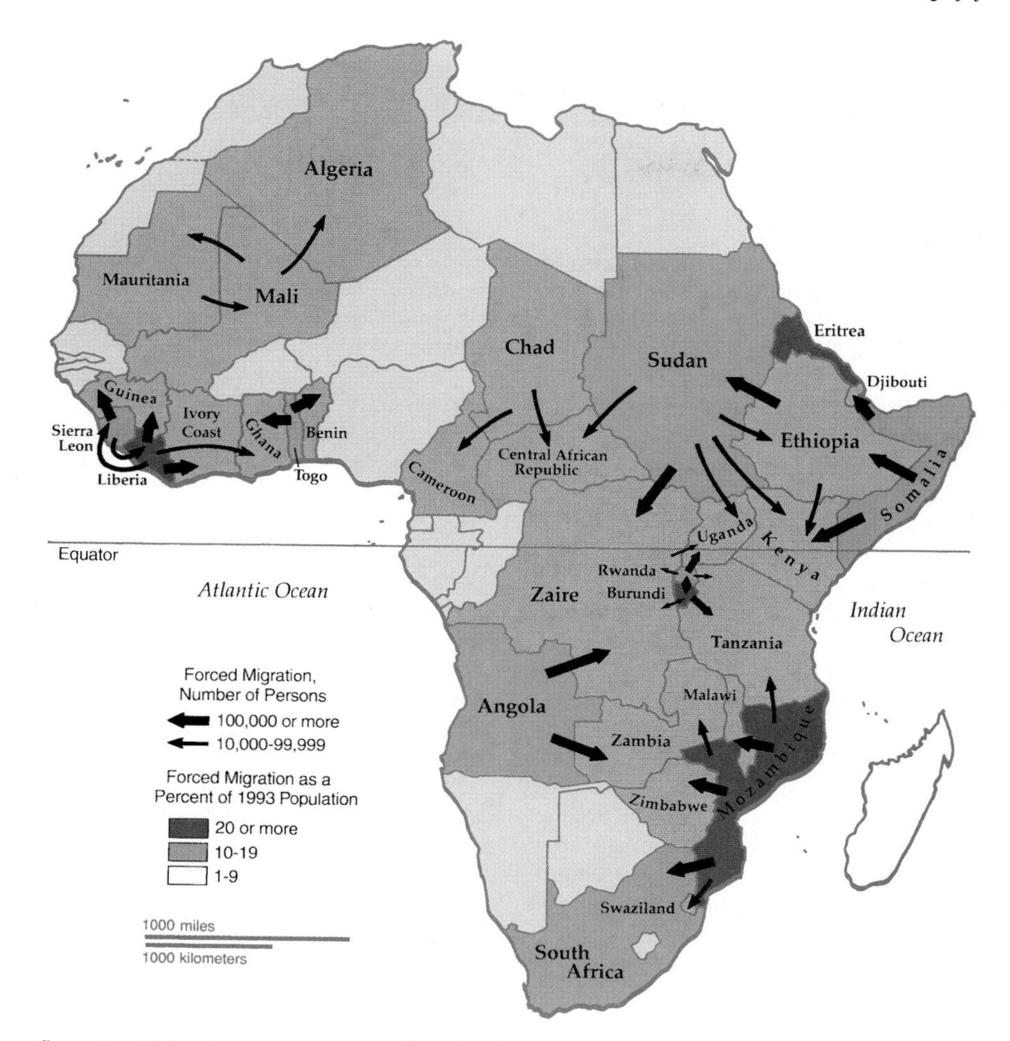

FIGURE 8.21 **Refugee flows in Africa.** In recent years, political upheavals have forced millions of Africans from their homes and across international borders.

Redrawn from W. Wood, "Forced Migration," *The Annals of the Association of American Geographers,* vol. 84, no. 4, Fig. 3, p. 619. Association of American Geographers, 1994.

large cities). They are particularly important to relatively affluent people seeking "the good life." Amenities help account for the attractiveness of the so-called Sunbelt states in this country for retirees; a similar movement to the southern coast has also been observed in countries such as the United Kingdom and France.

The significance of the various factors also varies according to the age, sex, education, and economic status of the migrants (see "Gender and Migration"). For the modern American, reasons to migrate have been summarized into a limited number of categories that are not mutually exclusive. They include:

1. changes in life course, such as getting married, having children, getting a divorce, or needing less dwelling space when the children leave home;
2. changes in career cycle, such as leaving college, getting a first job or a promotion, receiving a career transfer, or retiring;
3. forced migrations associated with urban development, construction projects, and the like;
4. neighborhood changes from which there is flight, perhaps pressures from new and unwelcome ethnic groups, building deterioration, street gangs, or similar rejected alterations in activity space;
5. changes of residence associated with individual personality (chronic mobility).

Some people simply tend to move often for no easily discernible reasons, whereas others, *stayers,* settle into a community permanently. Of course, for a country such as China, with its limitations on emigration and severe housing shortages, a totally different set of summary migration factors would be present.

The factors that contribute to mobility tend to change over time. However, in most societies one group has always been the most mobile: young adults. They are the members of society who are launching careers and making initial decisions about occupation and location. They have the fewest responsibilities of all adults; thus, they are not as strongly tied to family and institutions as older people. Most of the major voluntary migrations have been composed primarily

Gender and Migration

Gender is involved in migration at every level. In a household or family, women and men are likely to play different roles regarding decisions or responsibilities for activities such as child care. These differences, and the inequalities that underlie them, help determine who decides whether the household moves, which household members migrate, and the destination for the move. Outside the household, societal norms about women's mobility and independence often restrict their ability to migrate.

The economies of both sending and receiving areas play a role as well. If jobs are available for women in the receiving area, women have an incentive to migrate, and families are more likely to encourage the migration of women as necessary and beneficial. Thousands of women from East and Southeast Asia have migrated to the oil-rich countries of the Middle East, for example, to take service jobs.

The impact of migration is also likely to be different for women and men. Moving to a new economic or social setting can affect the regular relationships and processes that occur within a household or family. In some cases, women might remain subordinate to the men in their families. A study of Greek-Cypriot immigrant women in London and of Turkish immigrant women in the Netherlands found that although these women were working for wages in their new societies, these new economic roles did not affect their subordinate standing in the family in any fundamental way.

In other situations, however, migration can give women more power in the family. In Zaire, women in rural areas move to towns to take advantage of job opportunities there, and gain independence from men in the process.

One of the keys to understanding the role of gender in migration is to disentangle household decision-making processes. Many researchers see migration as a family decision or strategy, but some family members will have more sway than others, and some members will benefit more than others from those decisions.

For many years, men predominated in the migration streams flowing from Mexico to the United States. Women played an important role in this migration stream, even when they remained in Mexico. Mexican women influenced the migration decisions of other family members; they married migrants to gain the benefits from and opportunity for migration; and they resisted or accepted the new roles in their families that migration created.

In the 1980s, Mexican women began to migrate to the United States in increasing numbers. Economic crises in Mexico and an increase in the number of jobs available for women in the United States, especially in factories, domestic service, and service industries, have changed the backdrop of individual migration decisions. Now, women often initiate family moves or resettlement efforts.

Mexican women have begun to build their own migration networks, which are key to successful migration and resettlement in the United States. Networks provide migrants with information about jobs and places to live and have enabled many Mexican women to make independent decisions about migrating.

In immigrant communities in the United States, women are often the vital links to social institution services and to other immigrants. Thus, women have been instrumental in the way that Mexican immigrants have settled and become integrated into new communities.

of young people who suffered from a lack of opportunities in the home area and who were easily able to take advantage of opportunities elsewhere.

The concept of **place utility** helps us understand the decision-making process that potential migrants undergo. It is the value that an individual puts on each known, potential migration site. The decision to migrate is a reflection of the appraisal by a potential migrant of the current homesite, as opposed to other sites of which something is known. The individual may adjust to conditions at the homesite, and thus decide not to migrate.

In the evaluation of the place utility of each potential site, the decision maker considers not only present relative place utility but also expected place utility (see "Rational Decision Making: An Example"). The evaluations are matched with the individual's aspiration level, that is, the level of accomplishment or ambition that the individual sees for himself or herself. Aspirations tend to be adjusted to what an individual considers attainable. If one is satisfied with present circumstances, then search behavior is not initiated. If, on the other hand, dissatisfaction is felt, a utility is assigned to each possible new site. The utility is based on past or expected future rewards at the various sites. Because the new places are unfamiliar to the individual, the information received about them acts as a substitute for the personal experience of the homesite. The decision maker can do no more than sample information about the new sites, and, of course, there may be sampling errors.

One goal of the potential migrant is to minimize uncertainty. Most decision makers either elect not to migrate or postpone the decision unless uncertainty can be lowered sufficiently. Most migrants reduce uncertainty by imitating the successful procedures followed by others. We see that the decision to migrate is not a perfunctory, spur-of-the-moment reaction to information. It is usually a long, drawn-out process based on a great deal of sifting and evaluating data.

RATIONAL DECISION MAKING: AN EXAMPLE

An example of a structure developed for making a rational decision whether or not to migrate can be found in the case of the selection of a job by a hypothetical unmarried male who is a recent college graduate. This person partitioned his consideration into four major aspects: the nature of the work, travel requirements, geographic location, and monetary compensation. Each of these aspects was subdivided, and some were subdivided again, as illustrated in the table.

Each category was given a value on a scale from 0 to 1.0, according to the importance that the potential migrant placed on it. The restriction that the total of the values must add up to 1.0 enabled the graduate to multiply across any path to find the true relative worth of the 15 categories. In the table, we see that the graduate placed the greatest emphasis on starting salary and the kind of work to be done at the outset. Retirement and insurance benefits were least important.

Once the graduate had a clearer idea of his values, he could take each of his job opportunities and evaluate them according to the best information available about each subcategory. For each possible job, he assigned a value to every category according to the system 90–100 is excellent, 80–89 is good, 70–79 is fair, and so on. By multiplying each of these by the relative worth of each category and summing for each column, the graduate was able to make an objective decision. For example, for job opportunity No. 1, 95 × 0.040 = 3.80; 80 × 0.059 = 4.72; and so on. For job opportunity No. 1, the estimated score was 75.2; for No. 2, the value was 74.8; and for No. 3, it was 82.3.

This decision was based on the subject's own value system. Each person's categories and weighting systems would be different. In the case illustrated, the graduate finally ranked the local opportunity ahead of the national and regional opportunities. The ranking led to the decision not to migrate at the present time.

AN EXAMPLE OF JOB EVALUATION AND SELECTION

Aspect	Subdivision	Category	Relative Worth Placed on Each Category	Ratings of Job Opportunities: 1 Large National Company	2 Regional Office	3 Local Job
Nature of Work	Current and future work features	Management training program	0.040	× 95 = 3.80	80 = 3.20	50 = 2.00
		Variety of work	0.059	× 80 = 4.72	75 = 4.43	70 = 4.13
		Technical challenge	0.049	× 80 = 3.92	80 = 3.92	95 = 4.66
	Immediate work features		0.132	× 70 = 9.24	80 = 10.56	90 = 11.88
Travel Requirements	Long trips away from office	Trip lengths	0.082	× 90 = 7.38	70 = 5.74	60 = 4.92
		Proportion time away	0.054	× 80 = 4.32	70 = 3.78	60 = 3.24
	Daily commuting characteristics		0.034	× 50 = 1.70	70 = 2.38	100 = 3.40
Geographic location		Climate	0.034	× 80 = 2.72	70 = 2.38	60 = 2.04
		Proximity to leisure-time activities	0.068	× 90 = 6.12	80 = 5.44	60 = 4.08
		Proximity to relatives	0.068	× 60 = 4.08	70 = 4.76	100 = 6.80
Monetary compensation	Future salary prospects	Ten-year increase	0.035	× 90 = 3.15	80 = 2.80	80 = 2.80
		Three-year increase	0.064	× 60 = 3.84	70 = 4.48	80 = 5.12
	Immediate prospects — Fringe benefits	Retirement	0.009	× 95 = 0.86	70 = 0.63	60 = 0.54
		Insurance	0.014	× 95 = 1.33	70 = 0.98	60 = 0.84
	Immediate prospects — Starting salary		0.258	× 70 = 18.06	75 = 19.35	100 = 25.80
			1.000	75.24	74.83	82.25

An example of some of these observations and generalizations in action can be seen in the case of the large numbers of young Mexicans who have migrated both legally and illegally to the United States over the last 20 years (see Figure 8.22 and "Backlash"). Faced with rural poverty and overpopulation at home, they regard the place utility in Mexico as minimal. Their space-searching ability is, however, limited by both the lack of money and the lack of alternatives in the land of their birth. With a willingness to work and with aspirations for success—perhaps wealth—in the United States, they are eager listeners to friends and relatives who tell of numerous job opportunities in the United States. Hundreds of thousands, presented with the glittering prospect of a certain job, low-paying though it might be, quickly place high utility on perhaps a temporary relocation (maybe 5 or 10 years) to the United States. Many know that dangerous risks are involved if they attempt to enter the country illegally, but the rewards are worth the risk. Their arrival indicates their assignment of higher utility to the new site than to the old one.

In the 20th century, nearly all countries have experienced a movement of population into the cities from their agricultural areas. The migration has presumably paralleled the number of perceived opportunities within cities and convictions of absence of place utility in the rural districts. Perceptions, of course, do not necessarily accord with reality. The enormous influx of rural folk to major urban areas in developing countries, and the economic destitution of many of those in-migrants, suggest recurrent gross misperceptions, faulty information, and sampling error. In Chapter 12, we discuss the reasons for this phenomenon.

Barriers to Migration

Paralleling the incentives to migration is a set of disincentives, or barriers, to migration. They help to account for the fact that many people do not choose to move even when conditions are bad at home and are known to be better elsewhere. Migration depends on a knowledge of the opportunities in other areas. People with a limited knowledge of the opportunities elsewhere are less likely to migrate than those who are better-informed. Other barriers include physical features, the costs of moving, ties to individuals and institutions in the original activity space, and government regulations.

Physical barriers to travel include seas, mountains, swamps, deserts, and other natural features. In prehistoric times, physical barriers played an especially significant role in limiting movement. Thus, the spread of the ice sheets across most of Europe in Pleistocene times was a barrier to both migration and human habitation. Physical barriers to movement have probably assumed less importance only within the last 400 years. The developments that made possible the great age of exploration, beginning about A.D. 1500, and the technological developments associated with industrialization have enabled people to conquer space more easily. With industrialization came improved forms of transportation that made travel faster, easier, and cheaper. Still, as the account of conditions on the Overland Trail illustrated, only a century and a half ago, travel could be arduous, and in some parts of the world, it remains so today.

Economic barriers to migration include the cost of both travel and of establishing a residence elsewhere. Frequently, the additional expense of maintaining contact with those left behind is also a pertinent cost factor. Normally, all of these costs increase with the distance traveled and are a more significant barrier to travel for the poor than for the rich. Many immigrants to this country were married men who came alone; when they had acquired enough money, they sent for their families to join them. This phenomenon is still evident among recent immigrants from the Caribbean area to the United States, and among Turks, Yugoslavs, and West Indians who have settled in Northern and Western European countries.

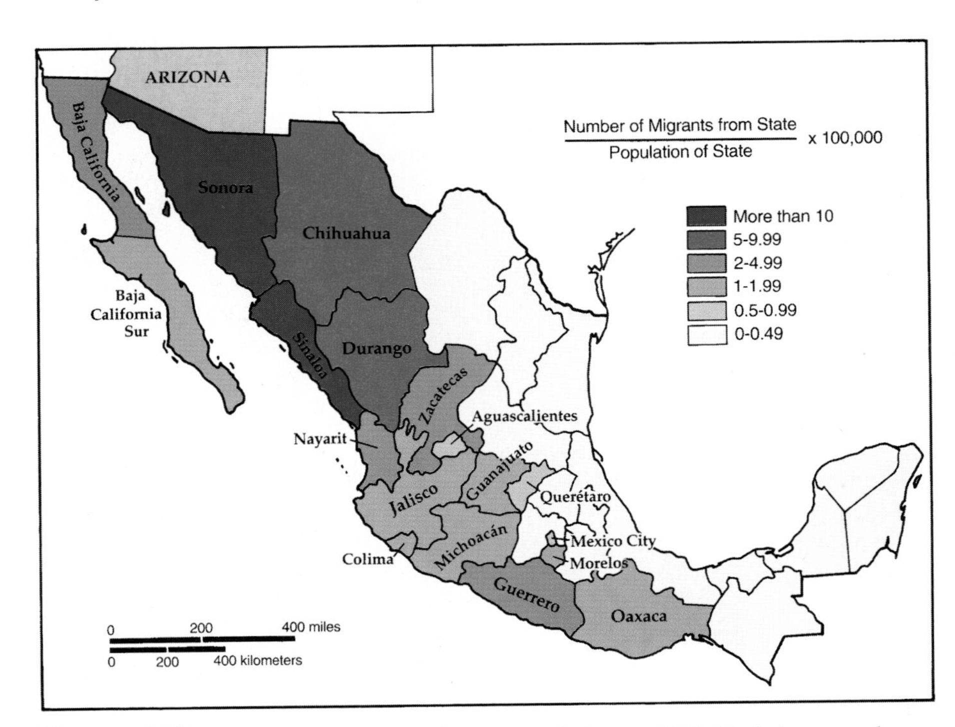

FIGURE 8.22 **Undocumented migration rate to Arizona; 1989–91.** Arizona and Sonora have historical ties that go back to the Gadsden Purchase in 1854. In many respects, the border with the United States cuts through a cultural region. Note that distance plays a large role, with over half of the migrants coming from three nearby states: Sonora, Sinaloa, and Chihuahua.

Redrawn from John P. Horner, "Continuity Amidst Change," *The Professional Geographer*, vol. 47, no. 4, Fig. 2, p. 403. Association of American Geographers, 1995.

Problems in the Sunbelt

Since 1970, there has been a net in-migration of more than 12 million people to the U.S. region known as the Sunbelt, the tier of southern states stretching from the Atlantic to the Pacific coast. This region captures the bulk of the nation's migrants and immigrants from other countries. Arizona's population grew to 3.7 million in 1990 from 1.8 million in 1970 and 2.7 million in 1980. Florida's growth rate for the 1980–1990 decade was more than 50%. Ironically, the rapid growth of the Sunbelt causes problems that threaten some of the very qualities that make the area attractive.

Low energy costs, low wages, and low taxes attracted businesses and industries from the North to the Sunbelt. Both blue- and white-collar workers from the North followed the companies, drawn by the promise of jobs and the particular attractions of the region: warmth and sunshine, unspoiled land, recreational opportunities, and other amenities. Although the Sunbelt continues to attract retirees, companies, and workers, its advantages are eroding rapidly, and growth itself is the reason.

With economic growth have come higher wages. In some parts of the region, taxes have risen already; taxes will continue to rise as states and cities try to cope with the services that population and economic growth demand. Across the Sunbelt, the rapid influx of people has strained the capacity of sewage treatment plants. Depending upon which city is considered, poor roads, inadequate fire and police protection, and limited mass transit are problems. Florida, California, and the Southwest face water shortages as well.

Many Sunbelt cities, although prosperous, exhibit the kinds of problems that we associate with northern and eastern cities: crowded freeways and traffic jams, noise, urban sprawl, rising crime rates, unemployment, poor inner cities, and air pollution. The clean air that attracted people with allergies, asthma, and other respiratory problems to Arizona is being polluted not only by cars and mining operations, but also by the foliage that migrants have planted to remind them of home. For example, the fast-growing mulberry tree provides good shade, but it produces pollen levels four times greater than other plants. Overall, Arizona has experienced a tenfold increase in airborne pollen since 1960. The problems caused by growth tend to inhibit further growth. The high crime rate in Florida, for example, hurts the huge tourist industry, and higher wages in the Southeast deter new companies from moving there.

A city does not create urban problems independently of the people who live there. When a substantial number of people move to a new and different city, the potential for the reappearance of the same problems, such as drugs and crime, that they hoped to leave behind is considerable.

The *cost factor* limits long-distance movement, but the larger the differential between present circumstances and perceived opportunities, the more individuals are willing to spend on moving. Figure 8.23 is an example of the effect of distance on relocation. For many, especially older people, the differentials must be extraordinarily high for movement to take place.

Cultural factors also contribute to decisions not to migrate. Family, religious, ethnic, and community relationships defy the principle of differential opportunities. Many people will not migrate under any but the most pressing of circumstances. The fear of change and human inertia—the fact that it is easier not to move than to do so—may be so great that people consider, but reject, a move. Ties to one's own country, cultural group, neighborhood, or family may be so strong as to compensate for the disadvantages of the home location. Returning migrants may convince potential leavers that the opportunities are not, in fact, better elsewhere—or, even if they are better, they are not worth entering an alien culture or sacrificing home and family.

FIGURE 8.23 Distance between old and new residences in the Asby area of Sweden. Notice how the number of movers decreased with increasing distance.

Source: Data from T. Hägerstrand, *Innovation Diffusion as a Spatial Process,* copyright © 1967 The University of Chicago Press, Chicago, IL.

Restrictions on immigration and emigration constitute *political barriers* to migration. Many governments frown on movements into or outside their own borders. Recognizing that immigration might be economically or politically disadvantageous, many nations restrict out-migration. These restrictions may make it impossible for potential migrants to leave, and they certainly limit the number who can do so.

On the other hand, countries suffering from an excess of workers often encourage emigration. The huge migration of people to the Americas in the late 19th and early 20th

Geography *and* Public Policy

Backlash

Migrants can enter a country legally—with a passport, visa, working permit, or other authorization—or illegally. Some aliens enter a country legally but on a temporary basis (as a student or tourist, for example), but then remain after their departure date. Recent years have seen a rising tide of emotion against the estimated 4 million people who reside illegally in the United States, a sentiment that has been reflected in a number of actions.

Greater efforts are being made to deter illegal crossings along the Mexican border by increasing the number of Border Patrol agents and by building steel fences near El Paso, Texas; Nogales, Arizona; and San Ysidro, California. Four states—Florida, Texas, Arizona, and California—are suing the federal government to win reimbursement for the costs of illegal immigration. The governor of California has proposed an amendment to the Constitution to deny citizenship to children born on American soil if their parents are not legal residents of the United States.

In the last several years, California voters have approved a trio of ballot initiatives aimed at curbing what their proponents see as privileges for immigrants, racial and ethnic minorities. Proposition 187, passed in November 1994, denies certain public services to illegal aliens. It prohibits state and local government agencies from providing publicly funded education, nonemergency health care, welfare benefits, and social services to any person whom they cannot verify is either a U.S. citizen or a person legally admitted to the country. The measure also requires state and local agencies to report suspected illegal immigrants to the Immigration and Naturalization Service and to certain state officials.

Proponents of Proposition 187 argued that California can no longer support the burden of high levels of immigration, especially if the immigrants cannot enter the more skilled professions. They contended that welfare, medical, and educational benefits are magnets that draw illegal aliens into the state. These unauthorized immigrants are estimated to cost California taxpayers more than $3.5 billion per year and result in overcrowded schools and public health clinics, and the reduction of services to legal residents. Why should the latter pay for benefits for people who are breaking the law, 187-supporters asked.

Those opposed to 187 contended that projected savings are illusory because the proposition collides with federal laws that

Reuters/Bettmann.

centuries is a good example of perceived opportunities for economic gain far greater than in the home country. Many European countries were overpopulated, and their political and economic systems stifled domestic economic opportunity at a time when people were needed by American entrepreneurs hoping to increase their wealth in untapped resource-rich areas.

The most developed countries where per capita incomes are high, or perceived as high, are generally the most desired international destinations. In order to protect themselves against overwhelming migration streams, such countries as the United States, Australia, France, and Germany have restrictive policies on immigration. In addition to absolute quotas on the number of immigrants (usually classified by country of origin), a country may impose other requirements, such as the possession of a labor permit or sponsorship by a recognized association.

Patterns of Migration

Several geographic concepts deal with patterns of migration. The first of these is the **migration field.** For any single place, the origin of its in-migrants and the destination of its out-migrants remain fairly stable spatially over time. Areas

guarantee access to public education for all children in the United States. It also violates federal Medicaid laws, so California will be in danger of losing all regular Medicaid funding. Forcing an estimated 300,000 children out of school and onto the streets will increase the risk of juvenile crime. Forbidding doctors from giving immunizations or basic medical care to anyone suspected of being an illegal immigrant will encourage the spread of communicable diseases throughout the state, putting everyone at risk. Educators, doctors, and other public service officials will be turned into immigration officers, a task for which they are ill-suited. Finally, opponents argued that the proposition will not stop the flow of illegal aliens because it does nothing to increase enforcement at the border or to punish employers who hire undocumented workers.

A week after the passage of Proposition 187, a federal judge issued a temporary restraining order blocking enforcement of most of its provisions pending the resolution of legal issues. Shortly thereafter, another federal judge struck down portions of the proposition, declaring them unconstitutional. "The state is powerless to enact its own scheme to regulate immigration or to devise immigration regulations which run parallel to or purport to supplement the federal immigration law," the judge wrote. Resolving the issues will require the involvement of the U.S. Supreme Court and is likely to take years.

Proposition 209, passed in November 1996, bans state and local government preferences based on race and gender in hiring and school admissions. No "positive" discrimination for racial minorities is allowed, and affirmative action programs are to be discontinued. Although opponents of the proposition appealed the initiative, the U.S. Supreme Court in 1997 declined to hear the appeal, allowing the ban on racial and gender preferences to stand.

Finally, in November 1998, California voters overwhelmingly approved Proposition 227, characterized by a spokesman for the Mexican American Legal Defense and Education Fund as the third in a row of anti-Latino measures. The proposition scraps the system of bilingual education, in which non-English-speaking children were taught in their native language until they learned English well enough to be mainstreamed into regular classrooms. Instead, students will receive one year of instruction in English. While opponents of the measure called it immigrant-bashing, its supporters argued that bilingual education has been a failure; few children graduate into English-speaking classes each year, and many leave school unable to speak, read, or write well in the language of their adopted country.

Questions to Consider

1. What do you think are the magnets that draw immigrants across the border: jobs or benefits? Is the denial of services likely to reduce illegal immigration?
2. Do you believe the federal government has an obligation to help states bear the costs of education, medical care, and incarceration for unauthorized immigrants? Why or why not?
3. Should the United States require citizens to have a national identification card? Why or why not?
4. If you had been able to vote on Proposition 187, how would you have voted? Why?
5. Is it good policy not to educate people or to give them basic medical care? If so, under what circumstances?

that dominate a locale's in- and out-migration patterns constitute the migration fields for the place in question, as shown in Figure 8.24. As would be expected, areas near the point of origin make up the largest part of the migration field. However, places far away, especially large cities, may also be prominent. These characteristics of migration fields are functions of the hierarchical movement to larger places (as we shall see) and the fact that so many people live in large metropolitan areas that one may expect some migration into and out of them from most areas within a country.

Migration fields do not conform exactly to the diffusion concepts mentioned earlier. As shown by Figure 8.25, some migration fields reveal a distinctly **channelized** pattern of flow. The channels link areas that are socially and economically tied to one another by past migration patterns, economic trade considerations, or by some other affinity. As a result, flows of migration along these channels are greater than would otherwise be the case. The movements of blacks from the southern United States to the North, of Scandinavians to Minnesota and Wisconsin, of Mexicans to such border states as California, Texas, and New Mexico, and of retirees to Florida and Arizona are all examples of channelized flows.

Sometimes the channelized migration is specific to occupational as well as ethnic groups. For example, nearly

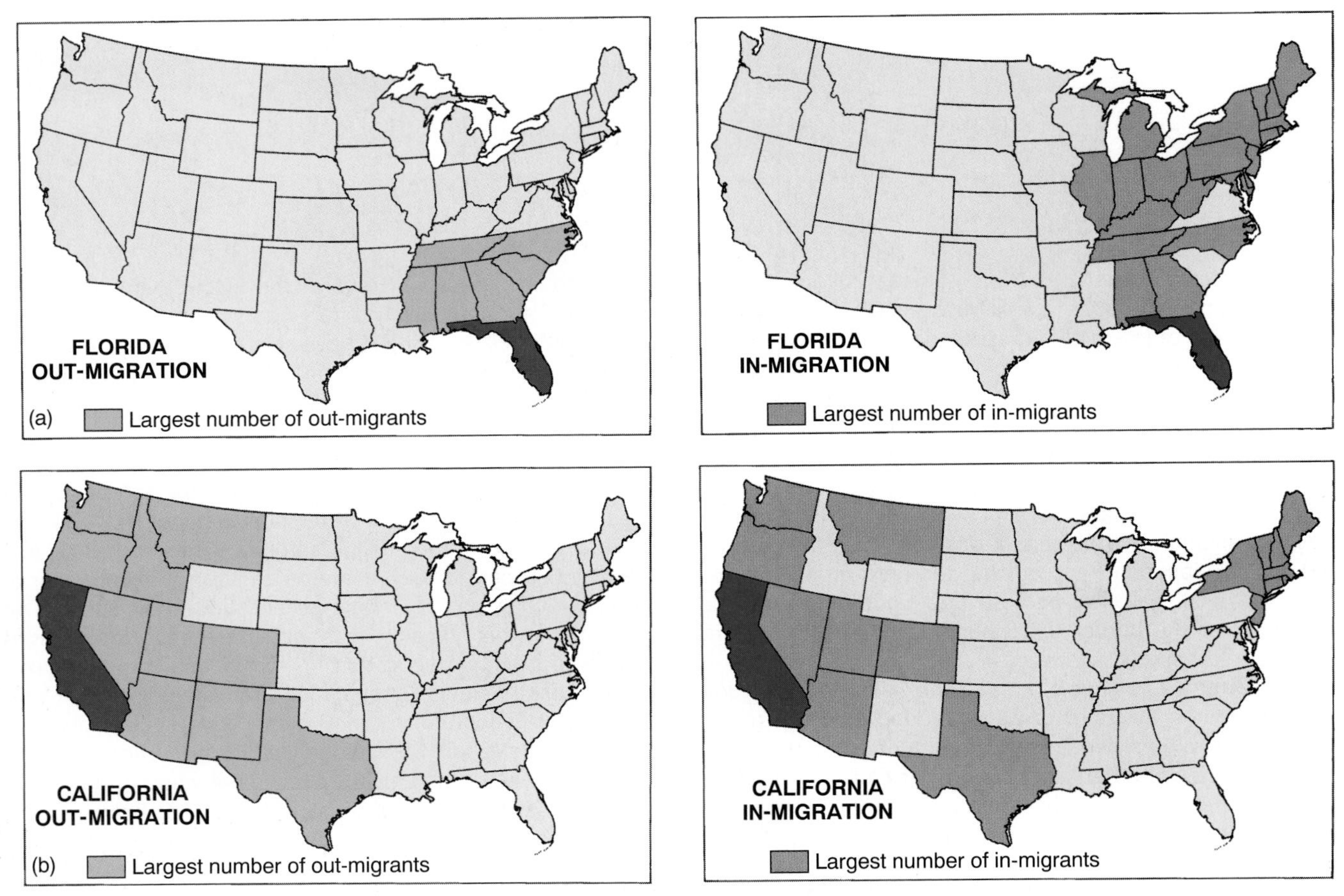

FIGURE 8.24 **The migration fields of Florida and California in 1980.** (a) For Florida, nearby southern states receive most out-migrants, but in-migrants, especially retirees, originate from much of eastern United States. (b) For California, the out-migration areas are western states. The in-migration areas include both western and heavily populated northeastern states.
From Kavita Pandit, "Differentiation between Subsystems and Typologies in the Analysis of Migration Regions," *The Professional Geographer,* vol. 46, no. 3, Figs. 5 & 6, pp. 342–3. Association of American Geographers, 1994.

all newspaper vendors in New Delhi, in the north of India, are reported to come from one small district in Tamil Nadu, in the south of India. Most construction workers in New Delhi come either from Orissa, in the east of India, or Rajasthan, in the northwest, and taxicab drivers originate in the Punjab area. The diamond trade of Bombay, India, is dominated by a network of about 250 related families who come from a small town several hundred miles to the north.

Return migration is a term used to refer to those who, soon after migrating, decide to return to their point of origin. If freedom of movement is not restricted, it is not unusual for as many as 25% of all migrants to return to their place of origin (Figure 8.26). Unsuccessful migration is sometimes due to an inability to adjust to the new environment. More often, it is the result of false expectations based on distorted mental images of the destination at the time of the move. Myths, secondhand and false information, and people's own exaggerations contribute to what turn out to be a mistaken decision to move. Although return migration often represents the unsuccessful adjustment of individuals to a new environment, it does not necessarily mean that negative information about a place returns with the migrant. It usually means a reinforcement of the channel, as communication lines between the unsuccessful migrant and would-be migrants take on added meaning and understanding.

In addition to channelization, the influence of large cities causes migration fields to deviate from the distance-decay pattern. The concept of **hierarchical migration** assists in understanding the nature of migration fields. Earlier we noted that sometimes information diffuses according to a hierarchical rule, that is, from city to city at the highest level in the hierarchy and then to lower levels. Hierarchical migration, in a sense, is a response to that flow. The tendency is for individuals in domestic relocations to move up the level in the hierarchy, from small places to larger ones. Very often, levels are skipped on the way up; only in periods of general economic decline is there considerable movement down the hierarchy. The suburbs of large cities are considered part of the metropolitan area, so the movement from a town to a small suburb would be considered moving up the hierarchy. From this pattern, we can envisage information flowing to a place via a hierarchical routing. Once the information is digested, some respond by migrating to the area from which the information came. In recent years, there has

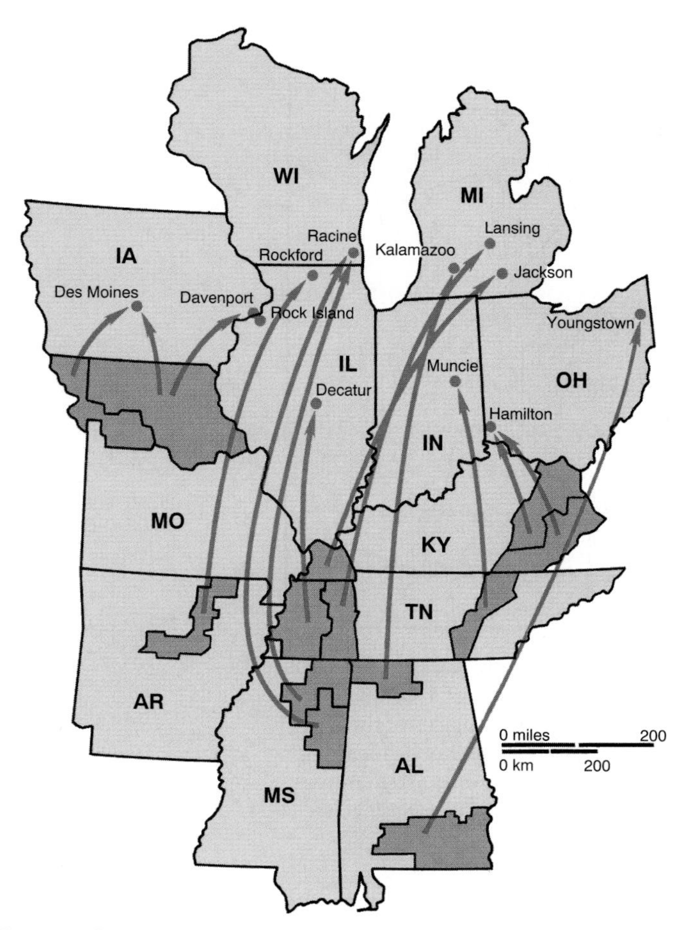

FIGURE 8.25 **Channelized migration flows from the rural South to Midwestern cities of medium size.** Distance is not necessarily the main determinant of flow direction. Perhaps through family and friendship links, the rural southern areas are tied to particular Midwestern destinations.

Redrawn by permission from *Proceedings of the Association of American Geographers,* C. C. Roseman, Vol. 3, p. 142. Association of American Geographers, 1971.

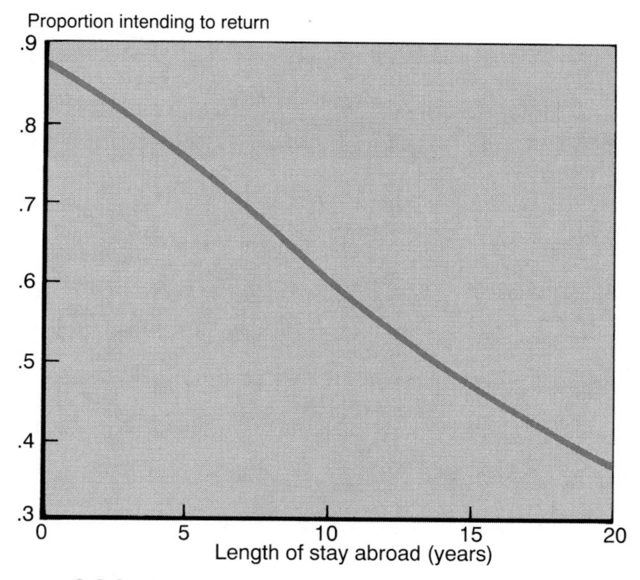

FIGURE 8.26 **Intended return migration of Yugoslavs from Germany.** As the length of stay in Germany increases, the proportion of Yugoslavs intending to return decreases.

Redrawn from B. Waldorf, "Determinants of International Return Migration Intentions," *The Professional Geographer,* vol. 47, no. 2, Fig. 2, p. 132. Association of American Geographers, 1995.

been a growing reversal of these trends; more and more people work at home, usually in computer-related occupations (telecommuting). Some even relocate to rural areas.

Spatial Search

Many of the ideas presented in this chapter are related to people's decision making in a geographic environment; that is, after evaluating alternatives, people come to a conclusion about where to satisfy their desires. Geographers call the process by which the alternatives are evaluated **spatial search.** The seeker may or may not actually travel to the various destinations; one may pursue the search by evaluating the information that comes to one's residence. For example, searching for a new residence may entail the reading of newspaper advertisements.

The quality of the decision is very much a function of the quality of available information. Some decisions, like buying a new home, are so important that individuals take months or even years to gather information. And, of course, in such circumstances, one is usually obliged to visit each possible homesite (Figure 8.27).

Decision making is based on a relationship between information availability and the preferences and motivations of the individual decision maker. Unfortunately, many decisions are made with inadequate or distorted information. Gathering information is a time-consuming activity. Many people are unable or unwilling to use their time for information gathering, and many disappointments are based on insufficient information. Often, even if the time were available and people were willing to gather information, the needed information might just be unavailable. Many of the return migrants previously discussed were responding to an inadequate understanding of the migration location.

People base decisions on their evaluation of the advantages and disadvantages of each possibility. They depend on their image—perhaps a mental map—of the thing or place being considered, and they use some type of scaling or preference criteria, keeping in mind the uncertainties that exist. Their motivations are based on their belief and value systems. For example, does the decision maker expect instant gratification or is she or he more inclined to expect gratification in several weeks, months, or years? Does one think that family satisfaction is more important than individual satisfaction? And so on.

In a study of the search for housing and residential choice in Toronto, many of the variables and factors involved in spatial search were evaluated. Among other things, it was found that people looking for apartments

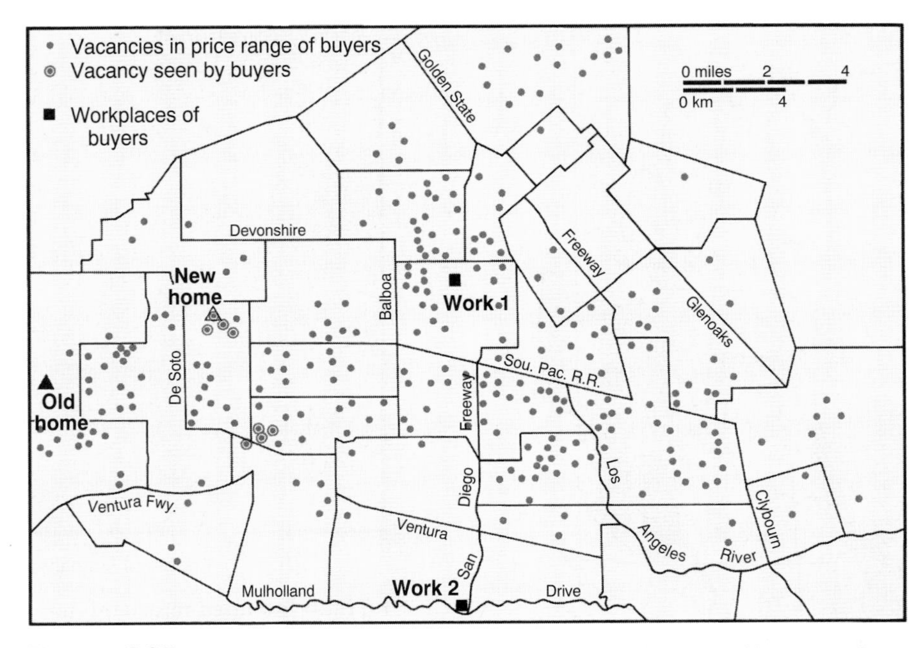

FIGURE 8.27 **An example of spatial search in the San Fernando Valley area of Los Angeles.** The dots represent the house vacancies in the price range of a sample family. Note (1) the relationship of the new house location to the workplaces of the married couple; (2) the relationship of the old house location to the new house location; and (3) the limited spatial search space. This example is typical of intraurban moves.

Redrawn by permission from J. O. Huff, *Annals of the Association of American Geographers,* Vol. 76, pp. 217–221. Association of American Geographers, 1986.

spent much less time looking than did potential home owners. The difference was ascribed to the importance people place on buying as opposed to renting. It was found, however, that apartment dwellers, even though they may complain about noise, inadequate space, and so on, are not much less satisfied with their decision than home owners. Most young apartment dwellers think of their quarters as temporary, so they set lower standards of comfort in order to take advantage of lower-cost housing, more convenience, and less responsibility.

In terms of the search itself, people spend a longer period of time deciding to look for new housing than they do in actually inspecting and choosing it. On average, home buyers look at seven or more housing units, compared to only about three for those who eventually choose apartments. A Los Angeles study concluded that those home buyers not leaving the metropolitan area altogether search for homes within a short distance of both their present home place and their workplace.

SUMMARY

In this discussion of the geography of spatial behavior, we have emphasized the factors that influence how individuals view and use space. At the beginning of the chapter, we posed a number of questions used as our theme. In considering how individuals view their environment, the concepts of spatial perception and spatial cognition were helpful. We spoke of the nature of the information available and of people's age, their past experiences, and their values. The question of differences in the extent of space used led us to the concept of activity space. The age of an individual, his or her degree of mobility, and the availability of opportunities all play a role in defining the limits of individual activity space. We saw, however, that as distance increases, familiarity with the environment decreases. The concept of critical distance identifies the distance beyond which the decrease in familiarity with places begins to be significant.

How space is used is a function of all of these factors, but certain factors having to do with the diffusion of innovations indicate what opportunities exist for individuals living in various places. Contagious and hierarchical diffusion influence the geographic direction that cultural change will take. Of course, barriers to diffusion, such as effort and cost, do exist. A special type of human interaction is migration. When strong enough, the various push and pull forces motivate a long-distance, permanent move. Migration fosters the spread of culture by means of a relocation diffusion process.

Finally, we examined how the ability of many to make well-reasoned, meaningful decisions is a function of the utility they assign to places and the opportunities at those places.

Geographers of spatial behavior view individual action in a special way. There is strong emphasis on information flow and the effect of distance decay. In many respects, it is an interdisciplinary field in that the combined ideas of geographers, psychologists, and other social scientists are crucial to an understanding of spatial behavior. Such behavior has a political component, and to that we next turn our attention.

Key Words

activity space 284
channelized migration 303
cognition 280
contagious diffusion 291
critical distance 287
diffusion 291
distance decay 287
environmental perception 280
hierarchical diffusion 293
hierarchical migration 304
mental map 280
migration 294
migration field 302
place utility 298
pull factor 295
push factor 295
return migration 304
spatial search 305
territoriality 284

For Review and Consideration

1. What is the difference in meaning between *environmental perception* and *cognition?* Give an example of each.
2. On a blank piece of paper, and without any maps to guide you, draw a map of the United States, putting in state boundaries wherever possible; this is your mental map of the nation. Compare it with a standard atlas map. What conclusions can you reach?
3. What is meant by *activity space?* What factors affect the areal extent of the activity space of an individual?
4. Recall the places you have visited in the past week. In your movements, were the distance-decay and critical-distance rules operative? What variables affect an individual's critical distance?
5. Briefly distinguish between *contagious diffusion* and *hierarchical diffusion.* In what ways, if any, were these forms of diffusion in operation in the culture hearths discussed in Chapter 7?
6. What considerations affect a decision to migrate? What is *place utility,* and how does its perception induce or inhibit migration?
7. What common barriers to migration exist? Why do most people migrate within their own country?
8. Define the term *migration field.* Some migration fields show a channelized flow of people. Select a particular channelized migration flow (such as the movement of Scandinavians to the United States, or people from the Great Plains to California, or southern blacks to the North) and explain why a channelized flow developed.

Selected References

Aitken, Stuart C. *Putting Children in Their Place.* Washington, D.C.: Association of American Geographers, 1994.

Boyle, Paul, and Keith Halfacre, eds. *Mirgration into Rural Areas: Theories and Issues.* London: John Wiley & Sons, 1998.

Brown, Lawrence A. *Innovation Diffusion: A New Perspective.* New York: Methuen, 1981.

Brunn, Stanley, and Thomas Leinbach. *Collapsing Space and Time: Geographic Aspects of Communication and Information.* London: Harper Collins Academic, 1991.

Castles, Stephen, and Mark J. Miller. *The Age of Migration: International Population Movements in the Modern World.* New York: Guilford Press, 1993.

Chisholm, Michael, and David M. Smith, eds. *Shared Space: Divided Space.* New York: Harper Collins Academic, 1990.

Clark, William A. V. *Human Migration.* Newbury Park, CA: Sage Publications, 1986.

Cutter, Susan. *Living with Risk: The Geography of Technological Hazard.* London: Edward Arnold, 1992.

Downs, Roger M., and David Stea, eds. *Image and Environment: Cognitive Mapping and Spatial Behavior.* Chicago: Aldine Publishing Co., 1973.

Gärling, Tommy, and Reginald G. Golledge, eds. *Behavior and Environment: Psychological and Geographical Approaches.* Amsterdam: Elsevier/North Holland, 1993.

Golledge, Reginald G., and Robert J. Stimson. *Decision Making and Spatial Behavior: A Geographic Perspective.* New York: Guilford Press, 1996.

Gould, P. R., and Rodney White. *Mental Maps.* 2d ed. Boston: Allen & Unwin, 1986.

Hägerstrand, Torsten. *Innovation Diffusion as a Spatial Process.* Chicago: University of Chicago Press, 1967.

Haines, David W., ed. *Refugees as Immigrants: Cambodians, Laotians, and Vietnamese in America.* Totowa, N.J.: Rowman & Littlefield Publishers, Inc., 1988.

Hewitt, Kenneth. *Regions of Risk: Hazards, Vulnerability and Disaster.* Harlow, Essex: Addison Wesley Longman, 1997.

King, Russel, ed. *Return Migration and Regional Economic Problems.* London: Croom Helm, 1986.

Long, Larry H. *Migration and Residential Mobility in the United States.* New York: Russell Sage Foundation, 1988.

Lynch, Kevin. *The Image of the City.* Cambridge, Mass.: MIT Press, 1960.

Morrill, Richard L., Gary L. Gaile, and Grant Ian Thrall. *Spatial Diffusion.* Scientific Geography Series; vol. 10. Newbury Park, Calif.: Sage Publications, 1988.

Palm, Risa. *Natural Hazards.* Baltimore, Md.: Johns Hopkins University Press, 1990.

Pooley, Colin G., and Ian D. Whyte, eds. *Migrants, Emigrants and Immigrants: A Social History of Migration.* New York: Routledge, 1991.

Roseman, Curtis C. "Channelization of Migration Flows from the Rural South to the Industrial Midwest." *Proceedings of the Association of American Geographers* 3 (1971): 140-146.

Thrift, Nigel. *Spatial Formations.* Thousand Oaks, CA: Sage Publications, 1996.

Walmsley, D. J., and G. J. Lewis, eds. *People and Environment: Behavioural Approaches in Human Geography.* 2d ed. New York: John Wiley & Sons, Inc., 1993.

Zonn, Leo E., ed. *Place Images in Media: Portrayal, Experience, and Meaning.* Savage, Md.: Rowman & Littlefield Publishers, Inc., 1990.

Don't forget about Dushkin's *Annual Editions Online: Geography* at http://dushkin.com/aeonline/. See preface for details.

CHAPTER

9

POLITICAL GEOGRAPHY

Uzbek soldiers guard the Friendship Bridge, which links Uzbekistan to Afghanistan at Hairaton.
©AP/Dimitri Messinis/Wide World Photos.

They met together in the cabin of the little ship on the day of the landfall. The journey from England had been long and stormy. Provisions ran out, a man had died, a boy had been born. Although they were grateful to have reached the calm waters off Cape Cod that November day of 1620, their gathering in the cramped cabin was not to offer prayers of thanksgiving but to create a political structure to govern the settlement they had come to establish (Figure 9.1). The Mayflower Compact was an agreement among themselves to "covenant and combine our selves together into a civill body politick . . . to enacte, constitute, and frame such just and equall lawes, ordinances, acts, constitutions, and offices . . . convenient for ye generall good of ye Colonie. . . ." They elected one of their company governor, and only after those political acts did they launch a boat and put a party ashore.

The land they sought to colonize had for more than 100 years been claimed by the England they had left. The New World voyage of John Cabot in 1497 had invested their sovereign with title to all of the land of North America and a recognized legal right to govern his subjects dwelling there. That right was delegated by royal patent to colonizers and their sponsors, conferring upon them title to a defined tract and the right to govern it. Although the Mayflower settlers were originally without a charter or patent, they recognized themselves as part of an established political system. They chose their governor and his executive department annually by vote of the General Court, a legislature composed of all freemen of the settlement.

As the population grew, new towns were established too distant for their voters to attend the General Court. By 1636 the larger towns were sending representatives to cooperate with the executive branch in making laws. Each town became a legal entity, with election of local officials and enactment of local ordinances the prime purpose of the town meetings that are still common in New England today.

The Mayflower Compact, signed by 41 freemen as their first act in a New World, was the first step in a continuing journey of political development for the settlement and for the larger territory of which it became a part. From company patent to crown colony to rebellious commonwealth under the Continental Congress to state in a new country, Massachusetts (and Plymouth Plantation) were part of a continuing process of the political organization of space.

FIGURE 9.1 **Signing the Mayflower Compact,** probably the first written plan for self-government in America. Forty-one adult male Pilgrims signed the Compact aboard the *Mayflower* before going ashore.
Courtesy of the Pilgrim Society, Plymouth, MA.

That process is as old as human history. From clans to kingdoms, human groups have laid claim to territory and have organized themselves and administered their affairs within it. Indeed, the political organizations of society are as fundamental an expression of culture and cultural differences as are forms of economy or religious beliefs. Geographers are interested in that structuring because it is both an expression of the human organization of space and is closely related to other spatial evidences of culture, such as religion, language, and ethnicity.

Political geography is the study of the organization and distribution of political phenomena in their areal expression. Nationality is a basic element in cultural variation among people, and political geography traditionally has had a primary interest in country units, or *states* (Figure 9.2). Of central concern have been spatial patterns that reflect the exercise of central governmental control, such as questions of boundary delimitation and effect. Increasingly, however, attention has shifted both upward and downward on the political scale. On the world scene, international alliances, regional compacts, and producer cartels have increased in prominence since World War II, representing new forms of spatial interaction. At the local level, voting patterns, constituency boundaries and districting rules, and political fragmentation have directed public attention to the significance of area in the domestic political process.

In this chapter, we discuss some of the characteristics of political entities, examine the problems involved in defining jurisdictions, seek the elements that lend cohesion to a political entity, explore the implications of partial surrender of sovereignty, and consider the significance of the fragmentation of

FIGURE 9.2 These flags, symbols of separate member states, grace the front of the United Nations building in New York City. Although central to political geographic interest, states are only one level of the political organization of space.
©Pictor/Uniphoto.

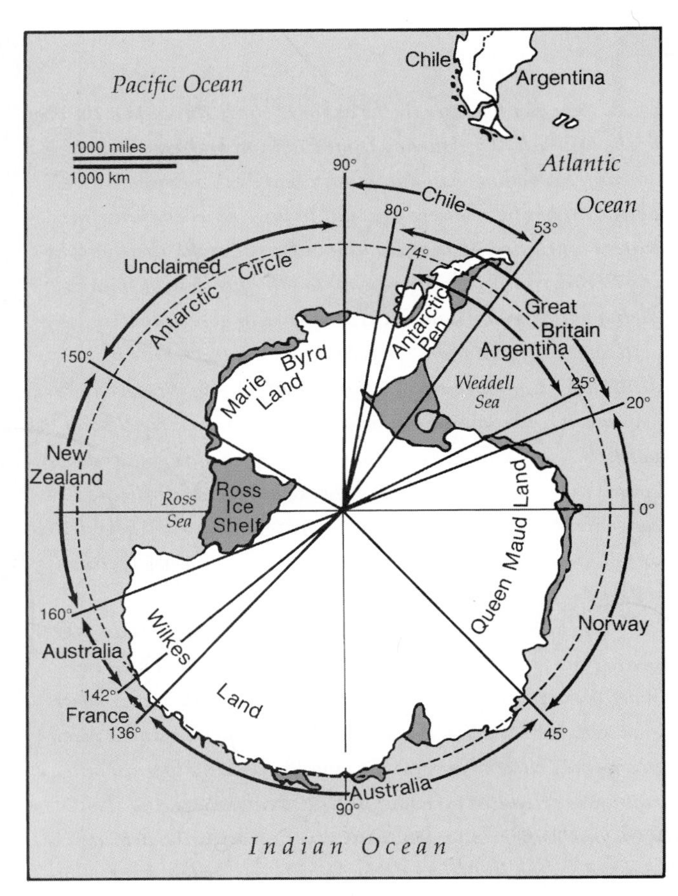

FIGURE 9.3 **Territorial claims in Antarctica.** Although seven countries claim sovereignty over portions of Antarctica, and three of the claims overlap, the continent has no permanent inhabitants or established local government.

political power. We begin with states and end with local political systems.

Emphasis here on political entities should not make us lose sight of the reality that states are rooted in the operations of the economy and society they represent, that social and economic disputes are as significant as border confrontations, and that in some regards transnational corporations and other nongovernmental agencies may exert more influence in international affairs than do the separate states in which they are housed or operate. Some of those expanded political considerations are alluded to in the discussions that follow; others are developed more fully in Chapter 10.

NATIONAL POLITICAL SYSTEMS

One of the most significant elements in cultural geography is the nearly complete division of the earth's land surface into separate national units, as shown on the Countries of the World map inside this book's cover. Even Antarctica is subject to the rival territorial claims of seven countries, although these claims have not been pressed because of the Antarctic Treaty of 1961 (Figure 9.3). A second element is that this division into country units is relatively recent. Although countries and empires have existed since the days of early Egypt and Mesopotamia, only in the last century has the world been almost completely divided into independent governing entities. Now, people everywhere accept the idea of the state, and its claim to sovereignty within its borders, as normal.

States, Nations, and Nation-States

Before we begin our consideration of political systems, we need to clarify some terminology. Geographers use the words *state* and *nation* somewhat differently than the way they are used in everyday speech; the confusion arises because each word has more than one meaning. A state can be defined as either (1) any of the political units forming a federal government (e.g., one of the United States) or as (2) an independent political unit holding sovereignty over a territory (e.g., the United States). In this latter sense, *state* is synonymous with *country* or *nation*. That is, a nation can also be defined as (1) an independent political unit holding sovereignty over a territory (e.g., a member of the United Nations). But it can also be used to describe (2) a community of people with a common culture and territory (e.g., the Kurdish nation). The second definition is *not* synonymous with state or country.

To avoid confusion, we shall define a **state** on the international level as an independent political unit occupying a defined, permanently populated territory and having full sovereign control over its internal and foreign affairs. We will use *country* as a synonym for the territorial and political concept of "state." Not all recognized territorial entities are states. Antarctica, for example, has neither established government nor permanent population; it is, therefore, not a state. Nor are *colonies* or *protectorates* recognized as states. Although they have defined extent, permanent inhabitants, and some degree of separate governmental structure, they lack full control over all of their internal and external affairs.

We use nation in its second sense, as a reference to people, not to political structure. A **nation** is a group of people with a common culture occupying a particular territory, bound together by a strong sense of unity arising from shared beliefs and customs. Language and religion may be unifying elements, but even more important are an emotional conviction of cultural distinctiveness and a sense of ethnocentrism.

The composite term **nation-state** properly refers to a state whose territorial extent coincides with that occupied by a distinct nation or people or, at least, whose population shares a general sense of cohesion and adherence to a set of common values (Figure 9.4a). That is, a nation-state is an entity whose members feel a natural connection with each other by virtue of sharing language, religion, or some other cultural characteristic strong enough both to bind them together and to give them a sense of distinction from all others outside the community. Although all countries strive for consensus values and loyalty to the state, few can claim to be ethnic nation-states. Iceland, Denmark, and Poland are often cited as acceptable European examples. Japan is an Asian illustration.

A *binational* or *multinational state* is one that contains more than one nation (Figure 9.4b). Often, no single ethnic group dominates the population. In the constitutional structure of the Soviet Union before 1988, one division of the legislative branch of the government was termed the Soviet of Nationalities. It was composed of representatives from civil divisions of the Soviet Union populated by groups of officially recognized "nations": Ukrainians, Kazakhs, Estonians, and others. In this instance, the concept of nationality was territorially less than the extent of the state.

Alternatively, a single nation may be dispersed across and be predominant in two or more states. This is the case with the *part-nation state* (Figure 9.4c). Here, a people's sense of nationality exceeds the areal limits of a single state. An example is the Arab nation, which dominates 17 states.

Finally, there is the special case of the *stateless nation*, a people without a state. The Kurds, for example, are a nation of approximately 20 million people divided among six states and dominant in none (Figure 9.4d). Kurdish nationalism has survived over the centuries, and many Kurds nurture a vision of an independent Kurdistan. Other

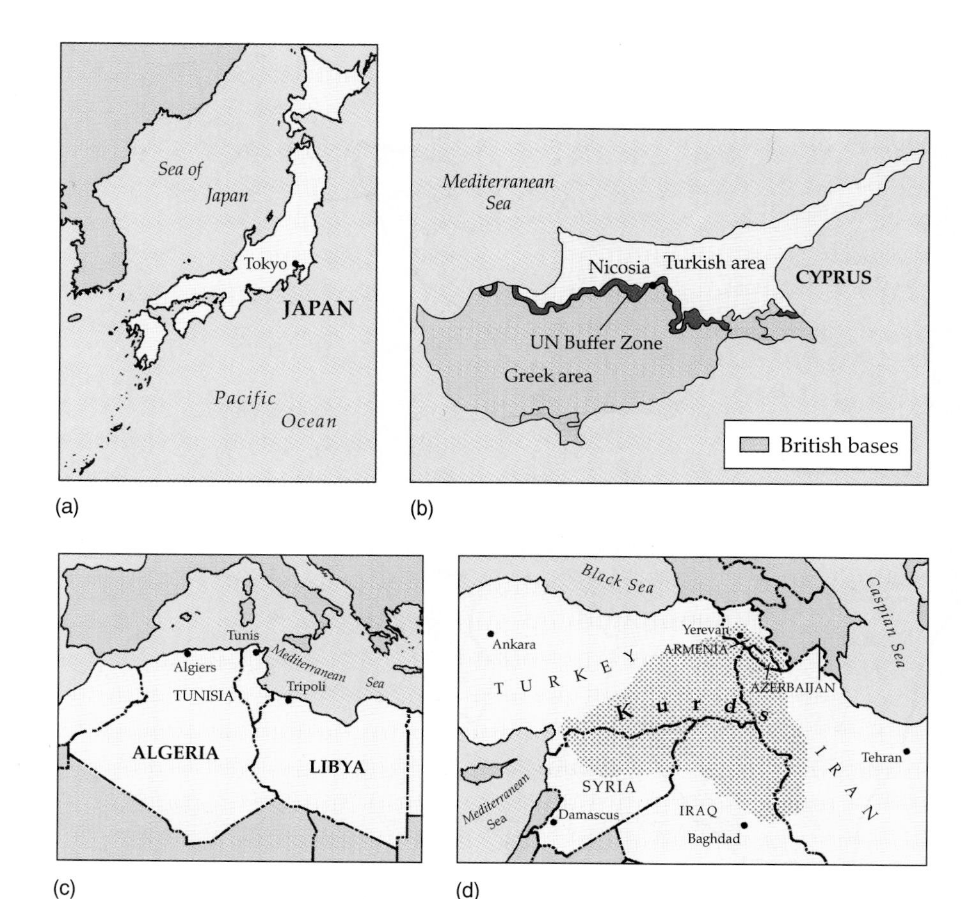

FIGURE 9.4 **Types of relationships between "states" and "nations."** (a) A **nation-state.** Japan is an example of a state occupied by a distinct nation, or people. (b) A **multinational state.** The island of Cyprus, in the eastern Mediterranean Sea, contains two distinct nations: Greeks and Turks. After Cyprus gained its independence from Britain in 1960, there was an upsurge of violence between the two groups, a military coup with the aim of uniting the island with Greece, and a retaliatory invasion by Turkey. Since 1974, Cyprus has been partitioned between the Turkish north and the Greek south. A buffer zone policed by United Nations troops separates the two sectors from one another. Britain retains sovereignty over two bases on the isalnd. (c) A **part-nation state.** The Arab nation extends across and dominates many states in northern Africa and the Middle East. (d) A **stateless nation.** An ancient group with a distinctive language, the Kurds are concentrated in Turkey, Iran, and Iraq. Smaller numbers live in Syria, Armenia, and Azerbaijan.

stateless nations are Macedonians, Basques, and Palestinians. The Palestine Liberation Organization (PLO) took over administrative control of the Gaza Strip and Jericho from Israel in 1994 and of additional towns and districts within the West Bank territory in 1995–96, steps that Palestinians hope will lead to eventual statehood.

The Evolution of the Modern State

The idea of the modern state was developed by European political philosophers in the 18th century. Their views advanced the concept that people owe allegiance to a state and the people it represents rather than to its leader, such as a king or a feudal lord. The new concept coincided in France with the French Revolution and spread over Western Europe, to England, Spain, and Germany.

Many states are the result of European expansion during the 17th, 18th, and 19th centuries, when much of Africa, Asia, and the Americas was divided into colonies. Usually these colonial claims were given fixed and described boundaries where none had earlier been formally defined. Of course, precolonial native populations had relatively fixed home areas of control within which there was recognized dominance and border defense and from which there were, perhaps, raids of plunder or conquest of neighboring "foreign" territories. Beyond understood tribal territories, great empires arose, again with recognized outer limits of influence or control: Mogul and Chinese; Benin and Zulu; Incan and Aztec. Upon them where they still existed, and upon the less formally organized spatial patterns of effective tribal control, European colonizers imposed their arbitrary new administrative divisions of the land. In fact, tribes that had little in common were often joined in the same colony (Figure 9.5). The new divisions, therefore, were not usually based on meaningful cultural or physical lines. Instead, the boundaries simply represented the limits of the colonizing empire's power.

FIGURE 9.5 **The discrepancies between tribal and national boundaries in Africa.** Tribal boundaries were ignored by European colonial powers. The result has been significant ethnic diversity in nearly all African countries.

Redrawn from *World Regional Geography: A Question of Place* by Paul Ward English, with James Andrew Miller. Copyright © 1977 Harper & Row. Used by permission of the author.

As these former colonies have gained political independence, they have retained the idea of the state. They have generally accepted—in the case of Africa, by a conscious decision to avoid precolonial territorial or ethnic claims that could lead to war—the borders established by their former European rulers. The problem that many of the new countries face is "nation-building"—developing feelings of loyalty to the state among their arbitrarily associated citizens. The Democratic Republic of the Congo ("Zaire" from 1971 to 1997), the former Belgian Congo, contains more than 250 frequently antagonistic tribes. Only if past tribal animosities can be converted into an overriding spirit of national cohesion will such countries truly be nation-states.

The end of the colonial era brought a rapid increase in the number of sovereign states. At the beginning of World War II, in 1939, there were about 70 independent countries. By 1970, the number had more than doubled, and by 1990, independent states totaled nearly 200. The number of independent states increased again following—among other political geographic developments—the disintegration of the USSR and Yugoslavia during the early 1990s (Figure 9.6).

Geographic Characteristics of States

Every state has certain geographic characteristics by which it can be described and that set it apart from all other states. A look at the world political map inside the cover of this book confirms that every state is unique. The size, shape, and location of any one state combine to differentiate it from all others. These characteristics are of more than academic interest, because they also affect the power and stability of states.

Size

The area that a state occupies may be large, as is true of China, or small, as is Liechtenstein. The world's largest country, Russia, occupies more than 17 million square kilometers (6.5 million sq mi), or some 11% of the land surface of the world. It is more than a million times as large as Nauru, one of the ministates found in all parts of the world (see "The Ministates").

An easy assumption would be that the larger a state's area, the greater is the chance that it will have resources, such as fertile soil and minerals, from which it can benefit. In general, that assumption is valid, but much depends upon accidents of location. Mineral resources are unevenly distributed, and size alone does not guarantee their presence within a state. And Australia, Canada, and Russia, though large, have relatively small areas capable of supporting productive agriculture. Great size, in fact, may be a disadvantage. A very large country may have vast areas that are inaccessible, sparsely populated, and hard to integrate into the mainstream of economy and society. Small states are more apt than large ones to have a culturally homogeneous population. They find it easier to develop transportation and communication systems to link the sections of the country, and, of course, they have shorter boundaries to defend against invasion. Size alone, then, is not critical in determining a country's stability and strength, but it is a contributing factor.

Shape

Like size, a country's shape can affect the well-being of a state by fostering or hindering effective organization. Assuming no major topographical barriers, the most efficient form would be a circle, with the capital located in the center. In such a country, all places could be reached from the center in a minimal amount of time and with the least expenditure for roads, railway lines, and so on. It would also have the shortest possible borders to defend. Uruguay and Poland have roughly circular shapes, forming a **compact state** (Figure 9.7).

FIGURE 9.6 By mid-1992, 15 newly independent countries had taken the place of the former USSR.

The Ministates

Totally or partially autonomous political units that are small in area and population pose some intriguing questions. Should size be a criterion for statehood? What is the potential of ministates to cause friction among the major powers? Under what conditions are they entitled to representation in international assemblies like the United Nations?

About half of the world's independent countries contain fewer than 5 million people. Of these, more than 40 have under 1 million, the population size adopted by the United Nations as the upper limit defining "small states," though not too small to be members of that organization. Nauru has about 10,000 inhabitants on its 21 square kilometers (8.2 sq mi). Other areally small states like Singapore, covering 580 square kilometers (224 sq mi), have populations (3.5 million) well above the UN criterion. Many are island territories located in the West Indies and the Pacific Ocean (such as Grenada and Tonga Islands), but Europe (Vatican City and Andorra), Asia (Bahrain and Macao), and Africa (Djibouti and Equatorial Guinea) have their share.

Many ministates are vestiges of colonial systems that no longer exist. Some of the small states of West Africa and the Arabian peninsula fall into this category. Others, such as Mauritius, served primarily as refueling stops on transoceanic voyages. However, some occupy strategic locations (such as Bahrain, Malta, and Singapore), and others contain valuable minerals (Kuwait, Nauru, and Trinidad). The possibility of claiming 370-kilometer-wide (200 nautical mile) zones of adjacent seas adds to the attraction of yet others.

Their strategic or economic value can expose small islands and territories to unwanted attention from larger neighbors. The 1982 war between Britain and Argentina over the Falkland Islands (called the Islas Malvinas by Argentina) and the Iraqi invasion of Kuwait in 1990 demonstrate the ability of such areas to bring major powers into conflict and to receive world attention that is out of proportion to their size and population.

The proliferation of tiny countries raises the question of their representation and their voting weight in international assemblies. Should there be a minimum size necessary for participation in such bodies? Should countries receive a vote proportional to their population? Within the United Nations, the Small Island Developing States (SIDS) recently have emerged as a significant power bloc, controlling more than one-fifth of UN General Assembly votes. Since forming themselves into the Alliance of Small Island States (AOSIS) in November, 1990, they have proved to have unexpected clout and played a key role in placing New Zealand on the Security Council in 1993.

The influence of the United States and other major powers in the United Nations has already been eroded by the small states. Although the United States pays 25% of the UN budget and has about one and one-half times the population of all the small countries combined, its vote can be balanced by that of any of them.

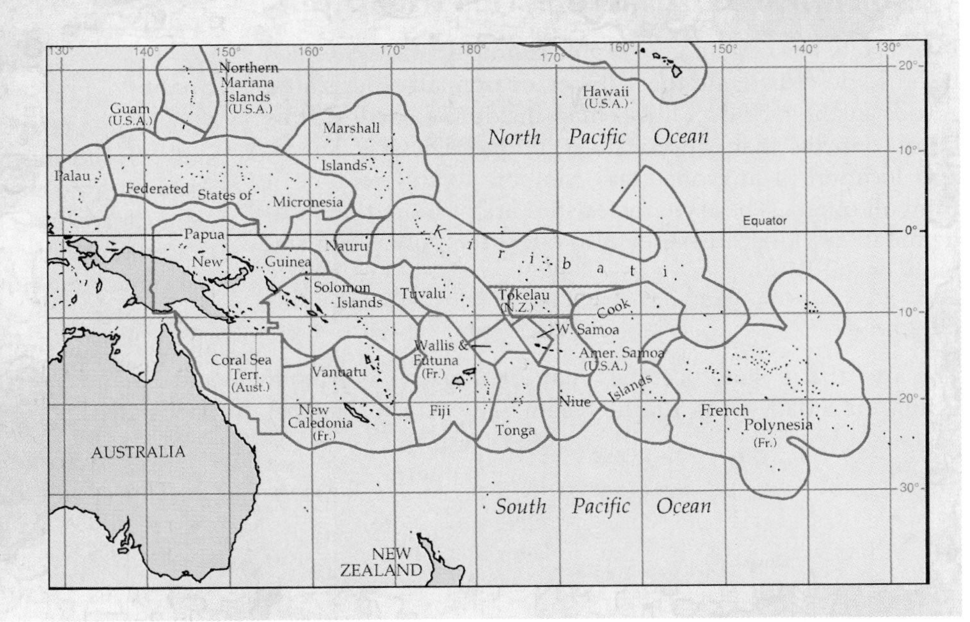

Prorupt states are nearly compact but possess one or sometimes two narrow extensions of territory. Proruption may simply reflect peninsular elongations of land area, as in the case of Myanmar and Thailand. In other cases, the extensions have an economic or strategic significance, having been designed to secure state access to resources or to establish a buffer zone between states that would otherwise adjoin. The proruptions of Afghanistan, Democratic Republic of the Congo, and Namibia fall into this category. The Caprivi Strip of Namibia, for example, which extends eastward from the main part of the country, was designed by the Germans to give what was then their colony of Southwest Africa access to the Zambezi River. Whatever their origin, proruptions tend to isolate a portion of a state.

The least efficient shape administratively is represented by countries like Norway and Chile, which are long and narrow. In such **elongated states,** the parts of the country far from the capital are likely to be isolated because great expenditures are required to link them to the core. These countries are also likely to encompass more diversity of climate, resources, and peoples than compact states, perhaps to the detriment of national cohesion or, perhaps, to the promotion of economic strength.

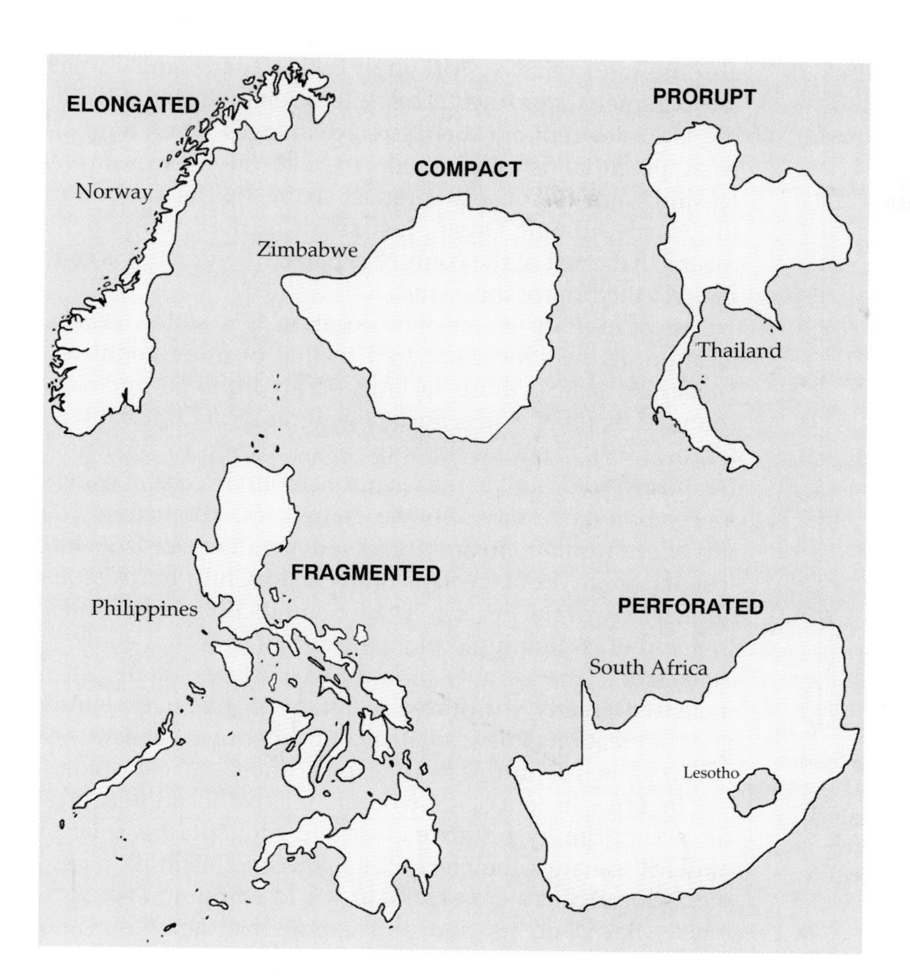

FIGURE 9.7 **Shapes of states.** The sizes of the countries should not be compared. Each is drawn on a different scale.

A fourth class of shapes, that of **fragmented states,** includes countries composed entirely of islands (e.g., the Philippines and Indonesia), countries that are partly on islands and partly on the mainland (Italy and Malaysia), and those that are chiefly on the mainland but whose territory is separated by another state (the United States). Fragmentation makes it harder for the state to impose centralized control over its territory, particularly when the parts of the state are far from one another. This is a problem in the Philippines and Indonesia, the latter made up of over 13,000 islands, stretched out along a 5100-kilometer (3200-mi) arc. Fragmentation helped lead to the disintegration of Pakistan. It was created in 1947 as a fragmented state, but East and West Pakistan were 1610 kilometers (1000 mi) from one another. That distance exacerbated economic and cultural differences between the two, and when the eastern part of the country seceded in 1971 and declared itself the independent state of Bangladesh, West Pakistan was unable to impose its control.

A special case of fragmentation occurs when a territorial outlier of one state, an **exclave,** is located within another state. Before German unification, West Berlin was an outlier of West Germany within East Germany (the German Democratic Republic). Europe has many such outlying bits of one country inside another. Kleinwalsertal, for example, is a piece of Austria accessible only from Germany. Baarle-Hertog is a fragment of Belgium inside Holland. Campione d'Italia is an Italian outlier in Switzerland, and Büsingen is a German one. Llivia is a Spanish town just inside France (Figure 9.8). Exclaves are not limited to Europe, of course. African examples include Cabinda, an exclave of Angola, and Melilla and Ceuta, two Spanish exclaves in Morocco.

The counterpart of an exclave, an **enclave,** helps to define the fifth class of shapes, the **perforated state.** A perforated state completely surrounds a territory that it does not rule, as, for example, the Republic of South Africa surrounds Lesotho. The enclave, the surrounded territory, may be independent or may be part of another state. Two of Europe's smallest independent states, San Marino and Vatican City, are enclaves that perforate Italy. As an *exclave* of West Germany, West Berlin perforated the national territory of former East Germany and was an *enclave* in it. The stability of the perforated state can be weakened if the enclave is occupied by people whose value systems differ from those of the surrounding country (see "The Gnarled Politics of an Enclave").

Location

The significance of size and shape as factors in national well-being can be modified by a state's location, both

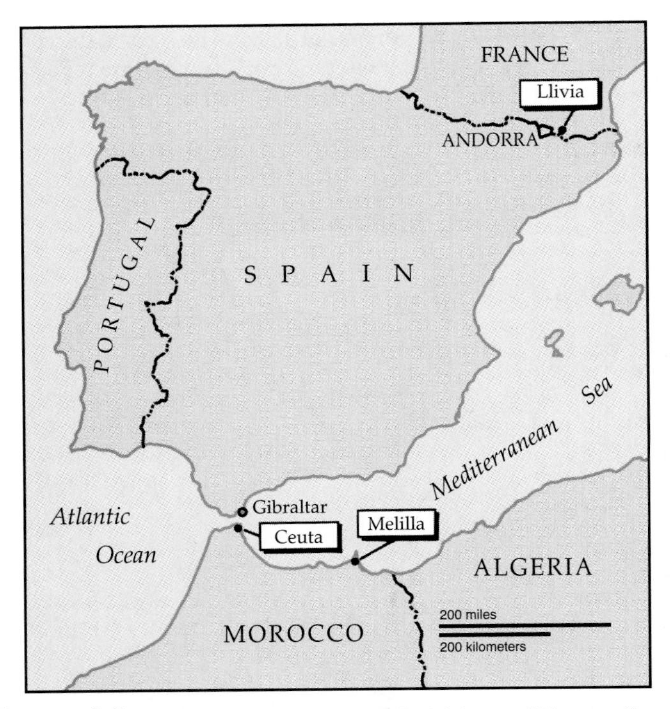

FIGURE 9.8 **Spanish exclaves in North Africa and France.** Spanish troops seized the garrison towns of Melilla and Ceuta almost 500 years ago after evicting the Moors from Spain proper. Llivia became an exclave in 1660 when Spain ceded the surrounding area to France in the Treaty of the Pyrenees. Gibraltar is a British colony, and Andorra is an independent ministate.

absolute and relative. Although both Canada and Russia are extremely large, their *absolute* location in the upper-middle latitudes reduces their size advantages when agricultural potential is considered. To take another example, Iceland has a reasonably compact shape, but its location in the North Atlantic Ocean, just south of the Arctic Circle, means that most of the country is barren. Settlement is confined to the rims of the island.

As important as absolute location is a state's *relative* location, its position compared to that of other countries. *Landlocked* states, those lacking ocean frontage and surrounded by other states, are at a geographical disadvantage (Figure 9.9). They lack both easy access to maritime (seaborne) trade and to the resources found in coastal waters and submerged lands. Bolivia gained 480 kilometers (300 mi) of sea frontier along with its independence in 1825, but lost its ocean frontage by conquest to Chile in 1879. Its annual Day of the Sea ceremony reminds Bolivians of their loss and of continuing diplomatic efforts to secure an alternate outlet. The number of landlocked states—about 40—increased greatly with the dissolution of the Soviet Union and the creation of new, smaller countries out of such former multinational countries as Yugoslavia and Czechoslovakia.

In a few instances, a favorable relative location constitutes the primary resource of a state. Singapore, a state of only 580 square kilometers (224 sq mi) and 3.5 million people, is located at a crossroads of world shipping and commerce. Based on its port and commercial activities, and

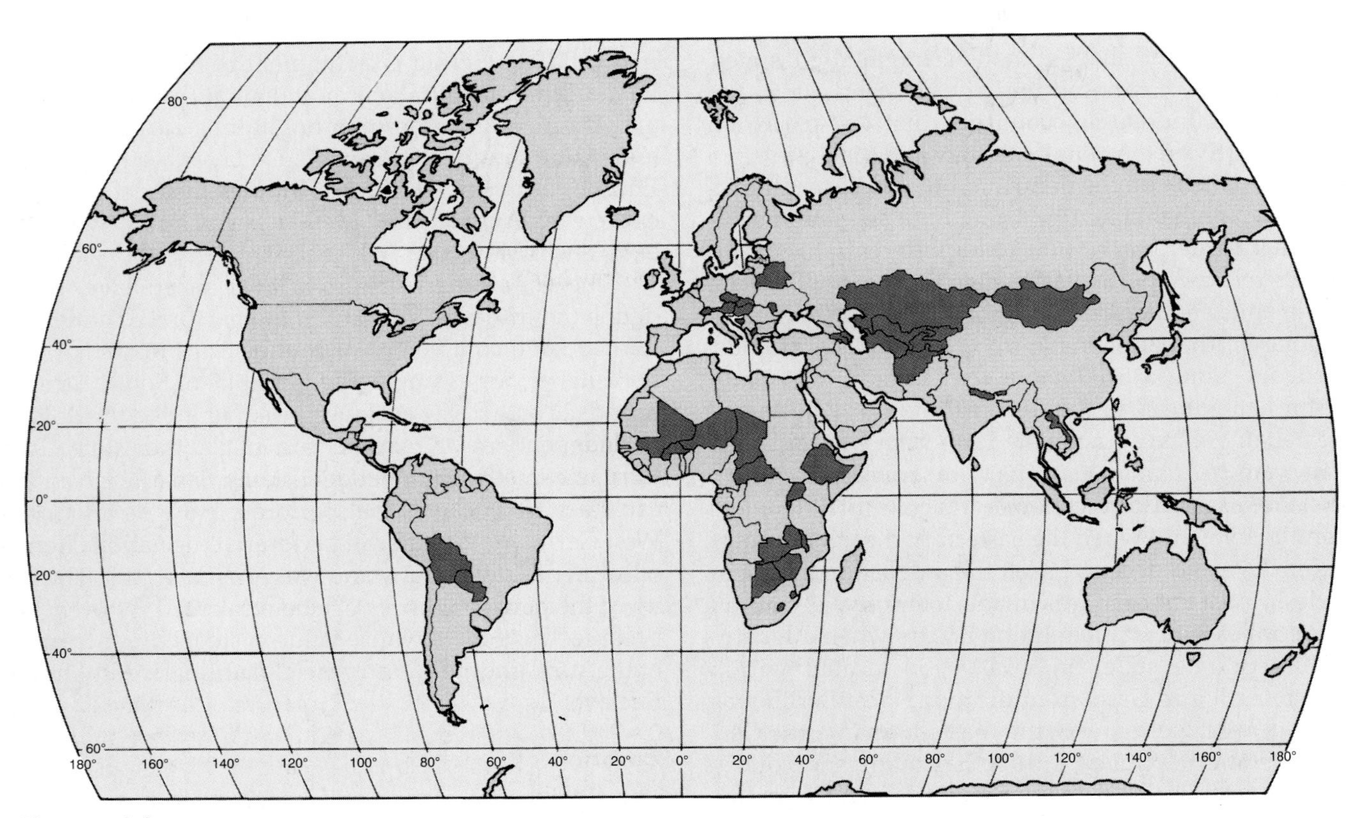

FIGURE 9.9 **Landlocked states.**

The Gnarled Politics of an Enclave

The collapse of the Soviet Union was accompanied by a resurgence of ethnic and nationalist feelings that had been long suppressed in its tightly controlled republics. Nowhere was this more evident than in Nagorno-Karabakh, a mountainous enclave in Azerbaijan that is 80% Armenian. Armenians are Christians and Azeris are Shiite Muslims, but religious differences are only partly to blame for the conflict that began to consume the region in 1988.

Both Armenia and Azerbaijan were incorporated into the Soviet Union during the early 1920s. Bowing to pressure from Turkey, which had a history of discord with its own Armenian minority group and did not want a larger than necessary Armenian presence along its border, Stalin awarded the region of Nagorno-Karabakh to Azerbaijan. Although it was accorded a separate status within that republic as an autonomous district, Armenians were resentful. The Azeris are ethnically Turkish, and Armenians claim that more than 1 million of their ethnic kin had been slaughtered by Turks during World War I.

The Soviets believed that nationalism would eventually disappear under communism, but that did not happen. Although vastly outnumbered by more than 7 million Azerbaijanis, the 180,000 Karabakh Armenians retained a strong sense of national identity buttressed by their own language and adherence to the Armenian Christian Church.

In 1987, encouraged by Mikhail Gorbachev's policy of liberalization and his stated willingness to rectify wrongs of the past, Karabakh Armenians agitated for union with Armenia. Banners displayed at mass demonstrations proclaimed, "Karabakh is a test case for *perestroika*" (restructuring). Rallies were held in Yerevan, the Armenian capital, to support Karabakh Armenians. There was talk of creating a corridor to link Nagorno-Karabakh with Armenia; less than 16 kilometers (10 mi) separate the two at the enclave's southwestern tip.

Azerbaijan responded by tightening its control of the enclave. Officials banned the teaching of Armenian history in the schools of Nagorno-Karabakh and discouraged the use of the Armenian language.

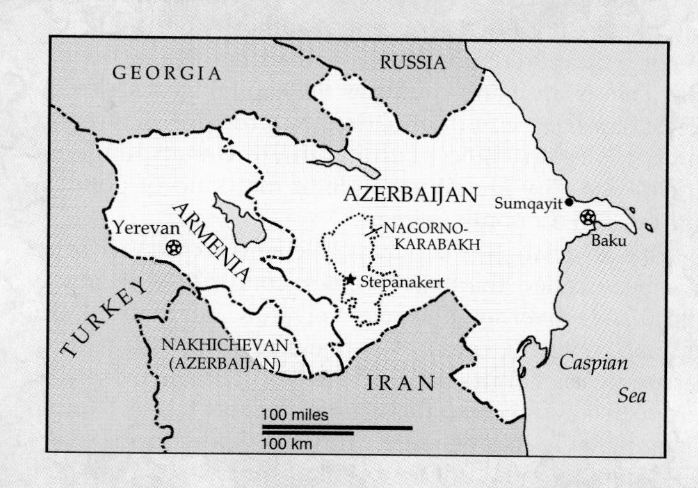

Violence against Armenians in the city of Sumqayit in February 1988 touched off an open warfare between Armenians and Azeris. Soviet troops were called in to stop the violence, but they were unable to restore order and were largely withdrawn by 1992, by which time both Armenia and Azerbaijan had become independent states. Thousands of people died; tens of thousands of both Azeris and Armenians became refugees, seeking sanctuary in their respective homelands.

Nagorno-Karabakh is now Azeri-free. Villages and towns once populated by Azeris have been resettled, mainly with Armenian refugees from Azerbaijan. Although the territory still is recognized internationally as part of Azerbaijan, it is controlled by its resident Armenians.

Like other enclaves, Nagorno-Karabakh tests the stability of the surrounding state. It demonstrates how boundary decisions made in the past may reverberate in the present, and how forces of disruption strengthen when central authority wanes. Finally, the conflict in the enclave raises the recurring political-geographic question of how peacefully to accommodate the demands of a minority when they conflict with the wishes of the majority.

buttressed by its more recent industrial development, Singapore has become a notable southeast Asian economic success. In general, history has shown that countries benefit from a location on major trade routes, not only from the economic advantages such a location carries, but also because they are exposed to the diffusion of new ideas and technologies.

Cores and Capitals

Many states have come to assume their present shape, and thus the location they occupy, as a result of growth over centuries. They grew outward from a central region, gradually expanding into surrounding territory. The original nucleus or **core area** of a state usually contains its densest population and largest cities, the most highly developed transportation system, the most developed economic base. All of these elements become less intense away from the national core. Urbanization ratios and city sizes decline, transport networks thin, and economic development is less intensive.

Easily recognized and unmistakably dominant national cores include the Paris Basin of France; London and southeastern England; Moscow and the major cities of European Russia; northeastern United States; and the Buenos Aires megalopolis in Argentina. Not all countries have such clearly defined cores—Chad, Mongolia, or Saudi Arabia, for example—and some may have two or more rival core areas. Ecuador, Nigeria, Democratic Republic of the Congo, and Vietnam are examples of multi-core states.

The capital city of a state is usually within its core region and frequently is the very focus of it, dominant not only because it is the seat of central authority but because of the concentration of population and economic functions as well. That is, in many countries the capital city is also the largest or *primate* city, dominating the structure of the entire country. Paris in France, London in the United Kingdom, and Mexico City are all example of that kind of political, cultural, and economic primacy.

This association of capital with core is common in what have been called the *unitary states,* countries with highly centralized governments, relatively few internal cultural contrasts, a strong sense of national identity, and borders that are clearly cultural as well as political boundaries. Most European cores and capitals are of this type. It is also found in many newly independent countries whose former colonial occupiers established a primary center of exploitation and administration and developed a functioning core in a region that lacked an urban structure or organized government. With independence, the new states retained the established infrastructure, added new functions to the capital, and, through lavish expenditures on governmental, public, and commercial buildings, sought to create a prestigious symbol of nationhood.

In *federal states,* associations of more or less equal provinces or states with strong regional governmental responsibilities, the capital city may have been newly created to serve as the administrative center. Although part of a generalized core region of the country, the capital was not its largest city and acquired few of the additional functions to make it so. Ottawa, Canada; Washington, D.C.; and Canberra, Australia, are examples (Figure 9.10).

All other things being equal, a capital located in the center of the country provides equal access to the government, facilitates communication to and from the political hub, and enables the government to exert its authority easily. Many capital cities, such as Washington, D.C., were centrally located when they were designated as seats of government, but lost their centrality as the state expanded.

Some capital cities have been relocated outside of peripheral national core regions, at least in part to achieve the presumed advantages of centrality. Two examples of such relocation are from Karachi inland to Islamabad in Pakistan, and from Istanbul to Ankara, in the center of Turkey's territory. A particular type of relocated capital is the *forward-thrust capital* city, one that has been deliberately sited in a state's frontier zone to signal the government's awareness of regions away from the core and its interest in encouraging more uniform development. In the late 1950s, Brazil moved its capital from Rio de Janeiro to the new city of Brasília to demonstrate its intent to develop the vast interior of the country. The West African country of Nigeria has been building the new capital of Abuja near its geographic center since the late 1970s, with the relocation there of government offices and foreign embassies in the early 1990s. The British colonial government relocated Canada's capital six times between 1841 and 1865, in part seeking centrality to the mid-19th century population pattern and in part seeking a location that bridged that country's cultural divide (Figure 9.11).

Boundaries: The Limits of the State

We noted earlier that no portion of the earth's land surface is outside the claimed control of a national unit, that even uninhabited Antarctica has had territorial claims imposed upon it (see Figure 9.3). Each of the world's states is separated from its neighbors by *international boundaries,* or lines that establish the limit of each state's jurisdiction and authority. Boundaries indicate where the sovereignty of one state ends and that of another begins.

Within its own bounded territory, a state administers laws, collects taxes, provides for defense, and performs other such governmental functions. Thus, the location of the boundary determines the kind of money people in a given area use, the legal code to which they are subject, the army they may be called upon to join, and the language and perhaps the religion children are taught in school. These examples suggest how boundaries serve as powerful reinforcers of cultural variation over the earth's surface.

Territorial claims of sovereignty, it should be noted, are three-dimensional. International boundaries mark not only the outer limits of a state's claim to land (or water) surface, but are also projected downward to the center of the earth in accordance with international consensus allocating rights to subsurface resources. States also project their sovereignty upward, but with less certainty because of a lack of agreement on the upper limits of territorial airspace. Properly viewed, then, an international boundary is a line without breadth; it is a vertical interface between adjacent state sovereignties.

Before boundaries were delimited, nations or empires were likely to be separated by *frontier zones,* ill-defined and fluctuating areas marking the effective end of a state's authority. Such zones were often uninhabited or only sparsely populated and were liable to change with shifting settlement patterns. Many present-day international boundaries lie in former frontier zones, and in that sense, the boundary line has replaced the broader frontier as a marker of a state's authority.

Classification of Boundaries

Geographers have traditionally distinguished between "natural" and "artificial" boundaries. **Natural** (or *physical*) **boundaries** are those based on recognizable physiographic features such as mountains, rivers, and lakes. Although they might seem to be attractive as borders because they actually exist in the landscape and are visible dividing elements, many natural boundaries have proved to be unsatisfactory. That is, they do not effectively separate states.

FIGURE 9.10 Canberra, the planned capital of Australia, was deliberately sited away from the country's two largest cities, Sydney and Melbourne. Planned capitals are often architectural showcases, providing a focus for national pride.
Australian Information Service.

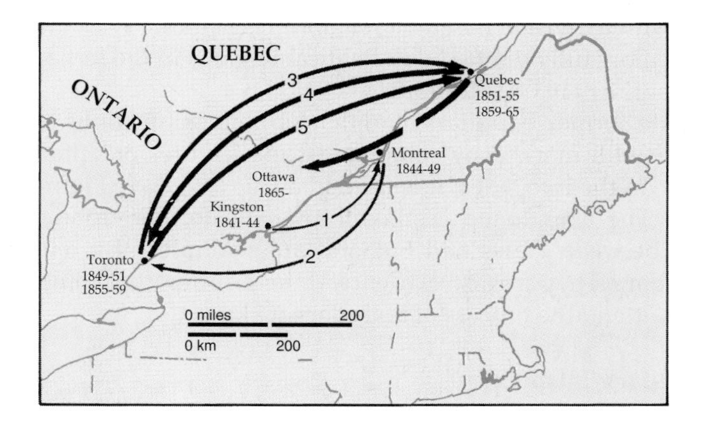

FIGURE 9.11 **Canada's migratory capital.** Kingston was chosen as the first capital of the united Province of Canada in preference to either Quebec, capital of Lower Canada, or Toronto, that of Upper Canada. In 1844, governmental functions were relocated to Montreal, where they remained until 1849, after which they shifted back and forth—as the map indicates—between Toronto and Quebec. An 1865 session of the provincial legislature was held in Ottawa, the city that became the capital of the Confederation of Canada in 1867.
Redrawn with permission from David B. Knight, *A Capital for Canada* (Chicago: University of Chicago, Department of Geography, Research Paper No. 182, 1977), Figure 1, p. vii.

Many international boundaries lie along mountain ranges, for example in the Alps, Himalayas, and Andes, but while some have proved to be stable, others have not. Mountains are rarely total barriers to interaction. Although they do not invite movement, they are crossed by passes, roads, and tunnels. High pastures may be used for seasonal grazing, and the mountain region may be the source of water for hydroelectric power. Nor is the definition of a boundary along a mountain range a simple matter. Should it follow the crests of the mountains or the *water divide* (the line dividing two drainage areas)? The two are not always the same. Border disputes between China and India are in part the result of the failure of mountain crests and headwaters of major streams to coincide (Figure 9.12).

FIGURE 9.12 Several international borders run through the jumble of the Himalayas. The mountain boundary between India and China has long been in dispute.
Fred Bavendam/Peter Arnold, Inc.

Rivers can be even less satisfactory as boundaries. In contrast to mountains, rivers foster interaction. River valleys are likely to be agriculturally or industrially productive, and to be densely populated. For example, for hundreds of miles the Rhine River serves as an international boundary in Western Europe. It is also a primary traffic route lined by chemical plants, factories, blast furnaces, and power stations, and dotted by the castles and cathedrals that make it one of Europe's major tourist attractions. It is more a common intensively used resource than a barrier in the lives of the states it borders.

With any river, it is not clear precisely where the boundary line should lie: along the right or left bank, along the center of the river, or perhaps along the middle of the navigable channel? Even an agreement in accordance with international custom that the boundary be drawn along the main channel may be impermanent if the river changes its course, floods, or dries up.

The alternative to natural boundaries are artificial or **geometric boundaries.** Frequently delimited as sections of parallels of latitude or meridians of longitude, they are found chiefly in Africa, Asia, and the Americas. The western portion of the United States–Canada border, which follows the 49th parallel, is an example of a geometric boundary. Many such boundaries were established when the areas in question were colonies, the land was only sparsely settled, and detailed geographic knowledge of the frontier region was lacking.

Boundaries can also be classified according to whether they were laid out before or after the principal features of the cultural landscape developed. An **antecedent boundary** is one drawn across an area before it is well populated, that is, before most of the cultural landscape features developed. To continue our earlier example, the western portion of the United States–Canada boundary is such an antecedent line, having been established by a treaty between the United States and Great Britain in 1846.

Boundaries drawn after the development of the cultural landscape are termed **subsequent.** One type of subsequent boundary is **consequent** (also called *ethnographic*), a border drawn to accommodate existing religious, linguistic, ethnic, or economic differences between countries. An example is the boundary drawn between Northern Ireland and Eire (Ireland). Subsequent **superimposed boundaries** may also be forced upon existing cultural landscapes, a country, or a people by a conquering or colonizing power that is unconcerned about preexisting cultural patterns. The colonial powers in 19th-century Africa superimposed boundaries upon established African cultures without regard to the tradition, language, religion, or tribal affiliation of those whom they divided (see Figure 9.5).

When Great Britain prepared to leave the Indian subcontinent after World War II, it was decided that two independent states would be established in the region: India and Pakistan. The boundary between the two countries, defined in the partition settlement of 1947, was thus both a *subsequent* and a *superimposed* line. As millions of Hindus migrated from the northwestern portion of the subcontinent to seek homes in India, millions of Muslims left what would become India for Pakistan. In a sense, they were attempting to insure that the boundary would be *consequent,* that is, that it would coincide with a division based on religion. This boundary example is more fully discussed in Political Regions in the Indian Subcontinent in Chapter 13.

If a former boundary line that no longer functions as such is still marked by some landscape features or differences on the two sides, it is termed a *relic boundary* (Figure 9.13). The abandoned castles dotting the former frontier zone between Wales and England are examples of a relic boundary. They are also evidence of the disputes that sometimes attend the process of boundary making.

Boundary Disputes

Boundaries create many possibilities and provocations for conflict. Since World War II, almost half of the world's sovereign states have been involved in border disputes with neighboring countries. Just like householders, states are far more likely to have disputes with their neighbors than with

FIGURE 9.13 Like Hadrian's Wall in the north of England or the Great Wall of China, The Berlin Wall was a demarcated boundary. Unlike them, it cut across a large city and disrupted established cultural patterns. The Berlin Wall, therefore, was a *subsequent superimposed* boundary. The dismantling of the wall in 1990 marked the reunification of Germany; any of it that remains standing as an historic monument is a *relic* boundary.
Stephanie Maze/National Geographic Image.

more distant parties. It follows that the more neighbors a state has, the greater the likelihood of conflict.

Although the causes of boundary disputes and open conflict are many and varied, they can reasonably be placed into four categories.

1. **Positional disputes** occur when states disagree about the interpretation of documents that define a boundary and/or the way the boundary was delimited. Such disputes typically arise when the boundary is antecedent, preceding effective human settlement in the border region. Once the area becomes populated and gains value, the exact location of the boundary becomes important.

   The boundary between Argentina and Chile, originally defined during Spanish colonial rule, was to follow the highest peaks of the southern Andes and the watershed divides between east- and west-flowing rivers. Because the terrain had not been adequately explored, it wasn't apparent that the two do not always coincide. In some places, the water divide is many miles east of the highest peaks, leaving a long, narrow area of several hundred square miles in dispute (Figure 9.14). During the late 1970s, Argentina and Chile nearly went to war over the disputed territory, whose significance had been increased by the discovery of oil and natural gas deposits.
2. **Territorial disputes** over the ownership of a region often, though not always, arise when a boundary that has been superimposed on the landscape divides an ethnically homogeneous population. Each of the two states then has some justification for claiming the territory inhabited by the ethnic group in question. We noted earlier that a single nation may be dispersed across several states (see Figure 9.4). Conflicts can arise if people of one state want to annex a territory whose population is ethnically related to that of the state but now subject to a foreign government. This type of expansionism is called **irredentism.** In the late 1930s, Hitler used the existence of German minorities in Czechoslovakia and Poland to justify Germany's occupation of those countries. More recently, Somalia has had many border clashes with Ethiopia over the rights of Somalis living in that country (Figure 9.15).
3. Closely related to territorial conflicts are **resource disputes.** Neighboring states are likely to covet the resources—whether they be valuable mineral deposits, fertile farmland, or rich fishing grounds—lying in border areas and to disagree over their use. In recent years, the United States has been involved in disputes with both its immediate neighbors: with Mexico over the shared resources of the Colorado River and Gulf of Mexico, and with Canada over the Georges Bank fishing grounds in the Atlantic Ocean.

   One of the causes of the 1990–1991 war in the Persian Gulf was the huge oil reservoir known as the Rumaila field (Figure 9.16). More than 80 kilometers (50 mi) long, the oil field lies mainly under Iraq with a small extension into Kuwait. Because the two countries had been unable to agree on percentages of ownership, or a formula for sharing production costs and revenues, Kuwait pumped oil from Rumaila without any agreement. Iraq helped justify its invasion of Kuwait by contending that the latter had been stealing Iraqi oil in what amounted to economic warfare.
4. **Functional disputes** arise when neighboring states disagree over policies to be applied along a boundary. Such policies may concern immigration, the movement of traditionally nomadic groups, customs regulations, or land use. U.S. relations with Mexico, for example, have been affected by the increasing number of illegal aliens and the flow of drugs entering the United States from Mexico (Figure 9.17a&b). In Central America, relations between Honduras and El Salvador, two countries that have long disputed their common boundary, worsened in the late 1970s, when Honduras expelled Salvadoran farmers who had illegally occupied available agricultural land in western Honduras.

Geopolitical Assessments

Many analysts have generalized about the relationship between national power and a variety of geographic elements including location, territorial extent, population, and resources. **Geopolitics** is a branch of political geography that considers the economic, political, and military value of space in order to present recommendations about courses of action in international relations best designed to advance

FIGURE 9.14 **Areas of international dispute in Latin America.** Among the countries still disputing the precise location of their boundaries are Venezuela and Guyana, and Honduras and El Salvador. Guatemala claims all of Belize, and in official government documents, Bolivia claims that it "has a right to a coastline" at Chile's expense. In 1998, Peru and Ecuador ended a border dispute that had lasted nearly 60 years. The two countries agreed to a demarcation of a disputed 68-kilometer (48-mi) stretch of land in the Condor mountain range. Also in 1998, Agentina and Chile signed an accord settling their last remaining Andean territorial dispute.

Redrawn from Ernst Griffin, "Latin America" in D. Gordon Bennett, *Tension Areas of the World,* 1982, Park Press, Champaign, IL. Used by permission of the author.

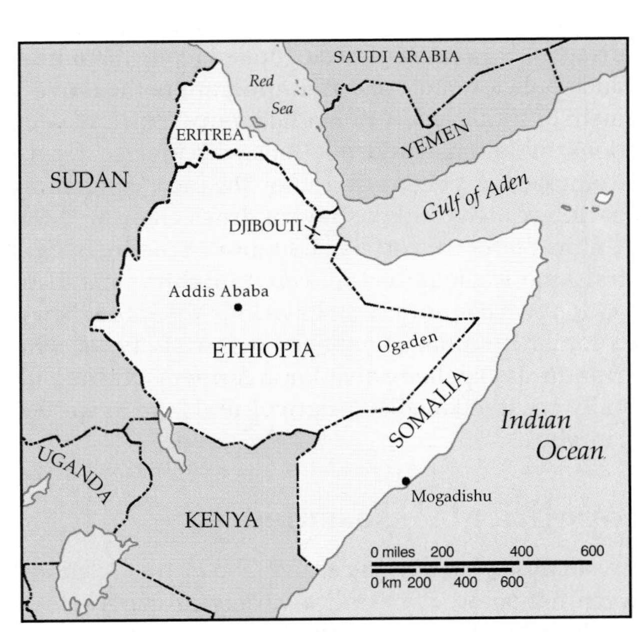

FIGURE 9.15 **Irredentism in East Africa.** In the 1970s, Somalia claimed the eastern portion of Ethiopia, Ogaden, because it is peopled mostly by Somalis. Although Somali army troops and ethnic Somali rebels in Ethiopia were defeated in 1978, guerrilla fighting in Ogaden continued for another decade. More than 1 million Somalis from Ogaden became refugees in Somalia.

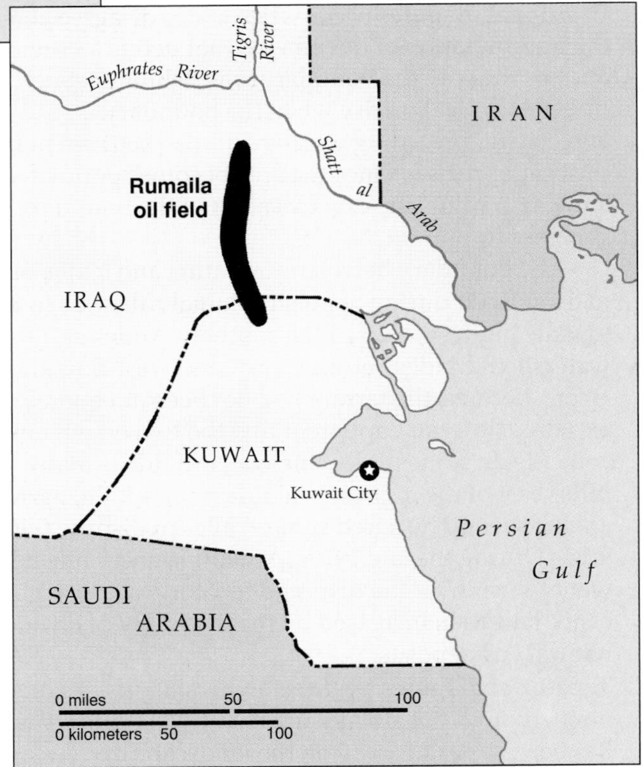

FIGURE 9.16 **The Rumaila oil field.** One of the world's largest oil reservoirs, Rumaila straddles the Iraq-Kuwait border. Iraqi grievances over Kuwaiti drilling were partly responsible for Iraq's invasion of Kuwait in 1990.

(a)

(b)

FIGURE 9.17 (a) To stem the flow of undocumented migrants entering California from Baja California, the United States, in 1993, constructed a fence 3 meters (10 ft) high along the border. (b) Protesters on the Mexican side of the border expressed their anger over the new barrier.
©San Diego Union Tribune/John Nelson.

the cause of national security, projection of power and influence, or territorial aggrandizement.

Modern geopolitics was rooted in the concern of an eminent English geographer, Halford Mackinder, with the balance of power in the world at the beginning of the 20th century. Believing that the major powers would be those that controlled the land, not the seas, he developed what came to be known as the **heartland theory.** The greatest land power, he argued, would be sited in Eurasia, the "World-Island" containing the world's largest landmass in both area and population. Its interior or heartland, he warned, would provide a base for world conquest, and Eastern Europe was the core of that heartland (Figure 9.18). Mackinder warned, "Who rules East Europe commands the Heartland, who rules the Heartland commands the World-Island, who rules the World-Island commands the World."[1]

Developed in a century that saw first Germany and then the Soviet Union dominate East Europe, and the decline of Britain as a superpower, Mackinder's theory impressed many. Near the end of World War II, the theory was modified by Nicholas Spykman, who agreed with Mackinder that Eurasia was the likely base for potential world domination, but argued that the coastal fringes of the landmass, not its heartland, were the key. The coastal margins, or rimland, contained dense populations, abundant resources, and had controlling access both to the seas and to the continental interior. Spykman's **rimland theory,** published in 1944, stated, "Who controls the Rimland rules Eurasia, who rules Eurasia controls the destinies of the world."[2] The rimland has tended throughout history to be politically fragmented, and Spykman believed that it would be to the advantage of both the United States and USSR if it remained that way.

By the end of World War II, the Heartland was equated in American eyes with the USSR. To prevent Soviet domination of the World-Island, U.S. foreign policy during the Cold War was based on the notion of **containment,** or confining the USSR within its borders by means of a string of regional alliances in the Rimland: the North Atlantic Treaty Organization (NATO) in Western Europe, the Central Treaty Organization (CENTO) in West Asia, and the Southeast Treaty Organization (SEATO). Military intervention was deemed necessary where communist expansion, whether Soviet or Chinese, was a threat—in Berlin, the Middle East, and Korea, for example.

A simple spatial model, the **domino theory,** was used as an adjunct to the policy of containment. According to this analogy, adjacent countries are lined up like dominoes; if one topples, the rest will fall. In the early 1960s, the domino theory was invoked to explain and justify U.S. intervention in Vietnam, and in the 1980s the theory was applied to involvement in Central America. The fear that war among the Serbs, Croatians, and Bosnians in Bosnia-Herzegovina would lead to the downfall of that state and spread into other parts of the former Yugoslavia led in 1995 to NATO airstrikes against the Serbs, a peace agreement forged with American help in Dayton, Ohio, and stationing of United Nations peacekeeping forces in Bosnia.

These and other models aimed at realistic assessments of national power and foreign policy stand in contrast to "organic

[1]Halford J. Mackinder, *Democratic Ideals and Reality* (London: Constable, 1919), p. 150.

[2]Nicholas J. Spykman, *The Geography of the Peace* (New York: Harcourt Brace, 1944), p. 43.

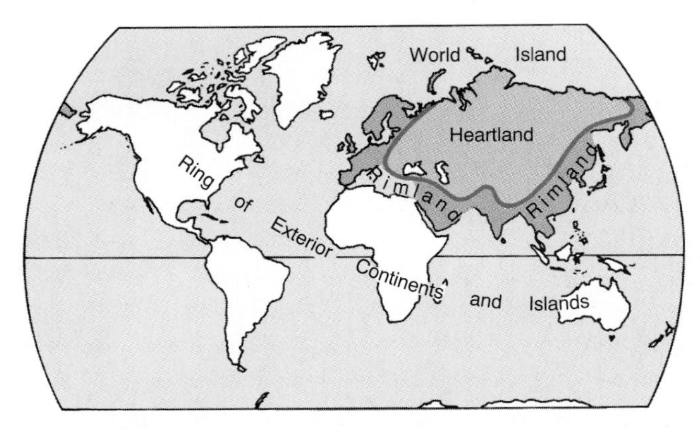

FIGURE 9.18 **Geopolitical viewpoints.** Both Mackinder and Spykman believed that Eurasia possessed strategic advantages, but they disagreed on whether its heartland or rimland provided the most likely base for world domination.

state theory" based on the 19th-century idea of the German geographer Friedrich Ratzel (1844–1904) that the state was an organism conforming to natural laws and forced to grow and expand into new territories (*Lebensraum*) in order to secure the resources needed for survival. Without that growth, the state would wither and die. These ideas, later expanded in the 1920s by the German Karl Haushofer (1869–1946) as *Geopolitik,* were used by the Nazi party as the presumed intellectual basis for wartime Germany's theories of race superiority and need for territorial conquest. Repudiated by events and Germany's defeat, *Geopolitik* for many years gave bad odor to any study of geopolitics, which only recently has again become a serious subfield of political geography.

Many analysts believe that the older geopolitical concepts no longer apply to the current world situation. A number of developments have rendered them obsolete: the dissolution of the USSR and the presumed end of the Cold War; the proliferation of nuclear technology; and the rise of Japan, China, and Western Europe to world power status. Geopolitical reality is now seen less in terms of military advantage and confrontation—the East-West rivalry of the Cold War era—and more as a reflection of two other forms of competition.

One is economic rivalry between economic core countries and emerging peripheral states. The contrasting terms *North* and *South* are commonly used to emphasize the distinctions between the rich, advanced, developed countries of the Northern Hemisphere (to which Australia and New Zealand in the Southern Hemisphere are added) and all the rest of the world, the less developed countries of the South.

The other is competition rooted in more fundamental conflicts between different "civilizations." It has been suggested that the world will increasingly be shaped by the interactions and conflicts among seven or eight major civilizations: Western, Confucian, Japanese, Islamic, Hindu, Slavic, Latin American, and possibly African. The differences between such civilizations, it is thought, are basic and antagonistic, rooted in enduring differences of history, language, culture, tradition, and religion. These differences, the argument runs, are less easily resolved than purely political and economic ones and underlie such recent clashes as Indian rivalries between Hindus and Muslims, those of Sri Lanka between Tamils and Buddhists, conflicts in former Yugoslavia and between Armenians and Azeris in the Caucasus, and between and within other states and areas where "civilizations" come in contact and competition.

Centripetal Forces: Promoting State Cohesion

At any moment in time, a state is characterized by forces that promote unity and national stability and by others that disrupt them. Political geographers refer to the former as **centripetal forces.** These are factors that bind together the people of a state, that enable it to function and give it strength. **Centrifugal forces,** on the other hand, destabilize and weaken a state. If centrifugal forces are stronger than those promoting unity, the very existence of the state will be threatened. In the sections that follow, we examine four forces—nationalism, unifying institutions, effective organization and administration of government, and systems of transportation and communication—to see how they can promote cohesion.

Nationalism

One of the most powerful of the centripetal forces is **nationalism,** an identification with the state and the acceptance of national goals. Nationalism is based on the concept of allegiance to a single country; it thus fosters a feeling of collective distinction from all other peoples and lands. It is an emotion that provides a sense of identity and loyalty and of collective distinction from all other peoples and lands.

States purposely try to instill feelings of allegiance in their constituents, for such feelings give the political system strength. People who have such allegiance are likely to accept the rules governing behavior in the area and to participate in the decision-making process establishing those rules. In addition, a sense of unity binding the people of a state together is necessary to overcome the divisive forces present in most societies. Not everyone, of course, will feel the same degree of commitment or loyalty. The important consideration is that the majority of a state's population accept its ideologies, adhere to its laws, and participate in its effective operation. For many countries, such acceptance and adherence has come only recently and partially; in some, it is frail and endangered.

We noted earlier that true nation-states are rare; in only a few countries do the territory occupied by the people of a particular nation and the territorial limits of the state coincide. Most countries have more than one culture group that considers itself separate in some important way from other citizens. In a multicultural society, nationalism helps integrate different groups into a unified population. This has occurred in countries like the United States and Switzerland, where different culture groups have joined together to create political entities commanding the loyalties of all of their citizens.

On-Line Political Geography

As the chapter content indicates, the range of topics of political geographic interest is great. The number of websites addressing those interests is also large, and no more than a few can be suggested here. Most of those cited below are either guides to a variety of those general or specialized political sites or have valuable links to other organizational home pages.

At the international level, the *Official WEB Site Locator for the United Nations System* is of obvious interest. It is a catalog of UN system websites linking directly to agencies by title, by subject matter interest, and by a search program. Links to other international organizations are also provided.

http://www.unsystem.org/

Another good place to start locating international agency and foreign governmental information is at the University of Colorado at Boulder *Government Publications Library* site. Although its connections to publications and agencies at the state and U.S. national level are extensive and valuable, the library's superb collection of links at the foreign country and international levels makes it an indispensable reference tool. For information about countries other than the United States, follow its options through "Foreign Resources" at

http://www-libraries.colorado.edu/ps/gov/for/foreign.htm

Access "International Resources" for links to various international agencies, such as the European Union, World Bank, and the United Nations and several of its organizations.

http://www-libraries.colorado.edu/ps/gov/int/internat.htm

The World-Wide Web Virtual Library page on *International Affairs Resources* provides extensive links by type (e.g., periodicals, statistical information); by source (e.g., primary sources, university centers); and by topic (foreign policy, international trade, peacekeeping, etc.).

http://www.pitt.edu/~ian/ianres.html

For information on more than 250 countries, check *The World Factbook 1997* published by the U.S. Central Intelligence Agency. Geographic, economic, demographic, government, and other data are provided for each country for the latest year available.

http://www.etown.edu/home/selchewa/international_studies/ firstpag.htm

In addition to the general sites listed above, the Internet has a number of special topic sites related to political geography.

The *Law of the Sea and Ocean Affairs* page maintained by the Council on Ocean Law contains an overview and the full text of the UN Convention on the Law of the Sea, a list of ratifying countries, and links to such related conferences as that on high seas fisheries.

http://www.oceanlaw.org/index.html

The North Atlantic Treaty Organization (NATO) maintains this official home page.

http://www.nato.int/

Political Boundary Links, an index of websites related to territorial disputes and political boundaries, is maintained by the International Boundaries Research Unit at Durham University.

http://www-ibru.dur.ac.uk/links.html

The World Trade Organization, established in 1995, is the principal agency of the world's multilateral trading system. Its home page includes access to documents discussing international conferences and agreements, reviewing its own publications, and summarizing the current state of world trade.

http://www.wto.org/

Elections and Electoral Systems by Country, a page of pointers to the latest worldwide election news, is maintained as part of the Political Science Resources page at the University of Keele in the United Kingdom. Information is available on more than 40 countries and can include election results, referendums, and electoral information.

http://www.psr.keele.ac.uk/election.htm

In addition, Keele's *Political Science Resources Guide* is among the better guides to worldwide political science resources.

http://www.psr.keele.ac.uk/htm

Most U.S. government agencies have their own home pages. The *FedWorld Information Network* is maintained by the National Technical Information Service to help people access federal government information online.

http://www.fedworld.gov/

The Department of State *Foreign Affairs Network* (*DOSFAN*) maintained by the University of Illinois at Chicago is an indispensable site for research into U.S. foreign policy.

http://dosfan.lib.uic.edu/index.html

Redistricting Rulings is an up-to-date collection of recent U.S. Supreme Court decisions on legislative redistricting.

http://www.nando.net/insider/redistrict/redistrict.html

The University of Colorado at Boulder Libraries site mentioned previously also provides extensive links to state government and municipality resources and pages.

http://www-libraries.colorado.edu/ps/gov/st/allstate.htm

The *Government of Canada* site provides links through its federal organizations listings to all departments and agencies of the national government.

http://Canada.gc.ca/

In addition, you can find an updated guide to Canadian government information by subject, province, and municipality from the *University of Waterloo Electronic Library.*

http://dsp-psd.pwgsc.gc.ca/dsp-psd/Reference/cgii_index-e.html

States promote nationalism in a number of ways. *Iconography* is the study of the symbols that help unite people. National anthems and other patriotic songs, flags, national flowers and animals, colors, and rituals and holidays are all developed as symbols of a state in order to promote nationalism and attract allegiance (Figure 9.19). By ensuring that all citizens, no matter how diverse the population may be, will have at least these symbols in common, they impart a sense of belonging to a political entity called, for example, Japan or Canada. In some countries, certain documents, such as the Magna Charta in England or the Declaration of Independence in the United States, serve the same purpose. Royalty may fill the need: in Sweden, Japan, and Great Britain, the monarchy functions as the symbolic focus of allegiance. Symbols such as these are significant insofar as ideologies and beliefs are an important aspect of every culture. When a culture is very heterogeneous, composed of people with different customs, religions, and languages, belief in the national unit can help weld them together.

FIGURE 9.19 The ritual of the Pledge of Allegiance is just one way in which schools in the United States seek to instill a sense of national identity in students.
© Michael Siluk.

Unifying Institutions

A number of institutions help to develop the sense of commitment and cohesiveness essential to the state. Schools, particularly elementary schools, are among the most important of these. Children learn the history of their own country and relatively little about other countries. Schools are expected to inculcate the society's goals, values, and traditions; allegiance to the state is accepted as the norm. As a rule, schools teach youngsters to identify with their country rather than with the world or with humanity as a whole.

Other institutions that promote nationalism are the armed forces and, sometimes, a state church. The armed forces are of necessity taught to identify with the state. They see themselves as protecting the state's welfare from what are perceived to be its enemies. In some countries, the religion of the majority of the people may be designated a state church. In such cases, the church sometimes becomes a force for cohesion, helping to unify the population. This is true of the Roman Catholic Church in the Republic of Ireland, Islam in Pakistan, and Judaism in Israel. In countries like these, the religion and the church are so identified with the state that belief in one is transferred to allegiance to the other.

The schools, the armed forces, and the church are just three of the institutions that teach people what it is like to be members of a state. As institutions, they operate primarily on the level of the sociological subsystem of culture, helping to structure the outlooks and behaviors of the society. But by themselves, they are not enough to give cohesion, and thus strength, to a state.

Organization and Administration

A further bonding force is public confidence in the effective organization of the state. Can it provide security from external aggression and internal conflict? Are its resources distributed and allocated in such a way as to be perceived to promote the economic welfare of all its citizens? Are all citizens afforded equal opportunity to participate in governmental affairs (see "Legislative Women")? Do institutions that encourage consultation and the peaceful settlement of disputes exist? How firmly established are the rule of law and the power of the courts? Is the system of decision making responsive to the people's needs?

The answers to the questions, and the relative importance of the answers, will vary from country to country, but they and similar ones are implicit in the expectation that the state will, in the words of the Constitution of the United States, "establish justice, insure domestic tranquility, provide for the common defense, (and) promote the general welfare. . . ." If those expectations are not fulfilled, the loyalties promoted by national symbols and unifying institutions may be weakened or lost.

Transportation and Communication

A state's transportation network fosters political integration by promoting interaction between areas and by joining them economically and socially. The role of a transportation network in uniting a country has been recognized since ancient times. The saying that all roads lead to Rome had its origin in the impressive system of roads that linked Rome to the rest of the empire. Centuries later, a similar network was built in France, linking Paris to the various departments of the country. Often the capital city is better connected to other cities than the outlying cities are to one another. In France, for example, it can take less time to travel from one city to another by way of Paris than by direct route.

Roads and railroads have played a historically significant role in promoting political integration. In the United States and Canada, they not only opened up new areas for settlement, but they increased interaction between rural and urban areas. Because transportation systems play a major role in a state's economic development, it follows that the more economically advanced a country is, the more extensive its transport network is likely to be. At the same

Legislative Women

Women, a majority of the world's population, in general get a raw deal in the allocation of such resources as primary and higher education, employment opportunities and income, and health care. That their lot is improving is encouraging. In every developing country, women have been closing the gender gap in literacy, school enrollment, and acceptance in the job market.

But in the political arena, where power ultimately lies, women's share of influence is increasing only slowly and selectively. In 1995, only nine countries out of a world total of some 200 had women as heads of government: presidents or prime ministers. Nor did they fare much better as members of parliaments. Women in early 1998 held just 12% of all the seats in the world's legislatures.

In only the 19 countries noted here did women occupy more than 20% of national legislative seats in 1998. Sweden was the most feminist, with 40% female members. In no country were women a legislative majority.

Asia
China (21%)
North Korea (20%)
Vietnam (26%)

Europe
Denmark (33%)
Finland (34%)
Norway (36%)
Sweden (40%)
Austria (25%)
Germany (26%)
Netherlands (28%)
Switzerland (20%)
Spain (20%)

Africa
Eritrea (21%)
Mozambique (25%)
South Africa (24%)

Oceania
Australia (21%)
New Zealand (29%)

North America
Canada (21%)

South America
Argentina (23%)

Nine of those 19 countries are in Europe, but in many more European countries, women occupy a tiny minority of legislative seats. This includes both established democracies of Northern and Western Europe and virtually all of the countries of Southern and Eastern Europe. Typical percentages are those for France (9%), the United Kingdom (12%), and Belgium (16%).

Many countries are witnessing increased discontent with the proportion of women in legislatures. In a play on words, banners carried by women protesting outside France's newly elected National Assembly in 1994 read, "You love us as mothers (*mères*) or whores; why not as mayors (*maires*) or deputies?" Political parties from Mexico to China have tried to correct female under-representation, usually by setting quotas for women candidates, and a few governments have tried to require their political parties to improve their balance. The French Socialists, for example, decided that 30% of their candidates in the 1998 elections would be women.

Quotas are controversial, however, and often are viewed with disfavor even by avowed feminists. Some argue that quotas imply that women cannot match men on merit alone. Others fear that other groups (e.g., homosexuals, religious groups, ethnic minorities) will seek quotas to assure their fair representation in legislatures.

time, the higher the level of development, the more money there is to be invested in building transport routes. In other words, the two reinforce one another.

Transportation and communication, while encouraged within a state, are frequently curtailed or at least controlled between states as a conscious device for promoting state cohesion through limitation on external spatial interaction (Figure 9.20). The mechanisms of control include restrictions on trade through tariffs or embargoes, legal barriers to immigration and emigration, and limitations on travel through passports and visa requirements.

Centrifugal Forces: Challenges to State Authority

State cohesion is not easily achieved or, once gained, invariably retained. Destabilizing centrifugal forces are ever-present, sowing internal discord and challenges to the state's authority. Transportation and communication may be hindered by a country's shape or great size, leaving some parts of the country not well integrated with the rest. A state that is not well organized or administered stands to lose the loyalty of its citizens. Institutions that in some states promote unity can be a divisive force in others. Organized religion, for example, may compete with the state for people's allegiance, it may oppose state policies or exacerbate tensions between the state and its neighbors.

We previously identified four types of relationships between states and nations: a nation-state, a multinational state, a part-nation state, and a stateless nation (see Figure 9.4). Only nation-states have cohesion between the nation and the state. That is, most of the residents belong to one nationality *and* almost all people of that nationality live in the state. Denmark and Japan were cited as examples. Compared to other types of countries, nation-states tend to lack many of the centrifugal forces that can weaken or destroy a state's unity and stability.

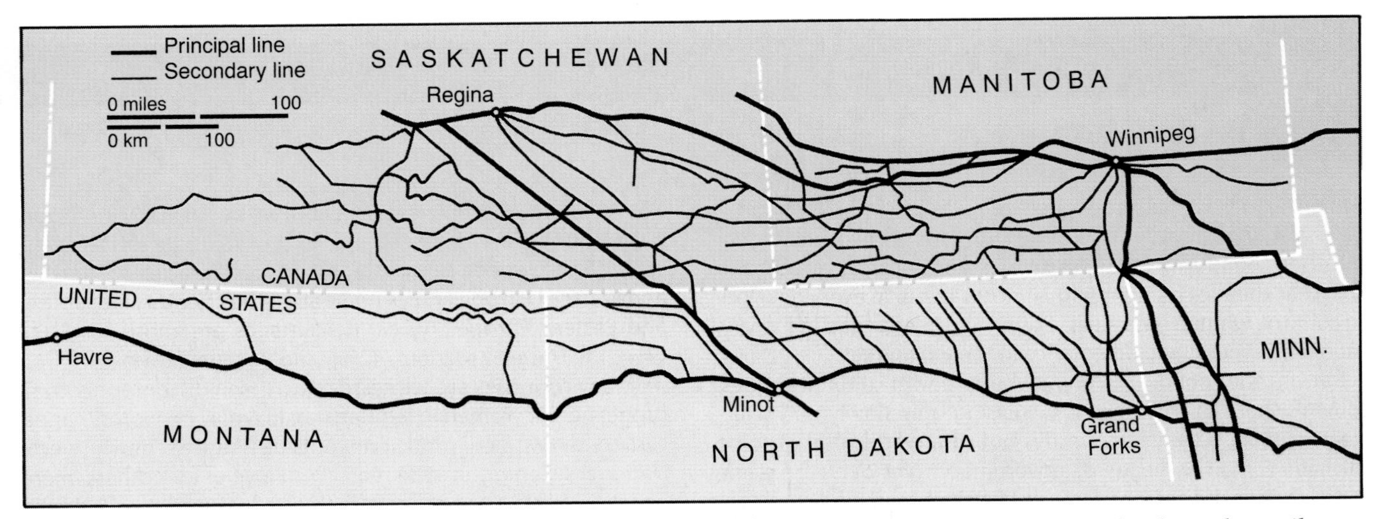

FIGURE 9.20 **Canadian–United States railroad discontinuity.** Canada and the United States developed independent railway systems connecting their respective prairie regions with their separate national cores. Despite extensive rail construction during the 19th and early 20th centuries, the pattern that emerged even before recent track abandonment was one of discontinuity at the border. Note how the political boundary restricted the ease of spatial interaction between adjacent territories. Many branch lines approached the border, but only eight crossed it. In fact, for more than 480 kilometers (300 mi), no railway bridged the boundary line. The international border—and the cultural separation it represents—inhibits other expected degrees of interaction. Telephone calls between Canadian and U.S. cities, for example, are far less frequent than would be expected if distance alone were the controlling factor.

Centrifugal forces are particularly strong in states containing two or more nationalities, each occupying a distinct territory. Such states might be characterized by any or all of the following: unassimilated minorities; racial or ethnic conflict; contrasting cultures; a multiplicity of languages and religions; and an unequal distribution of wealth.

A country whose population is bound not by a shared sense of nationalism but is split by several local primary allegiances suffers from **subnationalism.** That is, many people give their primary allegiance to traditional groups or nations that are smaller than the population of the entire state. Subnationalism can be a disruptive centrifugal force, particularly if a group believes that its right to **self-determination** has not been achieved. Self-determination is the concept that nationalities have the right to govern themselves in their own state or territory, a right to self-rule. They may try to carve out a new nation-state from portions of existing areas.

Any country that contains one or more important national minorities is susceptible to nationalist challenges from within its borders if the minority group has an explicit territorial identification. In its intense form, **regionalism**—a strong minority group self-awareness and identification with a region—can be expressed politically as a desire for more autonomy (self-government) or even separation from the rest of the country. It is prevalent in many parts of the world today and has created curents of unrest within many countries, even long-established ones.

Canada, for example, houses a powerful secessioinist movement in French-speaking Quebec, the country's largest province. In October 1995, a referendum to secede from Canada and become a sovereign country failed by a scant margin (49% yes, 51% no). Quebec's nationalism is fueled by strong feelings of collective identity and distinctiveness, and by a desire to protect its language and culture. Additionally, separatists believe that the province, which has ample resources and one of the highest standards of living in the industrialized world, could manage as well or better on its own than within Canada.

In Western Europe, five countries (the United Kingdom, France, Belgium, Italy, and Spain) contain political movements whose members reject total control by the existing sovereign state and who claim to be the core of a separate national entity (Figure 9.21). Some separatists would be satisfied with *regional autonomy,* usually in the form of self-government or "home rule"; others seek complete independence for their regions.

In an effort to accommodate these separatist movements, several European governments have moved in the direction of regional recognition and **devolution** (decentralization) of political control. Recognizing the need for administrative structures that reflect regional concerns, the central governments have granted a degree of political autonomy to recognized political subunits. France, for example, established 22 regional governments in 1986. Spain has a program of devolution for its 17 "autonomous communities," a program that Italy is beginning to emulate.

Nationalist challenges to state authority affect many countries outside of Western Europe, of course. The Basques of Spain and the Bretons of France have their counterparts in the Palestinians in Israel, the Sikhs in India, the Tamils in Sri Lanka, the Moros in the Philippines, and many others.

The countries of Eastern Europe and the republics of the former Soviet Union have recently seen an explosion of

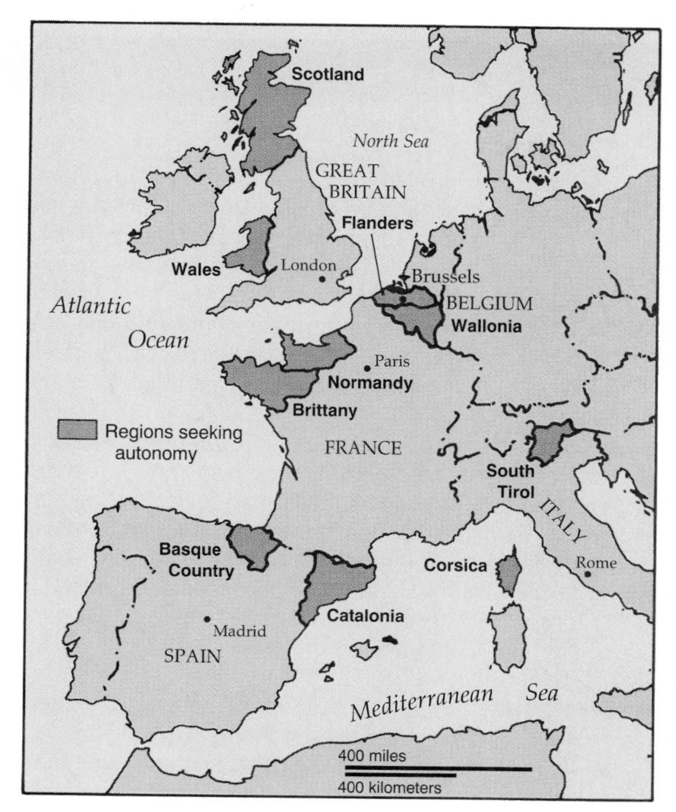

FIGURE 9.21 Regions in Western Europe seeking autonomy. Despite long-standing state attempts to culturally assimilate these historic nations, each contains a political movement that has recently sought or is currently seeking a degree of self-rule that recognizes its separate identity. Separatists on the island of Corsica, for example, want to secede from France. And despite having been granted a measure of self-rule for their regions, militant separatists in the Basque country and Catalonia demand independence from Spain. The desires of nationalist parties in both Wales and Scotland were partially accommodated by separate majority votes in 1997 for their own parliaments and a degree of regional autonomy, an outcome labeled "separation but not a divorce" from the United Kingdom.

nationalist feelings. Now that the forces of ethnicity, religion, and culture are no longer suppressed by communism, ancient rivalries are more evident than at any time since World War II. The end of the Cold War aroused hopes of decades of peace. Instead, the collapse of communism and the demise of the USSR have spawned many smaller wars. Numerous ethnic groups large and small are asserting their identities and their right to determine their own political status.

By mid-1992, the 15 former Soviet constituent republics had declared their status as fully independent states (refer back to Figure 9.6), but those declarations did not assure the satisfaction of all separatist movements within them. Many of the new individual countries are subject to strong destabilizing forces that challenge their territorial integrity and survival. The Russian Federation itself, the largest and most powerful remnant of the former USSR, has 89 components, including 21 "ethnic republics" and a number of other nationality regions. Many are rich in natural resources, have non-Russian majorities, and seek greater autonomy within the federation. Some, indeed, want total independence. One, the predominantly Muslim republic of Chechnya, in 1994 claimed the right of self-determination and attempted to secede from the federation, provoking a bloody war. If Chechnya's secession had succeeded, other republics might have tried to follow suit. Additional declarations of independence could plunge the federation into civil wars, chaos, and dictatorship.

As the USSR declined and eventually disbanded, it lost control of its communist satellites in Eastern Europe. That loss and resurgent nationalism led to a dramatic reordering of the region's political map. Some of the key developments are listed below.

- East Germany was reunified with West Germany in 1990.
- Yugoslavia broke into five pieces in 1991–92 (see "The Disintegration of Yugoslavia").
- More peacefully, in 1993 the people of Czechoslovakia agreed to split their country into two separate, ethnically-based states—the Czech Republic and Slovakia. The population of the former is largely Czech and Moravian. Slovakia is 86% Slovak, but the 11% of the population that is Hungarian displays some separatist yearnings.

It is too early to tell whether these new states and others will be viable political entities. We can, however, make some generalizations about nationalist challenges to state authority.

The two preconditions common to all separatist movements are *territory* and *nationality.* First, the group must be concentrated in a core region that it claims as a national homeland. It seeks to regain control of land and power that it believes were unjustly taken by the ruling party. Second, certain cultural characteristics must provide a basis for the group's perception of separateness, identity, and cultural unity. These might be language, religion, or distinctive group customs, which promote feelings of group identity at the same time that they foster exclusivity. Normally, these cultural differences have persisted over several generations and have survived despite strong pressures toward assimilation.

Other characteristics common to many separatist movements are a *peripheral location* and *social and economic inequality.* Troubled regions tend to be peripheral, often isolated in rural pockets, and their location away from the seat of central government engenders feelings of alienation, exclusion, and neglect. Second, the dominant culture group is often seen as an exploiting class that has suppressed the local language, controlled access to the civil service, and taken more than its share of wealth and power. Poorer regions complain that they have lower incomes and greater unemployment than prevail in the rest of the state, and that "outsiders" control key resources and industry. Separatists in relatively rich regions believe that they could exploit their resources for themselves and do better economically without the constraints imposed by the central state.

THE DISINTEGRATION OF YUGOSLAVIA

The history of Yugoslavia in this century demonstrates how difficult it is for a multinational state to be stable, how easy it can be for centrifugal forces to tear it apart. "I was born in one Yugoslavia, grew up in another, and now I'm a citizen of a third, and it is the smallest of the three," a woman commented as she watched the ceremony proclaiming a new Yugoslav state in 1992.

The first Yugoslavia was a kingdom created in 1918 when Serbia was joined by Croatia and Slovenia, both previously part of the Austro-Hungarian Empire. A combination of distinct groups, the new country was far from homogeneous. Slovenes and Croatians were primarily Roman Catholic and wrote in the Roman alphabet. Serbs, previously part of the Ottoman Empire, practiced the Eastern Orthodox religion and wrote in Cyrillic. Muslims were the largest ethnic group in Bosnia-Herzegovina.

The second Yugoslavia began life in 1946 as a centrally controlled communist state. It was a federation of six republics, but an uneasy federation; the two largest republics, Serbia and Croatia, had been on opposing sides during World War II. Indeed, thousands of Serbs had died in concentration camps run by the Nazi puppet government in Croatia.

In large part, this Yugoslavia was held together by the leadership of Josip Broz Tito, who had led the partisan army opposing the Nazis. He played down ethnic rivalries, brought a measure of economic prosperity to the country, and kept open ties to the West.

The federation began unraveling after Tito's death in 1980 and the decline of communism. As Yugoslavia's new political leaders appealed to nationalist feelings, ethnic, religious, and cultural differences came to the fore. Fearing domination by the Serbs (the largest ethnic group, with 36% of the population), four of the six republics seceded from the federation in 1991: Slovenia, Croatia, Bosnia, and Macedonia.

In its third incarnation, Yugoslavia consists only of Serbia and Montenegro. It has less than half the area and half the population of the old state.

By and large, the boundaries of the five republics do not match the territories occupied by nationalities. Only *Slovenia* fits the definition of a nation-state. That is, most of its inhabitants are Slovenes, and almost all Slovenes live in Slovenia. Compared to the other newly established republics, Slovenia is stable and peaceful. Each of the other republics contains significant minorities and has experienced civil strife as nationalities have fought to redefine the boundaries of their countries. One tactic used to transform a multinational region into one containing only one nation is **ethnic cleansing,** the killing or forcible relocation of less powerful minority nationalities. It has occurred in both Croatia and Bosnia.

After *Croatia* declared its independence from Yugoslavia in 1991, Serbs attempted to carve out their own state in eastern Croatia, where they outnumbered the Croats. Aided by Yugoslav armed forces, Serb militias bombarded hundreds of Croatian towns and villages and ethnically cleansed Croats from about one-third of their country. Croatian armies regained control of their territory in 1995, tens of thousands of Serbs fled the country, United Nations troops policed the region until 1998, and an uneasy peace now prevails.

An even bloodier war erupted in *Bosnia,* where the largest group of people (40%) is Muslim. Not wanting to live in a country with a Muslim plurality, Serbs and Croats attempted to unite the territories they inhabited with Serbia and Croatia. Bosnian Serbs gained control of much of Bosnia, killing Muslims or forcing them into detention camps. As in Croatia, an uneasy peace has prevailed in Bosnia since the Dayton accords of 1995.

Within the new Yugoslavia, civil strife broke out in the Kosovo region of southern *Serbia* after constitutional changes revoked the province's autonomy. Serbs alleged the changes were necessary to protect the Serbian minority in the province from domination by the predominantly ethnic Albanian population (90%) of the province. Led by the Kosovo Liberation Army, Albanian separatists are fighting for independence from Serbia.

That separtist movement is watched with interest by ethnic Albanians in *Macedonia.* Since gaining its independence in 1991, Macedonia has seen the emergence of a nationalist political movement that is hostile to the country's Albanian minority. Ethnic Macedonians, who are Slavs and Orthodox Christians, constitute about 70% of the population. The government has made Macedonian the sole official language and the Orthodox Church the state religion.

International Political Systems

The strivings of groups with distinct territorially based identities and a desire for fuller control of that territory are a reminder of the fragility of the modern state. In many ways, countries are now weaker than ever before. Many are economically frail, others are politically unstable, and some are both. Strategically, no state is safe from military attack, for technology now enables countries to shoot weapons halfway around the world. Some people believe that no national security is possible in the atomic age.

The recognition that a country cannot by itself guarantee either its prosperity or its own security has led to increased cooperation among states. In a sense, these cooperative ventures are replacing the empires of yesterday. They are proliferating quickly, and they involve countries everywhere.

The United Nations and Its Agencies

The United Nations (UN) is the only organization that tries to be universal. Its membership has expanded from 51 countries in 1945 to 185 in 1997. Switzerland is the only independent, sovereign state with a population of more than 1 million that is not a member of the UN.

The UN is the most ambitious attempt ever undertaken to bring together the world's countries in international assembly and to promote world peace. It is stronger and more representative than its predecessor, the League of Nations. It provides a forum where countries may discuss international problems and regional concerns and a mechanism, admittedly weak but still significant, for forestalling disputes or, when necessary, for ending wars (Figure 9.22). The United Nations also sponsors 40 programs and agencies aimed at fostering international cooperation with respect to specific goals. Among these are the World Health Organization (WHO), the Food and Agriculture Organization (FAO), and the United Nations Educational, Scientific, and Cultural Organization (UNESCO). Many other UN agencies and much of the UN budget are committed to assisting member states with matters of economic growth and development.

Member states have not surrendered sovereignty to the UN, and the world body is legally and effectively unable to make or enforce a world law. Nor is there a world police force. Although there is recognized international

FIGURE 9.22 **United Nations peacekeeping forces on duty in Bosnia-Herzegovina.** Under the auspices of the UN, soldiers from many different countries staff peacekeeping forces and military observer groups in many world regions in an effort to halt or mitigate conflicts. The demand for peacekeeping and observer operations is indicated by the recent deployment of UN forces in Angola, Bosnia, Cambodia, Cyprus, Haiti, Iraq, Israel, Lebanon, India, Pakistan, Rwanda, Somalia, and elsewhere.
©United Nations Photo Library.

law adjudicated by the International Court of Justice, rulings by this body are sought only by countries agreeing beforehand to abide by its arbitration. Finally, the United Nations has no authority over the military forces of individual countries.

A pronounced change both in the relatively passive role of the United Nations and in traditional ideas of international relations has begun to emerge, however. Long-established rules of total national sovereignty that allowed governments to act internally as they saw fit, free of outside interference, are fading as the United Nations increasingly applies a concept of "interventionism." The Persian Gulf War of 1991 was UN authorized under the old rules prohibiting one state (Iraq) from violating the sovereignty of another (Kuwait) by attacking it. After the war, the new interventionism sanctioned UN operations within Iraq to protect Kurds within that country. Later, the UN intervened with troops and relief agencies in Somalia, Bosnia, and elsewhere, invoking an "international jurisdiction over inalienable human rights" that prevails without regard to state frontiers or sovereignty considerations.

Whatever the long-term prospects for interventionism replacing absolute sovereignty, for the short term the UN remains the only institution where the vast majority of the world's countries can collectively discuss international political and economic concerns and attempt peacefully to resolve their differences. It has been particularly influential in formulating a law of the sea.

Maritime Boundaries

Boundaries define political jurisdictions and areas of resource control. However, we have considered only boundaries on land. As water covers about two-thirds of the earth's surface and people have used the seas since ancient times, it is not surprising that boundaries across water are also important. A basic question involves the right of states to control water and the resources it contains. The inland waters of a country, such as rivers and lakes, have traditionally been regarded as being within the sovereignty of that country. Oceans, however, are not within any country's borders. Are they, then, to be open to all states to use, or may a single country claim sovereignty and limit access and use by other countries?

For most of human history, the oceans remained effectively outside individual national control or international jurisdiction. The seas were a common highway for those daring enough to venture on them, an inexhaustible larder for fishermen, and a vast refuse pit for the muck of civilization. By the end of the 19th century, however, most coastal countries claimed sovereignty over a continuous belt 3 or 4 nautical miles wide (a *nautical mile,* or *nm,* equals 1.15 statute miles, or 1.85 kilometers). At the time, the 3-nm limit represented the farthest range of artillery and thus the effective limit of control by the coastal state. Though recognizing the rights of others to innocent passage, such sovereignty permitted the enforcement of quarantine and customs regulations, allowed national protection of coastal fisheries, and made claims of neutrality effective during other people's wars. The primary concern was with security and unrestricted commerce. No separately codified law of the sea existed, however, and none seemed to be needed until after World War I.

A League of Nations Conference for the Codification of International Law, convened in 1930, inconclusively discussed maritime legal matters and served to identify areas of concern that were to become increasingly pressing after World War II. Important among these was an emerging shift from interest in commerce and national security to a preoccupation with the resources of the seas, an interest fanned by the *Truman Proclamation* of 1945. Motivated by a desire to exploit offshore oil deposits, the U.S. federal government, under this doctrine, laid claim to all resources on the continental shelf contiguous to its coasts. Other states, many claiming even broader areas of control, hurried to annex marine resources. Within a few years, a quarter of the earth's surface was appropriated by individual coastal countries.

Unrestricted extensions of jurisdiction and territorial disputes over proliferating claims to maritime space and resources led to a series of United Nations conferences on the Law of the Sea. Meeting over a period of years, delegates from more than 150 countries attempted to achieve consensus on a treaty that would establish an internationally agreed-upon "convention dealing with all matters relating to the Law of the Sea." The meetings culminated in a draft treaty in 1982, the **United Nations Convention on the Law of the Sea.**

An International Law of the Sea

The convention delimits territorial boundaries and rights by defining four zones of diminishing control (Figure 9.23).

1. A *territorial sea* of up to 12 nm (19 km) in breadth, over which coastal states have sovereignty, including exclusive fishing rights. Vessels of all types normally have the right of innocent passage through the territorial sea, although under certain circumstances noncommercial vessels (primarily military and research) can be challenged.
2. A *contiguous zone* to 24 nm (38 km). Although a coastal state does not have complete sovereignty in this zone, it can enforce its customs, immigration, and sanitation laws and has the right of hot pursuit out of its territorial waters.
3. An **exclusive economic zone (EEZ)** of up to 200 nm (370 km) in which the state has recognized rights to explore, exploit, conserve, and manage the natural resources, both living and nonliving, of the seabed and waters (Figure 9.24). Countries have exclusive rights to the resources lying within the continental shelf when this extends farther, up to 350 nm (560 km) beyond

their coasts. The traditional freedoms of the high seas are to be maintained in this zone.

4. The *high seas* beyond the EEZ. Outside any national jurisdiction, they are open to all states, whether coastal or landlocked. Freedom of the high seas includes the right to sail ships, fish, fly over, lay submarine cables and pipelines, and pursue scientific research. Mineral resources in the international deep seabed area beyond national jurisdiction are declared the common heritage of humankind, to be managed for the benefit of all the peoples of the earth.

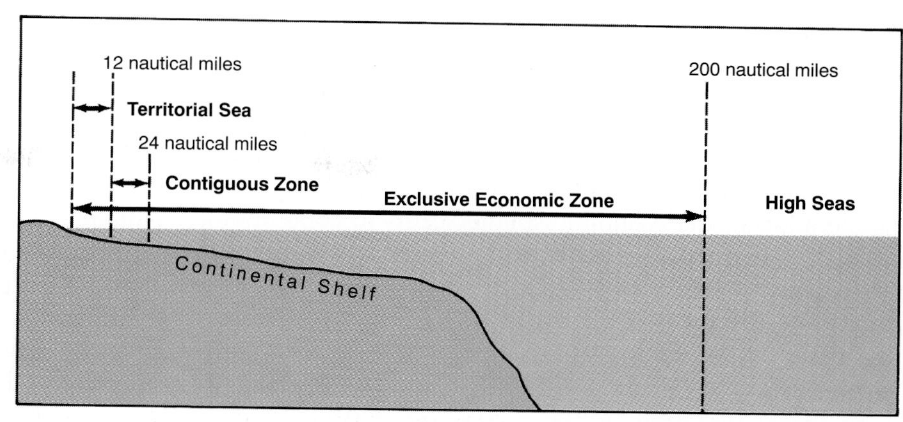

FIGURE 9.23 **Territorial claims permitted by the 1982 United Nations Convention on the Law of the Sea (UNCLOS).**

By the end of the 1980s, most coastal countries, including the United States, had used the UNCLOS provisions to proclaim and reciprocally recognize jurisdiction over 12-nm (19-km) territorial seas and 200-nm (370-km) economic zones. Despite reservations held by the United States and a few other industrial countries about the deep seabed mining provisions, the convention received the necessary ratification by 60 states and became international law in 1994.

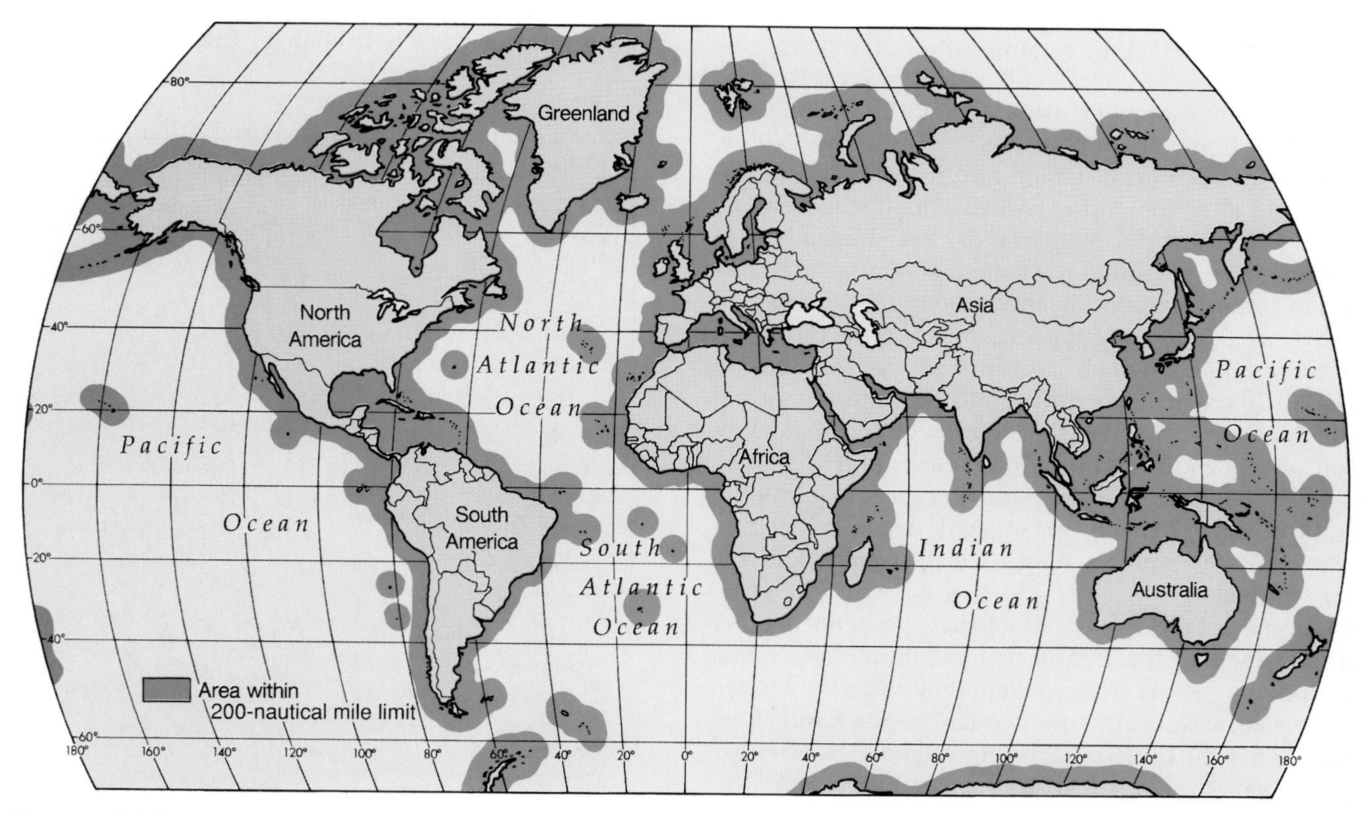

FIGURE 9.24 **The 200-nautical-mile exclusive economic zone (EEZ) claims of coastal states.** The provisions of the Law of the Sea Convention have in effect changed the maritime map of the world. Three important consequences flow from the 200-nm EEZ concept: (1) islands have gained a new significance; (2) countries have a host of new neighbors; and (3) the EEZ lines result in overlapping claims. EEZ lines are drawn around a country's possessions as well as around the country itself. Every island, no matter how small, has its own 200-nm EEZ. This means that while the United States shares continental borders only with Canada and Mexico, it has maritime boundaries with countries in Asia, South America, and Europe. All told, the United States may have to negotiate some 30 maritime boundaries, which is likely to take decades. Other countries, particularly those with many possessions, will have to engage in similar lengthy negotiations.

Regional Alliances

Countries have shown themselves to be willing to relinquish some of their independence to participate in smaller multinational systems. These groupings may be economic, military, or political, and many have been formed since 1945. Cooperation in the economic sphere seems to come more easily to states than does political or military cooperation.

Economic Alliances

The World Trade Organization (WTO), which came into existence in 1995, is charged with enforcing the global trade accord that grew out of years of international negotiations under the terms of the General Agreement on Tariffs and Trade (GATT). The basic principle behind the WTO is that the 100-plus member countries should work to cut tariffs, dismantle other barriers to trade, liberalize trade in services, and treat all other countries uniformly in matters of trade. Any preference granted to one should be available to all. Increasingly, however, regional rather than global trade agreements are being struck, and free-trade areas are proliferating. Such regional alliances (some 80 of them by the late 1990s), it can be argued, make world trade less free by scrapping tariffs on trade among member states but retaining them on exchanges with nonmembers.

Among the most powerful and far-reaching of the economic alliances are those that have evolved in Europe, particularly the European Union and its several forerunners. Shortly after the end of World War II, the Benelux countries (Belgium, the Netherlands, and Luxembourg) formed an economic union to create a common set of tariffs and to eliminate import licenses and quotas. Formed at about the same time were the Organization for European Cooperation (1948), which coordinated the distribution and use of Marshall Plan funds, and the European Coal and Steel Community (1952), which integrated the development of that industry in the member countries. A few years later, in 1957, the *European Economic Community (EEC)*, or *Common Market,* was created, composed at first of only six states: France, Italy, West Germany, and the Benelux countries.

To counteract these Inner Six, as they were called, other countries joined in the European Free Trade Association (EFTA). Known as the Outer Seven, they were the United Kingdom, Norway, Denmark, Sweden, Switzerland, Austria, and Portugal (Figure 9.25). Between 1973 and 1986, three members (the United Kingdom, Denmark, and Portugal) left EFTA for membership in the Common Market and were replaced by Iceland and Finland. Other Common Market additions were Greece in 1981, and Spain and Portugal in 1986. Austria, Finland, and Sweden became members in 1995.

Invitations to preliminary entry negotiations, conditional on continued economic restructuring, were issued to Poland, the Czech Republic, Hungary, Slovenia, and Estonia in mid-1997 (Figure 9.26). Five other European applicants (Bulgaria, Latvia, Lithuania, Romania, and Slovakia) were not included in the membership talks.

Over the years, members of the **European Union (EU)**, as the organization embracing the Common Market is now called, have taken many steps to integrate their economies and coordinate their policies in such areas as transportation, agriculture, and fisheries. A council of Ministers, a Commission, a European Parliament, and a Court of Justice give the European Union supranational institutions with effective ability to make and enforce laws. By January 1, 1993, the EU had abolished most remnant barriers to free trade and the free movement of capital and people among its members, creating a single European market. Plans call for monetary integration and a single European currency (the euro) for at least some countries by 1999.

We have traced this European development process, not because it is important to remember all the forerunners or the present structure of the European Union, but to illustrate the fluid process by which regional alliances are made. Countries come together in an association, some drop out, and others join. New treaties are made, and new coalitions emerge. It seems safe to predict that although the alliances themselves will change, the idea of economic

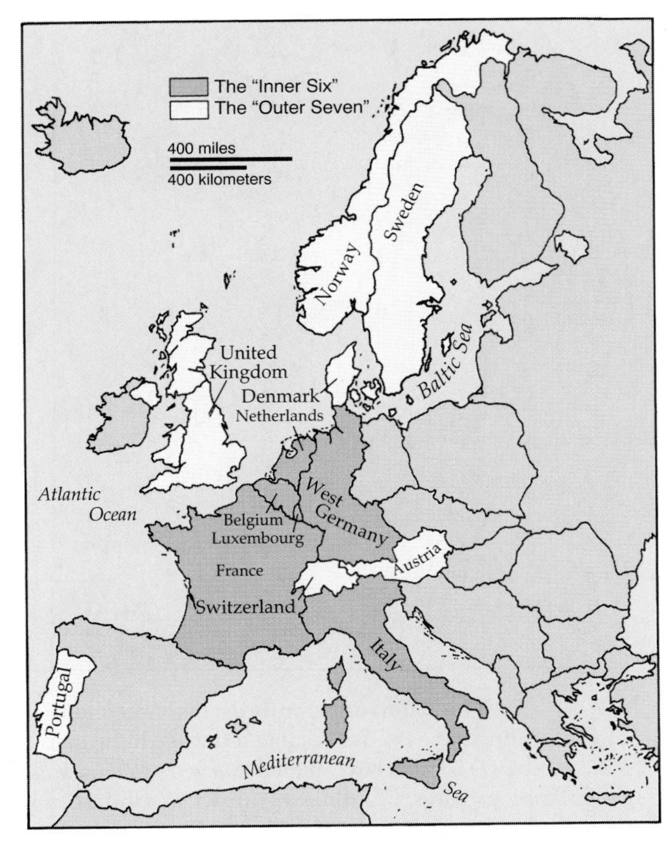

FIGURE 9.25 **The original "Inner Six" and "Outer Seven" of Europe.**

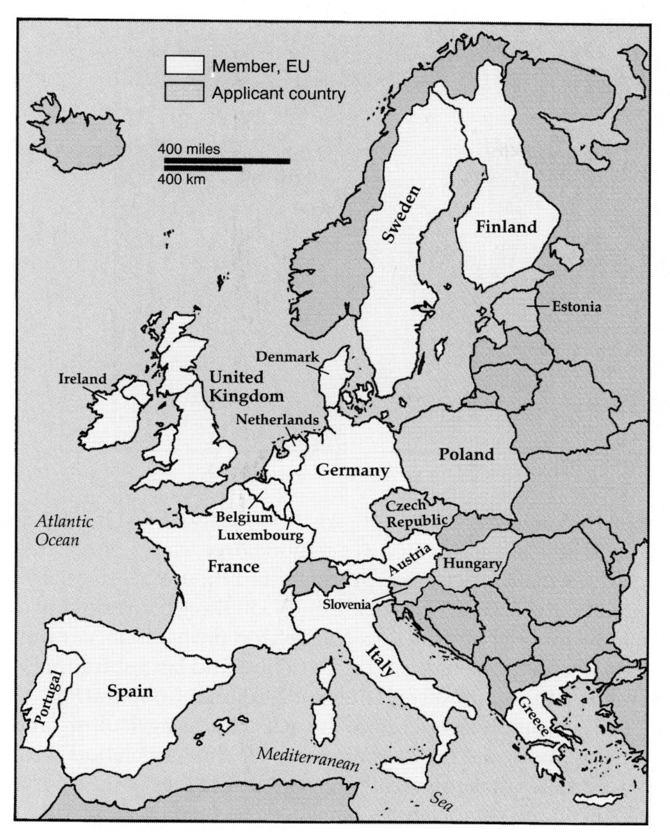

FIGURE 9.26 **The European Union (EU)** on January 1, 1995, expanded from 12 to 15 members, as Austria, Finland, and Sweden joined the organization. In 1998, the EU opened negotiations on membership with five Central and East European countries: Estonia, Poland, the Czech Republic, Slovenia, and Hungary. The EU has stipulated that in order to join, a country must have stable institutions guaranteeing democracy, the rule of law, human rights and protection of minorities; a functioning market economy; and the ability to accept the obligations of membership, including the aims of political, economic, and monetary union.

associations has been permanently added to that of political and military leagues, which are as old as nation-states themselves.

Three further points about economic unions are worth noting. The first, which also applies to military and political alliances, is that the formation of a coalition in one area often stimulates the creation of another alliance by countries left out of the first. Thus, the union of the Inner Six gave rise to the treaty among the Outer Seven. Similarly, a counterpart of the Common Market was the Council of Mutual Economic Assistance (CMEA), also known as Comecon, which linked the former communist countries of Eastern Europe and the USSR through trade agreements.

Second, the new economic unions tend to be composed of contiguous states (Figure 9.27a&b). This was not the case with the recently dissolved empires, which included far-flung territories. Contiguity facilitates the movement of people and goods. Communication and transportation are simpler and more effective among adjoining countries than among those far removed from one another.

Finally, it does not seem to matter whether countries are alike or distinctly different in their economies, as far as joining economic unions is concerned. There are examples of both. If the countries are dissimilar, they may complement each other. This was one basis for the European Common Market. Dairy products and furniture from Denmark are sold in France, freeing that country to specialize in the production of machinery and clothing. On the other hand, countries that produce the same raw materials hope that by joining together in an economic alliance, they might be able to enhance their control of markets and prices for their products. The Organization of Petroleum Exporting Countries (OPEC) is a case in point. Other attempts to form commodity cartels and price agreements between producing and consuming countries are represented by the International Tin Agreement, the International Coffee Agreement, and others.

Military and Political Alliances

Countries form alliances for other than economic reasons. Strategic, political, and cultural considerations may also foster cooperation. *Military alliances* are based on the principle that unity brings strength. Such pacts usually provide for mutual assistance in the case of aggression. Once again, action breeds reaction when such an association is created. The formation of the North Atlantic Treaty Organization (NATO), a defensive alliance of many European countries and the United States, was countered by the establishment of the Warsaw Treaty Organization, which joined the USSR and its satellite countries of Eastern Europe. Both pacts allowed the member states to base armed forces in one another's territories, a relinquishment of a certain degree of sovereignty unique to this century.

Military alliances depend on the perceived common interests and political goodwill of the countries involved. As political realities change, so, too, do the strategic alliances. NATO was created to defend Western Europe and North America against the Soviet military threat. When the dissolution of the USSR and Warsaw Pact removed that threat, the purpose of the NATO alliance became less clear and, during the 1990s, its relationships with Eastern European states and Russia were under review and change as most of those countries sought ways to foster cooperation with NATO. Three of them (Poland, The Czech Republic, and Hungary) were invited in 1997 to join the alliance (Figure 9.28).

All international alliances recognize communities of interest. In economic and military associations, common objectives are clearly seen and described, and joint actions are agreed upon with respect to the achievement of those

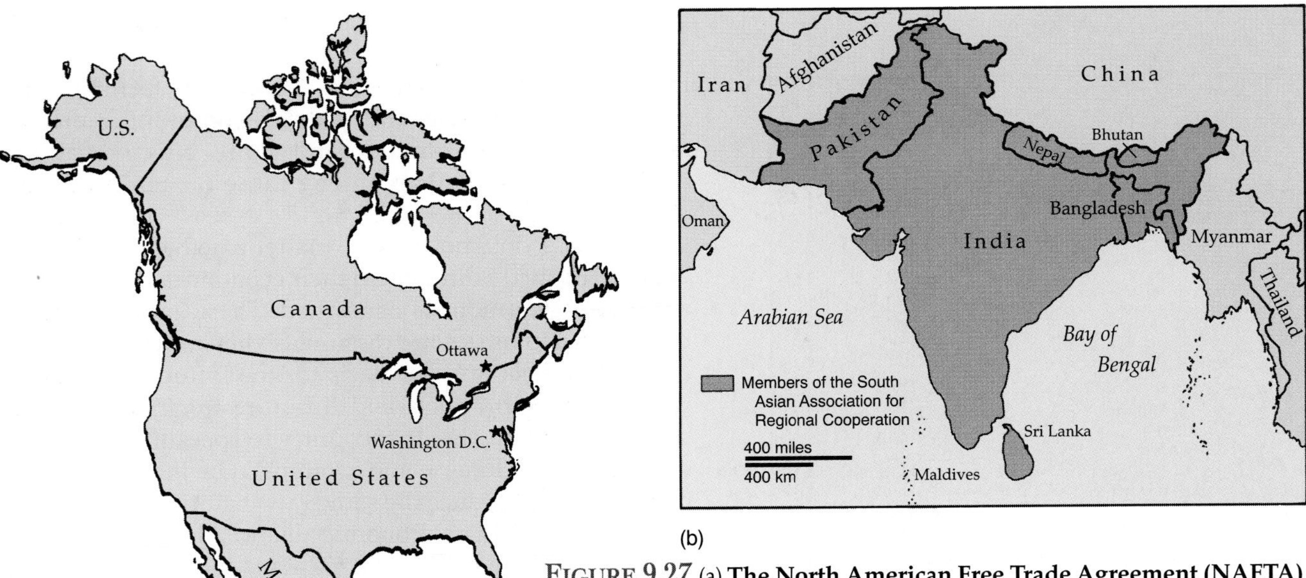

FIGURE 9.27 (a) **The North American Free Trade Agreement (NAFTA)** is intended to unite Canada, the United States, and Mexico in a regional free trade zone. Under the terms of the treaty, tariffs on all agricultural products and thousands of other goods are to be eliminated by the end of 1999. In addition, all three countries are to ease restrictions on the movement of business executives and professionals. If fully implemented, the treaty will create one of the world's richest and largest trading blocs. (b) **Members of the South Asian Association for Regional Cooperation.** Formed in December 1985 to promote cooperation in such areas as agriculture and rural development, transportation, and telecommunications, the Association is composed of seven neighboring states that contain about one-fifth of the world's population. Many other regional economic associations have been formed or proposed, including: the Arab Maghreb Union (of North African states); the Asean Free Trade Area (Southeast Asia); the Southern Cone Common Market (Mercosur) and the Andean Group (both in South America); the Caribbean Group and Common Market; and the Southern African Development Community.

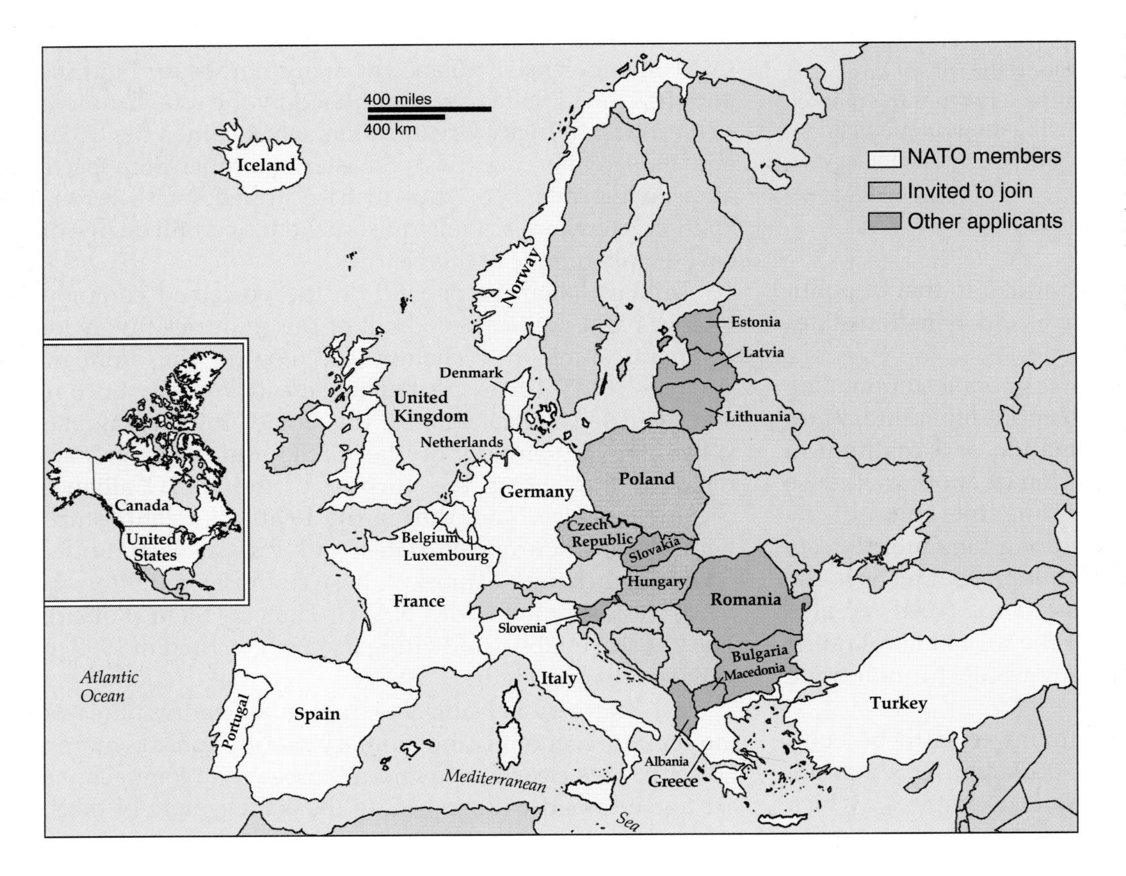

FIGURE 9.28 **The NATO military alliance** as of 1998 had 16 members. Three countries (Poland, the Czech Republic, and Hungary) have been invited to join the alliance, the first enlargement since Spain became a member in 1982. If the expansion is approved by all 16 NATO members, the three could enter the alliance in 1999 or 2000. Nine other countries have applied for membership; their applications will be considered at a special summit in 1999. Proponents of expansion argue that it is necessary in order to create a zone of stability and security throughout Europe. Opponents contend that enlargment is a divisive move that will cast a shadow over the future of relations with Russia, which is opposed to expansion so close to its borders.

objectives. More generalized common concerns or appeals to historical interest may be the basis for primarily *political alliances.* Such associations tend to be rather loose, not requiring their members to yield much power to the union. Examples are the Commonwealth of Nations (formerly the British Commonwealth), composed of many former British colonies and dominions, and the Organization of American States, both of which offer economic as well as political benefits.

There are many examples of abortive political unions that have foundered precisely because the individual countries could not agree on questions of policy and were unwilling to subordinate individual interests to make the union succeed. The United Arab Republic, the Central African Federation, the Federation of Malaysia and Singapore, and the Federation of the West Indies fall within this category.

Although many such political associations have failed, observers of the world scene speculate about the possibility that "superstates" will emerge from one or more of the international alliances that now exist. Will a "United States of Europe," for example, under a single common government, be the logical outcome of the successes of the EU? No one knows, but as long as the individual state is regarded as the highest form of political and social organization (as it is now), and as the body in which sovereignty rests, such total unification is unlikely.

Local and Regional Political Organization

The most profound contrasts in cultures tend to occur between, rather than within, states, one reason political geographers traditionally have been primarily interested in country units. The emphasis on the state, however, should not obscure the fact that for most of us it is at that local level that we find our most intimate and immediate contact with government and its influence on the administration of our affairs. In the United States, for example, an individual is subject to the decisions and regulations made by the local school board, the municipality, the county, the state, and perhaps, a host of special-purpose districts—all in addition to the laws and regulations issued by the federal government and its agencies. Among other things, local political entities determine where children go to school, the minimum size lot on which a person can build a house, and where one may legally park a car. Adjacent states of the United States may be characterized by sharply differing personal and business tax rates; differing controls on the sale of firearms, alcohol, and tobacco; variant administrative systems for public services; and different levels of expenditures for them (Figure 9.29).

All of these governmental entities are *spatial systems.* Because they operate within defined geographic areas, and because they make behavior-governing decisions, they are topics of interest to political geographers. In the concluding sections of this chapter, we examine two aspects of political organization at the local and regional level. Our emphasis will be on the United States and Canadian scene simply because their local political geography is familiar to most of us. We should remember, however, the North American structure of municipal governments, minor civil divisions, and special-purpose districts has counterparts in other regions of the world.

The Geography of Representation: The Districting Problem

There are more than 85,000 local governmental units in the United States. Slightly more than half of these are municipalities, townships, and counties. The remainder are school districts, water-control districts, airport authorities, sanitary districts, and other special-purpose bodies. Around each of these districts, boundaries have been drawn. Although the number of districts does not change greatly from year to year, many boundary lines are redrawn in any single year.

For example, the ruling of the U.S. Supreme Court in 1954 in *Brown v. Board of Education of Topeka, Kansas,* that the doctrine of "separate but equal" school systems for blacks and whites was unconstitutional, led to the redrawing of

FIGURE 9.29 **The Four Corners Monument,** marking the meeting of Utah, Colorado, New Mexico, and Arizona. Jurisdictional boundaries within countries may be precisely located but are usually not highly visible on the landscape. At the same time, those boundaries may be very significant in citizens' personal affairs and in the conduct of economic activities.
© Cameramann International, Ltd.

thousands of attendance boundaries of school districts. Likewise, the court's "one person, one vote" ruling in *Baker v. Carr* (1962) signified the end of overrepresentation of sparsely populated rural districts in state legislatures and led to the frequent adjustment of electoral districts within states and cities to attain roughly equal numbers of voters. Such *redistricting* or *reapportionment* is made necessary by shifts in population, as areas gain or lose people.

The analysis of how boundaries are drawn around voting districts is one aspect of **electoral geography,** which also addresses the spatial patterns yielded by election results and their relationship to the socioeconomic characteristics of voters. In a democracy, it might be assumed that election districts should contain roughly equal numbers of voters, that electoral districts should be reasonably compact, and that the proportion of elected representatives should correspond to the share of votes cast by members of a given political party. Problems arise because the way in which the boundary lines are drawn can maximize, minimize, or effectively nullify the power of a group of people.

Gerrymandering is the practice of drawing the boundaries of legislative districts so as to unfairly favor one political party over another, to fragment voting blocs, or to achieve other nondemocratic objectives (Figure 9.30). A number of strategies have been employed over the years for that purpose. *Stacked* gerrymandering involves drawing circuitous boundaries to enclose pockets of strength or weakness of the group in power; it is what we usually think of as gerrymandering. The *excess vote* technique concentrates the support of the opposition in a few districts, which they can win easily, but leaves them few potential seats elsewhere. Conversely, the *wasted vote* strategy dilutes the opposition's strength by dividing its votes among a number of districts.

Assume that X and O represent two groups with an equal number of voters but different policy preferences. Although there are equal numbers of Xs and Os, the way electoral districts are drawn affects voting results. In Figure 9.31a, the Xs are concentrated in one district and will probably elect only one representative of four. The power of the Xs is maximized in Figure 9.31b, where they may control three of the four districts. The voters are evenly divided in Figure 9.31c, where the Xs have the opportunity to elect two of the four representatives. Finally, Figure 9.31d shows how both political parties may agree to delimit the electoral districts to provide "safe seats" for incumbents. Such a partitioning offers little chance for change.

Figure 9.31 depicts a hypothetical district, compact in shape with an even population distribution and only two

FIGURE 9.30 **The original gerrymander.** The term *gerrymander* originated in 1811 from the shape of an electoral district formed in Massachusetts while Elbridge Gerry was governor. When an artist added certain animal features, the district resembled a salamander and quickly came to be called a gerrymander. The Bettmann Archive.

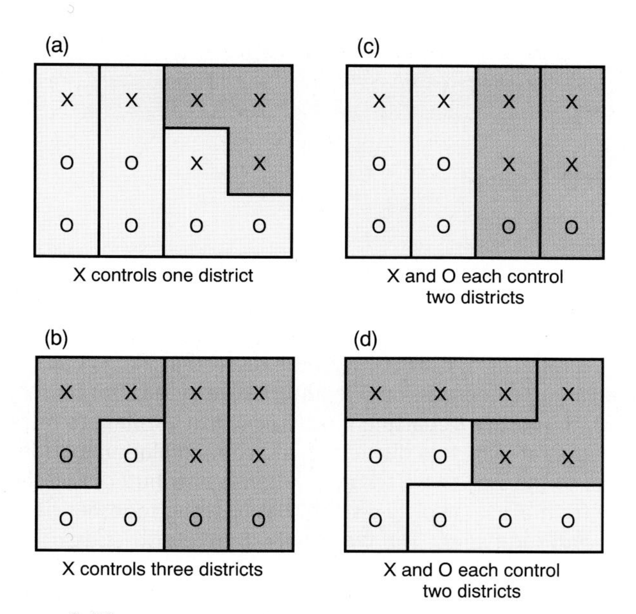

FIGURE 9.31 Alternative districting strategies. Xs and Os might represent Republicans and Democrats, urban and rural voters, blacks and whites, or any other distinctive groups.

groups competing for representation. In actuality, American municipal voting districts are often oddly shaped because of such factors as the city limits, historic settlement patterns, current population distribution, and transportation routes—as well as past gerrymandering. Further, in any large area, many groups vie for power. Each electoral interest group promotes its version of "fairness" in the way boundaries are delimited. Minorities seek representation in proportion to their numbers, so that they will be able to elect representatives who are concerned about and responsive to their needs (see "Voting Rights and Race").

We should note that gerrymandering is not automatically successful. First, a districting arrangement that appears to be unfair may be appealed to the courts. In addition, many factors other than political party affiliation influence voting decisions. Candidates usually receive more votes in their home areas, for example. Some candidates may have more workers or more money to spend on a campaign than do their opponents. A key issue may cut across party lines.

The Fragmentation of Political Power

Boundary drawing at any electoral level is never easy, particularly when political groups want to maximize their representation and minimize that of opposition groups. Furthermore, the boundaries that we may want for one set of districts may *not* be those that we want for another. For example, sewage districts must take natural drainage features into account, whereas police districts may be based on the distribution of the population or the number of miles of street to be patrolled. And school attendance zones must consider the numbers of school-aged children and the capacities of individual schools.

As these examples suggest, the United States is subdivided into great numbers of political administrative units whose areas of control are spatially limited. The 50 states are partitioned into more than 3000 counties ("parishes" in Louisiana), most of which are further subdivided into townships, each with a still lower level of governing power. This political fragmentation is further increased by the existence of nearly innumerable special-purpose districts whose boundaries rarely coincide with the standard major and minor civil divisions of the country, or even with each other (Figure 9.32). Each district represents a form of political allocation of territory to achieve a specific aim of local need or legislative intent (see "Too Many Governments").

Canada, a federation of 10 provinces and two territories (joined in April 1999 by a third, Nunavut), has a similar pattern of political subdivision. Each of the provinces contains minor civil divisions—municipalities—under provincial control, and all (cities, towns, villages, and rural municipalities) are governed by elected councils. Ontario and Quebec also have counties that group smaller municipal units for certain purposes. In general, municipalities are responsible for police and fire protection, local jails, roads and hospitals, water supply and sanitation, and schools, duties which are discharged either by elected agencies or appointed commissions.

Most North Americans live in large and small cities. In the United States, these, too, are subdivided, not only into wards or precincts for voting purposes but also into special districts for such functions as fire and police protection, water and electricity supply, education, recreation, and sanitation. These districts almost never coincide with one another, and the larger the urban area, the greater the proliferation of small, special-purpose governing and taxing units. Although no Canadian community has quite the multiplication of governmental entities as, for example, Chicago, Illinois, with well more than 1000 special- and general-purpose governments, major Canadian cities may find themselves with complex and growing systems of similar nature. Even before its major expansion on January 1, 1998, for example, metropolitan Toronto had more than 100 authorities that could be classified as "local governments."

The existence of such a great number of districts in metropolitan areas may cause inefficiency in public services and hinder the orderly use of space. *Zoning ordinances,* for example, are determined by each municipality. They are intended to allow the citizens to decide how land is to be used and, thus, are a clear example of the effect of political decisions on the division and development of space. Zoning policies dictate the areas where light and heavy industries may be located, the sites of parks and other

Geography *and* Public Policy

Voting Rights and Race

The irregularly shaped Congressional voting districts shown here represent a deliberate attempt to balance voting rights and race. They have been called extreme examples of racial gerrymandering, however, and at least in part ruled unconstitutional by the Supreme Court.

All of the districts shown contain a majority of black voters. They were created by state legislatures after the 1990 census to make minority representation in Congress more closely resemble minority presence in the state's total population. Specifically, they were intended to comply with the federal Voting Rights Act of 1965, which provides that members of racial minorities shall not have "less opportunity than other members of the electorate . . . to elect representatives of their choice." In the 1980s, the Justice Department prodded states to create districts designed to give black voters representation in rough proportion to their numbers in the population.

In North Carolina, for example, although 24% of the population of that state is black, past districting had divided black voters among a number of districts, with the result that blacks had not elected a single Congressional representative in the 20th century. In 1991, the Justice Department ordered North Carolina to redistrict so that at least two districts would contain black majorities. Because of the way the black population is distributed, the only way to form black-majority districts was to string together cities, towns, and rural areas. The two newly created districts had slim (53%) black majorities.

The redistricting in North Carolina and other states had immediate effects. Black membership in the House of Representatives increased from 26 in 1990 to 39 in 1992; blacks constituted nearly 9% of the House as against 12% of blacks in the total population. Within a year, those electoral gains were threatened as lawsuits challenging the redistricting were filed in a number of states. The

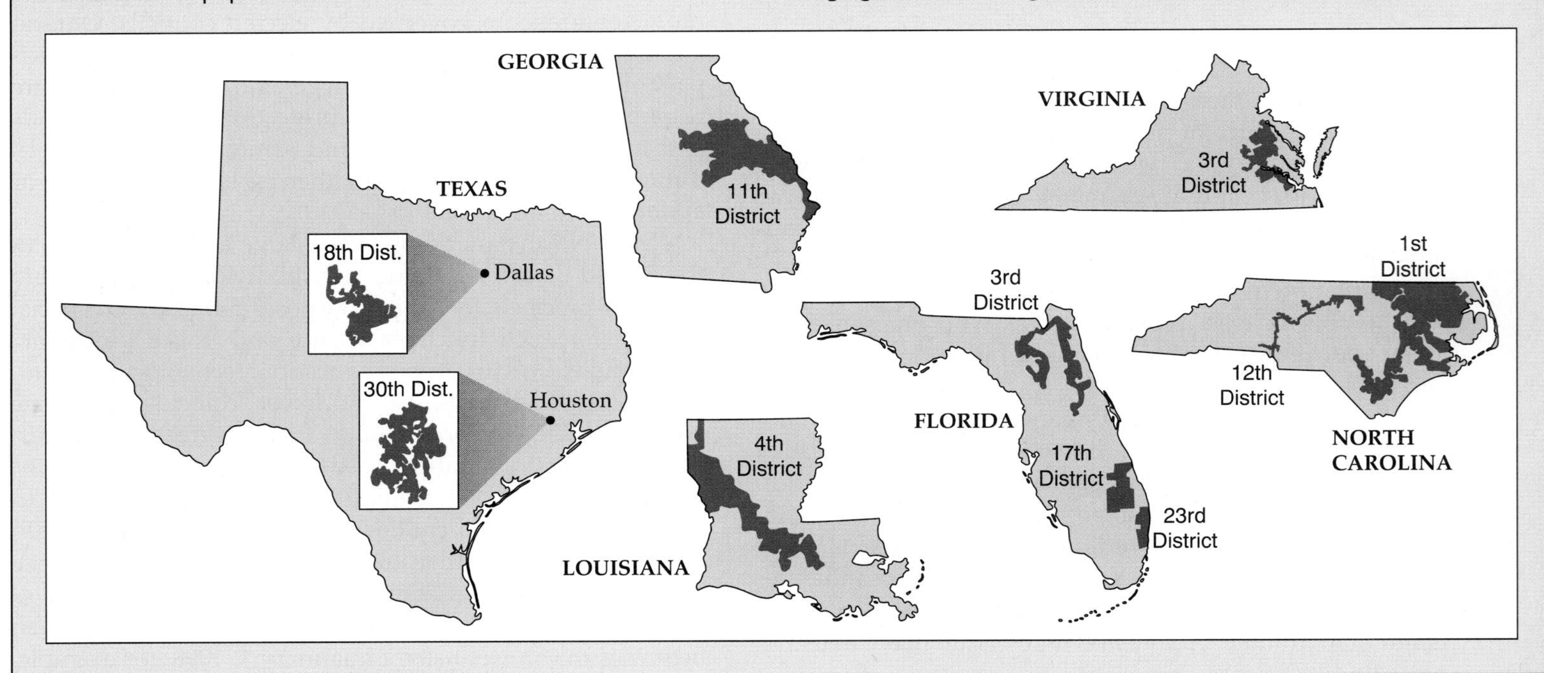

recreational areas, the location of business districts, and the types and location of housing. Unfortunately, in large urban areas, the efforts of one community may be hindered by the practices of neighboring communities. Thus, land zoned for an industrial park in one city may abut land zoned for single-family residences in an adjoining municipality. Each community pursues its own interests, which may not coincide with those of its neighbors or of the larger region.

Inefficiency and duplication of effort characterize not just zoning but many of the services provided by local governments. The efforts of one community to avert air and water pollution may be, and often are, counteracted by the rules and practices of other towns in the region, although state and national environmental protection standards are now reducing such potential conflicts. Social as well as physical problems spread beyond city boundaries. Thus, nearby suburban communities are affected when a central chief contention of the plaintiffs was that the irregular shapes of the districts were a product of racial gerrymandering and amounted to reverse discrimination against whites.

In June 1993, a sharply divided Supreme Court ruled in *Shaw v. Reno* that North Carolina's 12th Congressional District might violate the constitutional rights of white voters and ordered a district court to review the case. The 5–4 ruling gave evidence that the country had not yet reached agreement on how to comply with the Voting Rights Act. It raised a central question: should a state maximize the rights of racial minorities or not take racial status into consideration? A divided court provided answers in 1995, 1996, and 1997 rulings that rejected Congressional redistricting maps for Georgia, Texas, North Carolina, and Virginia on the grounds that "race cannot be the predominant factor" in drawing election district boundaries, nor can good-faith efforts to comply with the Voting Rights Act insulate redistricting plans from constitutional attack.

One way to eliminate gerrymandering and its distortion of voter representation is through a modified at-large voting system called *preference* or *cumulative voting.* Under this system, voters elect several candidates in a larger district rather than just one in a traditional district. North Carolina, for example, might contain three Congressional districts instead of its current 12, with the divisions along county lines. The districts would be assigned a number of representatives proportional to their population. If one district had four representatives, voters would have four votes to assign however they liked. They could cast all four for one candidate or spread them among two, three, or four candidates. By concentrating their votes, minority groups could increase the chances that their candidates would be elected. Preference voting is common in Europe and on corporate boards in the United States. Although it is used in a number of American communities, no state employs the system for electing members of Congress.

Questions to Consider

1. Do you believe that race should be a consideration in the electoral process? Why or why not? If so, should voting districts be drawn to increase the likelihood that representatives of racial or ethnic minorities will win elections? If not, how can one be certain that the voting power of minorities will not be diluted in white-majority districts?
2. With which of the following arguments in *Shaw v. Reno* do you agree? Why? ". . . Racial gerrymandering, even for remedial purposes, may balkanize us into competing racial factions; it threatens to carry us further from the goal of a political system in which race no longer matters." (Justice Sandra Day O'Connor) ". . . Legislators will have to take race into account in order to avoid dilution of minority voting strength." (Justice David Souter)
3. One of the candidates in North Carolina's 12th Congressional District said, "I love the district because I can drive down I-85 with both car doors open and hit every person in the district." Given a good transportation and communication network, how important is it that voting districts be compact?
4. Blacks face difficult hurdles in being elected in districts that do not have a black majority, as witnessed by their numerical underrepresentation in most legislative bodies. But critics of "racial gerrymandering" contend that blacks have been and can continue to be elected in white-majority districts and, further, that white politicians can and do adequately represent the needs of all, including black, voters in their districts. Do you agree? Why or why not?

city lacks the resources to maintain high-quality schools or to attack social ills. The provision of health care facilities, electricity and water, transportation, and recreational space affects the whole region and, many professionals think, should be under the control of a single unified metropolitan government.

The growth in the number and size of metropolitan areas has increased awareness of the problems of their administrative fragmentation. In response, new approaches to the integration of those areas have been proposed and adopted. The aims of all plans of metropolitan government are to create or preserve coherence in the management of areawide concerns and to assure that both the problems and the benefits of growth are shared without regard to their jurisdictional locations.

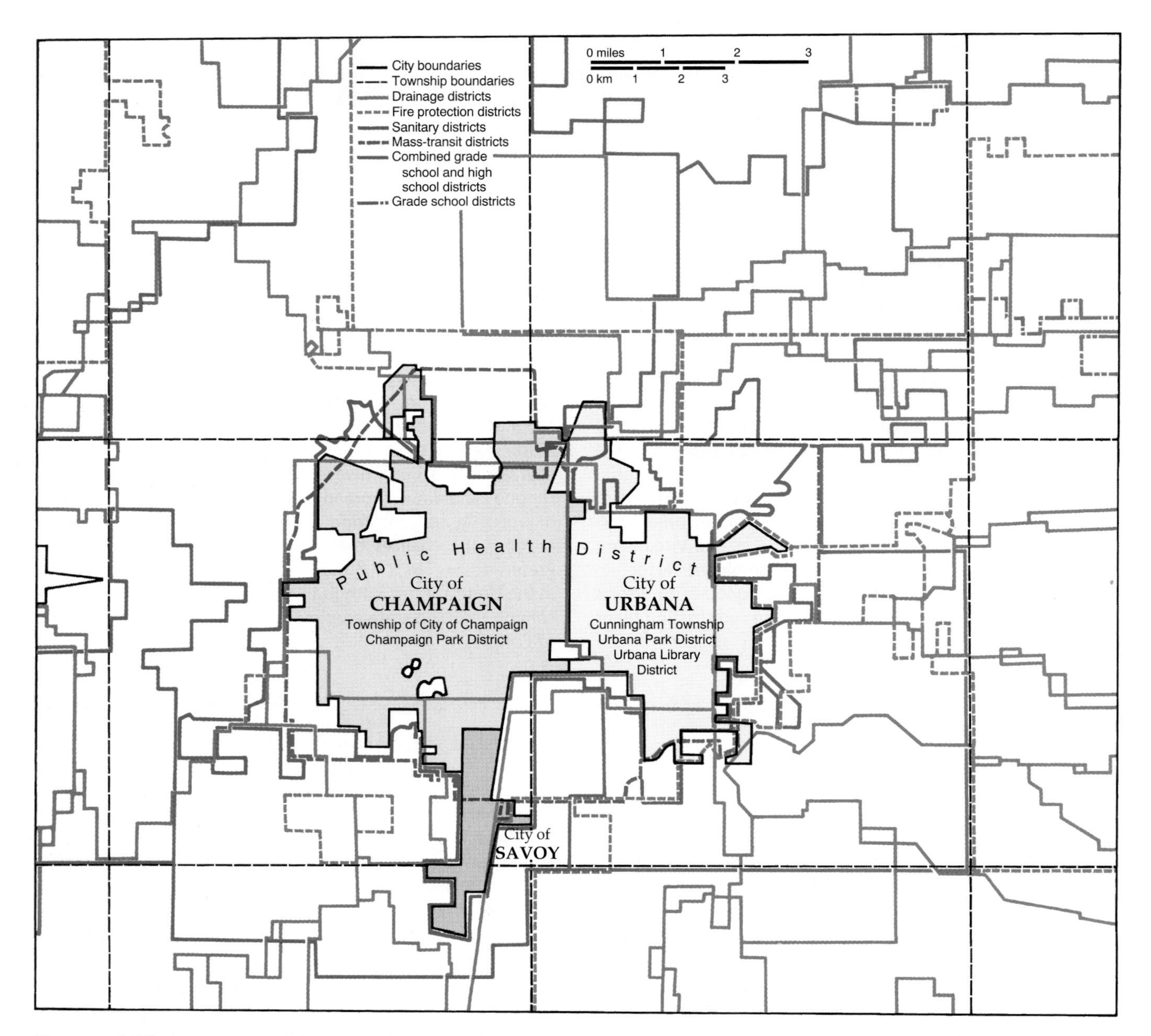

FIGURE 9.32 **Political fragmentation in Champaign County, Illinois.** The map shows a few of the independent administrative agencies with separate jurisdictions, responsibilities, and taxing powers in a portion of a single Illinois county. Among the other such agencies forming the fragmented political landscape are Champaign County itself, a forest preserve district, a public health district, a mental health district, the county housing authority, and a community college district.

SUMMARY

The sovereign state is the dominant entity in the political subdivision of the world. It constitutes an expression of cultural separation and identity as pervasive as that inherent in language, religion, or ethnicity. A product of 18th-century political philosophy, the idea of the state was diffused globally by colonizing European powers. In most instances, the colonial boundaries they established have been retained as their international boundaries by newly independent countries.

The greatly varying geographic characteristics of states contribute to national strength and stability. Size, shape, and relative location influence countries' economies and international roles, while national cores and capitals are the heartlands of states. Boundaries, the legal definition of a state's size and shape, determine the limits of its sovereignty. They may or may not reflect preexisting cultural

Too Many Governments

If you are a property owner in Wheeling Township in the city of Arlington Heights in Cook County, Illinois, here's who divvies up your taxes: the city, the county, the township, an elementary school district, a high school district, a junior college district, a fire protection district, a park district, a sanitary district, a forest preserve district, a library district, a tuberculosis sanitarium district, and a mosquito abatement district.

Lest you attribute this to population density or the Byzantine ways of Cook County politics, it's not that much different elsewhere in Illinois—home to more governmental units than any other state in the United States. According to late-1980s figures from the U.S. Bureau of the Census, there were 6626 local government units in Illinois. Second-place Pennsylvania had 4956 governments—1670 fewer than Illinois—and the average for all states was 1663.

Along with its 102 counties, Illinois has nearly 1300 municipalities, more than 1400 townships, and over 1000 school districts; but the biggest factor in the governmental unit total are single-function special districts. These were up from 2600 in 1982 to nearly 2800 in 1987, with no end to their increase in sight. Special districts range from Chicago's Metropolitan Sanitary District to the Caseyville Township Street Lighting District. Most of these governments have property-taxing power. Some also impose sales or utility taxes.

This proliferation is in part a historical by-product of good intentions. The framers of the state's 1870 constitution, wanting to prevent overtaxation, limited the borrowing and taxing power of local governments to 5% of the assessed value of properties in their jurisdictions. When this limit was reached and the need for government services continued to grow with population, voters and officials circumvented the constitutional proscription by creating new taxing bodies—special districts. Illinois' special districts grew because they could be fitted to service users without regard to city or county boundaries.

Critics say all these governments result in duplication of effort, inefficiencies, higher costs, and higher taxes. Supporters of special districts, townships, and small school districts argue that such units fulfill the ideal of a government close and responsive to its constituents.

landscapes and, in any given case, may or may not prove to be viable. Whatever their nature, boundaries are at the root of many international disputes. Maritime boundary claims, particularly as reflected in the UN Convention on the Law of the Sea, add a new dimension to traditional claims of territorial sovereignty.

State cohesiveness is promoted by a number of centripetal forces. Among these are national symbols, a variety of institutions, and confidence in the aims, organization, and administration of government. Also helping to foster political and economic integration are transportation and communication connections. Destabilizing centrifugal forces, particularly ethnically based separatist movements, threaten the cohesion and stability of many states.

Although the state remains central to the partitioning of the world, a broadening array of political entities affects people individually and collectively. Recent decades have seen a significant increase in the number and variety of global and regional alliances, to which states have surrendered some sovereign powers. At the other end of the spectrum, expanding Anglo American urban areas and governmental responsibilities raise questions of fairness in districting procedures and of effectiveness when political power is fragmented.

Key Words

antecedent boundary 320
centrifugal force 324
centripetal force 324
compact state 313
consequent boundary 320
containment 323
core area 317
devolution 328
domino theory 323
electoral geography 338
elongated state 314
enclave 315
ethnic cleansing 330
European Union (EU) 334
exclave 315
exclusive economic zone (EEZ) 332
fragmented state 315
functional dispute 321
geometric boundary 320
geopolitics 321
gerrymandering 338
heartland theory 323
irredentism 321
nation 311
nationalism 324
nation-state 311
natural boundary 318
perforated state 315
positional dispute 321
prorupt state 314
regionalism 328
resource dispute 321
rimland theory 323
self-determination 328
state 310
subnationalism 328
subsequent boundary 320
superimposed boundary 320
territorial dispute 321
United Nations Convention on the Law of the Sea 332

FOR REVIEW AND CONSIDERATION

1. What are the differences between a *state,* a *nation,* and a *nation-state?* Why is a colony not a state? How can one account for the rapid increase in the number of states since World War II?
2. What attributes differentiate states from one another? How do a country's size and shape affect its power and stability? Can a piece of land be both an *enclave* and an *exclave?*
3. How may boundaries be classified? How do they create opportunities for conflict? Describe and give examples of three types of border disputes.
4. How did Mackinder and Spykman differ in their assessments of Eurasia as a likely base for world conquest? What post-1945 developments suggest that there may be no enduring correlation between location and national power?
5. Distinguish between *centripetal* and *centrifugal* political forces. What are some of the ways national cohesion and identity are achieved?
6. What characteristics are common to all or most autonomist movements? Where are some of these movements active? Why do they tend to be located on the periphery rather than at the national core?
7. What types of international organizations and alliances can you name? What were the purposes of their establishment? What generalizations can you make regarding economic alliances?
8. How does the *United Nations Convention on the Law of the Sea* define zones of diminishing national control? What are the consequences of the concept of the 200-nm *exclusive economic zone?*
9. Why does it matter how boundaries are drawn around electoral districts? Theoretically, is it always possible to delimit boundaries "fairly"? Support your answer.
10. What reasons can you suggest for the great political fragmentation of the United States? What problems stem from such fragmentation?

SELECTED REFERENCES

Agnew, John. *Political Geography: A Reader.* New York: John Wiley & Sons, 1996.

Blake, Gerald H. *Maritime Boundaries.* World Boundaries Series, vol. 5. London: Routledge, 1994.

Boyd, Andrew. *An Atlas of World Affairs.* 10th ed. New York: Routledge, 1998.

Chinn, Jeff, and Robert Kaiser. *Russians as the New Minority: Ethnicity and Nationalism in the Soviet Successor States.* Boulder, Colo.: Westview Press, 1996.

Cohen, Saul B. "Global Geopolitical Change in the Post-Cold War Era." *Annals of the Association of American Geographers* 81, no. 4 (1991):551–580.

Demko, George J., and William B. Wood, eds. *Reordering the World: Geopolitical Perspectives on the Twenty-First Century.* Boulder, Colo.: Westview Press, 1994.

Gellner, Martin I. *Encounters with Nationalism.* Oxford, England: Basil Blackwell Ltd., 1994.

Gibb, Richard, and Mark Wise. *The European Union.* London: Edward Arnold, 1998.

Glassner, Martin I. *Political Geography.* 2d ed. New York: John Wiley & Sons, 1996.

Grofman, Bernard. *Political Gerrymandering and the Courts.* New York: Agathon Press, 1990.

Heffernan, Michael. *The Meaning of Europe.* London: Edward Arnold, 1998.

Hooson, David, ed. *Geography and National Identity.* Oxford, England: Basil Blackwell Ltd., 1994.

Johnston, Ronald J., David B. Knight, and E. Kofman, eds. *Nationalism, Self-Determination and Political Geography.* London and New York: Croom Helm, 1988.

Morrill, Richard L. "Gerrymandering." *Focus* 41, no. 3 (1991):23–27.

Newhouse, John. "Europe's Rising Regionalism," *Foreign Affairs* 76 (January/February 1997): 67–84.

O'Tuathail, Gearoid. *Critical Geopolitics.* London: Routledge, 1996.

Prescott, J. R. V. *Political Frontiers and Boundaries.* New York: Harper Collins Academic, 1990.

"The Rise of Europe's Little Nations." *The Wilson Quarterly* 18, no. 1 (1994):50–81.

Scholfield, Clive H., ed. *Global Boundaries.* World Boundaries Series, vol. 1. London: Routledge, 1994.

Shelley, Fred M., et al. *Political Geography of the United States.* New York: Guilford Press, 1996.

Short, John R. *An Introduction to Political Geography.* 2d ed. New York: Routledge, 1993.

Spencer, Metta, ed. *Separatism: Democracy and Disintegration.* Lanham, Md.: Rowman and Littlefield, 1997.

Taylor, Peter J. *Political Geography: World-Economy, Nation-State and Locality.* 3d ed. New York: John Wiley & Sons, 1993.

Williams, Allan M. *The European Community: The Contradiction of Integration.* 2d ed. Oxford, England: Basil Blackwell, 1994.

Don't forget about Dushkin's *Annual Editions Online: Geography* at http://dushkin.com/aeonline/. See preface for details.

PART THREE

THE LOCATIONAL TRADITION

Given, then, our population map, what has it to show us? Starting from the most generally known before proceeding towards the less familiar, observe first the mapping of London—here plainly shown, as it is properly known, as Greater London—with its vast population streaming out in all directions—east, west, north, south—flooding all the levels, flowing up the main Thames valley and all the minor ones, filling them up, crowded and dark, and leaving only the intervening patches of high ground pale. . . . This octopus of London, polypus rather, is something curious exceedingly, a vast irregular growth without previous parallel in the world of life—perhaps likest to the spreading of a great coral reef. Like this, it has a stony skeleton, and living polypes—call it, then, a "man-reef" if you will. Onward it grows, thinly at first, the pale tints spreading further and faster than the others, but the deeper tints of thicker population at every point steadily following on. Within lies a dark crowded area; of which, however, the daily pulsating centre calls on us to seek some fresh comparison to higher than coralline life.[1]

Thus, in the lavish prose of the early 20th century, did Patrick Geddes, "spokesman for man and the environment," capsulize the spread of people in the London region. The physical environment, the vibrating human life of cities, and the change in human settlement patterns are all implicit in his remarks, as they are in the locational tradition of geography.

A theme that has run like a thread through the first two parts of this book may aptly be termed the *theme of location.* Distributions of climates, landforms, culture hearths, and religions were points of interest in our examinations of physical and cultural landscapes. In Part III of our study, the analysis of location becomes our central concern rather than just one strand among many, and the locational tradition of geography is brought to the fore.

In the study of economic, resource, or urban geography, a central question has to do with *where* certain types of human activities take place, not just in absolute terms but also in relation to one another. In studying the distribution of a given activity, such as commercial grain farming, the concern is with identifying the locational pattern, analyzing it to see why it is arranged the way it is, and searching for underlying principles of location. Because the elements of culture are integrated, such a study often must take into consideration the ways in which the location of one element affects the locations of others. For example, certain characteristics of the labor pool helped alter the distribution of the textile industry in America, and that alteration had economic consequences for both New England and the South and affected the location of other economic activities as well.

In Chapter 10, "Economic Geography," our attention is directed to the location of economic activities as we seek to answer the question of why they are distributed as they are. What forces operate to make some regions extremely productive and others less so, or some enterprises successful and others not? The stages of economic activities from primary production to sophisticated technical services are examined.

The resources upon which humankind draws and depends for sustenance and development are not distributed uniformly in quantity or quality over the earth. Chapter 11, "The Geography of Natural Resources," centers on the relationship between the demands people place on the spatially varying environment—demands that are culture-bound—and the ability of the environment to sustain those demands. In an increasingly integrated economic and social world of growing consumption demands, both access to resources, particularly energy and mineral resources, and their wise and productive use loom ever more important in human affairs. The understanding of resources in both their distributions and exploitation characteristics is a theme of growing interest in the locational tradition of geography.

The increasing urbanization of the world's population represents another stage of the process of cultural development, economic change, and environmental control, aspects of which have been the topics of earlier separate chapters. In Chapter 12, urban areas themselves are viewed in two ways. First, we focus on cities as points in space with patterns of distribution and specializations of function that invite analysis. Second, we consider cities as landscape entities with specialized arrangements of land uses resulting from recognizable processes of urban growth and development. Within their confines, and by means of the interconnected functional systems that they form, cities summarize a present stage of economic patterns and resource use of humankind. The consideration of urban geography thus logically concludes our examination of the locational tradition of geography.

[1]From *Patrick Geddes: Spokesman for Man and the Environment,* ed. by Marshall Stalley. New Brunswick, N.J.: Rutgers University Press, 1972, p. 123.

A cruise ship stopping at Miami, Florida.

CHAPTER

10

ECONOMIC GEOGRAPHY

The fuels and petrochemicals produced in this Australian petroleum refinery and others elsewhere are essential to modern advanced economies.

The crop bloomed luxuriantly that summer of 1846. The disaster of the preceding year seemed over and the potato, the sole sustenance of some 8 million Irish peasants, would again yield in the bounty needed. Yet within a week, wrote Father Mathew, "I beheld one wide waste of putrefying vegetation. The wretched people were seated on the fences of their decaying gardens . . . bewailing bitterly the destruction that had left them foodless." Colonel Gore found that "every field was black," and an estate steward noted that "the fields . . . look as if fire has passed over them." The potato was irretrievably gone for a second year; famine and pestilence were inevitable.

Within five years, the settlement geography of the most densely populated country in Europe was forever altered. The United States received a million immigrants, who provided the cheap labor needed for the canals, railroads, and mines that it was creating in its rush to economic development. New patterns of commodity flows were initiated as American maize for the first time found an Anglo-Irish market—as part of Poor Relief—and then entered a wider European market that had also suffered general crop failure in that bitter year. Within days, a microscopic organism, the cause of the potato blight, had altered the economic and human geography of two continents.

That alteration resulted from a complex set of intertwined causes and effects that demonstrates once again our repeated observation that apparently separate physical and cultural geographic patterns are really interconnected parts of a single reality. Central among those patterns are the ones the economic geographer isolates for special study.

Simply stated, **economic geography** is the study of how people earn their living, how livelihood systems vary by area, and how economic activities are spatially interrelated and linked. Of course, we cannot really comprehend the totality of the economic pursuits of 6 billion human beings. We cannot examine the infinite variety of production and service activities found everywhere on the earth's surface; nor can we trace all their innumerable interrelationships, linkages, and flows. Even if that level of understanding were possible, it would be valid for only a fleeting instant of time, for economic activities are constantly undergoing change.

Economic geographers seek consistencies. They attempt to develop generalizations that will aid in the comprehension of the maze of economic variations characterizing human existence. From their studies emerges a deeper awareness of the dynamic, interlocking diversity of human enterprise, of the impact of economic activity on all other facets of human life and culture, and of the increasing interdependence of differing national and regional economic systems (see "Economic Regions" in Chapter 13). The potato blight, although it struck only one small island, ultimately affected the economies of continents. In like fashion, the depletion of America's natural resources and the "deindustrialization" of its economy are altering the relative wealth of countries, flows of international trade, domestic employment and income patterns, and more (Figure 10.1).

The Classification of Economic Activity and Economies

The search for understanding of livelihood patterns is made more difficult by the complex environmental and cultural realities controlling the economic activities of humans. Many production patterns are rooted in the spatially variable circumstances of the *physical environment.* The staple crops of the humid tropics, for example, are not part of the agricultural systems of the midlatitudes; livestock types that thrive in American feedlots or on western ranges are not adapted to the Arctic tundra or the margins of the Sahara desert. The unequal distribution of useful mineral deposits gives some regions and countries economic prospects and employment opportunities denied to others. Forestry and fishing depend on still other natural resources unequal in occurrence, type, and value.

Within the bounds of the environmentally possible, economic or production decisions may be conditioned by *cultural considerations.* For example, culturally based food preferences rather than environmental limitations may dictate the choice of crops or livestock. Maize is a preferred grain in Africa and the Americas, wheat in North America, Australia, Argentina, southern Europe, and Ukraine, and rice in much of Asia. Pigs are not produced in Muslim areas. Level of *technological development* of a culture will affect its recognition of resources or its ability to exploit them. Preindustrial societies do not know of, or need, iron ore or coking coal underlying their hunting, gathering, or gardening grounds. *Political decisions* may encourage or discourage—through subsidies, protective tariffs, or production restrictions—patterns of economic activity. And, ultimately, production is controlled by *economic factors* of demand, whether that demand is expressed through a free market mechanism, through government intervention, or through the consumption requirements of a single family producing for its own needs.

Categories of Activity

Such regionally varying environmental, cultural, technological, political, or market conditions add spatial details to more generalized ways of categorizing the world's productive work. One approach to that categorization is to view economic activity as ranged along a continuum of both increasing complexity of product or service and increasing distance from the natural environment. Seen from that perspective, a small number of distinctive stages of production and service activities may be distinguished (Figure 10.2).

FIGURE 10.1 These Japanese cars unloading at Seattle were forerunners of a continuing flow of imported goods capturing a share of the domestic market traditionally held by American manufacturers. Established patterns of production and exchange are constantly subject to change in a world of increasing economic and cultural interdependence and of changing relative competitive strengths.

Primary activities are those that harvest or extract something from the earth. They are at the beginning of the production cycle, where humans are in closest contact with the resources and potentialities of the environment. Such primary activities involve basic foodstuff and raw material production. Hunting and gathering, grazing, agriculture, fishing, forestry, and mining and quarrying are examples. **Secondary activities** are those that add value to materials by changing their form or combining them into more useful, and therefore more valuable, commodities. That provision of *form utility* may range from simple handicraft production of pottery or woodenware to the delicate assembly of electronic goods or space vehicles (Figure 10.3). Copper smelting, steel making, metal working, automobile production, textile and chemical industries—indeed, the full array of *manufacturing* and *processing industries*—are included in this phase of the production process. Also included are the production of *energy* (the "power company") and the *construction* industry.

Tertiary activities consist of those business and labor specializations that provide *services* to the primary and secondary sectors and *goods* and *services* to the general community and to the individual. They include professional, clerical, and personal services. They constitute the vital link between producer and consumer, for tertiary occupations importantly include the wholesale and retail *trade* activities necessary in highly interdependent societies.

In economically advanced societies, many individuals and some entire organizations are engaged in the processing and dissemination of information and in the administration and control of their own or other enterprises. The term **quaternary** is applied to this fourth class of economic activities, which is composed entirely of services rendered by white collar professionals working in education, government, management, information processing, and research. The distinctions between tertiary and quaternary activities are further developed later in this chapter under the section "Tertiary and Beyond."

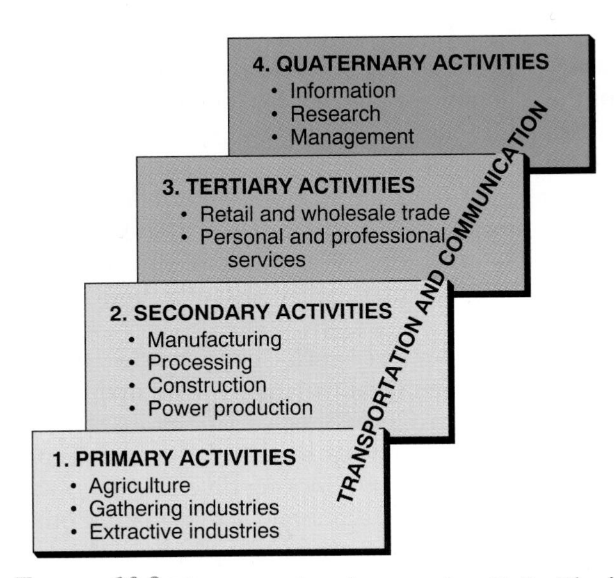

FIGURE 10.2 **The categories of economic activity.** The four main sectors of the economy do not stand alone. They are connected and integrated by transportation and communication services and facilities not assigned to any single sector but common to all.

FIGURE 10.3 These logs entering a lumber mill are products of *primary production.* Processing them into boards, plywood, or prefabricated houses is a *secondary activity* that increases their value by altering their form.
©Mark Gibson.

These categories of production and service activities help us to see an underlying structure to the near-infinite variety of things people do to earn a living and to sustain themselves. But they tell us little about the organization of the larger economy of which the individual worker or enterprise is a part. For that sort of organizational understanding of world and regional economies we look to *systems* rather than *components* of economies.

Types of Economic Systems

Broadly viewed, national economies of the last half of the 20th and start of the 21st century fall into one of three major types of system: subsistence, commercial, or planned. None of these economic systems is or has been "pure." That is, none exists in isolation in an increasingly interdependent world. Each, however, displays certain underlying characteristics based on its distinctive forms of resource management and economic control.

In a **subsistence economy,** goods and services are created for the use of the producers and their kinship groups. Therefore, there is little exchange of goods and only limited need for markets. In the **commercial economies** that have become dominant in nearly all culture areas, producers or their agents freely market their goods and services, the laws of supply and demand determine price and quantity, and market competition is the primary force shaping production decisions and distributions. In the extreme form of **planned economies** associated with the communist-controlled societies that have now collapsed in nearly every country where they were formerly created or imposed, producers or their agents disposed of goods and services through government agencies that controlled both supply and price. The quantities produced and the locational patterns of production were tightly programmed by central planning departments.

Rigidly planned economies no longer exist in their classical form; they have been modified or dismantled now in favor of free market structures or are only partially retained in the lesser degree of economic control associated with governmental supervision or ownership of selected sectors of increasingly market-oriented economies. Nevertheless, their landscape evidence lives on. The physical structures, patterns of production, and imposed regional interdependencies they created remain to influence or distort the economic decisions of successor societies.

In actuality, few people are members of only one of these systems, although one may be dominant. A farmer in India may produce rice and vegetables primarily for the family's consumption, but also save some of the produce to sell. In addition, members of the family may market cloth or other handicrafts they make. With the money derived from those sales, the Indian peasant is able to buy, among other things, clothes for the family, tools, or fuel. Thus, that Indian farmer is a member of at least two systems: subsistence and commercial.

In the United States, government controls on the production of various types of goods and services (such as growing wheat or tobacco, producing alcohol, constructing and operating nuclear power plants, or engaging in licensed personal and professional services) mean that the country does not have a purely commercial economy. To a limited extent, its citizens participate in a controlled and planned as well as in a free market environment. Many African, Asian, and Latin American market economies have been decisively shaped by government policies encouraging or demanding production of export commodities rather than domestic foodstuffs, or promoting through import restrictions the development of domestic industries not readily supported by the national market alone. Example after example would show that there are very few people in the world who are members of only one type of economic system (Figure 10.4).

Nonetheless, in a given country, one of the three systems tends to dominate, and it has in the past been relatively easy subjectively to classify countries by that dominance even while recognizing that some elements of the other two systems may exist within the controlling scheme. However, inevitably spatial patterns change, including those of economic systems and activities. For example, the commercial economies of Western European countries, some with sizable infusions of planned economy controls, are being restructured by both increased free market competition and supranational regulation under the World Trade Organization and the Common Market of the European Union (see p. 334). Many of the countries of Latin America, Africa, Asia, and the Middle East that traditionally were dominated by subsistence economies are adopting advanced technologies and simultaneously becoming either more commercial or more planned.

No matter what economic system may locally prevail, in all systems transportation is a key variable. No advanced economy can flourish without a well-connected transport network. All subsistence societies—or subsistence areas of

FIGURE 10.4 Independent street merchants, shop owners, and most farmers in modern China are members of both a planned and market system. The country's more than 15 million registered private businesses in the mid-1990s far exceeded the total number of private enterprises operating in 1949 when the communists took power. Since government price controls on most food items were removed in May 1985, free markets have multiplied. Increasingly, manufacturing, too, is being freed of central government control. In 1978, state-owned enterprises accounted for 78% of China's industrial output; by 1998, they produced only 34%, though they accounted for two-thirds of urban employment. In late 1997, the government announced its intention to sell off the majority of China's large and medium-sized state-owned industrial firms and most of its 300,000 smaller companies, converting them to shareholder ownership through publicly sold stock. Now, too, foreign investment is encouraged and capitalism with all its risks and rewards is substantially replacing the security and distortions of communism in the economic sphere. The photo shows a portion of the crowded produce market in Kunming, Yunnan Province.

developing countries—are characterized by their isolation from regional and world routeways (Figure 10.5), and that isolation restricts their progression to more advanced forms of economic structure.

Although the former sharp contrasts in economic organization are becoming blurred and national economic orientations are changing, the evident contrasts between subsistence, planned, and commercial systems have differently affected national patterns of livelihood, production, and economic decision making. Indeed, both approaches to economic classification—by types of activities and by organization of economies—help us to visualize and understand world economic geographic patterns. In this chapter, our path to that understanding leads through the successive categories of economic activity, from primary to quaternary, with emphasis on the technologies, spatial patterns, and organizational systems that are involved in each category.

Primary Activities: Agriculture

Before there was farming, *hunting and gathering* were the universal forms of primary production. These preagricultural pursuits are now practiced by at most a few thousand persons worldwide, primarily in isolated and remote pockets within the low latitudes and among the sparse populations of very high latitudes. The interior of New Guinea, rugged areas of interior Southeast Asia, diminishing segments of the Amazon rain forest, a few districts of tropical Africa and northern Australia, and parts of the Arctic regions still contain such preagricultural people. Their numbers are few and declining, and wherever they are brought into contact with more advanced cultures, their way of life is eroded or lost.

Agriculture, defined as the growing of crops and the tending of livestock, whether for the subsistence of the producers

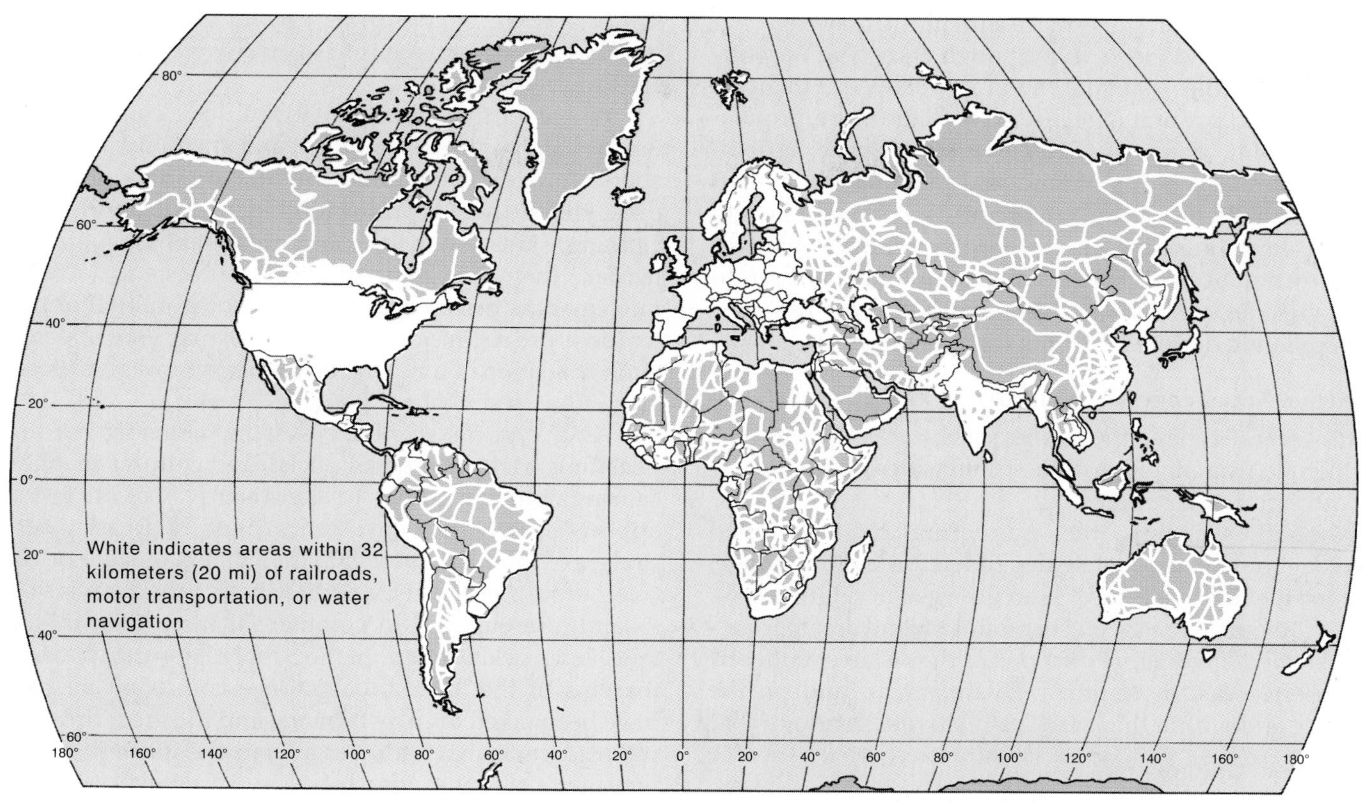

FIGURE 10.5 **Patterns of access and isolation.** Accessibility is a key measure of economic development and of the degree to which a world region can participate in interconnected market activities. Isolated areas of countries with advanced economies suffer a price disadvantage because of high transportation costs. Lack of accessibility in subsistence economic areas slows their modernization and hinders their participation in the world market.

Source: Copyright Permission: Hammond Incorporated, Maplewood, NJ 07040.

or for sale or exchange, has replaced hunting and gathering as economically the most significant of the primary activities. It is spatially the most widespread, found in all world regions where environmental circumstances permit. Crop farming alone covers some 15 million square kilometers (5.8 million sq mi) worldwide, about 10% of the earth's total land area. In many developing economies, at least three-fourths of the labor force is directly involved in farming and herding. In some, such as Nepal and Bhutan in Asia or Burundi in Africa, the figure is more than 90%.

In highly developed commercial economies, on the other hand, direct employment in agriculture involves only a small fraction of the labor force: less than 10% in most of Western Europe, 5% in Canada, and less than 3% in the United States (for the world pattern of the agricultural labor force, see Figure 7.6). Indeed, a declining number and proportion of farm workers, along with farm consolidation and increasing output, is typical in all present-day highly developed commercial agriculture systems.

It has been customary to classify agricultural societies on the twin bases of the importance of off-farm sales and the level of mechanization and technological advancement. *Subsistence, traditional* (or *intermediate*), and *advanced* (or *modern*) are usual terms employed to recognize both aspects. These are not mutually exclusive but rather are recognized stages along a continuum of farm economy variants. At one end lies production solely for family sustenance, using rudimentary tools and native plants. At the other is the specialized, highly capitalized, near-industrialized agriculture for off-farm delivery that marks advanced economies. Between these extremes is the middle ground of traditional agriculture, where farm production is in part destined for home consumption and in part oriented towards off-farm sale either locally or in national and international markets. We can most clearly see the variety of agricultural activities and the diversity of controls on their spatial patterns by examining the "subsistence" and "advanced" ends of the agricultural continuum.

Subsistence Agriculture

By definition, a *subsistence* economic system involves nearly total self-sufficiency on the part of its members. Production for exchange is minimal, and each family or close-knit social group relies on itself for its food and other most essential requirements. Farming for the immediate needs of the family is, even today, the predominant occupation of humankind. In most of Africa, much of Latin America, and most of southern and eastern Asia the majority of people are primarily concerned with feeding themselves from their own land and livestock.

Two chief types of subsistence agriculture may be recognized: *extensive* and *intensive.* Although each type has several variants, the essential contrast between them is realizable yield per unit of area used and, therefore, population-supporting potential. **Extensive subsistence agriculture** involves large areas of land and minimal labor input per hectare. Both product per land unit and population densities are low. **Intensive subsistence agriculture** involves the cultivation of small land holdings through the expenditure of great amounts of labor per acre. Yields per unit area and population densities are both high (Figure 10.6).

Extensive Subsistence Agriculture

Of the several types of *extensive subsistence* agriculture—varying one from another in their intensities of land use—two are of particular interest.

Nomadic herding, the wandering but controlled movement of livestock solely dependent on natural forage, is the most extensive type of land use system (Figure 10.6). That is, it requires the greatest amount of land area per person sustained. Over large portions of the Asian semidesert and desert areas, in certain highland areas, and on the fringes of and within the Sahara, a relatively small number of people graze animals for consumption by the herder group, not for market sale. Sheep, goats, and camels are most common, while cattle, horses, and yaks are locally important. The reindeer of Lapland were formerly part of the same system.

Whatever the animals involved, their common characteristics are hardiness, mobility, and an ability to subsist on sparse forage. The animals provide a variety of products: milk, cheese, and meat for food; hair, wool, and skins for clothing; skins for shelter; and excrement for fuel. For the herder, they represent primary subsistence. Nomadic movement is tied to sparse and seasonal rainfall or to cold temperature regimes and to the areally varying appearance and exhaustion of forage. Extended stays in a given location are neither desirable nor possible.

As a type of economic system, nomadic herding is declining. Many economic, social, and cultural changes are causing nomadic groups to alter their way of life or to disappear entirely. On the Arctic fringe of Russia, herders under communism were made members of state or collective herding enterprises. In northern Scandinavia, Lapps (Saami) are engaged in commercial more than in subsistence livestock farming. In the Sahel region of Africa on the margins of the Sahara, oases once controlled by herders have been taken over by farmers, and the great droughts of recent decades have forever altered the formerly nomadic way of life of thousands.

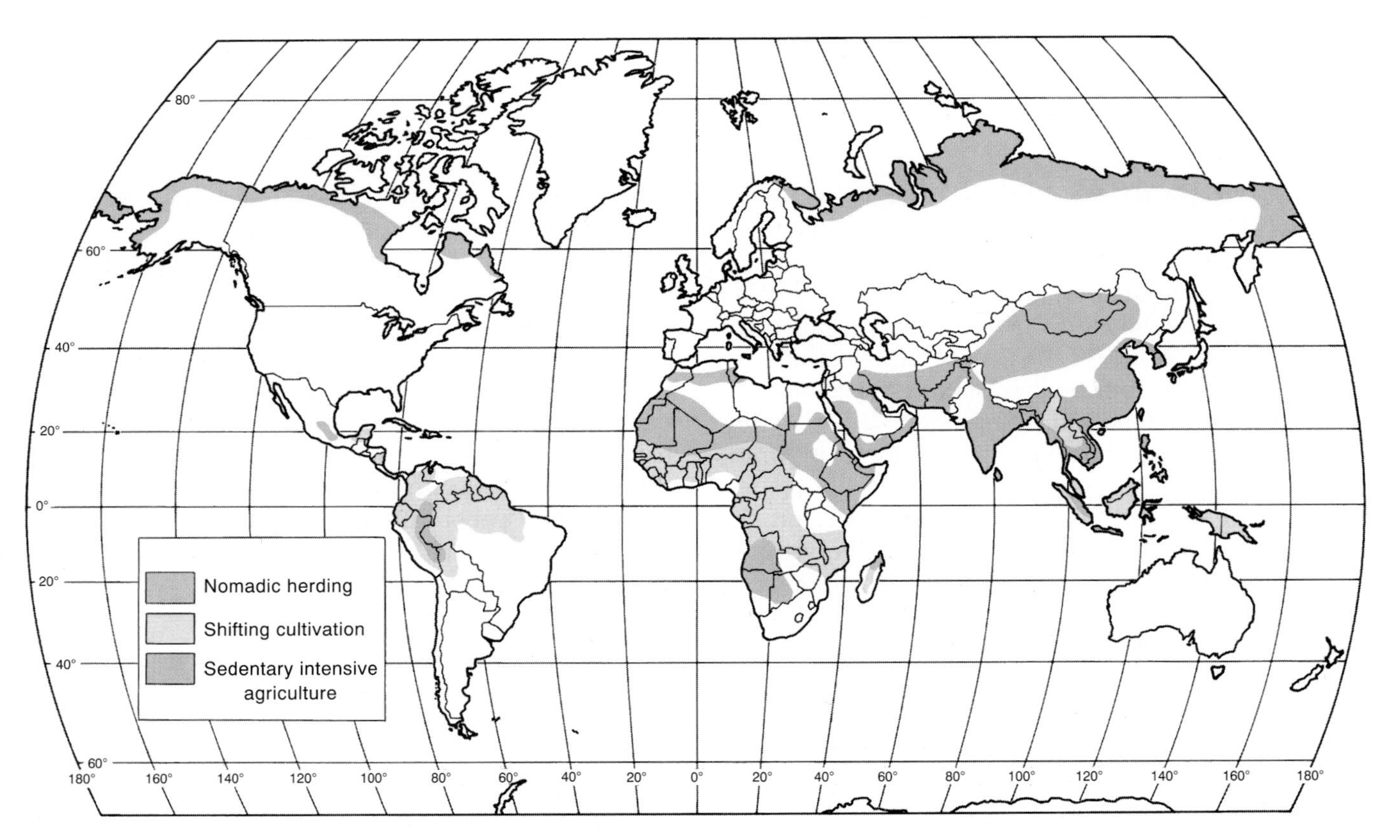

FIGURE 10.6 **Subsistence agricultural areas of the world.** Nomadic herding, supporting relatively few people, was the age-old way of life in large parts of the dry and cold world. Shifting or swidden agriculture maintains soil fertility by tested traditional practices in tropical wet and wet-and-dry climates. Large parts of Asia support millions of people engaged in sedentary intensive cultivation, with rice and wheat the chief crops.

On-Line Economic Geography

The range of subject matter in economic geography—and in this chapter—is broad and the number World Wide Web sites touching on some or all of it is great. Primary and secondary activities, domestic and international trade, developed and developing economies, economic theory, and the like, all are part of a vast web library of information. More than for most other subjects of this book, economic geography investigation will involve "surfing" the web, following leads, links, and interconnections in pursuit of the particular topics of interest to you. A few suggestions for getting started—tied to the main chapter topics—are offered here; nearly all provide extensive cross-references and links to related websites.

Primary Activities **Agriculture** sites worth checking include the *Food and Agriculture Organization of the United Nations*

http://www.fao.org/

and the *United States Department of Agriculture*

http://www.usda.gov/

Both provide guides to their own activities and agencies and links to other, related sites, as does the *Agriculture Canada* page.

http://www.agr.ca/

The *Iowa State University's Agriculture WWW Sites* page gives extensive links to agriculture-related sites on-line and is a good starting point for those with specific crop and livestock interests.

http//www.ag.iastate.edu/other.html

Other primary resources are considered on both international and purely domestic sites, including *Resources for the Future*

http://www.rff.org

and the U.S. Department of Interior's *Minerals Information* page.

http://minerals.er.usgs.gov/minerals

The *World Resources Institute* maintains websites for its areas of work in economics, forests, energy, sustainable agriculture, resources, and environmental management. Look for them through the Institute's home page.

http://www.wri.org/

Trade *The World Trade Organization,* established in 1995, is the principal agency of the world's multilateral trading system. Its home page includes access to documents discussing international conferences and agreements, reviewing its publications, and summarizing the current state of world trade.

http://www.wto.org/

Two other useful websites of broader topical interest, both yielding research results and statistical data for worldwide and national economies, are well worth checking. *The World Bank* is a leading source for country studies, research, and statistics covering all aspects of economic development and world trade; its site provides access to the contents of the bank's publications including its *World Development Report,* and to its research areas.

http://www.worldbank.org/

The Paris-based *Organization for Economic Cooperation and Development* site gives access to comparative statistics from 25 economically advanced countries and to the documents produced by the OECD itself.

www.oecd.org/

Secondary and Tertiary Activities The *U.S. Department of Commerce* is charged with promoting American business, manufacturing, and trade. Its home page connects with the websites of its constituent agencies.

http://www.doc.gov/

The *Bureau of Labor Statistics* page contains economic data, including unemployment rates, worker productivity, employment surveys and statistical summaries.

http://stats.bls.gov/blshome.html

A valuable Canadian counterpart is *Industry Canada,* with links to its own sectors, to other related branches of government, and to pertinent industry pages.

http://info.ic.gc.ca/

General "Economic" Sites *Econolink* advertises itself as featuring "the best web sites that have anything to do with economics." Its listings are selective with brief descriptions of site content; many referenced sites themselves have web links. Although more purely "economic" than "economic geographic," the page will help connect you to topics of particular interest to you.

http://www.progress.org/econolink

A much differently based and distributed form of extensive subsistence agriculture is found in all of the warm, moist, low-latitude areas of the world. There, many people engage in a kind of nomadic farming. Through clearing and use, the soils of those areas lose many of their nutrients (as soil chemicals are dissolved and removed by surface and groundwater or nutrients are removed from the land in the vegetables picked and eaten), and farmers cultivating them need to move on after harvesting several crops. In a sense, they rotate fields rather than crops to maintain productivity. This type of **shifting cultivation** has a number of names, the most common of which are *swidden* (an English localism for "burned clearing") and *slash-and-burn.*

Characteristically, the farmers hack down the natural vegetation, burn the cuttings, and then plant such crops as maize (corn), millet (a cereal grain), rice, manioc or cassava, yams, and sugarcane (Figure 10.7). Increasingly included in many of the crop combinations are such high value, labor-intensive commercial crops as coffee, providing the cash income that is evidence of the growing integration of all peoples into exchange economies. Initial yields—the first and second crops—may be very high, but they quickly become lower with each successive planting on the same plot. As that occurs, cropping ceases, native vegetation is allowed to reclaim the clearing, and gardening shifts to another newly prepared site. The first clearing will ideally not be used again for crops until, after many years, natural fallowing replenishes its fertility (see "Swidden Agriculture").

Nearly 5% of the world's people are still predominantly engaged in tropical shifting cultivation on some one-fifth of the world's land area (Figure 10.6). Since the essential characteristic of the system is the intermittent cultivation of the land, each family requires a total occupance area equivalent to the garden plot in current use plus all land left fallow for regeneration. Population densities are characteristically low, for much land is needed to support few people. It may be argued that shifting cultivation is a highly efficient cultural adaptation where land is abundant in relation to population, and levels of technology and capital availability are low. As those conditions change, the system becomes less viable.

Intensive Subsistence Agriculture

Nearly half of the people of the world are engaged in intensive subsistence agriculture, which predominates in areas shown in Figure 10.6. As a descriptive term, *intensive subsistence* is no longer fully applicable to a changing way of life and economy in which the distinction between subsistence and commercial is decreasingly valid. While families may still be fed primarily with the produce of their individual plots, the exchange of farm commodities within the system is considerable. Production of foodstuffs for sale in rapidly growing urban markets is increasingly vital for the rural economies of "subsistence farming" areas and for the sustenance of the growing proportion of national and regional populations no longer themselves engaged in farming. Nevertheless, hundreds of millions of Indians, Chinese, Pakistanis, Bangladeshis, and Indonesians, plus further millions in other Asian, African, and Latin American countries remain small-plot, mainly subsistence producers of rice, wheat, maize, millet, or pulses (peas, beans, and other legumes). Most live in monsoon Asia, and we will devote our attention to that area.

Intensive subsistence farmers are concentrated in such major river valleys and deltas as the Ganges and the Chang Jiang (Yangtze), and in smaller valleys close to coasts—level areas with fertile alluvial soils. These warm, moist districts are well suited to the production of rice, a crop that under ideal conditions can provide large amounts of food per unit of land. Rice also requires a great deal of time and attention, for planting rice shoots by hand in standing fresh water is a tedious art (Figure 10.8). In the cooler and drier portions of Asia north of 20°N, wheat is grown intensively, along with millet and, less commonly, upland rice.

Intensive subsistence farming is characterized by large inputs of labor per unit of land, by small plots, by the intensive use of fertilizers, mostly animal manure (see "The Economy of a Chinese Village"), and by the promise of high yields in good years. For food security and dietary custom, some other products are also grown. Vegetables and some livestock

FIGURE 10.7 An African swidden plot being fired. Stumps and trees left in the clearing will remain after the burn.
©Wolfgang Kaehler.

Swidden Agriculture

The following account describes shifting cultivation among the Hanunóo people of the Philippines. Nearly identical procedures are followed in all swidden farming regions.

When a garden site of about one-half hectare (a little over one acre) has been selected, the swidden farmer begins to remove unwanted vegetation. The first phase of this process consists of slashing and cutting the undergrowth and smaller trees with bush knives. The principal aim is to cover the entire site with highly inflammable dead vegetation so that the later stage of burning will be most effective. Because of the threat of soil erosion the ground must not be exposed directly to the elements at any time during the cutting stage. During the first months of the agricultural year, activities connected with cutting take priority over all others.

Once most of the undergrowth has been slashed, the larger trees must be felled or killed by girdling (cutting a complete ring of bark) so that unwanted shade will be removed. Some trees, however, are merely trimmed but not killed or cut, both to reduce labor and to leave trees to reseed the swidden during the subsequent fallow period.

The crucial and most important single event in the agricultural cycle is swidden burning. The main firing of a swidden is the culmination of many weeks of preparation in spreading and leveling chopped vegetation, preparing firebreaks to prevent flames escaping into the jungle, and allowing time for the drying process. An ideal burn rapidly consumes every bit of litter; in no more than an hour or an hour and a half, only smoldering remains are left.

Swidden farmers note the following as the benefits of a good burn: 1) removal of unwanted vegetation, resulting in a cleared swidden; 2) extermination of many animal and some weed pests; 3) preparation of the soil for dibble (any small hand tool or stick to make a hole) planting by making it softer and more friable; 4) provision of an evenly-distributed cover of wood ashes, good for young crop plants and protective of newly-planted grain seed. Within the first year of the swidden cycle, an average of between 40 and 50 different types of crop plants have been planted and harvested.

The most critical feature of swidden agriculture is the maintenance of soil fertility and structure. The solution is to pursue a system of rotation of one to three years in crop and ten to twenty in woody or bush fallow regeneration. When population pressures mandate a reduction in the length of fallow period, productivity tends to drop as soil fertility is lowered, marginal land is utilized, and environmental degradation occurs. The balance is delicate.

Adapted from *Hanunóo Agriculture*, by Harold C. Conklin (FAO Forestry Development Paper No. 12), 1957.

are part of the agricultural system, and fish may be reared in rice paddies and ponds. Cattle are a source of labor and of food. Food animals include swine, ducks, and chickens, but since Muslims eat no pork, hogs are absent in their areas of settlement. Because of poverty, tradition, and religion, Hindus eat relatively little meat. The large number of cattle in India are primarily important, indeed vital, for labor, as a source of milk, and as producers of fertilizer and fuel.

Not all of the world's subsistence farming is based in rural areas. Urban agriculture is a rapidly growing activity, providing a significant amount of the world's food, particularly to the swiftly expanding cities of developing countries. Using the garbage dumps of Jakarta, the rooftops of Mexico City, and meager dirt strips along roadways in Calcutta or Kinshasa, millions of people are feeding their own families and supplying local markets with vegetables, fruit, fish, and even meat—all produced within the cities themselves. Such urban agriculture, the United Nation reckons, accounts for one-seventh of the world's total food production. In China, cities produce 90% and more of the vegetables consumed; in Africa, a reported 20% of urban nutritional requirement is produced in the towns and cities. In all parts of the developing world, urban-origin foodstuffs have reduced the incidence of adult and child malnutrition in cities rapidly expanding by their own birth rates and by the growing influx of displaced rural folk.

Green Revolution

Population pressures and new agricultural technologies are forcing change upon traditional intensive subsistence farming practices and societies. **Green Revolution** is the shorthand reference to a complex of seed and management improvements adapted to the needs of intensive agriculture that have brought larger harvests from a given area of farmland. The "revolution," indeed, is basically part of a worldwide long transition to the industrialization of agriculture. Between 1967 and 1997, world grain production rose nearly 86%; more than three-quarters of that growth was due to increases in yields rather than expansions in cropland. For Asia as a whole, cereal yields grew by more than 40% between 1980 and 1996, accounted for largely by increases in China and India, and increased by over 35% in South America. These yield increases and the improved food supplies they represent have been particularly important in densely populated, subsistence farming areas heavily dependent on rice and wheat cultivation (Figure 10.9). In fact, despite rapidly growing population numbers over the period, the proportion of malnourished fell, the Food and

FIGURE 10.8 Transplanting rice seedlings requires arduous hand labor by all members of the family. The newly flooded diked fields, previously plowed and fertilized, will have their water level maintained until the grain is ripe. This photograph was taken in Indonesia; the scene is repeated wherever subsistence wet-rice agriculture is practiced.

©Sean Sprague/Impact Visuals.

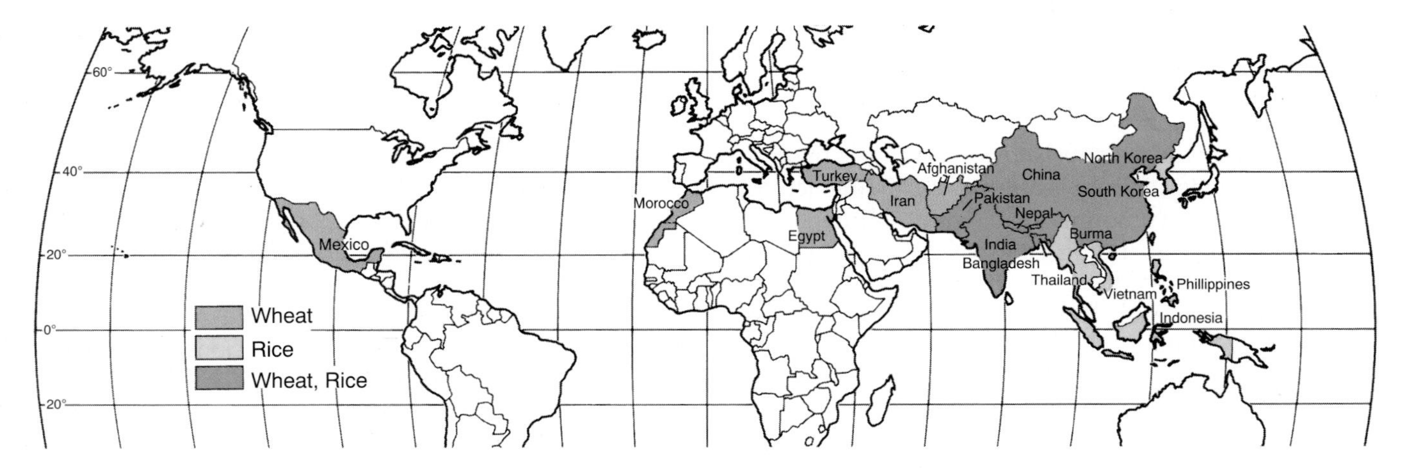

FIGURE 10.9 **Early beneficiaries of the Green Revolution.** In the 11 countries that were early adopters of new rice varieties and cropping techniques, average yields increased by 52% between 1965 and 1983. In the rest of the world, they actually dropped by 4% during the same period. Wheat yields increased 66% in the 9 reporting countries (excluding Mexico); in the rest of the world they grew only 29%. By the late 1990s in the developing countries as a group, new high-yielding crop varieties were being used on 55% of the farmland planted in wheat, 57% of land planted in rice, and 53% of maize land.

Redrawn with permission from Robert E. Huke, "The Green Revolution," *Journal of Geography,* vol. 84, no. 6, 1985. Copyright © 1985 National Council for Geographic Education, Indiana, PA.

The Economy of a Chinese Village

The village of Nanching is in subtropical southern China on the Zhu River delta near Guangzhou (Canton). Its subsistence agricultural regime was described by a field investigator, whose account is here condensed. The system is found in its essentials in other rice-oriented societies.

In this double-crop region, rice was planted in March and August and harvested in late June or July and again in November. March to November was the major farming season. Early in March the earth was turned with an iron-tipped wooden plow pulled by a water buffalo. The very poor who could not afford a buffalo used a large iron-tipped wooden hoe for the same purpose.

The plowed soil was raked smooth, fertilizer was applied, and water was let into the field, which was then ready for the transplanting of rice seedlings. Seedlings were raised in a seedbed, a tiny patch fenced off on the side or corner of the field. Beginning from the middle of March, the transplanting of seedlings took place. The whole family was on the scene. Each took the seedlings by the bunch, ten to fifteen plants, and pushed them into the soft inundated soil. For the first thirty or forty days the emerald green crop demanded little attention except keeping the water at a proper level. But after this period came the first weeding; the second weeding followed a month later. This was done by hand, and everyone old enough for such work participated. With the second weeding went the job of adding fertilizer. The grain was now allowed to stand to "draw starch" to fill the hull of the kernels. When the kernels had "drawn enough starch," water was let out of the field, and both the soil and the stalks were allowed to dry under the hot sun.

Then came the harvest, when all the rice plants were cut off a few inches above the ground with a sickle. Threshing was done on a threshing board. Then the grain and the stalks and leaves were taken home with a carrying pole on the peasant's shoulder. The plant was used as fuel at home.

As soon as the exhausting harvest work was done, no time could be lost before starting the chores of plowing, fertilizing, pumping water into the fields, and transplanting seedlings for the second crop. The slack season of the rice crop was taken up by chores required for the vegetables which demanded continuous attention, since every peasant family devoted a part of the farm to vegetable gardening. In the hot and damp period of late spring and summer, eggplant and several varieties of squash and beans were grown. The green-leafed vegetables thrived in the cooler and drier period of fall, winter, and early spring. Leeks grew the year round.

When one crop of vegetables was harvested, the soil was turned and the clods broken up by a digging hoe and leveled with an iron rake. Fertilizer was applied, and seeds or seedlings of a new crop were planted. Hand weeding was a constant job; watering with the long-handled wooden dipper had to be done an average of three times a day, and in the very hot season when evaporation was rapid, as frequently as six times a day. The soil had to be cultivated with the hoe frequently as the heavy tropical rains packed the earth continuously. Instead of the two applications of fertilizer common with the rice crop, fertilizing was much more frequent for vegetables. Besides the heavy fertilizing of the soil at the beginning of a crop, usually with city garbage, additional fertilizer, usually diluted urine or a mixture of diluted urine and excreta, was given every ten days or so to most vegetables.

Agriculture Organization (FAO) reports, from 36% to 20% of the population of developing countries between 1965 and 1995.

Expanded food production made possible through the Green Revolution has helped alleviate some of the shortages and famines predicted for intensive subsistence agricultural regions since the early 1960s. But a price has been paid. The Green Revolution is commercially oriented and demands high inputs of costly hybrid seeds, mechanization, irrigation, fertilizers, and pesticides. As it is adopted, traditional and subsistence agriculture are being displaced. Lost, too, are the food security that distinctive locally adapted native crop varieties provide and the nutritional diversity and balance that multiple-crop intensive gardening assure. Subsistence farming, wherever practiced, is oriented toward risk-minimization. Many differentially hardy varieties of a single crop guarantee some yield whatever adverse weather, disease, or pest problems might occur. Commercial agriculture, however, aims at profit-maximization, not minimal food security.

The presumed benefits of the Green Revolution are not available to all subsistence agricultural areas or advantageous to everyone engaged in farming (see "Women and the Green Revolution"). Africa is a case in point. Green Revolution crop improvements have concentrated on wheat, rice, and maize (corn). Of these, only maize is important in Africa, where principal food crops include millet, sorghum, cassava, manioc, yams, cowpeas, and peanuts. New varieties of maize resistant to the drought and acidic soils common in Africa were announced in the middle 1990s. However, both the belated research efforts directed to other African crops and the great range of growing conditions on the continent

Women and the Green Revolution

Women farmers grow at least half of the world's food and up to 80% in some African countries. They are responsible for an even larger share of food consumed by their own families: 80% in sub-Saharan Africa, 65% in Asia, and 45% in Latin America and the Caribbean. Further, women comprise between one-third and one-half of all agricultural laborers in developing countries. For example, African women perform about 90% of the work of processing food crops and 80% of the work of harvesting and marketing. Despite their fundamental role, however, women do not share equally with men in the rewards from agriculture, nor are they always beneficiaries of presumed improvements in agricultural technologies and practices. Often, they cannot own or inherit the land on which they work and frequently have difficulty in obtaining improved seeds or fertilizers available to male farmers.

As a rule, women farmers work longer hours and have lower incomes than do male farmers. This is not because they are less competent. Rather, it is due to restricting cultural and economic factors. First, most women farmers are involved in subsistence farming and food production for the local market that yields little cash return. Second, they have far less access than men to credit at bank or government-subsidized rates that would make it possible for them to acquire the Green Revolution technology, such as hybrid seeds and fertilizers. Third, in some cultures women cannot own land and so are excluded from agricultural improvement programs and projects aimed at landowners. For example, many African agricultural development programs are based on the conversion of communal land, to which women have access, to private holdings, from which they are excluded. In Asia, inheritance laws favor male over female heirs, and female-inherited land is managed by husbands; in Latin America, discrimination results from the more limited status held by women under the law.

At the same time, the Green Revolution and its greater commercialization of crops has generally required an increase in labor per hectare, particularly in tasks typically reserved for women, such as weeding, harvesting, and postharvest work. If women are provided no relief from their other daily tasks, the Green Revolution for them may be more burden than blessing. But when mechanization is added to the new farming system, women tend to be losers. Frequently, such predominantly female tasks as harvesting or dehusking and polishing of grain—all traditionally done by hand—are given over to machinery, displacing rather than employing women. Even the application of chemical fertilizers (a man's task) instead of cow dung (women's work) has reduced the female role in agricultural development programs. The loss of those traditional female wage jobs means that already poor rural women and their families have insufficient income to improve their diets even in the light of substantial increases in food availability through Green Revolution improvements.

If women are to benefit from the Green Revolution, new cultural norms—or culturally acceptable accommodations within traditional household, gender, and customary legal relations—will be required. These must permit or recognize women's landowning and other legal rights not now clearly theirs, access to credit at favorable rates, and admission on equal footing with males to government assistance programs. Recognition of those realities fostered the Food and Agriculture Organization of the United Nations' "FAO Plan of Action for Women in Development (1996–2001)" aimed at stimulating and facilitating efforts to enhance the role of women as contributors and beneficiaries of economic, social, and political development. Objectives of the plan include promoting gender-based equity in control of productive resources; enhancing women's participation in decision- and policy-making processes at all levels, local and national; and providing women with access to credit to enable them to engage as creators and owners of small-scale manufacturing, trade, or service businesses.

The model for that credit access is the Grameen Bank, established by a Bangladeshi economist in 1976. Based on the conviction that access to credit should be a basic human right, the bank extends "microcredit"—the average loan is U.S. $100—for "microenterprises," with women the primary borrowers. By 1998, the bank had made over 2 million loans in some 35,000 villages. More than 93% of the borrowers have been women and repayment rates reach 97%. The Grameen concept has spread from its Bangladesh origins to elsewhere in Asia and to Latin American and Africa. By 1998, over 10 million customers of microcredit institutions and small credit cooperatives were operating worldwide. These represent, however, only some 2% of the estimated 500 million women who still have virtually no access to credit or to the economic, social, or educational benefits that come from its availability.

suggest that the dramatic regionwide increases in food production experienced with rice in Southeast Asia will be delayed or perhaps never realized in the African context.

Commercial Agriculture

Few people or areas still retain the isolation and self-containment characteristic of pure subsistence economies. Nearly all have been touched by a modern world of trade and exchange and, in response, have adjusted their traditional economies. Modifications of subsistence agricultural systems have inevitably made them more complex by imparting to them at least some of the diversity and linkages of activity that mark the advanced economic systems of the more developed world. Farmers in those systems produce not for their own subsistence but primarily for a market off the farm itself. They are part of integrated exchange economies in which agriculture is but one element in a complex structure that includes employment in mining, manufacturing, processing, and the service activities of the tertiary and quaternary sectors. In those economies, agricultural patterns presumably reflect production responses to market demand expressed through price, and are related to the consumption requirements of the larger society rather than to the immediate needs of farmers themselves.

Production Controls

Agriculture within modern, developed economies is characterized by specialization—by enterprise (farm), by area, and even by country; by *off-farm sale* rather than subsistence production; and by *interdependence* of producers and buyers linked through markets. Farmers in a free market economy supposedly produce those crops that their estimates of market price and production cost indicate will yield the greatest return. Theoretically, farm products in short supply will command an increased market price. That, in turn, should induce increased production to meet the demand with a consequent reduction of market price to a level of equilibrium with production costs.

Supply, demand, and the market price mechanism are the presumed controls on agricultural production in commercial economies. In reality, they are joined by a number of nonmarket governmental influences that may be as decisive as market forces in shaping farmers' options and spatial production patterns. If there is a glut of wheat on the market, for example, the price per ton will come down and the area sown to it should diminish. It will also diminish regardless of supply if governments, responding to economic or political considerations, impose acreage controls.

Distortions of market control may also be introduced to favor certain crops or commodities through subsidies, price supports, market protections, and the like. The political power of farmers in the European Union, for example, has secured generous product subsidies and, for the Union, immense unsold stores of butter, wine, and grains until 1992, when reforms began to reduce the unsold stockpiles even while increasing total farm spending. In Japan, the home market for rice is largely protected and reserved for Japanese rice farmers even though their production efficiencies are low and their selling price is high by world market standards. In the United States, programs of farm price supports, acreage controls, financial assistance, and other governmental involvements in agriculture have been of recurring and equally distorting effect (Figure 10.10).

A Model of Agricultural Location

Early in the 19th century, before such governmental influences were the norm, Johann Heinrich von Thünen (1783–1850) observed that lands of apparently identical physical properties were used for different agricultural purposes. Around each major urban market center, he noted, developed a set of concentric rings of different farm products (Figure 10.11). The ring closest to the market specialized in perishable commodities that were both expensive to ship and in high demand. The high prices they could command in the urban market made their production an appropriate use of high-valued land near the city. Rings of farmlands farther away from the city were used for less perishable commodities with lower transport costs, reduced demand, and lower market prices. General farming and grain farming

FIGURE 10.10 Open storage of 1 million bushels of Iowa corn. In the world of commercial agriculture, supply and demand are not always in balance. Both the bounty of nature in favorable crop years and the intervention of governmental programs that distort production decisions can create surpluses for which no market is readily available.
©Bill Gillette/Stock, Boston.

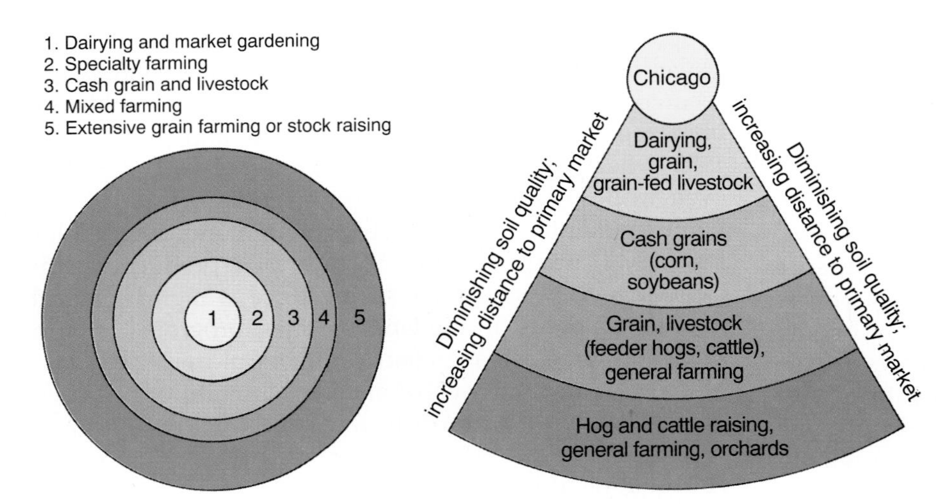

FIGURE 10.11 (a) **von Thünen's model.** Recognizing that as distance from the market increases, the value of land decreases, von Thünen developed a descriptive model of intensity of land use that holds up reasonably well in practice. The most intensively produced agricultural crops are found on land close to the market; the less intensively produced commodities are located at more distant points. The numbered zones of the diagram represent modern equivalents of the theoretical land use sequence von Thünen suggested some 175 years ago. As the metropolitan area at the center increases in size, the agricultural specialty areas are displaced outward, but the relative position of each is retained. (b) **A schematic view of the von Thünen zones** in the sector south of Chicago. There, farmland quality decreases southward as the boundary of recent glaciation is passed and hill lands are encountered in southern Illinois. On the margins of the city near the market, dairying competes for space with livestock feeding and suburbanization. Southward into flat, fertile central Illinois, cash grains dominate. In southern Illinois, livestock rearing and fattening, general farming, and some orchard crops are the rule.

(b) Modified with permission from Bernd Andreae, *Farming, Development and Space: A World Agricultural Geography*, trans. Howard F. Gregor (Berlin; Hawthorne, N.Y.: Walter de Gruyter Publishers, 1981).

replaced the market gardening of the inner ring. At the outer margins of profitable agriculture, farthest from the single central market, livestock grazing and similar extensive land uses were found.

To explain why this should be so, von Thünen proposed a formal spatial model, perhaps the first developed to analyze human activity patterns. He concluded that the uses to which parcels were put were a function of the differing "rent" values placed on seemingly identical lands. Those differences, he claimed, reflected the cost of overcoming the distance separating a given farm from a central market town ("A portion of each crop is eaten by the wheels," he observed). The greater the distance, the higher was the operating cost to the farmer, since transport charges had to be added to other expenses. When a commodity's production costs plus its transport costs just equaled its value at the market, a farmer was at the economic margin of its cultivation. A simple exchange relationship ensued: the greater the transportation cost, the lower the rent that could be paid for land if the crop produced was to remain competitive in the market.

Since in the simplest form of the model transport costs are the only variable, the relationship between land rent and distance from market can be easily calculated by reference to each competing crop's *transport gradient.* Perishable commodities such as fruits and vegetables would encounter high transport rates per unit of distance; other items such as grain would have lower rates. Land rent for any farm commodity decreases with increasing distance from the central market, and the rate of decline is determined by the transport gradient for that commodity. Crops that have both the highest market price and the highest transport costs will be grown nearest to the market. Less perishable crops with lower production and transport costs will be grown at greater distances away (Figure 10.12). Since in this model transport costs are uniform in all directions away from the center, the concentric zonal pattern of land use called the **von Thünen rings** results.

The von Thünen model may be modified by introducing ideas of differential transport costs, variations in topography or soil fertility, or changes in commodity demand and market price. With or without such modifications, von Thünen's analysis helps explain the changing crop patterns and farm sizes evident on the landscape at increasing distance from major cities, particularly in regions dominantly agricultural in economy. Farmland close to markets takes on high value, is used *intensively* for high-value crops, and is subdivided into relatively small units. Land far from markets is used *extensively* and in larger units.

In dominantly industrial and postindustrial economies, it has been suggested, the basic forces determining agricultural land use near cities are those associated with urban expansion itself, and von Thünen regularities are less certain. Rather, irregularities and uncertainties of peripheral city growth, the encroachment on agricultural land by expansion from two or more cities, and the withholding of land from farming in anticipation of subdivision may locally reverse the von Thünen intensity rings. Where those urbanizing forces dominate, the agricultural pattern often may be one of increasing—rather than decreasing—intensity with distance from the city.

Intensive Commercial Agriculture

Farmers producing for off-farm sales who apply large amounts of capital (for machinery and fertilizers, for example) and/or labor per unit of land engage in **intensive commercial agriculture.** The crops that justify such costly inputs are characterized by high yields and high market value per unit of land. They include fruits, vegetables, and

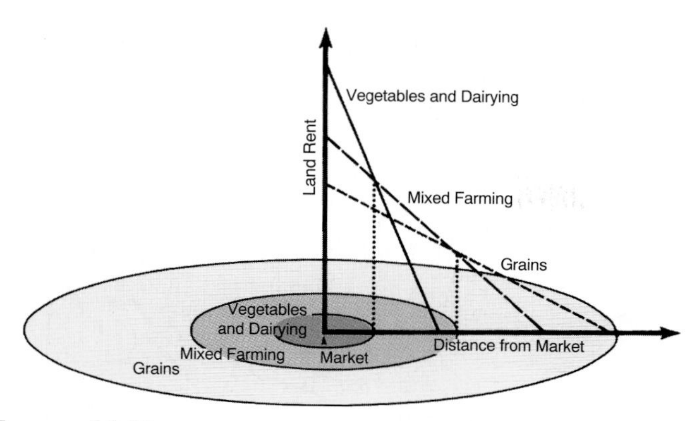

FIGURE 10.12 **Transport gradients and agricultural zones.**

dairy products, all of which are highly perishable. Dairy farms and *truck farms* (horticultural or "market garden" farms that produce a wide range of vegetables and fruits) are found near most medium-sized and large cities. Since their products are perishable, transport costs increase because of the special handling that is needed, such as use of refrigerated trucks and custom packaging. This is another reason for locations close to market. Note the distribution of truck and fruit farming in Figure 10.13, which also suggests the importance of climatic conditions in fruit and vegetable growing.

Livestock-grain farming involves the growing of grain to be fed on the producing farm to livestock, which constitute the farm's cash product. In Western Europe, three-fourths of crop land is devoted to production for animal consumption; in Denmark, 90% of all grains are fed to livestock for conversion into meat, butter, cheese, and milk. Although livestock-grain farmers work their land intensively, the value of their product per unit of land is usually less than that of the truck farm. Consequently, in North America at least, livestock-grain farms are farther from the main markets than are horticultural and dairy farms. In general, the livestock-grain belts of the world are close to the great coastal and industrial zone markets. The "corn belt" of the United States and the livestock region of Western Europe are two examples.

Extensive Commercial Agriculture

Farther from the market, on less expensive land, there is less need to use the land intensively. Cheaper land gives rise to larger farm units. **Extensive commercial agriculture** is typified by large wheat farms and livestock ranching.

Large-scale wheat farming requires sizable capital inputs for planting and harvesting machinery, but the inputs per unit of land are low; wheat farms are very large. Nearly half the farms in Saskatchewan, for example, are more than 400 hectares (1000 acres). The average farm in Kansas is over 400 hectares, and in North Dakota more than 525 (1300 acres). In North America, the spring wheat (planted in spring, harvested in autumn) region includes the Dakotas, eastern Montana, and the southern parts of the Prairie Provinces of Canada. The winter wheat (planted in fall, harvested in midsummer) belt focuses on Kansas and includes adjacent sections of neighboring states (Figure 10.13). Argentina is the only South American country to have comparable large-scale wheat farming. In the Eastern Hemisphere, the system is fully developed only east of the Volga River in northern Kazakhstan and the southern part of Western Siberia, and in southeastern and western Australia.

Livestock ranching differs significantly from livestock-grain farming and, by its commercial orientation and distribution, from the nomadism it superficially resembles. A product of the 19th-century growth of urban markets for beef and wool in Western Europe and northeastern United States, ranching has been primarily confined to areas of European settlement. It is found in western United States and adjacent sections of Mexico and Canada (Figure 10.13); the grasslands of Argentina, Brazil, Uruguay and Venezuela; the interior of Australia; the uplands of South Island, New Zealand; and the Karoo and adjacent areas of South Africa (Figure 10.14). All except New Zealand and the humid pampas of South America have semiarid climates. All, even the most remote from markets, were a product of improvements in transportation by land and sea, refrigeration of carriers, and of meat-canning technology. In all, introduced beef cattle or sheep replaced original native fauna, such as bison on North America's Great Plains, almost always with eventual severe environmental deterioration. More recently, the midlatitude demand for beef has been blamed for expanded cattle production and extensive destruction of tropical rain forests in Central America and the Amazon Basin.

In all of the ranching regions, livestock range (and the area exclusively in ranching) has been reduced as crop farming has encroached on its more humid margins, as pasture improvement has replaced less nutritious native grasses, and as grain fattening has supplemented traditional grazing. Since ranching can be an economic activity only where alternative land uses are nonexistent and land quality is low, ranching regions of the world characteristically have low population densities, low capitalization per land unit, and relatively low labor requirements.

Special Crops

Proximity to the market does not guarantee the intensive production of high-value crops, should terrain or climatic circumstances hinder it. Nor does great distance from the market inevitably determine that extensive farming on low-priced land will be the sole agricultural option. Special circumstances, most often climatic, make some places far from markets intensively developed agricultural areas. Two special cases are agriculture in Mediterranean climates and in plantation areas (Figure 10.14).

Most of the arable land in the Mediterranean basin itself is planted to grains, and much of the agricultural area is used for grazing. *Mediterranean agriculture* as a specialized farming economy, however, is known for grapes, olives,

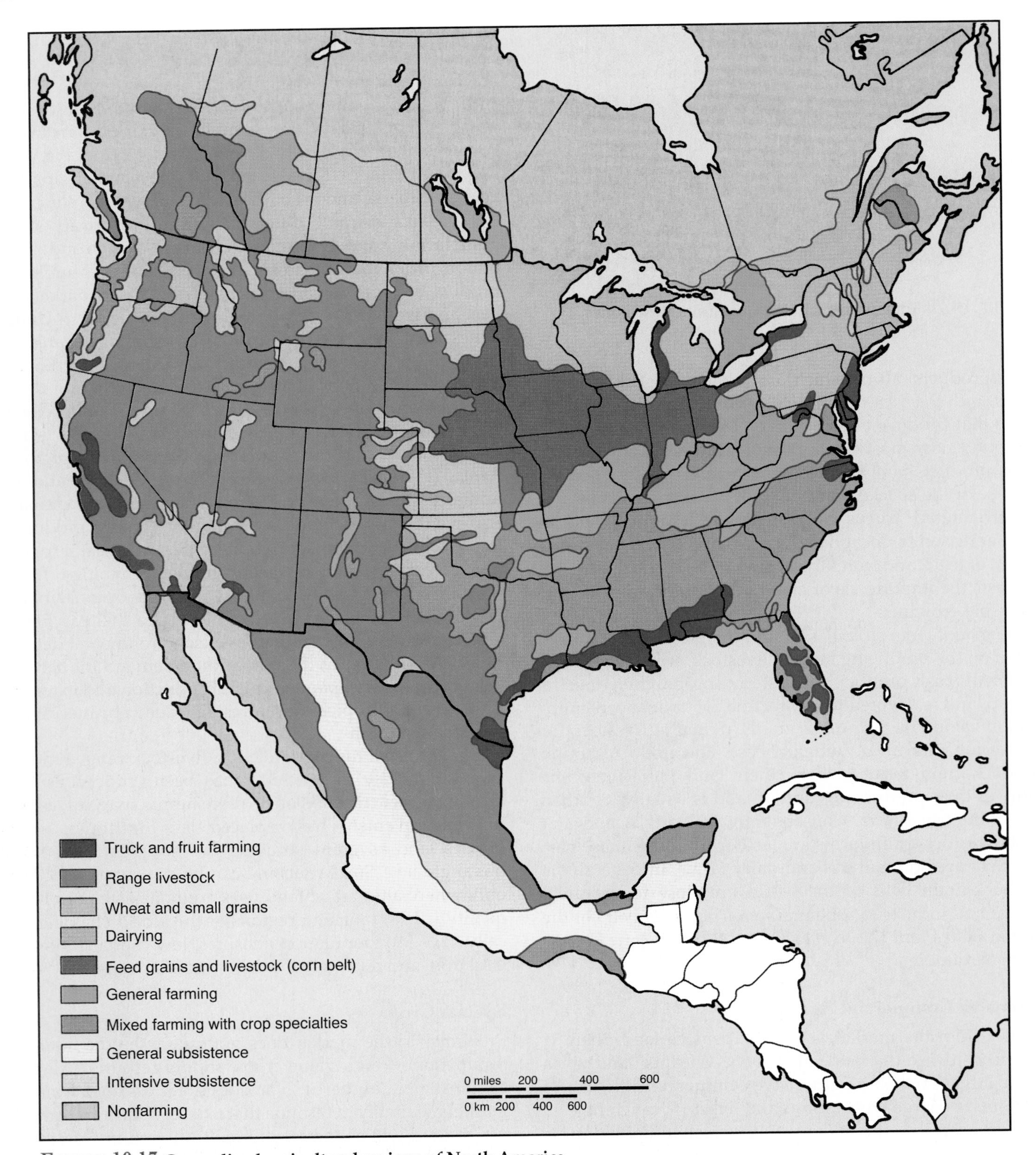

FIGURE 10.13 Generalized agricultural regions of North America.
Sources: U.S. Bureau of Agricultural Economics; Agriculture Canada; Mexico, Secretaría de Agricultura y Recursos Hidráulicos.

oranges, figs, vegetables, and similar commodities. These crops need warm temperatures all year round and a great deal of sunshine in the summer. The Mediterranean agricultural lands indicated in Figure 10.14 are among the most productive agricultural lands in the world. The precipitation regime of Mediterranean climate areas—winter rain and summer drought—lends itself to the controlled use of water. Of course, much capital must be spent for the irrigation systems, another reason for the intensive use of the land for high-value crops that are, for the most part, destined for

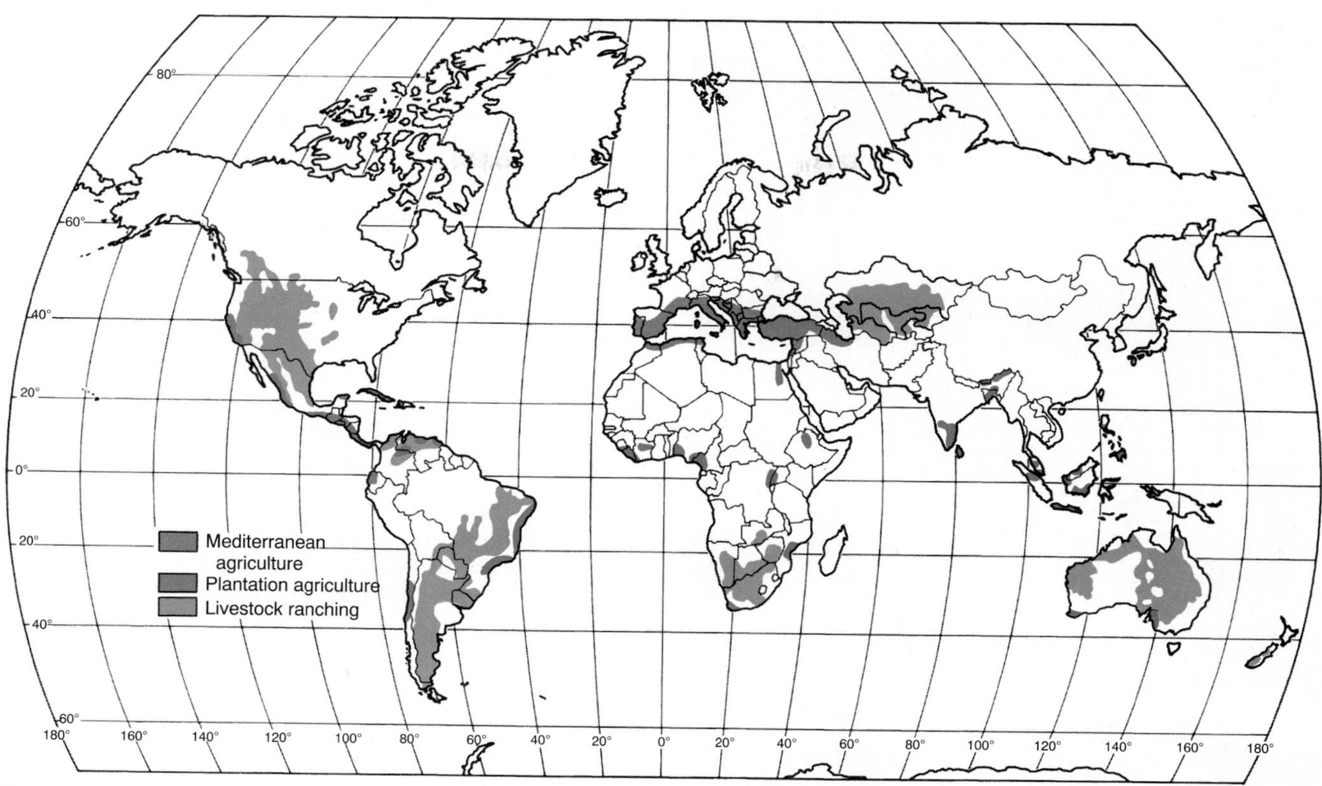

FIGURE 10.14 **Livestock ranching and special crop agriculture.** Livestock ranching is primarily a midlatitude enterprise catering to the urban markets of industrialized countries. Mediterranean and plantation agriculture are similarly oriented to the markets provided by advanced economies of Western Europe and North America. Areas of Mediterranean agriculture—all of roughly comparable climatic conditions—specialize in similar commodities, such as grapes, oranges, olives, peaches, and vegetables. The specialized crops of plantation agriculture are influenced by both physical geographic conditions and present or, particularly, former colonial control of areas.

export to industrialized countries or areas outside the Mediterranean climatic zone and even, in the case of Southern Hemisphere locations, to markets north of the equator.

Climate is also considered the vital element in the production of what are commonly but imprecisely known as *plantation crops.* The implication of **plantation** is the introduction of a foreign element—investment, management, and marketing—into an indigenous culture and economy, often employing an introduced alien labor force. The plantation itself is an estate whose resident workers produce one or two specialized crops. Those crops, although native to the tropics, were frequently foreign to the areas of plantation establishment: African coffee and Asian sugar in the Western Hemisphere and American cacao, tobacco, and rubber in Southeast Asia and Africa are examples (Figure 10.15). Entrepreneurs in western countries such as England, France, the Netherlands, and the United States became interested in the tropics partly because they afforded them the opportunity to satisfy a demand in temperate lands for agricultural commodities not producible in the market areas. Custom and convenience usually retain the term "plantation" even where native producers of local crops dominate, as they do in cola nut production in Guinea, spice growing in India or Sri Lanka, or sisal production in the Yucatán.

The major plantation crops and the areas where they are produced include tea (India and Sri Lanka); jute (India and Bangladesh); rubber (Malaysia and Indonesia); cacao (Ghana and Nigeria); cane sugar (Cuba and the Caribbean area, Brazil, Mexico, India, and the Philippines); coffee (Brazil and Colombia); and bananas (Central America). As Figure 10.14 suggests, for ease of access to shipping most plantation crops are cultivated along or near coasts since production for export rather than for local consumption is the rule.

Agriculture in Planned Economies

As their name implies, planned economies have a degree of centrally directed control of resources and of key sectors of the economy that permits the pursuit of governmentally determined objectives. When that control is extended to the agricultural sector—as it was during particularly the latter part of the 20th century in communist-controlled Soviet Union, Eastern Europe, mainland China, and elsewhere—state and collective farms and agricultural communes replace private farms or subsistence gardens, crop production is divorced from market control or family need, and prices are established by plan rather than by demand or production cost.

FIGURE 10.15 An Indonesian rubber plantation worker collects latex in a small cup attached to the tree and cuts a new tap just above the previous one. The scene typifies classical plantation agriculture in general. The plantation was established by foreign capital (Dutch) to produce a nonnative (American) commercial crop for a distant, midlatitude market using nonnative (Chinese) labor supervised by foreign (Dutch) managers. Present-day ownership, management, and labor may have changed, but the nature and market orientation of the enterprise remains.
©Wolfgang Kaehler.

Such extremes of rural control have in recent years been relaxed or abandoned in most formerly strictly planned economies. Wherever past centralized control of agriculture was imposed and long endured, however, traditional rural landscapes were altered and the organization of rural society was disrupted. The programs decreed by Stalin and his successors in the former Soviet Union, for example, fundamentally restructured the geography of agriculture of that country, transforming the Soviet countryside from millions of small farm holdings to a consolidated pattern of fewer than 50,000 centrally controlled operating units. Reestablishment of private agriculture was undertaken quickly in Russia following USSR collapse in late 1991. By July of 1992, 120,000 newly created family farms marked the first phase of privatization of the gigantic communal farms of the Soviet period. Nonetheless, even by the late 1990s more than 80% of Russian agriculture was still under the control of inefficient and undercapitalized remnants of the Soviet farm system.

A similar progression from private and peasant agriculture, through collectivization, and back to what is virtually a private farming system has taken place in the planned economy of the People's Republic of China. After its assumption of power in 1949, the communist regime redistributed all farmlands to some 350 million peasants in inefficiently small (0.2 hectare, or 0.5 acre) subsistence holdings that were totally inadequate for the growing food needs of the country. By the end of 1957, some 90% of peasant households had been collectivized into about 700,000 communes, a number further reduced in the 1970s to 50,000 communes averaging some 13,000 members.

After the death of Chairman Mao in 1976, what became effectively a private farming system was reintroduced when 180 million new farms were allocated for unrestricted use to peasant families under rent-free leases. Most staple crops are still sold under enforced contracts at fixed prices to government purchasers, but increasingly vegetables and meat are sold on the free market (see Figure 10.4) and a slow shift from grain production to vegetables and industrial crops is occurring. Total agricultural output increased by over 50% between 1978 and the late-1990s, and even allowing for population growth, per capita food production gains were impressive: up by about one-third.

OTHER PRIMARY ACTIVITIES

In addition to agriculture, primary economic activities include fishing, forestry, and the mining and quarrying of minerals. These industries involve the direct exploitation of natural resources that are unequally available in the environment and differentially evaluated by different societies. Their development, therefore, depends on the occurrence of perceived resources, the technology to exploit their natural availability, and the cultural awareness of their value. (The definition, perception, and utilization of resources are explored in depth in Chapter 11.)

Two of them—fishing and forestry—are **gathering industries** based on harvesting the natural bounty of renewable resources, though ones in serious danger of depletion through overexploitation. Mining and quarrying are **extractive industries,** removing nonrenewable metallic and nonmetallic minerals, including the mineral fuels, from the earth's crust. They are the initial raw material phase of modern industrial economies.

Fishing and Forestry

Fish as a food resource and forests as a source for building materials, cellulose, and fuel are heavily exploited renewable resources. Livelihoods based on both resources are spatially widespread and parts of both subsistence and advanced economies. In both fishing and forestry, evidence is mounting that at least locally, their *maximum sustainable yield* is actually or potentially being exceeded. **Maximum sustainable yield** of a resource is the largest volume or rate of use that will not impair its ability to be renewed or to maintain the same future productivity. For fishing or forestry, that level is marked by a catch or harvest equal to the net growth of the replacement stock. When sustainable yields are exceeded, stocks may be depleted below recovery levels and livelihoods and economies based on the resource are endangered or destroyed. Both fishery and forestry resources and exploitations are discussed in Chapter 11.

Mining and Quarrying

Societies at all stages of economic development can and do engage in agriculture, fishing, and forestry. The extractive industries—mining and drilling for nonrenewable mineral wealth—emerged only when cultural advancement and economic necessity made possible a broader understanding of the earth's resources. Now those extractive industries provide the raw material and energy base for the way of life experienced by people in the advanced economies and are the basis for an important part of the international trade connecting the developed and developing countries of the world.

Extractive industries depend on the exploitation of minerals unevenly distributed in amounts and concentrations determined by past geologic events, not by contemporary market demand. Because usable deposits are the result of geologic accident, there is no necessary relationship between the size and wealth of a country and the resources its territory contains. Although larger states are more likely to be the beneficiaries of such "accidents," many smaller developing countries are major sources of one or more critical raw materials and, therefore, important participants in the growing international trade in minerals.

Transportation costs play a major role in determining where *low-value minerals* will be mined. Materials such as gravel, limestone for cement, and aggregate are found in such abundance that they have value only when they are near the site where they are to be used (Figure 10.16). For example, gravel for road building has value if it is at the road-building site, not otherwise. Transporting gravel hundreds of miles is an unprofitable activity.

The production of other minerals, especially *metallic minerals* such as copper, lead, and iron ore, is affected by a balance of three forces: the quantity available, the richness of the ore, and the distance to markets. A fourth factor, land acquisition and royalty costs, may equal or exceed other considerations in mine development decisions (see "Public Land, Private Profit"). Even if these conditions are favorable, mines may not be developed or even remain operating if supplies from competing sources are more cheaply available in the market. In the 1980s, more than 25 million tons of iron ore-producing capacity was permanently shut down in the United States and Canada as a result of such price competition. Similar declines occurred in North American copper, nickel, zinc, lead, and molybdenum mining as market prices fell below domestic production costs. Of course, increases in mineral prices may be reflected in opening or reopening mines that, at lower returns, were deemed

FIGURE 10.16 The Vancouver, British Columbia, municipal gravel quarry and storage yard. Proximity to market gives utility to low-value minerals unable to bear high transportation charges.
©Thomas Kitchin/Valan Photos.

Geography and Public Policy

Public Land, Private Profit

When President Ulysses S. Grant signed the Mining Act of 1872, the presidential and congressional goal was to encourage Western settlement and development by allowing any "hard-rock" miners (including prospectors for silver, gold, copper, and other metals) to mine federally owned land without royalty payment. It further permitted mining companies to gain clear title to publicly owned land and all subsurface minerals for no more than $12 a hectare ($5 an acre). Under those liberal provisions, mining firms have bought 1.3 million hectares (3.2 million acres) of federal land since 1872 and each year remove some $1.2 billion worth of minerals from government property. In contrast to the royalty-free extraction privileges granted to metal miners, oil, gas, and coal companies pay royalties of as much as 12.5% of their gross revenues for exploiting federal lands.

Whatever the merits of the 1872 law in encouraging economic development of lands otherwise unattractive to homesteaders, modern-day mining companies throughout the Western states have secured enormous actual and potential profits from the law's generous provisions. In Montana, a company claim to 810 hectares (2000 acres) of land would cost it less than $10,000 for an estimated $4 billion worth of platinum and palladium; in California, a gold mining company in 1994 sought title to 93 hectares (230 acres) of federal land containing a potential of $320 million of gold for less than $1200. Foreign as well as domestic firms may be beneficiaries of the 1872 law. In 1994, a South African firm arranged to buy 411 hectares (1016 acres) of Nevada land with a prospective $1.1 billion in gold from the government for $5100. And a Canadian firm in 1994 received title to 800 hectares (nearly 2000 acres) near Elko, Nevada, that cover a likely $10 billion worth of gold—a transfer that Interior Secretary Bruce Babbitt dubbed "the biggest gold heist since the days of Butch Cassidy." And in 1995, Mr. Babbitt conveyed about $1 billion worth of travertine (a mineral used in whitening paper) under 45 hectares (110 acres) of Idaho to a Danish-owned company for $275.

The "gold heist" characterization summarized a growing administration and congressional feeling that what was good in 1872 and today for metal mining companies was not necessarily beneficial to the American public that owns the land. In part, that feeling results from the fact that mining companies commit environmental sins that require public funding to repair or public tolerance to accept. The mining firms may destroy whole mountains to gain access to low-grade ores and leave toxic mine tailings, surface water contamination, and open-pit scarring of the landscape as they move on or disappear. Projected public costs of cleaning up 56 of the most damaged abandoned mining sites are estimated at $32 billion.

A congressional proposal introduced in 1993 would require mining companies to pay royalties of 8% on gross revenues for all hard-rock ores extracted and prohibit them from outright purchase of federal land. The royalty provision alone would have yielded nearly $100 million annually at 1994 levels of company income. Mining firms claim that imposition of royalties might well destroy America's mining industry. They stress both the high levels of investment they must make to extract and process frequently low-grade ores and the large number of high-wage jobs they provide as their sufficient contribution to the nation. The Canadian company involved in the Elko site, for example, reports that since it acquired the claims in 1987 from their previous owner, it has expended more than $1 billion, plus making additional donations for town sewer lines and schools and creating 1700 jobs. The American Mining Congress estimates the proposed 8% royalty charge would cost 47,000 jobs out of 140,000, and even the U.S. Bureau of Mines assumes a loss of 1100 jobs.

Questions to Consider

1. Do you believe the 1872 Mining Law should be repealed or amended? If not, what are your reasons for arguing for retention?

   If so, would you advocate the imposition of royalties on mining company revenues? At what levels, if any, should royalties be assessed? Should hard-rock and energy companies be treated equally for access to public land resources? Why or why not?
2. Would you propose to prohibit outright land sales to mining companies? If not, should sales prices be determined by surface value of the land or by the estimated (but unrealized) value of mineral deposits it contains?
3. Do you think that cleanup and other charges now borne by the public are acceptable in view of the capital investments and job creation of hard-rock companies? Do you accept the industry's claim that imposition of royalties would destroy American metal mining? Why or why not?

unprofitable. However, the developed industrial countries of commercial economies, whatever their former or even present mineral endowment, frequently find themselves at a competitive disadvantage against producers in developing countries with lower cost labor and state-owned mines with abundant, rich reserves.

When the ore is rich in metallic content, it is profitable to ship it directly to the market for refining. But, of course, the highest-grade ores tend to be mined first. Consequently, the demand for low-grade ores has been increasing in recent years as richer deposits have been depleted. Low-grade ores are often upgraded by various types of separation treatments at the mine site to avoid the cost of transporting waste materials not wanted at the market. Concentration of copper is nearly always mine-oriented (Figure 10.17); refining takes place near areas of consumption.

The large amount of waste in copper (98–99% or more of the ore) and in most other industrially significant ores should not be considered the mark of an unattractive deposit. Indeed, the opposite may be true. Many higher-content ores are left unexploited—because of the cost of extraction or the smallness of the reserves—in favor of the utilization of large deposits of even very low-grade ore. The attraction of the latter is a size of reserve sufficient to justify the long-term commitment of development capital and, simultaneously, to assure a long-term source of supply. At one time, high-grade magnetite iron ore was mined and shipped from the Mesabi area of Minnesota. Those deposits are now exhausted. Yet immense amounts of capital have been invested in the mining and processing into high-grade iron ore pellets of the virtually unlimited supplies of low-grade iron-bearing rock (taconite) still remaining.

Such investments do not assure the profitable exploitation of the resource. The metals market is highly volatile. Rapidly and widely fluctuating prices can quickly change profitable mining and refining ventures to losing undertakings. Marginal gold and silver deposits are opened or closed in reaction to trends in precious metals prices. Taconite beneficiation (waste material removal) in the Lake Superior region has virtually ceased in response to the decline of the U.S. steel industry and the price advantage of imported ores. In commercial economies, cost and market controls dominate economic decisions. In planned economies, cost may be a less important consideration than are such other concerns as goals of national development and resource independence.

The advanced economies have reached that status through their control and use of energy. Domestic supplies of mineral fuels, therefore, are often considered basic to national strength and independence. When those supplies are absent, developed countries are concerned with the availability and price of coal, oil, and natural gas in international trade. Mineral fuels and other energy sources are given extended discussion in Chapter 11.

FIGURE 10.17 Copper ore concentrating and smelting facilities at the Phelps-Dodge mine in Morenci, Arizona. Concentrating mills crush the ore, separating copper-bearing material from the rocky mass containing it. The great volume of waste material removed assures that most concentrating operations are found near the ore bodies. Smelters separate concentrated copper from other, unwanted minerals such as oxygen and sulfur. Because smelting is also a "weight-reducing" activity, it is frequently—though not invariably—located close to the mine as well.

TRADE IN PRIMARY PRODUCTS

International trade expanded by more than 6% a year between 1950 and 1996, a rate over 50% faster than growth in total gross world product. Primary commodities—agricultural goods, minerals, and fuels—make up a declining share of world trade, but still account for nearly one-fifth of the total dollar value of that international movement. The world distribution of supply and demand for those primary items in general results in an understandable pattern of commodity flow: from the producers located in less developed countries to the processors, manufacturers, and consumers of the more developed ones (Figure 10.18).

The reverse stream carries manufactured goods processed in the industrialized states back to the developing

FIGURE 10.18 Sugar being loaded for export at the port of Cebu in the Philippines. Much of the developing world depends on exports of mineral and agricultural products to the developed countries for the major portion of its income. Fluctuations in market demand and price of some of those commodities can have serious and unexpected consequences.
©United Nations Photo Library.

countries, a flow that accounted for 25% of total industrial country exports in the late 1990s. The trade benefits the developed countries by providing access to a continuing supply of industrial raw materials and foodstuffs not available domestically. It provides less developed states with capital to invest in their own development or to expend on the importation of manufactured goods, food supplies, or commodities—such as petroleum—they do not themselves produce. Increasingly, of course, low-wage developing countries are creating their own export manufacturing plants, often as subsidiaries or branch plants of European, North American, or Japanese owners. Where, however, their dependency is still on primary rather than manufactured goods trade, the terms of that exchange have been criticized as unequal and potentially damaging to commodity exporting countries.

Many developing countries rely greatly on the export of primary commodities—for example, on plantation agriculture crops such as coffee, cocoa, rubber, or sugar, or on unprocessed ores and metals, such as bauxite and copper. At times during the 1990s, some 90% of Zambia's exports were as copper, and 90% of those of Mauritius were sugar; early in the decade, 60% of Jamaica's exports were either bauxite or alumina. Many developing states have tried to break this degree of dependency as did, successfully, Zambia, Mauritius, and Jamaica by the end of the 1990s. Many others increasingly have emphasized the expansion of their industrial capabilities and export of manufactured goods, but with mixed and incomplete results. Although primary products declined as a share of exports of developing countries as a group from 80% in 1980 to 40% in the late 1990s, in Latin America, primary commodities often constitute more than two-thirds of country exports, and in about half of the African states they account for 75% or more.

Such dependency has left those countries exposed and weakened. Because commodity prices are volatile, rising sharply in periods of product shortage or international economic growth and, likely, falling abruptly with supply glut or international recession, commodity exporting states risk disruptive fluctuations in levels of income. This, in turn, can create serious difficulties in economic planning and debt repayment. Average commodity prices dropped by more than half in real terms between 1980 and 1995, representing an annual income loss to developing countries of $100 billion. Over the longer term, it has been argued, depletion of their natural capital through export of nonrenewable or destructively exploited resources endangers the future economic well-being of countries dependent wholly or largely on those natural resources for jobs and national income.

While prices paid for commodities tend to be low, prices charged for the manufactured goods offered in exchange by the developed countries tend to be high. In reaction to those perceived trade inequities, developing states promoted, in 1964, the establishment of the United Nations Conference on Trade and Development (UNCTAD). Its central constituency, the "Group of 77," now expanded to some 130 developing states, continues to press for a new world economic order based in part on an increase in the prices and values of exports from developing countries, a system of import preferences for their manufactured goods, and a restructuring of international cooperation to stress trade promotion and recognition of the special needs of poor countries.

Secondary Activities: Manufacturing

Although primary industries are locationally tied to the natural resources they gather or exploit, secondary and later stages of economic activity are less concerned with conditions of the physical environment. For them, enterprise location is more closely related to cultural and economic than to physical circumstances. They are movable rather than spatially tied, and are assumed to respond to recurring locational requirements and controls.

Those controls are rooted in observations about human spatial behavior in general and economic behavior in particular. We have already explored some of those assumptions in earlier discussions. We noted, for example, that the intensity of spatial interaction decreases with increasing separation of points—distance decay, we called it. Recall that von Thünen's model of agricultural land use was rooted in conjectures about transportation cost and land value relationships.

Such simplifying assumptions help us understand a presumed common set of controls and motivations guiding human economic behavior. We assume, for example, that people are *economically rational;* that is, given the information at their disposal, they make locational, production, or purchasing decisions in light of a perception of what is most cost-effective and advantageous. From the standpoint of producers or sellers of goods or services, it is assumed each is intent on *maximizing profit.* To reach that objective, a host of production and marketing costs and political, competitive, and other limiting factors may be considered, but the ultimate goal of profit-seeking remains clear. Finally, in commercial economies it is assumed that the best measure of the correctness of economic decisions is afforded by the market mechanism and the equilibrium between supply and demand that market prices establish (Figure 10.19).

Industrial Locational Models

When market principles are controlling, entrepreneurs seek to maximize profits by locating manufacturing activities at sites of lowest total input costs (and high revenue yields). In order to assess the advantages of one location over another, industrialists must evaluate the most important **variable costs.** They subdivide their total costs into categories and note how each cost will vary from place to place. In different industries, transportation charges, labor rates, power costs, plant construction or operation expenses, the interest rate of money, or the price of raw materials may be the major variable cost. The industrialist must look at each of these and, by a process of elimination, eventually select the lowest-cost site. If the producer then determines that a large enough market can be reached cheaply enough, the location promises to be profitable.

In the economic world, nothing remains constant. Because of a changing mix of input costs, production techniques, and marketing activities, many initially profitable locations do not remain advantageous. Migrations of population, technological advances, and changes in the demand for products affect industrialists and industrial locations greatly. The abandoned mills and factories of New England or of the steel towns of Pennsylvania, even the "deindustrialization" of America itself in the face of foreign competition, are testimony to the impermanence of the "best" locations.

The concern with variable costs as a determinant in industrial location decisions has inspired an extensive theoretical literature. Much of it is based on and extends the **least cost theory** proposed by the German location economist Alfred Weber (1868–1958), and sometimes called *Weberian analysis.* Weber explained the optimum location of a manufacturing establishment in terms of minimization of three basic expenses: relative transport costs, labor costs, and agglomeration costs. **Agglomeration** refers to the clustering of productive activities and people for mutual advantage. Such clustering can produce *agglomeration economies* through shared facilities and services. Diseconomies such as higher rents or wage levels resulting from competition for these localized resources may also occur.

Weber concluded that transport costs were the major consideration determining location. That is, the optimum

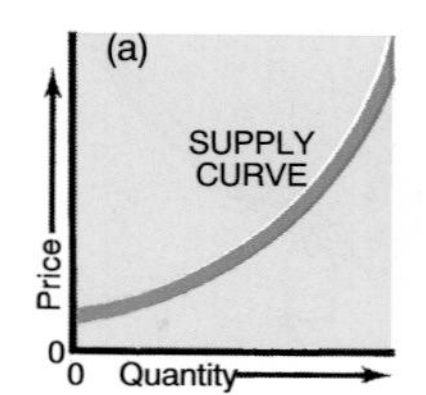

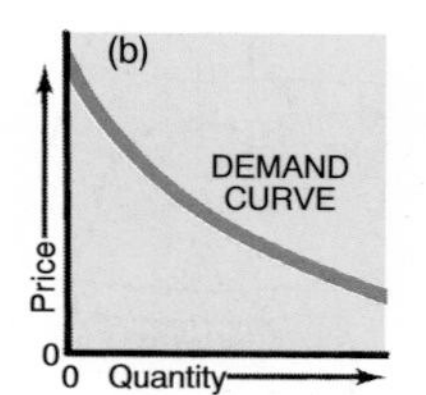

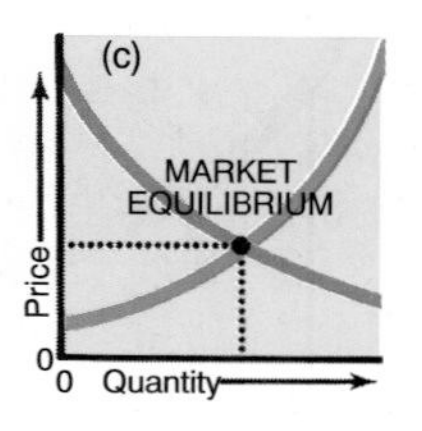

FIGURE 10.19 **Supply, demand, and market equilibrium.** The regulating mechanism of the market may be visualized graphically. (a) The *supply curve* tells us that as the price of a good increases, more of that good will be made available for sale. Countering any tendency for prices to rise to infinity is the market reality that the higher the price, the smaller the demand as potential customers find other purchases or products more cost-effective. (b) The *demand curve* shows how the market will expand as prices drop and the good becomes more affordable and attractive to more customers. (c) *Market equilibrium* is marked by the point of intersection of the supply and demand curves and determines the price of the good, the total demand, and the quantity bought and sold.

location would be found where the costs of transporting raw materials to the factory and finished goods to the market were at their lowest (Figure 10.20). He noted, however, if variations in labor or agglomeration costs were sufficiently great, a location determined solely on the basis of transportation costs might not in fact be the optimum one.

Assuming, however, transportation costs determine the "balance point," optimum location will depend on distances, the respective weights of the raw material inputs, and the final weight of the finished product. It may be either *material oriented* or *market oriented.* Material orientation reflects a sizable weight loss during the production process; market orientation indicates a weight gain (Figure 10.21).

For some theorists, Weber's least cost analysis is unnecessarily rigid and restrictive. They propose instead a *substitution principle* that recognizes that in many industrial processes it is possible to replace a declining amount of one input (e.g., labor) with an increase in another (e.g., capital for automated equipment) or to increase transportation costs while simultaneously reducing land rent. With substitution, a number of different points may be optimal manufacturing locations. Further, they suggest, a whole series of points may exist where total revenue of an enterprise just equals its total cost of producing a given output. These points, connected, mark the *spatial margin of profitability* and define the area within which profitable operation is possible (Figure 10.22). Location anywhere within the margin assures some profit and tolerates both imperfect knowledge and personal (rather than economic) considerations.

Other Locational Considerations

The behavior of individual firms seeking specific production sites under competitive conditions forms the basis of most industrial location theory. But such theory does not fully explain world or regional patterns of industrial localization or specialization, nor does it account for locational behavior that is uncontrolled by objective "factors," or that is directed by noncapitalistic planning goals.

Transport Characteristics

For example, within both national and international economies, the type and efficiency of transport media, as well as transportation costs, are central to the spatial patterning of production, explaining the location of a large variety of economic activities. Water-borne transportation is nearly always cheaper than any other mode of conveyance, and the enormous amount of commercial activity that takes place on coasts or on navigable rivers leading to

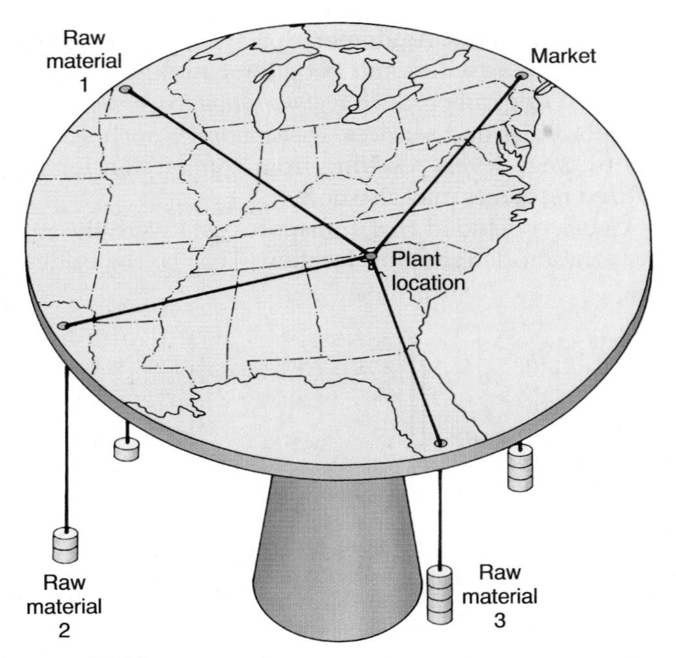

FIGURE 10.20 Plane table solution to a plant location problem. This mechanical model, suggested by Alfred Weber, uses weights to demonstrate the least transport cost point where there are several sources of raw materials. When a weight is allowed to represent the "pull" of raw material and market locations, an equilibrium point is found on the plane table. That point is the location at which all forces balance each other and represents the least-cost plant location.

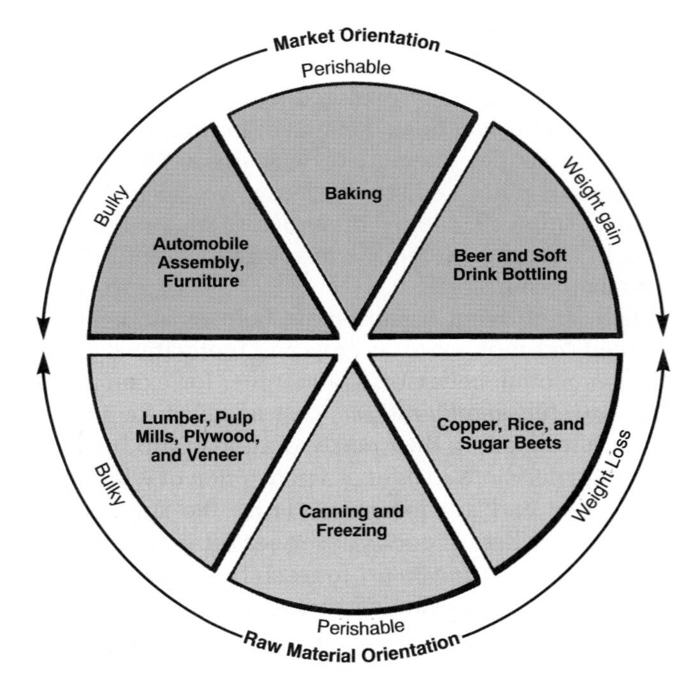

FIGURE 10.21 Spatial orientation tendencies. *Raw material orientation* is presumed to exist when there are limited alternative material sources, when the material is perishable, or when, in its natural state, it contains a large proportion of impurities or nonmarketable components. *Market orientation* represents the least-cost solution when manufacturing uses commonly available materials that add weight to the finished product, when the manufacturing process produces a commodity much bulkier or more expensive to ship than its separate components, or when the perishable nature of the product demands processing at individual market points.

Redrawn with permission from Truman A. Hartshorn, *Interpreting the City.* Copyright © 1980 John Wiley & Sons, Inc., New York, N.Y.

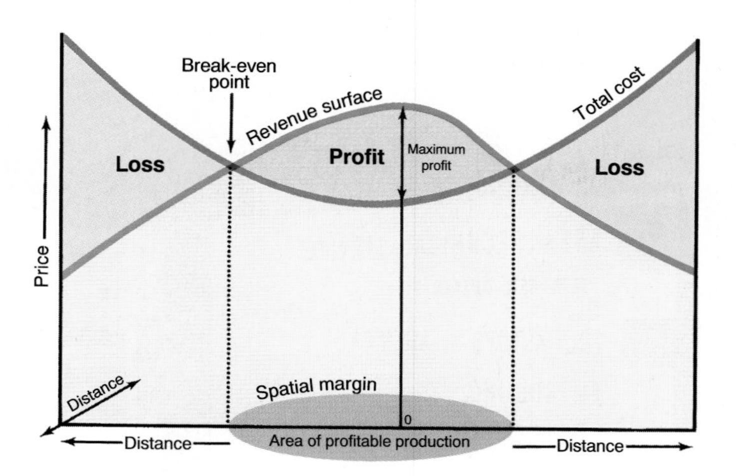

FIGURE 10.22 **The spatial margin of profitability.** In the diagram, **0** is the single optimal profit-maximizing location, but location anywhere within the area defined by the intersects of the total cost and total revenue surfaces will permit profitable operation. Some industries will have wide margins; others will be more spatially constricted. Skilled entrepreneurs may be able to expand the margins farther than less able industrialists. Importantly, a *satisficing* location may be selected by reasonable estimate even in the absence of the totality of information required for an *optimal* decision.

coasts is an indication of that cost advantage. When railroads were developed and the commercial exploitation of inland areas could begin, coastal sites continued to be important as more and more goods were transferred there between low cost water and land media. The advent of highway transportation vastly increased the number of potential "satisficing" manufacturing locations by freeing the locational decision from fixed-route production sites. Every change in carrier mode, efficiency, or cost structure has direct implications for locations of economic activity.

In the rare instance when transportation costs become a negligible factor in production and marketing, an economic activity is said to be *footloose.* Some manufacturing is located without reference to raw materials; for example, the raw materials for electronic products such as computers are so valuable, light, and compact that transportation costs have little bearing on where production takes place. Others are inseparable from the markets they serve and are so widely distributed that they are known as *ubiquitous industries.* Newspaper publishing, bakeries, and dairies, all of which produce a highly perishable commodity designed for immediate consumption, are examples.

Agglomeration Economies

The cumulative and reinforcing attractions of industrial concentration and urban growth are recognized locational factors, but ones not easily quantified. Both cost-minimizing and profit-maximizing theories make provision for *agglomeration,* the spatial concentration of people and activities for mutual benefit. That is, both recognize that areal grouping of industrial activities may produce benefits for individual firms that they could not experience in isolation. Those benefits—**external economies** or *agglomeration economies*—accrue in the form of savings from shared transport facilities, social services, public utilities, communication facilities, and the like. Collectively, these and other installations and services needed to facilitate industrial and other forms of economic development are called *infrastructure.* Areal concentration may also create pools of skilled and ordinary labor, of capital, ancillary business services, and, of course, a market built of other industries and urban populations. New firms, particularly, may find significant advantages in locating near other firms engaged in the same activity, for labor specializations and support services specific to that activity are already in place. Some may find profit in being near other firms with which they are linked either as customers or suppliers.

A concentration of capital, labor, management skills, customer base, and all that is implied by the term *infrastructure* will tend to attract still more industries from other locations to the agglomeration. In Weber's terms, that is, economies of association distort or alter locational decisions that otherwise would be based solely on transportation and labor costs, and once in existence agglomerations will tend to grow (Figure 10.23). Through a *multiplier effect,* each new firm added to the agglomeration will lead to the further development of infrastructure and linkages. As we shall see in Chapter 12, the "multiplier effect" also implies total (urban) population growth and thus the expansion of the labor pool and the localized market that are part of agglomeration economies.

Comparative Advantage

A third consideration affecting industrial location and specialization is the principle of **comparative advantage.** It tells us that areas tend to specialize in the production of those items for which they have the greatest relative advantage over other areas, or for which they have the least relative disadvantage, as long as free trade exists. This principle, basic to the understanding of regional specializations, applies as long as areas have different relative advantages for two or more goods.

Assume that two countries both have a need for, and are able to successfully produce domestically, two commodities. Further assume that there is no transport cost consideration. No matter what its cost of production of either commodity, Country A will choose to specialize in only one of them if, by that specialization and through exchange with Country B for the other, Country A stands to gain more than it loses. The key to comparative advantage is the utilization of resources in such a fashion as to gain, by specialization, a volume of production and a selling price that permit exchange for a needed commodity at a cost level that is below that of the domestic production of both.

At first glance, the concept of comparative advantage may at times seem to defy logic. For example, Japan may be

FIGURE 10.23 On a small scale, the planned industrial park furnishes its tenants external agglomeration economies similar to those offered by large urban concentrations to industry in general. An industrial park provides a subdivided tract of land developed according to a comprehensive plan for the use of (frequently) otherwise unconnected firms. Since the park developers, whether private companies or public agencies, supply the basic infrastructure of streets, water, sewage disposal, power, transport facilities, and, perhaps, private police and fire protection, park tenants are spared the additional cost of providing these services themselves. In some instances, factory buildings are available for rent, still further reducing firm development outlays. Counterparts of industrial parks for manufacturers are the office parks, research parks, science parks, and the like for "high-tech" firms and for enterprises in tertiary and quaternary services.

able to produce airplanes and home appliances more cheaply than the United States, thereby giving it an apparent advantage in both goods. But it benefits both countries if they specialize in the good in which they have a comparative advantage. In this instance, Japan's manufacturing cost structure makes it more profitable for Japan to specialize in the volume production of appliances and to buy airplanes from the United States, where large civilian and military markets encourage aircraft manufacturing specialization and efficiency.

When other countries' comparative advantages reflect lower labor, land, raw material and capital costs, manufacturing activities may voluntarily relocate from higher cost market locations to lower cost foreign production sites. Such voluntary *outsourcing*—producing parts or products abroad for domestic sale—by American manufacturers has employment and areal economic consequences no different from those resulting from successful competition by foreign companies or from industrial locational decisions favoring one section of the country over others.

A North American case in point is found along the northern border of Mexico. In the 1960s, Mexico enacted legislation permitting foreign (specifically, American) companies to establish "sister" plants, called *maquiladoras,* within 20 kilometers (12 miles) of the U.S. border for the duty-free assembly of products destined for reexport. By the late 1990s, more than 2500 such assembly and manufacturing plants had been established to produce a diversity of goods including electronic products, textiles, furniture, leather goods, toys, and automotive parts. The plants generated direct employment for more than one million Mexican workers (Figure 10.24) and a large number of U.S.

citizens, employees of growing numbers of American-side *maquila* suppliers and of diverse service-oriented business spawned by the "multiplier effect." The North American Free Trade Agreement (NAFTA) creating a single Canadian–United States–Mexican production and marketing community promises to turn "outsourcing" in the North American context from a search abroad for low-cost production sites to a review of best locations within a broadened unified economic environment (Figure 9.27a). The United States also benefits from outsourcing by other countries. Japan, Germany, Britain, Sweden, and others have established automobile and other manufacturing and processing plants to take advantage of American lower production and labor costs, with at least part of the product reexported to other markets.

FIGURE 10.24 American manufacturers, seeking lower labor costs, began to establish component manufacturing and assembly operations in the 1960s along the international border in Mexico. United States laws allowed finished or semifinished products to be brought into the country duty-free, as they are from the Converse Sport Shoe factory at Reynosa. *Outsourcing* has moved a large proportion of American electronics, small appliance, and garment industries to offshore subsidiaries or contractors in Asia and Latin America.

©Sharon Stewart/Impact Visuals.

Imposed Considerations

Locational theories dictate that in a pure, competitive economy, the costs of material, transportation, labor, and plant should be dominant in locational decisions. Obviously, neither in the United States nor in any other market economy do the idealized conditions exist. Other constraints—some representing cost considerations, others political or social impositions—also affect, perhaps decisively, the locational decision process. Land use and zoning controls, environmental quality standards, governmental area-development inducements, local tax abatement provisions or developmental bond authorizations, noneconomic pressures on quasi-governmental corporations, and other considerations constitute attractions or repulsions for industry outside of the context and consideration of pure theory (see "Contests and Bribery"). If these noneconomic forces become compelling, the assumptions of the commercial economy classification no longer obtain, and locational controls reminiscent of those imposed by current or former centrally planned economies become determining.

Transnational Corporations (TNCs)

Outsourcing is but one small expression of the growing international structure of modern manufacturing and service enterprises. Business and industry are increasingly stateless and economies borderless as giant **transnational corporations**—private firms that have established branch operations in nations foreign to their headquarters country—become ever-more important in the world space economy. By the late 1990s there were some 40,000 transnational (or multinational) companies controlling some 215,000 foreign affiliates. They varied greatly in size and power with the top 100 multinationals (excluding those in finance and banking) accounting for some $3.5 trillion in global assets and for about half of all assets held by transnationals outside of their home countries. Together, multinationals annually produce in sales through their foreign affiliates an amount greater than the world's total exports. Of the world's 100 largest economies, 51 are corporations, not countries.

TNCs are increasingly international in origin. In 1970, of the some 7000 multinational companies then identified by the United Nations, over half were from only two countries: the United States and Britain. By the mid-1990s, of the world's 100 largest TNCs—all based in 15 developed countries—more than half were in ten European countries, a third in Anglo America, 16 in Japan, and one each in Australia and New Zealand. Those same leading 100 multinationals are reckoned to control some 16% of the world's productive assets, with the top 300 together controlling about one-quarter.

Most transnational corporations operate in only a few industries: computers, electronics, petroleum and mining, motor vehicles, chemicals, and pharmaceuticals. Their worldwide impact is broader than those company specializations suggest. Some dominate the marketing and distribution of basic commodities; a few TNCs, for example,

Geography and Public Policy

Contests and Bribery

In 1985 it cost Kentucky over $140 million in incentives—some $47,000 a job—to induce Toyota to locate an automobile assembly plant in Georgetown, Kentucky. That was cheap. By 1993 Alabama spent $200,000 per job to lure Mercedes-Benz to that state, and Kentucky bid $350,000 per job in tax credits to bring a Canadian steel mill there. The spirited auction for jobs is not confined to manufacturing. A University of Minnesota economist calculates that his state is to spend $500,000 for each of the 1500 or more permanent jobs to be created by Northwest Airlines at two new maintenance facilities. For some, the bidding between states and locales to attract new employers and employment gets too fierce. Kentucky withdrew from competition for a United Airlines maintenance facility, letting Indianapolis have it when Indiana's offered package exceeded $300 million.

Inducements to lure companies are not just in cash and loans—though both figure in some offers. For manufacturers, incentives may include workforce training, property tax abatement, subsidized costs of land and building or their outright gifts, below-market financing of bonds, and the like. Similar offers are regularly made by states, counties, and cities to wholesalers, retailers, major office-worker and other service activity employers. The objective, of course, is not just to secure the new jobs represented by the attracted firm but to benefit from the general economic stimulus and employment growth that those jobs—and their companies—generate. Auto parts manufacturers are presumably attracted to new assembly plant locations; cities grow and service industries of all kinds—doctors, department stores, restaurants, food stores—prosper from the investments made to attract new basic employment.

Not everyone is convinced that those investments are wise, however. A poll of Minnesotans showed a majority opposed the generous offer made by the state to Northwest Airlines. In the late 1980s, the governor of Indiana, a candidate for Kentucky's governorship, and the mayor of Flat Rock, Michigan, were all defeated by challengers who charged that too much had been spent in luring the Suburu-Isuzu, Toyota, and Mazda plants, respectively. Established businesses resent what often seems neglect of their interests in favor of spending their tax money on favors to newcomers. The Council for Urban Economic Development, surveying the escalating bidding wars, has actively lobbied against incentives, and many academic observers note that industrial attraction amounts to a zero sum game: unless the attracted newcomer is a foreign firm, whatever one state achieves in attracting an expanding U.S. company comes at the expense of another state.

Some doubt that inducements matter much anyway. Although, sensibly, companies seeking new locations will shop around and solicit the lowest-cost, best deal possible, their site choices are apt to be determined by more realistic business considerations: access to labor, suppliers, and markets; transportation and utility costs; weather; the nature of the workforce; and overall costs of living. Only when two or more similarly attractive locations have essentially equal cost structures might such special inducements as tax reductions or abatements be determining in a locational decision.

Questions to Consider

1. As citizen and taxpayer, do you think it is appropriate to spend public money to attract new employment to your state or community?
2. If not, why not? If yes, what kinds of inducements and what total amount offered per job seem appropriate to you? What reasons support your opinion?
3. If you believe that "best locations" for the economy as a whole are those determined by pure location theory, what arguments would you propose to discourage locales and states from making financial offers designed to circumvent decisions clearly justified on abstract theoretical grounds?

account for 85% or more of world trade in wheat, maize, coffee, cotton, iron ore, and timber. As a group, TNCs directly employ some 75 million persons at home and abroad, or about 10% of worldwide nonagricultural paid employment. But because of their outsourced purchases of raw materials, parts and components, and services, the total number of worldwide jobs associated with TNCs in the late 1990s reached 150 million or more.

On average, a transnational corporation produces more than two-thirds of its output and hires two-thirds of its employees in its home country; employees working for the companies' foreign affiliates make up just over 1% of the world's workforce. Even that foreign impact is limited to a relatively few developing countries and regions. Majority inflows of foreign direct investment (FDI) are concentrated in 10 to 15 countries, mainly in South, Southeast, and East Asia (China is the largest recipient), and in Latin America and the Caribbean. The least-developed countries as a group—including nearly all African states—have received little FDI. Overall, the United States is not only the largest source of foreign direct investment, but its largest recipient as well.

Because they are international in operation with multiple markets, plants, and raw material sources, the TNCs are active and effective practitioners of the principle of comparative advantage. They produce in that country or region where costs of materials, labor, or other production inputs are minimized, while maintaining operational control and declaring taxes in localities where the economic climate is most favorable. Indeed, one-third of all trade consists of intrafirm transactions. From a locational perspective, TNCs have internationalized the plant-siting decision process and carried it further than single-country firms by multiplying the number of locationally separate operations seeking least-cost locations. Increasingly, service activities as well as manufacturing companies are international in scope, frequently under the control of "transnational integral conglomerates," which span a large spectrum of both service and industrial sectors.

Industrial Location in Planned Economies

The theoretical controls on plant location decisions that apply in commercial economies were not, by definition, determinant in the centrally planned Marxist economies of Eastern Europe and the former Soviet Union. In those economies, plant locational decisions were made by government agencies rather than by individual firms. Bureaucratic rather than company decision making did not mean that location assessments based on factor cost were ignored; it meant that central planners were more concerned with other than purely economic considerations in the creation of new industrial plants and concentrations. Since major capital investments are relatively permanent additions to the landscape, the results of their often noneconomic political or philosophical decisions are fixed and will long remain to influence industrial regionalism and competitive efficiencies into the post-communist present and future.

Important in the former Soviet Union, for example, was a controlling policy of the *rationalization of industry* through full development of the resources of the country wherever they were found and without regard to the cost or competitiveness of such development. Rationalization reflected a Soviet application of the Marxist-Leninist call for *industrial diversification* and *regional self-sufficiency.* Those developmental goals required each section of the country to contain a variety of industrial types (and, if possible, a local agricultural base) to assure its independent operation even if its connections to other sections were to be severed.

The *territorial production complex* was the planning mechanism created to achieve such economic development. As an areally based organizational form, it was assigned responsibility to develop necessary regional infrastructure, to facilitate the specialization of individual enterprises, and to promote overall regional economic growth and integration. Inevitably, although the factors of industrial production are identical in capitalist and noncapitalist economies, the philosophies and patterns of industrial location and areal development will differ between them.

World Manufacturing Patterns and Trends

Whether locational decisions are made by private entrepreneurs or central planners—and on whatever considerations those decisions are based—the results over many years have produced a distinctive world pattern of manufacturing. Figure 10.25 suggests the striking prominence of a relatively small number of major industrial concentrations localized within relatively few countries primarily but not exclusively parts of the "industrialized" or "developed" world. These may be roughly grouped into four commonly recognized major manufacturing regions: *Eastern Anglo America, Western and Central Europe, Eastern Europe,* and *Eastern Asia.* Together, the industrial plants within these established concentrations account for an estimated two-thirds of the world's manufacturing output by volume and value.

Their continuing dominance on the regional and international industrial scene is by no means assured. The first three—those of Anglo America and Europe—were the beneficiaries of an earlier phase in the development and spread of manufacturing following the Industrial Revolution of the 18th century and lasting until after World War II. The countries within them now are increasingly developing "postindustrial" economies in which traditional manufacturing and processing are of decreasing relative importance.

The fourth—the East Asian industrial region—is a part and forerunner of the wider, new pattern of world industrialization that has emerged in recent years, the result of international cultural convergence and technology transfers in the latter half of the 20th century, as briefly reviewed in Chapter 7 (page 235). The older rigid industrial "North–South" split between the developed and developing worlds depicted on Figure 7.8 has rapidly weakened as the full range of industrial activities from primary metal processing (e.g., the iron and steel industries) through advanced electronic assembly has been dispersed from, or separately established within, an ever-expanding list of countries.

Such states as Mexico, Brazil, China, and others of the developing world have created industrial regions of international significance, and the contribution to world manufacturing activity of the smaller newly industrializing countries (NICs) has been growing significantly. Even economies that until recently were overwhelmingly subsistence or dominated by agricultural or mineral exports have become important players in the changing world manufacturing scene. Foreign branch plant investment in low-wage Asian, African, and Latin American states has not only created there an industrial infrastructure but has as well increased their gross national products and per capita incomes sufficiently to permit expanded production for growing domestic—not just export—markets.

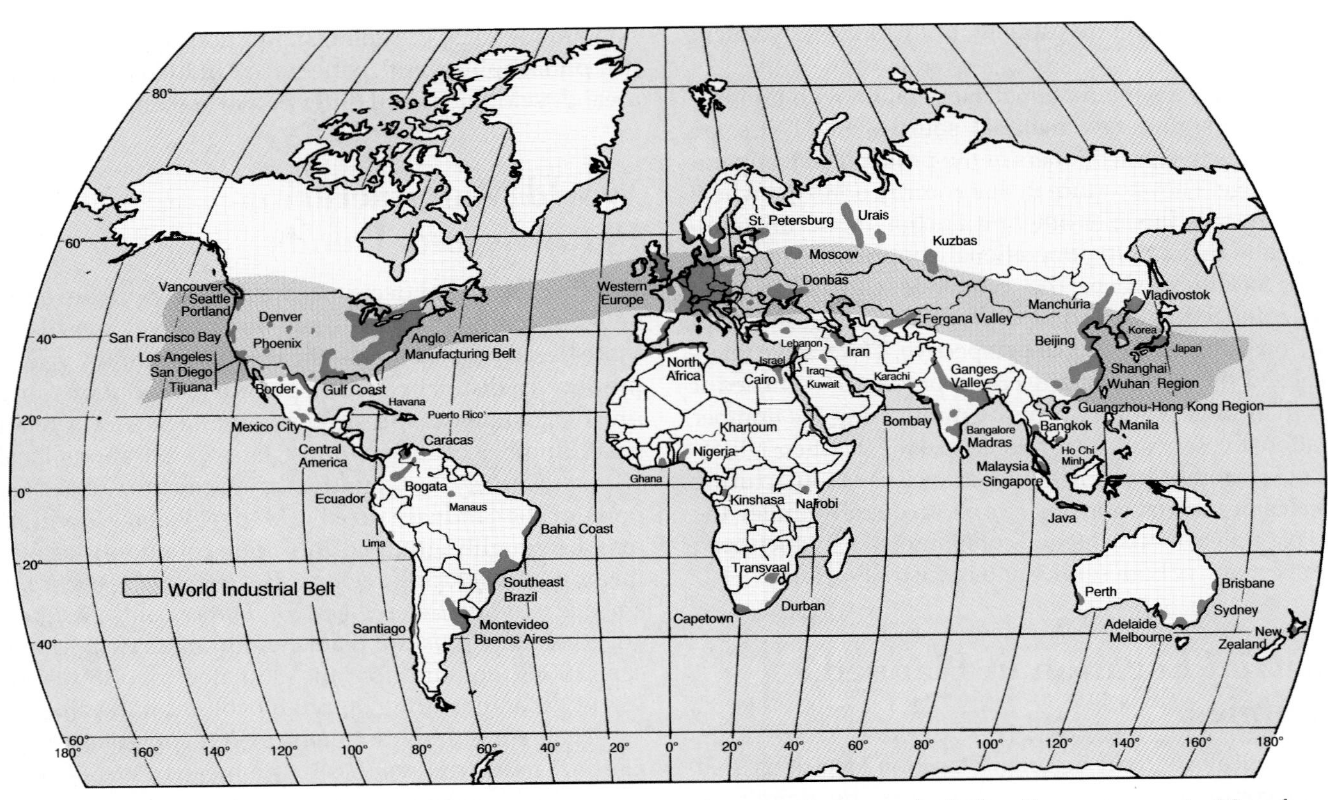

FIGURE 10.25 **World industrial regions.** Industrial districts are not as continuous or "solid" as the map suggests. Manufacturing is a relatively minor user of land even in areas of greatest concentration. There is a loose spatial association of major industrial districts in an "industrial belt" extending from Western Europe eastward to the Ural Mountains and, through outliers in Siberia, to the Far East; the belt picks up again on the west coast of North America, though its major Anglo American concentration lies east of the Mississippi River. The former overwhelming production dominance of that belt is increasingly being eroded by the industrialization of countries throughout the developing world.

In Malaysia, for example—one of the rapidly growing southeast Asian economies—agriculture's share of gross domestic product was cut almost in half between 1980 and 1996 and the share of merchandise exports from manufacturing rose from 19% to 76%. For Pakistan in South Asia, manufactured goods jumped from less than half to 84% of export values in that period, and for the Dominican Republic in Central America manufacturing grew from less than 25% to over three-quarters of merchandise exports over the same 16 years. Collectively, the "developing world" countries account now for some 65% of *gross world product;* increasingly, their share of that output is comprised of the manufactured goods that formerly were nearly exclusively the product of the "advanced world" of the North (Figure 7.8).

High-Tech Patterns

Major industrial districts of the world developed over time as entrepreneurs and planners established traditional secondary industries according to the pulls and orientations predicted by classical location theories. Those theories are less applicable in explaining the location of the latest generation of manufacturing activities: the high technology—or *high-tech*—processing and production that is increasingly part of the advanced economies. For these firms, new and different patterns of locational orientation and advantage have emerged, based on other than the traditional regional and site attractions.

High technology is more a concept than a definition. It probably is best understood as the application of intensive research and development efforts to the creation and manufacture of products of an advanced scientific and engineering character. Professional—"white collar"—workers make up a large share of the total workforce. The impact of high-tech industries on patterns of economic geography is expressed in at least two different ways.

First, high-tech activities are becoming major factors in employment growth in the advanced economies. In the United States, for example, between 1972 and 1997, the high-tech industries added over 2 million new jobs to the secondary sector of the economy, replacing thousands of other workers who lost jobs to outsourcing, foreign competition, and changing markets. England, Germany, Japan, and other advanced countries—though not yet those of Eastern Europe—have had similar employment shifts.

Second, high-tech industries have tended to become regionally concentrated in their countries of development, and within those regions they frequently form self-sustaining, highly specialized agglomerations. California, for example, has a share of United States high-tech employment far in excess of its share of American population. Along with California, the Pacific Northwest including British Columbia, New England, New Jersey, Texas, and Delaware all have proportions of their workers in high-tech industries above the national average. And within these and other states or regions of high-tech concentration, specific locales have achieved prominence: "Silicon Valley" near San Francisco, the "Silicon Forest" near Seattle, North Carolina's Research Triangle, Utah's "Software Valley," Route 128 around Boston, Ottawa, Canada's "Silicon Valley North," or the Canadian Technology Triangle west of Toronto are familiar Anglo American examples. Scotland's "Silicon Glen" or England's "Sunrise Strip" are other examples of industrial landscapes characterized by low, modern, dispersed office-plant-laboratory buildings rather than by massive factories, mills, or assembly structures, and storage areas.

The new distributional patterns of high-tech industries suggest they respond to different localizing forces than those controlling older generation industries. Important locational considerations appear to include: (1) proximity to major universities and large pools of scientific and technical labor skills; (2) avoidance of areas with strong labor unionization; (3) locally available venture capital and entrepreneurial daring; (4) location in regions and major metropolitan areas with favorable "quality of life" reputations and an employment base sufficiently large to supply needed workers and provide job opportunities for professionally trained spouses; (5) availability of first-quality communication and transportation facilities.

Agglomerating forces are also important. New firm formation is frequent and rapid in industries where discoveries are constant and innovation is continuous. Since many are "spin-off" firms founded by employees leaving established local companies, areas of existing high-tech concentration tend to spawn new entrants and to provide necessary labor skills. Agglomeration, therefore, is both a product and a cause of spatial associations.

Tertiary and Beyond

Primary activities, you will recall, gather, extract, or grow things. Secondary industries give *form utility* to the products of primary industry through manufacturing and processing efforts. A major and growing segment of both domestic and international economic activity, however, involves *services* rather than the production of commodities. Such activities consist of those business and labor specializations that provide services to the primary and secondary sectors, to the general community, and to the individual. They imply pursuits other than the actual production of tangible commodities.

As we saw earlier in this chapter, regional and national economies undergo fundamental changes in emphasis in the course of development. Subsistence societies exclusively dependent on primary industries may progress to secondary stage processing and manufacturing activities. In that progression, the importance of agriculture, for example, as an employer of labor or contributor to national income declines as that of manufacturing expands. As economic growth continues, secondary activities in their turn are replaced by service, or tertiary, functions as the main support of the economy. Advanced economies that have made that transition are often referred to as "postindustrial" because of the dominance of their service sectors and the significant decline of manufacturing as a generator of employment and national income.

Perhaps more than any other economy, the United States has reached postindustrial status (Figure 10.26). Its primary sector component has fallen from 66% of the labor force in 1850 to 4% in the late 1990s and the service sector has risen from 18% to near 80%. Of the about 35 million new jobs created in the United States between 1980 and 1999, nearly all of them, after discounting for job losses in other employment sectors, occurred in services. Comparable changes are found in other countries. By the late 1990s, between 65% and 80% of jobs in such developed economies as Japan, Canada, Australia, Israel, and all major Western European countries were also in the service sector; Russia and Eastern Europe average rather less.

"Service," however, is a broad concept that covers a range of activities from neighborhood barber to World Bank president. Collectively, services accounted for some 65% of gross world product in the late 1990s, up from barely half in 1980. That growth encompasses not only traditional low-order personal and retail services but also most importantly higher-order knowledge-based professional services performed primarily for other businesses, not for individual consumption.

Logically, the service category should be subdivided to distinguish between those activities answering to the daily living and support needs of individuals and local communities and those involving professional, administrative, or financial management tasks at regional, national, and international scales. Those differing levels of activity and scope

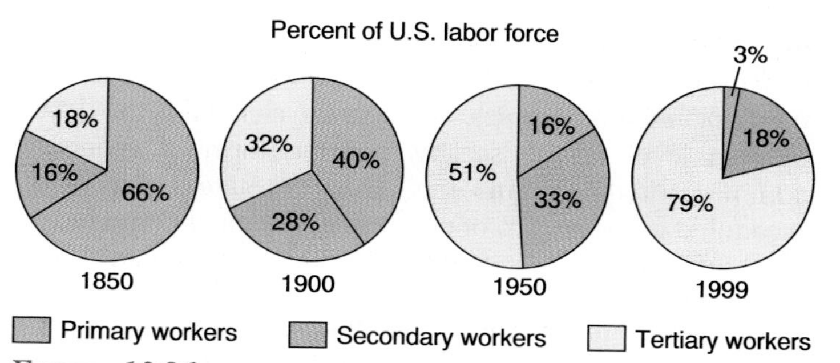

FIGURE 10.26 The changing sectoral allocation of the U.S. labor force is a measure of the economic development of the country. Its progression from a largely agricultural to postindustrial status is clearly evident.

represent different locational principles and quite different roles in their contribution to domestic and world economies. They differ as well in their import in establishing and maintaining local versus international patterns of connectivity and exchange.

To recognize such fundamental contrasts, we may usefully restrict the term "tertiary" specifically to those lower level activities largely related to day-to-day needs of people and to the usual range of functions found in smaller towns and cities worldwide. We can then relegate higher level, more specialized information, research, and management activities to a distinctive "quaternary" classification (Figure 10.2) with quite different and distinctive characteristics and significance.

Tertiary Sector

Some essential services are concerned with the wholesaling or retailing of goods, providing what economists call *place utility* to items produced elsewhere. They fulfill the exchange function of advanced economies and provide the market transactions necessary in highly interdependent societies. In commercial economies, tertiary activities also provide vitally needed information about market demand without which economically justifiable production decisions are impossible. Most tertiary activities, however, are concerned with personal and business services performed in shops and offices that, typically, cluster in cities large and small.

The supply of those kinds of low level services of necessity must be identical to the spatial distribution of *effective demand*—that is, wants made meaningful through purchasing power. Retail and personal service activities are localized by their markets, for the production of the service and its consumption are simultaneous occurrences. Retailers and personal service providers tend to locate, therefore, where market density is greatest and multiple service demands are concentrated (Figure 10.27). Their locational patterns and the employment support they imply are important aspects of urban economic structure and are dealt with in Chapter 12.

Traditional retailing and personal service activities represent a mark of contrast between advanced and subsistence societies, and the greater their proliferation, the greater the interdependence of the society of which they are a part. Their expansion has been great in both the commercial and the formerly planned economies. The decisive element is the development level of the society, not the form of economic administration. Within the United States, "services" accounted for some 80% of nonfarm employment (and nearly three-quarters of gross domestic product) in the late 1990s; manufacturing provided about 15% of employment (Table 10.1) and less than 20% of gross domestic product. In 1960, services were far less dominant and "goods production" occupied almost 38% of the nonfarm employed labor force.

In all of the world's increasingly interdependent postindustrial societies, the growth of the service component

FIGURE 10.27 Service activities are an increasing part of all economies, developing as well as advanced, as this garment repairman in an Ecuador city marketplace attests. Such "informal sector" employment—street vendors, odd-job handymen, open-air dispensers of such personal services as barbering, shoe shining, clothes mending, and the like—accounts for some 30% of all jobs in Latin America and 60% of those in Africa.

TABLE 10.1

U.S. EMPLOYMENT PATTERNS, 1960–1999[1]

	Percent of Nonfarm Jobs		Percent Increase in Jobs
	1960	**1999**	**1960–1999**
In services:			
Miscellaneous services[2]	13.6	29.4	+398
Retail trade	15.2	18.0	+173
State and local government	11.2	13.8	+182
Finance, insurance	4.8	5.8	+176
Wholesale trade	5.8	5.5	+116
Transport and utilities	7.5	5.3	+64
Federal government	4.2	2.1	+18
Total	**62.3**	**79.9**	**+195**
In goods production:			
Manufacturing	31.0	15.0	+11
Construction	5.4	4.7	+99
Mining	1.3	0.4	–21
Total	**37.7**	**20.1**	**+21**
Nonfarm jobs (total)			**+126**

[1]Payroll employees only; self-employed not included
[2]Includes education, health, food and lodgings, etc.

Source: Bureau of Labor Statistics, *Employment and Earnings.*

reflects not only the development of increasingly complex social, economic, and administrative structures. It also indicates changes made possible by growing personal incomes or possible and necessary by alterations in family structure and individual lifestyle. For example, in subsistence economies families produce, prepare, and consume food within the household. Urbanizing industrial societies have increasing dependence on specialized farmers growing food and wholesalers and retailers selling food to households that largely prepare and consume it at home. Postindustrial America increasingly opts to purchase prepared foods in restaurants, fastfood, or carry-out establishments with accelerating growth of the tertiary foodservice workers that change demands. People are still fed, but the employment structure has changed.

Part of the growth in the tertiary component is statistical, rather than functional. We saw in our discussion of modern industry that "outsourcing" was increasingly employed as a device to reduce costs and enhance manufacturing and assembly efficiencies. In the same way, outsourcing of services formerly provided in-house is also characteristic of current business practice. Cleaning and maintenance of factories, shops, and offices—formerly done by the company itself as part of internal operations—now is subcontracted to specialized service providers. The job is still done, perhaps even by the same personnel, but worker status has changed from "secondary" (as employees of a manufacturing plant, for example) to "tertiary" (as employees of a service company).

Whatever the origins of tertiary employment growth, the social and structural consequences are comparable. The process of development leads to increasing labor specialization and economic interdependence within a country. That has been true during the latter 20th century for all economies, as Table 10.2 attests. Carried to the postindustrial stage of advanced technology-based economies and high per capita income, the service component of both the employed labor force and the gross domestic product rises to dominance.

TABLE 10.2

CONTRIBUTION OF THE SERVICES SECTOR TO GROSS DOMESTIC PRODUCT

	Percentage of GDP		
Country Group	**1960**	**1980**	**1996**
Low income	32	29	35
Middle income	47	48	52
High income	54	61	66
United States	58	63	72

Source: World Bank, *World Development Report.*

Quaternary Sector

Available statistics unfortunately do not always permit a clear distinction between *tertiary* service employment that is a reflection of daily lifestyle or corporate structural changes and the more specialized, higher-level *quaternary* activities. The quaternary sector may be seen realistically as an advanced form of services involving specialized knowledge, technical skills, communication ability, or administrative competence. These are the activities carried on in office buildings, elementary and university classrooms, hospitals and doctors' offices, theaters, television stations, and the like. With the explosive growth in demand for and consumption of information-based services—mutual fund managers, tax consultants, Internet and software developers, research institutes, and more at near-infinite length—the quaternary sector in the most highly developed economies has unmistakably replaced manufacturing and all primary and secondary employment as the basis for economic growth.

Domestically, quaternary activities performed for other business organizations often embody "externalization" of specialized services similar to the outsourcing of low-level tertiary functions. The distinction between them lies in the fact that knowledge and skill-based free-standing quaternary service establishments can be spatially divorced from their clients; they are not tied to resources, affected by the environment, or necessarily localized by market. They can realize cost reductions through serving multiple clients in highly technical areas, and permit client firms to utilize specialized skills and efficiencies to achieve competitive advantage without the expense of adding to their own labor force.

In addition to serving as external suppliers of traditional business needs, quaternary specialists have increasingly developed new services attractive to an expanding client base. National accounting firms, for example, have developed management consulting services used by companies that may also use local accounting companies, and advertising agencies provide market analyses and marketing programs in addition to creating advertising copy. Such multiplication of specialized skills has, by multiplier effect, greatly accelerated quaternary sector employment within advanced economies.

The often close functional association of client and service firms within a country suggests that quaternary establishment location and employment patterns might well closely resemble that of the headquarters distribution of the primary and secondary industries served when high-level personal contacts are required. But the transportability of quaternary services also means that many quaternary activities can be spatially isolated from their client base. In the United States, at least, these combined trends have resulted both in the concentration of certain specialized services—merchant banking or bond underwriting, for example—in major metropolitan areas and, as well, in a regional diffusion of the quaternary sector to accompany the growing regional deconcentration of the client firm base, the dispersion of service needs to branch plants, and the growing use of service professionals by even the smallest primary, secondary, and tertiary companies. Similar locational tendencies have been noted even for the spatially more restricted advanced economies of, for example, England and France.

Information, administration, and the "knowledge" activities in their broadest sense are dependent on communication. Their spatial dispersion, therefore, has been abetted by the underlying technological base of most quaternary activities: electronic digital processing and telecommunication transfer of data. That technology permits many "back-office" tasks to be spatially far distant from the home office locations of either the service or client firms. Insurance claims, credit card charges, mutual fund and stock market transactions, and the like, are more efficiently and economically recorded or processed in low-rent, low-labor cost locations—often in suburbs or small towns and in rural states—than in the financial districts of major cities. Production and consumption of such services can be spatially separated in a way not feasible for tertiary, face-to-face activities (Figure 10.28).

Just as quaternary activities have been major engines of growth within the most advanced states, so too have they become an increasing factor in international trade flows and economic interdependence. Between 1980 and 1999, services increased from 15% of total world trade to about 25%. The fastest growing segment of that increase was in such private services as financial, brokerage, and leasing activities, which grew to 50% of all commercial services trade by the late 1990s. As in the domestic arena, rapid advances in information technology and electronic data transmission have been central elements in the internationalization of services. Many services considered nontradable even during the 1980s are now actively traded at long distance.

Developing countries have been particular beneficiaries of the new technologies. Their exports of services—already valued at $180 billion in the mid-1990s—grew at an annual 12% rate in that decade, twice as fast as service exports from

FIGURE 10.28 "Office parks" such as this complex at Tysons Corner, Virginia, outside of Washington, D.C., are increasingly familiar concentrations of quaternary workers in the outlying zones of major metropolitan centers. In the late 1990s, more than 80,000 office workers were employed in this "suburban downtown."
©Llewellyn.

industrial regions. The increasing tradability of services has expanded the international comparative advantage of developing states in relatively labor-intensive long-distance service activities such as mass data processing, computer-software development, and the like. At the same time, they have benefited from increased access to efficient, state-of-the-art equipment and techniques transferred from advanced economies. The concentration of computer-software development around Banglaore has made India a major world player in software innovation, for example, while elsewhere in that country increasing volumes of back-office work for Western insurance companies and airlines is being performed. Claims processing for life and health insurance firms have become concentrated in English-speaking Caribbean states to take advantage of lower wages and availability of a large pool of educated workers there. In all such cases, the result is an acceleration in the transfer rate of technology in such expanding areas as information and telecommunications services and an increase in the rate of developing-country integration in the world economy.

Most of the current developing-country gains in international quaternary services are the result of increased foreign direct investment (FDI) in the services sector. Those flows accounted for three-fifths of all FDI at the end of the 1990s. The majority of such investment, however, is transferred within the advanced countries themselves rather than between industrial and developing states. In either case, as transnational corporations employ mainframe computers around the clock for data processing, they can exploit or eliminate time zone differences between home office countries and host countries of their affiliates. Such cross-border intrafirm service transactions are not usually recorded in balance of payment or trade statistics, but materially increase the volume of international services flows.

The same cost and skill advantages that enhance the growth and service range of quaternary firms on the domestic scene also operate internationally. Principal banks of all advanced countries have established foreign branches, and the world's leading banks have become major presences in the primary financial capitals. In turn, a relatively few world cities have emerged as international business and financial centers whose operations and influences are continuous and borderless, while a host of offshore banking havens have emerged to exploit gaps in regulatory controls and tax laws (Figure 10.29). Accounting firms, advertising agencies, management consulting companies, and similar quaternary sector establishments of North American or European origin primarily have increasingly established their international presence, with main branches located in principal business centers worldwide.

The list of tertiary and quaternary employment is long. Its diversity and familiarity remind us of the complexity of modern life and of how far removed we are from the subsistence economies. As societies advance economically, the shares of employment and national income generated by the primary, secondary, tertiary, and quaternary sectors continually change, and spatial patterns of human activity reflect those changes. The shift is steadily away from primary production and secondary processing, and toward the trade, personal, and professional services of the tertiary sector and the information and control activities of the quaternary. The transition is recognized by the now-familiar term "postindustrial."

Summary

How people earn their living and how the diversified resources of the earth are employed by different peoples and cultures are fundamental concerns of the locational tradition of geography. In seeking spatial and activity regularities we can recognize three types of economic systems: *subsistence, commercial,* and *planned.* The first is concerned with production for the immediate consumption of individual producers and family members. In the second, economic decisions ideally respond to impersonal market forces and reasoned assessments of monetary gain. In the third, at least some nonmonetary social or political goals influence production decisions.

We can further classify economic activities according to the stages of production and the degree of specialization they represent. In that way, we make distinctions between *primary* activities (food and raw material production), *secondary* industries (processing and manufacturing), *tertiary* activities (distribution and general professional and personal service), and the administrative, informational, and technical specializations (*quaternary* activities) that mark highly advanced societies of either planned or commercial systems.

Agriculture, the most extensively practiced of the primary industries, is part of the spatial economy of both subsistence and advanced societies. In the first instance, it is responsive to the immediate consumption needs of the producer group and reflective of the environmental conditions under which it is practiced. In the second, agriculture reacts to consumer demand expressed through free or controlled markets. Its spatial expression reflects assessments of profitability and the dictates of social and economic planning.

Manufacturing is the dominant form of secondary activity, and is evidence of economic advancement beyond the subsistence level. Location theories help explain observed patterns of industrial development. Those theories are based on simplifying assumptions about fixed and variable costs of production and distribution, including costs of raw materials, power, labor, market accessibility, and transportation. *Weberian analysis* argues that least-cost locations are optimal and are strongly or exclusively influenced by transportation charges. Less rigid locational theory admits the possibility of multiple acceptable locations within a *spatial margin of profitability.* Agglomeration economies and the multiplier effect may make attractive locations not otherwise predicted for individual firms,

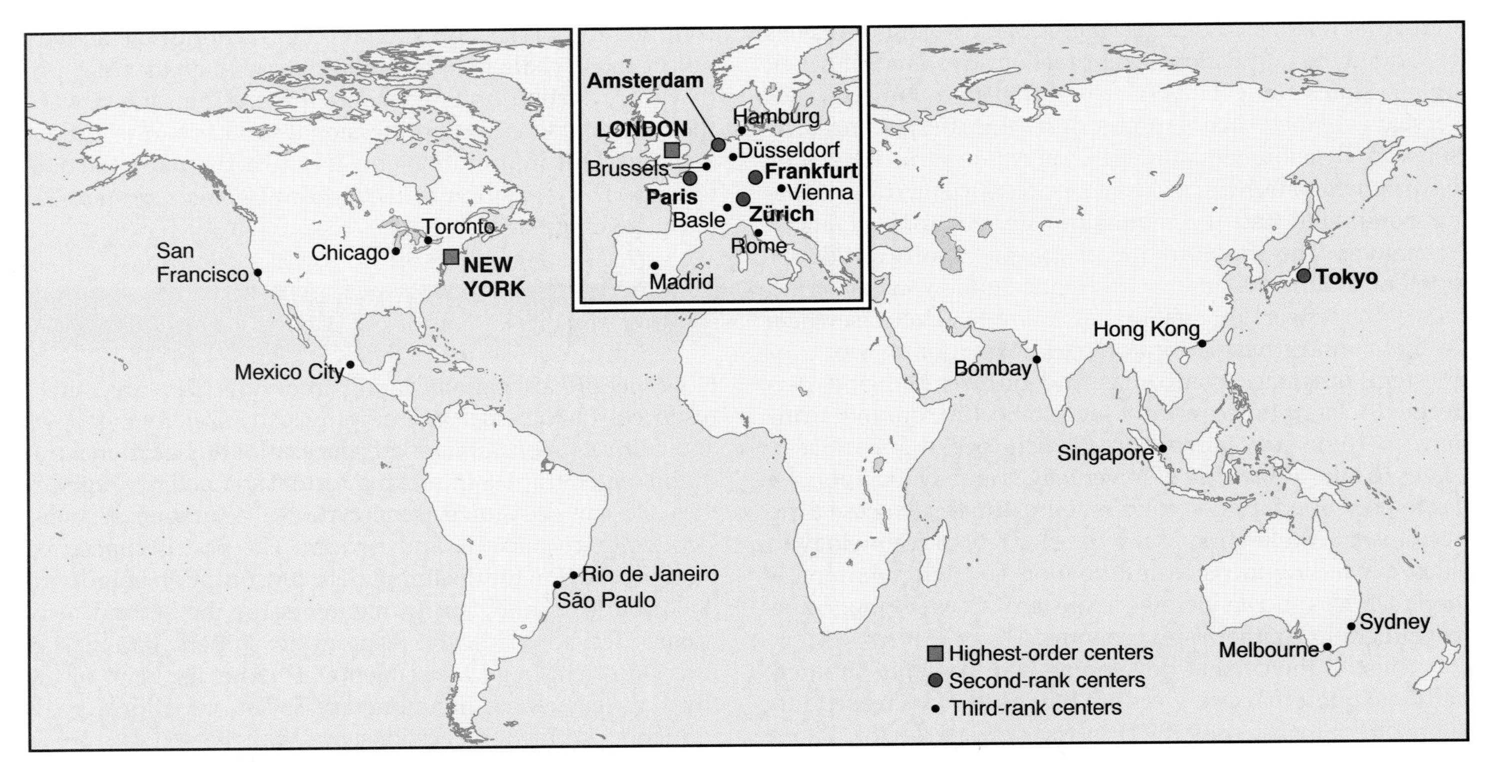

(a)

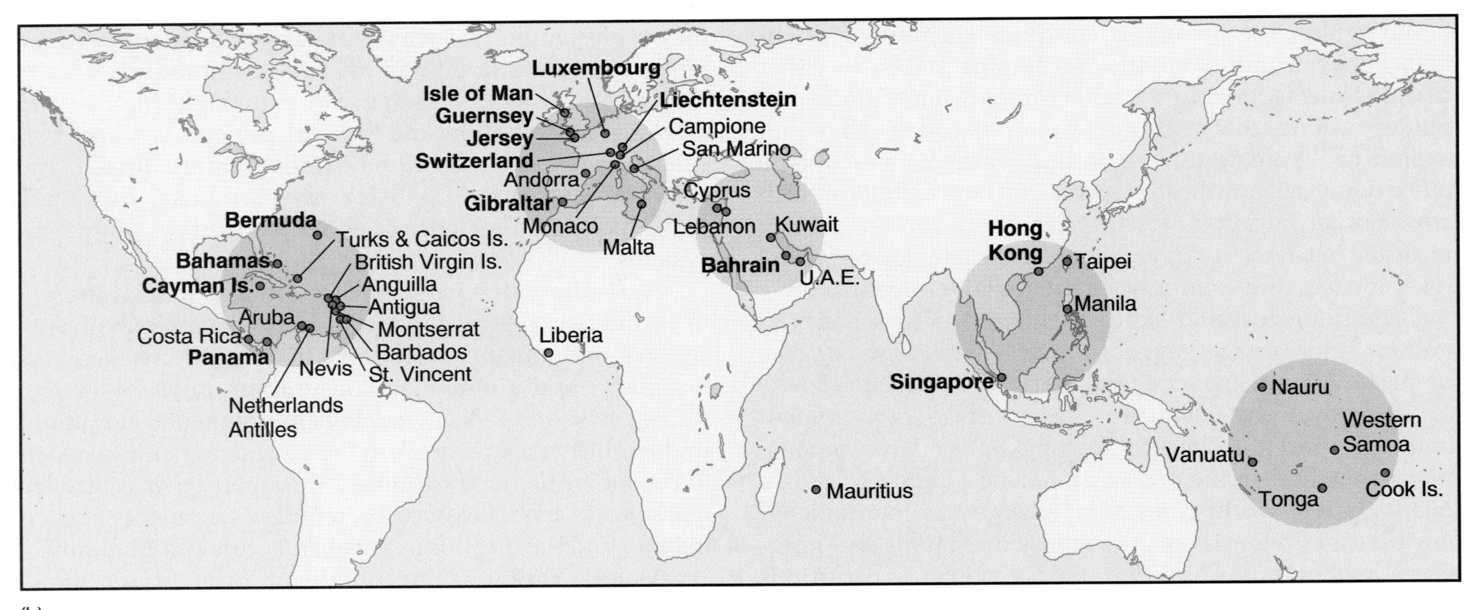

(b)

FIGURE 10.29 (a) **The hierarchy of international financial centers,** topped by New York and London, indicates the tendency of highest-order quaternary activities to concentrate in a few world and national centers. (b) At the same time, the multiplication of off-shore locations where "furtive money" avoiding regulatory control and national taxes finds refuge, suggests that dispersed convenience sites also serve the international financial community.

Source: Peter Dicken, *Global Shift.* 3rd ed. Guilford Press, 1998, Figures 12.9 and 12.10.

while comparative advantage may influence production decisions of entrepreneurs. The advanced rigidly planned economies did not necessarily ignore the locational constraints implicit in market economy decisions. They did, however, show the application of additional—frequently contradictory—bases for locational decisions uncontrolled by market forces.

A major share of global secondary manufacturing activity is found within a relatively small number of major industrial concentrations and multinational regions. The most advanced countries within those regions, however, are undergoing deindustrialization as newly industrializing countries with more favorable cost structures compete for markets. In the advanced economies, tertiary and quaternary activities become more important as secondary sector employment and share of gross national product declines. The new high-tech and postindustrial spatial patterns are not necessarily identical to those developed in response to theoretical and practical determinants of manufacturing success.

One final reminder is needed: an economy in isolation no longer exists. The world pattern of economic and cultural integration is too complete to allow totally separate national economies. Events affecting one ramify to affect all. The potato blight in an isolated corner of Europe a century and a half ago still holds its message. Despite differences in language, culture, or ideology, we are inseparably a single people economically.

Key Words

agglomeration 371
commercial economy 351
comparative advantage 373
economic geography 349
extensive commercial agriculture 363
extensive subsistence agriculture 354
external economies 373
extractive industries 366
gathering industries 366
Green Revolution 357
intensive commercial agriculture 362
intensive subsistence agriculture 354
least cost theory 371
maximum sustainable yield 367
nomadic herding 354
planned economy 351
plantation 365
primary activity 350
quaternary activity 350
secondary activity 350
shifting cultivation 355
subsistence economy 351
tertiary activity 350
transnational corporation (TNC) 375
variable costs 371
von Thünen rings 362

For Review and Consideration

1. What are the distinguishing characteristics of the economic systems labeled *subsistence, commercial,* and *planned?* Are they mutually exclusive, or can they coexist within a single political unit?
2. How is *intensive subsistence* agriculture distinguished from *extensive subsistence* cropping? Why, in your opinion, have such different land use forms developed in separate areas of the warm, moist tropics?
3. Briefly summarize the assumptions and dictates of von Thünen's agricultural model. How might the land use patterns predicted by the model be altered by an increase in the market price of a single crop? A decrease in the transportation costs of one crop but not of all crops?
4. What economic or ecological problems can you cite that do or might affect the *gathering industries* of forestry and fishing? What is *maximum sustainable yield?* Is that concept related to the problems you discerned?
5. What simplifying assumptions did Weber make in his theory of plant location? In what ways does the Weberian search for the *least cost location* differ from the recognition of the *spatial margin of profitability?*
6. How, in your opinion, do the concepts or practices of *comparative advantage* and *outsourcing* affect the industrial structure of advanced and developing countries?
7. As high-tech industries and *quaternary* employment become more important in the economic structure of advanced countries, what consequences for economic geographic patterns do you anticipate? Explain.

Selected References

Berry, Brian J. L., Edgar C. Conkling, and D. Michael Ray. *The Global Economy in Transition.* Upper Saddle River, N.J.: Prentice-Hall, 1997.

Chapman, Keith, and David Walker. *Industrial Location.* 2d ed. Cambridge, Mass.: Basil Blackwell, 1991.

Corbridge, Stuart, ed. *World Economy.* The Illustrated Encyclopedia of World Geography. New York: Oxford University Press, 1993.

Dicken, Peter. *Global Shift: Transforming the World Economy.* 3d ed. New York: Guilford Press, 1998.

Dicken, Peter, and Peter E. Lloyd. *Location in Space: Theoretical Perspectives in Economic Geography.* 3d ed. New York: Harper & Row, 1990.

Grigg, David. *An Introduction to Agricultural Geography.* 2d ed. New York: Routledge, 1995.

Hamilton, Ian, ed. *Resources and Industry.* The Illustrated Encyclopedia of World Geography. New York: Oxford University Press, 1992.

Hanink, Dean M. *Principles and Applications of Economic Geography.* New York: John Wiley & Sons, 1997.

Harrington, J. W., and Barney Warf. *Industrial Location.* London and New York: Routledge, 1995.

International Bank for Reconstruction and Development/The World Bank. *World Development Report.* Published annually for the World Bank by Oxford University Press, New York.

Malecki, Edward J. "Industrial Location and Corporate Organization in High Technology Industries." *Economic Geography* 61, no. 4 (1985):345–369.

Peters, William J., and Leon F. Neuenschwander. *Slash and Burn: Farming in the Third World Forest.* Moscow: University of Idaho Press, 1988.

South, Robert B. "Transnational 'Maquiladora' Location." *Annals of the Association of American Geographers* 80, no. 4 (1990):549–570.

Tarrant, John., ed. *Farming and Food.* The Illustrated Encyclopedia of World Geography. New York: Oxford University Press, 1991.

Turner, B. L. II, and Stephen B. Brush, eds. *Comparative Farming Systems.* New York: The Guilford Press, 1987.

United Nations Conference on Trade and Development (UNCTAD). *Transnational Corporations: Employment and the Workplace.* World Investment Report 1994. New York and Geneva: United Nations, 1994.

Weber, Alfred. *Theory of the Location of Industries.* Translated by Carl J. Friedrich. Chicago: University of Chicago Press, 1929. Reissue. New York: Russell & Russell, 1971.

Wheeler, James O., Peter Muller, Grant Thrall, and Timothy Fik. *Economic Geography.* 3d ed. New York: John Wiley & Sons, 1998.

World Resources . . . A Report by the World Resources Institute in collaboration with the United Nations Environment Programme and the United Nations Development Programme. New York: Oxford University Press, annual or biennial beginning 1986.

Young, John E. *Mining the Earth.* Worldwatch Paper 109. Washington, D.C.: Worldwatch Institute, 1992.

Don't forget about Dushkin's *Annual Editions Online: Geography* at http: / / dushkin.com / aeonline / . See preface for details.

CHAPTER

11

THE GEOGRAPHY OF NATURAL RESOURCES

A Norwegian oil production platform in the North Sea.
©Georg Gerster/Photo Researchers, Inc.

If present trends continue, the world in 2000 will be more crowded, more polluted, less stable ecologically and more vulnerable to disruption than the world we live in now. Serious stresses involving population, resources and environment are clearly visible ahead. Despite greater material output, the world's people will be poorer in many ways than they are today."

The warning above is from The Global 2000 Report to the President. Published in 1980, the report was the result of a request made by President Jimmy Carter to 13 federal agencies, asking them to participate in an exhaustive study of likely long-term changes in the world's population, natural resources, and environment.

Alarmed by the pessimism evident in Global 2000, Herman Kahn, then head of the Hudson Institute, and Julian Simon, a senior fellow at the Heritage Foundation, asked a number of experts to undertake their own study of the future. Modeling their language on the earlier report, the contributors to The Resourceful Earth predicted:

"If present trends continue, the world in 2000 will be less crowded (though more populated), less polluted, more stable ecologically, and less vulnerable to resource-supply disruption than the world we live in now. Stresses involving population, resources, and environment will be less in the future than now. The world's people will be richer in most ways than they are today. The outlook for food and other necessities of life will be better. Life for most people on earth will be less precarious economically than it is now."

These disparate views highlight the relationship between numbers of people, natural resources, and the environment. They show that thoughtful people can come to quite opposite conclusions regarding future scenarios and warn, in a sense, against believing without question both bleak and rosy assessments of our prospects. As Chapters 6 and 10 revealed, growing population numbers and economic development have magnified the extent and the intensity of human depletion of the treasures of the earth. Resources of land, of ores, and of most forms of energy are finite, but the resource demands of an expanding, economically advancing population appear to be limitless. This imbalance between resource availability and use has been a concern for more than a century, at least since the time of Malthus and Darwin, but it wasn't until the 1970s that the rate of resource depletion and the environmental degradation associated with it became a major and controversial issue.

Because resources are unevenly distributed in kind, amount, and quality, and do not match uneven distributions of population and demand, a consideration of natural resources falls within the locational distributional tradition of the discipline of geography. In this chapter, we survey the natural resources on which societies depend, their patterns of production and consumption, and the problem of managing those resources in light of growing demands and shrinking reserves. We do not attempt to predict which of the views expressed above is more likely to prove accurate, or whether some other scenario will occur, but our hope is that the chapter content will enable readers to make their own assessment.

We begin our discussion by defining some commonly employed terms.

Resource Terminology

A **resource** is a naturally occurring, exploitable material that a society perceives to be useful to its economic and material well-being. Willing, healthy, and skilled workers constitute a valuable resource, but without access to materials like fertile soil or petroleum, human resources are limited in their effectiveness. In this chapter, we devote our attention to physically occurring resources, or, as they are more commonly called, *natural* resources.

The availability of natural resources is a function of two things: the physical characteristics of the resources themselves, and human economic and technological conditions. The physical processes that govern the formation, distribution, and occurrence of natural resources are determined by physical laws over which people have no direct control. We take what nature gives us. To be considered a resource, however, a given substance must be *understood* to be a resource. This is a cultural, not purely a physical, circumstance. Native Americans may have viewed the resource base of Pennsylvania as composed of forests for shelter and fuel, and as the habitat of the game animals on which they depended for food. European settlers viewed the forests as the unwanted covering of the resource that they perceived to be of value: soil for agriculture. Still later, industrialists appraised the underlying coal deposits, ignored or unrecognized as a resource by earlier occupants, as the item of value for exploitation (Figure 11.1)

Natural resources are usually recognized as falling into one of two broad classes: renewable and nonrenewable.

Renewable Resources

Renewable resources are materials that can be regenerated in nature as fast or faster than they are exploited by society. They can be used over and over again; the supplies are not depleted. A distinction can be made, however, between those that are perpetual and those that are renewable only if carefully managed (Figure 11.2a). **Perpetual resources** are inexhaustible. They include the sun, wind, running water, waves, tides, and geothermal energy. The hydrologic cycle (see Figure 5.4) assures that water, no matter how often used or how much abused, will return over and over to the land for further exploitation.

FIGURE 11.1 The original hardwood forest covering these Virginia hills was removed by settlers who saw greater resource value in the underlying soils. The soils, in turn, were stripped away for access to the still more valuable coal deposits below. Resources are as a culture perceives them, though exploitation may consume them and destroy the potential of an area for alternate uses.
©Richard Pasley/Stock, Boston.

Potentially renewable resources are renewable if left to nature but can be destroyed if people use them carelessly. These include groundwater, soil, plants, and animals. If the rate of exploitation exceeds that of regeneration, these renewable resources can be depleted. Groundwater extracted beyond the replacement rate in arid areas may be as permanently dissipated as if it were a nonrenewable ore. Soils can be lost by mismanagement that leads to total erosion. Forests are a renewable resource only if people are planting at least as many trees as are being cut.

Nonrenewable Resources

In their original forms, **nonrenewable resources** are generated in nature so slowly that for all practical purposes they exist in finite amounts. They include the fossil fuels (coal, crude oil, and natural gas), the nuclear fuels (uranium and thorium), and a variety of nonfuel minerals, both metallic and nonmetallic. Although the elements of which these resources are composed cannot be destroyed, they can be altered to less useful or available forms, and they are subject to depletion. The energy stored in a unit volume of the fossil fuels may have taken eons to concentrate in usable form; it can be converted to heat in an instant and be effectively lost forever.

Fortunately, many minerals can be *reused* even though they cannot be *replaced.* If they are not chemically destroyed—that is, if they retain their original chemical composition—they are potentially reusable. Aluminum, lead, zinc, and other metallic resources, plus many of the nonmetallics, such as diamonds and petroleum by-products, can be used time and time again. However, many of these materials are used in small amounts in any given object, so that recouping them is economically unfeasible. In addition, many materials are now being used in manufactured products, so that they are unavailable for recycling unless the product is destroyed. Consequently, the term *reusable resource* must be used carefully. At present, all mineral resources are being mined much faster than they are being recycled.

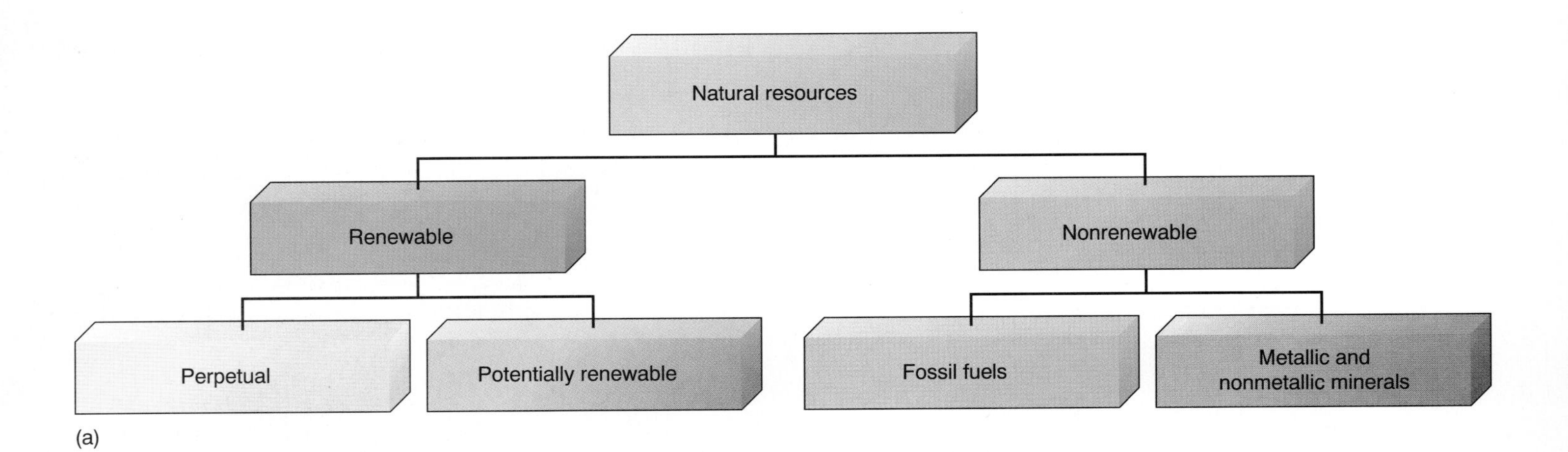

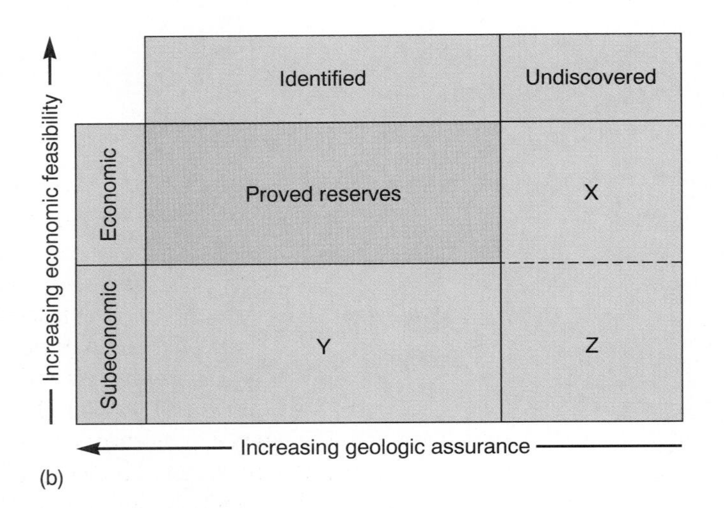

FIGURE 11.2 (a) **A classification of natural resources.** Renewable resources can be depleted if the rate of use exceeds that of regeneration. (b) **Proved reserves** consist of amounts that have been identified and can be recovered at current prices and with current technology. X denotes amounts that would be attractive economically but have not yet been discovered. Identified but not economically attractive amounts are labeled Y, and Z represents undiscovered amounts that would not be attractive now even if they were discovered.

Source: General classification of mineral resources by the U.S. Geological Survey.

Resource Reserves

Some regions contain many resources, others relatively few. No industrialized country, however, has all the resources it needs to sustain itself. The United States has abundant deposits of many minerals, but it depends on other countries for such items as tin and manganese. The actual or potential scarcity of key nonrenewable resources makes it desirable to predict their availability in the future. We want to know, for example, how much petroleum remains in the earth and how long we will be able to continue using it.

Any answer will be only an estimate, and for a variety of reasons such estimates are difficult to make. Exploration has revealed the existence of certain deposits, but we have no sure way of knowing how many remain undiscovered. Further, our definition of what constitutes a usable resource depends on current *economic* and *technological* conditions. If they change—if, for example, it becomes possible to extract and process ores more efficiently—our estimate of reserves also changes. Finally, the answer depends in part on the rate at which the resource is being used, but it is impossible to predict future rates of use with any certainty. The current rate could drop if a substitute for the resource in question is discovered, or increase if population growth or industrialization places greater demands on it.

A useful way of viewing reserves is illustrated by Figure 11.2b. Assume that the large rectangle includes the total stock of a particular resource, all that exists of it in or on the earth. Some deposits of that resource have been discovered; they are shown in the left-hand column as "identified amounts." Deposits that have not been located are called "undiscovered amounts." Deposits that are economically recoverable with current technology are at the top of the diagram, while those labeled "subeconomic" are not attractive for any of a number of reasons (the concentration is not rich enough, it would require expensive treatment after mining, it is not accessible, and so on).

We can properly term **proved reserves**—quantities of a resource that can be extracted profitably from known deposits—only the portion of the rectangle indicated by the pink tint. These are the amounts that have been identified and that can be recovered under existing economic and operating conditions. If new deposits of the resource are discovered, the reserve category will shift to the right; improved technology or increased prices for the product can shift the reserve boundary downward. An ore that was not considered a reserve in 1950, for example, may become a reserve in 2000 if ways are found to extract it economically.

ENERGY RESOURCES AND INDUSTRIALIZATION

Although people depend on a wide range of resources contained in the biosphere, energy resources are the "master" natural resources. We use energy to make all other resources available (see "What Is Energy?"). Without the energy resources, all other natural resources would remain in place,

What Is Energy?

People have built their advanced societies by using inanimate energy resources. **Energy**—the ability to do work—exists in two forms: potential and kinetic. *Potential* energy is stored energy; when released, it is in a form that can be harnessed to do work. *Kinetic* energy is the energy of motion; all moving bodies possess kinetic energy.

Assume that a reservoir contains a large amount of stored water. The water is a storehouse of *potential* energy. When the gates of the dam holding back the water are opened, water rushes out. Potential energy has become *kinetic* energy, which can be harnessed to do such work as driving a generator of electricity. No energy has been lost, it has simply been converted from one form to another.

Unfortunately, energy conversions are never complete. Not all of the potential energy of the water can be converted into electrical energy. Some potential energy is always converted to heat and dissipated to the surroundings. *Efficiency* is the measure of how well we can convert one form of energy into another without waste.

unable to be mined, processed, and distributed. When water becomes scarce, we use energy to pump groundwater from greater depths, or to divert rivers and build aqueducts. Likewise, we increase crop yields in the face of poor soil management by investing energy in fertilizers, herbicides, farm implements, and so on. By the application of energy, the conversion of materials into commodities and the performance of services far beyond the capabilities of any single individual are made possible. Further, the application of energy can overcome deficiencies in the material world that humans exploit. High-quality iron ore may be depleted, but by massive applications of energy, the iron contained in rocks of very low iron content can be extracted and concentrated for industrial use.

Energy can be extracted in a number of different ways. Humans themselves are energy converters, acquiring their fuel from the energy contained in food. Our food is derived from the solar energy stored in plants. In fact, nearly all energy sources are really storehouses for energy originally derived from the sun. Among them are wood, water, the ocean tides, the wind, and the fossil fuels. People have harnessed each of these energy sources, to a greater or lesser degree. Preagricultural societies depended chiefly on the energy stored in wild plants and animals for food, although people developed certain tools (such as spears) and customs to exploit the energy base. For example, they added to their own energy resources by using fire for heating, cooking, or clearing forest land.

Sedentary agricultural societies developed the technology to harness increasing amounts of energy. The domestication of plants and animals, the use of wind to power ships and windmills, and of water for waterwheels all expanded the energy base. For most of human history, wood was the predominant source of fuel, and even today at least half the world's people depend largely on fuel wood for cooking and heating.

However, it was the shift from renewable resources to those derived from nonrenewable minerals, chiefly fossil fuels, that sparked the Industrial Revolution, made possible the population increases discussed in Chapter 6, and gave population-supporting capacity to areas far in excess of what would be possible without inanimate energy sources. The enormous increase in individual and national wealth in industrialized countries has been built in large measure on an economic base of coal, oil, and natural gas. They are used to provide heat, to generate electricity, and to run engines.

Energy consumption goes hand in hand with industrial production and increases in per capita income (Figure 11.3). In general, the greater the level of energy consumption, the higher the gross national product per capita. This correlation of energy consumption with economic development points up a basic conflict between societies. Countries that can afford to consume great amounts of energy continue to expand their economies and to increase their levels of living. Those without access to energy, or those unable to afford it, see the gap between their economic prospects and those of the developed countries growing ever greater.

Nonrenewable Energy Resources

Crude oil, natural gas, and coal have formed the basis of industrialization. Figure 11.4 shows past energy consumption patterns in the United States. Burning wood supplied most energy needs until about 1885, by which time coal had risen to prominence. The proportion of energy needs satisfied by burning coal peaked about 1910; from then on, oil and natural gas were increasingly substituted for coal. The graph shows the absolute dominance of the fossil fuels as energy sources during the last 100 years. In 1990, they accounted for about 90% of national energy consumption.

Crude Oil

Today, crude oil and its by-products account for almost 40% of the commercial energy (excluding wood and other traditional fuels) consumed in the world. Some world regions and industrial countries have a far higher dependency. Figure 11.5 shows the main producers of crude oil.

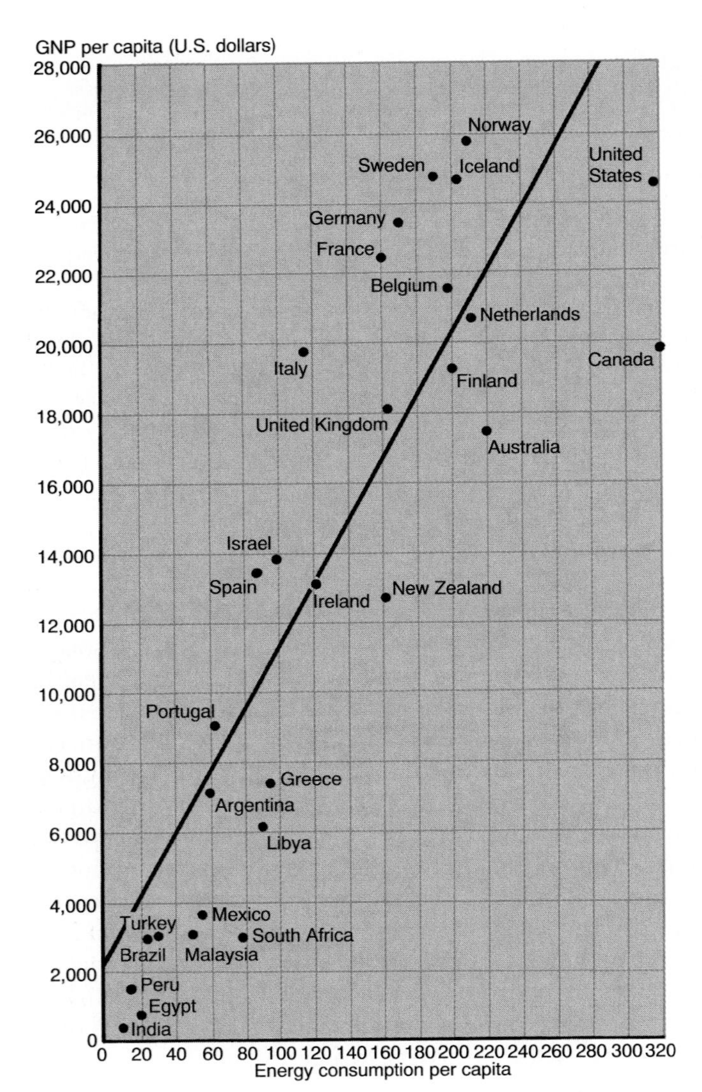

FIGURE 11.3 **Energy consumption rises with increasing gross national product.** Because the internal combustion engine accounts for a large share of national energy consumption, this graph reflects both economic development and the roles of mass transportation, automotive efficiency, and mechanization in different national economies.

Source: Data from World Resources Institute, 1996.

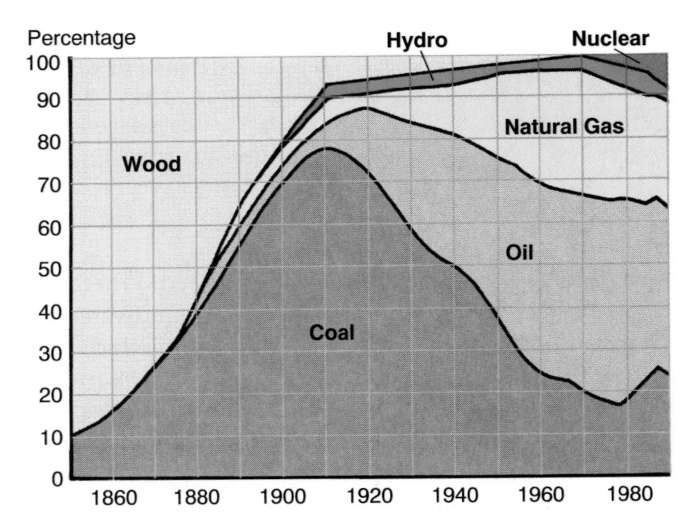

FIGURE 11.4 **Sources of energy in the United States, 1850–1990.** The fossil fuels provided about 90% of the energy supply in 1990.

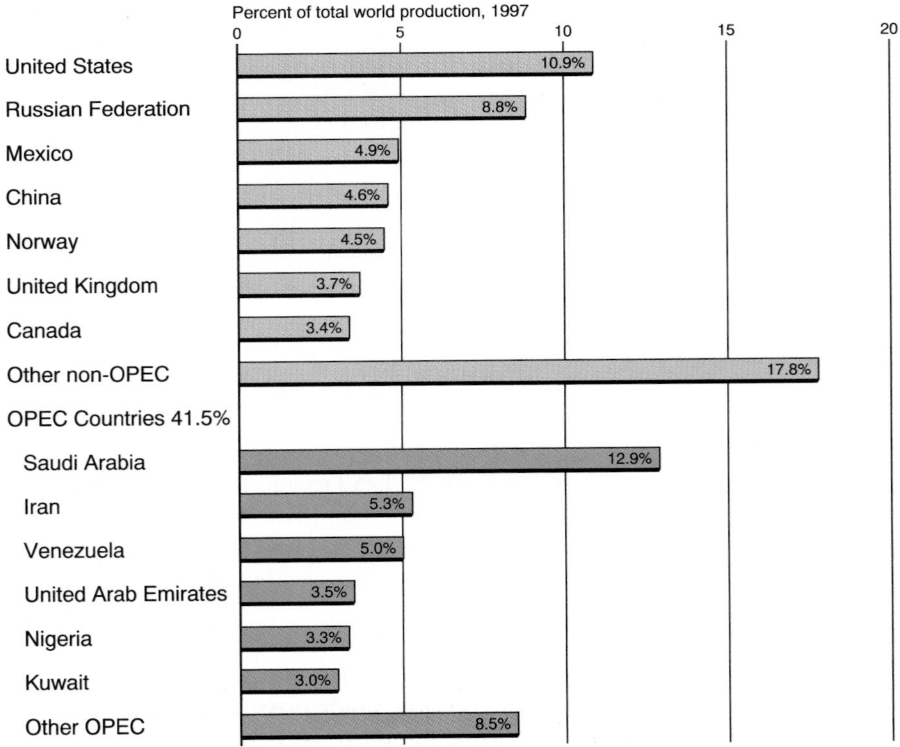

FIGURE 11.5 **Share of total international crude-oil production, 1997.** In that year, Saudi Arabia was the world's largest oil *producer,* followed by the United States. The largest *consumers* of oil were the United States and Japan, taking 25% and 8%, respectively, of the world total.

Source: Data from British Petroleum Company, *The BP Statistical Review of World Energy 1998.*

After it is extracted from the ground, crude oil must be refined. The hydrocarbon compounds are separated and distilled into waxes and tars (for lubricants, asphalt, and many other products) and various fuels. Petroleum rose to importance because of its combustion characteristics and its adaptability as a concentrated energy source for powering moving vehicles. Although there are thousands of oil-based products, fuels such as home heating oil, diesel and jet fuels, and gasoline are the major output of refineries. In the United States, transport fuels account for two-thirds of all oil consumption.

As Figure 11.6 shows, oil from a variety of production centers flows, primarily by water, to the industrially advanced countries. Note that the United States imports oil from a number of regions. The other major importers, Western Europe and Japan, import chiefly Middle Eastern oil.

The efficiency of pipelines, supertankers, and other modes of transport and the low cost of oil helped to create a world dependence on that fuel even though coal was still generally and cheaply available. The pattern is aptly illustrated by American reliance on foreign oil. For many years, United States oil production had remained at about the same level, 8 to 9 million barrels per day. Between 1970 and 1977, however, as domestic supplies became much more expensive to extract, consumption of oil from foreign sources increased dramatically, until almost half of the oil consumed nationally was imported. The dependence of the United States and other advanced industrial economies on imported oil gave the oil-exporting countries tremendous power, reflected in the soaring price of oil in the 1970s. During that decade, oil prices rose dramatically, largely as a result of the strong market position of the Organization of Petroleum Exporting Countries (OPEC).

Among the side effects of the oil "shocks" of 1973–1974 and 1979–1980 were worldwide recessions, large net trade deficits for oil importers, a reorientation of world capital flows, and a depreciation of the U.S. dollar against many other currencies. On the positive side, the soaring oil prices of the 1970s triggered oil exploration in non-OPEC countries, improvements in oil-drilling technology, and a search for alternative energy sources. Perhaps most important, for a number of years, they diminished total energy demand, partly because of the recession and partly because the high prices fostered conservation. Industrial countries have learned to use much less oil for each unit of output. Cars, planes, and other machines are more energy efficient than they were in the 1970s, as are industries and buildings constructed in recent years.

Since 1985, however, both global production and consumption of oil have increased steadily. And the United States, which satisfied 69% of its oil needs by domestic production in each year between 1982 and 1986, has since then

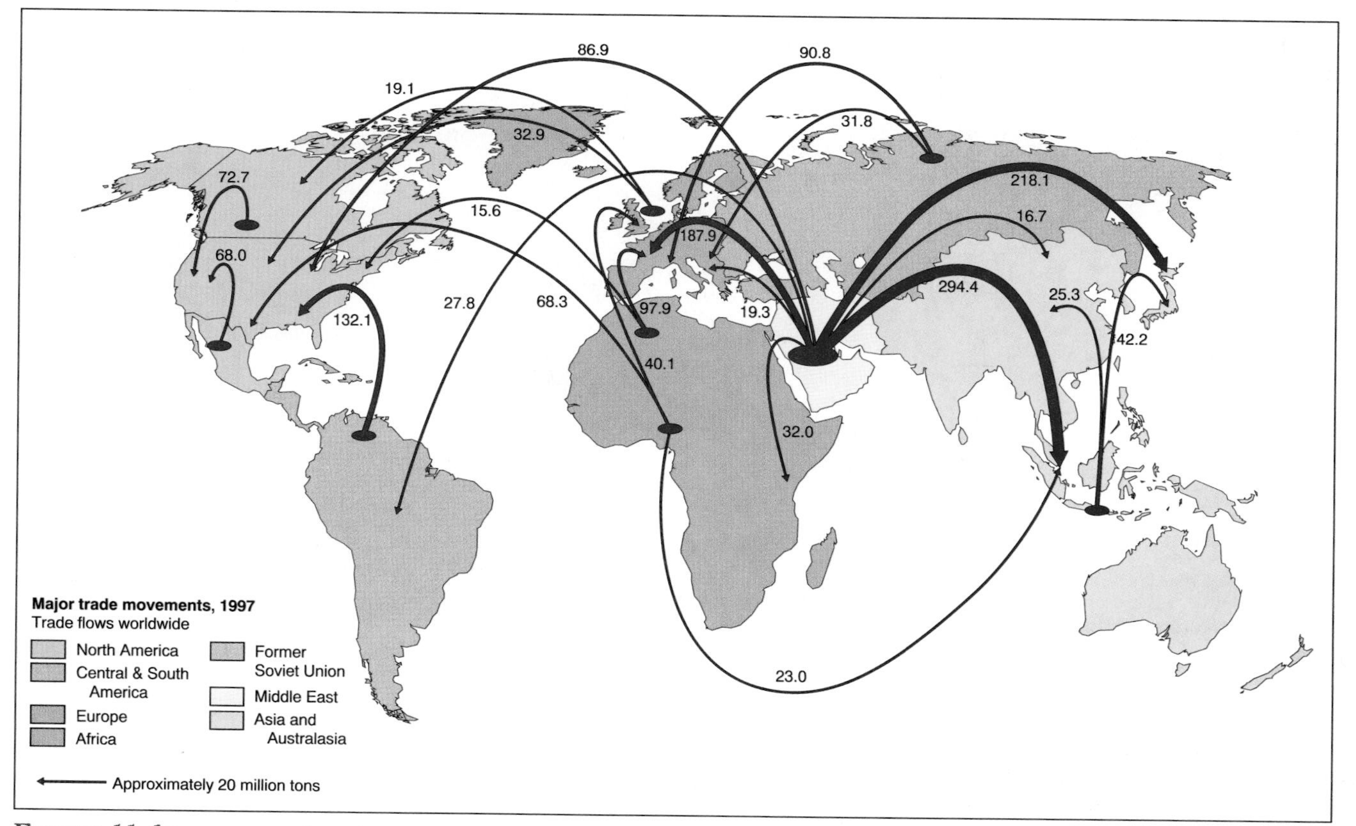

FIGURE 11.6 **International crude-oil flow by sea, 1997.** Note the dominant position of the Middle East in terms of oil exports. The arrows indicate origin and destination, not specific routes. The line widths are proportional to the volume of movement. The United States imports slightly more than half of the oil it uses.

Source: The *BP Statistical Review of World Energy 1998*. Used with permission.

relied increasingly on imports. By 1998, the United States depended on foreign sources for 53% of the oil it consumed annually (see "A Costly Habit").

It is particularly difficult to estimate the size of oil reserves. Not only are estimates constantly revised as oil is extracted and new reserves are located, but many governments tend to maintain some secrecy about the sizes of reserves, understating official national estimates. Nonetheless, it is clear that oil is a scarce resource and that oil reserves are very unevenly distributed among the world's countries (Figure 11.7). Slightly more than 1000 billion barrels are classified as proved reserves, and another 900 billion are thought to exist in undiscovered reservoirs. If all the oil could be extracted from the known reserves, and if the current rate of production holds, the proved reserves would last only about 40 years. The ratio of production to reserves, however, gives some Middle Eastern countries more than a century of pumping at current rates before their oil fields run dry.

The realization that oil supplies are finite and that some countries are likely to deplete their reserves in the foreseeable future has sparked renewed interest in another fossil fuel, coal.

Coal

Coal was the fuel basis of the Industrial Revolution. From 1850 to 1910, the proportion of U.S. energy supplied by coal rose from 10% to almost 80%. Although the consumption of coal declined as the use of petroleum expanded, coal remained the single most important domestic energy source until 1950 (see Figure 11.4).

Although coal is a nonrenewable resource, world supplies are so great that its resource life expectance may be measured in centuries, not in the decades usually cited for oil and natural gas. The United States alone possesses more than 250 billion tons of coal considered potentially mineable on an economic basis with existing technology. At current production levels, these demonstrated reserves would be sufficient to meet the domestic demand for coal for another 2.5 centuries.

Worldwide, the most extensive deposits are concentrated in the middle latitudes of the Northern Hemisphere, as shown in Figure 11.8. Two countries have dominated world coal production in recent years, accounting for more than half of the coal produced in the world: China and the United States. Since 1990, the use of coal worldwide has

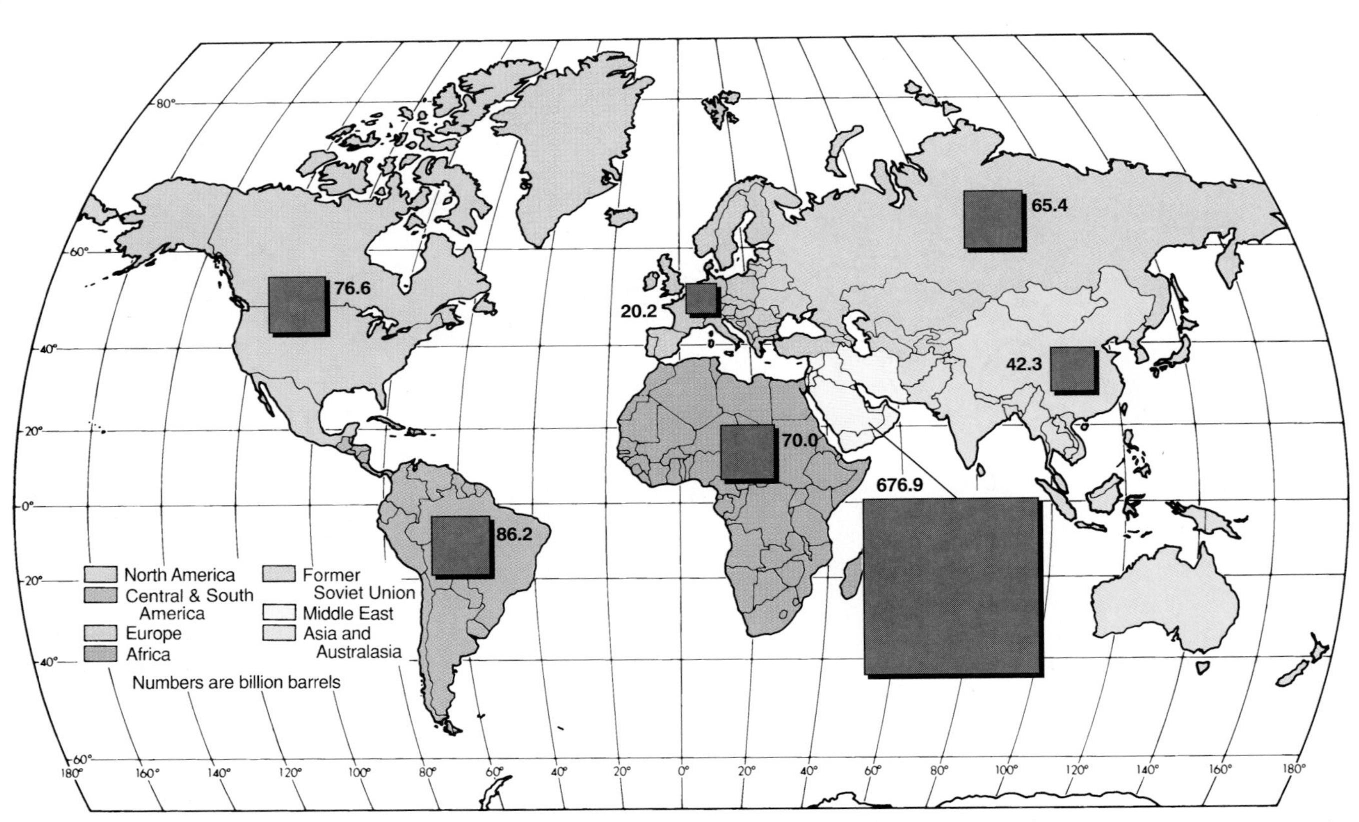

FIGURE 11.7 **Regional shares of proved oil reserves, in billions of barrels, January 1, 1998.** Middle Eastern countries contain about two-thirds of the proved reserves. At the beginning of 1998, the world's proved oil reserves were estimated to be 1038 billion barrels, a figure that tends to rise over time with new methods of locating and extracting oil deposits, and changes in price. In the late 1990s, global consumption of oil was about 26 billion barrels per year.

Source: Data from British Petroleum Company, *The BP Statistical Review of World Energy 1998.*

A Costly Habit

The United States is a crude-oil junkie, dependent on daily fixes of petroleum. On average, Americans consume more than 17 million barrels of oil a day. That is equivalent to 2.7 gallons per person per day, or about 1000 gallons per person per year. What are some of the implications of this dependence?
Consider the data in the table.

Country	1997 Proved Reserves (billion barrels)	1997 Production (billion bls.)	1997 Consumption (billion bls.)
United States	29.8	3.0	6.5
Canada	6.8	.9	.6
Mexico	40.0	1.2	.6

Notice the imbalance between production and consumption for the United States. In contrast to its hemispheric neighbors, Canada and Mexico, the United States consumes far more oil than it produces. At current rates of consumption, and assuming no imports, the proved reserves would meet domestic demand for only five years. Americans continue to drive their cars, and their manufacturing plants continue to turn out a wide range of petroleum-based products, only because the country imports almost 10 million barrels of oil a day. That is, the United States relies on foreign sources to meet slightly more than half of its crude-oil needs.

American dependence on imports should ease, and U.S. proved reserves increase significantly, as deepwater fields in the Gulf of Mexico are tapped. They are estimated to hold about 15 billion barrels of oil, considerably more oil than the giant Prudhoe Bay fields in Alaska, which are currently one of the largest sources of domestic oil. The fields lie beyond the continental shelf and, until recently were considered too deep—deeper than 1500 feet (456 meters)—to reach economically.

Several petroleum companies are building oil production platforms that will be anchored one-half mile (.8 km) or more below the surface of the water. One company, Shell Oil, has begun producing oil from the first two of these platforms, in its Auger and Mars fields. Two more fields—Ram-Powell and Ursa—are scheduled for production in 1997 and 1999, respectively. Unlike oil rigs in shallower water, which are rigid towers built on fixed platforms attached to the sea bottom, the new platforms float on the surface, tethered by steel tendons to enormous anchors. Each platform costs about $1 billion, reflecting the high price of dependence on oil.

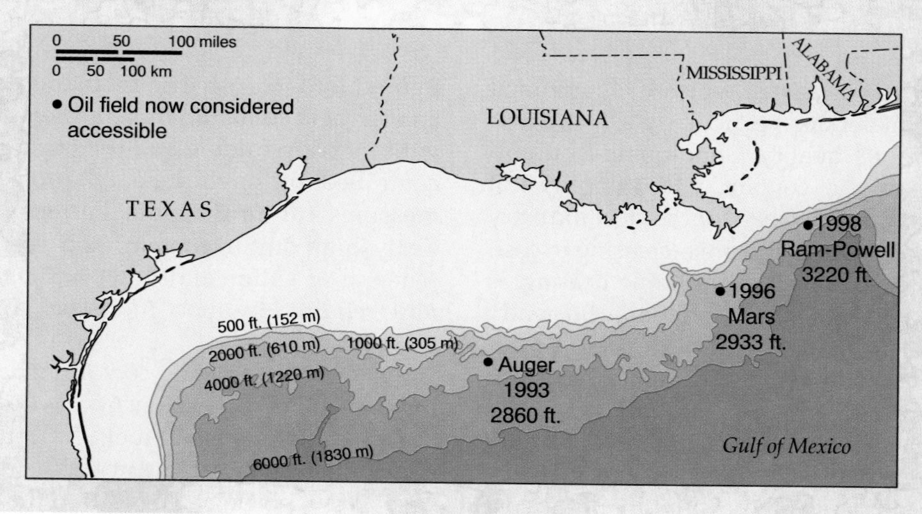

remained steady, but distinct regional variations in its use are evident. Coal production has risen in the United States but fallen in Europe and the countries of the former Soviet Union, as governments have discontinued subsidies to the industry. The use of coal continues to grow in developing countries, particularly China and India. In the United States and other industrialized countries, coal is used chiefly for electric power generation and to make coke for steel production. In less developed countries, coal is widely used for home heating and cooking, as well as to generate electricity and fuel factories.

Coal is not a resource of constant quality. It ranges from lignite (barely compacted from the original peat) through bituminous coal (soft coal) to anthracite (hard coal), each *rank* reflecting the degree to which organic material has been transformed. Anthracite has a fixed carbon content of about 90% and contains very little moisture. Conversely, lignite has the highest moisture content and the lowest amount of elemental carbon, and thus the lowest heat value. About half of the demonstrated reserve base in the United States is bituminous coal, concentrated primarily in the states east of the Mississippi River.

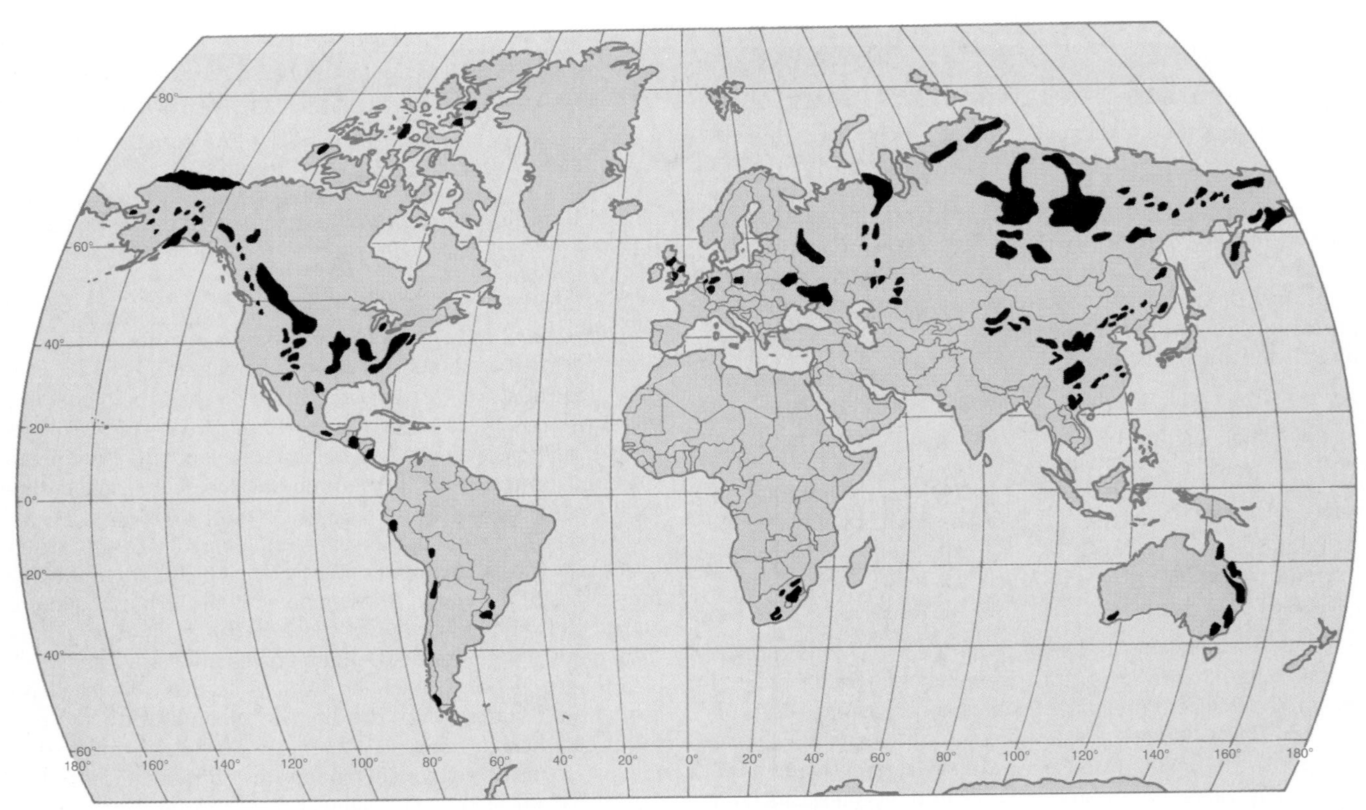

FIGURE 11.8 **Major coal basins** are concentrated in the Northern Hemisphere. China is the world's largest coal producer and consumer. The primary source of energy in China, coal supplies about 75% of the country's fuel needs.

Besides rank, the *grade* of a coal, which is determined by its content of waste materials (particularly ash and sulfur), helps to determine its quality. Good-quality bituminous coals with the caloric content and the physical properties suitable for producing coke for the steel industry are decreasingly available readily and are increasing in cost. Anthracite, formerly a dominant fuel for home heating, is now much more expensive to mine and finds no ready industrial market. The Schuylkill anthracite deposits of eastern Pennsylvania are discussed as a special type of resource region in Chapter 13.

The value of a given coal deposit is determined not only by its rank and grade but also by its accessibility, which depends on the thickness, depth, and continuity of the coal seam and its inclination to the surface. Much coal can be mined relatively cheaply by surface techniques, in which huge shovels strip off surface material and remove the exposed seams. Much, however, is available only by expensive and more dangerous shaft mining, as in Appalachia and most of Europe. In spite of their generally lower heating value, western United States coals are now attractive because of their low sulfur content. They do, however, require expensive transportation to markets or high-cost transmission lines if they are used to generate electricity for distant consumers (Figure 11.9).

The ecological, health, and safety problems associated with the mining and the combustion of coal must also be figured into its cost. The mutilation of the original surface and the acid contamination of lakes and streams associated with the strip mining and the burning of coal are partially controlled by environmental protection laws, but these measures add to the costs. Eastern U.S. coals have a relatively high sulfur content, and costly techniques for the removal of sulfur and other wastes from stack gases are now required by most industrial countries, including the United States.

The cost of moving coal influences its patterns of production and consumption. Coal is bulky and is not as easily transported as nonsolid fuels. As a rule, coal is usually consumed in the general vicinity of the mines. Indeed, the high cost of transporting coal induced the development of major heavy industrial centers directly on coalfields—for example, Pittsburgh, the Ruhr, the English Midlands, and the Donets district of Ukraine.

Natural Gas

Coal is the most abundant fossil fuel, but natural gas has been called the nearly perfect energy source. It is a highly efficient, versatile fuel that requires little processing and is environmentally benign. Of the fossil fuels, natural gas (which is mainly methane) has the least impact on the environment. It burns cleanly. The chemical products of burned methane are carbon dioxide and water vapor, which are not

FIGURE 11.9 Long-distance transportation adds significantly to the cost of low-sulfur western coals because they are remote from eastern U.S. markets. To minimize these costs, unit trains carrying only coal engage in a continuous shuttle movement between western strip mines and eastern utility companies.
© Jim Shippee/Unicorn Stock Photos.

pollutants, although the methane does add to the rising carbon dioxide level of the atmosphere.

As Figure 11.4 indicates, this century has seen an appreciable growth in the proportion of U.S. energy supplied by gas. In 1900, it accounted for about 3% of the national energy supply. By 1980, the figure had risen to 30%, but then declined to about 27% by 1998. The trend in the rest of the world has been in the opposite direction. Global production increased significantly after the oil shock of 1973–1974 and by 1998 had nearly doubled.

Most gas is used directly for industrial and residential heating. In fact, gas has overtaken both coal and oil as a house-heating fuel, and more than half of the homes in the United States are now heated by gas. A portion is also used in electricity-generating plants, and some is chemically processed into products as diverse as motor fuels, plastics, synthetic fibers, and insecticides.

Very large natural gas fields were discovered in Texas and Louisiana as early as 1916. Later, additional large deposits were found in the Kansas-Oklahoma-New Mexico region. At that time, the south-central United States was too sparsely settled to make use of the gas, and in any case, it was oil, not gas, that was being sought. Many wells that produced only gas were capped. Gas found in conjunction with oil was vented or burned at the wellhead as an unwanted by-product of the oil industry. The situation changed only in the 1930s, when pipelines were built to link the southern gas wells with customers in Chicago, Minneapolis, and other northern cities.

Like oil, natural gas flows easily and cheaply by pipeline. Unlike oil, however, gas does not move as freely in international trade (Figure 11.10a). Transoceanic shipment involves costly equipment for liquefaction and for vessels that can contain the liquid under appropriate temperature conditions. **Liquefied natural gas (LNG)** is extremely hazardous because the mixture of methane and air is explosive. Although the United States has imported some LNG, chiefly from Algeria, most gas is transferred by pipeline. In the United States, the pipeline system is more than 1.6 million kilometers (1 million mi) long (Figure 11.10b).

Like other fossil fuels, natural gas is nonrenewable; its supply is finite. Estimates of reserves are difficult to make because they depend on what customers are willing to spend for the fuel, and estimates have risen as the price of gas has increased. Further complicating the estimate of supplies is uncertainty about potential gas resources in unusual types of geologic formations. These include tight sandstone formations, deep (below 6000 meters, or 20,000 ft) geologic basins, and shale and coal beds.

Worldwide, two regions contain about 70% of the proved gas reserves: Russia and the Middle East (Figure 11.11). The remaining 30% is divided roughly equally among North America, Western Europe, Africa, Asia, and Latin America, each of which has from 4% to 7% of the total. The gas in these reserves would last about 65 more years at current production rates, but developing countries, particularly in South and Southeast Asia, may well have undiscovered deposits that could add significantly to the life expectance of world reserves if they were developed.

In the United States, the Texas-Louisiana and Kansas-Oklahoma-New Mexico regions account for about 90% of the domestic natural gas output, but there are thought to be gas deposits beneath almost all states. In addition, many offshore areas are known to contain gas. Potential Alaskan reserves are estimated to be at least twice as large as today's proved reserves in the rest of the country, containing enough gas to heat all of the houses in the country for a decade. Estimates of U.S. proved reserves indicate that at the current rate of consumption, there is enough gas to last anywhere from 10 to 20 years. But if the technology necessary to produce gas from unconventional sources is developed, gas reserves may be sufficient for another century. Of course, these less accessible supplies will be more costly to develop, and hence more expensive.

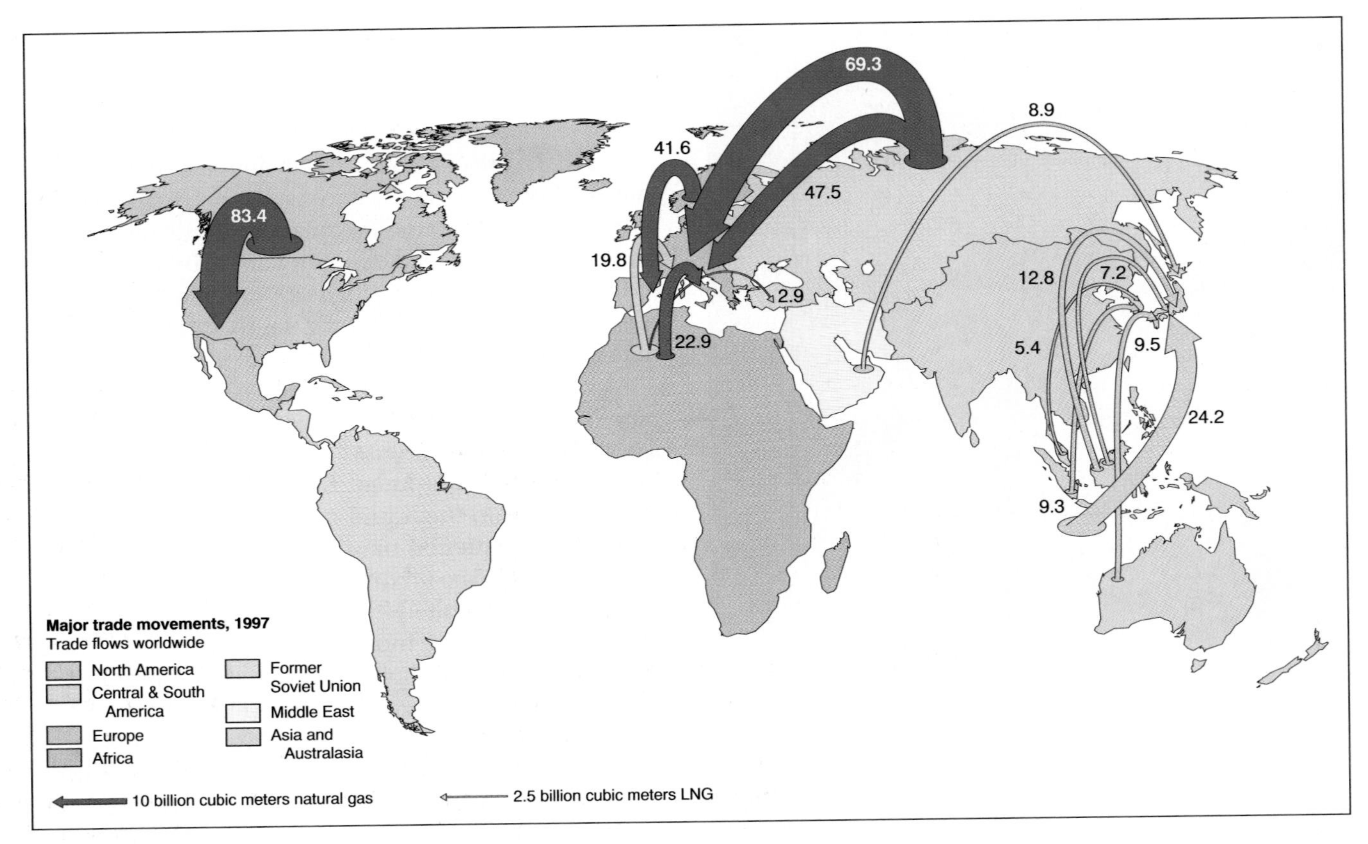

(a)

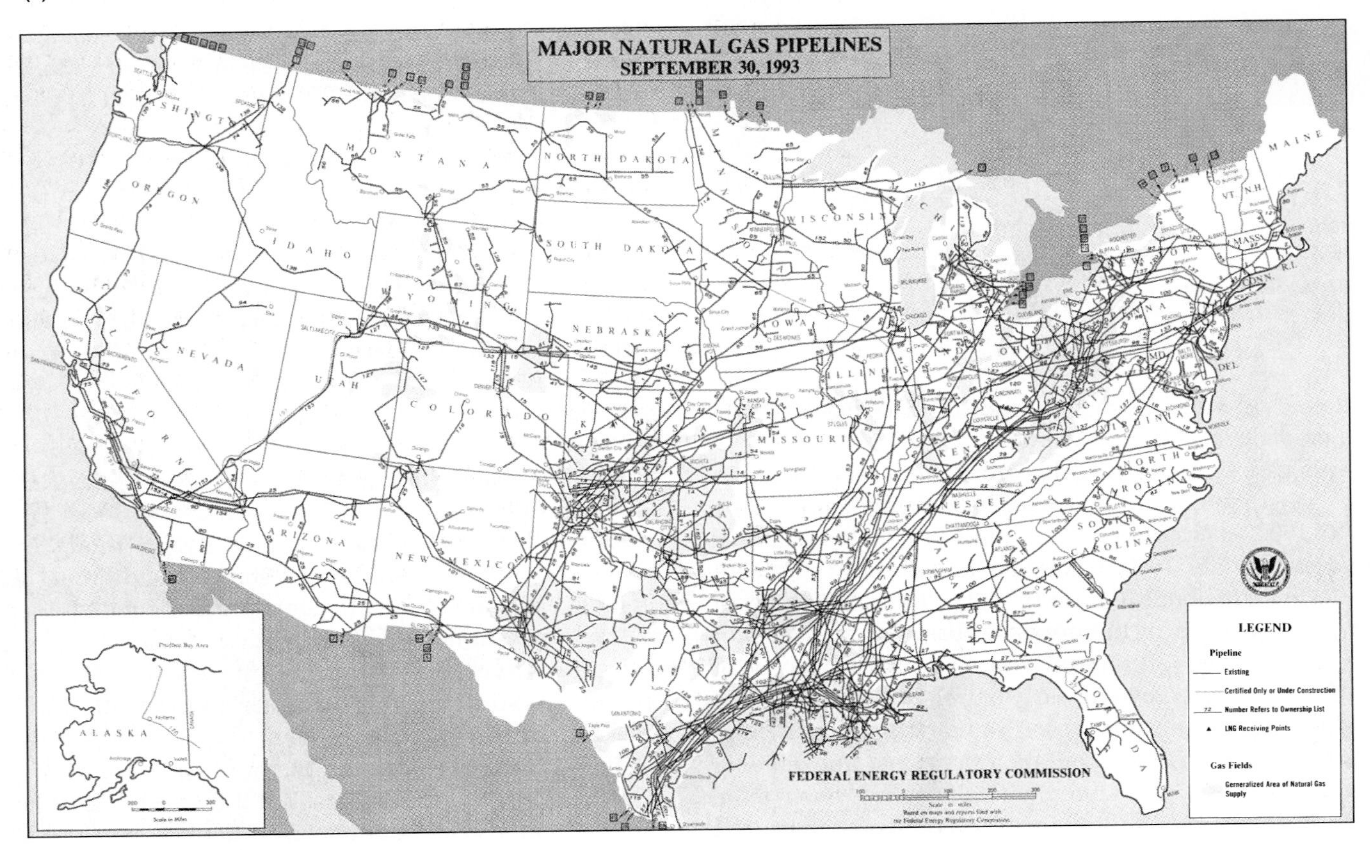

(b)

FIGURE 11.10 (a) **Worldwide trade flows of natural gas, 1997.** Most natural gas flows by pipeline. Notice that Japan is the only country that imports large amounts of liquefied natural gas, though Spain, France, and Belgium import some LNG from Algeria. (b) **Major natural gas pipelines in the United States, 1993.**

(a) Source: Data from British Petroleum Company, *The BP Statistical Review of World Energy 1998*.

(b) Source:U.S. Department of Energy.

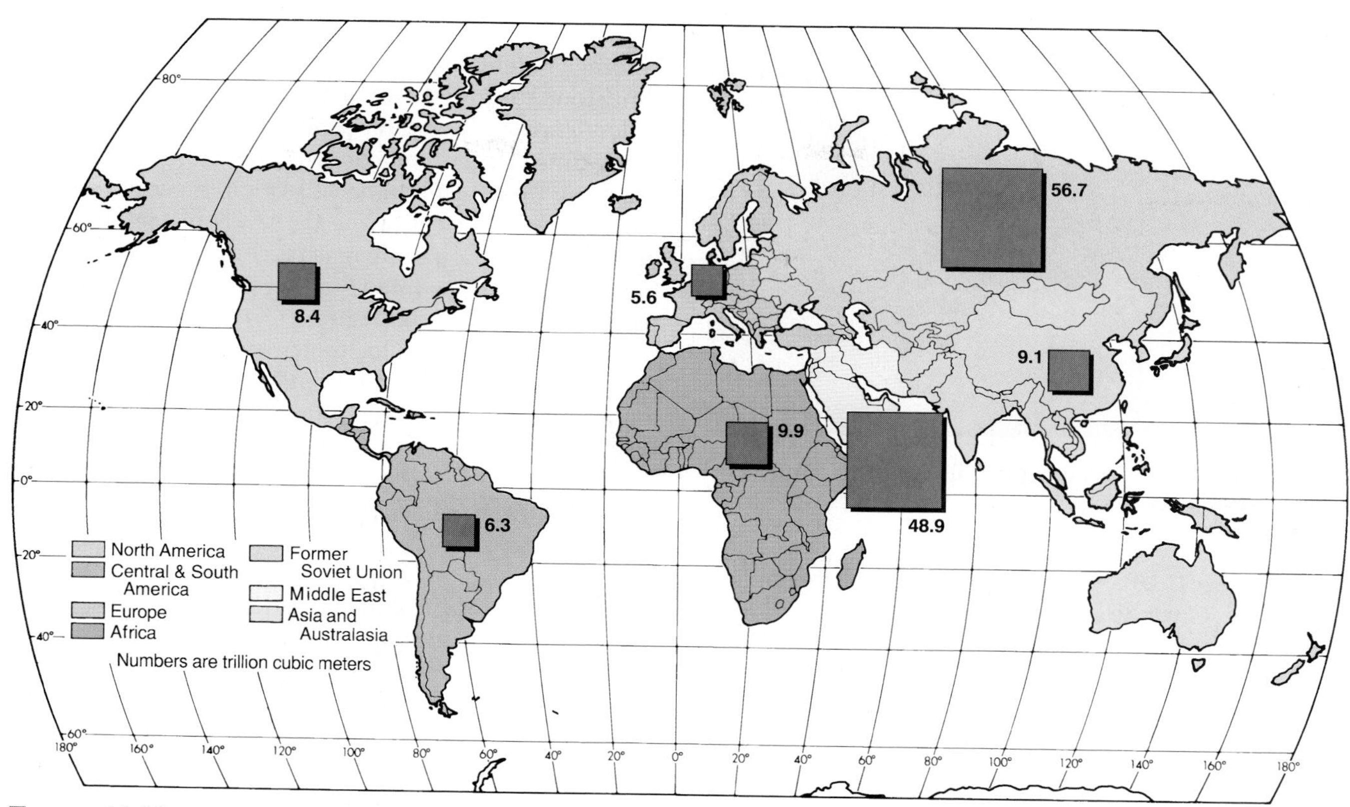

FIGURE 11.11 **Proved natural gas reserves, January 1, 1998.** Russia contains the largest natural gas reserves, about one-third of the world total (and more than 10 times the size of U.S. reserves). Major reserves also exist in the countries of the Middle East.
Source: Data from British Petroleum Company, *The BP Statistical Review of World Energy 1998.*

Synthetic Fuels

The rising energy prices of the 1970s stimulated interest in developing substitutes for oil and natural gas from underused resources. **Synthetic fuels** ("synfuels") are naturally occurring organic products that can be converted into synthetic petroleum and natural gas. The most promising synfuels are those produced from coal, oil shale, and tar sands.

Coal Conversion

Two broad categories of coal conversion techniques are gasification and liquefaction. **Coal gasification** is a process whereby coal is burned under high pressure in the presence of steam and either air or oxygen to produce either a high- or low-heating-value gas. With additional processing, the product can be converted into a gas of the quality of methane. Although gasification techniques have been employed on a small scale for several decades, the product generally has been too expensive to compete with natural gas.

Coal liquefaction processes are used to convert coal into liquid products comparable to petroleum. One type of liquefaction involves heating coal at relatively low temperatures and under pressure to extract the volatile (easily vaporized) matter. Volatiles are recovered as liquids and gases; the liquids are treated for the removal of organic sulfur, nitrogen, and oxygen. The resulting product, a synthetic crude oil, can be refined into gasoline and fuel oils comparable in quality to petroleum-based products. Liquefaction processes appear to be more complex and costly than gasification techniques, and the synthetic petroleum is not as clean a fuel as methane.

Producing energy from either gasified or liquefied coal is considerably more expensive than burning the coal directly and requires large capital outlays to construct the conversion facilities (Figure 11.12). Furthermore, roughly one-third of the energy in the coal is consumed in the conversion process. While synthetic gas and petroleum from coal may someday compete economically with natural gas and oil, the cost of the synfuels will be high.

Oil Shale

A similar situation affects the prospects of the extraction of oil from **oil shale,** fine-grained rock containing organic material called *kerogen.* A tremendous potential reserve of hydrocarbon energy, the rocks involved are not shales but calcium and magnesium carbonates more similar to limestone than to shale, and the hydrocarbon, kerogen, is not oil but a waxy, tarlike substance that adheres to the grains of carbonate material. The crushed rock is heated to a temperature high enough (more than 480°C, or 900°F) to decompose the kerogen, releasing a liquid oil product, *shale oil.*

World reserves of oil shale are enormous. Known deposits estimated to contain billions of barrels of shale oil are found in the United States, Brazil, Russia, China, and

FIGURE 11.12 **A coal-gasification plant in Daggett, California.**
Vince Streano/Tony Stone Images.

Australia (Figure 11.13). The richest deposits in the United States are in the Green River Formation near the junction of Colorado, Utah, and Wyoming. They contain enough oil to supply the needs of the United States for another century, and in the 1970s were thought to be the answer to national energy self-sufficiency. Several billion dollars were invested in oil shale research and development operations in the Piceance Basin near Grand Junction, Colorado, but as oil prices fell in the 1980s, interest in the projects waned. The last plant, that at Parachute Creek, Colorado, was abandoned in 1991.

Tar Sands

Another potential source of petroleum liquids is **tar sand,** sandstone saturated with a viscous high-carbon petroleum called *bitumen.* Global tar sand resources are thought to be many times larger than conventional oil resources, containing more than 2 trillion barrels of oil, most of it in Canada and Venezuela. Most of the tar sands in the United States are in Utah. Although there is no commercial production of tar-sand oil in the United States, oil is produced from similar deposits in Canada (Figure 11.14).

Like other synfuel technologies, producing oil from tar sands requires high capital outlays and carries substantial environmental costs. Nevertheless, because they exist in vast amounts, coal, oil shale, and tar sands could be plentiful sources of gaseous and liquid fuels. Unlike nuclear energy or most renewable energy sources (such as solar or hydroelectric power), they could provide the gasoline, jet, and other fuels on which industrialized societies depend. The lowering of oil and gas prices in the 1980s and the decline in federal government support put a temporary halt to efforts to make synfuel products economically competitive with oil imports. The prospect of depletion of oil and natural gas resources within the next century, however, indicates that at some point their prices are bound to rise. When that occurs, countries are likely to try to turn to these more unconventional sources of fuel. In the meantime, technologies employing nuclear energy as a source of nonfuel power are already developed.

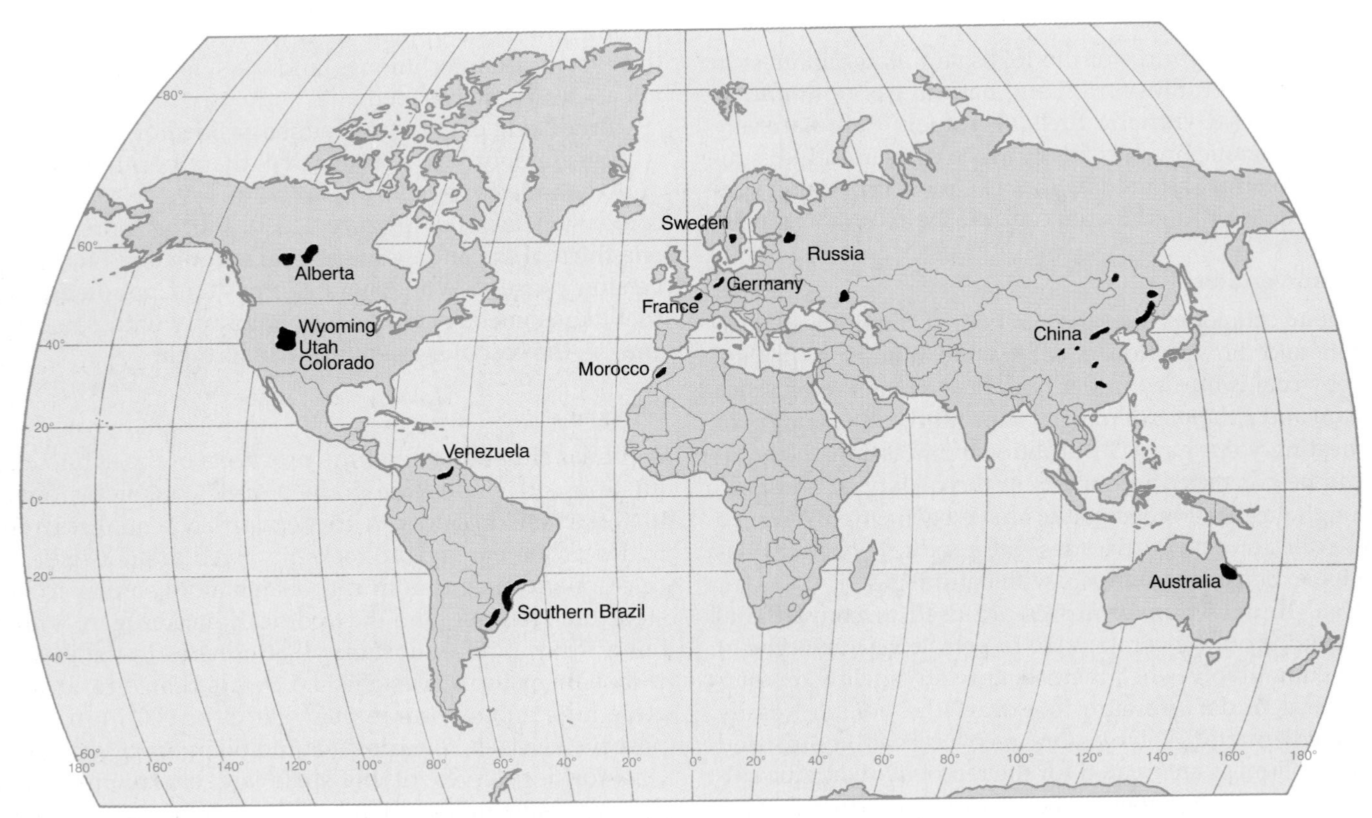

FIGURE 11.13 **Oil shale deposits.** The Piceance Basin in Colorado contains one of the richest hydrocarbon deposits in the world.

built to feed three large generating stations. Even farther into the future are plans to divert the Nottaway and Rupert Rivers into the Broadback, although growing opposition to the entire James Bay project makes completion of that phase uncertain.

Critics of the project include the region's indigenous population, the Cree and Inuit. The latter would be affected by James Bay II, but much of the Crees' ancestral burial and hunting grounds has already been submerged. Environmental experts charge that further flooding would not only threaten the Crees' source of livelihood but also endanger many species of wildlife that depend on the area's forests and wetlands. Mercury from submerged vegetation has already leached into reservoirs, and the level of mercury in the region's freshwater fish has increased sixfold since completion of the first phase. The fish are a dietary staple of the Cree, many of whom now have mercury levels that exceed the safety standards set by the World Health Organization.

The James Bay project is already more than 20 years old. In 1994, the first concrete was poured for what is intended to be the world's largest hydroelectric project, the Three Gorges Dam. The enormous structure is to block the Yangtze River in China's scenic Three Gorges area, where the river flows through a series of sheer chasms. Although the dam itself would not be the world's biggest, its hydroelectric output would be more than 18,000 megawatts, by far the largest in the world. The project raises so many questions, however, that some observers believe that construction could still be halted or at least postponed.

Scheduled to be completed in 2009, the Three Gorges Dam will have significant ecological and social costs. Upstream, the water will slow down and drop an enormous amount of silt on the bottom of the 565-kilometer-long (350-mi) reservoir; as it fills with sediment, upstream flooding is possible. The reduced water flow during the dry season will disrupt the Yangtze's estuary, causing sea water to move upstream. There will be a loss of nutrients to river life because the trapped sediment will contain organic matter that nourishes downriver food webs. Some analysts predict a dramatic reduction in the number of fish because the lower water temperature and the curtailing of spring floods will affect spawning behavior. Social costs of the project will be borne by the 1.9 million people whose houses will be submerged by the dam's huge reservoir.

Questions to Consider

1. The advantages of hydropower are manifold. The fuel (running water) is free, operating costs tend to be low, it does not pollute the air or contribute to global warming, and many suitable sites exist in less developed countries. Further, dams can control floods and provide a regulated flow of water for irrigation. How would you balance these advantages against the environmental and social costs of a project such as James Bay?
2. If you were the Minister of Water Resources in China, would you recommend that the Three Gorges Dam be completed? What questions would you want answered in order to assess the likely benefits and costs of the dam? How would you respond to critics who argue that China could more quickly and cheaply meet its energy and flood-control needs by reinforcing dikes and building smaller dams along the Yangtze River and its tributaries?
3. About 100,000 dams regulate American rivers, yet according to the U.S. Geological Survey, flooding continues to be the most destructive and costly type of natural disaster in the country, there has been no reduction in the average number of flood deaths each year, and even when adjusted for inflation, flood damage to property has almost tripled since 1951. Can you think of some reasons why this might be so?

They can turn turbines directly, do not use any fuel, and can be built and erected rather quickly. They need only strong, steady winds to operate, and these exist at many sites. Furthermore, wind turbine generators do not pollute the air or water and do not deplete scarce natural resources.

About 20% of the world's wind generation capacity is in three areas of California, where some 16,000 turbines provide about 1.5% of the state's electricity (Figure 11.22). California's dominance is due less to favorable wind energy conditions than to such factors as the state's commitment to developing renewable resources and the support of utility companies. Other states with *wind farms* include Oregon, Maine, New Hampshire, and Vermont. A wind farm is a cluster of wind-powered turbines in a favorable geographic location, that is, one where wind speeds average 14 to 20 miles per hour.

In recent years, governmental incentives have stimulated the growth of wind power installations in many European countries. In some, such as Denmark and Germany, turbines tend to be sited singly or in small clusters on farms. Other

FIGURE 11.20 **Parabolic trough reflectors at a solar thermal energy plant in the Mojave Desert near Daggett, California.** The facility uses sunlight to produce steam to generate electricity. Guided by computers, the parabolic reflectors follow the sun, focusing solar energy onto a steel tube filled with heat transfer fluid.
©Cameramann International, Ltd.

FIGURE 11.21 **The largest wet steam geothermal electric power plant in the world is located in Wairakei, New Zealand.** Magma radiates heat through the rock above it, heating water in underground reservoirs. Drilled wells tap the steam and bring it to the surface, where it is piped to the power plant. As the photograph suggests, one drawback of geothermal plants is that they release gases into the atmosphere, although scrubbers like those used on coal stacks can reduce gaseous emissions to an acceptable level.
©Nicholas DeVore / Tony Stone Images.

European countries, including Spain and the United Kingdom, are investing in large wind farms. Denmark has the highest per capita output of wind energy in the world, with wind power generating 7% of the country's electricity in 1997.

The chief disadvantage of wind power is that it is unreliable and intermittent; because its energy cannot be easily stored, it requires a backup system. Detractors point out that it takes thousands of wind turbines to produce the same amount of electricity as a single nuclear power plant. Furthermore, wind farms are very visible, often covering entire hillsides and dominating the landscape. Finally, local residents sometimes complain of the noise associated with the turbines, although the newest machines are scarcely audible.

FIGURE 11.22 **A "wind farm" in the Tehachapi Mountains Wind Resource Area of California.** Other wind farms—clusters of wind turbines—are located in Altamont Pass east of San Francisco and at the San Gorgonio Pass near Palm Springs. The wind turbine generators harness wind power to produce electricity. California's wind farms produce enough electricity to supply the residential power demands of nearly one million people.
Courtesy of Doug Sherman.

NONFUEL MINERAL RESOURCES

The mineral resources already discussed provide the energy that enables people to do their work. Equally important to our economic well-being are the *nonfuel* minerals, for they can be processed into steel, aluminum, and other metals, and into glass, cement, and other products. Our buildings, tools, and weapons are chiefly mineral in origin.

Virtually all the resources we deem essential, including metals, nonmetallic minerals, rocks, and the fuels, are contained in the earth's crust, the thin outer skin of the planet. Of the 92 natural elements, just eight account for more than 98% of the mass of the earth's crust (Figure 11.23). They can be thought of as geologically abundant, and all others as geologically scarce.

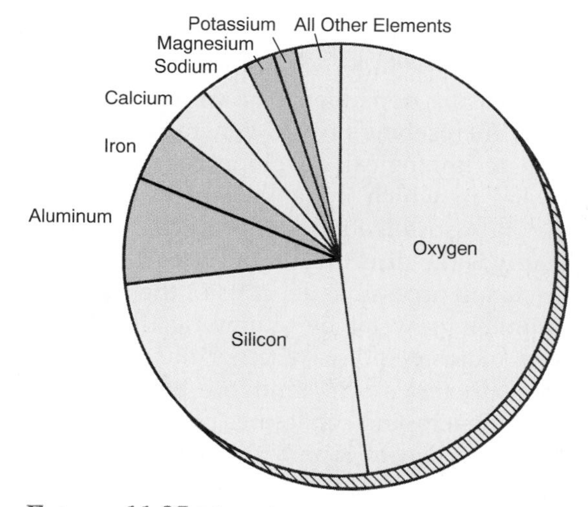

FIGURE 11.23 **The relative abundance, by weight, of elements in the earth's crust.** Only four of the economically important elements are geologically abundant, accounting for more than 1% of the total weight of the earth's crust. Fortunately, these and other commercially valuable minerals have been concentrated in specific areas within the crust. Were they uniformly disseminated throughout the crust, their exploitation would not be feasible.

Exploitation of a mineral resource typically involves four steps:

1. *exploration* (finding concentrated deposits of the material);
2. *extraction* (removing it from the earth);
3. *concentration* (separating the desired material from the ore);
4. *smelting* and / or *refining* (breaking down the mineral to desired pure material).

Each step requires inputs of energy and materials.

Natural processes produce minerals so slowly that they fall into the category of nonrenewable resources, existing in finite deposits. The supplies of some, however, are so abundant that a ready supply will exist far into the future. These include coal, sand and gravel, potash and magnesium. The supply of others, such as tin and mercury, is small and getting smaller as industrial societies place ever greater demands on them. Table 11.1 gives one estimate of "years remaining" for some important metals. It should be taken as suggestive rather than definitive because mineral reserves are difficult to estimate. As we noted in the case of fossil fuels, such estimates are based on economic and technologic conditions, and we cannot predict either future prices for minerals or improvements in technology.

Although human societies began to use metals as early as 3500 B.C. world demand remained small until the

TABLE 11.1

WORLD MINERAL RESERVES

MINERAL	YEARS REMAINING
Zinc	20
Lead	23
Copper	33
Tin	41
Mercury	45
Nickel	59
Iron ore	152
Bauxite	207

Note: These figures reflect the approximate number of years the world's identified reserves of selected minerals will last, based on current rates of production. Such figures are only suggestive because reserve totals and consumption rates fluctuate over time. A decline in the costs of exploitation, increases in value of the material, and/or new discoveries of deposits will extend the lifetimes shown here.

Source: Data from *World Resources 1996–97,* Oxford University Press, 1996.

Industrial Revolution. It was not until after World War II that increasing shortages and rising prices (and in the United States, increasing dependence on foreign sources) began to impress themselves on the general consciousness. Worldwide technological development has established ways of life in which minerals are the essential constituent. That industrialization has proceeded so rapidly and so cheaply is the direct result of the earlier ready availability of rich and accessible deposits of the requisite materials. Economies grew fat by skimming the cream. The question, yet unanswered, is whether the remaining supplies of scarce minerals will limit the expansion of industrialized and developing economies or whether, and how, people will find a way to cope with shortages.

The Distribution of Resources

Because the distribution of resources is the result of long-term geologic processes that concentrated certain elements into commercially exploitable deposits, it follows that the larger the country, the more likely it is to contain such deposits. And in fact, Russia, Canada, the United States, and Australia possess abundant and diverse mineral resources. As Figure 11.24 indicates, these are the leading mining countries. They contain roughly half of the nonfuel mineral resources and produce the bulk of the metals (e.g., iron, manganese, and nickel) and nonmetals (e.g., potash and sulfur).

Many types of nonfuel minerals are concentrated in a small number of countries, and some scarce elements occur in just a few regions of the world. Thus, extensive deposits of cobalt and diamonds are largely confined to Russia and central-southern Africa. Some countries contain only one or two exploitable minerals—Morocco has phosphates, for example, and New Caledonia, nickel. Several countries with large populations are at a disadvantage with respect to mineral reserves. They include industrialized countries like France and Japan, which are able to import the resources, as well as developing countries like Nigeria and Bangladesh, which are less able to afford imports.

It is important to note that no country contains all of the economically important mineral resources. Some, like the United States, which were bountifully supplied by nature, have spent much of their assets and now depend on foreign sources. Although the United States was virtually self-sufficient in mineral supplies in the 1940s and 1950s, it is not today. Because of its past history of use of domestic reserves and its continually expanding economy, the United States now depends on other countries for more than 50% of its supply of a number of essential minerals (Figure 11.25).

The increasing costs and the declining availability of metals encourage the search for substitutes. The fact that industrial chemists and metallurgists have been so successful in the search for new materials that substitute for the traditional resources has tended to allay fears of possible resource depletion. But it must be understood that no adequate replacements have been found for some minerals, such as cobalt and chromium. Other substitutes are frequently synthetics, often employing increasingly scarce and costly hydrocarbons in their production. Many, in their use or disposal, constitute environmental hazards, and all have their own high and increasing price tags.

Copper: A Case Study

Table 11.1 indicates that the world reserves of copper will last only another 33 years or so, based on current rates of production and assuming that no new extractable reserves appear. Copper is a relatively scarce mineral, and its importance to industrialized societies is evidenced by the fact that more copper is mined annually than any other nonferrous metal except aluminum.

Three properties make copper desirable: it conducts both heat and electricity extremely well, it can be hammered or drawn into thin films or wires, and it resists corrosion. More than half of all copper produced is used in electrical equipment and supplies, mainly as wire. Most of the remainder is used in construction, in industrial and farm machinery, for transportation, and for plumbing pipes. In alloys with other metals, copper is used to make bronze and brass.

Like most minerals, copper is unevenly distributed in the earth's crust. The largest copper deposits are found at convergent tectonic margins in western North America, western South America, and Australia. Copper deposits in sedimentary basins include those extending across northern Europe, from England to Poland, and the copper belt of central Africa (Zambia and Democratic Republic of Congo). Of an estimated 310 million metric tons of copper reserves,

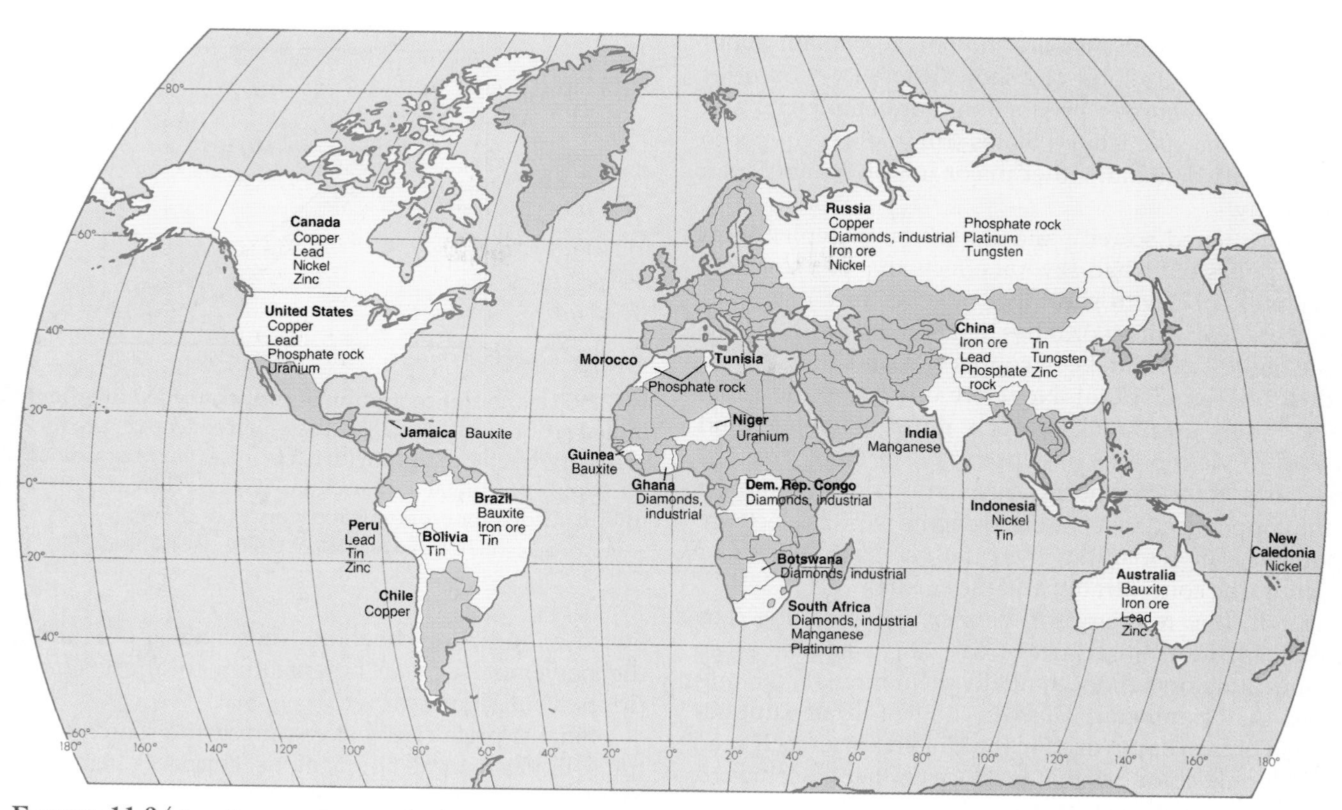

FIGURE 11.24 **Leading producers of selected minerals.** The countries shown are not necessarily those with the largest deposits. India, for example, contains reserves of bauxite, China of manganese, and South Africa of phosphates, but none is yet a major producer of those materials.

Data from World Resources Institute, 1996.

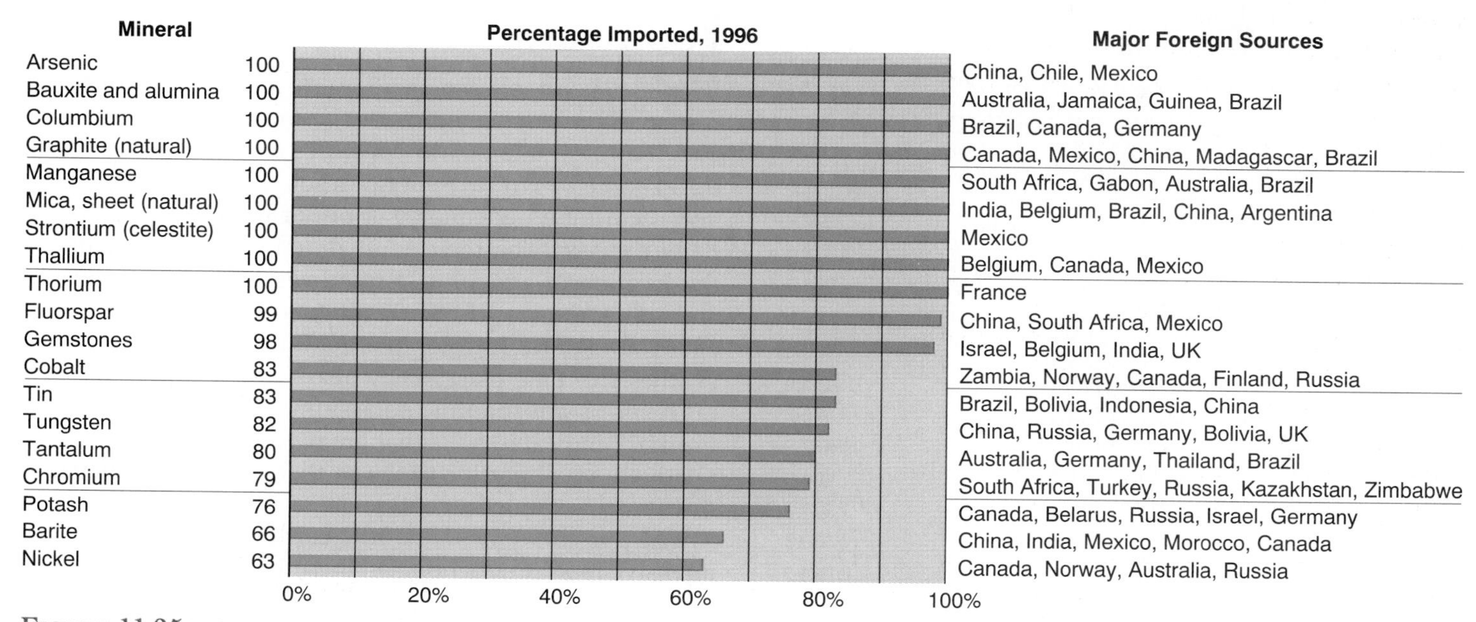

FIGURE 11.25 **U.S. reliance on foreign supplies of minerals.** Like other industrially developed countries, the United States is a leading consumer of the world's mineral resources. Imports account for 50% or more of national consumption of the minerals indicated.

Source: Data from U.S. Geological Survey, U.S. Dept. of the Interior.

Chile and the United States are thought to contain about 140, or almost half. As Table 11.2 shows, those two countries lead the world in copper production. For commercial and economic reasons, the United States still imports copper to meet part of its demand. Other major importers are Japan and Germany.

The threatened scarcity and the ultimate depletion of copper supplies have had several effects that suggest how societies will cope with shortages of other raw materials. First, the grade of mined ores has decreased steadily. Those with the highest percentage of copper (2% and above) were mined early (Figure 11.26). Now, ore of 0.5% grade is the average. Thus, 1000 tons of rock must be mined and processed to yield 5 tons of copper—or, in more practical terms, 3 tons of rock are necessary to equip one automobile with the copper used in its radiator and its various electrical components. The remaining 2.985 tons is waste, generated at the mine, the concentrator, and the smelter.

Second, the recovery of copper by recycling has increased. In the United States, old scrap and new scrap from fabricating operations annually yield more than a million tons of the mineral. The recycling of scrap supplies nearly half of the copper used in the United States each year.

Developing countries that produce copper have moved to gain control over their own resources and have banded together in a cartel. Copper mines in Chile, Peru, Zambia, and Democratic Republic of Congo, once privately owned, have all been nationalized. Those countries are all members of the Intergovernmental Council of Copper Exporting Countries (CIPEC). Should world demand for copper increase, and should serious shortages appear imminent, the power of such an organization to affect supply and prices would likely grow.

Finally, price rises have spurred the search for substitutes. In many of its applications, copper is being replaced by other, less expensive materials. Aluminum is replacing copper in some electrical applications and in heat exchangers. Plastics are supplanting copper in plumbing pipes and building materials. Glass fibers are employed in many telephone transmission lines, and steel can be used in shell casings and coinage. Due to its increased use in motors and electronic equipment, however, the amount of copper used in automobiles actually increased by 40% since 1980.

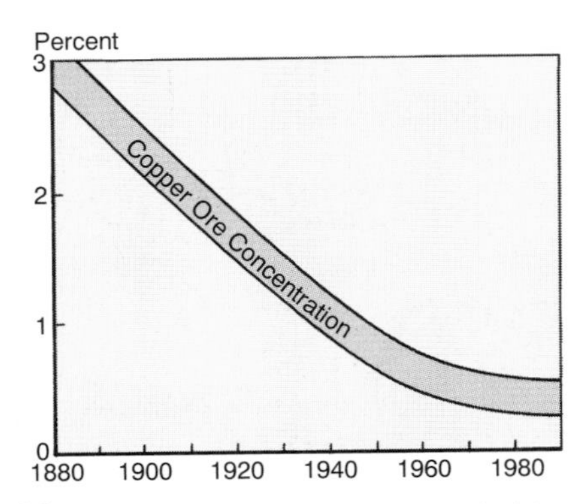

FIGURE 11.26 Concentration of copper needed in order to be mined economically. In 1880, three-percent copper ore rock was necessary, but today, rock with 0.5% or less copper is mined. As the supply of a metal decreases and its price increases, the concentration needed for economic recovery goes down.

Source: Data from the U.S. Bureau of Mines, U.S. Dept. of the Interior.

TABLE 11.2

THE TEN LEADING PRODUCERS AND CONSUMERS OF COPPER, 1994 (FIGURES IN MILLIONS OF METRIC TONS)

Production		Consumption	
Chile	2219.9	United States	2674.3
United States	1795.4	Japan	1374.9
Canada	617.3	Germany	983.1
Former Soviet Union	540.0	China	745.7
China	432.1	Former Soviet Union	560.0
Australia	415.6	France	495.0
Zambia	384.4	Korea, Rep.	476.2
Poland	376.8	Italy	467.9
Peru	359.9	Belgium	404.9
Indonesia	333.8	United Kingdom	377.3
Subtotal	7475.2	Subtotal	8559.3
World total	9522.6	World total	11084.2
World reserves	310,000		
World reserves life index	33 years		

Source: *World Resources 1996–97*, Oxford University Press, 1996.

FOOD RESOURCES

People depend on the fossil fuels and nonfuel mineral resources because they provide the energy to perform work and the raw materials to construct nearly everything we use. People's own raw energy resource, however, is food, and securing an adequate supply of food is a paramount daily concern. The three most important determinants of the location of agricultural resources are sunshine, water, and soil. Plants must receive an adequate amount of solar energy and of fresh water in order for the process of photosynthesis to take place. They also need soils rich in such nutrients as phosphorus, potassium, and nitrogen. As Figure 11.27 indicates, soils that are naturally fertile occupy a relatively small portion of the earth, and they are unevenly distributed among the continents.

At present, only 11% of the earth's surface is used for intensive food production. On these 1.5 billion hectares (3.7 billion acres) must be grown all of the crops on which the planet depends, and rapid population growth has placed increasing pressure on them. The world's population has more than doubled since 1950. To date, the output of the

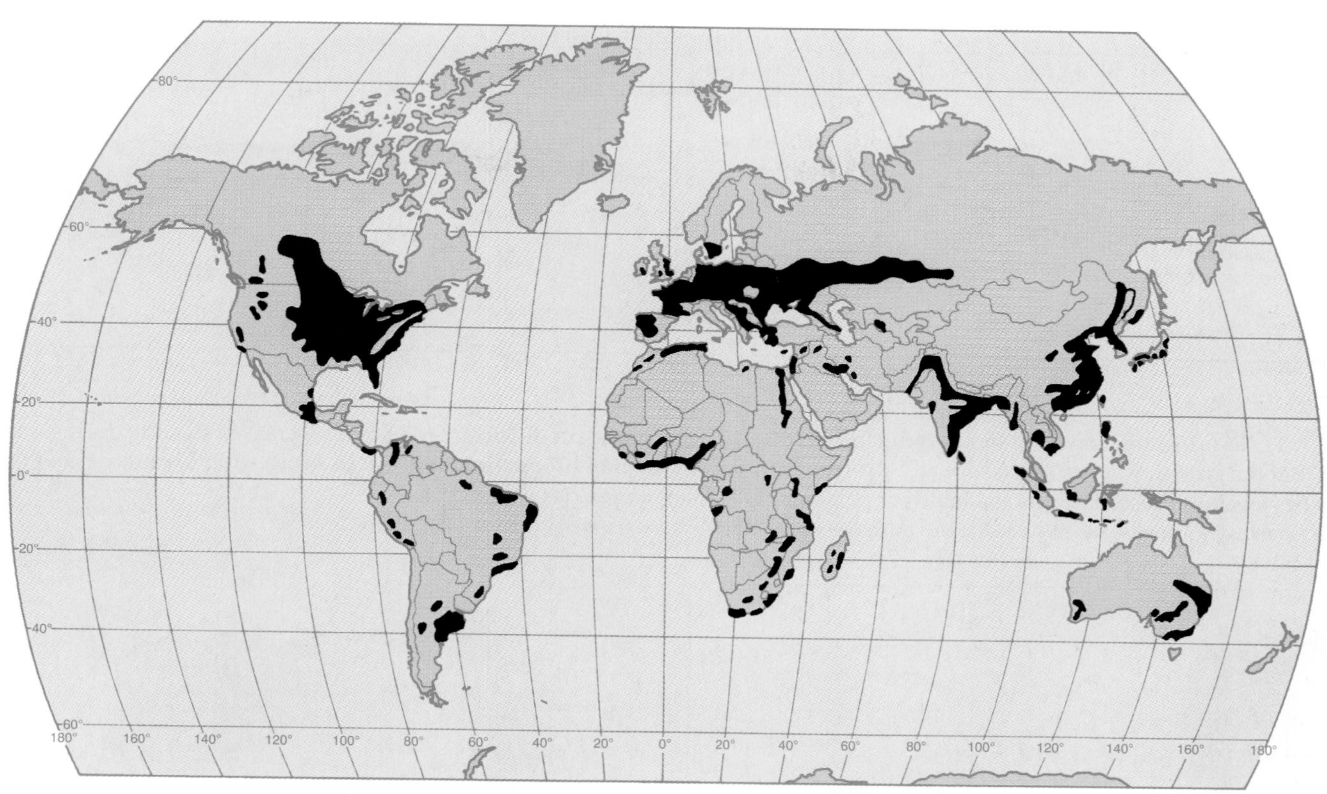

FIGURE 11.27 **Areas with naturally fertile soils** account for much of the world's grain production. They include the corn and wheat belts of North America and Ukraine and the rice-producing regions of India and Southeast Asia.

world's farms has kept pace with demand—that is, the amount of food produced since 1950 has also more than doubled (Figure 11.28). This increase in food production, which has occurred both in developed and developing countries, has been due chiefly to the expansion of land area under cultivation and to increasing productivity.

If foodstuffs were evenly distributed, everyone would receive several pounds of food per day. That is, the total supply of food produced is sufficient to meet world demand. At the same time, however, more people than ever before are malnourished. By some estimates, one-eighth of the world's people—chiefly small children and pregnant women—suffer from clinically defined malnutrition (Figure 11.29).

This seeming contradiction between ample food supplies and widespread malnutrition results from inequalities in population growth rates, lack of access to fertile soils, local climatic catastrophes, the inadequacy of storage facilities and road systems in many countries, and problems of credits and trade imbalances that make it difficult to import food supplies. A number of countries in Latin America (e.g., Peru and Haiti), Africa (e.g., Somalia and Ethiopia), and Asia (e.g., Bangladesh and Laos) have faced serious food shortages in recent years, but on a regional basis the greatest discrepancy between population growth and food production has been in Africa (Figure 11.30).

The secondary effects of chronic or periodic shortages are of concern both to individual countries and to the international community. When malnutrition is the rule, not just starvation but low resistance to infectious diseases, high child mortality rates, mental damage, social disorder, and political unrest or upheaval are likely consequences. The interconnections of the world's peoples mean that food supply problems are not simply domestic concerns in the seemingly remote areas of their occurrence, but in some form have an impact on all societies.

In the next ten years, the world's population is expected to grow by about one billion. That is, each year, on average, another 100 million people will be added, and each year, food supplies will have to increase accordingly. Indeed, for everyone in the world to have an adequate diet, food production should actually exceed population growth in order to provide the grain reserves needed to improve diets above subsistence levels and to compensate for variations in crop yields from year to year. In the following sections, we consider methods of expanding food supplies.

Expansion of Cultivated Areas

One way to increase food production is to expand the areas being cultivated. Approximately 70% of the world's land is not suitable for intensive human use, either because it is too cold, too dry, too steep, or infertile. Essentially all activities must be concentrated on the remaining 30% of land, or about 4.5 billion hectares (11 billion acres). Of those, only one-third (1.5 billion hectares, or 3.7 billion

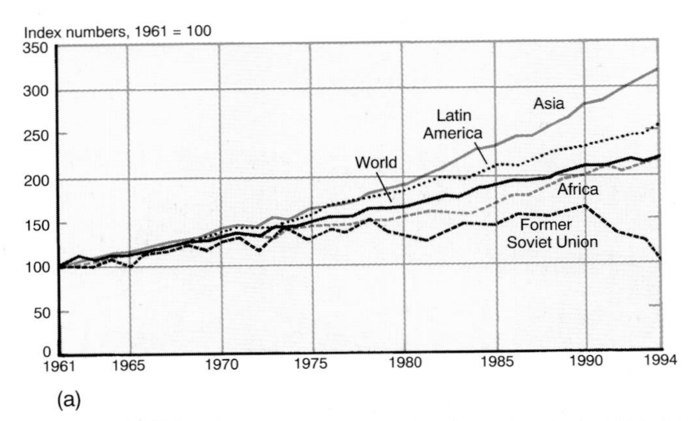

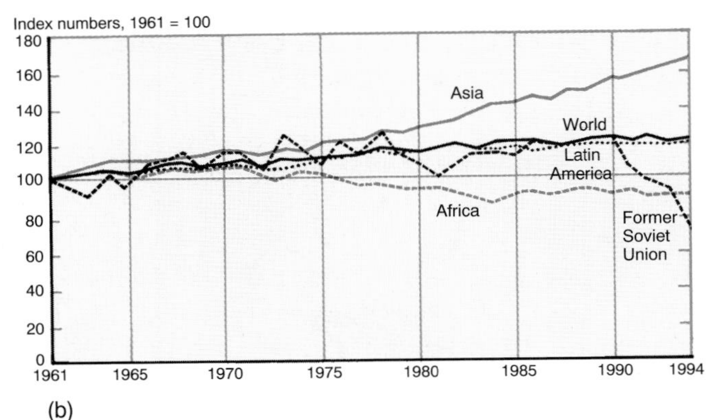

FIGURE 11.28 (a) **Trends in world food production, 1961–94.** Globally, the production of food crops increased steadily during this period, but the growth was not shared equally. (b) **Trends in per capita food production, 1961-94.** In Africa and the countries of the former Soviet Union, less food was available per person in 1994 than in previous years.

Source: Food and Agriculture Organization of the United Nations as reprinted in *World Resources 1996–97*, the World Resources Institute.

FIGURE 11.29 **Malnourished children in Somalia.** The world's food supply is unequally distributed among and within countries, and sometimes within households. Approximately 840 million people worldwide, including 200 million children, do not get enough to eat. Malnutrition takes its heaviest toll on young children. Thousands of them die every day, partly because malnutrition makes them susceptible to infections and parasites.

©Schiller/The Image Works.

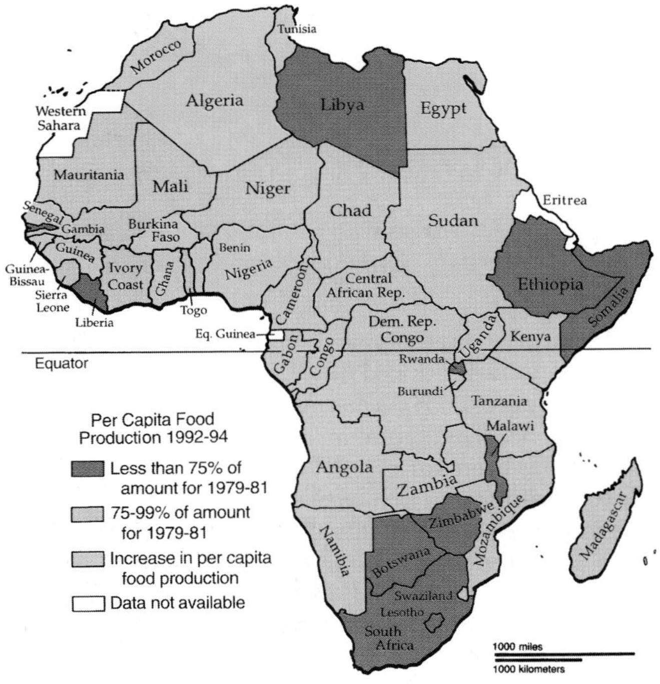

FIGURE 11.30 **Per capita food production 1992–94 compared to 1979–81.** Although the total amount of food produced in African countries has doubled over the last 30 years, the population has grown even faster, canceling the gains in production. In most African countries, less food is available per person now than it was a generation ago, and food shortages are now common in much of sub-Saharan Africa. Since 1980, both the absolute number and percentage of people suffering from malnutrition have increased in sub-Saharan Africa.

Data from World Resources Institute, 1996.

acres) are actually used in intensive food production. The United Nations Food and Agricultural Organization estimates that, at most, another 1 billion hectares (2.5 billion acres) are potentially suitable for agricultural use, depending on definitions of "suitable" and "agricultural use."

Clearly, some land is more suitable than other land for regular, sustained food production, and experts generally agree that most of the land in the world that is well suited for agriculture is already being cultivated. An estimated 82% of potential cropland is already under production in Asia. Few reserves of arable land remain in the temperate regions of Europe and North America. Indeed, in the United States, more land is regularly taken *out* of production than is added. Each year more than 80,000 hectares (2 million acres), almost all of it prime agricultural land, are lost to expanding cities, highways, and other nonagricultural uses. A similar loss of cropland to urbanization occurs in Japan, the densely populated, rapidly industrializing countries of East Asia, and many of the West European countries. Worldwide, additional millions of hectares are lost annually through soil erosion and salinization and to the spread of deserts by overgrazing and deforestation. They may remove as much land from production as is added each year.

Most of the land that is deemed potentially suitable for cultivation *and* that receives enough rainfall for intensive agriculture is in the rain forests of Africa and the Amazon Basin of South America. Although food needs are great in those areas, soils (oxisols) in the tropical rain forest are delicate, low in nutrients and humus content, poor at holding water, and need time to recover their fertility after cropping.

As described in Chapter 10, farmers traditionally adapted successfully to these limitations by practicing slash-and-burn agriculture. Increasing population pressures now threaten the viability of the system, particularly in the Amazon Basin. There, some 2% of the forest is being cleared for agriculture every year, cleared areas are becoming more closely spaced, and insufficient time is being allowed for vegetation to regenerate. Within just a few years' time, nutrients are leached from the soil, soil erosion increases, and yields drop.

There is some evidence that crop yields can be sustained in tropical forests if fertilizers are used to add nutrients to the soil, and that a given plot can support three crops per year—producing more food than does slash-and-burn agriculture. Such continuous cultivation will become successful only if a number of conditions are met. Fertilizers will have to be available at affordable prices, soil erosion controlled, and crops rotated in order to return nutrients to the soil and maintain yields. Most important, farmers will have to be trained in the new techniques.

In summary, expanding the amount of cultivated land cannot be viewed as the solution to increasing world food supplies and may, indeed, render land that is improperly added to the agricultural base unfit for any use.

Increasing Yields

A greater potential for expanding agricultural output lies in increasing the yields on land that is already being cultivated. Corn yields, for example, can vary from 12 tons per hectare (200 bushels per acre) in commercial agricultural societies to less than 1.5 tons per hectare (24 bushels per acre) in countries relying on traditional farming techniques. Rice yields on irrigated land vary from 1 ton per hectare (16 bushels per acre) to ten times that amount. Much of the growth in the world food supply throughout the world since 1950 has been due to such improved yields (Figure 11.31).

These have occurred in both developed and developing countries. Corn yields in the United States have nearly quadrupled since 1950, a result chiefly of the heavy application of nitrogen fertilizer to hybrid corn strains. In developing countries, changes in farming technology are known collectively as the Green Revolution (see Chapter 10). Wheat, corn, and rice—the three crops that account for roughly half of the world's food production by weight—have responded to applications of fertilizers and pesticides, to irrigation and farm mechanization, and to techniques of genetic engineering that produce high-yielding strains. The question is whether the dramatic gains in yields achieved since 1950 can be sustained or whether the gains most easily obtained have already been exhausted.

The techniques for increasing yields include:

1. use of high-yielding varieties of grains (mainly rice and wheat) that make more efficient use of fertilizers and, often, take less time to mature than do native strains;
2. expansion of irrigated acreage, which is disproportionately important to global food production;

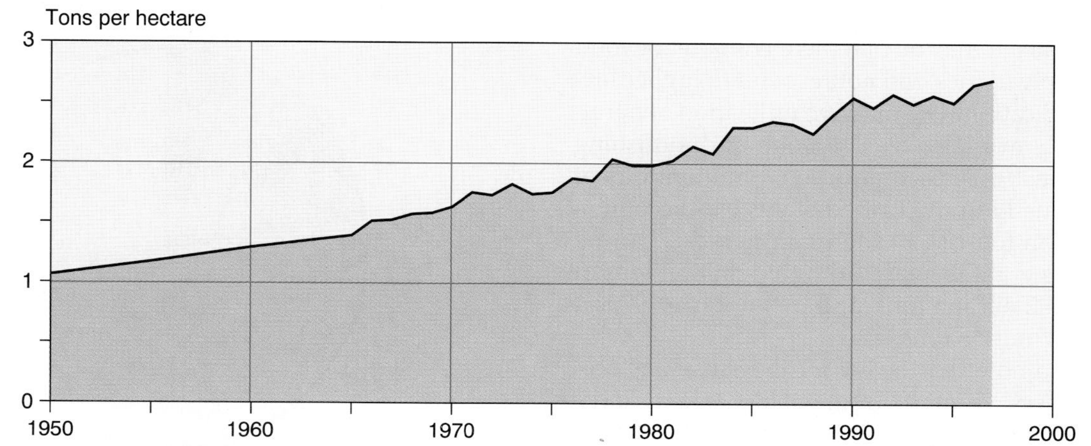

FIGURE 11.31 **World grain yield per hectare, 1950–1997.** Improved yields have contributed significantly to the growth in food production since 1950. Although grain production expanded nearly 3% annually between 1950 and 1984, however, it has dropped since then to about 1% annual growth. It is not clear whether the slowdown in growth since 1984 indicates that the greatest gains of the Green Revolution already have been realized.

Source: U.S. Department of Agriculture, as reported in Lester R. Brown, et al., *Vital Signs 1998.* W. W. Norton & Co. and Worldwatch Institute, 1998.

3. increased use of fertilizers, both organic (crop residues and animal manure) and inorganic (e.g., phosphorus, potassium, and ammonia);
4. multiple-cropping—growing two or three crops in rotation in one year on a single piece of land;
5. control of pests, which include thousands of species of rodents, insects, bacteria, and fungi that are estimated to destroy about 20% of the world's harvested food each year;
6. farm mechanization.

As attractive as it seems, the yield-increase strategy poses a number of problems. First, the most dramatic production increments are potentially obtainable in the less-developed countries, which are least able financially to make the conversion from labor-intensive to capital-intensive agriculture. High-yield agriculture is capital intensive, requiring great expenditures for equipment, fertilizers, pesticides, and energy. Requisite amounts of capital simply are not available in many developing countries. In addition, heavy farm machines are not well suited for use on small farms or in areas where holdings are fragmented.

Second, it is not clear how much irrigated acreage can expand. Irrigated land is vitally important to agricultural productivity, supplying some 36% of the world's food from only 16% of its cropland. Between 1950 and 1980 there was a rapid growth in the amount of land that is irrigated. The growth slowed appreciably after that, and since 1990, little increase in irrigated acreage has occurred. A number of reasons have been advanced for the decline: the number of appropriate sites for new dams has dwindled; in many regions, pumping from rivers and aquifers is approaching the sustainable yield; reservoirs have filled with sediment; some irrigated acreage has been lost to salinization.

Fertilizer use has exhibited a similar trend. From 1950–89, productivity soared as farmers expanded their use of fertilizers tenfold, from 14 million to 146 million tons. Since 1989, however, the number of tons of fertilizers used has declined annually. It may be that there is a point beyond which existing crop varieties will not respond to higher levels of fertilizer application. The massive utilization of herbicides and pesticides may also reach a point of diminishing returns. During the 1950s, each additional million tons of fertilizer used annually in the United States was accompanied by a 10-million-ton rise in the grain harvest. During the early 1960s, the grain increase from that million tons of input declined to 8.2 million tons, and by the early 1970s it had declined to 5.8 million tons.

We should also note that heavy fertilizer use has ecological consequences. Runoff from fields where large amounts of fertilizers have been used pollutes surface waters, and nitrogen fertilizers consume great quantities of natural gas in their manufacture, using a scarce resource and contributing to atmospheric heat. Since it takes about five tons of mostly fossil fuels to produce one ton of fertilizer, the consequences of increasing yields by heavy fertilizer applications are serious.

Finally, the Green Revolution endangers populations by encouraging dependence upon *monoculture,* total emphasis upon a single crop, which is inherently dangerous. It demands total dependence for human existence in marginally fed countries upon a single crop with a wide variety of essential inputs. Should the crop fail from disease, insect damage, or input shortage, disaster is unavoidable. Perhaps unwittingly, but wisely, subsistence agriculture is characterized by the production of a great variety of foodstuffs. Diversity is the subsistence farmer's insurance policy.

Increasing Fish Consumption

The pressing nutritional need in much of the world is for protein. The amino acids contained in protein are the essential building blocks of the body. The oceans have long been viewed as a seemingly unlimited source of protein, and increasing people's consumption of fish and shellfish is a third means by which global food supplies might be augmented. Although fish and shellfish account for less than 20% of all human protein consumption worldwide, an estimated 1 billion people depend on fish as their primary source of protein. Reliance on fish is greatest in the developing countries of eastern and southeastern Asia, Africa, and parts of Latin America. Fish are also very important in the diets of some advanced economies with well-developed fishing industries—Norway, Iceland, and Japan, for example.

The annual fish supply comes from three sources:

1. the *inland catch,* from ponds, lakes, and rivers;
2. *aquaculture,* where fish are produced in a controlled environment;
3. the *marine catch,* all fish harvested in coastal waters or on the high seas.

As Figure 11.32 indicates, inland waters supply only 17% of the global fish catch. The other 83% of the harvest comes

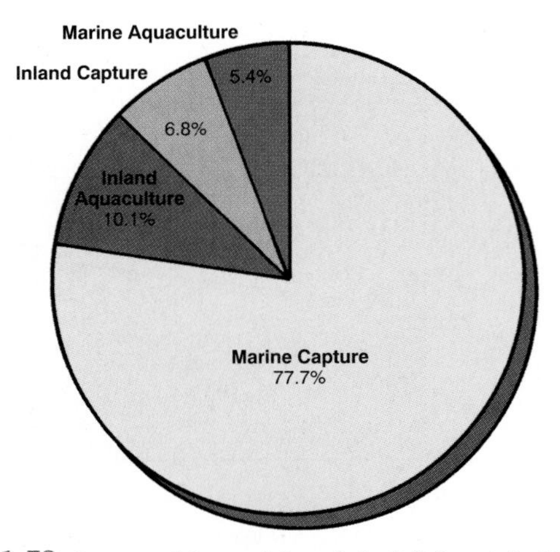

FIGURE 11.32 **Composition of the global fish catch.** The world's oceans supply most of the fish catches, but overfishing threatens the productivity and viability of marine ecosystems.
Source: Food and Agriculture Organization of the United Nations (FAO), *Global Fish and Shellfish Production in 1993* (FAO Fisheries Dept., Fisheries Information, Data and Statistics Services, Rome, March 1995). Table 1, p. 2.

On-Line The Geography of Natural Resources

Many federal government departments and agencies that are involved with natural resources have their own home pages, a few of which are noted here.

Energy Information Administration is a logical place to begin a search for information about energy resources. This home page provides information and links to petroleum, natural gas, coal, nuclear, electric, and renewable energy resources.

http://www.eia.doe.gov/

The Department of Energy home page can be accessed at:

http://www.doe.gov/

The *USDA Forest Service Home Page* is at:

http://www.fs.fed.us/

U.S. Fish and Wildlife Service Endangered Species Program web pages are highlighted by a list of endangered species in the U.S. This list is available by state and type of animal or plant. There is also a foreign species index. Links from selected species lead to more information about those species. The site also contains a copy of the Endangered Species Act of 1973, selected policy documents, and tables of contents and selected articles from the "Endangered Species Bulletin," a bimonthly publication of the USFWS.

http://www.fws.gov/r9endspp/endspp.html

National Wetlands Inventory: Part of the U.S. Fish and Wildlife Service, the National Wetlands Inventory plans, directs, and monitors the gathering, analysis, dissemination, and evaluation of information relating to the location, quantity, and ecological importance of America's wetlands. Its WWW server gives access to wetlands maps, a database, and links to the Wetlands Inventory files.

http://www.nwi.fws.gov/

Savannah River Site Home Page: The Savannah River Site (SRS) is a nuclear facility in South Carolina operated by Westinghouse Electric Corporation for the U.S. Department of Energy. This Web page contains information about nuclear power, site maps, historical highlights, future plans, and nuclear waste management, as well as environmental restoration technologies, initiatives, and activities.

http://www.srs.gov/

Natural Resources Canada Homepage contains general information about the department, services to the general public, and links to sectors such as the Canadian Forest Service, Energy, and Mining. There are additional links to Geodetic Survey of Canada, Pacific Forestry Centre, and the Canada Centre for Mapping.

http://www.nrcan.gc.ca/

Sustainable Development Dimensions, provided by the Food and Agriculture Organization (FAO) of the United Nations, is a gateway to information about sustainable development. The site is divided into four major categories: people, institutions, knowledge, and environment.

http://www.fao.org/sd/

In addition to government sources, a variety of private organizations can be accessed for information about natural resources. They include those noted below.

Solstice is a file server maintained by the Center for Renewable Energy and Sustainable Technology in Washington, D.C. It contains state-of-the-art information on renewable energy, energy efficiency, the environment, and sustainable community development.

http://solstice.crest.org/

The World Wildlife Fund's *World Forest Maps—WWF* contains maps of forest coverage and protected areas at several different scales for sections of North, Central, and South America, Europe, Africa, Asia, and Oceania.

http://www.panda.org/forests4life/ffl_pickmap.htm

The full text of *World Resources 1996–97: A Guide to the Global Environment,* produced by the World Resources Institute, United Nations Environment Programme, United Nations Development Programme, and World Bank, is available via the Web. This eleventh edition of "World Resources" contains two major sections: Global Conditions and Trends and Data Tables, and The Urban Environment.

http://www.wri.org/wri/wr-98-99/index.html

Rainforest Action Network (*RAN*), founded in 1985, "works to protect the Earth's rainforests and support the rights of their inhabitants through education, grassroots organizing, and non-violent direct action." RAN's website includes information about rainforests, the organization, and its recent campaigns (including Amazon protection and wood reduction campaigns, among others).

http://www.ran.org/

Oilworld is an oil newsletter with daily prices, petroleum futures, leases, and other information pertaining to the oil and gas industry.

http://oilworld.com/

Electric Utility Home Page provides links to the home pages of electric utilities in the United States.

http://www.electricrates.com/erholink.htm (A–I)

http://www.electricrates.com/erlnk2.htm (J–Z)

from the world's oceans, and most of that catch is made in coastal wetlands and relatively shallow coastal waters above the continental shelf. Near shore, shallow embayments and marshes provide spawning grounds, and river waters supply nutrients to an environment highly productive of fish. Offshore, ocean currents and upwelling water move great amounts of nutritive salts from the ocean floor through the sunlit surface waters, nourishing *plankton*—minute plant and animal life forming the base of the marine food chain.

Commercial fishing is largely concentrated in northern waters, where warm and cold currents join and mix, and where such familiar food species as herring, cod, mackerel, haddock, and flounder congregate on the broad continental shelves favorable for fish production. Two of the most heavily fished regions are the North Pacific and North Atlantic, which together supply more than half of the fish catch (Figure 11.33). Tropical fish species tend not to school and, because of their high oil content and unfamiliarity, seem to be less acceptable in the commercial market. They are, however, of great importance for local consumption.

If the sea is to be a reservoir for extended exploitation, it must be more wisely handled than in the past. As Figure 11.34a indicates, between 1950 and 1989, the world fish catch more than quadrupled as modern technology was applied to harvesting food fish. This technology included the use of sonar, radar, helicopters, and satellite communications to locate schools of fish; more efficient nets and tackle; and factory trawlers to follow the fishing fleets to prepare and freeze the catch (Figure 11.34b). The rapid rate of increase led to inflated projections of probable fisheries productivity and to the feeling that the resources of the oceans were inexhaustible. Quite the opposite has proved to be true.

In recent years, the productivity of fishing areas has declined as *overfishing* (catches above reproduction rates) and pollution of coastal waters seriously endangered the supplies of traditional and desired food species. Since 1989, the annual catch has never exceeded 95 million tons, and because world population has continued to grow, there has been an actual decline in the average fish catch per person. According to the UN Food and Agriculture Organization, all 17 major oceanic fishing areas in the world are now being fished at or beyond capacity; 13 are in decline. The plundering of United States coastal waters has imperiled a number of the most desirable fish species, including haddock, flounder, and cod in New England waters; Spanish mackerel, grouper, and red snapper off the Gulf of Mexico; halibut and striped bass off California; and salmon and steelhead in the Pacific Northwest.

Overfishing is partly the result of the accepted view that the world's oceans are common property, a resource open to anyone's use with no one responsible for its maintenance, protection, or improvement. The result of this "open seas" principle is just one expression of the so-called

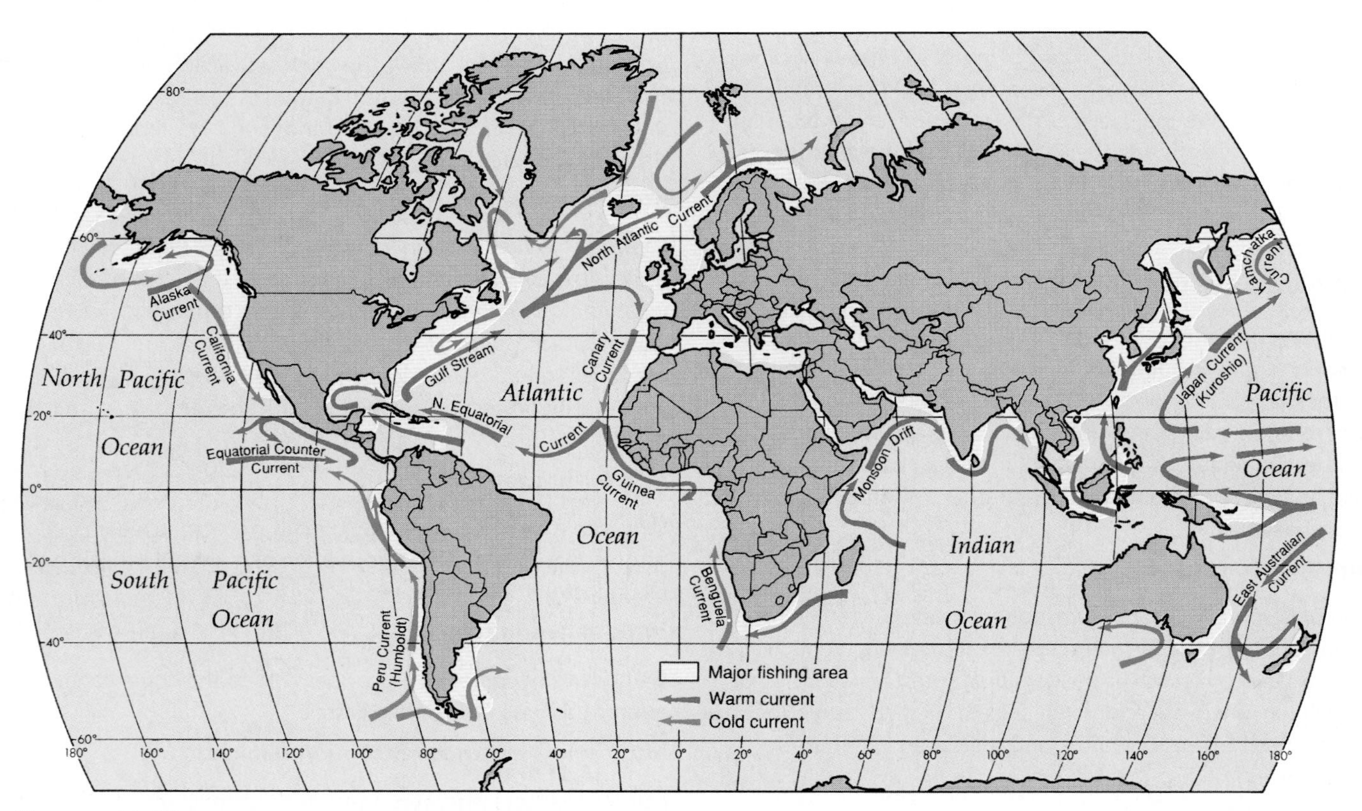

FIGURE 11.33 **The major commercial marine fisheries of the world.** The waters within 325 kilometers (200 mi) of the United States coastline account for almost one-fifth of the world's annual fish and shellfish harvests. Overfishing, urban development, and the contamination of bays, estuaries, and wetlands have contributed to the depletion of fish stocks in those coastal waters.

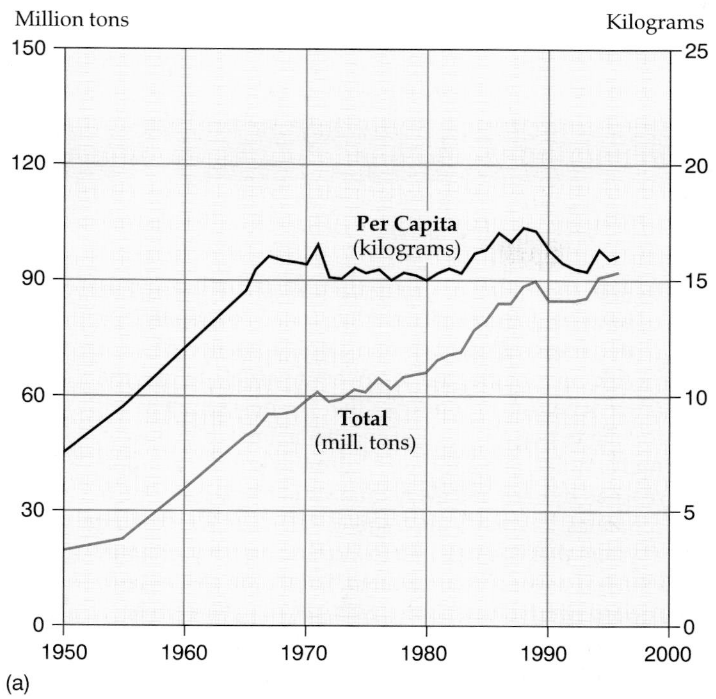

FIGURE 11.34 (a) **Annual fish harvests** rose from 22 million tons in 1950 to 100 million tons in 1989. After years of steady growth, the total fish catch has risen only slightly since 1989. Because of population growth, the amount of fish available per person—the per capita fish catch—has actually declined since 1989. (b) **A Japanese factory ship in the Bering Sea.** Factory ships prepare and freeze the catch of smaller vessels. This ship can process 25 to 30 tons of fish per hour. The increased efficiency of commercial fishing has led to serious depletion of a food source once thought to be inexhaustible.

(a) Source: FAO as reported in Lester Brown, et al., *Vital Signs 1998.* W. W. Norton & Co. and Worldwatch Institute, 1998.

(b) © William McCloskey.

tragedy of the commons—the economic reality that when a resource is available to all, each user thinks he or she is best served by exploiting the resource to the maximum even though this means its eventual depletion (see "The Tragedy of the Commons").

One approach to increasing the fish harvest is through **aquaculture,** the breeding of fish in freshwater ponds, lakes, and canals or in fenced-off coastal bays and estuaries (Figure 11.35). Fish farming has long been practiced in Asia, where fish is a major source of protein, but now takes place on every continent. Fish farms in the United States produce significant amounts of catfish, crawfish, trout, and oysters. Aquaculture offers several advantages over conventional fishing. There is no shortage of land suitable for aquaculture, the technology is relatively simple and requires little or no fuel, farmers can choose the species they want to raise, and yields are high.

Until recently, aquaculture was small-scale and localized. Much of it still is, but the recent development of large-scale, mechanized farms has contributed significantly to increased production. Aquaculture yielded 6.5 million tons of fish and shellfish in 1984. That figure more than doubled in a decade, and reached 23 million tons in 1996, accounting for more than 20% of the total world fish harvest. About two-thirds of the production is from inland culture; the rest is coastal.

LAND RESOURCES

As we have seen, cropland occupies only 11% of the world's total land area. Other types of land resources include forests, rangelands, parks, and wilderness areas. In this final section of the chapter, we examine the distribution and status of two of those resources: coastal wetlands, which comprise a minute portion of the land area, and forests, which cover about a third of the earth.

Coastal Wetlands

The part of the sea lying above the continental shelf if the most productive of all ocean waters, supporting the major commercial marine fisheries. Because it is not very deep, this *neritic zone* is penetrated and warmed by sunlight. It

The Tragedy of the Commons

In a classic essay entitled "The Tragedy of the Commons," the biologist Garrett Hardin outlined the basic problem inherent in the common ownership of resources. The original "commons" were the open land of a village available to anyone who wished to graze cattle.

> "The tragedy of the commons develops in this way. Picture a pasture open to all. It is to be expected that each herdsman will try to keep as many cattle as possible on the commons. Such an arrangement may work reasonably satisfactorily for centuries because tribal wars, poaching, and disease keep the numbers of both man and beast well below the carrying capacity of the land. Finally, however, comes the day of reckoning, that is, the day when the long-desired goal of social stability becomes a reality. At this point, the inherent logic of the commons remorselessly generates tragedy.
>
> As a rational being, each herdsman seeks to maximize his gain. Explicitly or implicitly, more or less consciously, he asks, 'What is the utility to me of adding one more animal to my herd?' This utility has one negative and one positive component.
>
> 1. The positive component is a function of the increment of one animal. Since the herdsman receives all the proceeds from the sale of the additional animal, the positive utility is nearly +1.
> 2. The negative component is a function of the additional overgrazing created by one more animal. Since, however, the effects of overgrazing are shared by all the herdsmen, the negative utility for any particular decision-making herdsman is only a fraction of –1.
>
> Adding together the component partial utilities, the rational herdsman concludes that the only sensible course for him to pursue is to add another animal to his herd. And another; and another. . . . But this is the conclusion reached by each and every rational herdsman sharing a commons. Therein is the tragedy. Each man is locked into a system that compels him to increase his herd without limit—in a world that is limited. Ruin is the destination toward which all men rush, each pursuing his own best interest in a society that believes in the freedom of the commons. Freedom in a commons brings ruin to all."

We can think of the ecosphere as one vast commons, stocked with resources not owned by anyone in particular and available to all: air and water, forests and grasslands, wild animals and fish. Because they are "free" goods, Hardin argues, people tend to overexploit them, even if it means the degradation, depletion, or eventual exhaustion of the resource. Air and water become polluted because it costs little for individuals, industries, or cities to dispose of wastes by burning or dumping them. Fishermen attempt to harvest as much fish from oceans as possible, even though this means depletion of the resource on which they depend.

also receives the nutrients flowing into oceans from streams and rivers so that vegetation and a great variety of aquatic life can flourish. However, the neritic zone depends to a considerable extent on the continued functioning of the **estuarine zone,** the relatively narrow area of wetlands along coastlines where salt water and fresh water meet and mix (Figure 11.36).

Coastal wetlands take a variety of forms, including marshes, swamps, tidal flats, and estuaries. Extremely valuable ecological systems, they have a number of vital functions. Trapping the silt and other organic matter that rivers bring downstream, they provide shelter and food for a variety of fish and shellfish, and indeed are essential to the survival of many species by serving as their spawning grounds. Coastal wetlands are also breeding, feeding, nesting, and wintering grounds for many types of birds (Figure 11.37). Not only are these areas extraordinarily productive themselves, but they also contribute to the productivity of the neritic zone, where fish feed on the life that flows from wetlands into the sea. In addition, the wetlands absorb floodwaters, provide barriers to coastal erosion, and remove pollutants from surface waters.

In much of the world, coastal wetlands are in danger. They have been drained, dredged, built upon, mined for phosphate, and used as garbage dumps. They are polluted by chemicals, excess nutrients, and other water-borne wastes. Natural shorelines have been bulldozed, and artificial levees and breakwaters interfere with the flooding that nurtures wetlands with fresh infusions of sediment and water. Habitat alteration or destruction inevitably disrupts the intricate ecosystems of the wetlands. Scientists estimate that as much as half of the estuarine zone in the United States has been damaged, drastically altered, or destroyed. At present, the only way to halt this disruption seems to be to preserve the wetlands as they are, as ecological reserves. Should we fail to do so, there will be negative consequences not only for the water life they sustain, but also for that in the adjoining oceans.

Forest Resources

Coastal wetlands are only one of the renewable resources in danger of irreparable damage by human action. In many parts of the world, forests are similarly endangered.

FIGURE 11.35 **Harvesting fish at an aquaculture farm in Thailand.** Fish farming, or aquaculture, is one of the fastest-growing sectors in world food production. Asian countries supply about three-quarters of the total aquaculture harvest.

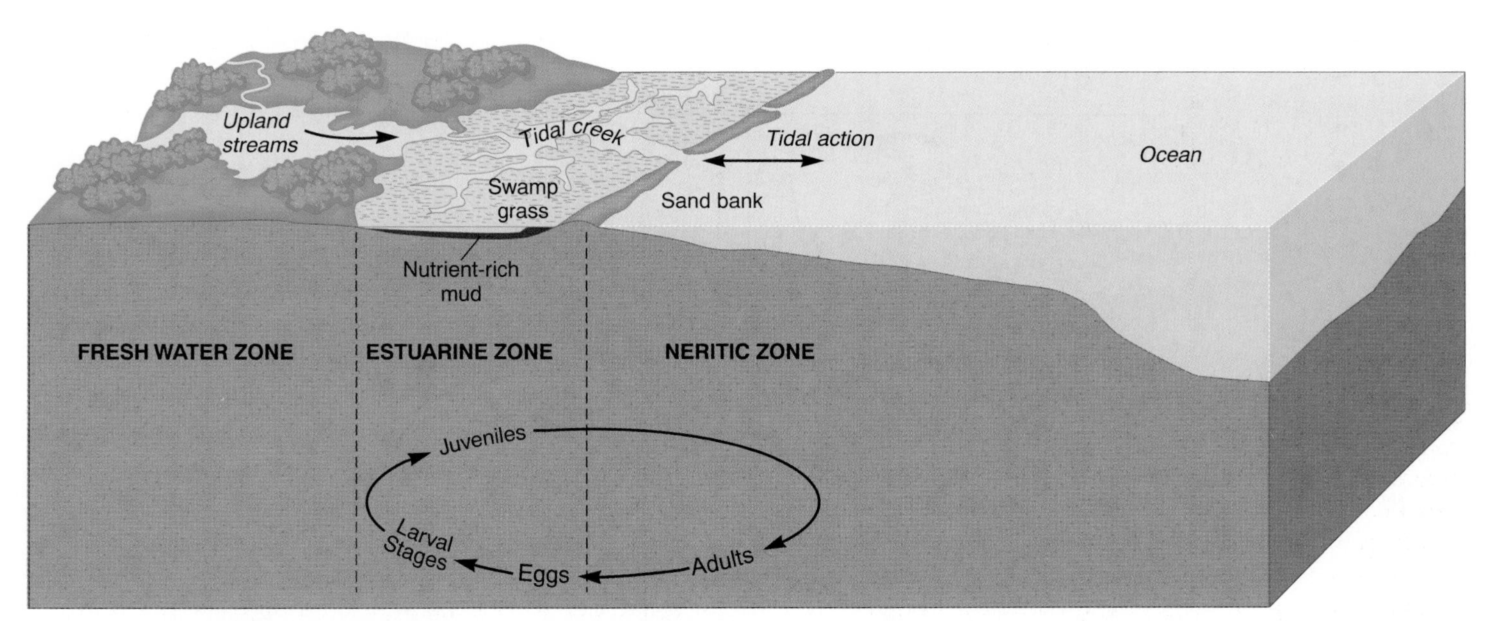

FIGURE 11.36 **The estuarine zone.** The outflow of fresh water from streams and the action of tides and wind serve to mix deep ocean waters with surface waters in estuaries, contributing to their biological productivity. The saline content of estuaries is lower than that of the open sea. Many fish and shellfish require water of low salinity at some point in their life cycles.

Before the rise of agriculture some 11,000 years ago, the world's forests probably covered some 45% of the earth's land area exclusive of Antarctica. Even after millennia of land clearance for agriculture and, more recently, commercial lumbering, cattle ranching, and fuelwood gathering, forests still cover about one-third of the world's land area. As an industrial raw material source, however, forests are more restricted in area. *Commercial forests* are restricted to two very large global belts (Figure 11.38). One, nearly continuous, occupies the upper-middle latitudes of the Northern Hemisphere. The second straddles the equatorial zones of South and Central America, Central Africa, and Southeast Asia.

The northern coniferous, or softwood, forest is the largest and most continuous stand. Its pine, spruce, fir, and

FIGURE 11.37 A salt marsh in Louisiana. Tidal marshlands have been subjected to dredging and filling for residential and industrial development. The loss of such areas reduces the essential habitat of waterfowl, fish, crustaceans, and mollusks. Many waterfowl breed and feed in coastal marshes and use them for rest during long migrations.

other species are used for construction lumber and to produce pulp for paper, rayon, and other cellulose products. To its south are found the deciduous hardwoods: oak, hickory, maple, birch, and the like. These and the trees of the mixed forest lying between the hardwood and softwood belts have been much reduced in areal extent by centuries of agricultural and urban settlement and development, though they still are commercially important for hardwood applications: furniture, veneers, railroad ties, and so on. The tropical lowland hardwood forests are exploited primarily for fuelwood and charcoal, although an increasing quantity of special quality woods is cut for export as specialty lumber.

The adage about not being able to see the forest for the trees is applicable to those who view forests only for the commercial value of the trees they contain. Forests are more than trees, and timbering is only one purpose that forests serve. Chief among the other purposes are soil and watershed conservation, the provision of a habitat for wildlife, and recreation. Forests also play a vital role in the global recycling of water, carbon, and oxygen.

Because forests serve a variety of purposes, the kind of management techniques employed in any one area depend on the particular use(s) to be emphasized. Thus, if the goal is to maintain a diversity of native plant species in order to provide a maximum number of ecological niches for

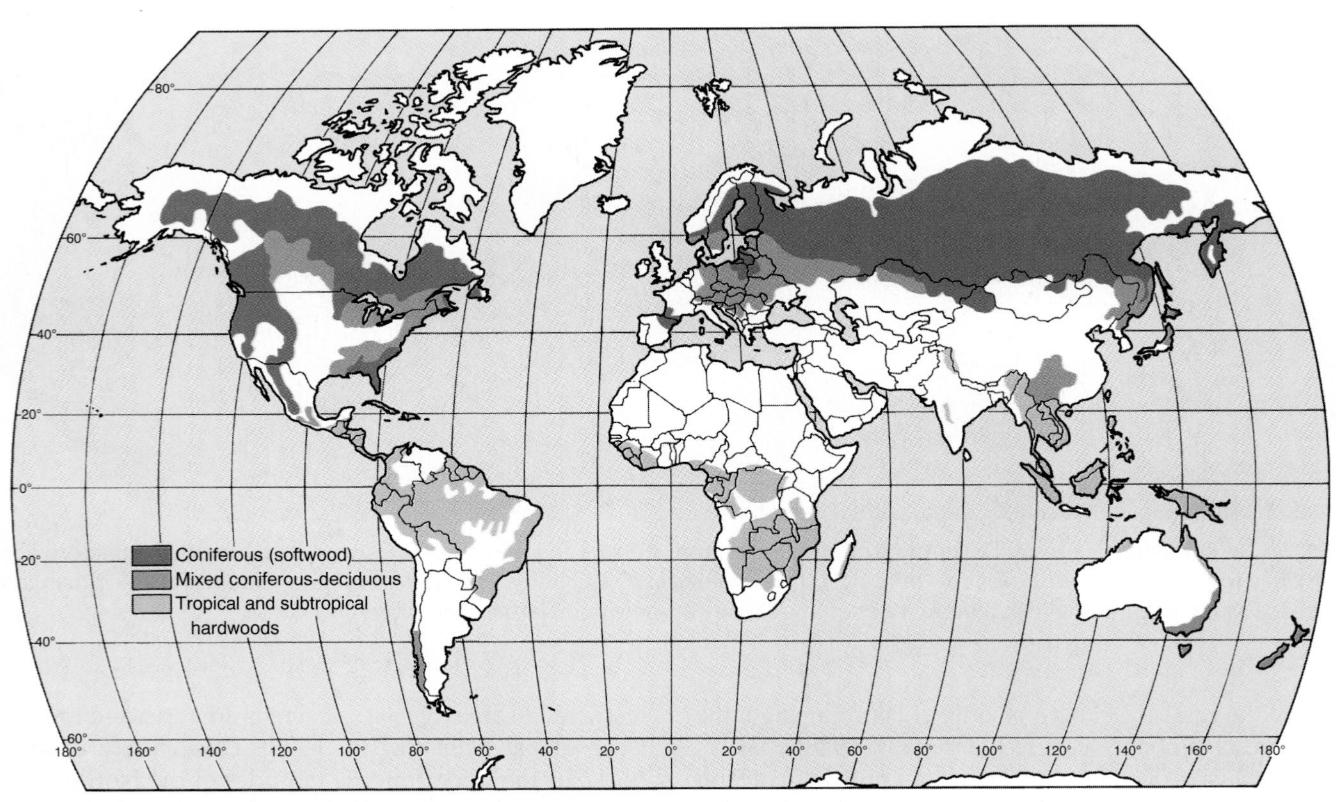

FIGURE 11.38 Major commercial forest regions. Much of the original forest, particularly in midlatitude regions, has been cut over. Many treed landscapes that remain do not contain commercial stands. Significant portions of the northern forests are not readily accessible and at current prices cannot be considered commercial.

wildlife, the forest will be managed differently than if it is designed for public recreation or the protection of watersheds. Even if the use to be emphasized is timber production, different management approaches may be taken. Logging techniques for the production of plywood or wood chips, for example, differ from those used for the production of high-quality lumber.

Commercial forests can be considered a renewable resource only if sustained-yield techniques are practiced—that is, if harvesting is balanced by new growth (see the term *maximum sustainable yield* in Chapter 10). Timber companies employ a number of different methods of tree harvesting and regeneration. Two quite different practices, selective cutting and clear cutting, illustrate the diversity of such approaches (Figure 11.39).

Selective cutting is practiced in mixed-forest stands containing trees of varying ages, sizes, and species. Medium and large trees are cut either singly or in small groups, encouraging the growth of younger trees that will be harvested later. Over time, the forest will regenerate itself. With *clear cutting,* as the name implies, all the trees are removed from a given area at one time. The site is then left to regenerate naturally or is replanted, often with fast-growing seedlings of a single species. Excessive clear cutting destroys wildlife habitats, accelerates soil erosion and water pollution, replaces a mixed forest with a wood plantation of no great genetic diversity, and reduces or destroys the recreational value of the area.

U.S. National Forests

Roughly one-third of the United States is forested, the same proportion for the world as a whole. Only some 40% of those forests provide the annual harvest of commercial timber. The remaining forests either do not contain economically valuable species, they are in small, fragmented holdings, are inaccessible, or are in protected areas. Of that 40% of commercial forestland, almost half is in 171 national forests owned by the public and managed by the U.S. Forest Service (Figure 11.40). Logging by private companies is permitted; timber companies pay for the right to cut designated amounts of timber. Currently, the Forest Service is at the center of debates over how the forests should be managed. Among the issues are methods of harvesting, the cutting of very old tree stands, road building, and rates of reforestation.

By law, the national forests are to be managed under the principle of *multiple use,* which is intended to balance the needs of recreation and wildlife with those of such developmental activities as logging, mining, and drilling for oil and gas. Although no use is to be particularly favored over others, conservationists charge that the Forest Service increasingly supports commercial logging and that the forests are being cut at an unprecedented rate.

In recent years, billions of board feet of timber have been taken from the national forests. Environmentalists are especially concerned that nearly half of this has come from national forests in Oregon and Washington, most of

(a)

(b)

FIGURE 11.39 (a) **Selective cutting in eastern Canada.** Older, mature specimens are removed at first cutting. Younger trees are left for later harvesting. (b) **Clear cutting in the Gifford Pinchot National Forest, Washington.** Cutting every tree, regardless of species or size, drives out wildlife, damages watersheds, disrupts natural regeneration, and removes protective ground cover, exposing slopes to erosion.

(a) ©R. Moller/Valan Photos.

(b) ©Milton Rand/Tom Stack & Associates.

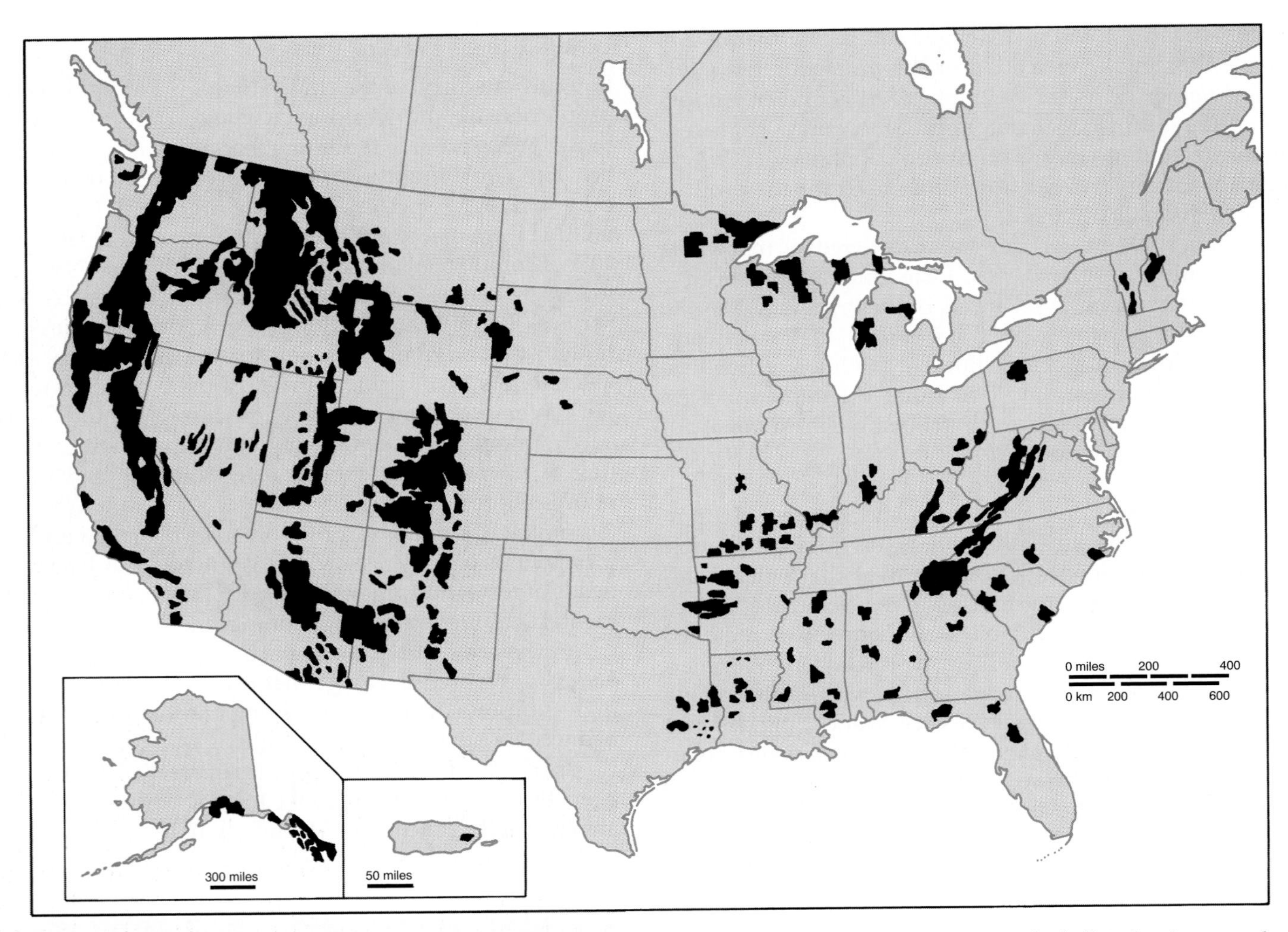

FIGURE 11.40 National forests of the United States. Trees cut from national forests supply about 15% of all timber harvested in the United States.

it irreplaceable "old growth." These virgin forests contain trees that are among the tallest and oldest in the world, indeed that were alive when Pilgrims set foot on Plymouth Rock.

Old-growth forests include trees of every age and size, both living and dead. Some ancient trees are immense, capable of growing 90 meters (300 ft) high, and they may live for more than 1000 years. They include the Douglas fir, Western red cedar, sequoia, and redwood. Tons of dead and decaying logs carpet the forest floor, where, sodden with moisture, they help control erosion and protect the forest from fire. As they decay, the logs release nutrients back into the soil. Such forests provide a habitat for hundreds of types of insects and animals, some of them threatened or endangered species.

The only large expanses of old-growth forest remaining in the United States are in the Pacific Northwest, most of them owned by the federal government. These ancient forests once covered about 60% of the forested areas between the Cascade Mountains and the Pacific Ocean, stretching 3200 kilometers (2000 mi) from California to Alaska. Today, only 10% of the old-growth forests remain, and they are being logged at the rate of 25,000 hectares (60,000 acres) per year. If logging continues at the present rate, they will be gone in about 20 years, ceasing to exist as they are today. Although companies plant new seedlings to replace those they cut, timber is being harvested twice as fast as new trees can replace it. Further, traditional management practices, including clear cutting, road building, and harvesting after decades, not centuries, of regeneration prevent the development of a true old-growth forest ecosystem.

It is ironic that many Americans condemn the burning of the tropical rain forests while at the same time, the U.S. government not only permits the destruction of forests just as ecologically precious but in fact subsidizes that destruction. The federal government annually loses hundreds of millions of dollars on timber sales, because building and maintaining the logging roads costs far more than the timber companies pay for the wood.

Perhaps the worst abuse has occurred in the Tongass National Forest, North America's largest temperate rain forest, which covers much of Alaska's southern panhandle (Figure 11.41). In the last decade, the government has spent

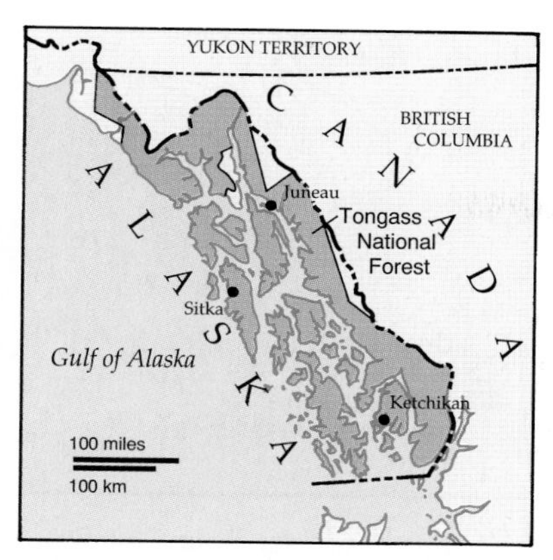

FIGURE 11.41 **The Tongass National Forest,** the largest national forest in America, covers approximately 7 million hectares (17 million acres). Since the late 1950s, the federal government has subsidized the timber industry by promising companies a long-term supply of cheap wood from the Tongass. Conservationists contend that old-growth trees are not needed for forest products and that the Tongass should be managed for wildlife, fishing, and tourism, as well as logging.

more than $250 million to build access roads and promote commercial timbering and received in return less than $3 million from the sales. Five-hundred-year-old trees 3 meters (10 ft) in diameter and more than 32 meters (105 ft) tall have been sold for $3 each and ground into pulp that is sent overseas and converted into products such as rayon and chopsticks.

Tropical Forests

It is not only in the United States that government economic policies accelerate the rate of forest destruction. Much of the deforestation occurring in tropical areas also has governmental sanction. Brazil, Indonesia, and the Philippines are among the countries where governments subsidize projects aimed at converting forests to other uses, such as farming, cattle ranching, and industry. Of course, their economic policies are driven by the continual demand for resources that poverty and a growing population place on a society.

The tropical forests extend across parts of Asia, Latin America, and Africa (Figure 11.42). We noted in Chapter 5 that some 100,000 square kilometers (40,000 sq mi) are being completely cleared every year, and that almost half of their original expanse has already been either cleared or degraded. Nearly half of Asia's natural forest is gone. Seventy percent of the tropical forests of Central America and some 40% of those of South America have disappeared. Africa has lost more than half of its original forest, especially in the east and south. While parts of central and western Africa are still heavily forested, the United Nations estimates that at current rates of logging, Nigeria, the Ivory Coast, and other West African countries will have no forests left in another ten years.

Why should North Americans care what happens to the tropical rain forests? Their destruction raises three principal global concerns and a host of local ones. First, all forests play a major role in maintaining the oxygen and carbon balance of the earth. People and their industries consume oxygen; vegetation both extracts the carbon from atmospheric carbon dioxide and releases oxygen back into the atmosphere. Indeed, the forests of the Amazon have been called the "lungs of the world" for their major contribution to the oxygen breathed by humankind. When the tropical forest is cleared, its role both as a carbon "sink" and oxygen-replenisher is lost.

A second global concern is the contribution of forest clearing to air pollution and climate change. Deforestation by burning releases vast quantities of carbon dioxide into the atmosphere. Brazilian scientists estimate that the thousands of fires that are set to clear the Amazon forest account for one-tenth of the global production of carbon dioxide (Figure 11.43). The destruction thus contributes to the warming of the atmosphere and, ultimately, to the greenhouse effect. In addition, the fires generate gases (nitrogen oxides and methane) that create acid rain and contribute to the depletion of the ozone layer (see Chapter 5).

Finally, the eradication of tropical forests is already leading to the loss of a major part of the biological diversity of the planet. The forests are one component in an intricate ecosystem that has developed over millions of years. The trees, vines, flowering plants, animals, and insects depend on one another for survival. The destruction of the habitat by clearance annually causes the extinction of thousands of plant and animal species that exist nowhere else. Of the estimated 5–30 million plant and animal species believed to exist on earth, a minimum of 40% are native to the tropical rain forest. Many of the plants have become important world staple food crops, among them rice, corn, cassava, squash, banana, pineapple, and sugarcane. Unknown

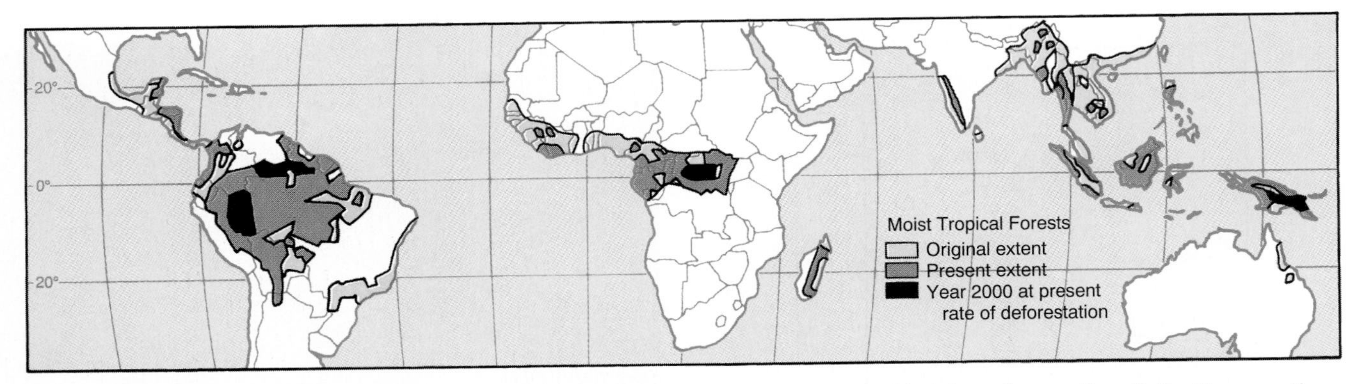

FIGURE 11.42 **Areas of tropical rain forest.** Three countries contain more than half of these forests: Brazil, the Democratic Republic of Congo, and Indonesia. Every second, an area of forest the size of a football field is destroyed to make way for farming, cattle ranching, commercial logging, and development projects. At current rates of deforestation, only four big blocks of tropical forest will be left in another decade.

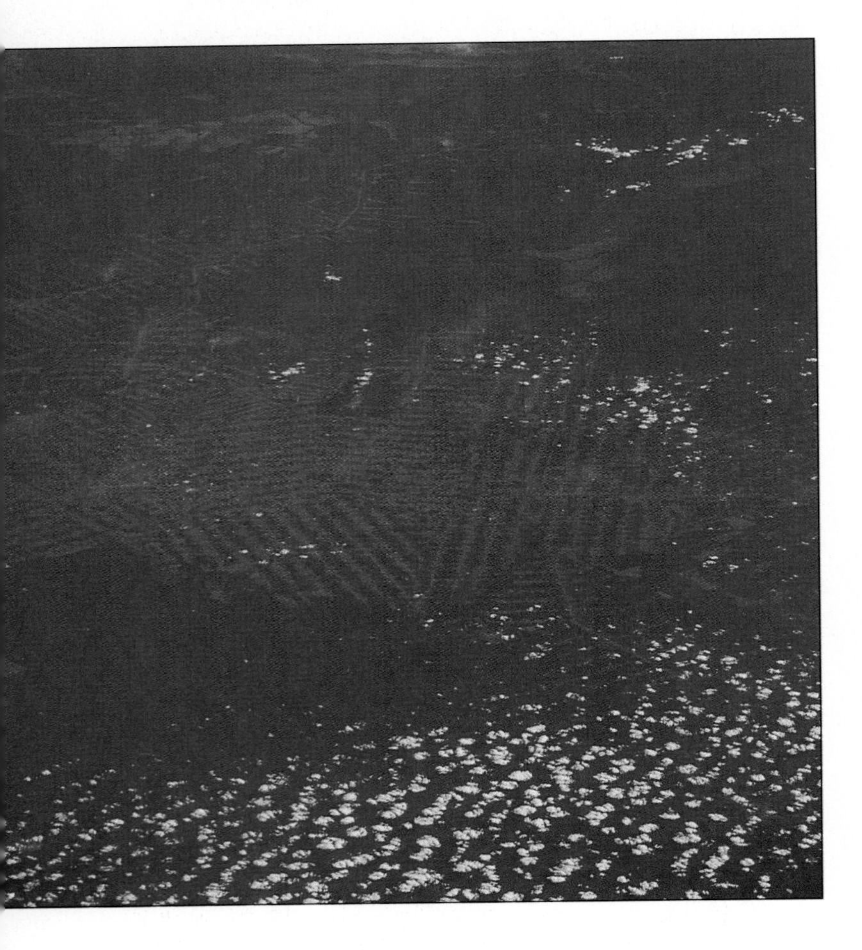

FIGURE 11.43 **Deforestation in Rondônia, Brazil,** has been concentrated near roads, yielding the "fishbone" pattern evident in this satellite image. The fastest and cheapest way to clear areas for farms and ranches is to burn them. About the size of Oregon, Rondônia is the Brazilian state with the largest percentage of its forest cover destroyed by fire, although the states of Mato Grosso and Pará also show significant deforestation. The World Wildlife Fund estimates that from 12 to 15% of the Amazon forest has been destroyed since record-keeping began in the early 1980s. Likened by some to an environmental holocaust, the fires generate hundreds of millions of tons of gases that contribute to global warming and depletion of the earth's protective ozone layer.
NASA

additional potential food species remain as yet unexploited. In addition, the tropical forests yield an abundance of industrial products (oils, gums, latexes, and turpentines) and are the world's main storehouse of medicinal plants (see "Tropical Forests and Medical Resources").

Deforestation also incurs heavy environmental, economic and social costs on a more local basis. All forests anchor topsoil and absorb excess moisture. In a vicious cycle, forest clearance accelerates soil erosion and siltation of streams and irrigation channels, leaving the area vulnerable to flooding and drought, and leading in turn to future shortages of food and fuelwood. Within a matter of years, land that has been cleared for agriculture can become unsuitable for that use. In the Himalayan watershed, in the Ethiopian highlands, and in numerous other places, deforestation, erosion, and rainfall runoff have aggravated floods that have killed tens of thousands of people and left millions of others homeless.

Tropical Forests and Medical Resources

Tropical forests are biological cornucopias, possessing a stunning array of plant and animal life. Costa Rica, about the size of South Carolina, contains as many bird species as all of North America, more species of insects, and nearly half the number of plant species. One stand of rain forest in Kalimantan (formerly Borneo) contains more than 700 species of tree, as many as exist in North America. Forty-three species of ant inhabit a single tree species in Peru, dependent on it for food and shelter and providing in return protection from other insects.

The tropical forests yield an abundance of chemical products used to manufacture alkaloids, steroids, anesthetics, and other medicinal agents. Indeed, half of all modern drugs, including strychnine, quinine, curare, and ipecac, come from the tropical forests. A single flower, the Madagascar periwinkle, produces two drugs used to treat leukemia and Hodgkin's disease.

As significant as these and other modern drugs derived from tropical plants are, scientists believe that the medical potential of the tropical forests remains virtually untapped. They fear that deforestation will eradicate medicinal plants and traditional formulas before their uses become known, depriving humans of untold potential benefits that may never be realized. Tribal peoples make free use of plants of the rain forest for such purposes as treating stings and snakebites, relieving burns and skin fungi, reducing fevers and curing earaches. Yet botanists have only recently begun to identify tropical plants and study traditional herbal medicines to discover which of them might contain medically important compounds.

A second concern is that forest destruction will create shortages of drugs already derived from those plants. According to some experts, as many as 60,000 plants with valuable medical properties are likely to become extinct by 2050.

Resource Management

The destruction of the rain forests is a tragedy that yields no long-term benefits. The world is approaching the end of a period in which resources were cheap, readily available, and lavishly used. Over the centuries, the earth has been viewed as an almost inexhaustible storehouse of resources for humans to exploit and, simultaneously, as a vast repository for the waste products of society. Now there is a growing realization that resources can be depleted, even renewable ones like forests, that many have lifespans measured only in decades, and that the air, water, and soil—also resources—cannot absorb massive amounts of pollutants and yet retain their life-supporting abilities. That realization was reflected in June 1992, at the Earth Summit in Rio de Janeiro, when the world's governments agreed to form the UN Commission on Sustainable Development. Since that time, more than 70 countries, including the United States, have launched efforts to chart a path toward sustainable development.

An *environmentally sustainable economy* is one that satisfies current needs without jeopardizing the prospects of future generations. The principles of sustainability are straightforward. Over the long term,

- soil erosion cannot exceed soil formation;
- forest destruction cannot exceed forest regeneration;
- species extinction cannot exceed species evolution;
- fish catches cannot exceed the regenerative capacity of fisheries; and
- pollutants cannot exceed the capacity of the system to absorb them.

A society can violate the principles of sustainability in the short run but not in the long run and still endure.

The wise management of resources entails three strategies: conservation, reuse, and substitution. By *conservation* we mean the careful use of resources so that future generations can obtain as many benefits from them as we now enjoy. It includes decreasing our consumption of resources, avoiding their wasteful use, and preserving their quality. Thus soils can be conserved and their fertility maintained by contour plowing, crop rotation, and a variety of other practices. Properly managed, forests can be preserved even as their resources are tapped.

Opportunities to reduce the consumption of energy resources are many and varied. Nearly everything can be made more energy efficient. Motor vehicles use a significant portion of the world's oil output. Doubling their fuel efficiency by reducing vehicle weight and using more efficient engines and tires would save at least 20% of the world's total annual oil output. Industries have an enormous potential for saving energy by using more efficient equipment and processes. The Japanese steel industry, for example, uses one-third less energy to produce a ton of steel than does that industry in most other countries. Energy used for heating, cooling, and lighting homes and office buildings could be reduced by half if they were properly constructed.

The *reuse* of materials also reduces the consumption of resources. Instead of being buried in landfills, waste can be burned or decomposed and fermented to provide energy. Recycling of steel, aluminum, copper, glass, and other materials can be greatly increased, not only to recover the materials themselves but also to recoup the energy invested in their production (Figure 11.44). It has been estimated that

FIGURE 11.44 **Aluminum cans await recycling.** It takes only 5% as much electricity to make aluminum from scrap as from raw materials. In other words, manufacturers can make 20 cans out of recycled material with the same energy it takes to make a single can out of new material. Another good reason to recycle is that an aluminum can thrown away in the trash will still be in the landfill 100 years from now.

throwing away an aluminum soft drink container wastes as much energy as filling the can half full of gasoline and pouring it on the ground.

Finally, the *substitution* of other energy sources for gas and oil, and the substitution of other materials for nonfuel minerals in short supply can be more actively pursued. If the technology to exploit them economically can be developed, coal and oil shale could supply fuel needs well into the future. In addition, the renewable energy resources such as biomass, solar, and geothermal power, are virtually infinite in their amount and variety. While no single renewable source is likely to be as important as oil or gas, collectively they could make a significant contribution to energy needs.

Summary

Our economic and material well-being depend on our use of natural resources. Renewable resources like soil and plants are those that can be regenerated in nature as fast as or faster than societies exploit them, although even renewable resources can be depleted if the rate of use exceeds that of regeneration. Nonrenewable resources—the fossil fuels and nonfuel minerals—are generated so slowly that they are thought of as existing in finite amounts. The proved reserves of a resource are the amounts that have been identified and that can be extracted profitably.

The Industrial Revolution was characterized by a shift from societal dependence on renewable resources to those derived from nonrenewable minerals, chiefly fossil fuels. The industrially advanced countries depend for about 40% of their commercial energy on crude oil, a scarce and unevenly distributed resource whose known reserves are likely to be exhausted before the middle of the next century. Only two regions, Russia and the Middle East, have large reserves of natural gas, another resource likely to approach depletion by the mid-21st century. Synthetic petroleum and gas can be produced from a number of sources, but at the present time all synfuels have high economic and environmental costs. Although nuclear power plants produce less than one-third of the world's electricity as a whole, the pattern is one of uneven dependence. Some countries receive more than half of their electricity from such plants, while others have none.

Renewable natural resources are more widely and evenly distributed than the nonrenewable ones. Wood and other forms of biomass are the primary source of energy for more than half of the world's people. Hydropower is a major source of electricity in many countries. Other renewable resources, including geothermal energy and wind power, make a more localized and limited contribution to energy needs.

The earth's crust contains a variety of nonfuel mineral resources from which people fashion metals, glass, stone, and other products. All are nonrenewable. Some exist in vast amounts, others in relatively small quantities; some are widely distributed, others concentrated in just a few locations.

About one-tenth of the earth's surface is used for the intensive production of food. Although food production has increased as fast as world population, more people than ever before are malnourished. Methods of expanding food supplies include the expansion of cultivated areas, increasing yields on land already under cultivation, and increasing the production of fish.

Human activities have had and continue to have a severe impact on two types of land resources, coastal wetlands and forests. Both play vital ecological roles, yet the wetlands have been degraded in many parts of the world, and forests are being destroyed faster than they can regenerate. The clearance of tropical forests is a matter of global, not just local, concern.

The growing demand for resources, induced by population increases and economic development, strains the earth's supply of raw materials. We began this chapter by presenting two significantly different views of the future. Because the shape of the future will be determined by the way the present generation of people thinks and acts, the wise and careful management of natural resources of all types is essential for all the world's economies, regardless of their stage of development.

Key Words

aquaculture 419
biomass 403
coal gasification 399
coal liquefaction 399
energy 391
estuarine zone 420
geothermal energy 405
hydropower 404
liquefied natural gas (LNG) 397
nonrenewable resource 389
nuclear fission 401
nuclear fusion 403
oil shale 399
perpetual resource 388
photovoltaic cell (PV) 405
potentially renewable resource 389
proved reserves 390
renewable resource 388
resource 388
solar energy 405
synthetic fuel 399
tar sand 400
tragedy of the commons 419

For Review and Consideration

1. What is the basic distinction between a *renewable* and a *nonrenewable* resource? Why do estimates of *proved reserves* vary over time?
2. Why are energy resources called the "master" natural resources? What is the relationship between energy consumption and industrial production? Briefly describe historical energy consumption patterns in the United States.
3. Why has oil become the dominant form of commercial energy? Which countries are the main producers of crude oil? How long are proved reserves of oil likely to last?
4. Why has the proportion of U.S. energy supplied by coal increased since 1961? What ecological and social problems are associated with the use of coal?
5. Review some of the techniques by which synthetic fuels may be produced. Why are synfuels not yet economically competitive with the other fossil fuels?
6. What are the different methods of generating nuclear energy? Why is there public opposition to nuclear power?
7. Which are the most widely used ways of using renewable resources to generate energy? What are the advantages of using such resources?
8. What, in general, are the leading mining countries? What role do developing countries play in the production of critical raw materials? How have producing countries reacted to the threatened scarcity of copper?
9. Since food resources are considered renewable, why is there concern about their exploitation? Discuss three ways of increasing food production. What problems do they pose?
10. What is the *estuarine zone?* What role does it play in the maintenance of marine life? In what ways is the ecology of the zone being assaulted?
11. What vital ecological functions do forests perform? Where are the tropical rain forests located, and what concerns are raised by their destruction?
12. Discuss three ways of reducing demands on resources.

Selected References

Bender, William, and Margaret Smith. *Population, Food, and Nutrition.* Washington, D.C.: Population Reference Bureau, Inc. 1997.

Castillon, David A. *Conservation of Natural Resources: A Management Approach.* 2d ed. Dubuque, Iowa: Wm. C. Brown, 1997.

Coull, James R. *World Fisheries Resources.* New York: Routledge, 1993.

Cutter, Susan, Hilary L. Renwick, and William H. Renwick. *Exploitation, Conservation, Preservation: A Geographic Perspective on Natural Resource Use.* 2d ed. New York: John Wiley & Sons, 1991.

Dower, Roger, et al. *A Sustainable Future for the United States.* Washington, D.C.: World Resources Institute, 1996.

Elliott, David. *Energy, Society and Environment.* New York: Routledge, 1997.

Energy for Tomorrow's World. The World Energy Council. New York: St. Martin's Press, 1993.

Grigg, David. *The World Food Problem.* 2d ed. Cambridge, Mass.: Basil Blackwell, 1996.

Hinrichs, Roger A. *Energy.* Philadelphia: Saunders College Publishing, 1991.

Kesler, Stephen E. *Mineral Resources and the Environment.* New York: Macmillan Publishing Co., 1994.

Mather, A. S., and K. Chapman. *Environmental Resources.* New York: John Wiley & Sons, 1996.

McNeely, Jeffrey, et al. *Conserving the World's Biological Diversity.* New York: World Resources Institute, 1990.

Miller, G. Tyler, Jr. *Resource Conservation and Management.* Belmont, Calif.: Wadsworth Publishing Co., 1990.

Miller, Kenton, and Laura Tangley. *Trees of Life: Saving Tropical Forests and Their Biological Wealth.* Washington, D.C.: World Resources Institute, 1991.

Mounfield, Peter R. *World Nuclear Power.* New York: Routledge, 1991.

Park, Chris C. *Tropical Rainforests.* New York: Routledge, 1992.

Rees, Judith. *Natural Resources: Allocation, Economics, Policy.* 2d ed. New York: Routledge, 1990.

Simmons, I. G. *Earth, Air and Water: Resources and Environment in the Late 20th Century.* London: Edward Arnold, 1991.

World Resources Institute and the International Institute for Environment and Development. *World Resources.* New York: Oxford University Press. Annual or biennial.

World Watch, a bimonthly magazine published by the Worldwatch Institute, Washington, D.C.

Worldwatch Institute. *State of the World.* New York: W. W. Norton & Co. Annual.

Worldwatch Institute. *Worldwatch Papers* issued several times a year provide in-depth analysis of a variety of resource issues. Washington, D.C.

Don't forget about Dushkin's *Annual Editions Online: Geography* at http://dushkin.com/aeonline/. See preface for details.

CHAPTER

12

URBAN GEOGRAPHY

View over Tower Bridge, the Thames River, and London, England.
©Jason Hawkes/Tony Stone Images.

In the 1930s, Mexico City was described as perhaps the handsomest city in North America and the most exotic capital city of the hemisphere, essentially unchanged over the years and timeless in its atmosphere. It was praised as beautifully laid out, with wide streets and avenues, still the "city of palaces" as described by Baron von Humboldt in the 19th century. The 70-meter-wide (200-ft-wide) Paseo de la Reforma, often noted as "one of the most beautiful avenues in the world," was shaded by a double row of trees and lined with luxurious residences.

By the 1950s, with a population of more than 2 million and an area of 52 square kilometers (20 sq mi), Mexico City was no longer unchanged. The old, rich families who formerly resided along the Paseo de la Reforma had fled from the noise and crowding. Their "palaces" were being replaced by tall blocks of apartments and hotels. Industry was expanding and multiplying, tens of thousands of peasants were flocking in from the countryside every year. By the late 1990s, with its population estimated at 16 million and its area at over 1350 square kilometers (522 sq mi), metropolitan Mexico City was among the world's largest urban complexes.

The toll exacted by that growth is heavy. Each year the city pours more than 5 million tons of pollutants into the air, 80% coming from unburned gas leaked from residents' stoves and heaters and from the exhausts of their estimated 4 million motor vehicles; the rest is produced by nearly 35,000 industrial plants. More than 5 million people citywide have no access to tap water; in some squatter neighborhoods less than half do. Some 4 million residents have no access to the sewage system. Approximately one-third of all families—and they average five people—live in just a single room, and that room generally is in a hovel in one of the largest slums in the world.

The changes in Mexico City since the 1930s have been profound. Already one of the world's most populous metropolitan areas, Mexico City is a worst-case scenario of an urban explosion that sees an increasing proportion of the world's population housed within a growing number of immense cities (Figure 12.1).

Figure 12.2 presents evidence that the growth of major metropolitan areas has been astounding in this century. More than 285 metropolitan areas had a population with more than 1 million people in 1992; at the beginning of the century there were only about a dozen such areas. By 2015, there will be 516 cities with populations over 1 million. Fifteen metropolises (Tokyo, São Paulo, New York, Mexico City, Shanghai, Bombay, Los Angeles, Beijing, Calcutta, Seoul, Jakarta, Buenos Aires, Tianjin, Osaka, Lagos) have over 10 million people (Figure 12.3). In 1900, none were of that size. Of course, as we saw in Chapter 6, it follows that since the world's population has greatly increased, so, too, would the urban component increase. But the fact remains that urbanization and metropolitanization have increased more rapidly than the growth of total population. The amount of urban growth differs from continent to continent and from country to country (Figure 12.4), but all countries have one thing in common: the proportion of the people living in cities is rising.

Table 12.1 shows world urban population by region. Note that the most industrialized parts of the world, North America and Western Europe, are the most urbanized in terms of percentage of people living in cities, while Asia and Africa have lower proportions of urban population. What is evident is that the industrialization process gives rise to increasing urbanization. As the world continues to industrialize, especially in developing countries, one can expect further large increases in urbanization. It is interesting to note in Table 12.1 that even though China, Southeast Asia (Vietnam, Indonesia, etc.), and South Asia (India, Pakistan, and Bangladesh) have relatively low proportions of people in urban regions, the absolute number of people in urban areas is among the highest in the world. Given the huge populations in Asia, and the relatively heavy emphasis on agriculture (excluding Japan and Korea), it sometimes escapes us that many large cities throughout parts of the world exist where subsistence agriculture still engages most people.

In this chapter, our first objective is to consider the major factors responsible for the size and location of urban areas. The second goal is to identify the nature of land use patterns within those areas. Third, we attempt to differentiate cities around the world by a review of the factors that help explain their special nature.

TABLE 12.1

Estimated Urban Share of Total Population, 1950 and 1990, with Projections to 2025

Region	1950	1990	2025
North America	64%	74%	85%
Europe	56	75	83
Russia	45	74	82
East Asia (except China)	43	73	87
China	12	23	55
Southeast Asia	—	27	49
South Asia	15	26	49
Latin America	41	69	84
Africa	15	31	52
Oceania	61	70	78
World	29	47	61

Sources: United Nations and authors.

FIGURE 12.1 **Sprawling Mexico City.** With more than 15 million people, the Mexico City metropolitan area is one of the largest in the world. Ringed by mountains, the area frequently experiences temperature inversions resulting in the smog visible in this photograph.

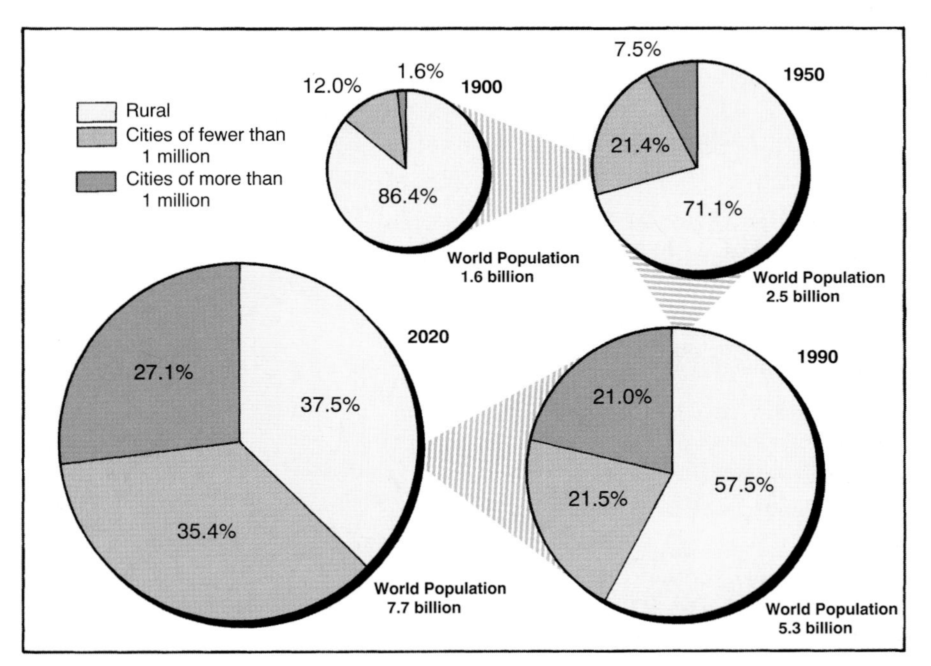

FIGURE 12.2 **Trends in world urbanization.**
Source: Estimates and projections from Population Reference Bureau and other sources.

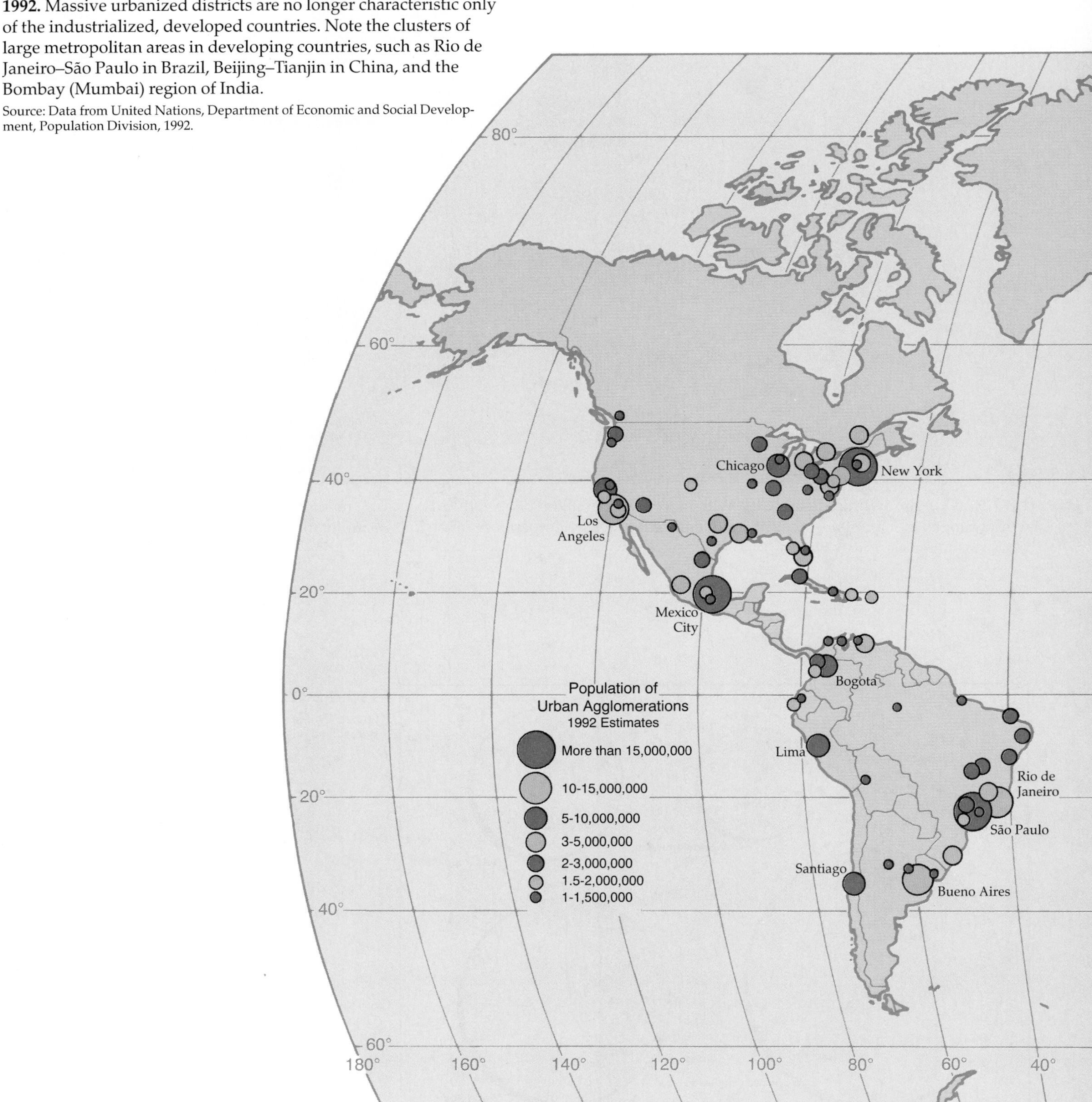

FIGURE 12.3 **World metropolitan areas of 1 million or more in 1992.** Massive urbanized districts are no longer characteristic only of the industrialized, developed countries. Note the clusters of large metropolitan areas in developing countries, such as Rio de Janeiro–São Paulo in Brazil, Beijing–Tianjin in China, and the Bombay (Mumbai) region of India.

Source: Data from United Nations, Department of Economic and Social Development, Population Division, 1992.

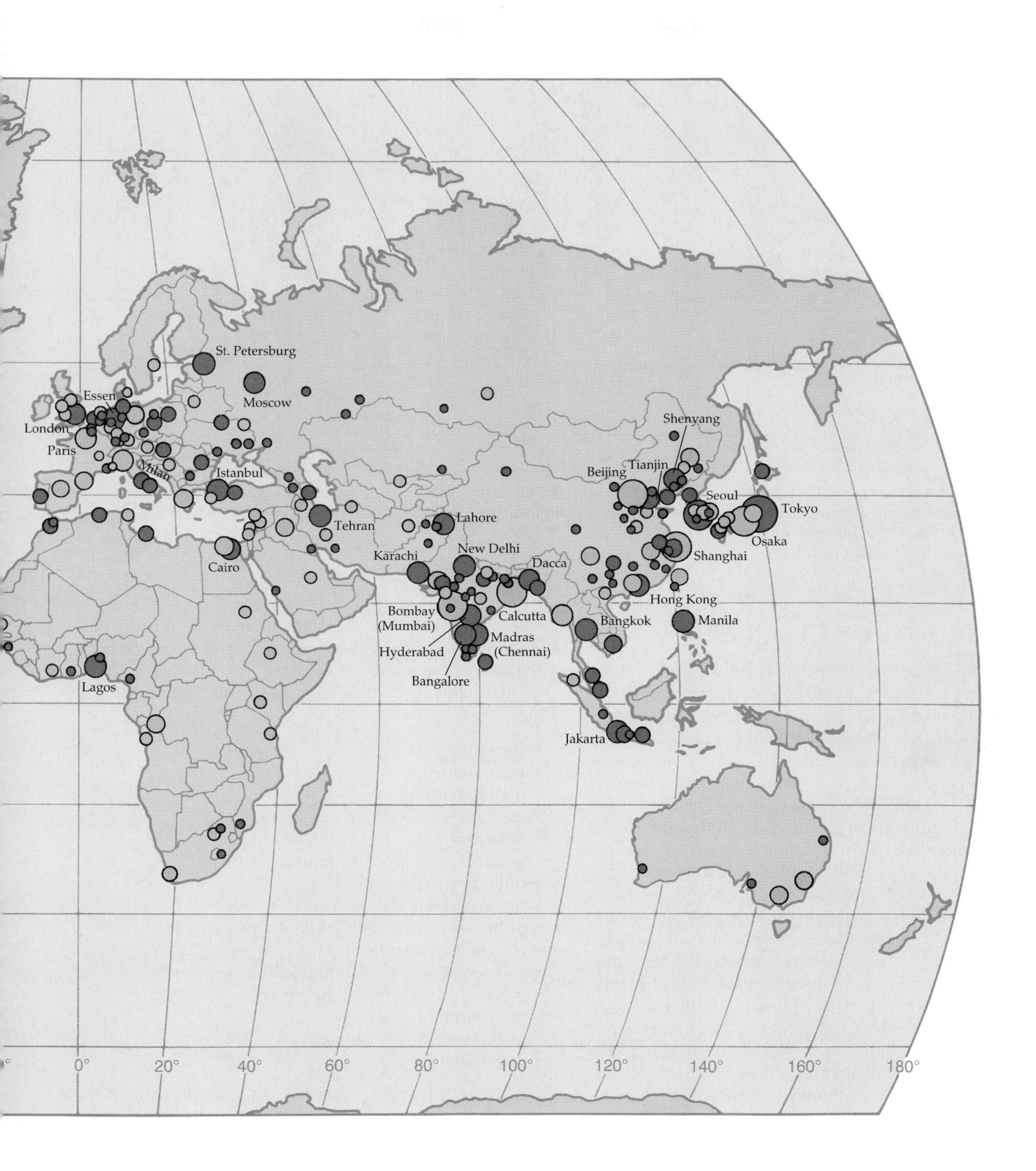
St. Petersburg
Moscow
Essen
London
Paris
Milan
Istanbul
Cairo
Lagos
Tehran
Lahore
Karachi
New Delhi
Dacca
Bombay
(Mumbai)
Calcutta
Hyderabad
Madras
(Chennai)
Bangalore
Bangkok
Jakarta
Manila
Hong Kong
Shanghai
Beijing
Tianjin
Shenyang
Seoul
Tokyo
Osaka
0°
20°
40°
60°
80°
100°
120°
140°
160°
180°

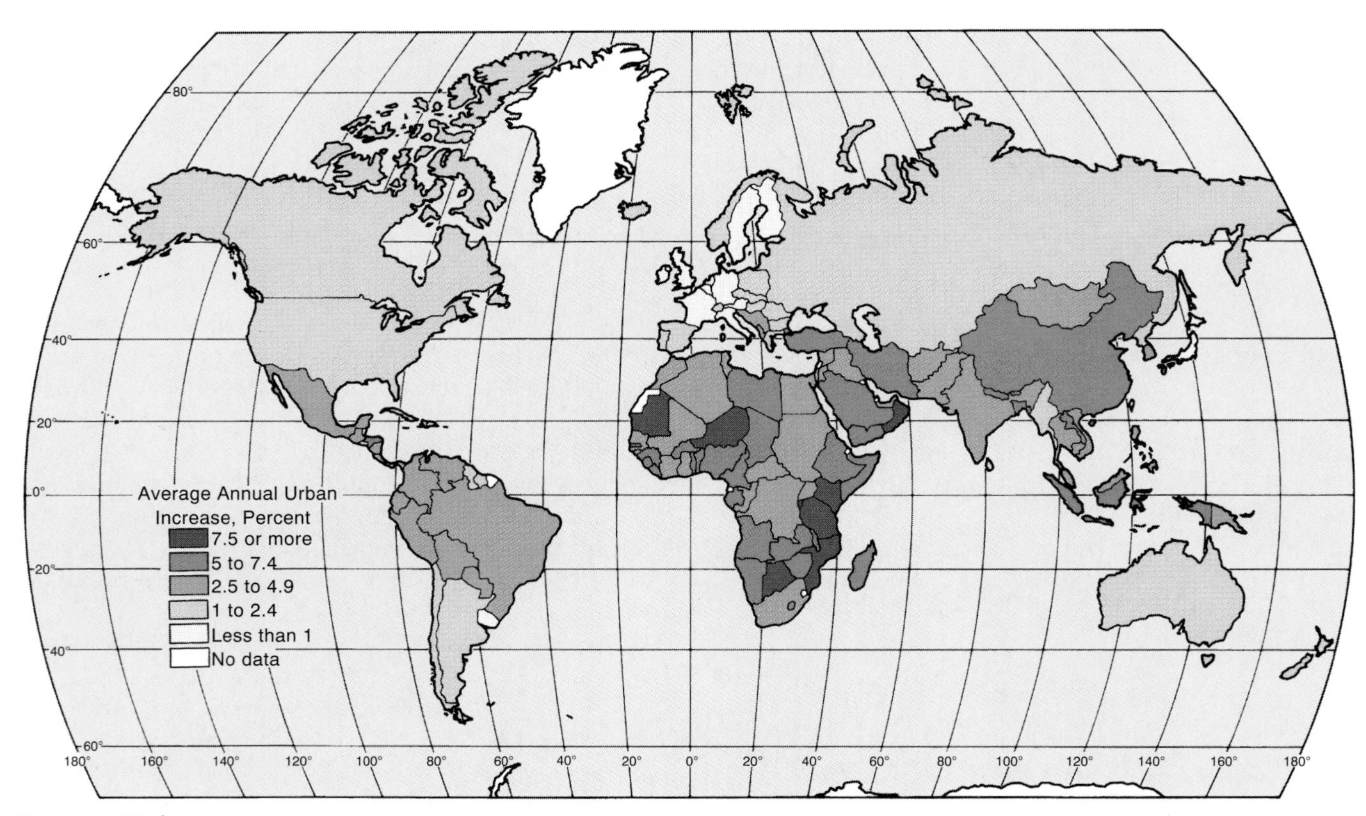

FIGURE 12.4 **Average annual urban population growth rates, 1980–1990.** In general, developing countries show the highest percentage increases in their urban populations, and the already highly urbanized and industrialized countries have the lowest—less than 1% per year in most of Europe.

Sources: Data from World Bank, *World Development Report* 1992; and United Nations.

The Functions of Urban Areas

Except for the occasional recluse or hermit, people gather together to form couples, families, groups, organizations, towns, and so forth. This desire to be near one another, however, is more than just a function of the need to socialize. Our human support systems are based on the flow of information, goods and services, and cooperation among people who are located at convenient places relative to one another. Unless individuals can produce all that they need themselves, and relatively few can, they must depend on shipments of food and supplies to their home place or convenient outlet centers. Non-subsistence groups establish stores, places of worship, repair centers, and production sites as close to their home places as is possible and reasonable. The result is the establishment of towns. These may grow to the size of a Tokyo metropolitan area (about 26 million people) or a Mexico City area (about 16 million).

Whether they are villages, towns, or cities, urban settlements exist for the efficient performance of functions required by the society that creates them. They reflect the saving of time, energy, and money that the agglomeration of people and activities implies. The more accessible the producer to the consumer, the worker to the workplace, the citizen to the town hall, the worshiper to the church, or the lawyer or doctor to the client, the more efficient is the performance of their separate activities, and the more effective is the integration of urban functions.

Urban areas provide all or some of the following types of functions:

- retailing;
- wholesaling;
- manufacturing;
- business service;
- entertainment;
- political and official administration;
- military defensive needs;
- social and religious service;
- public service including sanitation and police;
- educational service;
- transportation and communication service;
- meeting place activity;
- recreation;
- visitor service;
- residential areas.

Because all urban functions and people cannot be located at a single point, cities themselves must take up space, and land uses and populations must have room within them. Because interconnection is essential, the nature of the transportation system will have an enormous bearing on the total number of services that can be performed and the efficiency with which the functions can be carried out (Figure 12.5). The totality of people and functions of a city constitutes a distinctive cultural landscape whose similarities and differences from place to place are the subjects for urban geographic analysis.

Some Definitions

Urban areas are not of a single type, structure, or size. Urban areas are alike in that they are nucleated, nonagricultural settlements. At one end of the size scale, urban areas are small towns with perhaps a single main street of shops; at the opposite end, they are complex, multifunctional metropolitan areas or supercities (Figure 12.6). The word *urban* is often used in place of such terms as town, city, suburb, and metropolitan area, but it is a general term, not used to specify a particular type of settlement. Although the terms designating the different types of urban settlements, like city, are employed in common speech, they are not uniformly applied by all users. What is recognized as a city by a resident of rural Vermont or West Virginia may not at all be afforded that name and status by an inhabitant of California or New Jersey. It is necessary in this chapter to agree on the meanings of terms commonly employed but varyingly interpreted.

The words **city** and **town** denote nucleated settlements, multifunctional in character, including an established central business district and both residential and nonresidential land uses. **Towns** are smaller in size and have less functional complexity than cities, but they still have a nuclear business concentration. **Suburb** denotes a subsidiary area, a functionally specialized segment of a large urban complex. It may be dominantly or exclusively residential, industrial, or commercial, but by the specialization

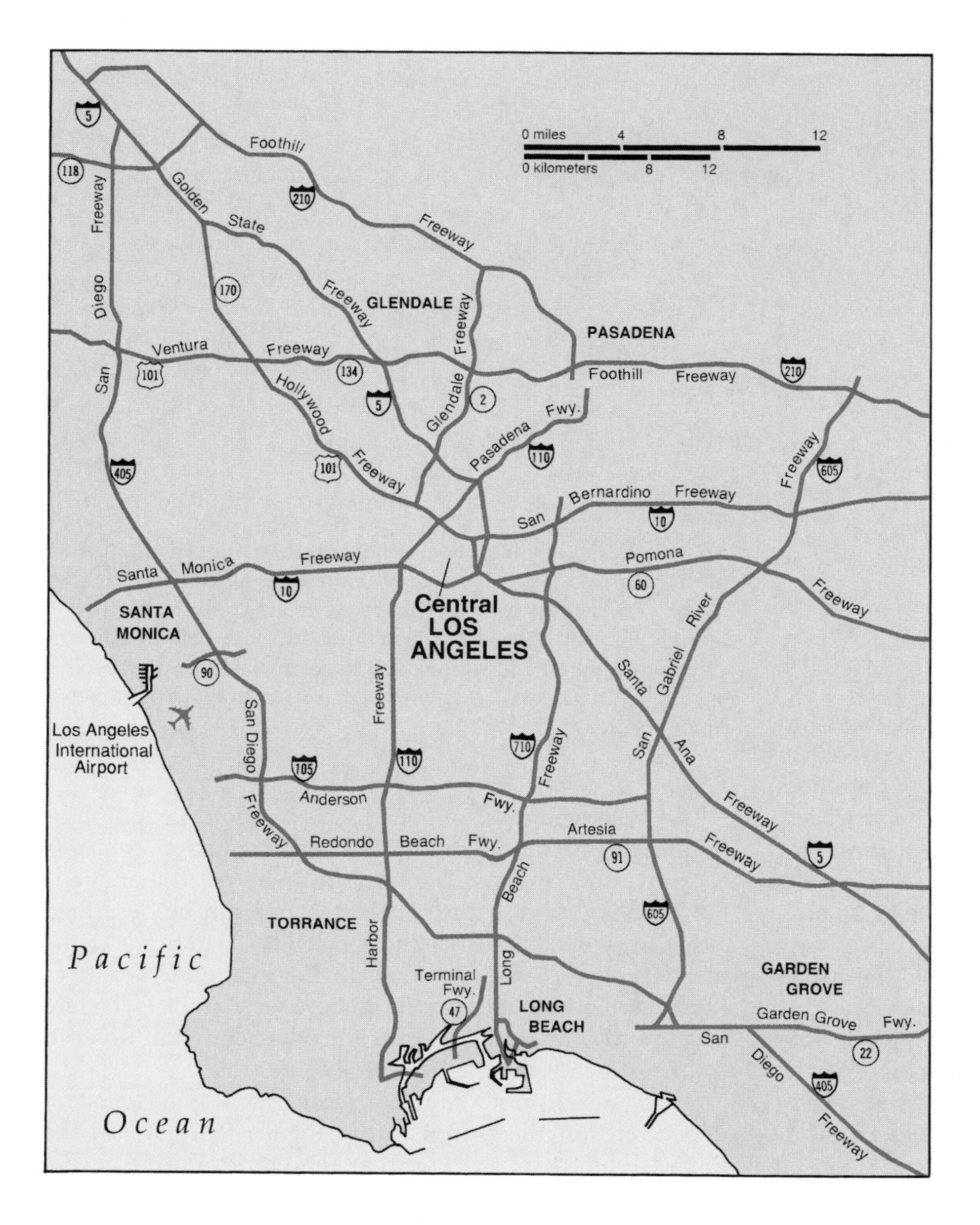

FIGURE 12.5 **The Los Angeles freeway system.** This is the world's most extensive system of high-speed highways for a metropolitan area. It fostered the growth of the Los Angeles metropolitan area, but in recent years the system has become congested.

(a)

(b)

FIGURE 12.6 The differences in size, density, and land use complexity are immediately apparent between (a) New York City and (b) a small town. One is a city, one is a town, but both are urban areas.
(a) ©Carl Purcell.
(b) ©Susan Reisenweaver.

of its land uses and functions, a suburb depends on urban areas outside of its boundaries. Suburbs, however, can be independent political entities. For large cities having many suburbs, it is common to call that part of the urban area contained within the official boundaries of the main city around which the suburbs have been built the **central city.**

The **urbanized area** refers to a continuously built-up landscape defined by building and population densities with no reference to political boundaries. It may be viewed as the physical city and may contain a central city with many contiguous cities, towns, suburbs, and other urban tracts. A **metropolitan area,** on the other hand, refers to a large-scale functional entity, perhaps containing several urbanized areas, discontinuously built up, but nonetheless operating as an integrated economic whole (Figure 12.7) (see "The Definition of *Metropolitan* in the United States"). A list of the large U.S. metropolitan areas is given in Table 12.2.

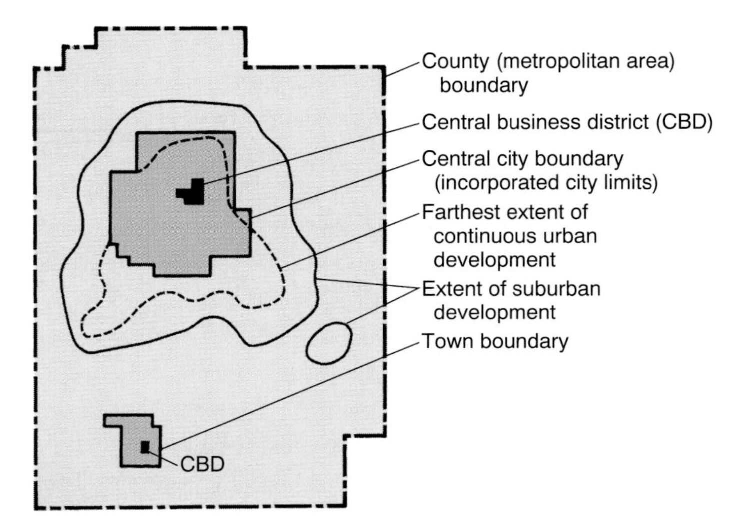

FIGURE 12.7 **A hypothetical spatial arrangement of urban units within a metropolitan area.** Sometimes official limits of the central city are very extensive and contain areas commonly thought of as suburban. On the other hand, older eastern U.S. cities and San Francisco in the west have official limits that contain only half or less of the population associated with their metropolitan areas.

The Location of Urban Settlements

Urban centers are functionally connected to other cities and rural areas. In fact, the reason for the existence of an urban center is not only to provide services for itself, but for others outside of it. The urban area is a consumer of food, a processor of materials, and an accumulator and dispenser of goods and services, but it must rely on outside areas for its supplies and as a market for its activities. In order to adequately perform the tasks that support it and to add new functions as demanded by the larger economy, the urban unit must be efficiently located. That efficiency may be marked by centrality to the area served. It may derive from the physical characteristics of its site, or placement may be related to the resources, productive regions, and transportation network of the country, so that the effective performance of a wide array of activities is possible.

In discussing urban settlement location, geographers frequently differentiate between site and situation, concepts already introduced in Chapter 1. You will recall that **site** is the exact location of the settlement and can be described either in terms of latitude and longitude, or in terms of the physical characteristics of the site. For example, the site of Philadelphia is an area bordering and west of the Delaware River north of the intersection with the Schuylkill River in southeast Pennsylvania (Figure 12.8).

The description can be more or less exhaustive depending on the purpose it is meant to serve. In the Philadelphia case, the fact that the city is partly on the

THE DEFINITION OF *METROPOLITAN* IN THE UNITED STATES

Definitions of various types of urban areas must be clear if proper accounting is to be made by governmental authorities. The United States Bureau of the Census has refined and redefined the concept of "metropolitan" from time to time to summarize the realities of the changing population, physical size, and functions of urban regions.

Until 1983, the *Standard Metropolitan Statistical Area (SMSA)* was recognized. It was made up of one or more functionally integrated counties focusing upon a central city of at least 50,000 inhabitants. Now, the minimum size requirement for central cities has been dropped and central city status is determined by other qualities—whether, for example, a city is an employment center surrounded by bedroom community-type suburbs. Automatically, the number of central cities (and metropolitan areas) increased. The statistical structure of urban America has been altered as individual communities exercised their rights to withdraw from former metropolitan affiliations or opted to join with neighboring cities in new ones.

In the mid-1980s, old and new metropolitan areas were redefined into *Metropolitan Statistical Areas* (MSAs are economically integrated urbanized areas in one or more contiguous counties), *Primary Metropolitan Statistical Areas* (PMSAs are those counties that are part of MSAs that have less than 50% resident workers working in a different county), and *Consolidated Metropolitan Statistical Areas* (an MSA becomes a CMSA if it contains 1 million or more people and is composed of PMSAs).

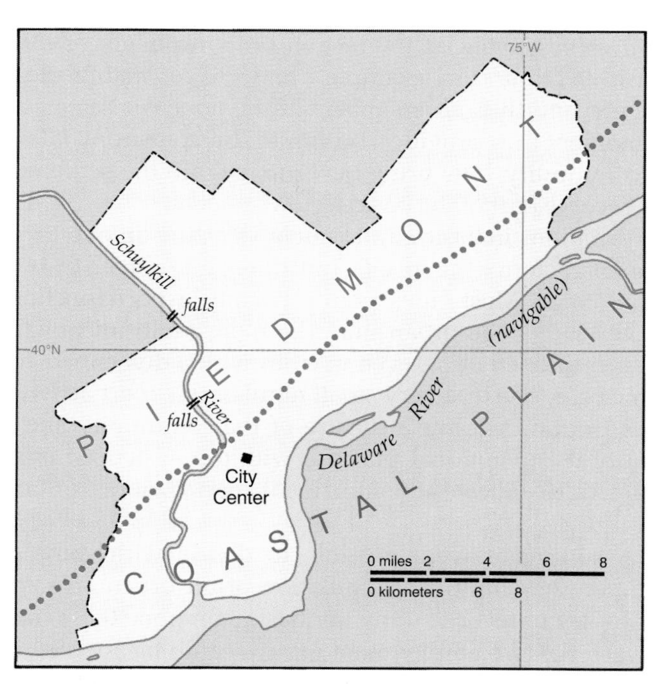

FIGURE 12.8 **The site of Philadelphia.**

Atlantic coastal plain, partly in the piedmont (foothills), and is served by a navigable river is important if one is interested in the development of the city during the Industrial Revolution. As Figure 12.9 suggests, water transportation was an important localizing factor when the major American cities were established.

If site suggests absolute location, **situation** indicates relative location. The relative location places a settlement in relation to the physical and human characteristics of the surrounding areas. Very often it is important to know what kinds of possibilities and activities exist in the area near a settlement, such as the distribution of raw materials, market areas, agricultural regions, mountains, and oceans.

The site of central Chicago is 41°52′N, 87°40′W, on a lake plain. More important, however, is its situation close to the deepest penetration of the Great Lakes system into the interior of the country, astride the Great Lakes–Mississippi waterways, and near the western margin of the manufacturing belt, the northern boundary of the corn belt, and the southeastern reaches of a major dairy region. References to railroads, coal deposits, and ore fields would amplify its situational characteristics (Figure 12.10). As a gateway to the west from the east and *vice versa,* Chicago's O'Hare International Airport is one of the busiest in the United States. From this description of Chicago's situation, implications relating to market, to raw materials, and to transportation centrality can be drawn.

The site or situation that originally gave rise to an urban unit may not remain the essential ingredient for its growth and development for very long. Agglomerations, originally successful for whatever reason may, by their success, attract people and activities totally unrelated to the initial localizing forces. By what has been called a process of "circular and cumulative causation," a successful urban unit may acquire new populations and functions attracted by the already existing markets, labor force, and urban facilities.

THE ECONOMIC BASE

When one or more urban settlements within a well-linked system increases its productivity, perhaps because of an increase in demand for the special goods or services that it produces, all members of the system are likely to benefit. The concept of the *economic base* shows how settlements are affected by changes in economic conditions. In the

TABLE 12.2

United States Metropolitan Area Populations with More than One Million, 1996

Rank	Metropolitan Area (Principal Cities)	Population (In Thousands)
1	New York	19,938
2	Los Angeles	15,495
3	Chicago	8600
4	Washington-Baltimore	7165
5	San Francisco-Oakland	6608
6	Philadelphia	5973
7	Boston	5563
8	Detroit	5284
9	Dallas-Fort Worth	4575
10	Houston	4253
11	Atlanta	3541
12	Miami	3514
13	Seattle	3321
14	Cleveland	2913
15	Minneapolis-St. Paul	2765
16	Phoenix	2747
17	San Diego	2655
18	St. Louis	2548
19	Pittsburgh	2379
20	Denver	2277
21	Tampa	2199
22	Portland	2078
23	Cincinnati	1921
24	Kansas City	1690
25	Milwaukee	1643
26	Sacramento	1632
27	Norfolk	1540
28	Indianapolis	1492
29	San Antonio	1490
30	Columbus	1447
31	Orlando	1417
32	Charlotte	1321
33	New Orleans	1313
34	Salt Lake City	1218
35	Las Vegas	1201
36	Buffalo	1175
37	Hartford	1145
38	Greensboro	1141
39	Providence	1124
40	Nashville	1117
41	Rochester	1088
42	Memphis	1078
43	Austin	1041
44	Oklahoma City	1027
45	Raleigh-Durham	1025
46	Grand Rapids	1015
47	Jacksonville	1009

Source: United States Bureau of the Census.

discussion that follows, we concentrate on changes within urban areas.

Part of the employed population of an urban unit is engaged in either the production of goods or in the performance of services for areas and people outside that urban area. They include workers engaged in "export" activities, whose efforts result in money flowing into the community. Collectively, they constitute the **basic sector** of the urban area's total economic structure. Other workers support themselves by producing goods for residents of the urban unit itself. Their efforts, necessary to the well-being and the successful operation of the settlement, do not generate new money for it but comprise a **nonbasic sector** of its economy. These people are responsible for the internal functioning of the urban unit. They are crucial to the continued operation of its stores, professional offices, city government, local transit, and school systems.

The total economic structure of an urban area equals the sum of its basic and nonbasic activities. In actuality, it is the rare urbanite who can be classified as belonging entirely to one sector or another. Some part of the work of most people involves financial interaction with residents of other areas. Doctors, for example, may have mainly local patients, and thus are members of the nonbasic sector, but the moment they provide a service to someone from outside the community, they bring new money into the settlement and become part of the basic sector.

Variations in basic employment structure among urban areas characterize the specific functional role played by individual cities. Most cities perform many export functions, and the larger the urban unit, the more multifunctional it becomes. Nonetheless, even in cities with a diversified economic base, one or a very small number of export activities tends to dominate the structure of the community and to identify its operational purpose within a system of cities. Figure 12.11 indicates the functional specializations of large U.S. cities.

Assuming it were possible to divide with complete accuracy the employed population of an urban area into totally separate basic and nonbasic components, a ratio between the two employment groups could be established. With exception for some high-income communities, this basic/nonbasic ratio is roughly similar for cities of similar size irrespective of their functional specializations. Further, as a settlement increases in size, the number of nonbasic personnel grows faster than the number of new basic workers. Thus, in cities with a population of 1 million, the ratio is about two nonbasic workers for every basic worker; the addition of ten new basic employees implies the expansion of the labor force by 30 (10 basic, 20 nonbasic) and an increase in total population equal to the added workers plus their dependents. A **multiplier effect** thus exists, associated with economic growth. The term multiplier effect implies the addition of nonbasic workers and dependents to a settlement's total employment and population as a supplement of new basic employment; the size of the effect is

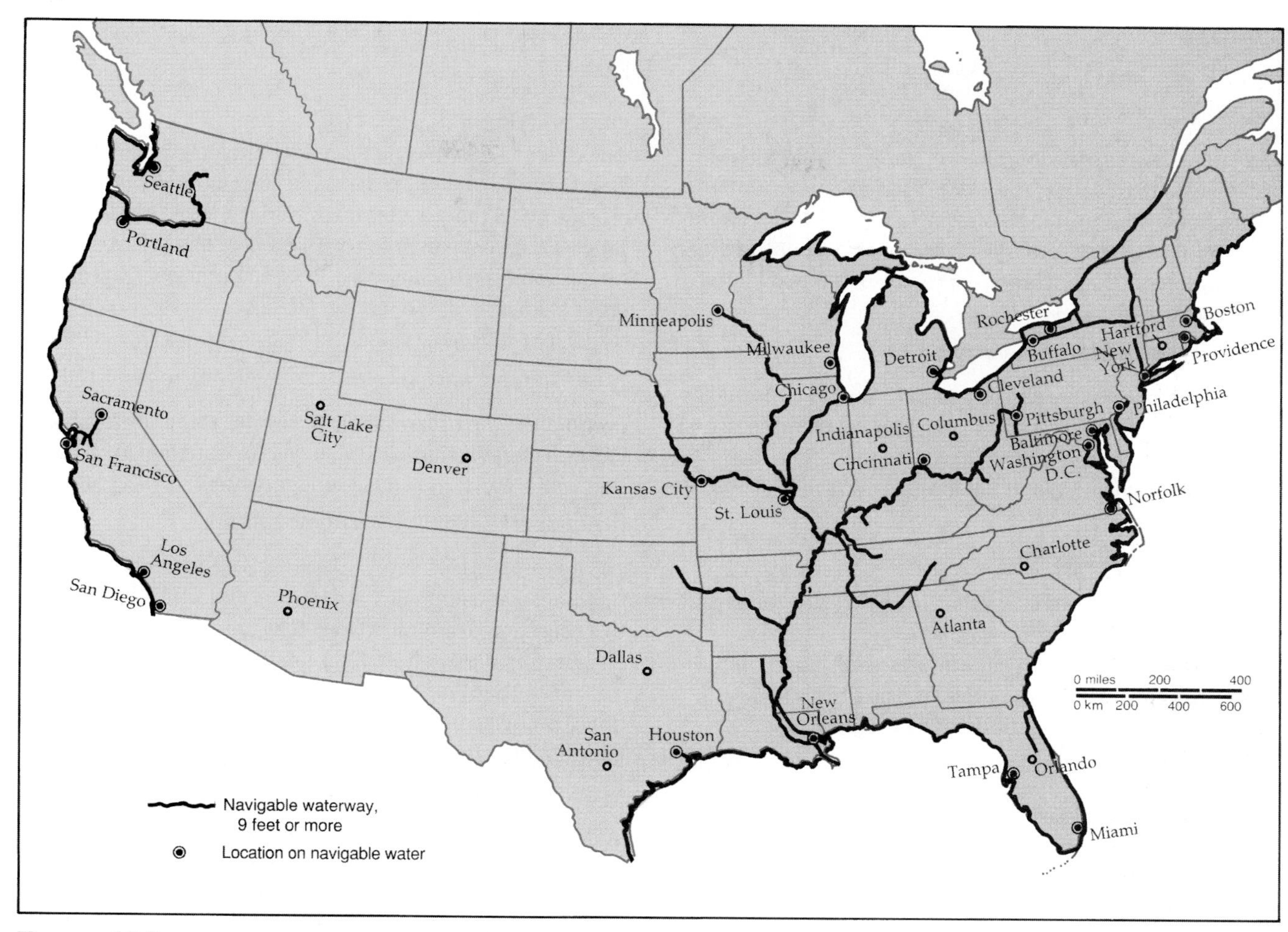

FIGURE 12.9 **Metropolitan areas in the United States with 1 million or more residents as of February, 1991.** Notice the association of principal cities and navigable water. Before the advent of railroads in the middle of the 19th century, all major cities were associated with waterways.

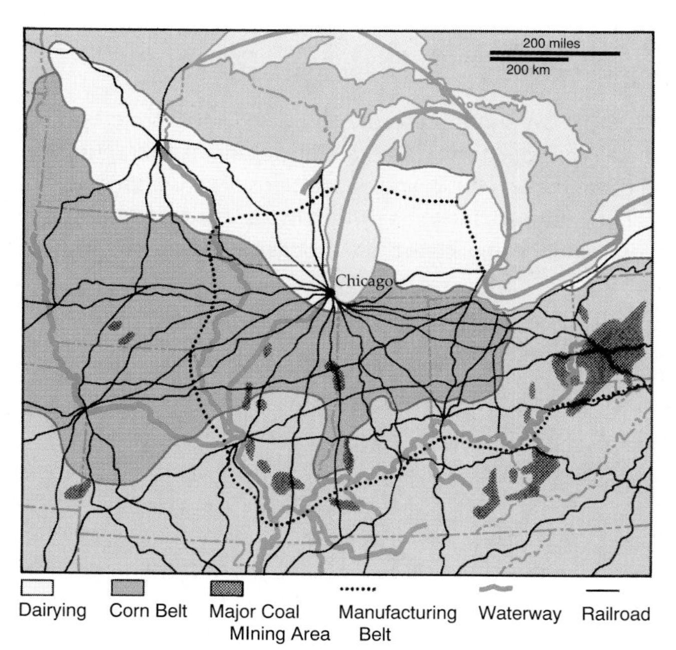

FIGURE 12.10 **The situation of Chicago** helps to suggest the reasons for its functional diversity and size.

determined by the community's basic/nonbasic ratio (Figure 12.12).

The changing numerical relationships shown in Figure 12.12 are understandable when we consider how settlements add functions and grow in population. A new industry selling services to other communities requires new workers, who thus increase the basic workforce. These new employees, in turn, demand certain goods and services, such as clothing, food, and medical assistance, which are provided locally by the nonbasic workforce. Those who perform such services must themselves have services available to them. For example, a grocery clerk must also buy groceries. The more nonbasic workers an urban area has, the more nonbasic workers are needed to support them, and the application of the multiplier effect becomes obvious.

The growth of cities may be self-generating—"circular and cumulative"—in a way related not to the development of industries that specialize in the production of material objects for export, like automobiles and paper products, but to the attraction of what would be classified as *service* industry. Banking and legal services, a sizable market, a diversified labor force, extensive public services, and the like may

On-Line Urban Geography

United Nations Population Information Network (*POPIN*) gives access to a wealth of data at the world, regional, and country levels. Available are statistical tables (including the UN's "World Urbanization Prospects" country counts and projections), official documentation of UN population conferences, full-text journals, and other materials.

http://www.undp.org/popin/popin.htm

Two other UN sites are more directly urban. The *UN Centre for Human Settlements* (*Habitat*) home page provides entree to a variety of documents and reports, including the 1997 Urban Environment Forum and World Habitat Day and the 1996 Habitat II conference.

http://www.unhabitat.org/

The UN's education program *Cities of Today, Cities of Tomorrow,* although designed for grades 5–12, has a useful set of background papers for those of any age interested in world cities, their history, and their current problems. The site also gives access to a set of brief international "City Profiles."

http://www.un.org/Pubs/CyberSchoolBus/special/habitat

The *Best Practices for Human Settlements Database* contains proven solutions to common urban problems facing the world's cities today. First presented at the United Nations Habitat II City Summit in 1996, this "knowledge base" identifies ways in which shared solutions can address urban issues such as poverty, access to land, clean water, population, shelter, and transportation.

http://www.bestpractices.org/

Urban Geography on the Web is the official home page of the Urban Geography Specialty Group of the Association of American Geographers. It contains information on the subdiscipline as well as links to websites of interest.

http://www.geog-nt.geog.buffalo.edu/ugsg/

Two other professional groups concerned with urban matters are *Canadian Urban Institute* at

http://www.canurb.com/

and the *Urban Land Institute* at

http://www.uli.org/

The *Urban Morphology Research Group* (*UMRG*), founded in 1974 in the School of Geography at the University of Birmingham, specializes in the study of urban form. This home page provides an annotated guide to web resources for those interested in that topic. Detailed links are arranged alphabetically, giving the country of origin for each site.

http://www.bham.ac.uk/geography/umrg/

One example of changing urban morphology is offered by *Mapping Urban Sprawl in the San Francisco Bay Region.* Historical records, USGS topographic maps, aerial photos, and Landsat satellite data were used to compile a data base of urbanization for the San Francisco Bay area. The computer-generated time-series animation of the resulting data provides a visualization of settlement-related land use and land cover changes since 1850.

http://geo.arc.nasa.gov/esdstaff/william/urban.html

Evolution of the Urban System: This site, at the University of Western Ontario, Canada, combines the notion of urban evolution with a short icon animation sequence.

http://sparky.sscl.uwo.ca/Demo2.html

The Electronic Map Library, provided by the Department of Geography at California State University/Northridge, gives access to a collection of digital atlases. They are available for New York City, Boston, Washington, D.C., San Diego, Los Angeles, Sacramento, and San Francisco. Atlases for each are divided into four principal topics: population and race, income, poverty, and adult educational attainment.

http://geogdata.csun.edu/library.html

State of the Nation's Cities: The Center for Urban Policy Research at Rutgers University has assembled a comprehensive database on 77 American cities and suburbs. The database brings together information on over 3000 variables from a wide variety of sources, allowing easy comparability of indicators on employment and economic development, demographic measures, housing and land use, income and poverty, fiscal conditions, and a host of other health, social, and environmental indicators.

http://www.policy.rutgers.edu/cupr/SoNC.htm

ESPROMUD: Urbanization model: The ESPROMUD project is primarily concerned with the extent, type, and nature of urban land cover, the way it is changing, patterns of consumption of materials in the urbanization process, waste production, waste disposal, and the impact all those actions have on earth surface processes.

http://cchp3.unican.es/ESPROMUD/espromud53.html

The USA CityLink Project is a comprehensive listing of WWW home pages of individual U.S. cities.

http://usacitylink.com//

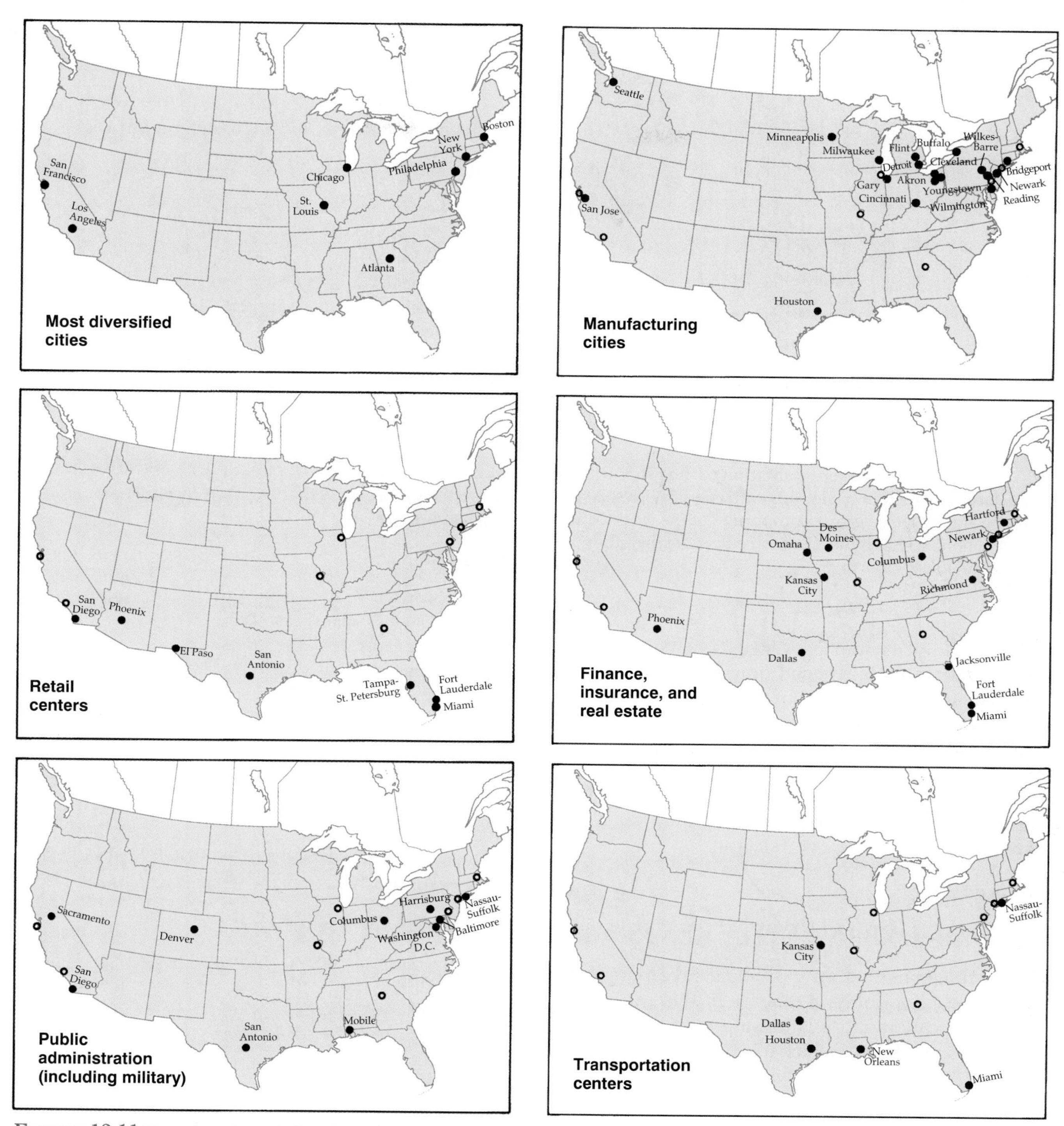

FIGURE 12.11 **Functional specialization of U.S. metropolitan areas.** Five categories of employment were selected to show patterns of specialization for some U.S. metropolitan areas. In addition, the category "Most Diversified" represents those areas that have a generally balanced employment distribution. Since these metropolitan areas specialize in the other categories, they are included as open circles on the specialization maps. Note that the most diversified urban areas tend to be the largest.

generate additions to the labor force not basic by definition, but nonbasic. In recent years, service industries have developed to the point where new service industries serve the older service industries. For example, computer systems firms aid banks on developing more efficient computer-driven banking systems.

In much the same way as settlements grow in size and complexity, so do they decline. When the demand for the goods and services of an urban unit falls, there is an obvious need for fewer workers, and thus both the basic and the nonbasic components of a settlement system are affected. There is, however, a resistance to decline that impedes the

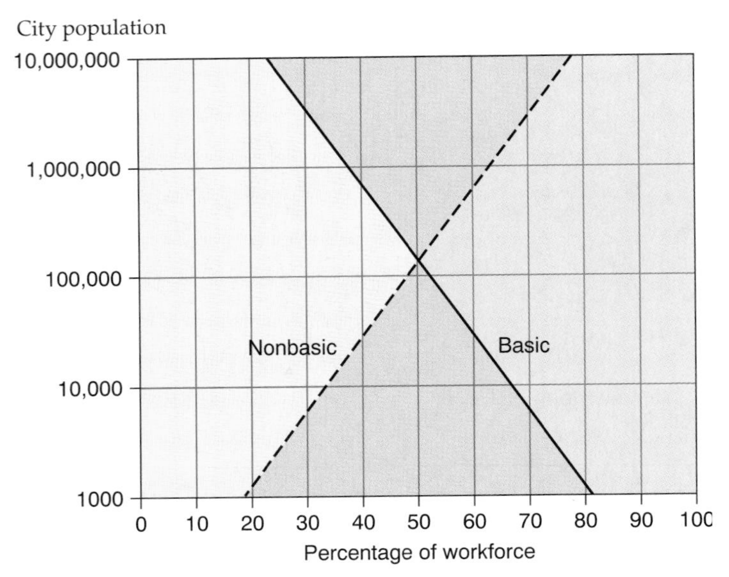

FIGURE 12.12 **A generalized representation of the proportion of the workforce engaged in basic and nonbasic activities.** As settlements become larger, a greater proportion of the workforce is employed in nonbasic activities. Larger centers are therefore more self-contained.

TABLE 12.3

FASTEST GROWING U.S. METROPOLITAN STATISTICAL AREAS 1990–1996, WITH MORE THAN 600,000 POPULATION

RANK	MSA	% GROWTH
1	Las Vegas, NV-AZ	40.9
2	Austin-San Marcos, TX	23.1
3	Phoenix-Mesa, AZ	22.7
4	Atlanta, GA	19.7
5	Raleigh-Durham-Chapel Hill, NC	19.4
6	Portland-Salem, OR-WA	15.9
7	Orlando, FL	15.7
8	El Paso, TX	15.7
9	Tucson, AZ	15.1
10	Denver-Boulder-Greeley, CO	15.0

SLOWEST GROWING OR DECLINING U.S. METROPOLITAN STATISTICAL AREAS, 1990–1996 WITH MORE THAN 600,000 POPULATION

RANK	MSA	% CHANGE
1	Scranton-Wilkes Barre, PA	–1.6
2	Buffalo-Niagara Falls, NY	–1.2
3	Hartford, CT	–1.1
4	Providence-Fall River-Warwick, RI-MA	–0.9
5	Pittsburgh, PA	–0.6
6	Toledo, OH	–0.4
7	Dayton-Springfield, OH	–0.1
8	Syracuse, NY	0.5
9	Philadelphia-Wilmington, PA-DE-NJ	1.4
10	Detroit-Ann Arbor-Flint, MI	1.9

BIGGEST POPULATION GAINS, 1990–1996 U.S. MSAS AND CONSOLIDATED METROPOLITAN STATISTICAL AREAS

RANK	MSA/CMSA	GAIN
1	Los Angeles-Riverside-Orange County	963,626
2	Atlanta	581,730
3	Dallas-Fort Worth	537,279
4	Houston-Galveston-Brazoria	522,399
5	Phoenix-Mesa	508,205
6	Washington-Baltimore	438,124
7	New York-Northern New Jersey-Long Island	388,843
8	Chicago-Gary-Kenosha	359,954
9	San Francisco-Oakland-San Jose	355,547
10	Seattle-Tacoma-Bremerton	350,529

Source: U.S. Bureau of the Census.

process and delays its impact. Whereas settlements can grow rapidly as migrants respond quickly to the need for more workers, under conditions of decline those that have developed roots in the community are hesitant to leave or may be financially unable to move to another locale. Figure 12.13 and Table 12.3 show that in recent years urban areas in the South and West of the United States have been growing, while decline is evident in the Northeast and the North Central regions.

SYSTEMS OF URBAN SETTLEMENTS

The various functions that an individual urban area performs are reflected not only in the size of that settlement, but also in its location and relationship with other urban units in the larger system of which it is part.

The Urban Hierarchy

Perhaps the most effective way to recognize how systems of cities are organized is to consider the **urban hierarchy.** Urban areas can be divided into size classes on the basis of their functional complexity. One can measure the numbers and kinds of services each city or metropolitan area provides. The hierarchy is then like a pyramid; a few large and complex cities are at the top and many smaller cities are at the bottom. There are always more smaller cities than larger ones.

A functional hierarchy, as shown in Figure 12.14, divides U.S. metropolitan areas into a spatial system that includes world centers and various levels of regional metropolitan areas. Goods, services, communication lines, and people flow up and down the hierarchy. The few high-level metropolitan areas provide specialized functions for large regions in the United States and the rest of the world, while the smaller cities have smaller regions that they serve. These cities serve the areas around them, but since cities of the same level provide roughly the same services, cities of the same

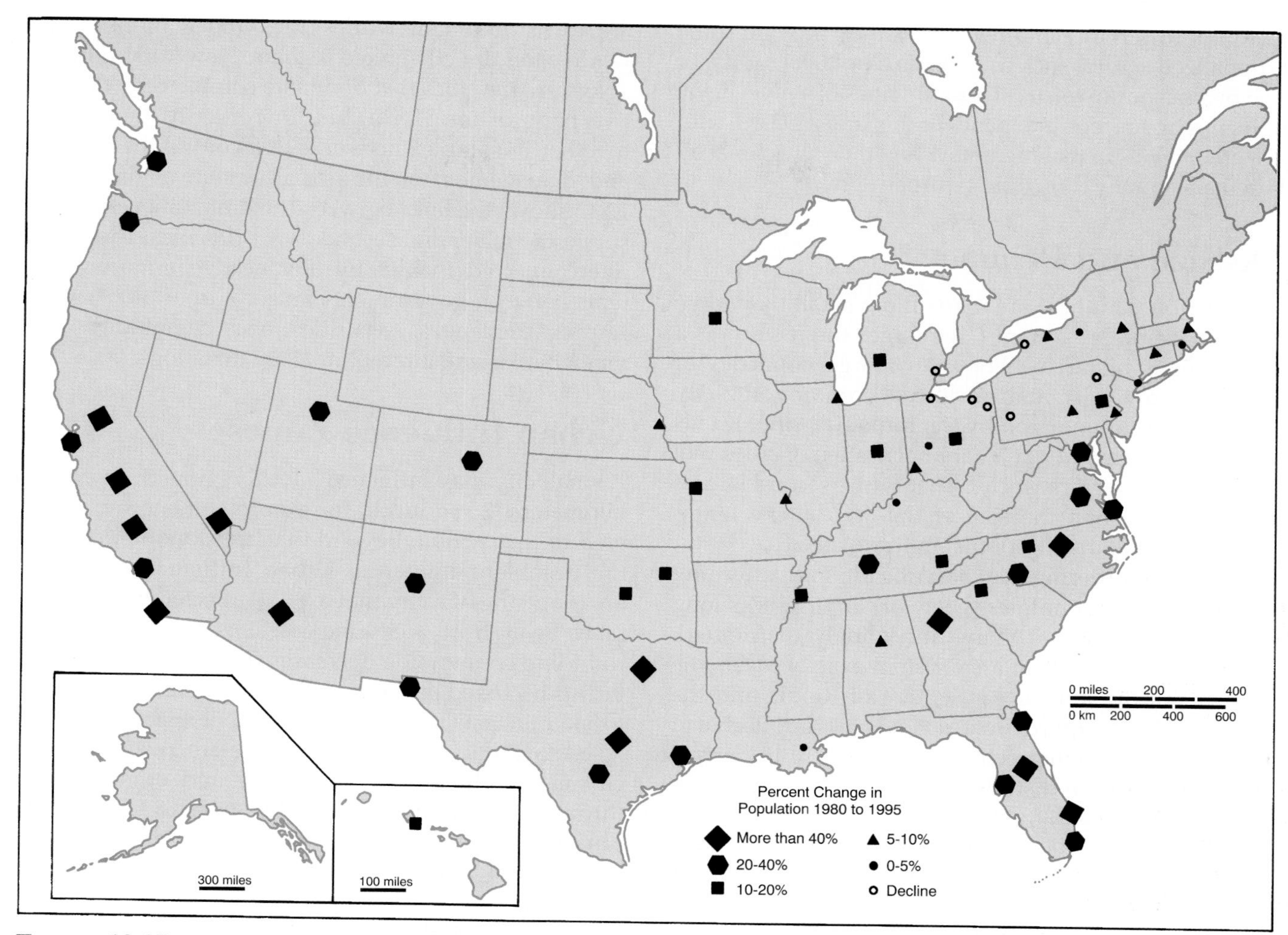

FIGURE 12.13 The pattern of metropolitan growth and decline in the United States, 1980–1995. Shown are metropolitan areas with 600,000 or more population in 1990.

Source: Data from U.S. Bureau of the Census.

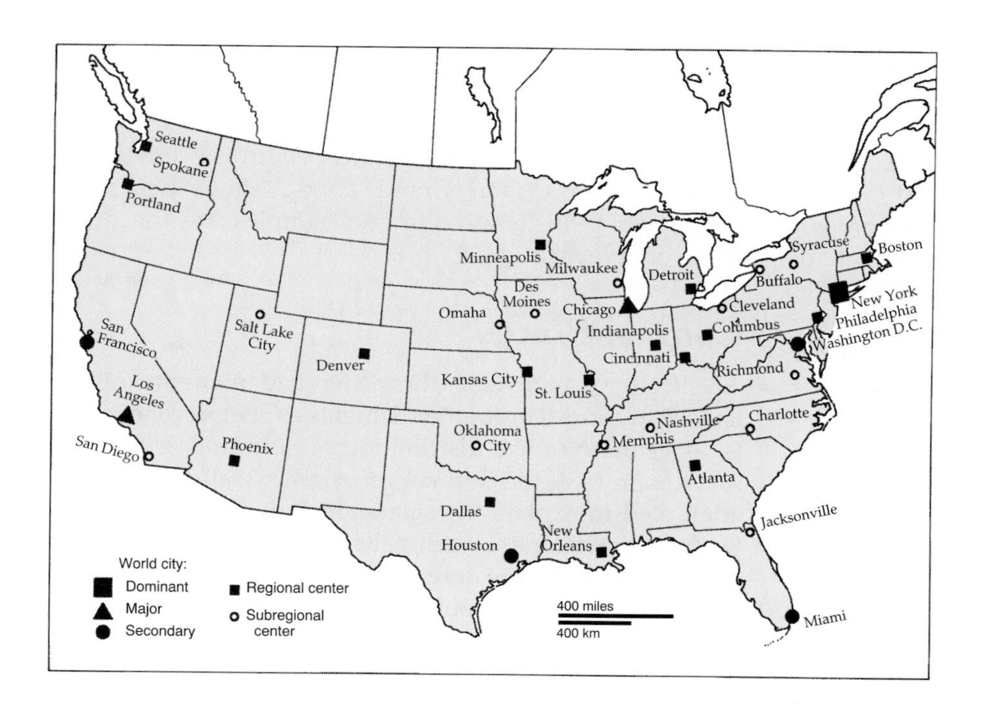

FIGURE 12.14 Functional hierarchy of U.S. metropolitan regions. Only the major metropolitan areas are shown. The hierarchy includes smaller urban areas (not shown) that depend on or serve the larger centers.

Redrawn from P. L. Knox, ed., *The United States. A Contemporary Human Geography.* Harlow, England: Longman, 1988, Fig. 5.5, p. 144.

size tend not to serve each other unless they provide some very specialized service such as housing a political capital of a region or a major university. Thus, the cities of a given level in the hierarchy are not independent, but interrelated with cities of other levels in that hierarchy. Together, all cities at all levels in the hierarchy constitute an urban system.

Rank-Size and Primacy

The observation that there are many more small than large cities within an urban system ("the larger the fewer") is a statement about hierarchy. For many large countries of great regional diversity and advanced economy, the city size hierarchy is summarized by the **rank-size rule.** It tells us that the *n*th largest city of a national system of cities will be $1/n$ the size of the largest city. That is, the second largest settlement will be half the size of the largest, the tenth-biggest will be 1/10 the size of the first-ranked city.

The rank-size ordering may describe the size patterning of cities in complex economies where urban history is long and urbanizing forces are many and widely distributed. Although no national urban area system exactly meets the requirements of the rank-size rule, that of Russia and the United States closely approximates it. It is less applicable to countries with developing economies or where the urban size hierarchy has been distorted through concentration of functions in a single, paramount center.

In some countries the urban system is dominated by a **primate city,** one that is far more than twice the size of the second-ranked city. In fact, there may be no obvious "second city" at all, for a characteristic of a primate city hierarchy is one very large city, few or no intermediate-sized cities, and many subordinate smaller settlements. For example, Seoul contains nearly 30% of the population of South Korea. The capital cities of many developing countries display this kind of overwhelming primacy. In part, their primate city pattern is a heritage of their colonial past, when economic development, colonial administration, and transportation and trade activities were concentrated at a single point. Kenya (Nairobi is the primate city), and many other African nations are examples. In other instances—Egypt (Cairo) or Mexico (Mexico City), for example—development and population growth have tended to concentrate disproportionately in a capital city whose very size attracts further development and growth. Many European countries also show a primate structure (United Kingdom and France), often ascribed to the former concentration of economic and political power around the royal court in a capital city that was, perhaps, also the administrative and trade center of a larger colonial empire.

World Cities

Standing at the top of national systems of cities are a relatively few cities that may be called **world cities.** These are urban centers that are control points for international production and marketing and for international finance. There are three dominant world cities and a number of other world cities directly linked to them. New York, London, and Tokyo are the cities that dominate commerce in their respective parts of the world, but, importantly, they are bound together in complex networks that control the organization and management of the global system of finance. Figure 12.15 shows the links between the dominant centers and the suggested major and secondary world cities. These cities are interconnected mainly by advanced communication systems between governments, major corporations, stock and futures exchanges, securities and commodity markets, major banks, and international organizations.

Urban Influence Zones

A small city may influence a local region of, say, 65 square kilometers (25 sq mi) if, for example, its newspaper delivered to that region. Beyond that area, another city may be the dominant influence. **Urban influence zones** are the areas outside of a city that are still affected by it. As the distance away from a city increases, its influence on the surrounding countryside decreases (recall the idea of distance decay discussed in Chapter 8). The sphere of influence of an urban unit is usually proportional to its size.

A large city located 100 kilometers away from a small city may influence the small city and other small cities through its banking services, TV stations, and large shopping malls. There is an overlapping hierarchical arrangement, and the influence of the largest cities is felt over the widest areas.

Intricate relationships and hierarchies are common. Consider Grand Forks, North Dakota, which for local market purposes dominates the rural area immediately surrounding it. However, Grand Forks is influenced by political decisions made in the state capital, Bismarck. For a variety of cultural, commercial, and banking activities, Grand Forks is influenced by Minneapolis. As the center of wheat production, Grand Forks and Minneapolis are subordinate to the grain market in Chicago. Of course, the pervasive agricultural and other political controls exerted from Washington, D.C., on Grand Forks, Minneapolis, and Chicago indicate how large and complex are the urban zones of influence.

Central Places

An effective way to realize how cities and towns are interrelated is to consider urban settlements as **central places,** that is, as centers for the distribution of economic goods and services. In 1933, the German geographer Walter Christaller attempted to explain the size and location of settlements. He developed a framework, called **central place theory,** for understanding urban interdependence. Christaller recognized that his theory would best be developed in rather idealized circumstances. He assumed that the following propositions were true.

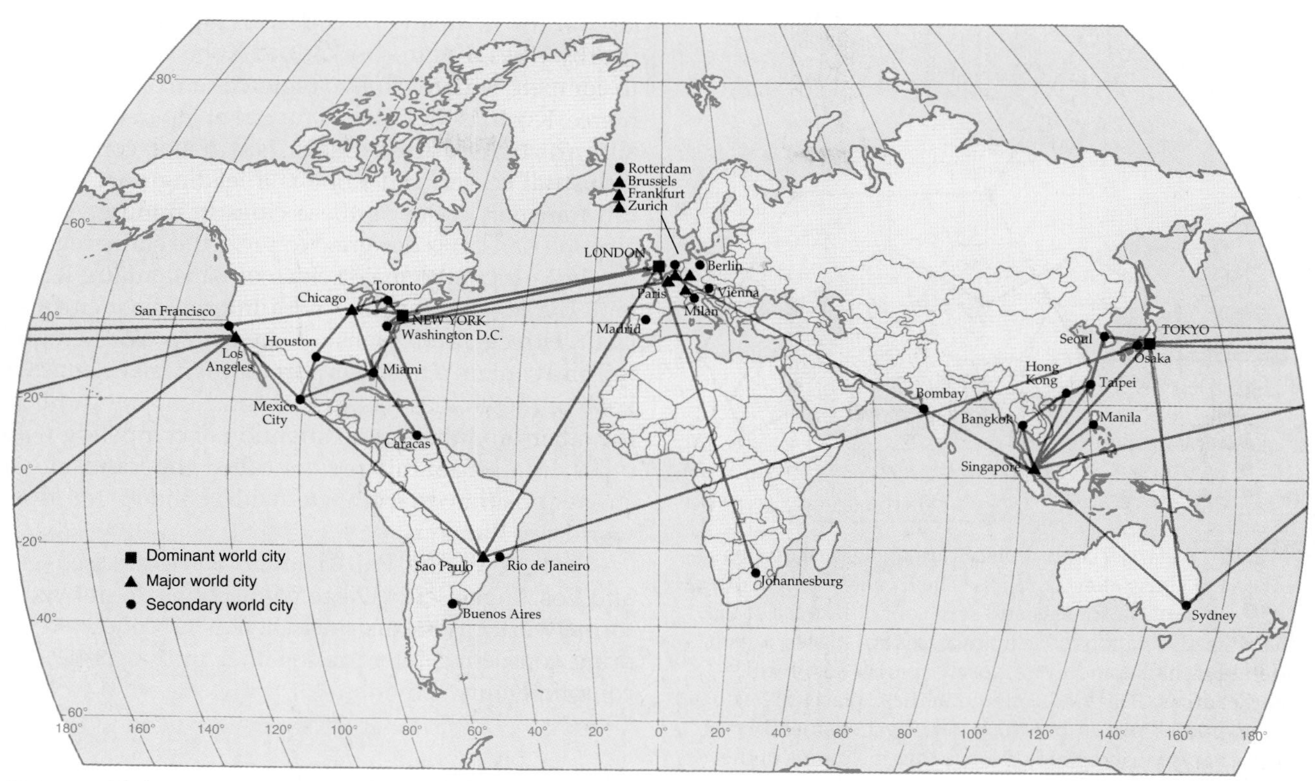

FIGURE 12.15 **A classification of world cities.** The ties between cities represent the flow of financial and economic information.

1. Towns that provide the surrounding countryside with such fundamental goods as groceries and clothing would develop in a plain where farmers specialized in commercial agricultural production.
2. The farm population would be dispersed in an even pattern.
3. The characteristics of the people would be uniform; that is, they would possess similar tastes, demands, and incomes.
4. Each kind of product or service available to the dispersed population would have its own *threshold,* or minimum number of consumers needed to support its supply. Because such goods as diamonds or fur coats are either expensive or not in great demand, they would have a high threshold, while a fewer number of consumers would be required to support a store that sells bread.
5. Consumers would purchase goods and services from the nearest opportunity (store).

When all of the assumptions are considered simultaneously, they yield the following results.

1. The agricultural plain will be divided into noncompeting markets where each entrepreneur has exclusive rights to the sale of a particular product.
2. A series of hexagonal market areas that cover the entire plain will emerge, as shown in Figure 12.16.
3. There will be a central place at the center of each of the hexagonal market areas.
4. The largest central places will supply all of the goods and services the consumers in that area demand and can afford.
5. The size of the market area of a central place will be proportional to the number of goods and services offered from that central place.
6. Contained within or at the edge of the largest market areas are central places serving a smaller population and offering fewer goods and services.

In addition, Christaller reached two important conclusions. First, towns of the same size will be evenly spaced, and larger towns will be farther apart than smaller ones. This means that many more small than large towns will exist. In Figure 12.16, the ratio of the number of small towns to towns of the next larger size is 3 to 1. This distinct, steplike series of towns in size classes differentiated by both size and function is called a *hierarchy of central places.*

Second, the system of towns is interdependent. If one town were eliminated, the entire system would have to readjust. Consumers need a variety of products, each of which has a different minimum number of customers required to support it. The towns containing many goods and services become regional retailing centers, and the small central places serve just the people immediately in their vicinity. The higher the threshold of a desired product, the farther, on average, the consumer must travel to purchase it.

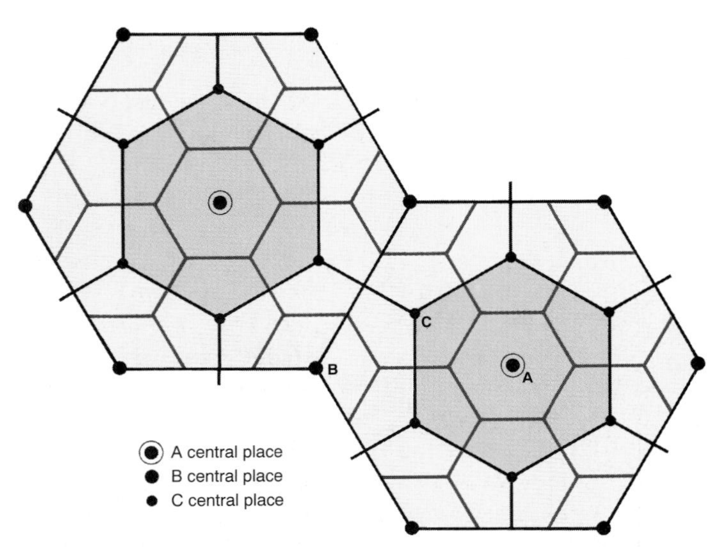

FIGURE 12.16 The two A central places are the largest on this diagram of one of Christaller's models. The B central places offer fewer goods and services for sale and serve only the areas of the intermediate-sized hexagons. The many C central places, which are considerably smaller and more closely spaced, serve still smaller market areas. The goods offered in the C places are also offered in the A and B places, but the latter offer considerably more and more specialized goods. Notice that the places of the same size are equally spaced.

From Arthur Getis and Judith Getis, "Christaller's Central Place Theory," *Journal of Geography*, 1966. Used with permission of the National Council for Geographic Education, Indiana, PA.

These conclusions have been shown to be generally valid in widely differing areas within the commercial world. When varying incomes, cultures, landscapes, and transportation systems are taken into consideration, the results, although altered to some extent, hold up rather well. They are particularly applicable to agricultural areas, especially with regard to the size and spacing of cities and towns, as Figure 12.17 suggests. One has to stretch things a bit to see the model operating in highly industrialized areas, where cities are more than just retailing centers. However, if we combine a Christaller-type approach with the ideas that help us understand industrial location and transportation alignments (see Chapter 10), we have a fairly good understanding of the location of the majority of cities and towns.

Network Cities

In recent years, a new kind of urban spatial pattern has begun to appear. A **network city** evolves when two or more previously independent nearby cities, potentially complementary in function, strive to cooperate by developing between them high speed transportation corridors and communications infrastructure. For example, with Hong Kong joining China, an infrastructure is being built that will integrate Hong Kong with Guangzhou, the huge Chinese city on the mainland of China. In Japan, three distinctive, nearby cities, Kyoto, Osaka, and Kobe are joining together to compete with the Tokyo region as a major center of commerce. Kyoto is the cultural capital of Japan with its temples and artistic treasures, Osaka is a major commercial and industrial center, and Kobe is a leading port. High speed rail transport connects these cities in minutes, and a new airport (Kansai) is designed to serve the entire region.

In Europe, the major cities of Amsterdam, Rotterdam, and The Hague, together with intermediate cities such as Delft, Utrecht, and Zaanstad, are connected by high speed rail and contain a major airport that serves the entire region. Each of these cities has special functions not duplicated in the others and there is no intention of competing with each other. In a sense, this region, called the Randstad, is in a strong position to compete with London for dominant world city status.

The New York-Philadelphia, San Francisco-Oakland, and Los Angeles-San Diego connections do not yet qualify for network city status since there is no concerted effort to bring competing interests together into a single center of complimentary activities.

Inside the City

The structure, patterns, and spatial interactions of systems of cities make up only half the story of urban settlements. The other half involves the distinctive cultural landscapes that are the cities themselves. An understanding of the nature of cities is incomplete without a knowledge of their internal characteristics. So far, we have explored the location, size, and growth and decline tendencies of cities

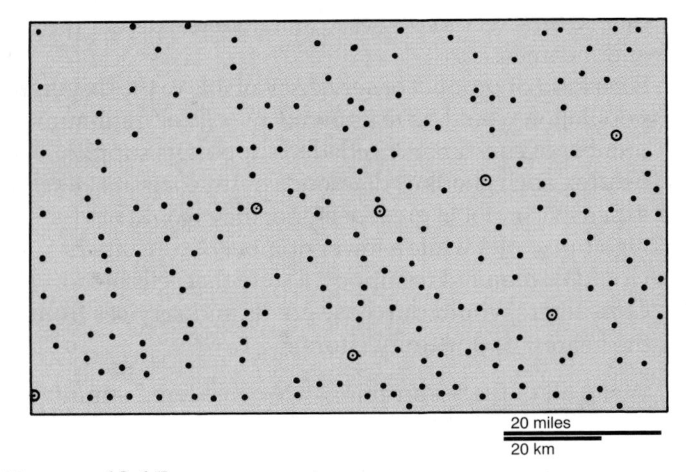

FIGURE 12.17 **The pattern of hamlets, towns, and cities in a portion of Indiana.** This map represents an area 132 by 82 kilometers (82 by 51 mi) just north of Indianapolis. Cities containing more than 10,000 people are circled. Notice that the pattern is remarkably even and includes a number of linear arrangements that correspond to highways and railroads.

within hierarchical urban systems. Now we look into the city itself in order to better understand how land uses are distributed, how social areas are formed, and how institutional controls, such as zoning regulations, affect its structure. This discussion will primarily relate to cities in the United States, although most cities of the world have been formed in a somewhat similar manner.

It is a common observation that a recurring pattern of land use arrangements and population densities exists within urban areas. There is a certain sameness to the way cities are internally organized, especially within one particular culture sphere like North America or Western Europe. The major variables responsible for shaping internal land use patterns are: accessibility, controls on the market in land, and the transportation technologies available during the periods of urban growth. These variables will be discussed together in the following sections.

The Competitive Bidding for Land

For its effective operation, the city requires close spatial association of its functions and people. As long as these functions were few and the population small, pedestrian movement and pack-animal haulage were sufficient for the effective integration of the urban community. With the addition of large-scale manufacturing and the accelerated urbanization of the economy during the 19th century, however, functions and populations—and therefore city area—grew beyond the interaction capabilities of pedestrian movement alone. Increasingly efficient and costly, mass-transit systems were installed. Even with their introduction, however, only land within walking distance of the mass-transit routes or terminals could successfully be incorporated into the expanding urban structure.

Usable land, therefore, was a scarce commodity, and by its scarcity, it assumed high market value and demanded intensive, high-density utilization. Because of its limited supply of usable land, the industrial city of the mass-transit era was compact, was characterized by high residential and structural densities, and showed a sharp break on its margins between urban and nonurban uses. The older central cities of, particularly, the northeastern United States and southeastern Canada were of that vintage and pattern.

Within the city, parcels of land were allocated among alternate potential users on the basis of the relative ability of those users to outbid their competitors for a chosen site. There was, in gross generalization, a continuous open auction in land in which users would locate, relocate, or be displaced in accordance with "rent-paying" ability. The attractiveness of a parcel, and therefore the price that it could command, was a function of its accessibility. Ideally, the most desirable and efficient location for all of the functions and the people of a city would be at a single point at which the maximum possible interchange could be achieved. Such total coalescence of activity is obviously impossible.

Because uses must therefore arrange themselves spatially, the attractiveness of a parcel is rated by its relative accessibility to all other land uses of the city. Store owners wish to locate where they can easily be reached by potential customers; factories need a convenient assembly of their workers and materials; residents desire easy connection with jobs, stores, and schools; and so forth. Within the older central city, the radiating mass-transit lines established the elements of the urban land use structure by freezing in the landscape a clear-cut pattern of differential accessibility. The convergence of that system on the city core gave that location the highest accessibility, the highest desirability, and, hence, the highest land values of the entire built-up area. Similarly, transit junction points were more accessible to larger segments of the city than locations along single traffic routes; the latter were more desirable than parcels lying between the radiating lines (Figure 12.18).

Society deems certain functions desirable without regard to their economic competitiveness. Schools, parks, and public buildings are assigned space without being participants in the auction for land. Other uses, through the process of that auction, are assigned spaces by market forces. The merchants with the highest-order goods and the largest threshold requirements bid most for, and occupy, parcels within the **central business district (CBD),** which is localized at the convergence of mass-transit lines. The successful bidders for slightly less accessible CBD parcels are the developers of the tall office buildings (*skyscrapers*) of major cities, the principal hotels, and similar land uses that help to produce the *skyline* of the commercial city. Table 12.4 lists the world's tallest office buildings; very often one is seen as the apex of an inverted V that makes up the CBD skyline.

Comparable, but lower-order, commercial aggregations develop at the outlying intersections—transfer points—of the mass-transit system. Industry takes control of parcels adjacent to essential cargo routes: rail lines, waterfronts, rivers, or canals. Strings of stores, light industries, and high-density apartment structures can afford and benefit from location along high-volume transit routes. The least accessible locations within the city are left for the least-competitive users: low-density residences. A diagrammatic summary of this repetitive allocation of space among competitors for urban sites is shown in Figure 12.19. Compare it to the generalized land use map of Calgary, Alberta, Canada (Figure 12.20).

Land Values and Population Density

Theoretically, the open land auction should yield two separate although related, distance-decay patterns, one related to land values and the other to population density (as distance increases away from the CBD, population density decreases). If one views the land value surface of the central

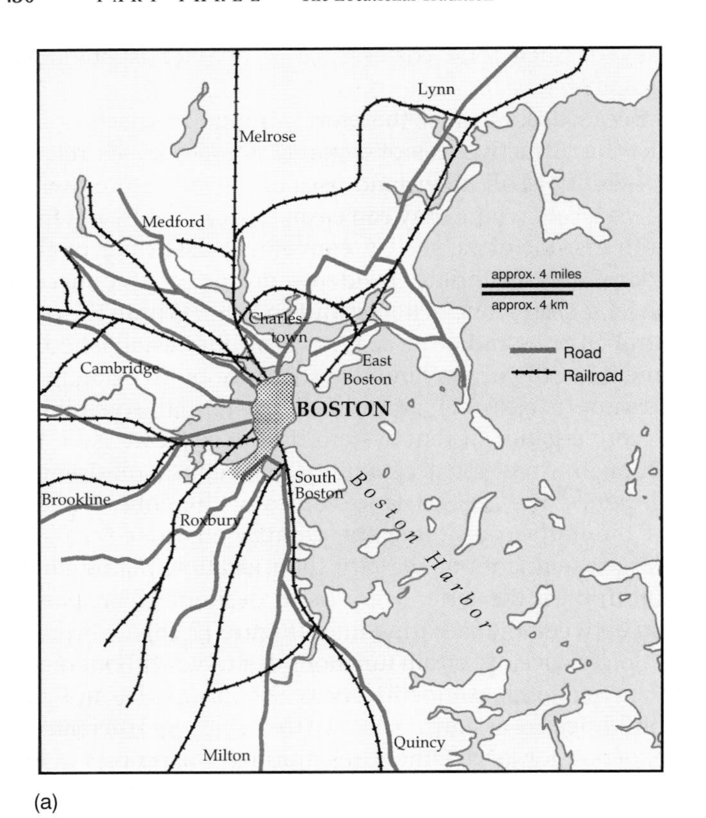

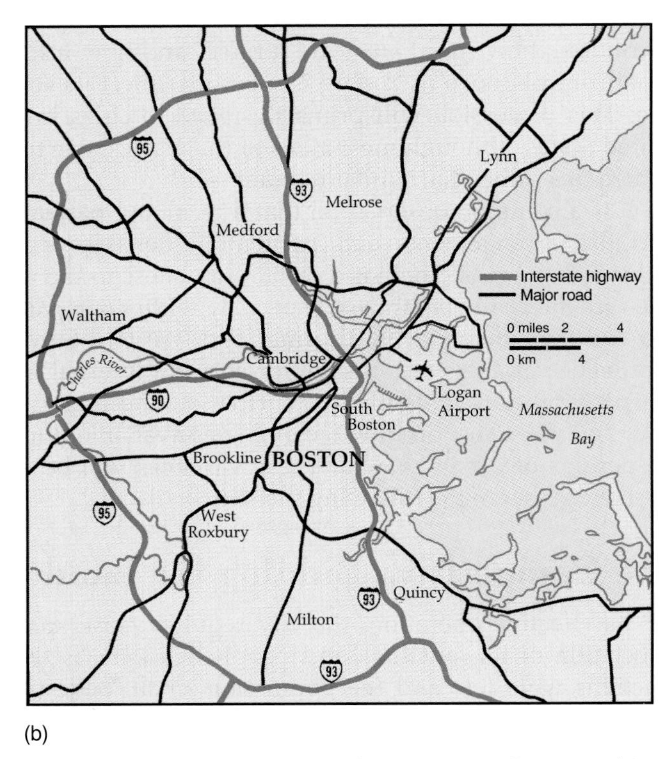

FIGURE 12.18 **Major transit lines in Boston in 1872 and 1994.** (a) Notice how the lines converge on the city center. Compare this map with the late 20th-century freeway patterns shown in (b) the Boston of 1994 and Los Angeles (see Figure 12.5).

TABLE 12.4

THE WORLD'S TALLEST OFFICE BUILDINGS

RANK	BUILDING NAME	CITY	COUNTRY	HEIGHT IN METERS (FEET)	STORIES
1	Petronas Tower I	Kuala Lumpur	Malaysia	450 (1483)	88
2	Petronas Tower II	Kuala Lumpur	Malaysia	450 (1483)	88
3	Sears Tower	Chicago	USA	443 (1454)	110
4	*Jin Mao	Shanghai	China	421 (1379)	88
5	One World Trade Center	New York	USA	417 (1368)	110
6	Two World Trade Center	New York	USA	415 (1362)	110
7	Empire State	New York	USA	381 (1250)	102
8	Central Plaza	Hong Kong	China	374 (1227)	78
9	Bank of China	Hong Kong	China	368 (1209)	70
10	*T&C Tower	Kaoshiung	Taiwan	348 (1140)	85
11	Amoco	Chicago	USA	346 (1136)	80
12	John Hancock	Chicago	USA	344 (1127)	100
13	First Interstate	Los Angeles	USA	338 (1107)	72
14	*Sky Central Plaza	Guangzhou	China	322 (1056)	80
15	*Baiyoke Tower II	Bangkok	Thailand	320 (1050)	90
16	Chrysler	New York	USA	319 (1046)	77
17	Shenzhen Avic Plaza	Shenzhen	China	313 (1025)	63
18	NationsBank	Atlanta	USA	312 (1023)	55
19	*Stratosphere Tower	Las Vegas	USA	308 (1010)	114
20	Texas Commerce Tower	Houston	USA	306 (1002)	75

*Under construction

Source: *The World Almanac and Book of Facts, 1997.*

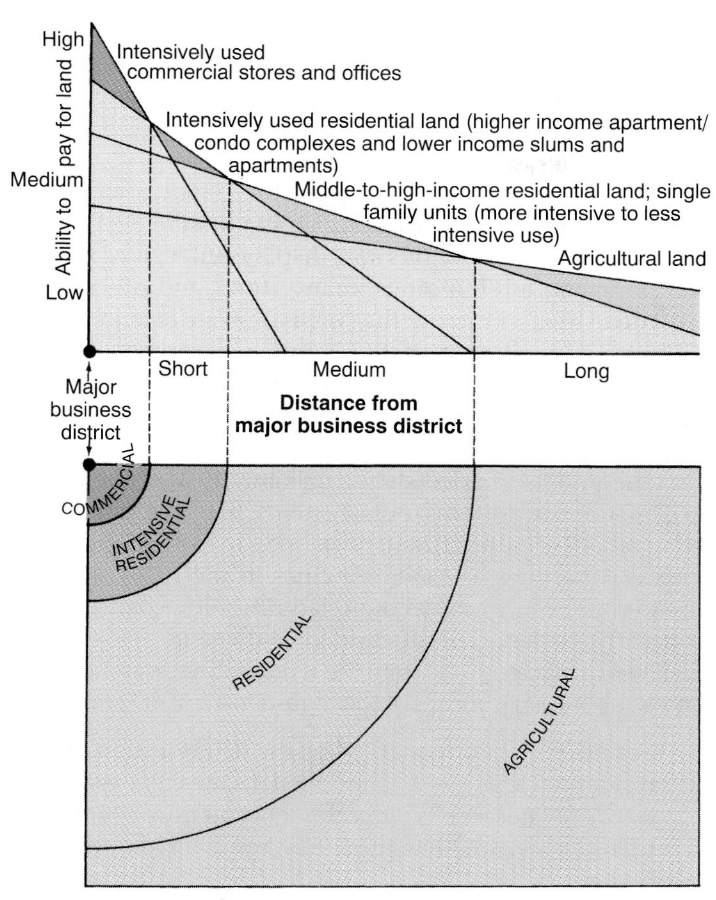

FIGURE 12.19 **Generalized urban land use pattern.** The model depicts the location of various land uses in an idealized urban area where the highest bidder gets the most accessible land.

city as a topographic map (Figure 12.21), with hills representing high valuations and depressions showing low prices, a series of peaks, ridges, and valleys would reflect the differentials in accessibility marked by the pattern of mass-transit lines, their intersections, and the unserved interstitial areas. Dominating these local variations, however, is an overall decline of valuations with increasing distance away from the *peak value intersection,* the most accessible and costly parcel of the central business district. As would be expected in a distance-decay pattern, the drop in valuation is precipitous within a short linear distance from that point, and then the valuation declines at a lesser rate to the margins of that built-up area.

With one important variation, the population density pattern of the central city shows a comparable distance-decay arrangement, as suggested by Figure 12.22. The exception is the tendency to form a hollow *at the center,* the CBD, which represents the inability of all but the most costly apartment houses to compete for space against alternative occupants desiring supremely accessible parcels. Yet accessibility is attractive to a number of residential users and brings its penalty in high land prices. The result is the high-density residential occupancy of parcels *near the center* of the city—by those who are too poor to afford a long-distance journey to work; are consigned by their poverty to the high-density, obsolescent slum tenements near the heart of the inner city; or are self-selected occupants of the high-density, high-rent apartments made necessary by the price of land. Other urbanites, if financially able, may opt to trade off higher commuting costs for lower-priced land and may reside on larger parcels away from high-accessibility, high-congestion

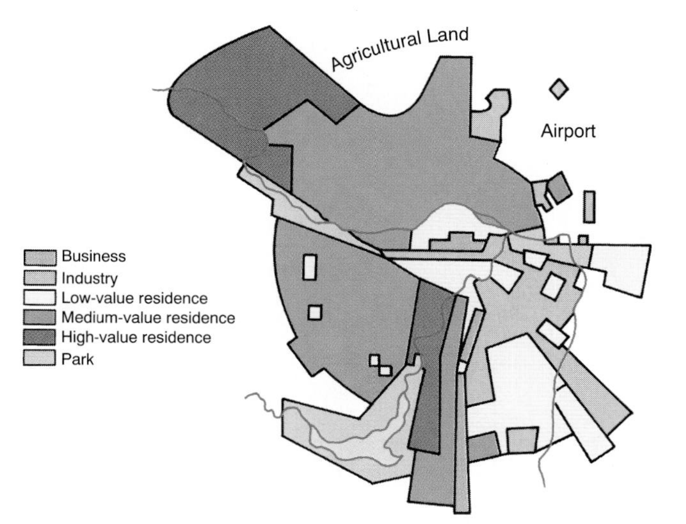

FIGURE 12.20 **The land use pattern in and around Calgary, Alberta, Canada, in 1981.** Physical and cultural barriers and the evolution of urban areas over time tend to result in a sectoral pattern of similar land uses. Calgary's central business district is the focus for many of the sectors.

Redrawn with permission from P. J. Smith, "Calgary: A Study in Urban Patterns," *Economic Geography,* Vol. 38, No. 4, p. 328. Copyright © 1962 Clark University, Worcester, MA.

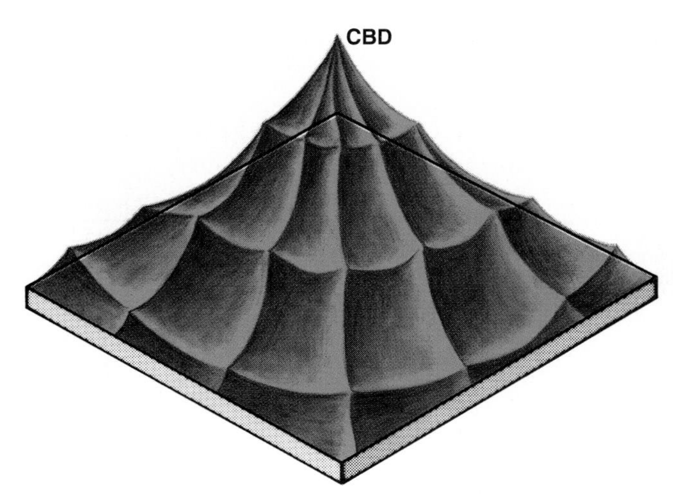

FIGURE 12.21 **Hypothetical pattern of land values.** The "topography" represented on this diagram shows the major land-value peak of the CBD. The ridges radiating from the CBD represent the higher land values along transit lines and major business thoroughfares. Regularly occurring minor peaks indicate commercial agglomerations at major transit intersections.

Redrawn with permission from B. J. L. Berry, *Commercial Structure and Commercial Blight,* Research Paper 85, Department of Geography Research Series, The University of Chicago, 1963.

locations. Residential density declines with increasing distance from the city center as this option is exercised.

As a city grows in population, the peak densities no longer increase, and the pattern of population distribution becomes more uniform. Secondary centers begin to compete with the CBD for customers and industry, and the residential areas become less associated with the city center and more dependent on high-speed transportation arteries. Peak densities in the inner city decline, and peripheral areas increase in population concentration.

The validity of these generalizations may be seen on Figure 12.23, a time series graph of population density patterns for Cleveland, Ohio, over a 50-year period. The peak density was 2.8 miles from the CBD in 1940, but by 1990 it was at 5.8 miles. As the city expanded, density decreased close to the center, but beyond 7.7 miles from the center, population density increased.

Models of Urban Land Use Structure

Generalized models of urban growth and land use patterns were proposed during the 1920s and 1930s describing the results of these controls on the observed structure of the central city. The models were simplified graphic summaries of United States mass-transit city growth processes as interpreted by different observers. Although the culture, society, economy, and technology they summarized have now been superseded, the physical patterns they explained or summarized still remain as vestiges and controls on the current landscape. A review of their propositions and conclusions still helps our understanding of the modern U.S. urban complex.

The common starting point of the classical models is the distinctive central business district found in every older central city. The core of this area displays intensive land use development: tall buildings, many stores and offices, and crowded streets. Framing the core is a fringe area of wholesaling activities, transportation terminals, warehouses, new car dealers, furniture stores, and even light industries. Just beyond the central business district frame is the beginning of residential land uses.

The land use models shown in Figure 12.24 differ in their explanation of patterns outside the CBD. The **concentric zone model** (Figure 12.24a), developed to explain the sociological patterning of American cities in the 1920s, sees the urban community as a set of nested rings. It recognizes four concentric circles of mostly residential diversity at increasing distance in all directions from the wholesaling, warehousing, and light industry border of the high-density CBD core:

- a zone of transition marked by the deterioration of old residential structures abandoned, as the city expanded, by the former wealthier occupants and now containing high-density, low-income slums, rooming houses, and perhaps ethnic ghettos;

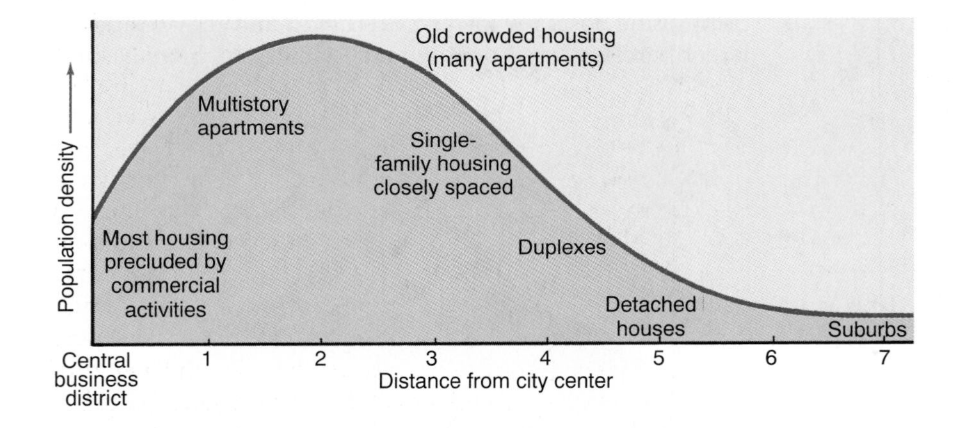

FIGURE 12.22 **A generalized population density curve.** As distance from the area of multistory apartment buildings increases, the population density declines.

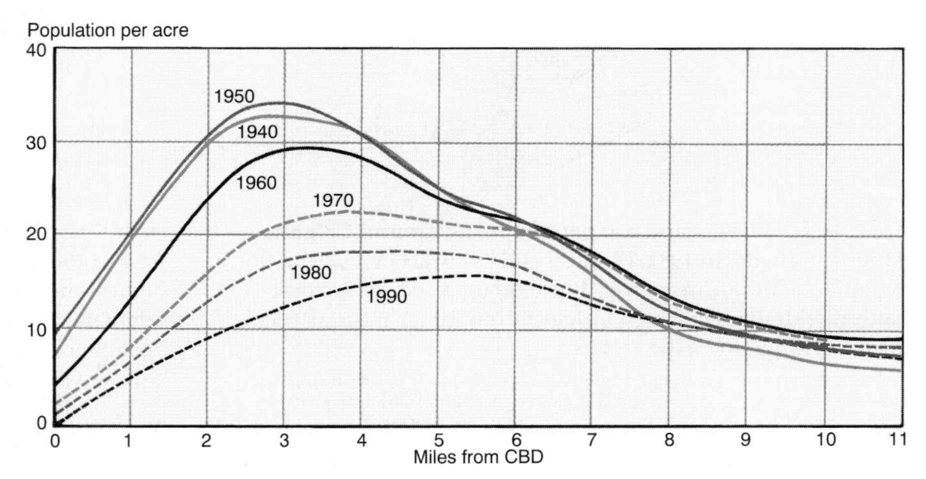

FIGURE 12.23 **Population density gradients for Cleveland, Ohio, 1940–1990.** The progressive depopulation of the central core and flattening of the density gradient over time to the margin of the city are clearly seen as Cleveland passed from mass transit to automobile domination.

Source: Anupa Mukhopadhyay and Ashok K. Dutt, "Population Density Gradient Changes for a Postindustrial City—Cleveland, Ohio 1940–1990," *GeoJournal* 34:517, no. 4, 1994. Redrawn by permission of Kluwer Academic Publishers and Ashok K. Dutt.

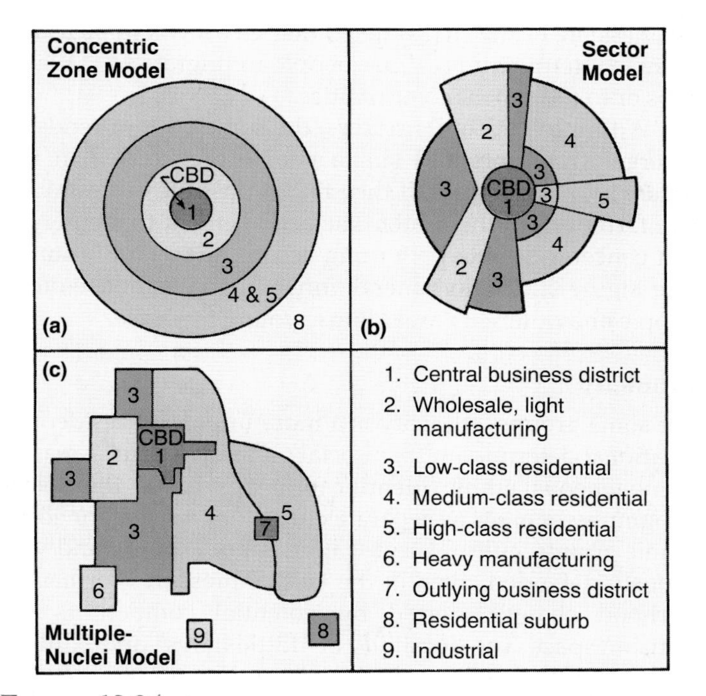

FIGURE 12.24 **Three classic models of the internal structure of cities.**

Redrawn from "The Nature of Cities" by C. D. Harris and E. L. Ullman in volume no. 242 of *The Annals of the American Academy of Political and Social Science.* Copyright © 1945 The American Academy of Political and Social Science, Philadelphia, PA. Used by permission of the publisher and authors.

- a zone of "independent working people's homes" occupied by industrial workers, perhaps second-generation Americans able to afford modest but older homes on small lots;
- a zone of better residences, single-family homes, or high-rent apartments occupied by those wealthy enough to exercise choice in housing location and to afford the longer, more costly journey to CBD employment;
- a commuters' zone of low-density, isolated residential suburbs, just beginning to emerge when this model was proposed.

The model is dynamic; it imagines the continuous expansion of inner zones at the expense of the next outer developed circles and suggests a ceaseless process of invasion and succession that yields a restructured land use pattern and population segregation by income level.

The **sector model** (Figure 12.24b) also concerns itself with patterns of housing and wealth, but it arrives at the conclusion that high-rent residential areas are dominant in city expansion and grow outward from the center of the city along major arterials. New housing for the wealthy, the model concludes, is added in an outward extension of existing high-rent axes as the city grows. Middle-income housing sectors lie adjacent to the high-rent areas, and low-income residents occupy the remaining sectors of growth. There tends to be a filtering-down process as older areas are abandoned by the outward movement of their original inhabitants, with the lowest-income populations (closest to the center of the city and farthest from the current location of the wealthy) becoming the dubious beneficiaries of the least desirable vacated areas. The accordance of the sector model with the actual pattern that developed in Calgary, Canada, is suggested in Figure 12.20 and for Chicago in Figure 12.26.

The concentric circle and sector models assume urban growth and development outward from a single central core, the site of original urban settlement that later developed into the central business district. These "single-node" models are countered by a **multiple-nuclei model** (Figure 12.24c), which maintains that large cities develop by peripheral spread from several nodes of growth, not just one. Individual nodes of special function—commercial, industrial, port, residential—are originally developed in response to the benefits accruing from the spatial association of like activities. Peripheral expansion of the separate nuclei eventually leads to coalescence and the meeting of incompatible land uses along the lines of juncture. The urban land use pattern, therefore, is not regularly structured from a single center in a sequence of circles or a series of sectors but based on separately expanding clusters of contrasting activities.

Social Areas of Cities

The larger and more economically and socially complex cities are, the stronger is the tendency for city residents to segregate themselves into groups based on *social status, family status,* and *ethnicity.* In a large metropolitan region, this territorial behavior may be a defense against the unknown or unwanted; a desire to be among similar kinds of people; a response to income constraints; or a result of social and institutional barriers. Most people feel more secure when they are near those with whom they can easily identify. In traditional societies, these groups are the families and tribes. In modern society, people group according to income or occupation (social status), stage in the life cycle (family status), and language or race (ethnic characteristics). Many of these groupings are fostered by the size and value of available housing. Land developers, especially in cities, produce homes of similar quality in specific areas. The sorting process, then, takes place in relation to existing land uses. Of course, as time elapses, there is a change in the quality of houses, land uses change, and new groups may replace old groups. In any case, neighborhoods of similar social characteristics evolve.

Social Status

The social status of an individual or family is determined by income, education, occupation, and home value. In the United States, high income, a college education, a professional or managerial position, and high home value constitute high status. High home value can mean an expensive rented apartment as well as a large house with extensive grounds.

A good housing indicator of social status is persons per room. A low number of persons per room tends to indicate high status. Low status characterizes people with low-income jobs living in low-value housing. There are many levels of status, and people tend to filter out into neighborhoods where most of the heads of households are of similar rank.

In most cities, people of similar social status are grouped in sectors which fan out from the innermost urban residential areas (Figure 12.25). The pattern in Chicago is illustrated in Figure 12.26. If the number of people within a given social group increases, they tend to move away from the central city along an arterial connecting them with the old neighborhood. Major transport routes leading to the city center are the usual migration routes out from the center. Social-status patterning agrees with the sector model.

Family Status

As the distance from the city center increases, the average age of the head of the household declines, or the size of the family increases, or both. Within a particular sector—say, that of high status—older people whose children do not live with them, or young professionals without families, tend to live close to the city center. Between these are the older families who lived at the outskirts of the city in an earlier period. The young families seek space for child rearing, and older people covet more the accessibility of the cultural and business life of the city. Where inner-city life is unpleasant, there is a tendency for older people to migrate to the suburbs or to retirement communities.

Within lower-status sectors, the same pattern tends to emerge. Transients and single people are housed in the inner city, and families, if they find it possible or desirable, live farther from the center. The arrangement that emerges is a concentric-circle patterning according to family status (see Figure 12.25). In general, inner-city areas house older people and outer-city areas house younger people.

Ethnicity

For some groups, ethnicity is a more important residential locational determinant than social or family status. Areas of homogeneous ethnic identification appear in the social geography of cities as separate clusters or nuclei. For some ethnic groups, cultural segregation is both sought and vigorously defended, even in the face of pressures for neighborhood change exerted by potential competitors for housing space. The durability of "Little Italys" and "Chinatowns," and of Polish, Greek, Armenian, and other ethnic neighborhoods in many American cities is evidence of the persistence of self-maintained segregation (see "Los Angeles in Flux").

Certain ethnic or racial groups, especially blacks, have been segregated in nuclear communities. Every city in the United States has one or more black areas which, in many respects, may be considered cities within a city. Figure 12.27 illustrates the concentration of blacks, Hispanics, and other ethnic groups in Los Angeles. The social and economic barriers to movement outside the area have always been high. In many American cities, the poorest residents are the blacks, who are often relegated to the lowest-quality housing in the least desirable areas of the city. Similar restrictions

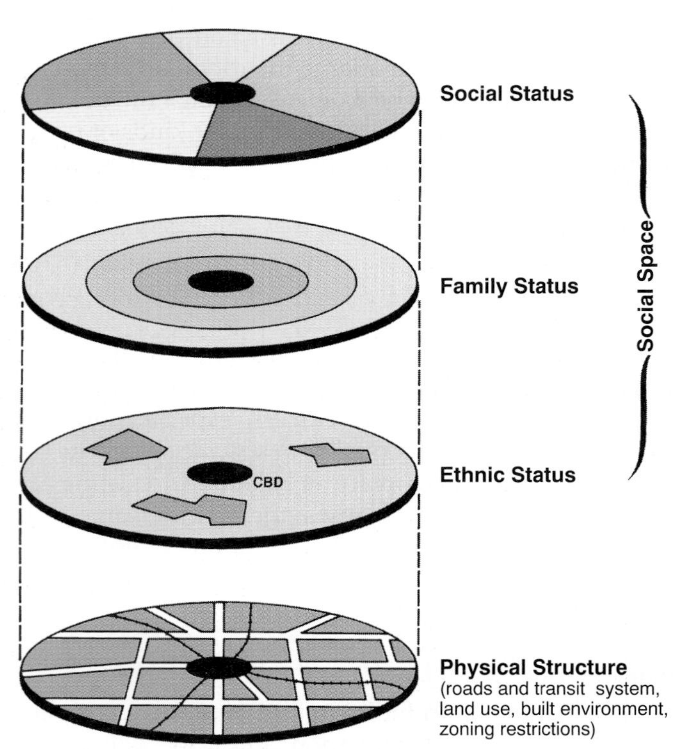

FIGURE 12.25 **The social geography of American and Canadian urban areas.**

Redrawn with permission from Robert A. Murdie, "*Factorial Ecology of Metropolitan Toronto*," Research Paper 116, Department of Geography Research Series, University of Chicago, 1969.

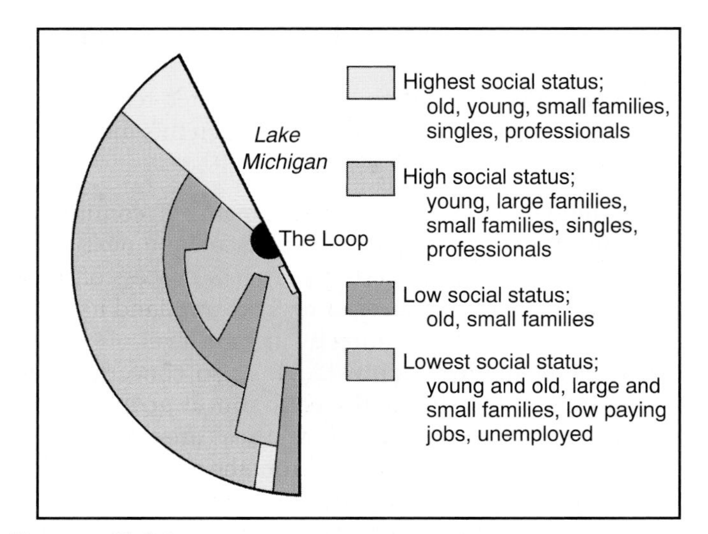

FIGURE 12.26 **A diagrammatic representation of the major social areas of the Chicago region.** The central business district is known as The Loop.

Redrawn with permission from Philip Rees, "*The Factorial Ecology of Metropolitan Chicago*," M. A. Thesis, University of Chicago, 1968.

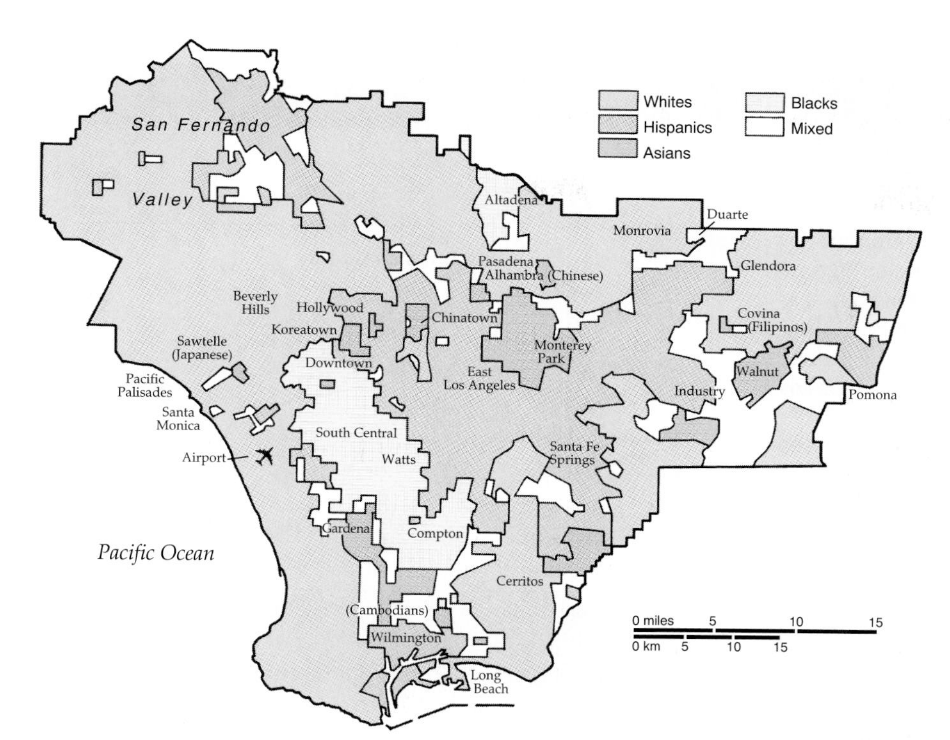

FIGURE 12.27 Ethnic patterns in Los Angeles, 1990.

Source: Data from Eugene Turner and James P. Allen, *An Atlas of Population Patterns in Metropolitan Los Angeles and Orange Counties, 1990,* Occasional Publication No. 8, Center for Geographical Studies, California State University, Northridge.

have been placed on Hispanics and other non-English speaking minorities. Note in Figure 12.27 the great diversity of ethnic groups in Los Angeles.

Of the three patterns, family status has undergone the most widespread change in recent years. Today, the suburbs house large numbers of singles and childless couples, as well as two-parent families. Areas near the central business district have become popular for young professionals. Much of this is a result of changes in family structure and the advent of large numbers of new jobs for professionals in the suburbs and the central business districts, but not in between. With more women in the workforce than ever before, and as a result of multiple-earner families, residential site selection has become a more complex undertaking.

Institutional Controls

Most governments have instituted innumerable laws to control all aspects of urban life, including the rules for using streets, the provision of sanitary services, and the use of land. In this section, we touch only upon land use.

Institutional or governmental controls have strongly influenced the land use arrangements and growth patterns of most cities in the world. Cities have adopted land use plans and enacted subdivision control regulations and zoning ordinances to realize those plans. They have adopted building, health, and safety codes to assure legally acceptable urban development and maintenance. All such controls are based on broad applications of the police powers of municipalities and their rights to assure public health, safety, and well-being even when private-property rights are infringed.

These nonmarket controls on land use are designed to minimize incompatibilities (residences adjacent to heavy industry, for example), provide for the creation in appropriate locations of public uses (the transportation system, waste disposal, government buildings, prisons, and parks) and private uses (colleges, shopping centers, and housing) needed for and conducive to a balanced, orderly community. In theory, such careful planning should preclude the emergence of slums, so often the result of undesirable adjacent uses, and should stabilize neighborhoods by reducing market-induced pressures for land use change.

Such controls in the United States, particularly zoning ordinances and subdivision regulations specifying large lot sizes for residential buildings and large house-floor areas, have been adopted as devices to exclude from upper-income areas, lower-income populations or those who would choose to build or occupy other forms of residences: apartments, special housing for the aged, and so forth. Bitter court battles have been waged, with mixed results, over "exclusionary" zoning practices that in the view of some, serve to separate rather than unify the total urban structure and to maintain or increase diseconomies of land use development. In addition, it is well known that some real estate agents "steer" people of certain racial and ethnic groups into neighborhoods that the agent thinks are appropriate.

In most of Asia there is no zoning and it is quite common to have small-scale industrial activities operating in residential areas. Even in Japan, a house may contain several people doing piecework for a local industry. In both Europe and Japan, neighborhoods have been built and rebuilt gradually over time, and it is quite common to have a wide variety of building types from various eras mixed together on the same street. In the United States and Canada, such mixing is much rarer and is often viewed as a temporary condition as areas are in transition toward total redevelopment. Perhaps the only exception to this in a large city is Houston, Texas, where there are no zoning regulations.

Los Angeles in Flux

Different ethnic and racial groups fit together in a large American metropolis much as separate pieces make up a giant jigsaw puzzle. Unlike an ordinary puzzle, however, the metropolis is often in a state of flux. Pieces change shape and move; more pieces may have to be squeezed within the borders of the puzzle. One of the urban areas that best exemplifies this situation is Los Angeles.

The population of the Los Angeles metropolitan area increased to 15.5 million in 1996 and continues to grow, though not nearly at the same rate as it did in the 1980s, when it increased by 26%. During that decade, millions of people came to Los Angeles from all parts of the United States and from Asia and Latin America.

In the early 1990s, a dramatic change in the industrial structure of the United States sent the economy of Los Angeles into a tailspin. The end of the Cold War led government leaders to reduce the United States budget for military equipment such as airplanes and missiles. The many defense-related industries in the Los Angeles area experienced massive layoffs. During the same period, the metropolitan area suffered a series of tragedies, including the Los Angeles riot (1992), devastating fires (1993), and the severe Northridge earthquake (1994).

As a result of these circumstances, an outpouring of people from Los Angeles occurred. Many people who left were able to sell their houses at a handsome profit and reinvest in areas where real estate prices were considerably lower. The out-migration has been primarily to other western states. Small towns in Idaho, Montana, Oregon, and Washington have been the recipients of Californians. In addition, the metropolitan areas of Seattle, Portland, Las Vegas, Salt Lake City, Phoenix, and Tucson have grown appreciably by the out-migration. In many cases, the Los Angelenos moved to get away from the troubles and congestion of a city that had grown faster than its service-sector was able to handle.

Despite the emigration, the population of the Los Angeles metropolitan area has continued to grow. Natural increase accounts for some of the growth, but immigrants from other countries streamed into California during the 1980s. That migration was mostly Hispanic and Asian. Mexicans represented the largest single immigrant group: some 860,000 arrived in Los Angeles during the 1980s. Large numbers also came from Central America, particularly El Salvador. As a result, Los Angeles has become one of the largest Spanish-speaking urban areas in the world. Hispanics constituted 38% of the population of Los Angeles County in 1990, up from 28% a decade earlier.

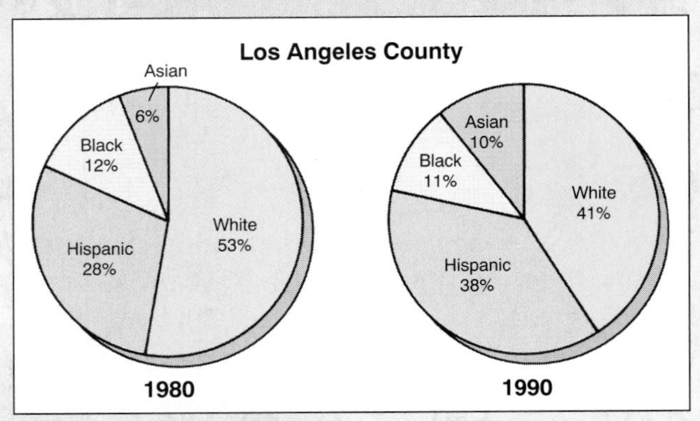

Source: Data from U.S. Bureau of the Census.

Asians also arrived in large numbers during the 1980s. They came chiefly from the Philippines, Vietnam, Korea, China (Hong Kong), and Japan. By 1990, the Asian population of Los Angeles County had risen to 10%, up from just 5.8% in 1980.

Today, one of every four people in Los Angeles County is foreign-born. Like migrants before them, these new arrivals have tended to cluster in certain areas. South Central Los Angeles, once predominantly black, is now 45% Hispanic. Many Vietnamese live or own businesses in the Westminster area (known as "Little Saigon"). Asians, especially Chinese, are now the largest group in both Alhambra and Monterey Park, communities that had negligible Asian populations only 20 years ago.

Suburbanization in the United States

The 20 years before World War II saw the creation of a technological, physical, and institutional structure that resulted, after that war, in a sudden and massive alteration of past urban forms. The improvement of the automobile increased its reliability, range, and convenience, freeing its owner from dependence on fixed-route public transit for access to home, work, or shopping. The new transport flexibility opened up vast new acreages of nonurban land to urban development. The acceptance of a maximum 40-hour work week guaranteed millions of Americans the time for a commuting journey not possible when the workdays of ten or more hours were common.

Finally, to stimulate the economy by the ripple effect associated with a potentially prosperous construction industry, the Federal Housing Administration was established as part of the New Deal programs under President Franklin D. Roosevelt. It guaranteed creditors the security of their mortgage loans, thus reducing down-payment requirements and lengthening mortgage repayment periods. In addition, veterans of World War II were granted generous terms on new housing.

Demands for housing, pent up by years of economic depression and wartime restrictions, were loosed in a flood after 1945, and a massive suburbanization of people and urban functions altered the existing pattern of urban America. Between 1950 and 1970, the two most prominent patterns of population growth were the metropolitanization of people and, within metropolitan areas, their suburbanization. During the 1960s, the interstate highway system was substantially completed, allowing sites 32 and 48 kilometers (20 and 30 mi) from workplaces to be within commuting distance from home places. Growth patterns for the Chicago area are shown in Figure 12.28. The high energy prices of the 1970s slowed the rush to the suburbs, but in the 1980s suburbanization again proceeded apace, although the tendency was as much for "filling-in" as it was for continued sprawl.

In the last 40 years, more and more industries of all types (first manufacturers, then service industries) moved to the suburbs. Their moves reflected the economies that began to accrue from modern single-story facilities with plenty of parking space for employees. Industries no longer needed to locate near rail facilities; freeways presented new opportunities for lower cost, more flexible truck transportation. Service industries took advantage of the large, well-educated labor force now living in the suburbs. In addition to the land that was developed around freeway intersections in the 1960s and along major commercial routeways intersecting freeways in the 1970s, was the residential land that filled the interstitial areas between major routeways. Now the major shopping districts of metropolitan areas, besides the central business district, are the shopping centers at freeway intersections, freeway frontage roads, and major connecting highways.

In time, in the United States, an established social and functional pattern of suburban land use emerged, giving evidence of a lower-density, more extensive repetition of the models of land use developed to describe the structure of the central city. Multiple nuclei of specialized land uses developed, expanded, and coalesced. Sectors of high-income residential use continued their outward extension beyond the central-city limits, usurping the most scenic and most desirable suburban areas and segregating them by price and zoning restrictions. As shown in Figure 12.29, middle-, lower-middle-, and lower-income groups found their own income-segregated portions of the fringe. Ethnic minorities were relegated to the inner city and some older industrial suburbs (see "The Dichotomous Metropolitan Region").

With increasing suburban sprawl and the rising costs implicit in the ever-greater spatial separation of the functional segments of the fringe, the limits of feasible expansion were reached, the supply of developable land was reduced (with corresponding increases in its price), and the intensity of land development grew. Changing lifestyles and cost constraints have resulted in a proliferation of suburban apartment complexes and gated communities and the disappearance of open land (see "The Gated Community").

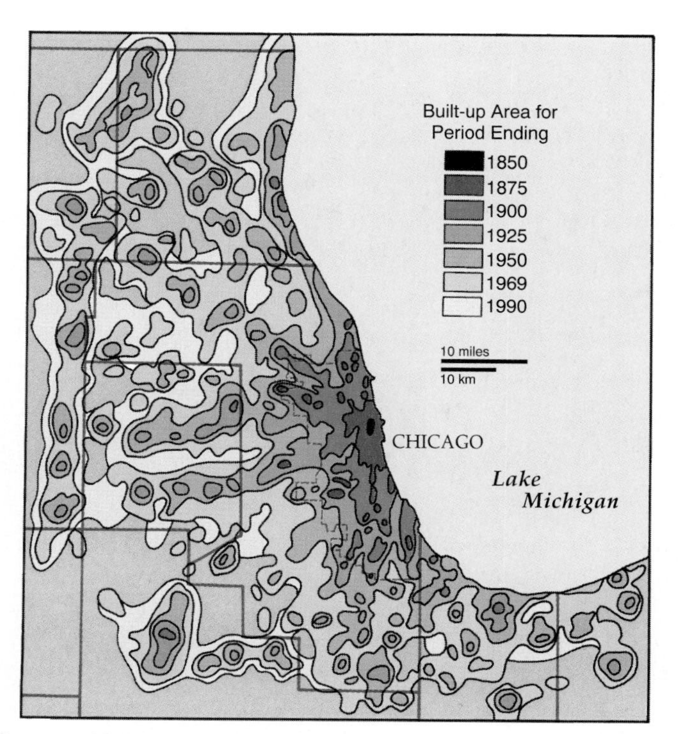

FIGURE 12.28 **Urban sprawl.** In Chicago, as in most large metropolitan areas, the size of the urbanized area has increased dramatically during the last 40 years.

Redrawn with permission from B. J. L. Berry, *Chicago: Transformation of an Urban System*, 1976.

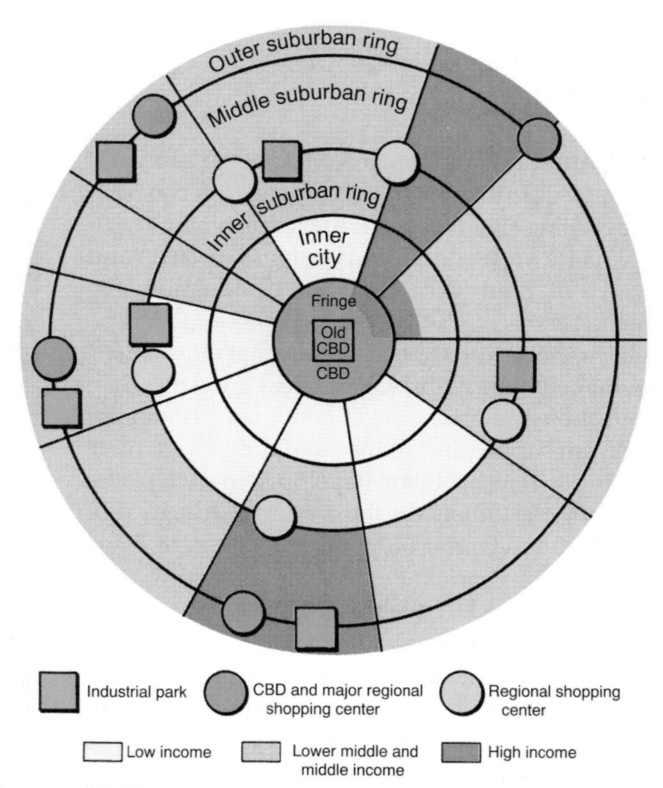

FIGURE 12.29 **A diagram of the present-day United States metropolitan area.** Note that aspects of the concentric, sector, and multiple-nuclei patterns are evident and carried out into the suburban fringe.

Figure 4.10, page 121 from *The North American City*, 4th Ed. by Maurice Yeates. Copyright © 1990 by Harper & Row, Publishers, Inc. Reprinted by permission of Addison-Wesley Educational Publishers Inc.

THE DICHOTOMOUS METROPOLITAN REGION

In the United States, the suburbanization movement that began at the end of World War II is still going strong today. As a result, one of the most important contrasts in the society is the difference between city life and suburban life. That difference is as much dependent on social, life cycle, and ethnic considerations as it is on industrial, commercial, and transportation differences. Without plan, the United States has developed dichotomous metropolitan areas. No other part of the world can claim this distinction. Some of the differences that have developed between city and suburb are noted in the following list.

Compared to the suburb, the city has:

- a better developed public transportation system, but a less developed system of highways;
- more people dependent on public transportation;
- a poorer population containing a large segment of the population that has not been drawn into its formal workforce;
- a large population that is dependent on municipal services and constitutes a drain on city resources;
- a slower growing population (a decline in many cases);
- more single people, fewer children, and more older people;
- more ethnic diversity with greater numbers of blacks, Hispanics, and immigrants;
- fewer women in the workforce;
- a less well-educated workforce, fewer jobs suitable for the inexperienced and poorly educated;
- a declining number of jobs in manufacturing and retailing, and only a modest development of industrial parks;
- more people dependent on temporary, part-time, and what is termed *casual* employment, that is, nonunion jobs where work is done in the home, on the street, and where no benefits are paid by employers;
- a higher proportion of in-commuters, and higher levels of wages paid, especially to the in-commuters;
- a more concentrated economic structure, with gains in employment in the low-wage jobs of the finance and insurance industries at the expense of all other categories of employment.

The suburbs have outgrown their former role as bedroom communities and have emerged as a chain of independent, multinucleated urban developments. Together, they are largely self-sufficient, divorced from the central city. Many, perhaps most, suburbanites have no connection with the core city, feel no ties to it, and satisfy almost all of their needs within the peripheral zone.

The maturation and coalescence of urban land uses have resulted in the emergence of coherent metropolitan-area cities that are suburban in traditional name only, containing business districts and the mix of land uses that the designation *city* implies.

In recent years, suburban areas have expanded to the point where metropolitan areas are coalescing. Within these suburban areas new major centers of growth are appearing. Office buildings and huge shopping centers are being developed in such places as Oak Brook, Illinois (in the western Chicago suburbs), Meadowlands, New Jersey (west of New York), King of Prussia, Pennsylvania (northwest of Philadelphia), and Costa Mesa, California (south of Los Angeles) (Figure 12.30). In fact, in the 1980s more office space was created in the suburbs than in the central cities of America. The Boston-to-Washington corridor, often called a **megalopolis,** is now a continuously built-up region with many new centers that compete with the business districts of Boston, Providence, New York, Philadelphia, Baltimore, and Washington. One summary of the new pattern is shown in Figure 12.31. A further analysis of the northeastern U.S. urban corridor appears in "Megalopolis" in Chapter 13.

FIGURE 12.30 Many business areas, such as Costa Mesa, California, south of Los Angeles, have become large centers comparable in size to the central business district of a moderate size city. These high-density, automobile-oriented, suburban regions exist outside the largest central cities.
©Tony Freeman/PhotoEdit.

The Gated Community

Associated with high-income suburbanites is a new type of community that appeared in America in the late 1970s, blossomed in the 1980s, and now flourishes in urban areas of the South and West. One might identify as its ancestors the walled towns of medieval Europe, for the distinguishing characteristic of the **gated community** is the residents' desire for security. A gated community is a fenced or walled residential area with check points staffed by security guards where access is limited to designated individuals.

From Miami to Houston to Los Angeles, people have fled the suburbs of the 1950s to reside in master-planned, upscale developments surrounded by walls or fences. Entry to these communities within communities is restricted; gates are manned by guards or are accessible only by computer key-card or telephone. Some of the communities hire private security forces to patrol the streets. Surveillance systems monitor common recreational areas, such as community swimming pools, tennis courts, and health clubs. Houses are commonly equipped with security systems. Troubled by the high crime rates, drug abuse, gangs, and drive-by shootings that characterize many urban areas, people seek safety within their walled enclaves.

The typical gated community has been built by a developer following a master plan. To preserve the upscale nature of the development and protect land values, self-governing community associations enact conditions and restrictions. Pervasive and detailed, they specify such things as the size, construction, and color of walls and fences, the size and permitted uses of rear and side yards, the design of lights and mailboxes. Some go so far as to tell residents what trees they can plant, what pets they may raise, and where they may park their boats or recreational vehicles.

© Richard B. Levine.

Central City Change

While the process of suburban spread continued, many central cities of the United States, especially in the Northeast and the Midwest, lost population. In the 1970s, there was a decided shift of population to cities in what are called the Sunbelt states. This trend was partly a result of the lower living costs (especially fuel costs) in the South and West, but it was chiefly a function of manufacturers seeking nonunionized workers or new plants unencumbered by high social and transportation costs. Perhaps as important as that trend was the movement of people out of metropolitan areas to nearby nonmetropolitan districts. In a sense, this latter trend represented a further sprawl of the metropolitan districts. This trend, however, came to an end in the 1980s, while suburbanization has continued.

The economic base and the financial stability of the central city have been grievously damaged by the process of suburbanization. In earlier periods of growth, as new settlement areas developed beyond the political margins of the

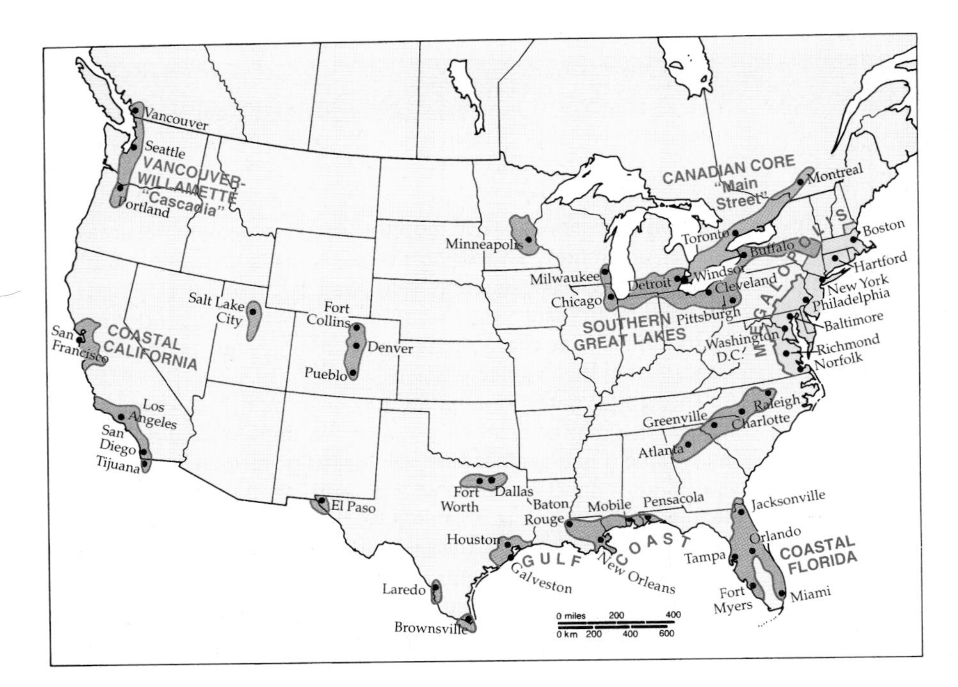

FIGURE 12.31 **The northeastern U.S. Boston-to-Washington megalopolis is the original and largest megalopolis.** The map depicts the various candidates for megalopolis status in the United States and Canada. These are areas where two or more metropolitan areas have coalesced to form a nearly continuously built-up region.

city, annexation absorbed new growth within the corporate boundaries of the expanding older city. The additional tax base and employment centers became part of the municipal whole. But in states that recognized the right of separate incorporation of the new growth area—particularly in the eastern part of the United States—the ability of the city to continue to expand was restricted. Lower suburban residential densities made septic tanks and private wells adequate substitutes for expensive city-sponsored sewers and water mains; lower structural densities and lower crime rates meant a less felt need for high-efficiency city fire and police protection. Where possible, suburbanites opted for a separation from the central city and for aloofness from the costs, deterioration, and adversities associated with it. Their homes, jobs, shopping, schools, and recreation all existed outside the confines of the city from which they had divorced themselves.

The redistribution of population caused by suburbanization resulted not only in the spatial but also in the political segregation of social groups of the metropolitan area. The upwardly mobile resident of the city—younger, wealthier, and better educated—took advantage of the automobile and the freeway to leave the central city. The poorer, older, least advantaged urbanites were left behind. The central cities and suburbs are becoming increasingly differentiated. Large areas within the cities now contain only the poor and minority groups, a population barely able to pay the rising costs of social services that their numbers, neighborhoods, and condition require.

The services needed to support the poor include welfare payments, social workers, extra police and fire protection, health delivery systems, and subsidized housing. Central cities, by themselves, are unable to raise the taxes needed to support such an array and intensity of social services since they have lost the tax bases represented by suburbanized commerce, industry, and upper-income residential uses. Lost, too, are the job opportunities that were formerly a part of the central-city structure. Increasingly, the poor and minorities are trapped in a central city without the possibility of nearby employment and are isolated by distance, immobility, and unfamiliarity from the few remaining low-skill jobs, which are now largely in the suburbs. This unfortunate circumstance is often called a *spatial mismatch.*

Abandonment of the central city by people and functions has nearly destroyed the traditional active, open auction of urban land, which led to the replacement of obsolescent uses and inefficient structures in a continuing process of urban modernization (Figure 12.32). In the vacuum left by the departure of private investors, the federal government, particularly since the landmark Housing Act of 1949, has initiated urban renewal programs with or without provisions for a partnership with private housing and redevelopment investment. Under a wide array of programs instituted and funded since the late 1940s, slum areas have been cleared, public housing has been built, cleared land has been conveyed at subsidized cost to private developers for middle-income housing construction, cultural complexes and industrial parks have been created, and city centers have been reconstructed (Figure 12.33).

With the continuing erosion of the urban economic base and the disadvantageous restructuring of the central-city population base, the hard-fought governmental battle to maintain or revive the central city is frequently judged to be a losing one. In recent years, the central city has been the destination of thousands of homeless people (see "The Homeless"). Many live in public parks, in doorways, by street level warm air exhausts of subway trains, and in subway stations. The urban economies, with their high land values, limited job opportunities for the unskilled, and inadequate resources for social services, have relegated many to a homeless existence.

There are two trends, however, that do give a new flavor to the central city. One is gentrification and the other is the vitality of the CBD. **Gentrification** is the movement of middle-class people to deteriorated portions of the inner city (Figure 12.34). The reasons for the upsurge of interest in

FIGURE 12.32 A derelict slum in the Maxwell Street area of Chicago.
© Tom Stack/Tom Stack & Associates.

urban housing reflect, to some extent, the recent changes in American family structure and in the employment structure of central business districts. Just 20 years ago, the suburbs, with their green spaces, were a powerful attraction for young married couples who considered single-family houses with ample yards as ideal places to raise children.

Now that the proportion of single people (whether never or formerly married) and childless couples in the American population has increased, the attraction of nearby jobs and social and recreational activities has become an important residential location factor. Many of the new jobs in the central business district are designed for professionals in the fields of banking, insurance, and financial services. These jobs are replacing the manufacturing jobs of an earlier period. The gentrification process has been a positive force in the renewal of some of the depressed housing areas in neighborhoods surrounding the central business district.

New office buildings are springing up in most large central cities (Figure 12.35). These represent a shift in employment emphasis from manufacturing to more service-oriented activities such as headquarters for industries, financial services, and business services in general. The number of jobs in the central city has not increased, but the

FIGURE 12.33 **Many elaborate—and massive—public housing projects have been failures.** Chicago's Robert Taylor Homes, shown here—with 4400 apartments in 28 16-story buildings, the largest public housing unit in the world—have been partly abandoned, a victim of soaring vandalism and crime rates. The first of several of its buildings to be razed was demolished in May, 1997; sixteen are slated for demolition by 2006. The growing awareness that public high-rise developments intended to revive the central city do not meet the housing and social needs of their inhabitants led to the razing of nearly 100,000 of the more than 1.3 million public housing units in cities around the country during the 1990s, many replaced by low-rise apartment or mixed-use development.
©Mark Pokempner/Tony Stone Images.

Geography *and* Public Policy

The Homeless

In the 1980s, the number of homeless people in the United States rose dramatically. Now every large city is apt to have hundreds or even thousands of people who lack homes of their own. One sees them pushing shopping carts containing their earthly goods, lined up at soup kitchens or rescue missions, and sleeping in parks or doorways. Reliable estimates of their numbers simply do not exist: official counts place the number of homeless Americans anywhere between 600,000 and 3 million.

The existence of the homeless raises a multitude of questions—the answers, however, are yet to be agreed upon by Americans. Who are the homeless, and why did their numbers increase? Who should be responsible for coping with the problems they raise? Are there ways to eliminate homelessness?

Some people believe the homeless are primarily the impoverished victims of a rich and uncaring society. They view them as people like us, but ones who have had a bad break and been forced from their homes by job loss, divorce, domestic violence, or incapacitating illness. They point to the increasing numbers of families and children numbered among the homeless. The women and children are less visible than the "loners" (primarily men) because they tend to live in cars, emergency shelters, or doubled up in substandard buildings. The advocates of the homeless argue that government policies of the 1980s are partly to blame for this phenomenon because they led to a dire shortage of affordable housing. The federal budget for building low-income and subsidized housing was more than $30 billion in 1980. By the end of the decade it had shrunk by three-quarters, to $7.5 billion. During the same period, city governments pursued policies aimed at the destruction of low-income housing, especially single-room occupancy hotels, and fostered gentrification. In addition, many states reduced funding for mental hospitals, casting institutionalized people out onto the streets.

At the other end of the spectrum are those who see the homeless chiefly as people responsible for their own plight, not unlike the skid row denizens of former years. In the words of one commentator, they are "deranged, pathological predators who spoil neighborhoods, terrorize passersby, and threaten the commonweal." They point to studies showing that nationally between 66% and 85% of all homeless suffer from alcoholism, drug abuse, or mental illness, and argue that people are responsible for the alcohol and drugs they ingest; they are not helpless victims of a disease.

A homeless man finds shelter on a bench near the White House in Washington, D.C.
UPI/Bettmann.

Communities have tried a number of strategies to cope with their homeless populations. Some set up temporary shelters, especially in cold weather; some subsidize permanent housing and/or group homes. They encourage private, nonprofit groups to establish soup kitchens and food banks. Others attempt to drive the homeless out of town, or at least to parts of town where they will not be visible. They forbid loitering in city parks or on beaches after midnight, install sleep-proof seats on park benches and bus stations, and outlaw aggressive panhandling.

Neither type of solution appeals to those who believe that homelessness is more than simply a lack of shelter, that it is a matter of a mostly disturbed population with severe problems that requires help getting off the streets and into treatment. What the homeless need, they say, is a "continuum of care"—an entire range of services that includes education; treatment for drug and alcohol abuse and mental illness; and job training.

Questions to Consider

1. What is the nature of the homeless problem in the community where you live or with which you are most familiar?
2. Where should responsibility for the homeless lie: at the federal, state, or local governmental level? Or is it best left to private groups such as churches and charities? Why?
3. Some people argue that giving money, food, or housing but no therapy to street people makes one an "enabler" or accomplice of addicts. Do you agree? Why or why not?
4. One columnist has proposed quarantining male street people on military bases and compelling them to accept medical treatment. Those who resist would be charged with crimes of violence and turned over to the criminal justice system. Do you believe the homeless should be forced into treatment programs or institutionalized against their will? If so, under what conditions?

FIGURE 12.34 Gentrified housing in the Georgetown section of Washington, D.C. Gentrification is especially noticeable in the major urban centers of the eastern United States, from Boston south along the Atlantic Coast to Charleston, South Carolina, and Savannah, Georgia.
©Carl Purcell.

FIGURE 12.35 An example of a dynamic central business district: the Dallas skyline.
©Susan Van Etten/PhotoEdit.

change favors the professionals over blue-collar workers. Many of the workers in these office buildings now live within the gentrified neighborhoods of the central city, displacing the poor and forcing them to find higher priced dwelling units elsewhere.

World Urban Diversity

The city, as Figure 12.3 reminds us, is a global phenomenon. It is also a regional and cultural variable. The descriptions and models we have used to study the functions, land use arrangements, suburbanization trends, and other aspects of the United States city would not, in all—or even many—instances, help us understand the structures and patterns of cities in other parts of the world. Those cities have been created under different historical, cultural, and technological circumstances. They have developed different functional and structural patterns, some so radically different from our United States model that we would find them unfamiliar and uncharted landscapes indeed. The city is universal; its characteristics are cultural and regional.

The United States and Canadian City

Even within the seemingly homogeneous culture realm of the United States and Canada, the city shows subtle, but significant, differences—not only between older eastern and newer western U.S. cities, but between cities of Canada and those of the United States. Although the urban expression is similar in the two countries, it is not identical. The Canadian city, for example, is more compact than its U.S. counterpart of equal population size, with a higher density of buildings and people and a lesser degree of suburbanization of populations and functions (Figure 12.36).

Space-saving and multiple-family housing units are more the rule in Canada, so a similar population size is housed on a smaller land area with much higher densities,

FIGURE 12.36 A bird's-eye view of Toronto, showing clusters of tall buildings in the central business district and at important intersections.

on average, within the central area of cities. The Canadian city is better served by and more dependent on mass transportation than is the U.S. city. Since Canadian metropolitan areas have only one-quarter the number of miles of expressway lanes per capita as U.S. metropolitan areas—and at least as much resistance to constructing more—suburbanization of peoples and functions is less extensive north of the border than south.

In social as well as physical structure, Canadian-United States contrasts are apparent. While cities in both countries are ethnically diverse (Canadian communities, in fact, have a higher proportion of foreign born), U.S. central cities exhibit far greater internal distinctions in race, income, and social status, and more pronounced contrasts between central-city and suburban residents. That is, there has been much less "flight to the suburbs" by middle-income Canadians. As a result, the Canadian city shows greater social stability, higher per capita average income, more retention of shopping facilities, and more employment opportunities and urban amenities than its U.S. central-city counterpart. In particular, it does not have the rivalry from well-defined competitive "outer cities" of suburbia that so spread and fragment United States metropolitan complexes.

The West European City

If such significant urban differences are found even within the tightly knit North American region, we can only expect still greater divergences from the U.S. model at greater linear and cultural distance, and in countries with long urban traditions and mature cities of their own. The political history of France, for example, has given Paris an overwhelmingly primate position in its system of cities. Political, economic, and colonial history has done the same for London in the United Kingdom. On the other hand, Germany and Italy came late to nationhood, and no overwhelmingly dominant cities developed in their systems.

Nonetheless, a generally common heritage of medieval origins, Renaissance restructurings, and industrial period extensions has given the cities of Western Europe features distinctly different from those of cities in other regions founded and settled by European immigrants. Despite wartime destructions and postwar redevelopments, many still bear the impress of past occupants and technologies, even back to Roman times in some cases. An irregular system of narrow streets may be retained from the random street pattern developed in medieval times of pedestrian and pack-animal movement. Main streets radiating from the city center and cut by circumferential "ring roads" tell us the location of high roads leading into town through the gates in city walls now gone and replaced by circular boulevards. Broad thoroughfares, public parks, and plazas mark Renaissance ideals of city beautification and the aesthetic need felt for processional avenues and promenades.

Although each is unique historically and culturally, West European cities as a group share certain common features that set them off from the United States model, though they are less far removed from the Canadian norm. Cities of Western Europe have, for example, a much more compact form and occupy less total area than American cities of comparable population; most of their residents are apartment dwellers. Residential streets of the older sections tend to be narrow, and front, side, or rear yards or gardens are rare.

European cities were developed for pedestrians and still retain the compactness appropriate to walking distances. The sprawl of American peripheral or suburban zones is generally absent. At the same time, compactness and high density do not mean skyscraper skylines. Much of urban Europe predates the steel-frame building and the elevator. City skylines tend to be low, three to five stories in height, sometimes (as in central Paris) held down by building ordinance or by prohibitions on private structures exceeding the height of a major public building, often the central cathedral (Figure 12.37).

Compactness, high densities, and apartment dwelling encouraged the development and continued importance of public transportation, including well-developed subway systems. The private automobile has become much more common of late, though most central-city areas have not yet been significantly restructured with wider streets and parking facilities to accommodate it. The automobile is not the universal need in Europe that it has become in American cities. Home and work are generally more closely spaced in Europe—often within walking or bicycling distance—while most sections of towns have first-floor retail and business establishments (below upper-story apartments), bringing both places of employment and retail shops within convenient distance of residences.

A generalized model of the social geography of the West European city has been proposed (Figure 12.38). Its exact counterpart can be found nowhere, but many of its general features are part of the spatial social structure of most major European cities. In the historic core, now increasingly gentrified, residential units for the middle class, the self-employed, and the older generation of skilled artisans share limited space with preserved historic buildings, monuments, and tourist attractions.

The old city fortifications may mark the boundary between the core and the surrounding transitional zone of substandard housing, 19th-century industry, and recent immigrants. The waterfront has similar older industry; newer plants are found on the periphery. Public housing and some immigrant concentrations may be near that newer industry, while other urban socioeconomic groups aggregate themselves in distinctive social areas within the body of the city.

The West European city is not characterized by inner-city deterioration and out-migration. Its core areas tend to be stable in population and attract, rather than repel, the successful middle class and the upwardly mobile, conditions far different from comparable sections of older American central cities.

FIGURE 12.37 Even in their central areas, many European urban centers show a low profile, like that of Paris seen here from the Eiffel Tower. Although taller buildings—20, 30, and even 50 or more stories in height—have become more common in major urban areas since World War II, they are not the universal mark of the central business districts they have become in the United States, nor the generally welcomed symbols of progress and pride.
©IPA / The Image Works.

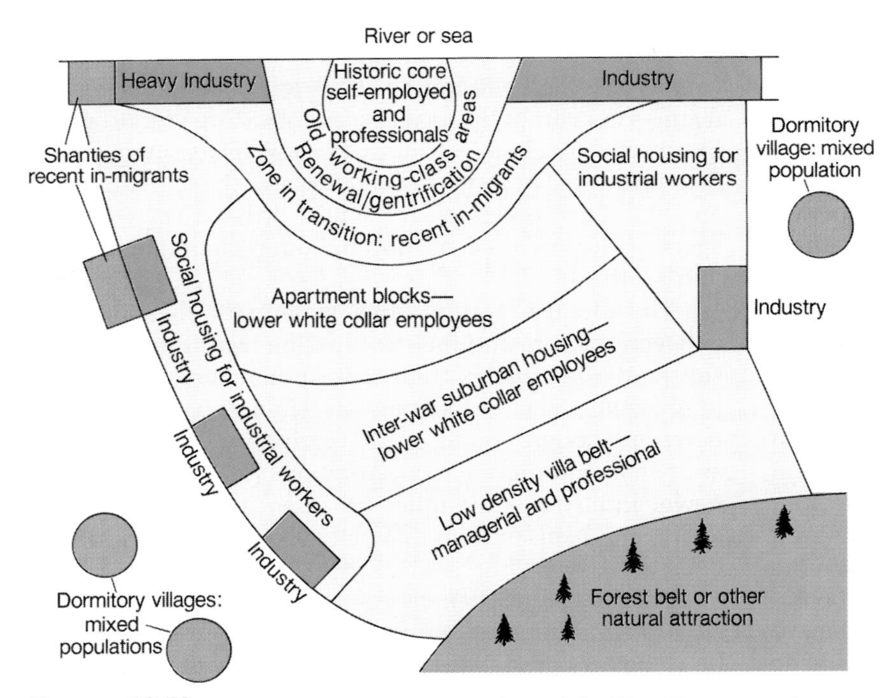

FIGURE 12.38 **A diagrammatic representation of the West European city.**
Redrawn from Paul White, *The West European City: A Social Geography*. Copyright © 1984 Longman Group UK Limited. Used by permission.

The East European City

Cities of Eastern Europe and the former European republics of the Soviet Union make up a separate urban class—the East European city. It is an urban form that shares many of the traditions and practices of West European cities, but it differs from them in the centrally administered planning principles that were, in the communist period (1945–1990), designed to shape and control both new and older settlements. The particular concerns were, first, limitation on size of cities to avoid supercity growth and metropolitan sprawl; second, assurance of an internal structure of neighborhood equality and self-sufficiency; and third, strict land use segregation. The East European city has fully achieved none of these objectives, but by attempting them it has emerged as a distinctive urban form.

The city is compact, with relatively high building and population densities reflecting the nearly universal apartment dwelling, and with a sharp break between urban and rural

The metropolitan area of **Shanghai** has a population of about 14 million. The city is extremely crowded, not unlike the teeming Lower East Side of New York in the 1920s. Most people walk or bicycle to work daily. The city depends on an overworked bus system to assist in moving millions of people to and from work daily. Work hours are staggered in order to alleviate the pressure on the transport system, but the streets are crowded with commuters and shoppers from dawn to after dark six days a week.

Shanghai may be the most crowded city in the world. The average population density is approximately 100,000 people per square mile, about three times more crowded than Tokyo and five times more crowded than Paris. The average living space is about 2 meters by 2 meters (6 ft by 6 ft) per person. This means that 10 or 11 people reside in an average size Shanghai apartment of 33 square meters (350 sq ft). The vast majority of people live in small two- and three-story houses that were converted into apartments long ago.

Crowding can be further understood by comparing the number of people per 9 square meters (100 sq ft) of living space (about the size of a typical room) by country. In the United States, the average is 0.5 persons. In France, the figure is 1.3. For Shanghai, a conservative estimate would be 3.0. Most amazing is that there are few, if any, squatter settlements. Many people live in temporary quarters, such as hallways and attics, but the use of land, which is publicly held, is tightly controlled by a government that discourages unplanned growth.

One major government strategy to alleviate the housing problem was to develop satellite towns around Shanghai. Twelve such satellites containing 300,000 jobs ring the city to a distance of 24 kilometers (15 mi), but until recently, only 45,000 people lived in them. Many are reluctant to leave cosmopolitan Shanghai for the sterile suburbs. As a result, the already overburdened bus system must deliver thousands of reverse commuters to jobs in the satellites. Commuting time for many of those workers is between two and three hours daily.

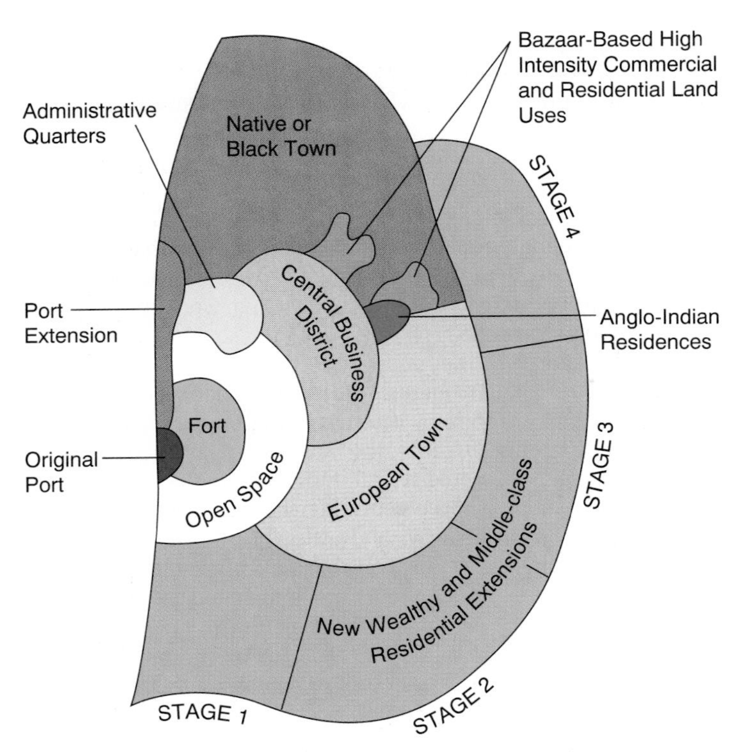

FIGURE 12.43 **A diagrammatic representation of the colonial-based South Asian city.**

Figure 9.7 on page 338 from Ashok K. Dutt, "Cities of South Asia," in *Cities of the World: World Regional Urban Development* by Stanley D. Brunn and Jack F. Williams. Copyright © 1983 by Harper Collins Publishers. Reprinted by permission of Addison-Wesley Educational Publishers, Inc.

SUMMARY

The urban area is an essential activity center that marks any society advanced beyond the level of subsistence. Although cities are among the oldest forms of civilization, only in this century have urban areas become home to the majority of people in industrialized countries and the commercial crossroads for uncounted millions in the developing world. At the global scale, four major world urban regions have emerged: Western Europe, South Asia, East Asia, and North America.

All urban areas growing beyond their village origins take on functions uniting them to the countryside and to a larger system of settlements. As they grow, they become functionally complex. Their economic base, composed of both *basic* and *nonbasic* activities, may become diverse. Basic activities represent the functions performed for the larger economy and urban system, while nonbasic activities satisfy the needs of the urban residents themselves. Functional classifications distinguish the economic roles of urban centers. Hierarchies of cities, central place theory, and network cities help us realize how urban areas are interrelated.

As North American urban centers expanded in population size and diversity, they developed structured land use and social patterns based on market allocations of urban space, channelization of traffic, and socioeconomic aggregation. Social areas can be identified based on social status, family status, and ethnicity. Since 1945, land use patterns and social areas have been modified by the suburbanization of people and functions that has led to the creation of new and complex outer urban areas and to the deterioration of the older central city itself.

Urbanization is a global phenomenon, and the North American models of city systems, land use, and social area patterns are not necessarily or usually applicable to other cultural contexts. In Europe, stringent land use regulations have brought about a compact urban form ringed by greenbelts. Although changing rapidly, the Eastern European urban area still shows a pattern of density and land use reflecting recent communist principles of city structure. Models descriptive of cities in the developing world do little to convey the fact that those settlements are currently growing faster than it is possible to provide employment, housing, safe water, sanitation, and other minimally essential services and facilities.

KEY WORDS

FOR REVIEW AND CONSIDERATION

1. Consider the city or town in which you live, attend school, or with which you are most familiar. In a brief paragraph, discuss that community's site and situation; point out the connection, if any, between its site and situation and the basic functions that it performed earlier or now performs.
2. Describe the *multiplier effect* as it relates to the population growth of urban units.
3. What area does a *central place* serve, and what kinds of functions does it perform? If an urban system were composed solely of central places, what summary statements could you make about the spatial distribution and the urban size hierarchy of that system?
4. Is there a hierarchy of retailing activities in the community with which you are most familiar? Of how many

and of what kinds of levels is that hierarchy composed? What localizing forces affect the distributional pattern of retailing within that community?
5. In what ways do social status, family status, and ethnicity affect the residential choices of households? What expected distributional patterns of urban social areas are associated with each? Does the social geography of your community conform to the predicted pattern?
6. How has suburbanization damaged the economic base and the financial stability of the central city?
7. In what ways does the Canadian city differ from the pattern of its United States counterpart?
8. Why are metropolitan areas in developing countries expected to grow larger than many Western metropolises by the year 2000? What do you expect the population density profile for Mexico City to look like in the year 2000?
9. What are *primate cities?* Why are primate cities in the developing countries overburdened? What can be done to alleviate the difficulties?
10. What are the significant differences in the generalized pattern of land uses of North American, West European, East European, Latin American, Asian, and African cities?

Selected References

Bourne, Larry S., and David F. Ley, eds. *The Changing Social Geography of Canadian Cities.* Montreal: McGill-Queen's University Press, 1993.

Brunn, Stanley D., and Jack F. Williams. *Cities of the World: World Regional Development.* 2d ed. New York: Harper & Row, 1993.

Carter, Harold. *The Study of Urban Geography,* 4th ed., New York: John Wiley & Sons, Inc., 1996.

Clark, David. *Urban World/Global City.* New York: Routledge, 1996.

Dutt, Ashok K., et al., eds. *The Asian City: Processes of Development, Characteristic and Planning.* Dordrecht / Boston / London: Kluwer Academic Publishers, 1994.

Ford, Larry R. *Cities and Buildings: Skyscrapers, Skid Rows, and Suburbs.* Baltimore: Johns Hopkins University Press, 1994.

Garreau, Joel. *Edge City: Life on the Frontier.* New York: Doubleday, 1991.

Goodwin, Mark. *Reshaping the City.* London: Edward Arnold, 1998.

Graham, Stephen, and Simon Marvin. *Telecommunications and the City.* New York: Routledge, 1996.

Gugler, Josef, ed. *Cities in the Developing World: Issues, Theory, and Policy.* Oxford: Oxford University Press, 1997.

Hanson, Susan, ed. *The Geography of Urban Transportation.* 2d ed. New York: Guilford, 1995.

Harvey, David. *Social Justice and the City.* London: Edward Arnold, 1973.

Herbert, David T., and Colin J. Thomas. *Cities in Space, City as Place.* New York: John Wiley & Sons, Inc., 1997.

Johnston, Ron, and Paul Knox, eds. *World Cities Series.* Includes books on (author's name in parenthesis): London (Michael Herbert), Tokyo (Roman Cybriwsky), Mexico City (Peter M. Ward), Tehran (Ali Madanipour), Beijing (Victor F.S. Sit), Buenos Aires (David Keeling), Vienna (Elisabeth Lichtenberger), Glasgow (Michael Pacione), Harare (Carole Rakodi), Havana (Roberto Segre et al.), Lagos (M. Peli), Paris (Daniel Noin and Paul White), Rome (J. Agnew), Seoul (Joochul Kim and Sang-Chuei Choe), Singapore (Martin Perry and Brenda Yeoh), Taipei (Roger M. Selya), and Dublin (Andrew MacLaran). New York: John Wiley & Sons, Inc., 1993 to 1998.

King, Leslie J. *Central Place Theory.* Newbury Park, Calif.: Sage, 1984.

Kivell, Philip. *Land and the City: Patterns and Processes of Urban Change.* London: Routledge, 1993.

Knox, Paul L. *Urbanization: An Introduction to Urban Geography.* Englewood Cliffs, N.J.: Prentice Hall, 1994.

Ley, David. *The New Middle Class and the Remaking of the Central City.* Oxford: Clarendon Press, 1996.

Lowder, Stella. *The Geography of Third World Cities.* Totowa, N.J.: Barnes and Noble, 1986.

Roseman, Curtis C., Hans Dieter Laux, and Gunter Thieme, eds. *EthniCity: Geographic Perspectives on Ethnic Change in Modern Cities.* Lanham, Md.: Rowan and Littlefield Publishers, Inc., 1996.

Scott, Allen J. *Metropolis: From the Division of Labor to Urban Form.* Berkeley: University of California Press, 1988.

Smith, Neil, and Peter Williams, eds. *Gentrification of the City.* Winchester, Mass.: Allen and Unwin, 1986.

Stimpson, C. R., E. Dixler, M. J. Nelson, and K. B. Yatrakis. *Women and the American City.* Chicago: University of Chicago Press, 1981.

Vance, James E. *The Continuing City: Urban Morphology in Western Civilization.* 2d ed. Baltimore, Md.: Johns Hopkins University Press, 1990.

Ward, David. *Cities and Immigrants.* New York: Oxford University Press, 1971.

Don't forget about Dushkin's *Annual Editions Online: Geography* at http: / / dushkin.com / aeonline / . See preface for details.

PART FOUR

The Area Analysis Tradition

Julius Caesar began his account of his transalpine campaigns by observing that all of Gaul was divided into three parts. With that spatial summary, he gave to every schoolchild an example of geography in action.

Caesar's report to the Romans demanded that he convey to an uninformed audience a workable mental picture of place. He was able to achieve that aim by aggregating spatial data, by selecting and emphasizing what was important to his purpose, and by submerging or ignoring what was not. He was pursuing the geographic tradition of area analysis, a tradition that is at the heart of the discipline and focuses on the recognition of spatial uniformities and the examination of their significance.

The tradition of area analysis is commonly associated with the term *regional geography,* the study of particular portions of the earth's surface. As did Caesar, the regional geographer attempts to view a particular area and to summarize what is spatially significant about it. One cannot, of course, possibly know everything about a region, nor would "everything" contribute to our understanding of its essential nature. Regional geographers, however, approach a preselected earth space—a continent, a country, or some other division—with the intent of making it as fully understood as possible in as many facets of its nature as they deem practicable. It is from this school of area analysis that "regional geographies" of, for example, Africa, the United States, or the Pacific Northwest derive.

Because the scope of their inquiries is so broad, regional geographers must become thoroughly versed in all of the topical subfields of the discipline, such as those discussed in this book. Only in that way can area specialists select those phenomena that give insight into the essential unity and diversity of their region of study. In their approach, and based on their topical knowledge, regional geographers frequently seek to delimit and study regions defined by one or a limited number of criteria.

The three topical traditions already discussed—earth science, culture–environment, and locational—have often misleadingly been contrasted to the tradition of area analysis. In actuality, practitioners of both topical and regional geography inevitably work within the tradition of area analysis. Both seek an organized view of earth space. The one asks what are the regional units that evolve from the consideration of a particular set of preselected phenomena; the other asks how, in the study of a region, its varied content may best be summarized and clarified.

Chapter 13 is devoted to the area analysis tradition. Its introductory pages explore the nature of regions and methods of regional analysis. The body of the chapter consists of a series of regional vignettes, each based on a theme introduced in one of the preceding chapters of the book. Since those chapters were topically organized, the separate studies that follow demonstrate the tradition of area analysis in the context of topical (sometimes called "systematic") geography. They are examples of the ultimate regionalizing objectives of geographers who seek an understanding of their data's spatial expression. The step from these limited topical studies to the broader, composite understandings sought by the area analyst is both short and intellectually satisfying.

Tea fields in Japan.
©Travelpix/FPG.

CHAPTER 13

THE REGIONAL CONCEPT

The distinctive rural landscape of Inishmaan, one of the Aran Islands off the entrance to Ireland's Galway Bay.
©Tim Thompson/Tony Stone Images.

The questions geographers ask, we saw in Chapter 1 of this book, ultimately focus on matters of location and character of place. We asked how things are distributed over the surface of the earth; how physical and cultural features of areas are alike or different from place to place; how the varying content of different places came about; and what all these differences and similarities mean for people.

The Nature of Regions

In the earlier chapters, we examined some of the physical and cultural content of areas and some spatially important aspects of human behavior. We looked at physical earth processes that lead to differences from place to place in the environment. We studied ways in which humans organize their actions in earth space—through political institutions, by economic systems and practices, and by cultural and social processes that influence spatial behavior and interaction. Population and settlement patterns and areal differences in human use and misuse of earth resources all were considered as part of the mission of geography.

For each of the topics we studied, from landforms to cities, we found spatial regularities. We discovered that things are not irrationally distributed over the surface of the earth, but reflect an underlying spatial order based on understandable physical and cultural processes. We found, in short, that although no two places are *exactly* the same, it is possible and useful to recognize segments of the total world that are internally similar in some important characteristic and distinct in that feature from surrounding areas.

These regions of significant uniformity of content are the geographer's spatial equivalent of the historian's "eras" or "ages," and are assigned brief summary names to indicate they are different in some important way from adjacent or distant territories. The **region,** then, is a device of areal generalization. It is an attempt to separate into recognizable component parts the otherwise overwhelming diversity and complexity of the earth's surface.

All of us have a general idea of the meaning of region, and all of us refer to regions in everyday speech and action. We visit "the old neighborhood" or "go downtown"; we plan to vacation or retire in the "Sunbelt," or we speculate on the effects of weather conditions in the "Northern Plains" or the "Corn Belt" on grain supplies or next year's food prices. In each instance we have mental images of the areas mentioned. Those images are based on place characteristics and areal generalizations that seem useful to us and recognizable to our listeners. We have, in short, engaged in an informal place classification to pass along quite complex spatial, organizational, or content ideas. We have applied the **regional concept** to bring order to the immense diversity of the earth's surface.

What we do informally as individuals, geography attempts to do formally as a discipline—define and explain regions (Figure 13.1). The purpose is clear: to make the infinitely varying world around us understandable through spatial summaries. That world is only rarely subdivided into neat, unmistakable "packages" of uniformity. Neither the environment nor human areal actions present us with a compartmentalized order, any more than the sweep of human history has predetermined "eras," or all plant specimens come labeled in nature with species names. We all must classify to understand, and the geographer classifies in regional terms.

Regions are spatial expressions of ideas or summaries useful to the analysis of the problem at hand. Although as many possible regions exist as there are physical, cultural, or organizational attributes of area, the geographer selects for study only those areal variables that contribute to the understanding of a specific topical or spatial problem. All others are disregarded as irrelevant.

In our reference to the Corn Belt, we delimit a portion of the United States showing common characteristics of

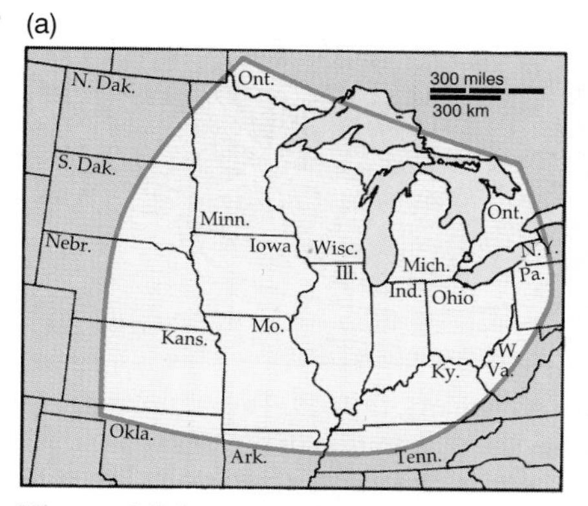

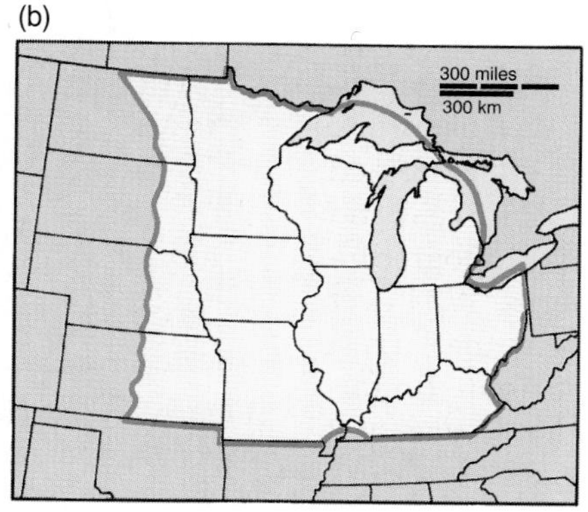

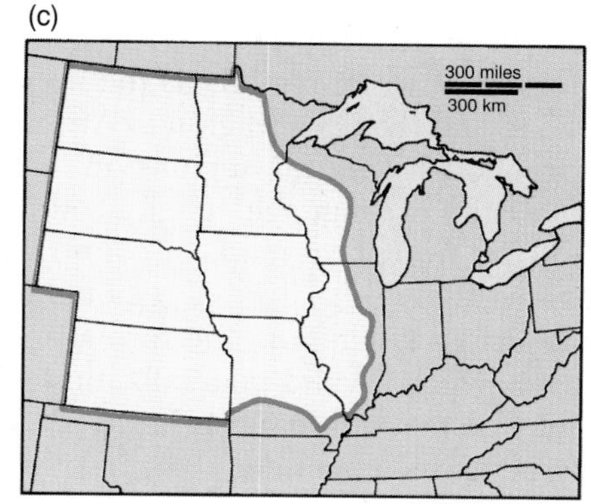

Figure 13.1 **The Midwest as seen by different professional geographers.** Agreement on the need to recognize spatial order and to define regional units does not imply unanimity in the selection of boundary criteria. All of the sources concur in the significance of the Midwest as a regional entity in the spatial structure of the United States and agree on its core area. They differ, however, in their assessments of its limiting characteristics.

Sources: (a) John H. Garland, ed., *The North American Midwest* (New York: John Wiley & Sons, 1955); (b) John R. Borchert and Jane McGuigan, *Geography of the New World* (Chicago: Rand McNally, 1961); and (c) Otis P. Starkey and J. Lewis Robinson, *The Anglo-American Realm* (New York: McGraw-Hill, 1969).

farming economy and marketing practices. We dismiss—at that level of generalization—differences within the region based on slope, soil type, state borders, or characteristics of population. The boundaries of the Corn Belt are assumed to be marked where the region's internal unifying characteristics change so materially that different agricultural economies become dominant and different regional summaries are required. The content of the region suggests its definition and determines the basis of its delimitation.

Although regions may vary greatly, they all share certain common characteristics related to earth space:

- **Regions have *location,*** often expressed in the regional name selected, such as the Midwest, the Near East, North Africa, and the like. This form of regional name underscores the importance of *relative location* (see Chapter 1, p. 8).
- **Regions have *spatial extent,*** recognized by territories over which defined characteristics or generalizations about physical, cultural, or organizational content are constant.
- **Regions have *boundaries*** based on the areal spread of the features selected for study. Since regions are the recognition of the features defining them, their boundaries are drawn where those features no longer occur or dominate. Regional boundaries are rarely as sharply defined as those suggested by Figure 13.2 or by the regional maps in this and other geography texts. More frequently, there exist broad zones of transition from one distinctive core area to another as the dominance of the defining regional features gradually diminishes outward from the core to the regional periphery. Linear boundaries are arbitrary divisions made necessary by the scale of world regional maps and by the summary character of most regional discussions.
- **Regions may be either *formal or functional,*** as we saw in Chapter 1 (p. 15).

Formal regions are areas of essential uniformity throughout in one or a limited combination of physical or cultural features. In previous chapters, we encountered such formal physical regions as the humid subtropical climate zone and the Sahel region of Africa, as well as formal (homogeneous) cultural regions in which standardized characteristics of language, religion, ethnicity, or livelihood existed. The frontsheet maps of countries and topographical regions show other formal regional patterns. Whatever the basis of its definition, the formal region is the largest area over which a valid generalization of attribute uniformity may be made. Whatever is stated about one part of it holds true for its remainder.

The **functional region,** in contrast, is a spatial system defined by the interactions and connections that give it a dynamic, organizational basis. Its boundaries remain constant only as long as the interchanges establishing it remain unaltered. The commuting region shown in Figure 1.10 retained its shape and size only as long as the road pattern and residential neighborhoods on which it was based remained as shown.

- **Regions are *hierarchically arranged.*** Although regions vary in scale, type, and degree of generalization, none stands alone as the ultimate key to areal understanding. Each defines only a part of spatial reality.

On a formal regional scale of size progression, the Delmarva Peninsula of the eastern United States may be seen as part of the Atlantic Coastal Plain, which is in turn a portion of the eastern North American humid continental climatic region, a hierarchy that changes the basis of regional recognition as the level and purpose of generalization alter (Figure 13.3). The central business district of Chicago is one land use complex in the functional regional hierarchy that describes the spatial influences of the city of Chicago and the metropolitan region of which it is the core. Each recognized regional entity in such progressions may stand alone and at the same time exist as a part of a larger, equally valid, territorial unit.

Such progressions may also reflect gradations in the intensity in spatial dominance of the phenomenon giving definition to the region. With particular attention to regions defined by the distribution of culture groups—but with application to regions delimited by other criteria—Donald Meinig suggested the term *core* to mean a centralized zone of concentration and greatest homogeneity of regional character. Areas

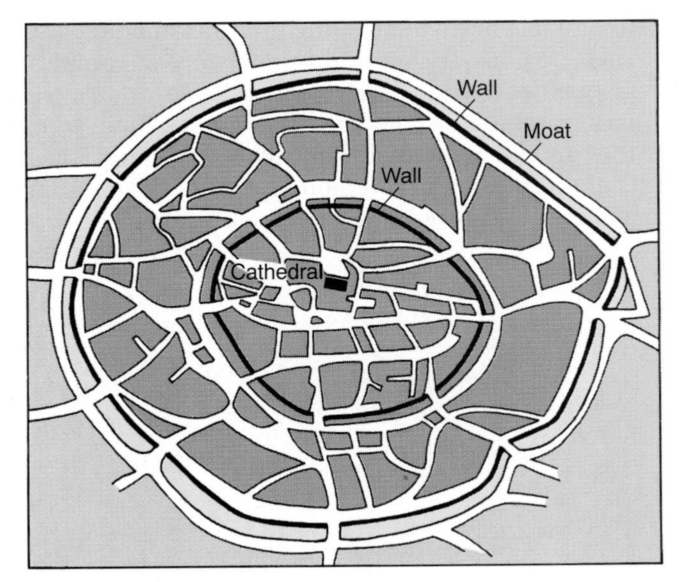

Figure 13.2 **Aachen, Germany, in 1649.** The acceptance of regional extent implies the recognition of regional boundaries. At some defined point, *urban* is replaced by *nonurban,* the Midwest ends and the Plains begin, or the rain forest ceases and the savanna emerges. Regional boundaries are, of course, rarely as precisely and visibly marked as were the limits of the walled medieval city. Its sprawling modern counterpart may be more difficult to define, but the boundary significance of the concept of *urban* remains.

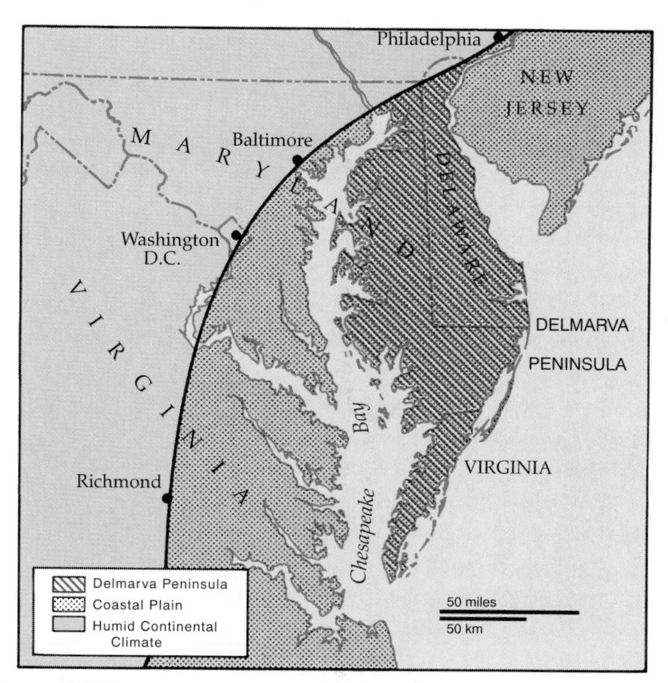

Figure 13.3 **A hierarchy of regions.** One possible nesting of regions within a *regional hierarchy* defined by differing criteria. Each regional unit has internal coherence. The recognition of its constituent parts aids in understanding the composite areal unit.

in which the culture (or other defining feature) is dominant, but with less intensity and totality of development, he labeled *domains*. Finally, Meinig proposed the term *sphere* to recognize the zone of broadest but least-intensive expression of the regional character, where some of its defining traits are encountered but where it is no longer spatially dominant.

These generalizations about the nature of regions and the regional concept are meant to instill firmly the understanding that regions are human intellectual creations designed to serve a purpose. Regions focus our attention on spatial uniformities. They bring clarity to the seeming confusion of the observable physical and cultural features of the world we inhabit. Regions provide the framework for the purposeful organization of spatial data.

The Structure of This Chapter

The remainder of the chapter contains examples of how geographers have organized regionally their observations about the physical and cultural world. Each study or vignette explores a different aspect of regional reality. Each organizes its data in ways appropriate to its subject matter and objective, but in each, some or all of the common characteristics of regional delimitation and structure may be recognized. Each of the regional examples is based on the content of one of the earlier chapters of this book, and the page numbers preceding the different selections refer to the material within those chapters most closely associated with, or explained further by, the regional case study. As an aid to visualizing the application of the regional concept, the examples are introduced by reference to the traditions of geography that unite their subject matter and approach.

Part I: Regions in the Earth Science Tradition

The simplest of all regions to define, and generally the easiest to recognize, is the formal region based on a single readily apparent component or characteristic. The island is land, not water, and its unmistakable boundary is naturally given where the one element passes to the other. The terminal moraine may mark the transition from the rich, black soils of recent formation to the parti-colored clays of earlier generation. The dense forest may break dramatically upon the glade or the open prairies. The nature of change is singular and apparent.

The physical geographer, although concerned with all of the earth sciences that explain the natural environment, deals at the outset with *single factor* formal regions. Many of the earth features of concern to physical geography, of course, do not exist in simple, clearly defined units. They must be arbitrarily "regionalized" by the application of boundary definitions. A stated amount of received precipitation, the presence of certain important soil characteristics, the dominance in nature of particular plant associations—all must be decided on as regional limits, and all such limits are subject to change through time or by purpose of the regional geographer.

Landforms as Regions

(*See "Stream Landscapes in Humid Areas," p. 85*)

The landform region exists in a more sharply defined fashion than such transitional physical features as soil, climate, or vegetation. For these latter areas, the boundaries depend on definitional decisions made (and defended) by the researcher. The landform region, on the other hand, arises—visibly and apparently unarguably—from nature itself, independent of human influence and unaffected by time on the human scale. Landforms constitute basic, naturally defined regions of physical geographic concern. The existence of major landform regions—mountains, lowlands, plateaus—is unquestioned in popular recognition or scientific definition. Their influence on climates, vegetational patterns, even on the primary economies of subsistence populations has been noted in earlier portions of this book. The following discussion of a distinct landform region, describing its constitution and its relationships to other physical and cultural features of the landscape, is adapted from a classic study by Wallace W. Atwood.

The Black Hills Province[1]

The Black Hills rise abruptly from the surrounding plains (Figure 13.4). The break in topography at the margin of this province is obvious to the most casual observer who visits that part of our country. Thus the boundaries of this area, based on contrasts in topography, are readily determined.

One who looks more deeply into the study of the natural environment may recognize that in the neighboring plains the rock formations lie in a nearly horizontal position. They are sandstones, shales, conglomerates, and limestones. In the foothills those same sedimentary formations are bent upward and at places stand in a nearly vertical position. Precisely where the change in topography occurs, we find a notable change in the geologic structure and thus discover an explanation for the variation in relief.

The Black Hills are due to a distinct upwarping, or doming, of the crustal portion of the earth. Subsequent removal by stream erosion of the higher portions of that dome and the dissection of the core rocks have produced the present relief features. As erosion has proceeded, more and more of a complex series of ancient metamorphic rocks has been uncovered. Associated with the very old rocks of the core and, at places, with the sedimentary strata, there are a number of later intrusions which have cooled and formed solid rock. They have produced minor domes about the northern margin of the Black Hills.

With the elevation of this part of our country there came an increase in rainfall in the area, and with the increase in elevation and rainfall came contrasts in relief, in soils, and in vegetation.

As we pass from the neighboring plains, where the surface is monotonously level, and climb into the Black Hills area, we enter a landscape having great variety in the relief. In the foothill belt, at the southwest, south, and east, there are hogback ridges interrupted in places by water gaps, *or gateways, that have been cut by streams radiating from the core of the Hills. Between the ridges there are roughly concentric valley lowlands. On the west side of the range, where the sedimentary mantle has not been removed, there is a plateau-like surface; hogback ridges are absent. Here erosion has not proceeded far enough to produce the landforms common to the east margin. In the heart of the range we find deep canyons, rugged intercanyon ridges, bold mountain forms, craggy knobs, and other picturesque features (Figure 13.5). The range has passed through several periods of mountain growth and several stages, or cycles, of erosion.*

The rainfall of the Black Hills area is somewhat greater than that of the brown, seared, semiarid plains regions, and evergreen trees survive among the hills. We leave a land of sagebrush and grasses to enter one of forests. The dark-colored evergreen trees suggested to early settlers the name Black Hills. As we enter the area, we pass from a land of cattle ranches and some seminomadic shepherds to a land where forestry, mining, general farming, and recreational activities give character to the life of the people. In color and form, in topography, climate, vegetation, and economic opportunities, the Black Hills stand out conspicuously as a distinct geographic unit.

[1]Adapted from *The Physiographic Provinces of North America* by Wallace W. Atwood. © Copyright, 1940, by Ginn and Company. Used by permission of Silver, Burdett & Ginn, Inc.

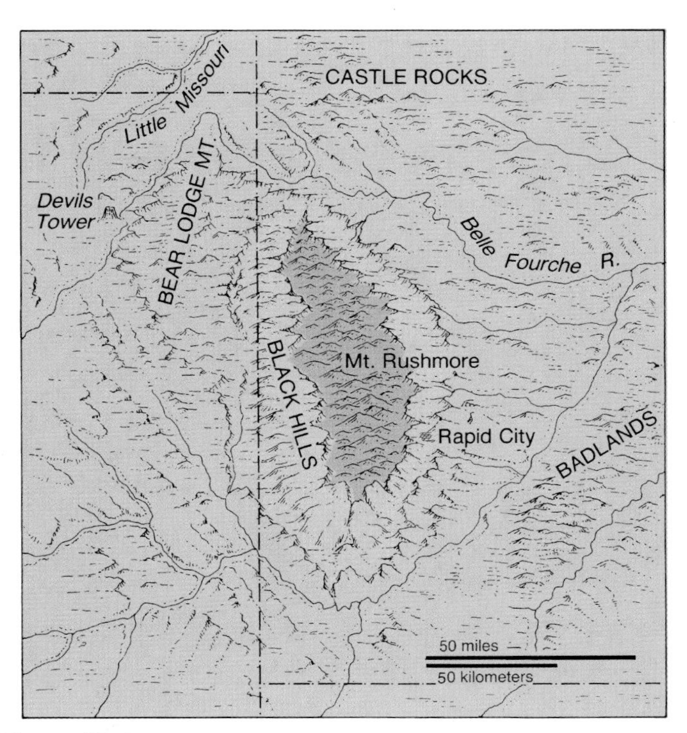

Figure 13.4 The Black Hills landform province.

Dynamic Regions in Weather and Climate

(See "Air Masses," p. 118)

The unmistakable clarity and durability of the Black Hills landform region and the precision with which its boundaries may be drawn are rarely echoed in other types of formal physical regions. Most of the natural environment, despite its appearance of permanence and certainty, is dynamic in nature. Vegetations, soils, and climates change through time by natural process or by the action of humans. Boundaries shift, perhaps abruptly, as witness the recent migration southward of the Sahara. The core characteristics of whole provinces change as marshes are drained or forests are replaced by cultivated fields.

That complex of physical conditions we recognize locally and briefly as weather and summarize as climate displays particularly clearly the temporary nature of much of the natural environment that surrounds us. Yet even in the turbulent change of the atmosphere, distinct regional entities exist with definable boundaries and internally consistent horizontal and vertical properties. "Air masses" and

Figure 13.5 The "Needles" of the Black Hills result from erosion along vertical cracks and crevices in granite.
©B. F. Molnia/Terraphotographics/BPS.

the consequences of their encounters constitute a major portion of contemporary weather analysis and prediction. Air masses further fulfill all the criteria of *multifactor formal regions,* though their dynamic quality and their patterns of movement obviously mark them as being of a nature distinctly different from such stable physical entities as landform regions, as the following extract from *Climatology and the World's Climates* by George R. Rumney makes clear.

Air Masses[2]

An air mass *is a portion of the atmosphere having a uniform horizontal distribution of certain physical characteristics, especially of temperature and humidity. These qualities are acquired when a mass of air stagnates or moves very slowly over a large and relatively unvaried surface of land or sea. Under these circumstances surface air gradually takes on properties of temperature and moisture approaching those of the underlying surface, and there then follows a steady, progressive transmission of properties to greater heights, resulting finally in a fairly clearly marked vertical transition of characteristics. Those parts of the earth where air masses acquire their distinguishing qualities are called* source regions.

The height to which an air mass is modified depends upon the length of time it remains in its source region and also upon the difference between the initial properties of the air when it first arrived and those of the underlying surface. If, for example, an invading flow of air is cooler than the surface beneath as it comes to virtual rest over a source region, it is warmed from below, and convective currents are formed, rapidly bearing aloft new characteristics of temperature and moisture to considerable heights. If, on the other hand, it is warmer than the surface of the source region, cooling of its surface layers takes place, vertical thermal currents do not develop, and the air is modified only in its lower portions. The process of modification may be accomplished in just a few days of slow horizontal drift, although it often takes longer, sometimes several weeks. Radiation, convection, turbulence, and advection are the chief means of bringing it about.

The prerequisite conditions for these developments are very slowly migrating, outward spreading, and diverging air and a very extensive surface beneath that is fairly uniform in nature. Light winds and relatively high barometric pressure characteristically prevail. Hence, most masses form within the great semipermanent anticyclonic regions of the general circulation, where calms, light variable winds, and overall subsidence of the atmosphere are typical.

Four major types of source regions are recognized: continental polar, maritime polar, continental tropical, and maritime tropical. Polar air masses are continental when they develop over land or ice surfaces in high latitudes; these are cold and dry. They are maritime when they form over the oceans in high latitudes. An air mass from these sources is cold and moist. Similarly, tropical air is continental when it originates along the Tropics of Cancer and Capricorn over northern Africa and northern Australia and is therefore warm and dry. It is maritime when it forms along the Tropics over the oceans, where it develops as a mass of warm, moist air. A single air mass usually covers thousands of square miles of the earth's surface when fully formed.

An air mass is recognizable chiefly because of the uniformity of its primary properties—temperature and humidity—and the vertical distribution of these. Secondary qualities, such as cloud types, precipitation, and visibility, are also taken

[2]George R. Rumney, *Climatology and the World's Climates,* Macmillan, 1968. Used by permission.

into account. These qualities are retained for a remarkably long time, often for several weeks, after an air mass has traveled far from its source region, and they are thus the means of distinguishing it from other masses of air.

The principal air masses of the Americas, their source regions, and paths of movement are shown in Figure 13.6.

Ecosystems as Regions

(*See "Ecosystems," p. 147*)

A traditional though oversimplified definition of geography as "the study of the areal variation of the surface of the earth" suggested that the discipline centered on the classification of areas and on the subdivision of the earth into its constituent regional parts. Considerations of organization and function were secondary and even unnecessary to the implied main purpose of regional study: the definition of cores and boundaries of areas uniform in physical or cultural properties.

Newer research approaches stress the need for the study of spatial relationships from the standpoint of *systems analysis,* which emphasizes the organization, structure, and functional dynamics within an area and provides for the quantification of the linkages between the things in space. The *ecosystem* or *biome,* introduced in Chapter 5, provides a systems-analytic concept of great flexibility that permits examination of the relationship between the environment and the biological realm. Since that relationship is structured, structure rather than spatial uniformity attracts attention and leads to new understanding of the region. The ecosystem concept, particularly, provides a point of view for investigating the complex consequences of human impact on the natural environment.

Note how, in the following description drawn from an article by William J. Schneider, the ecosystem concept is employed in the recognition and analysis of regions and subregions of varying size, complexity, and nature. Introduced, too, is the concept of *ecotone,* or zone of ecological stress, in this case induced by human pressures on the natural system.

The Everglades[3]

The Everglades is a river. Like the Hudson or the Mississippi, it is a channel through which water drains from higher to lower ground as it moves to the sea. It extends in a broad, sweeping arc from the southern end of Lake Okeechobee in central Florida to the tidal estuaries of the Gulf Coast and Florida Bay. As much as 70 miles (113 km) wide, but generally averaging 40 (64 km), it is a large, shallow slough that weaves tortuously through acres of saw grass and past "islands" of trees, its waters, even in the wet season, rarely deeper than 2 feet (0.6 meters). But, again like the Hudson or the Mississippi, the Everglades bears the significant imprint of civilization.

Since the close of the Pleistocene, 10,000 years ago, the Everglades has been the natural drainage course for the periodically abundant overflow of Lake Okeechobee. As the lake filled during the wet summer months or as hurricane winds blew and literally scooped the water out of the lake basin, excess water spilled over the lake's southern rim. This overflow, together with rainfall collected en route, drained slowly southward between Big Cypress Swamp and the sandy flatlands to the west and the Atlantic coastal ridge to the east, sliding finally into the brackish water of the coastal marshes (Figure 13.7).

Water has always been the key factor in the life of the Everglades. Three-fourths of the annual average 55 inches (140 cm) of rainfall occurs in the wet season, June through October, when water levels rise to cover 90% of the land area of the Everglades. In normal dry seasons in the past, water covered no more than 10% of the

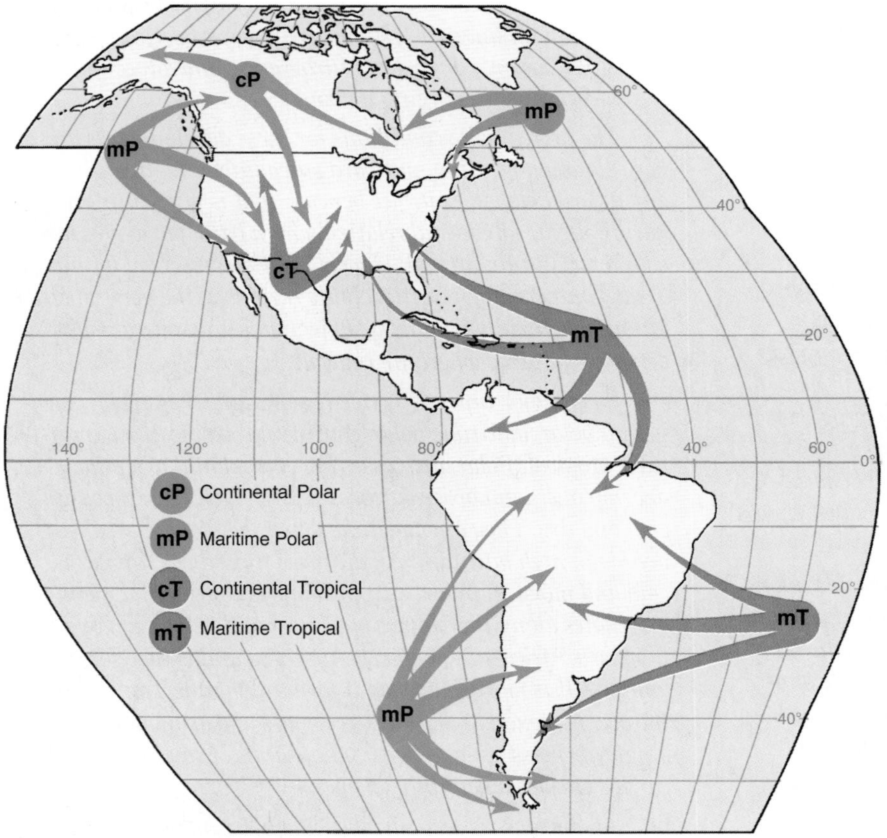

Figure 13.6 Air masses of North and South America, their source regions, and paths of movement.

[3] With permission from *Natural History,* November, 1966. Copyright American Museum of Natural History, 1966.

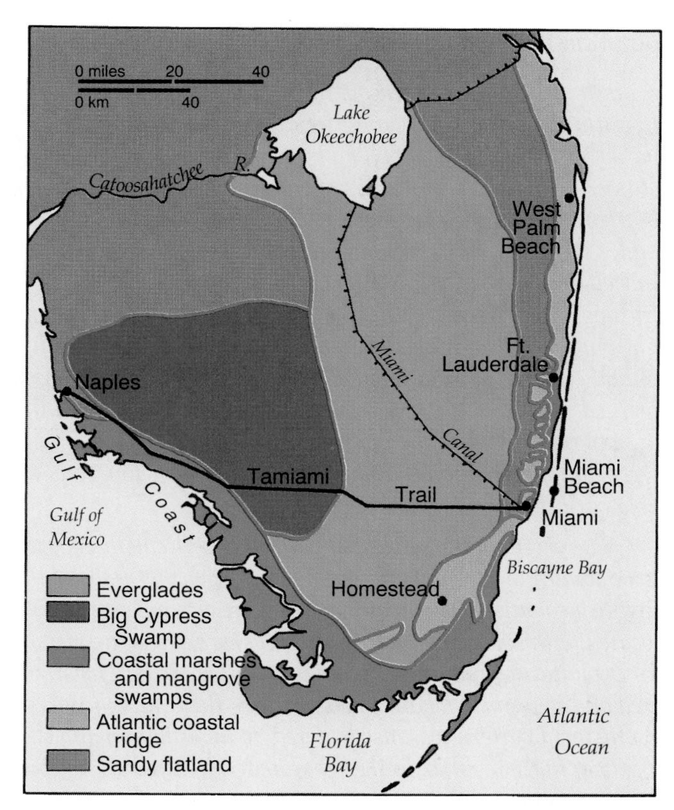

Figure 13.7 The Everglades is part of a complex of ecosystems stretching southward in Florida from Lake Okeechobee to the sea. Drainage and water-control systems have altered its natural condition.

land surface. Throughout much of the Everglades, and prior to recent engineering activities, this seasonal rain cycle caused fluctuations in water levels that averaged 3 feet (0.9 meters). Both occasional severe flooding and prolonged drought accompanied by fire imposed periodic stress upon the ecosystem. It may be that randomly occurring ecologic trauma is vital to the character of the Everglades.

Three dominant biological communities—open water, saw grass, and woody vegetation—reflect small, but consistent, differences in the surface elevation of peat soils that cover the Everglades (Figure 13.8). The open-water areas occur at the lower soil elevations; inundated much of the year, they contain both sparse, scattered marsh grasses and a mat of algae. The saw grass communities develop on a soil base only a few inches higher than that in the surrounding open glades. The soil base is thickest under the tree islands. The few inches' difference in soil depth apparently governs the species composition of these three communities.

Today, the Everglades is no longer precisely a natural river. Much of it has been altered by an extensive program of water management, including drainage, canalization, and the building of locks and dams. Large withdrawals of groundwater for municipal and industrial use have depleted the underlying aquifer and permitted the landward penetration of sea water through the aquifer and through the surface canals. Thousands of individual water-supply wells have been contaminated by encroaching saline water; large biotic changes have taken place in the former freshwater marshes south of Miami. Mangroves—indicators of salinity—have extended their habitat

Figure 13.8 Open water, saw grasses, and tree islands, all visible in this scene, constitute separate *biomes* of the Everglades, which is home also of a teeming animal life.

inland, and fires rage across areas that were formerly much wetter. The ecotone—the zone of stress between dissimilar adjacent ecosystems—is altering as a consequence of these human-induced modifications of the Everglades ecosystem.

The organization, structure, and functional dynamics of the Everglades ecosystems are thus undergoing change. The structured relationships of its components—in nature affected and formed by stress—are being subjected to distortions by humans in ways not yet fully comprehended.

Part II: Regions in the Culture–Environment Tradition

The earth science tradition of geography imposes certain distinctive limits on area analysis. However defined, the regions that may be drawn are based on nature and do not result from human action. The culture-environment tradition, however, introduces to regional geography the infinite variations of human occupation and organization of space. There is a corresponding multiplication of recognized regional types and of regional boundary decisions.

Despite the differing interests of physical and cultural geographers, one element of study is common to their concerns: that of process. The "becoming" of an ecosystem, of a cultural landscape, or of the pattern of exchanges in an economic system is an important open or implied part of nearly all geographic study. Evidence of the past as an aid to understanding of the present is involved in much geographical investigation, for present-day distributional patterns or qualities of regions mark a merely temporary stage in a continuing process of change.

Population as Regional Focus

(See "World Population Distribution," p. 213)

In no phase of geography are process and change more basic to regional understanding than in population studies. The human condition is dynamic and patterns of settlement are ever changing. Although these spatial distributions are related to the ways that people utilize the physical environment in which they are located, they are also conditioned by the purposes, patterns, and solutions of those who went before. In the following extract taken from the work of Glenn T. Trewartha, a dean of American population geographers, notice how population regionalization—used as a focal theme—ties together a number of threads of regional description and understanding. The aspirations of colonist-conquerors, past and contemporary transportation patterns, physical geographic conditions, political separatism, and the history and practice of agriculture and rural land holdings are all introduced to give understanding to population from a regional perspective.

Population Patterns of Latin America[4]

A distinctive feature of the spatial arrangement of population in Latin America is its strongly nucleated character; the pattern is one of striking clusters. Most of the population clusters remain distinct and are separated from other clusters by sparsely occupied territory. Such a pattern of isolated nodes of settlement is common in many pioneer regions; indeed, it was characteristic of early settlement in both Europe and eastern North America. In those regions, as population expanded, the scantily occupied areas between individual clusters gradually filled in with settlers, and the nodes merged. But in Latin America such an evolution generally did not occur, and so the nucleated pattern persists. Expectably, the individual population clusters show considerable variations in density.

The origin of the nucleated pattern of settlement is partly to be sought in the gold and missionary fever that imbued the Spanish colonists. Their settlements were characteristically located with some care, since only areas with precious metals for exploitation and large Indian populations to be Christianized and to provide laborers could satisfy their dual hungers. A clustered pattern was also fostered by the isolation and localism that prevailed in the separate territories and settlement areas of Latin America.

Almost invariably each of the distinct population clusters has a conspicuous urban nucleus. To an unusual degree the economic, political, and social life within a regional cluster centers on a single large primate city, which is also the focus of the local lines of transport.

The prevailing nucleated pattern of population distribution also bears a relationship to political boundaries. In some countries . . . a single population cluster represents the core area of the nation. In more instances, however, a population cluster forms the core of a major political subdivision of a nation-state, so that a country may contain more than one cluster. A consequence of this simple population distribution pattern and its relation to administrative subdivisions is that political boundaries ordinarily fall within the sparsely occupied territory separating individual clusters. In Latin America few national or provincial boundaries pass through nodes of relatively dense settlement (Figure 13.9).

Another feature arising out of the cluster pattern of population arrangement is that the total national territory *of a country is often very different from the* effective national territory, *since the latter includes only those populated parts that contribute to the country's economic support.*

A further consequence of the nucleated pattern is found in the nature of the transport routes and systems. Overland routes between population clusters are usually poorly

[4]From Glenn T. Trewartha, *The Less Developed Realm: A Geography of Its Population.* Copyright © 1972 by John Wiley & Sons, Inc. Reprinted by permission of John Wiley & Sons, Inc.

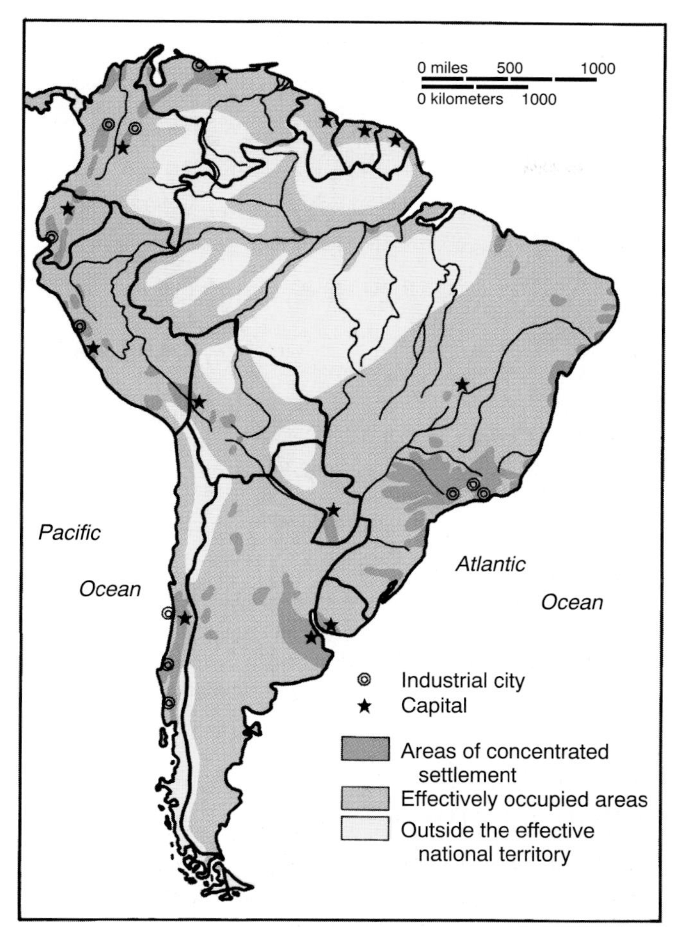

Figure 13.9 **Basic settlement patterns of South America.** Population clusters focused on urban cores and separated by sparsely populated rural areas were the traditional pattern in Latin American countries.

Used with permission of Simon & Schuster, Inc. from the Macmillan College Text *Introduction to Latin America,* by Preston E. James. Copyright © 1964 by Macmillan College Publishing Company, Inc.

developed, while a more efficient network ordinarily exists within *each cluster, with each such regional network joined by an overland route to the nearest port. Thus, the chief lines of transport connecting individual population clusters are often sea lanes rather than land routes. Gradually, as highways are developed and improved, intercluster overland traffic tends to increase.*

. . . The spectacular rates at which population numbers are currently soaring in Latin America are not matched by equivalent changes in their spatial redistributions. Any population map of Latin America reveals extensive areas of unused and underutilized land. Part of such land is highland and plagued by steep slopes, but by far the larger share of it is characterized by a moist tropical climate, either tropical wet or tropical wet and dry. Such a climatic environment, with its associated wild vegetation, soils, and drainage, admittedly presents many discouraging elements to the new settler of virgin lands. . . . [T]ropical climate alone is scarcely a sufficient explanation for the abundance of near-empty lands south of the Rio Grande. Cultural factors are involved as much as, if not more than, physical ones. One of the former is the unfortunate land-holding system that has been fastened on the continent, under which vast areas of potentially cultivable land are held out of active use by a small number of absentee landlords, who not only themselves make ineffective use of the land, but at the same time refuse to permit its cultivation by small operators. Because of the land-holding system, peasant proprietors are unable to secure their own lands, a situation that discourages new rural settlement. . . .

The changes now in progress in the spatial distributions accompanying the vast increase in numbers of people do not appear to involve any large-scale push of rural settlement into virgin territory. Only to a rather limited extent is new agricultural settlement taking place. Intercluster regions are not filling rapidly. The overwhelming tendency is for people to continue to pile up in the old centers of settlement in and around their cities, rather than to expand the frontier into new pioneer-settlement areas. . . .

Language as Region

(*See "Language," p. 248*)

The great culture realms of the world (outlined in Figure 7.3) are historically based composites of peoples. They are not closely identified with nation, language, religion, or technology, but with all these and more in varying combinations. Culture realms are therefore *multifactor regions* that obscure more than they clarify the distinctions between peoples that are so fundamental to the human mosaic of the earth. Basic to cultural geography is the recognition of small regions of single-factor homogeneity that give character to their areas of occurrence and that collectively provide a needed balance to the sweeping generalizations of the culture realm.

Language provides an example of such small area variation, one partially explored in Chapter 7. The language families shown on Figure 7.18 conceal the identity of and the distinctions among the different official tongues of separate countries. These, in turn, ignore or submerge the language forms of minority populations, who may base their own sense of proud identity upon their regional linguistic separateness. In scale and recognition even below these ethnically identified regional languages are those local speech variants frequently denied status as identifiable languages and cited as proof of the ignorance and the cultural deprivation of their speakers. Yet such a limited-area, limited-population tongue contains all the elements of the classic culturally based region. Its area is defined; its boundaries are easily drawn; it represents homogeneity and majority behavior among its members; and it summarizes, by a single cultural trait, a collection of areally distinctive outlooks.

Gullah as Language[5]

Isolation is a key element in the retention or the creation of distinctive and even externally unintelligible languages. The isolation of the ancestors of the some quarter-million present-day speakers of Gullah—themselves called Gullahs—was nearly complete. Held by the hundreds as slaves on the off-shore islands and in the nearly equally remote low country along the southeastern United States coast from South Carolina to the Florida border (Figure 13.10), they retained both the speech patterns of the African languages—Ewe, Fanti, Bambara, Twi, Wolof, Ibo, Malinke, Yoruba, Efik—native to the slave groups and over 4000 words drawn from them. Folk tales told in the Gullah creole can be heard and understood by Krio-speaking audiences in Sierra Leone today.

Forced to use English words for minimal communication with their white overseers, but modifying, distorting, and interjecting African-based substitute words into that unfamiliar language, the Gullahs kept intonations and word and idea order in their spoken common speech that made it unintelligible to white masters or to more completely integrated mainland slaves. Because the language was not understood, its speakers were considered ignorant, unable to master the niceties of English. Because ignorance was ascribed to them, the Gullahs learned to be ashamed of themselves, of their culture, and of their tongue, which even they themselves did not recognize as a highly structured and sophisticated separate language.

In common with many linguistic minorities, the Gullahs are losing their former sense of inferiority and gaining pride in their cultural heritage and in the distinctive tongue that represents it. Out of economic necessity, standard English is being taught to their schoolchildren, but an increasing scholarly and popular interest in the structure of their language and in the nature of their culture has caused Gullah to be rendered as a written language, studied as a second language, and translated into English.

In both the written and the spoken versions, Gullah betrays its African syntax patterns, particularly in its employment of terminal locator words: "Where you goin' at?" The same African origins are revealed by the absence in English translation of distinctive tenses: "I be tired" conveys the concept that "I have been tired for a period of time." Though tenses exist in the African root languages, they are noted more by inflection than by special words and structures.

"He en gut no morratater fer mak no pie wid" may be poor English, but it is good Gullah. Its translation—"He has no more sweet potatoes for making pie"—renders it intelligible to ears attuned to English but loses the musical lilt of original speech and, more importantly, obscures the cultural identity of the speaker, a member of a regionally compact group of distinctive Americans whose territorial extent is clearly defined by its linguistic dominance.

[5]By Jerome Fellmann.

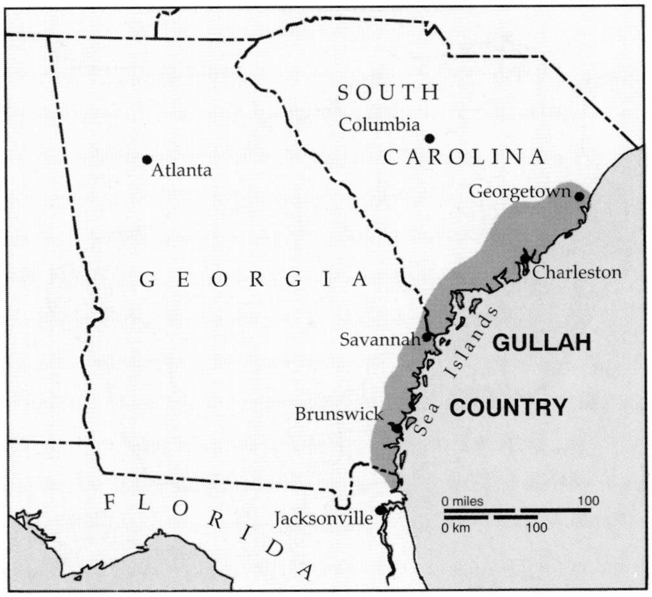

Figure 13.10 Gullah speakers are concentrated on the Sea Islands and the coastal mainland of South Carolina and Georgia. The isolation that promoted their linguistic distinction is now being eroded.

Mental Regions

(*See "Mental Maps," p. 280*)

The regional units so far used as examples, and the methods of regionalization they demonstrate, have a concrete reality. They are formal or functional regions of specified, measurable content. They have boundaries drawn by some objective measures of change or alteration of content, and they have location on an accurately measured global grid.

Individuals and whole cultures may operate, and operate successfully, with a much less formalized and less precise picture of the nature of the world and of the structure of its parts. The mental maps discussed in Chapter 8 represent personal views of regions and regionalization. The private world views they embody are, as we also saw, colored by the culture of which their holders are members.

Primitive societies, particularly, have distinctive world views by which they categorize what is familiar, and satisfactorily account for what is not. The Yurok Indians of the Klamath River area of northern California were no exception. Their geographic concepts were reported by T.T. Waterman, from whose paper, "Yurok Geography," the following summary is drawn.

The Yurok World View[6]

The Yurok imagines himself to be living on a flat extent of landscape, which is roughly circular and surrounded by ocean. By going far enough up the river, it is believed, "you

[6]Source: T. T. Waterman, "Yurok Geography," *University of California Publications in American Archaeology and Ethnology*, Vol. 16, No. 5, pp. 189–193, 1920.

come to salt water again." In other words, the Klamath River is considered, in a sense, to bisect the world. This whole earth mass, with its forests and mountains, its rivers and sea cliffs, is regarded as slowly rising and falling, with a gigantic but imperceptible rhythm, on the heaving, primeval flood. The vast size of the "earth" causes you not to notice this quiet heaving and settling. This earth, therefore, is not merely surrounded by the ocean but floats upon it. At about the central point of the "world" lies a place which the Yurok call qe'nek, on the southern bank of the Klamath, a few miles below the point where the Trinity River comes in from the south. In the Indian concept, this point seems to be accepted as the center of the world.

At this locality also the sky was made. Above the solid sky there is a sky-country, wo'noiyik, about the topography of which the Yurok's ideas are almost as definite as are his ideas of southern Mendocino County, for instance. Downstream from qe'nek, at a place called qe'nek-pul ("qe'nek-downstream"), is an invisible ladder leading up to the sky-country. The ladder is still thought to be there, though no one to my knowledge has been up it recently. The sky-vault is a very definite item in the Yurok's cosmic scheme. The structure consisting of the sky dome and the flat expanse of landscape and waters that it encloses is known to the Yurok as ki-we'sona (literally "that which exists"). This sky, then, together with its flooring of landscape, constitutes "our world." I used to be puzzled at the Yurok confusing earth and sky, telling me, for example, that a certain gigantic redwood tree "held up the world." Their ideas are of course perfectly logical, for the sky is as much a part of the "world" in their sense as the ground is.

The Yurok believe that passing under the sky edge and voyaging still outward you come again to solid land. This is not our world, and mortals ordinarily do not go there; but it is good, solid land. What are breakers over here are just little ripples over there. Yonder lie several regions. To the north (in our sense) lies pu'lekūk, downstream at the north end of creation. South of pu'lekūk lies tsī'k-tsīk-ol ("money lives") where the dentalium-shell, medium of exchange, has its mythical abode. Again, to the south there is a place called kowe'tsik, the mythical home of the salmon, where also all have a "house." About due west of the mouth of the Klamath lies rkrgr', where lives the culture-hero wo'xpa-ku-mä ("across-the-ocean that widower").

Still to the south of rkrgr' there lies a broad sea, kiolaaopa'a, which is half pitch—an Algonkian myth idea, by the way. All of these solid lands just mentioned lie on the margin, the absolute rim of things. Beyond them the Yurok does not go even in imagination. In the opposite direction, he names a place pe'tskuk ("up-river-at"), which is the upper "end" of the river but still in this world. He does not seem to concern himself much with the topography there.

The Yurok's conception of the world he lives in may be summed up in the accompanying diagram (Figure 13.11).

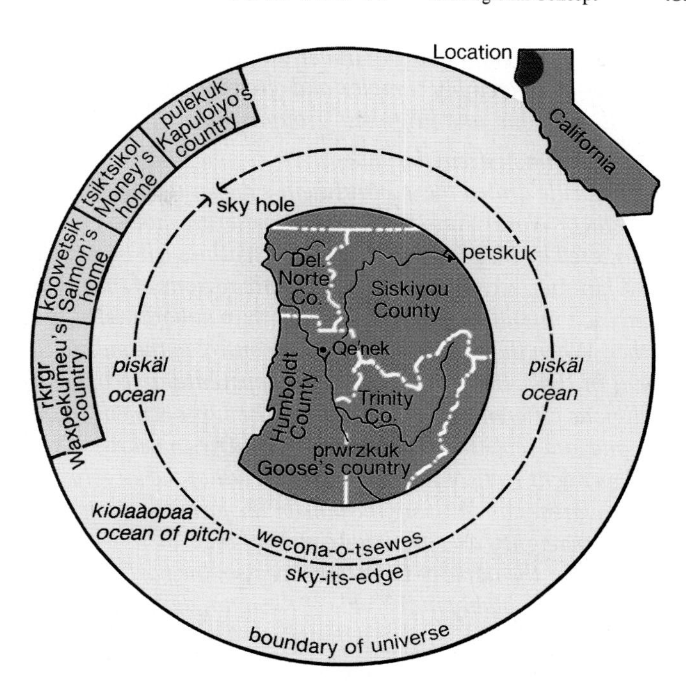

Figure 13.11 **The world view of the Yurok** as pieced together by T. T. Waterman during his anthropological study of the tribe. Qe'nek, in the center of the diagram, marks the center of the world in Indian belief.

Source: T. T. Waterman, "Yurok Geography," *University of California Publications in American Archaeology and Ethnology*, Vol. 16, No. 5, pp. 189–193, 1920.

Political Regions

(*See "Boundaries: the Limits of the State," p. 318*).

The most rigorously defined formal cultural region is the national state. Its boundaries are presumably carefully surveyed and are, perhaps, marked by fences and guard posts. There is no question of an arbitrarily divided transition zone or of lessening toward the borders of the basic quality of the regional core. This rigidity of a country's boundaries, its unmistakable placement in space, and the trappings—flag, anthem, government, army—that are uniquely its own give to the state an appearance of permanence and immutability not common in other, more fluid, cultural regions. But its stability is often more imagined than real. Political boundaries are not necessarily permanent. They are subject to change, sometimes violent change, as a result of internal and external pressures. The Indian subcontinent illustrates the point.

Political Regions in the Indian Subcontinent[7]

The history of the subcontinent since about 400 B.C. has been one of the alternating creation and dissolution of empires, of the extension of central control based upon the Ganges Basin, and of resistance to that centralization by the marginal territories of the peninsula. British India, created largely unintentionally by 1858, was only the last, though perhaps the most

[7]By Jerome Fellmann.

successful, attempt to bring under unified control the vast territory of incredibly complex and often implacably opposed racial, religious, and linguistic groupings.

A common desire for independence and freedom from British rule united the subcontinent's disparate populations at the end of World War II. That common desire, however, was countered by the mutual religious antipathies felt by Muslims and Hindus, each dominant in separate regions of the colony and each unwilling to be affiliated with or subordinated to the other. When the British surrendered control of the subcontinent in 1947, they recognized these apparently irreconcilable religious differences and partitioned the subcontinent into the second and seventh most populous countries on earth. The independent state of India was created out of the largely Hindu areas constituting the bulk of the former colony. Separate sovereignty was granted to most of the Muslim-majority area under the name of Pakistan. Even so, the partition left boundaries, notably in the Vale of Kashmir, dangerously undefined or in dispute.

An estimated 1 million people died in the religious riots that accompanied the partition decision. In perhaps the largest short-term mass migration in history, some 10 million Hindus moved from Pakistan to India, and 7.5 million Muslims left India for Pakistan, "The Land of the Pure."

Unfortunately, the purity resided only in common religious belief, not in spatial coherence or in shared language, ethnicity, customs, food, or economy. During its 23 years of existence as originally conceived, Pakistan was a sorely divided country. The partition decision created an eastern and a western component separated by more than 1600 kilometers (1000 miles) of foreign territory and united only by a common belief in Allah (Figure 13.12). West Pakistan, as large as Texas and Oklahoma combined, held 55 million largely light-skinned Punjabis with Urdu language and strong Middle Eastern cultural ties. Some 70 million Bengali-speakers, making up East Pakistan, were crammed into an Iowa-sized portion of the delta of the Ganges and Brahmaputra rivers. The western segment of the country was part of the semiarid world of western Asia; the eastern portion of Pakistan was joined to humid, rice-producing Southeast Asia.

Beyond the affinity of religion, little else united the awkwardly separated country. East Pakistan felt itself exploited by a dominating western minority that sought to impose its language and its economic development, administrative objectives, and military control. Rightly or wrongly, East Pakistanis saw themselves as aggrieved and abused. They complained of a per capita income level far below that of their western compatriots, claimed discrimination in the allocation of investment capital, found disparities in the pricing of imported foods, and asserted that their exports of raw materials—particularly jute—were supporting a national economy in which they did not share proportionally. They argued that their demands for regional autonomy, voiced since nationhood, had been denied.

When, in November of 1970, East Pakistan was struck by a cyclone and tidal wave that took an estimated 500,000 lives (see p. 2), the limit of eastern patience was reached. Resentful over what they saw as a totally inadequate West Pakistani effort of aid in the natural disaster that had befallen them, the East Pakistanis were further incensed by the refusal of the central government to convene on schedule a national assembly to which they had won an absolute majority of delegates. Civil war resulted, and the separate new state of Bangladesh was created. The sequence of political change in the subcontinent is traced in Figure 13.12.

The country and nation-state so ingrained in our consciousness and so firmly defined, as displayed on the map inside the cover of this book, is both a recent and an ephemeral creation of the cultural regional landscape. It rests upon a claim, more or less effectively enforced, of a monopoly of power and allegiance resident in a government and superior to the communal, linguistic, ethnic, or religious affiliations that

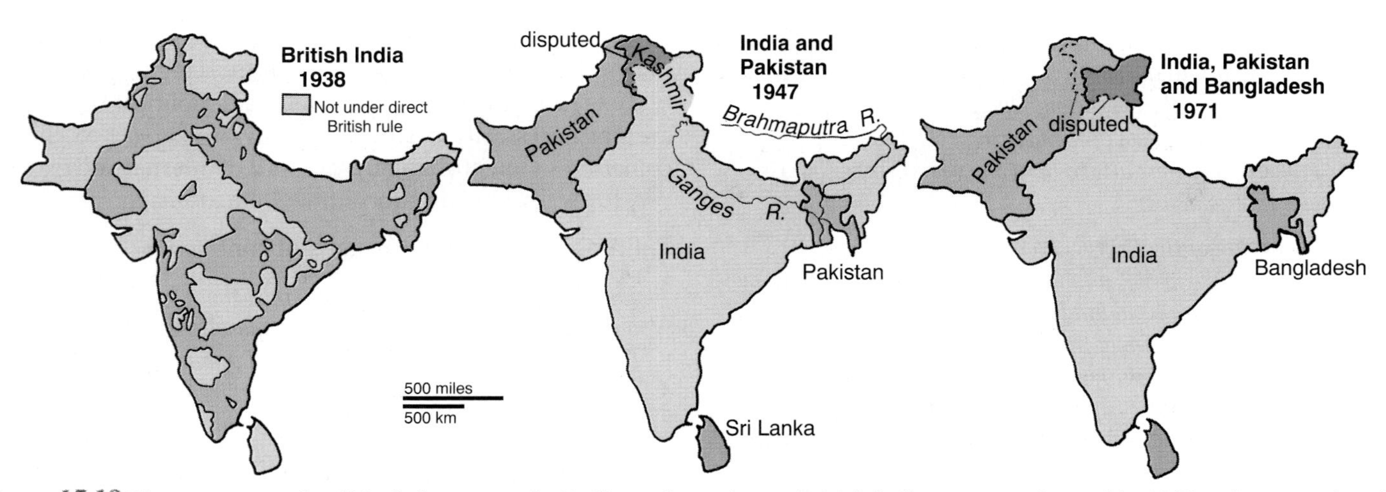

Figure 13.12 The sequence of political change on the Indian subcontinent. British India was transformed in 1947 to the countries of India and Pakistan, the latter a Muslim state with a western and an eastern component. In 1971 Pakistan was torn by civil war based on ethnic and political contrasts, and the eastern segment became the new country of Bangladesh.

preceded it or that claim loyalties overriding it. As the violent recent history of the Indian subcontinent demonstrates, nationalism may be sought, but its maintenance is not assured by the initiating motivations.

Part III: Regions in the Locational Tradition

While location, as we have seen, is a primary attribute of all regions, regionalization in the locational tradition of geography implies far more than a named delimitation of earth space. The central concern is with the distribution of human activities and of the resources on which those activities are based.

In this sense, world regionalization of agriculture and of the soils and climates with which it is related is within the locational tradition. For practical and accepted reasons, however, such underlying physical patterns have been included under the earth science tradition. But the point is made: the locational tradition emphasizes the "doing" in human affairs, and "doing" is not an abstract thing but an interrelation of life and the resources on which life depends.

The locational tradition, therefore, encourages the recognition and the definition of a far wider array of regional types than do the earth science or culture–environment traditions. Any single pattern of economic activity or of resource distribution invites the recognition of definable *formal regions.* The interchange of commodities, the control of urban market areas, the flows of capital, or the collection and distribution activities of ports are just a few examples of the infinite number of analytically useful *functional regions* that one may recognize.

Economic Regions

(*See p. 349*)

Economic regionalization is among the most frequent, familiar, and useful employments of the regional method. Through economic regions the geographer identifies activities and resources, maps the limits of their occurrence or use, and examines the interrelationships and flows that are part of the complexities of the contemporary world.

The economic region, examples of which were explored in Chapter 10, should be seen as potentially more than a device for recording what *is* in either a formal or a functional sense. It has increasingly become a device for examining what might or should be. The concept of the economic region as a tool for planning and a framework for the manipulation of the people, resources, and economic structure of a composite region first took root in the United States during the Great Depression years of the 1930s. The key element in the planning region is the public recognition of a major territorial unit in which economic change or decline is seen as the cause of a variety of interrelated problems including, for example, population out-migration, regional isolation, cultural deprivation, underdevelopment, and poverty.

Appalachia[8]

Until the early 1960s, "Appalachia" was for most a loose reference to the complex physiographic province of the eastern United States associated with the Appalachian mountain chain. If thought of at all, it was apt to be visualized as rural, isolated, and tree-covered; as an area of coal mining, hillbillies, and folksongs (Figure 13.13).

During the 1950s, however, the economic stagnation and the functional decline of the area became increasingly noticeable in the national context of economic growth, rising personal incomes, and growing concern with the elimination of the poverty and deprivation of every group of citizens. Less dramatically but just as decisively as the Dust Bowl or the Tennessee Valley of an earlier era, Appalachia became simultaneously a popularly recognized economic and cultural region and a governmentally determined planning region.

Evidences of poverty, underdevelopment, and social crisis were obvious to a country committed to recognize and eradicate such conditions within its own borders. By 1960, per capita income within Appalachia was $1400 when the national average was $1900. In the decade of the 1950s, mine employment fell 60% and farm jobs declined 52%; the rest of the country lost only 1% of mining jobs and 35% of agricultural employment. Rail employment fell with the drop in coal mining. Massive out-migration occurred among young adults, with such cities as Chicago, Detroit, Dayton, Cleveland, and Gary the targets. Even with these departures, unemployment among those who remained in Appalachia averaged 50% higher than the national rate. Because of the departures, the remnant population—only 47% of whom lived in or near cities in 1960, as against 70% for the entire United States—was distorted in age structure. The very young and the old were disproportionately represented; the productive working-age groups were, at least temporarily, emigrants.

When these and other socioeconomic indicators were plotted by counties and by state economic areas, an elongated but regionally coherent and clearly bounded Appalachia as newly understood was revealed by maps (Figure 13.14). It extended through 13 states from Mississippi to New York, covered some 505,000 square kilometers (195,000 sq mi) and contained 18 million people, 93% of them white.

By 1963, awareness of the problems of the area at federal and state levels passed beyond recognition of a multifactor economic region to the establishment of a planning region. A joint federal–state Appalachian Regional Commission was created to develop a program designed to meet the perceived needs of the entire area. The approach chosen was one of limited investment in a restricted number of highly localized developments, with the expectation that these would spark economic growth supported by private funds.

[8]By Jerome Fellmann.

Figure 13.13 This deserted eastern Kentucky cabin is a mute reminder of the economic and social changes in Appalachia. Increases in per capita income, expansion of urban job opportunities, and improved highway networks have reduced the poverty and isolation that has so long been associated with this region.

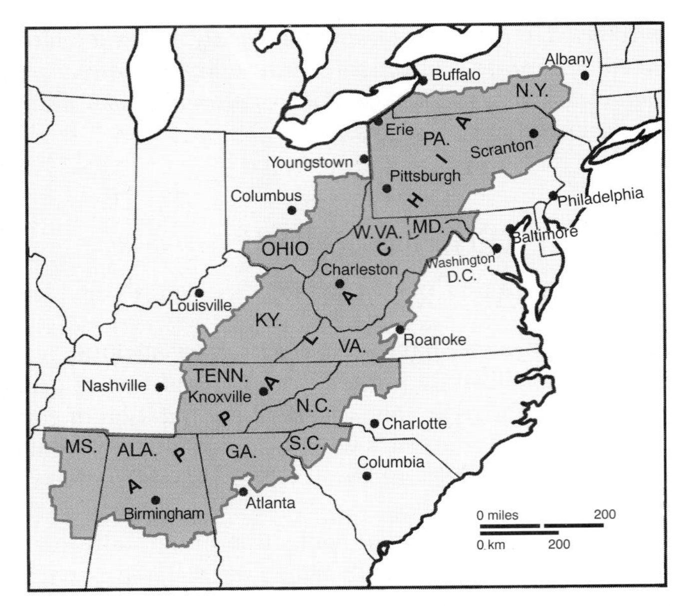

Figure 13.14 **The boundaries of "Appalachia"** as defined by the Appalachian Regional Commission were based on social and economic conditions influenced by political considerations, not topography.

In outline, the plan was (1) to ignore those areas of poverty and unemployment that were in isolated, inaccessible "hollows" throughout the region; (2) to designate "growth centers" where development potential was greatest and concentrate all spending for economic expansion there; the regional growth potential in targeted expenditure was deemed sufficient to overcome charges of aiding the prosperous and depriving the poor; and (3) to create a new network of roads so that the isolated jobless could commute to the new jobs expected to form in and near the favored growth centers. Road construction would also, of course, open inaccessible areas to tourism and strengthen the economic base of the entire planning region.

In the years since the Appalachian Regional Commission was established, and in ways not anticipated, the economic prospects of Appalachia have altered. New industrial jobs have multiplied as manufacturing has relocated to, or has been newly developed in, the Appalachia portion of the Sunbelt. The new factory employment opportunities have exceeded local labor pools, and out-migrants have returned home from cities outside the region, bringing again a more balanced population pyramid. At the same time, coal-mining jobs in the region have plummeted, and unemployment remains above the national average.

By 1994, the Appalachian Regional Commission had over the years committed more than $6 billion to the area, and over $10 billion more had come from other sources. The percentage of people living in poverty within the region declined from 31% in 1960 to 15% in 1990, and per capita income, which had been 79% of the national average in 1960, rose to 85% by the end of the 1980s. Population flows stabilized, with the number of people coming into Appalachia about equaling those leaving. More than 3500 kilometers (2200 miles) of road had been laid by the mid-1990s, though their construction did not give the total regional access that had been hoped. Smaller towns and their hinterlands have been bypassed, remain as isolated as ever, and have lost population. Some basic social services have, for the first time, reached essentially everyone in the region. Each of the 397 counties of the commission territory, for example, has been supplied with a clinic or a doctor.

Despite these evidences of progress, much of Appalachia still remained economically distressed in the 1990s. The mining and manufacturing recession of the earlier years of the decade was sorely felt within a region so dependent upon coal mining and with an industrial base of primary and fabricated metals, wood products, textiles, and apparel—all eroded by imports replacing domestic products.

Natural Resource Regions

(See "Coal," p. 394)

The unevenly distributed resources upon which people depend for existence are logical topics of interest within the locational tradition of geography. Resource regions are mapped, and raw material qualities and quantities are discussed. Areal relationships to industrial concentrations and the impacts of material extraction upon alternate uses of areas are typical interests in resource geography and in the definition of resource regions.

Those regions, however, are usually treated as if they were expressions of observable surface phenomena; as if, somehow, an oil field were as exposed and two-dimensional as a soil region or a manufacturing district. What is ignored is that most mineral resources are three-dimensional regions beneath the ground. In addition to the characteristics of an area that may form the basis for regional delimitations and descriptions of surface phenomena, regions beneath the surface add their own particularities to the problem of regional definition. They have, for example, upper and lower boundaries in addition to the circumferential bounds of surface features. They may have an internal topography divorced from the visible landscape. Subsurface relationships—for example, mineral distribution and accessibility in relation to its enclosing rock or to groundwater amounts and movement—may be critical in understanding these specialized, but real, regions. An illustration drawn from the Schuylkill field of the anthracite region of northeastern Pennsylvania helps illustrate the nature of regions beneath the surface.

The Schuylkill Anthracite Region[9]

Nothing in the wild surface terrain of the anthracite country suggested the existence of an equally rugged subterranean topography of coal beds and interstratified rock, slate, and fire clays forming a total vertical depth of 900 meters (3000 feet) at greatest development. Yet the creation of the surface landscape was an essential determinant of the areal extent of the Schuylkill district, of the contortions of its bedding, and of the nature of its coal content. A county history of the area reports, "The physical features of the anthracite country are wild. Its area exhibits an extraordinary series of parallel ridges and deep valleys, like long, rolling lines of surf which break upon a flat shore." Both the surface and the subsurface topographies reflect the strong folding of strata after the coal seams were deposited; the anthracite (hard) coal resulted from metamorphic carbonization of the original bituminous beds. Subsequent river and glacial erosion removed as much as 95% of the original anthracite deposits and gave to those that remained discontinuous existence in sharply bounded fields like the Schuylkill (Figure 13.15), a discrete areal entity of 470 square kilometers (181 sq mi).

The irregular topography of the underground Schuylkill region means that the interbedded coal seams, the most steeply inclined of all the anthracite regions (Figure 13.16), outcrop visibly at the surface on hillsides and along stream valleys. The outcrops made the presence of coal known as early as 1770, but not until 1795 did Schuylkill anthracite find its first use by local blacksmiths. Reviled as "stone coal" or "black stone" that would not ignite, anthracite found no ready commercial market, although it was used in wire and rolling mills located along the Schuylkill River before 1815 and to generate steam in the same area by 1830.

The resources of the subterranean Schuylkill region affected human patterns of surface regions only after the Schuylkill Navigation Canal was completed in 1825 (Figure 13.17), providing a passage to rapidly expanding external markets for the fuel and for the output of industry newly located atop the region. Growing demand induced a boom in coal exploitation, an exhaustion of the easily available outcrop coal, and the beginning of the more arduous and dangerous underground mining.

The early methods of mining were simple: merely quarrying the coal from exposed outcrops, usually driving on a slight incline to permit natural drainage. Deep shafts were unnecessary and, indeed, not thought of, since the presence of anthracite at depth was not suspected. Later, when it was no

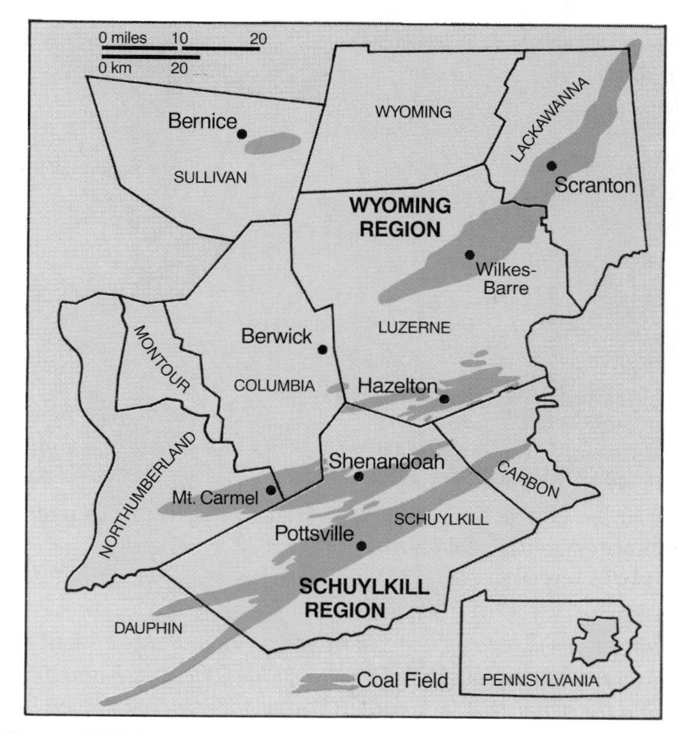

Figure 13.15 The anthracite regions of northeastern Pennsylvania are sharply defined by the geologic events that created them.

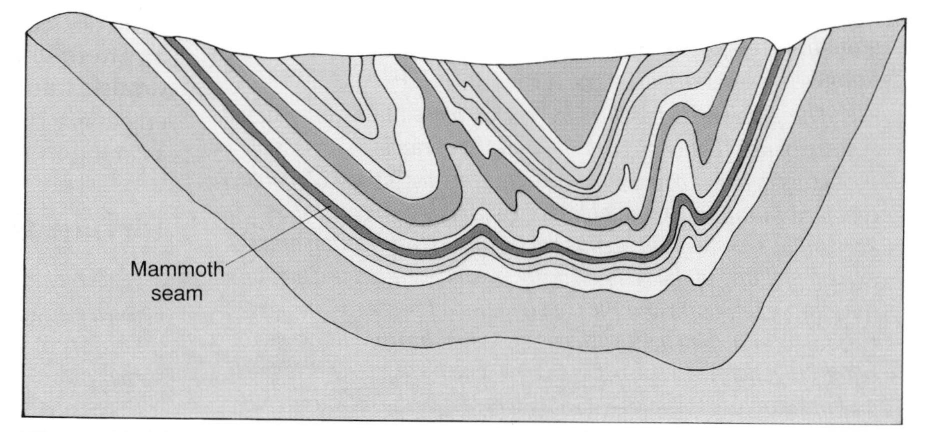

Figure 13.16 The deep folding of the Schuylkill coal seams made them costly to exploit. The Mammoth seam runs as deep as 450 to 600 meters (1500 to 2000 ft) below the surface.

[9]By Jerome Fellmann.

Figure 13.17 The Schuylkill Navigation Canal, like its counterpart Lehigh Canal shown here, provided an outlet to market for the anthracite region's coal resources after 1825.

longer possible to secure coal from a given outcrop, a small pit was sunk to a depth of 9–12 meters (30–40 feet); when the coal and the water that accumulated in the pit could no longer safely be brought to the surface by windlass, the pit was abandoned and a new one was started. Shaft mining, in which a vertical opening from the surface provides penetration to one or several coal beds, eventually became a necessity; with it came awareness of the complex interrelationships between seam thickness, the nature of interstratified rock and clays, the presence of gases, and the movement of subsurface water.

The subterranean Schuylkill region has a three-dimensional pattern of use. The configuration and the variable thickness of the seams demand concentration of mining activities. Minable coal is not uniformly available along any possible vertical or horizontal cross section because of the interstratification and the extreme folding of the beds. Mining is concentrated further by the location of shafts and the construction of passages, in their turn determined by both patterns of ownership and thickness of seam. In general, no seam less than 0.6 meters (2 feet) is worked, and the absolute thickness—15 meters (50 feet)—is found only in the Mammoth seam of the Schuylkill region.

Friable interstratified rock increases the danger of coal extraction and raises the costs of cave-in prevention. Although the Schuylkill mines are not gassy, the possibility of gas release from the collapse of coal pillars left as mine supports makes necessary systems of ventilation even more elaborate than those minimally required to provide adequate air to miners. Water is ever-present in the anthracite workings, and constant pumping or draining is necessary for mine operation. The collapse of strata underlying a river may result in sudden disastrous flooding.

The Schuylkill subterranean anthracite region presents a pattern of complexity in distribution of physical and cultural features and of interrelationships between phenomena every bit as great and inviting of geographical analysis as any purely surficial region.

Urban Regions

(See "Megalopolis," p. 458)

Urban geography represents a climax stage in the locational tradition of geography. Modern integrated, interdependent society on a world basis is urban-centered. Cities are the indispensable functional focuses of production, exchange, and administration. They exist individually as essential elements in interlocked hierarchical systems of cities. Internally, they display complex but repetitive spatial patterns of land uses and functions.

Because of the many-sided character of urbanism, cities are particularly good subjects for regional study. They are themselves, of course, formal regions. In the aggregate, their distributions give substance to formal regions of urban concentration. Cities are also the cores of functional regions of varying types and hierarchical orders. Their internal diversity of functional, land use, and socioeconomic patterns invite regional analysis. The employment of that approach in both its formal and functional modes is clearly displayed by Jean Gottmann, who, in examining the data and landscapes of the eastern United States at midcentury, recognized and analyzed Megalopolis. The following is taken from his study.

Megalopolis[10]

The northeastern seaboard of the United States is today the site of a remarkable development—an almost continuous stretch of urban and suburban areas from southern New Hampshire to northern Virginia and from the Atlantic shore to the Appalachian foothills (Figure 13.18). The processes of urbanization, rooted deep in the American past, have worked steadily here, endowing the region with unique ways of life and of land use. No other section of the United States has such a large concentration of population, with such a high average density, spread over such a large area. And no other section has a comparable role within the country or a comparable importance in the world. Here has been developed a kind of supremacy, in politics, in economics, and possibly even in cultural activities, seldom before attained by an area of this size.

Great, then, is the importance of this section of the United States and of the processes now at work within it. And yet it is difficult to single this area out from the surrounding areas, for its limits cut across established historical divisions, such as New England and the Middle Atlantic states, and across political entities, since it includes some states entirely and others only partially. A special name is needed, therefore, to identify this special geographical area.

This particular type of region is new, but it is the result of age-old processes, such as the growth of cities, the division of labor within a civilized society, the development of world resources. The name applied to it should, therefore, be new as a place name but old as a symbol of the long tradition of human aspirations and endeavor underlying the situations and problems now found here. Hence the choice of the term Megalopolis, *used in this study.*

As one follows the main highways or railroads between Boston and Washington, D.C., one hardly loses sight of built-up areas, tightly woven residential communities, or powerful concentrations of manufacturing plants. Flying this same route one discovers, on the other hand, that behind the ribbons of densely occupied land along the principal arteries of traffic, and in between the clusters of suburbs around the old urban centers, there still remain large areas covered with woods and brush alternating with some carefully cultivated patches of farmland (Figure 13.19). These green spaces, however, when inspected at closer range, appear stuffed with a loose but immense scattering of buildings, most of them residential but some of industrial character. That is, many of these sections that look rural actually function largely as suburbs in the orbit of some city's downtown. Even the farms, which occupy the larger tilled patches, are seldom worked by people whose only occupation and income are properly agricultural.

Thus the old distinctions between rural and urban do not apply here anymore. Even a quick look at the vast area of Megalopolis reveals a revolution in land use. Most of the people living in the so-called rural areas, and still classified as "rural population" by recent censuses, have very little, if anything, to do with agriculture. In terms of their interests and work they are what used to be classified as "city folks," but their way of life and the landscapes around their residences do not fit the old meaning of urban.

In this area, then, we must abandon the idea of the city as a tightly settled and organized unit in which people, activities, and riches are crowded into a very small area clearly separated from its nonurban surroundings. Every city in this region spreads out far and wide around its original nucleus; it grows amidst an irregularly colloidal mixture of rural and suburban landscapes; it melds on broad fronts with other mixtures, of somewhat similar though different texture, belonging to the suburban neighborhoods of other cities.

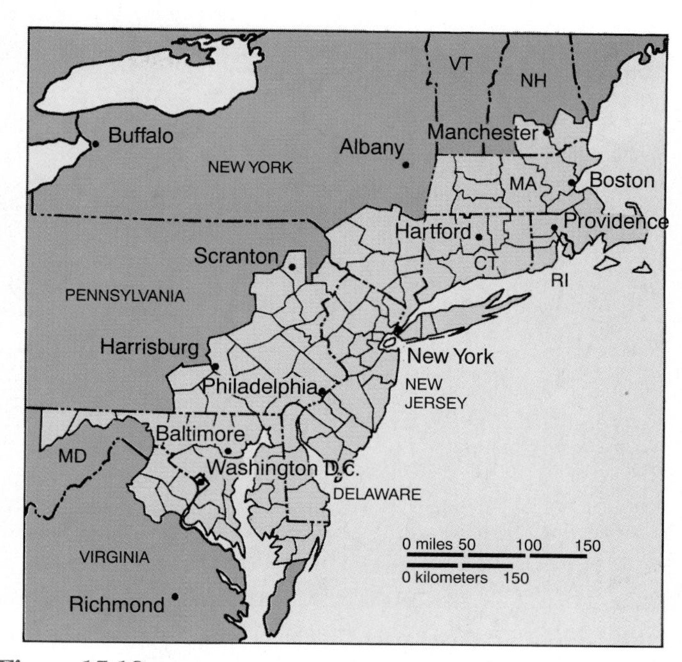

Figure 13.18 **Megalopolis in 1960.** The region was then composed of counties that, by United States census definition, were "urban" in population and economic characteristics. Much of the area is still distinctly "rural" in land use.

[10]Jean Gottmann, *Megalopolis: The Urbanized Northeastern Seaboard of the United States.* Copyright © 1961. Twentieth Century Fund, New York. Reprinted with permission.

Figure 13.19 The "Pine Barrens" of New Jersey is still a largely undisturbed natural enclave in the heart of Megalopolis. Its preservation is a subject of dispute between environmentalists and developers.
©Elizabeth J. Leppman.

> *Thus an almost continuous system of deeply interwoven urban and suburban areas, with a total population of about 37 million people in 1960, has been erected along the Northeastern Atlantic seaboard. It straddles state boundaries, stretches across wide estuaries and bays, and encompasses many regional differences. In fact, the landscapes of Megalopolis offer such variety that the average observer may well doubt the unity of the region. And it may seem to him that the main urban nuclei of the seaboard are little related to one another. Six of its great cities would be great individual metropolises in their own right if they were located elsewhere. This region indeed reminds one of Aristotle's saying that cities such as Babylon had "the compass of a nation rather than a city."*

The description of Megalopolis in 1960 represents, as does any regional study, a captured moment on the continuum of areal change. When that moment is lucidly described and the threads of areal character clearly delineated, the regional study serves as both a summary of the present and an augury of a future whose roots may be discerned in it. In the years since Gottmann described it, Megalopolis continued to develop along the lines he summarized. Urbanization proceeded, physically as well as functionally, to encroach upon the rural landscapes without regard for state boundaries or even the metropolitan cores that dominated in 1960. As then, new growth centers—becoming equivalents of central cities in their own right—were linked to the expanding transportation corridors, though now the lines of importance are increasingly expressways rather than railroads.

In the south, the corridors around Washington, D.C. are the Capital Beltway in Fairfax County and an extension from it westward to Loudoun County and Dulles Airport in Virginia and, to the north into Montgomery County, Maryland, the I-270 corridor. The Virginia suburbs, focusing on Tysons Corner (Figure 10.28) specialized in defense-related industries, but vast office building complexes and commercial centers are rapidly converting rural land to general urban uses. The Maryland suburbs emphasized health, space, and communications interests in their own new office, industrial, and commercial "parks" and complexes. Still farther to the north during the 1980s, in the "Princeton Corridor," a 42-kilometer (26-mile) stretch along Route 1 from New Brunswick to Trenton, New Jersey, huge corporate parks provided office and research space to companies relocating from the New York area and to new technology firms attracted by proximity to Princeton University. Morristown, New Jersey, and White Plains, in Westchester County, New York, were other similar concentrations of commercial, industrial, and office developments north and west of New York City. To the east, Stamford, Connecticut, with 150,000 daily in-commuters became by the late 1980s the headquarters center for major national corporations and had emerged as a truly major central city new on the scene in its present form since Megalopolis was first described.

Summary

The region is a mental construct, a created entity whose sole function is the purposeful organization of spatial data. The scheme of that organization, the selection of data to be analyzed, and the region resulting from these decisions are reflections of the intellectual problem posed.

This chapter has not attempted to explore all aspects of regionalism and of the regional method. It has tried, by example, only to document its basic theme: the geographer's regions are arbitrary but deliberately conceived devices for the isolation of things, patterns, interrelations, and flows that invite geographic analysis. In this sense, all geographers are regional geographers, and the regional examples of this chapter may logically complete our survey of the four traditions of geography.

Key Words

formal region 478
functional region 478
region 477
regional concept 477

For Review and Consideration

1. What do geographers seek to achieve when they recognize or define regions? On what basis are regional boundaries drawn? Are regions concrete entities whose dimensions and characteristics are agreed upon by all who study the same general segment of earth space? Ask three fellow students who are not participants in this course for their definition of the "South." If their answers differed, what implicit or explicit criteria of regional delimitation were they employing?
2. What are the spatial elements or the identifying qualities shared by all regions?
3. What is the identifying characteristic of a *formal region?* How are its boundaries determined? Name three different examples of formal regions drawn from any earlier chapters of this book. How was each defined and what was the purpose of its recognition?
4. How are *functional regions* defined? What is the nature of their bounding criteria? Give three or four examples of functional regions that were defined earlier in the text.
5. The ecosystem was suggested as a viable device for regional delimitation. What regional geographic concepts are suggested in ecosystem recognition? Is an ecosystem identical to a formal region? Why or why not?
6. National, linguistic, historical, planning, and other regions have been recognized in this chapter. With what other regional entities do you have acquaintance in your daily affairs? Are fire protection districts, police or voting precincts, or zoning districts regional units identifiable with geographers' regions as discussed in this chapter or elsewhere in this book? How influenced are you in your private life by your, or others', regional delimitations?

Selected References

Borchert, John R. *Megalopolis: Washington, D.C., to Boston.* Series: *Touring North America.* New Brunswick, N.J.: Rutgers University Press, 1992.

Freeman, T. W. *A Hundred Years of Geography.* Chapter 2, "The Regional Approach." Chicago: Aldine Publishing Co., 1961.

Gottmann, Jean. *Megalopolis Revisited: 25 Years Later.* College Park: Institute for Urban Studies, University of Maryland, 1987.

Johnston, R. J., J. Hauer, and G. A. Koekveld, eds. *Regional Geography: Current Developments and Future Prospects.* New York: Routledge, 1990.

McDonald, James R. "The Region: Its Conception, Design and Limitations." *Annals of the Association of American Geographers* 56 (1966):516–528.

Meinig, Donald. "The Mormon Culture Region: Strategies and Patterns in the Geography of the American West, 1847–1964." *Annals of the Association of American Geographers* 55 (1965): 191–220 (esp. 213–216).

Minshull, Roger. *Regional Geography.* Chicago: Aldine Publishing Co., 1967.

Moore, Tyrel G. "Core-Periphery Models, Regional Planning Theory, and Appalachian Development." *Professional Geographer* 46, no. 3 (1994):317–331.

Murphey, Rhoads. *The Scope of Geography.* 3d ed. Chapter 2, "The Region." London and New York: Methuen, Ltd., 1982.

Whittlesey, Derwent. "The Regional Concept and the Regional Method," in Preston E. James and Clarence F. Jones, eds. *American Geography: Inventory and Prospect.* Syracuse, N.Y.: Syracuse University Press (for the Association of American Geographers), 1954.

Don't forget about Dushkin's *Annual Editions Online: Geography* at http://dushkin.com/aeonline/
See preface for details.

APPENDIX 1998 WORLD POPULATION DATA

	Population mid-1998 (millions)	Births per 1,000 pop.	Deaths per 1,000 pop.	Natural Increase (annual, %)	"Doubling Time" in Year: at Current Rate	Projected Population for 2025 (millions)	Infant Mortality Rate[a]	Total Fertility Rate[b]	Percent of Population of Age <15/65+	Life Expectancy at Birth (years) Total	Percent Urban	Adult Literacy Rate (%)	% with Access to Safe Water	Energy Consumption per Capita[c]	Per Capita GNP 1997 (US$)[d]
WORLD	5,926	23	9	1.4	49	8,082	58	2.9	32/7	66	44	78	71	1,474	5,130
MORE DEVELOPED	1,178	11	10	0.1	548	1,240	8	1.6	19/14	75	73	99	+	—	(20,240)
LESS DEVELOPED	4,748	26	9	1.7	40	6,842	64	3.3	35/5	63	36	70	70	—	(1,230)
LESS DEVELOPED (EXCL. CHINA)	3,505	29	10	2.0	35	5,280	70	3.8	37/4	61	39	—	—	—	(1,420)
AFRICA	763	40	15	2.5	27	1,288	91	5.6	44/3	52	31	—	—	—	(650)
SUB-SAHARAN AFRICA	624	42	16	2.6	27	1,076	96	6.0	45/3	48	27	57	49	251	500
NORTHERN AFRICA	167	29	8	2.2	32	259	59	4.0	39/4	65	47	—	—	—	(1,280)
Algeria	30.2	31	7	2.4	29	47.3	44	4.4	39/4	67	56	62	78	866	1,490
Egypt	65.5	28	6	2.2	32	95.8	63	3.6	39/4	67	43	51	87	596	1,180
Libya	5.7	45	8	3.7	19	14.2	60	6.3	50/2	65	86	76	97	2,499	—
Morocco	27.7	24	7	1.8	40	41.2	62	3.3	36/5	72	52	44	65	311	1,250
Sudan	28.5	35	14	2.1	33	46.9	70	5.0	43/3	51	27	46	50	66	—
Tunisia	9.5	26	6	1.9	36	13.5	35	3.2	35/5	68	61	67	98	591	2,090
Western Sahara	0.2	47	18	2.9	24	0.4	150	6.9	—/—	47	—	—	—	—	—
WESTERN AFRICA	228	45	16	2.9	24	409	92	6.4	46/3	50	24	—	—	—	(310)
Benin	6.0	45	14	3.2	22	12.4	94	6.3	49/3	54	36	37	50	20	380
Burkina Faso	11.3	47	18	2.9	24	21.4	94	6.9	49/3	47	15	19	42	16	240
Cape Verde	0.4	36	8	2.9	24	0.5	52	5.3	45/6	70	44	72	51	307	1,090
Côte d'Ivoire	15.6	39	13	2.6	27	26.1	89	5.7	45/3	52	46	40	42	97	690
Gambia	1.2	43	19	2.4	29	2.0	90	5.9	44/3	45	37	39	48	56	350
Ghana	18.9	40	12	2.9	24	36.3	66	5.5	45/3	56	35	65	65	92	370
Guinea	7.5	43	19	2.4	29	13.1	153	5.7	47/3	45	29	36	46	64	570
Guinea-Bissau	1.1	42	22	2.1	34	1.9	141	5.8	42/4	43	22	55	59	37	240
Liberia	2.8	43	12	3.1	22	6.5	108	6.2	42/4	59	45	38	97	—	—
Mali	10.1	51	20	3.1	23	22.6	123	6.7	47/3	46	26	31	66	21	260
Mauritania	2.5	40	14	2.5	27	4.4	101	5.4	43/3	52	54	37	74	102	450
Niger	10.1	53	19	3.4	21	22.4	123	7.4	48/2	47	15	14	48	37	200
Nigeria	121.8	45	15	3.0	23	203.4	84	6.5	46/3	50	16	57	50	165	260

	Population mid-1998 (millions)	Births per 1,000 pop.	Deaths per 1,000 pop.	Natural Increase (annual, %)	"Doubling Time" in Year: at Current Rate	Projected Population for 2025 (millions)	Infant Mortality Rate[a]	Total Fertility Rate[b]	Percent of Population of Age <15/65+	Life Expectancy at Birth (years) Total	Percent Urban	Adult Literacy Rate (%)	% with Access to Safe Water	Energy Consumption Per Capital[c]	Per Capita GNP 1997 (US$)[d]
WESTERN AFRICA (continued)															
Senegal	9.0	43	16	2.7	26	17.0	68	5.7	45/3	49	42	33	63	104	550
Sierra Leone	4.6	49	30	1.9	36	8.2	195	6.5	44/3	34	36	31	34	72	200
Togo	4.9	46	11	3.6	19	11.0	84	6.8	46/3	58	31	52	55	45	330
EASTERN AFRICA	233	42	18	2.4	29	379	104	6.0	46/3	44	21	—	—	—	(230)
Burundi	5.5	43	18	2.5	28	10.5	105	6.6	47/3	46	5	35	52	23	180
Comoros	0.5	37	10	2.7	25	1.2	77	5.1	47/3	59	29	57	53	37	400
Djibouti	0.7	39	16	2.3	30	1.1	115	5.8	41/3	48	81	46	90	909	—
Eritrea	3.8	43	13	3.0	23	8.4	82	6.1	44/3	54	16	—	22	—	210
Ethiopia	58.4	46	21	2.5	28	98.8	128	7.0	46/3	42	16	36	25	21	110
Kenya	28.3	33	13	2.0	35	34.8	62	4.5	46/3	49	27	78	53	109	330
Madagascar	14.0	44	14	3.0	23	28.4	96	6.0	47/3	52	22	46	34	36	250
Malawi	9.8	42	24	1.7	41	10.9	140	5.9	48/3	36	20	56	37	38	220
Mauritius	1.2	17	7	1.0	69	1.4	21	2.0	27/6	70	43	83	98	388	3,800
Mozambique	18.6	41	19	2.2	32	33.3	134	5.6	46/2	44	28	40	63	38	90
Reunion	0.7	21	5	1.6	44	1.0	9	2.3	30/6	74	73	—	—	—	—
Rwanda	8.0	39	18	2.1	33	12.2	114	6.0	47/2	43	5	61	—	33	210
Seychelles	0.1	21	7	1.4	51	0.1	7	2.1	31/7	70	59	84	—	1,691	6,880
Somalia	10.7	50	19	3.2	22	23.7	122	7.0	48/3	47	24	24	31	—	—
Tanzania	30.6	42	17	2.5	28	50.7	100	5.7	46/3	47	21	68	38	32	210
Uganda	21.0	48	21	2.7	26	33.5	81	6.9	47/3	40	14	62	46	22	320
Zambia	9.5	42	23	1.9	36	16.2	109	6.1	45/3	37	39	78	27	145	380
Zimbabwe	11.0	35	20	1.5	46	12.4	53	4.4	44/3	40	31	85	79	424	750
MIDDLE AFRICA	90	46	16	3.1	23	187	103	6.5	46/3	49	34	—	—	—	(300)
Angola	12.0	51	19	3.2	22	25.5	124	7.2	48/3	47	42	42	32	89	340
Cameroon	14.3	41	13	2.8	25	28.5	65	5.9	44/4	55	44	63	50	117	650
Central African Republic	3.4	38	17	2.1	33	5.5	97	5.1	42/4	46	39	60	38	29	320
Chad	7.4	50	17	3.3	21	14.4	110	6.6	44/4	48	22	48	24	16	240
Congo	2.7	39	17	2.3	31	4.2	107	5.1	46/3	47	58	75	34	139	660
Congo, Dem. Rep. of (Zaire)	49.0	48	16	3.2	22	105.7	106	6.6	47/3	49	29	77	42	47	110
Equatorial Guinea	0.4	44	18	2.6	27	0.8	117	5.9	43/4	48	37	79	95	80	1,050
Gabon	1.2	35	15	2.0	35	2.1	94	5.0	38/6	54	73	63	68	587	4,230
Sao Tome and Principe	0.2	43	9	3.4	20	0.3	51	6.2	47/4	64	46	57	82	184	270
SOUTHERN AFRICA	45	28	12	1.7	42	54	55	3.5	36/4	56	53	—	—	—	(3,280)
Botswana	1.4	33	21	1.2	56	1.6	60	4.3	43/5	41	48	70	93	387	3,260
Lesotho	2.1	33	12	2.1	33	2.7	80	4.3	42/4	56	16	71	62	—	670
Namibia	1.6	36	20	1.7	42	2.3	68	5.1	42/4	42	27	76	57	—	2,220
South Africa	38.9	27	11	1.6	43	45.3	52	3.3	35/4	58	57	82	99	2,405	3,400
Swaziland	1.0	43	10	3.3	21	1.6	72	5.6	49/2	39	22	77	60	264	1,440
NORTH AMERICA	301	14	8	0.6	117	376	7	2.0	21/13	76	75	+	+	—	(27,100)
Canada	30.6	12	7	0.5	136	40.3	6.3	1.6	20/12	78	77	+	+	7,879	19,290
United States	270.2	15	9	0.6	116	335.1	7.0	2.0	22/13	76	75	+	+	7,905	28,740

	Population mid-1998 (millions)	Births per 1,000 pop.	Deaths per 1,000 pop.	Natural Increase (annual, %)	"Doubling Time" in Year: at Current Rate	Projected Population for 2025 (millions)	Infant Mortality Rate[a]	Total Fertility Rate[b]	Percent of Population of Age < 15/65+	Life Expectancy at Birth (years) Total	Percent Urban	Adult Literacy Rate (%)	% with Access to Safe Water	Energy Consumption Per Capital[c]	Per Capita GNP 1997 (US$)[d]
LATIN AMERICA & THE CARIBBEAN	500	25	7	1.8	38	697	36	3.0	34/5	69	72	87	77	969	3,880
CENTRAL AMERICA	132	29	5	2.3	30	197	32	3.4	37/4	71	66	—	—	—	(3,090)
Belize	0.2	33	6	2.7	26	0.4	34	4.1	42/5	72	51	87	89	417	2,740
Costa Rica	3.5	23	4	1.9	36	5.6	11.8	2.8	33/5	76	44	+	96	584	2,640
El Salvador	5.8	29	4	2.5	28	8.8	41	3.9	39/5	69	50	72	69	410	1,810
Guatemala	11.6	38	7	3.1	23	19.8	51	5.1	44/3	65	38	56	77	206	1,500
Honduras	5.9	33	6	2.8	25	9.7	42	4.4	42/3	68	44	73	87	236	700
Mexico	97.5	27	5	2.2	32	140.0	28	3.1	36/4	72	74	90	83	1,456	3,680
Nicaragua	4.8	38	6	3.2	22	8.5	46	4.6	44/3	66	63	66	61	265	410
Panama	2.8	23	5	1.8	39	3.7	22	2.7	33/5	74	55	91	93	678	3,080
CARIBBEAN	37	22	8	1.4	48	48	40	2.8	31/7	68	60	—	—	—	—
Antigua and Barbuda	0.1	17	5	1.2	59	0.1	18	1.7	30/6	74	36	+	+	2,017	7,380
Bahamas	0.3	23	6	1.7	42	0.4	19.0	2.0	32/5	72	86	+	+	6,864	11,830
Barbados	0.3	14	9	0.5	130	0.3	14.2	1.7	24/11	75	38	+	+	1,375	6,590
Cuba	11.1	14	7	0.6	107	11.8	7.2	1.4	22/9	75	74	+	93	923	—
Dominica	0.1	19	8	1.1	61	0.1	16.2	2.0	—/—	—	—	—	96	290	3,120
Dominican Republic	8.3	27	6	2.1	32	11.3	47	3.2	36/4	70	62	82	65	486	1,670
Grenada	0.1	29	6	2.3	30	0.2	12	3.8	38/6	71	32	+	—	293	3,000
Guadeloupe	0.4	18	6	1.2	58	0.5	7.9	2.0	26/9	77	99	—	—	—	—
Haiti	7.5	34	13	2.1	33	12.5	74	4.8	40/4	51	33	45	37	50	330
Jamaica	2.6	23	6	1.8	40	3.2	16	3.0	32/7	71	50	85	86	1,191	—
Martinique	0.4	15	6	0.9	75	0.5	6	1.7	24/11	78	81	—	—	—	—
Netherlands Antilles	0.2	19	7	1.2	59	0.3	6.3	2.2	27/7	75	90	—	—	—	—
Puerto Rico	3.9	18	8	1.0	71	4.4	11.5	2.1	27/10	74	71	—	—	—	7,010
St. Kitts-Nevis	0.04	19	9	1.0	69	0.1	25	2.6	34/10	67	43	90	+	486	6,160
Saint Lucia	0.1	25	6	1.9	36	0.2	18.0	2.7	37/6	71	48	—	—	338	3,620
St. Vincent & the Grenadines	0.1	22	7	1.6	44	0.2	19	2.4	37/6	73	25	82	89	199	2,500
Trinidad and Tobago	1.3	15	7	0.8	86	1.5	17.1	1.9	29/6	71	65	+	97	5,436	2,090
SOUTH AMERICA	331	24	7	1.7	42	453	37	2.8	33/6	69	76	—	—	—	(4,110)
Argentina	36.1	19	8	1.1	62	47.2	22.2	2.5	29/9	72	89	+	71	1,525	8,570
Bolivia	8.0	36	10	2.6	27	13.2	75	4.8	41/4	60	58	83	63	396	950
Brazil	162.1	22	8	1.4	48	208.2	43	2.5	32/6	67	76	83	76	772	4,720
Chile	14.8	19	6	1.4	50	19.5	11.1	2.4	29/7	75	85	+	95	1,065	5,020
Colombia	38.6	27	6	2.1	33	58.3	28	3.0	33/4	69	71	91	85	655	2,280
Ecuador	12.2	28	6	2.2	31	17.8	40	3.6	36/4	69	61	90	68	553	1,590
French Guiana	0.2	30	4	2.5	27	0.3	14	3.7	36/4	74	—	—	—	—	—
Guyana	0.7	24	7	1.7	40	0.7	63	2.7	35/4	66	36	+	61	350	800
Paraguay	5.2	32	6	2.7	26	9.4	27	4.4	41/4	69	52	92	60	308	2,010
Peru	26.1	28	6	2.2	32	39.2	43	3.5	35/4	69	71	89	67	421	2,460
Suriname	0.4	24	6	1.8	39	0.5	29	2.6	34/5	70	70	93	—	1,926	1,240
Uruguay	3.2	18	10	0.8	88	3.8	19.6	2.4	25/12	75	90	+	—	2,043	6,020
Venezuela	23.3	26	5	2.1	33	34.8	20.9	3.1	38/4	72	86	91	79	2,158	3,450

	Population mid-1998 (millions)	Births per 1,000 pop.	Deaths per 1,000 pop.	Natural Increase (annual, %)	"Doubling Time" in Year: at Current Rate	Projected Population for 2025 (millions)	Infant Mortality Rate[a]	Total Fertility Rate[b]	Percent of Population of Age <15/65+	Life Expectancy at Birth (years) Total	Percent Urban	Adult Literacy Rate (%)	% with Access to Safe Water	Energy Consumption Per Capita[c]	Per Capita GNP 1997 (US$)[d]
OCEANIA	30	18	7	1.1	63	40	28	2.4	27/10	73	71	—	—	—	(15,430)
Australia	18.7	14	7	0.7	101	23.5	5.3	1.8	21/12	78	85	+	+	5,215	20,540
Fed. States of Micronesia	0.1	33	8	2.6	27	0.2	46	4.7	44/4	66	27	81	100	—	1,980
Fiji	0.8	24	6	1.8	39	1.6	17	2.8	38/3	63	46	92	100	527	2,470
French Polynesia	0.2	23	5	1.8	39	0.4	11	3.1	36/3	70	54	—	—	—	—
Guam	0.2	28	4	2.4	29	0.2	9.1	3.4	30/4	74	38	—	—	—	—
Marshall Islands	0.1	43	7	3.6	19	0.2	26	6.7	51/3	62	65	91	74	—	1,770
New Caledonia	0.2	22	5	1.7	41	0.3	8	2.8	31/5	72	71	—	—	—	—
New Zealand	3.8	15	7	0.8	87	4.3	6.7	2.0	23/12	72	85	+	+	4,290	16,480
Palau	0.02	23	7	1.6	43	0.02	25	2.5	30/5	67	69	+	88	—	—
Papua-New Guinea	4.3	34	10	2.4	29	7.7	77	4.8	40/2	56	15	72	28	232	940
Solomon Islands	0.4	37	4	3.2	21	0.9	28	5.4	47/3	70	13	62	61	159	900
Vanuatu	0.2	35	7	2.8	25	0.3	41	4.7	46/4	63	18	64	87	279	1,310
Western Samoa	0.2	29	5	2.4	29	0.3	21	4.2	41/4	65	21	+	82	433	1,150
ASIA	3,604	23	8	1.5	46	4,965	57	2.8	32/6	65	34	—	—	—	(2,490)
ASIA (Excl. China)	2,361	26	8	1.8	39	3,404	66	3.3	35/5	63	36	—	—	—	3,500
WESTERN ASIA	182	28	7	2.2	32	303	54	4.0	37/5	67	64	—	—	—	—
Armenia	3.8	13	7	0.6	112	4.1	14	1.6	28/8	73	67	+	+	444	530
Azerbaijan	7.7	17	6	1.1	62	9.7	19	2.1	35/6	70	52	+	+	1,735	510
Bahrain	0.6	23	3	2.0	35	0.9	14	3.2	31/2	69	88	85	—	10,268	7,820
Cyprus	0.7	15	8	0.7	96	1.0	8	2.1	25/11	78	68	94	—	2,701	14,930
Gaza	1.1	52	6	4.6	15	3.0	33	7.4	50/3	72	—	—	—	—	—
Georgia	5.4	11	7	0.4	173	5.1	17	1.6	24/10	73	56	+	+	342	840
Iraq	21.8	38	10	2.8	25	41.6	127	5.7	43/3	59	70	58	78	1,213	—
Israel	6.0	21	7	1.5	47	8.1	6.7	2.9	30/9	78	90	+	+	3,003	15,810
Jordan	4.6	30	5	2.5	28	10.0	34	4.4	41/2	68	78	87	98	1,031	1,570
Kuwait	1.9	25	2	2.3	30	3.0	10	3.2	29/1	72	100	79	+	8,622	22,110
Lebanon	4.1	23	7	1.6	42	5.6	34	2.3	34/6	70	87	92	94	1,120	3,350
Oman	2.5	44	5	3.9	18	6.5	27	7.1	47/4	70	72	59	82	5,439	4,950
Qatar	0.5	19	2	1.7	42	0.7	12	4.1	27/1	71	91	79	—	12,597	11,570
Saudi Arabia	20.2	35	5	3.1	23	42.6	29	6.4	42/4	70	80	63	95	4,360	6,790
Syria	15.6	33	6	2.8	25	26.3	35	4.6	45/3	67	51	71	86	1,001	1,150
Turkey	64.8	22	7	1.6	45	88.0	42	2.6	31/5	68	64	82	49	1,009	3,130
United Arab Emirates	2.7	24	2	2.2	32	3.8	11	4.9	30/2	74	82	79	95	11,567	17,360
West Bank	1.8	40	6	3.4	21	3.9	27	5.4	45/4	72	—	—	—	—	—
Yemen	15.8	44	11	3.3	21	39.0	77	7.3	47/4	58	25	33	61	192	270
SOUTH CENTRAL ASIA	1,442	28	9	1.9	36	2,155	74	3.6	37/4	59	27	—	—	—	(410)
Afghanistan	24.8	43	18	2.5	28	48.0	150	6.1	41/3	46	18	32	12	—	—
Bangladesh	123.4	27	8	1.8	38	165.6	82	3.3	43/3	59	16	38	97	67	270
Bhutan	0.8	40	9	3.1	22	1.5	71	5.6	43/2	66	15	42	58	33	400
India	988.7	27	9	1.9	37	1,441.2	72	3.4	36/5	59	26	52	81	260	390
Iran	64.1	24	6	1.8	38	92.5	35	3.0	40/4	67	61	69	90	1,505	1,780
Kazakstan	15.6	15	10	0.5	133	18.7	25	1.9	30/7	65	56	+	93	3,3337	1,340

	Population mid-1998 (millions)	Births per 1,000 pop.	Deaths per 1,000 pop.	Natural Increase (annual, %)	"Doubling Time" in Year: at Current Rate	Projected Population for 2025 (millions)	Infant Mortality Rate[a]	Total Fertility Rate[b]	Percent of Population of Age <15/65+	Life Expectancy at Birth (years) Total	Percent Urban	Adult Literacy Rate (%)	% with Access to Safe Water	Energy Consumption Per Capital[c]	Per Capita GNP 1997 (US$)[d]
SOUTH CENTRAL ASIA (continued)															
Kyrgyzstan	4.7	22	8	1.5	48	7.0	28	2.8	37/6	67	34	+	71	513	440
Maldives	0.3	42	9	3.3	21	0.6	30	6.4	46/3	62	25	31	96	139	1,150
Nepal	23.7	33	11	2.2	32	39.5	79	4.6	43/4	55	10	28	63	33	210
Pakistan	141.9	39	11	2.8	25	258.1	91	5.6	41/4	58	28	38	74	243	490
Sri Lanka	18.9	19	6	1.3	53	24.1	16.5	2.2	35/4	72	22	90	57	136	800
Tajikistan	6.1	22	5	1.7	41	9.3	32	2.9	40/4	68	28	+	60	563	330
Turkmenistan	4.7	24	7	1.7	41	6.1	42	2.9	39/4	66	45	+	74	3,047	630
Uzbekistan	24.1	26	6	2.0	35	42.3	26	3.2	41/4	70	38	+	62	2,043	1,010
SOUTHEAST ASIA	512	24	8	1.6	42	709	55	2.9	35/4	64	35	—	—	—	(1,580)
Brunei	0.3	25	3	2.2	32	0.5	8.4	3.4	34/3	71	67	88	—	10,839	25,090
Cambodia	10.8	38	14	2.4	29	17.0	116	5.2	44/4	52	14	65	36	52	300
Indonesia	207.4	24	8	1.5	45	275.2	66	2.7	34/4	62	37	84	62	442	1,110
Laos	5.3	42	14	2.8	25	9.8	97	5.9	45/3	54	19	57	44	40	400
Malaysia	22.2	26	5	2.1	33	37.0	10	3.2	35/4	72	57	84	78	1,655	4,680
Myanmar	47.1	30	10	2.0	35	67.8	83	3.8	36/4	61	25	83	60	49	—
Philippines	75.3	30	7	2.3	30	116.8	34	3.7	38/4	66	47	+	84	307	1,220
Singapore	3.9	15	5	1.1	65	4.2	3.8	1.7	23/7	77	100	91	100	7,162	32,940
Thailand	61.1	17	7	1.1	64	71.6	25	2.0	27/5	69	31	94	89	878	2,800
Vietnam	78.5	19	6	1.2	57	109.5	38	2.3	40/5	67	20	94	43	104	320
EAST ASIA	1,469	16	7	0.9	74	1,798	29	1.8	25/7	72	37	—	—	—	(4,750)
China	1,242.5	17	7	1.0	69	1,561.4	31	1.8	26/6	71	30	82	67	707	860
China, Hong Kong SAR	6.7	9	5	0.4	161	7.8	4.0	1.1	18/10	79	—	—	+	2,185	25,280
Japan	126.4	10	7	0.2	330	120.9	3.8	1.4	15/16	80	78	+	+	3,964	37,850
Korea, North	22.2	18	9	0.9	75	26.1	39	1.9	28/6	66	59	—	81	1,129	—
Korea, South	46.4	16	6	1.0	68	52.7	11	1.7	22/6	74	79	+	93	3,225	10,550
Macao	0.5	13	3	1.0	71	0.6	5	1.5	25/7	80	97	—	—	—	—
Mongolia	2.4	24	7	1.6	42	3.3	49	3.1	36/4	57	57	83	40	1,045	390
Taiwan	21.7	15	7	1.0	73	25.4	6.7	1.7	23/8	75	75	+	+	—	—
EUROPE	728	10	11	-0.1	—	715	10	1.4	19/14	73	71	—	—	—	(13,710)
NORTHERN EUROPE	95	12	11	0.1	535	99	6	1.7	19/16	76	84	+	+	—	(20,320)
Denmark	5.3	13	11	0.2	431	5.6	5.8	1.8	18/16	75	85	+	+	3,918	32,500
Estonia	1.4	9	13	-0.4	—	1.2	10	1.3	20/14	68	70	+	+	3,454	3,330
Finland	5.2	12	10	0.2	377	5.2	3.5	1.7	19/17	77	65	+	+	5,613	24,080
Iceland	0.3	15	7	0.9	82	0.3	5.5	2.0	24/11	78	92	+	+	7,932	27,580
Ireland	3.7	14	9	0.5	133	3.8	5.5	1.9	23/11	75	57	+	+	3,196	18,280
Latvia	2.4	8	14	-0.6	—	2.0	16	1.2	20/14	70	69	+	+	1,471	2,430
Lithuania	3.7	11	12	-0.1	—	3.5	10	1.4	21/12	71	68	+	+	2,291	2,230
Norway	4.4	14	10	0.3	201	4.9	4.0	1.8	20/16	78	74	+	+	5,439	36,090
Sweden	8.9	10	11	-0.0	—	9.3	3.9	1.6	19/17	79	83	+	+	5,736	26,220
United Kingdom	59.1	13	11	0.2	433	62.6	6.1	1.7	19/16	77	90	+	+	3,786	20,710
WESTERN EUROPE	183	11	10	0.1	517	184	5	1.5	18/15	77	79	+	+	—	(28,250)
Austria	8.1	11	10	0.1	990	8.3	4.8	1.4	17/15	77	65	+	+	3,279	27,980

	Population mid-1998 (millions)	Births per 1,000 pop.	Deaths per 1,000 pop.	Natural Increase (annual, %)	"Doubling Time" in Year: at Current Rate	Projected Population for 2025 (millions)	Infant Mortality Rate[a]	Total Fertility Rate[b]	Percent of Population of Age <15/65+	Life Expectancy at Birth (years) Total	Percent Urban	Adult Literacy Rate (%)	% with Access to Safe Water	Energy Consumption Per Capital[c]	Per Capita GNP 1997 (US$)[d]
WESTERN EUROPE (continued)															
Belgium	10.2	12	10	0.1	529	10.3	5.8	1.6	18/16	77	97	+	+	5,167	26,420
France	58.8	12	9	0.3	210	64.2	5.1	1.7	19/16	78	74	+	+	4,150	26,050
Germany	82.3	10	10	-0.1	—	76.1	4.9	1.3	16/15	77	85	+	+	4,156	28,260
Liechtenstein	0.03	13	7	0.6	124	0.04	7.4	1.5	19/10	72	—	+	+	—	—
Luxembourg	0.4	14	9	0.4	161	0.5	4.9	1.8	19/14	76	86	+	+	9,361	45,330
Monaco	0.03	20	17	0.3	239	0.04	3.9	—	—/—	—	—	+	+	—	—
Netherlands	15.7	12	9	0.3	210	17.3	5.7	1.5	18/13	78	61	+	+	4,741	25,820
Switzerland	7.1	12	9	0.3	248	7.5	4.7	1.5	18/16	79	68	+	+	3,571	44,320
EASTERN EUROPE	307	9	13	-0.4	—	290	15	1.3	20/13	68	68	+	—	—	(2,350)
Belarus	10.2	9	13	-0.4	—	9.8	12	1.3	21/13	68	69	+	—	2,305	2,150
Bulgaria	8.3	9	14	-0.5	—	7.9	15.6	1.2	17/15	71	68	+	—	2,724	1,140
Czech Republic	10.3	9	11	-0.2	—	10.2	5.9	1.2	18/14	74	77	+	—	3,776	5,200
Hungary	10.1	10	14	-0.4	—	9.3	10.0	1.4	18/14	70	63	+	—	2,454	4,430
Moldova	4.2	12	12	0.1	1,386	4.9	20	1.8	26/9	66	46	+	—	963	540
Poland	38.7	11	10	0.1	630	40.8	12.2	1.6	22/11	72	62	+	—	2,448	3,590
Romania	22.5	10	12	-0.2	—	19.7	22.6	1.3	19/13	69	55	+	—	1,941	1,420
Russia	146.9	9	14	-0.5	—	134.6	17	1.2	20/12	67	73	+	—	4,079	2,740
Slovakia	5.4	11	10	0.2	408	5.3	10.2	1.5	22/11	73	57	+	—	3,272	3,700
Ukraine	50.3	9	15	-0.6	—	47.0	14	1.3	20/14	68	68	+	97	3,136	1,040
SOUTHERN EUROPE	144	10	9	0.1	853	143	8	1.3	17/15	77	61	—	—	—	(15,290)
Albania	3.3	17	5	1.2	58	4.6	20.4	2.0	34/6	72	37	85	—	314	750
Andorra	0.1	11	3	0.8	89	0.1	2.9	1.7	16/11	79	63	+	—	—	—
Bosnia-Herzegovina	4.0	13	7	0.6	124	4.3	—	1.5	23/7	72	40	+	—	348	—
Croatia	4.2	12	11	0.1	990	4.2	8.0	1.6	19/12	72	54	+	96	1,435	4,610
Greece	10.5	10	10	0.0	6,931	10.2	8.1	1.3	16/16	78	59	+	—	2,266	12,010
Italy	57.7	9	9	-0.0	—	54.8	5.8	1.2	15/17	78	67	+	+	2,821	20,120
Macedonia	2.0	16	8	0.8	90	2.3	16.4	2.1	26/9	71	60	—	—	1,308	1,090
Malta	0.4	13	7	0.6	120	0.4	10.7	2.1	22/12	77	89	91	—	2,511	8,630
Portugal	10.0	11	11	0.0	2,310	9.4	6.9	1.4	17/15	75	48	90	—	1,939	10,450
San Marino	0.03	11	7	0.4	161	0.03	10.6	1.3	15/15	76	89	+	—	—	—
Slovenia	2.0	9	9	0.0	—	2.0	4.7	1.3	18/13	75	50	+	—	2,806	9,680
Spain	39.4	9	9	0.0	1,733	39.0	4.7	1.2	16/16	77	64	+	99	2,639	14,510
Yugoslavia	10.6	13	11	0.2	289	11.4	14	1.8	22/12	72	51	+	—	1,110	—

[a]Infant deaths per 1,000 live births.

[b]Average number of children born to a woman during her lifetime.

[c]Commercial energy only, in kilograms of oil equivalent.

[d]Regional summaries in ( ) are 1996 data.

A dash (—) indicates data unavailable or inapplicable.

A plus sign (+) indicates that according to UNESCO, adult literacy is 95% or more. For countries of the industrialized "North," the "+" also implies essentially 100% access to safe water.

Urban population data are the percentage of the total population living in areas termed urban by that country.

Table modified from the **1998 World Population Data Sheet** of the Population Reference Bureau. Data for safe water supply are based on World Health Organization reports. Data on adult literacy are based on UNESCO sources. GNP Per Capita data are from the World Bank.

GLOSSARY

A

absolute direction Direction with respect to cardinal east, west, north, and south reference points

absolute location (*syn:* mathematical location) The exact position of an object or place stated in spatial coordinates of a grid system designed for locational purposes. In geography, the reference system is the global grid of parallels of latitude north or south of the equator and of meridians of longitude east or west of a prime meridian.

accelerated eutrophication The overnourishment of a water body with nutrients stemming from human activities such as agriculture, industry, and urbanization.

accessibility The relative ease with which a destination may be reached from other locations; the relative opportunity for spatial interaction. May be measured in geometric, social, or economic terms.

acculturation The cultural modification or change resulting from one culture group or individual adopting traits of a more advanced or dominant society; cultural development through "borrowing."

acid rain Precipitation that is unusually acidic; created when oxides of sulfur and nitrogen change chemically as they dissolve in water vapor in the atmosphere and return to earth as acidic rain, snow, fog, or dry particles.

activity space The area within which people move freely on their rounds of regular activity.

adaptation A presumed modification of heritable traits through response to environmental stimuli.

agglomeration The spatial grouping of people or activities for mutual benefit.

agglomeration economies (*syn:* external economies) The savings to an individual enterprise that result from spatial association with other similar economic activities.

agriculture Cultivating the soil, producing crops, and raising livestock; farming.

air mass A large body of air with little horizontal variation in temperature, pressure, and humidity.

air pressure The weight of the atmosphere as measured at a point on the earth's surface.

alluvial fan A fan-shaped accumulation of alluvium deposited by a stream at the base of a hill or mountain.

alluvium The sediment carried by a stream and deposited in a floodplain or delta.

amalgamation theory In human geography, the concept that multiethnic societies become a merger of the culture traits of their member groups.

anaerobic digestion The process by which organic waste is decomposed in an oxygen-free environment to produce methane gas (biogas).

anecumene *See* nonecumene.

animism A belief that natural objects may be the abode of dead people, spirits, or gods who occasionally give the objects the appearance of life.

antecedent boundary A boundary line established before the area in question is well populated.

aquaculture The breeding of fish in freshwater ponds, lakes, and canals or in fenced-off coastal bays and estuaries; fish farming.

aquifer Underground porous and permeable rock that is capable of holding ground water, especially rock that supplies economically significant quantities of water to wells and springs.

arable land Land that is or can be cultivated.

Arctic haze Air pollution resulting from the transport by air currents of combustion-based pollutants to the area north of the Arctic Circle.

area analysis tradition One of the four traditions of geography, that of regional geography.

arithmetic density *See* crude density.

arroyo A steep-sided, flat-bottomed gully, usually dry, carved out of desert land by rapidly flowing water.

artifacts The material manifestations of culture, including tools, housing, systems of land use, clothing, and the like. Elements in the technological subsystem of culture.

artificial boundary *See* geometric boundary.

assimilation The social process of merging into a composite culture, losing separate ethnic or social identity and becoming culturally homogenized.

asthenosphere A partially molten, plastic layer above the core and lower mantle of the earth.

atmosphere The gaseous mass surrounding the earth.

atoll A near-circular low coral reef formed in shallow water enclosing a central lagoon; most common in the central and western Pacific Ocean.

azimuthal projection A map projection that shows true directions from one central point to all other points.

B

barchan A crescent-shaped sand dune; the horns of the crescent point downwind.

basic sector Those products of an urban unit that are exported outside the city itself, earning income for the community.

bench mark A surveyor's mark indicating the position and elevation of some stationary object; used as a reference point in surveying and mapping.

biocide A chemical used to kill plant and animal pests and disease organisms. *See also* herbicide, pesticide.

biological magnification The accumulation of a chemical in the fatty tissue of an organism and its concentration at progressively higher levels in the food chain.

biomass Living matter, plant and animal, in any form.

biomass fuels The combustible and/or fermentable material of plant or animal origin, such as wood or corncobs, that can be used as a source of energy.

biome The total assemblage of living organisms in a single major ecological region.

biosphere (*syn:* ecosphere) The thin film of air, water, and earth within which we live, including the atmosphere, surrounding and subsurface waters, and the upper reaches of the earth's crust.

birth rate (*syn:* crude birth rate) The ratio of the number of live births during one year to the total population, usually at the midpoint of the same year, expressed as the number of births per year per 1000 population.

blizzard A heavy snowstorm accompanied by high winds.

boundary A line separating one political unit from another.

boundary definition A general agreement between two states about the allocation of territory between them.

boundary delimitation The plotting of a boundary line on maps or aerial photographs.

boundary demarcation The actual marking of a boundary line on the ground; the final stage in boundary development.

butte A small, flat-topped, isolated hill with steep sides, common in dry climate regions.

C

carcinogen A substance that produces or incites cancerous growth.

carrying capacity The numbers of any population that can be adequately supported by the available resources upon which that population subsists; for humans, the numbers supportable by the known and utilized resources—usually agricultural—of an area.

cartogram A map that has been simplified to present a single idea in a diagrammatic way; the base is not normally true to scale.

caste One of the hereditary social classes in Hinduism that determines one's occupation and position in society.

central business district (CBD) The center or "downtown" of an urban unit, where retail stores, offices, and cultural activities are concentrated and where land values are high.

central city That part of the urban area contained within the boundaries of the main city around which suburbs have developed.

central place A nodal point for the distribution of goods and services to a surrounding hinterland population.

central place theory A deductive theory formulated by Walter Christaller (1893–1969) to explain the size and distribution of settlements through reference to competitive supply of goods and services to dispersed rural populations.

centrifugal force In political geography, a factor that destabilizes and weakens a state.

centripetal force In political geography, a factor that promotes unity and national identity.

CFCs *See* chlorofluorocarbons.

channelization The modification of a stream channel; specifically, the straightening of meanders or dredging of the stream channel to deepen it.

channelized migration The tendency for migration to flow between areas that are socially and economically allied by past migration patterns, by economic trade considerations, or by some other affinity.

chemical weathering The decomposition of earth materials due to chemical reactions that include oxidation, hydration, and carbonation.

chlorofluorocarbons (CFCs) A family of synthetic chemicals that have significant commercial applications but whose emissions are contributing to the depletion of the ozone layer.

choropleth map A map that depicts quantities for areal units by varying pattern and/or color.

circumpolar vortex High-altitude winds circling the poles from west to east.

city A multifunctional nucleated settlement with a central business district and both residential and nonresidential land uses.

climate The long-term average weather conditions in a place or region.

climograph A bar and line graph used to depict average monthly temperatures and precipitation.

coal gasification A process by which crushed coal is burned in the presence of steam or oxygen to produce a synthetic gas.

coal liquefaction A process whereby coal is heated to produce a variety of liquid products that can be used as fuels.

coal slurry A mixture of finely ground coal and water that is moved by pipeline.

coastal wetland Land along a coastline that is occasionally or permanently covered with salt water, such as a marsh, swamp, or tidal flat.

cogeneration The simultaneous use of a single fuel for the generation of electricity and low grade central heat.

cognition The process by which an individual gives mental meaning to information.

cohort A population group unified by a specific common characteristic, such as age, and who are treated as a statistical unit during their lifetimes.

commercial economy The production of goods and services for exchange in competitive markets where price and availability are determined by supply and demand forces.

commercial energy Commercially traded fuels such as coal, oil, or natural gas and excluding wood, vegetable or animal wastes, or other biomass.

Common Market *See* European Union.

compact state A state whose territory is nearly circular.

comparative advantage A region's profit potential for a productive activity compared to alternate areas of production of the same good or to alternate uses of the region's resources.

concentric zone model A model describing urban land uses as a series of circular belts or rings around a core central business district, each ring housing a distinct type of land use.

conformal projection A map projection on which the shapes of small areas are accurately portrayed.

conic projection Any of several map projections based on the projection of the grid system onto a cone.

connectivity The directness of routes linking pairs of places; all of the tangible and intangible means of connection and communication between places.

consequent boundary (*syn:* ethnographic boundary) A boundary line that coincides with some cultural divide, such as religion or language.

conservation The wise use or preservation of natural resources so as to maintain supplies and qualities at levels sufficient to meet present and future needs.

contagious diffusion The spread of a concept, practice, or article from one area to others through contact and/or the exchange of information.

containment A guiding principle of U.S. foreign policy during the Cold War, that the USSR should be confined within its borders.

continental drift The hypothesis that an original single landmass (Pangaea) broke apart and that the continents have moved very slowly over the asthenosphere to their present locations.

contour interval The vertical distance separating two adjacent contour lines.

contour line A map line along which all points are of equal elevation above or below a datum plane, usually mean sea level.

conurbation A large metropolitan complex formed by the coalescence of two or more urban areas.

convection The circulatory movement of rising warm air and descending cool air.

convectional precipitation Rain produced when heated, moisture-laden air rises and then cools below the dew point.

Convention on the Law of the Sea *See* Law of the Sea Convention.

coral reef A rocklike landform in shallow tropical water composed chiefly of compacted coral and other organic material.

core The nucleus of a region or country, the main center of its industry, commerce, population, political, and intellectual life; in urban geography, that part of the central business district characterized by intensive land development.

core area The nucleus of a state, containing its most developed area, greatest wealth, densest populations, and clearest national identity.

Coriolis effect A fictitious force used to describe motion relative to a rotating earth; specifically, the force that tends to deflect a moving object or fluid to the right (clockwise) in the Northern Hemisphere and to the left (counterclockwise) in the Southern Hemisphere.

country *See* state.

creole A language developed from a pidgin to become the native tongue of a society.

critical distance The distance beyond which cost, effort, and/or means play an overriding role in the willingness of people to travel.

crude birth rate (CBR) *See* birth rate.

crude death rate (CDR) *See* death rate.

crude density (*syn:* arithmetic density, population density) The number of people per unit area of land.

crude oil A mixture of hydrocarbons that exists in a liquid state in underground reservoirs; petroleum as it occurs naturally, as it comes from an oil well, or after extraneous substances have been removed.

cultural convergence The tendency for cultures to become more alike as they increasingly share technology and organization structures in a modern world united by improved transportation and communication.

cultural divergence The likelihood or tendency for isolated cultures to become increasingly dissimilar with the passage of time.

cultural ecology The study of the interactions between societies and the natural environments they occupy.

cultural integration The observation that all aspects of a culture are interconnected; no part can be altered without impact upon other culture traits.

cultural lag The retention of established culture traits despite changing circumstances rendering them inappropriate.

cultural landscape The natural landscape as modified by human activities and bearing the imprint of a culture group or society; the built environment.

culture The totality of learned behaviors and attitudes transmitted within a society to succeeding generations by imitation, instruction, and example.

culture complex An integrated assemblage of culture traits descriptive of one aspect of a society's behavior or activity.

culture–environment tradition One of the four traditions of geography; in this text, identified with population, cultural, political, and behavioral geography.

culture hearth A nuclear area within which an advanced and distinctive set of culture traits develops and from which there is diffusion of distinctive technologies and ways of life.

culture realm A collective of culture regions sharing related culture systems; a major world area having sufficient distinctiveness to be perceived as set apart from other realms in its cultural characteristics and complexes.

culture region A formal or functional region within which common cultural characteristics prevail. It may be based on single culture traits, on culture complexes, or on political, social, or economic integration.

culture system A generalization suggesting shared, identifying traits uniting two or more culture complexes.

culture trait A single distinguishing feature of regular occurrence within a culture, such as the use of chopsticks or the observance of a particular caste system. A single element of learned behavior.

cyclone A type of atmospheric disturbance in which masses of air circulate rapidly about a region of low atmospheric pressure.

cyclonic precipitation (*syn:* frontal precipitation) The rain or snow that is produced when moist air of one air mass is forced to rise over the edge of another air mass.

cylindrical projection Any of several map projections based on the projection of the globe grid onto a cylinder.

D

data base *See* geographic data base.

DDT A chlorinated hydrocarbon that is among the most persistent of the biocides in general use.

death rate (*syn:* crude death rate, mortality rate) A mortality index usually calculated as the number of deaths per year per 1000 population.

decomposers Microorganisms and bacteria that feed on dead organisms, causing their chemical disintegration.

deforestation The clearing of land through total removal of forest cover.

delta A triangular-shaped deposit of mud, silt, or gravel created by a stream where it flows into a body of standing water.

demographic equation A mathematical expression that summarizes the contribution of different demographic processes to the population change of a given area during a specified time period.

demographic momentum *See* population momentum.

demographic transition A model of the effect of economic development on population growth. A first stage involves both high birth and death rates; the second phase displays high birth rates and falling mortality rates and population increases. Phase three shows reduction in population growth as birth rates decline to the level of death rates. The final, fourth, stage implies again a population stable in size but larger in numbers than at the start of the transition cycle.

demography The scientific study of population, with particular emphasis upon quantitative aspects.

density of population *See* population density.

dependency ratio The number of dependents, old or young, that each 100 persons in the productive years must support.

deposition The process by which silt, sand, and rock particles accumulate and create landforms such as stream deltas and talus slopes.

desertification The extension of desertlike landscapes as a result of climatic change or human activities, such as overgrazing or deforestation, usually in arid and semiarid regions.

developable surface A geometric form, such as a cylinder or cone, that may be flattened without tearing or stretching.

devolution Decentralization of political control.

dew point The temperature at which condensation forms, if the air is cooled sufficiently.

dialect A regional or socioeconomic variation of a more widely spoken language.

diastrophism The earth force that folds, faults, twists, and compresses rock.

dibble Any small hand tool or stick used to make a hole for planting.

diffusion The spread or movement of a concept, practice, article, or population from one point of origin to other areas.

distance The amount of separation between two objects, areas, or points; an extent of areal or linear measure.

distance decay The exponential decline of an activity or function with increasing distance from its point of origin.

domestication The successful transformation of plant or animal species from a wild state to a condition of dependency upon human management, usually with distinct physical change from wild forebears.

domino theory The geopolitical theory that if one state falls under the control of a rival power then its neighbors will follow in sequence into the enemy camp.

doubling time The time period required for any beginning total, experiencing a compounding growth, to double in size.

dune A wavelike desert landform created by windblown sand.

E

earthquake The movement of the earth along a geologic fault or at some other point of weakness at or near the earth's surface.

earth science tradition One of the four traditions of geography, identified with physical geography in general.

ecology The scientific study of how living creatures affect each other and what determines their distribution and abundance.

economic base The mix of manufacturing and service activities performed by the labor force of a city to satisfy demands both inside and outside the city and earn income to support the urban population.

economic geography The study of how people earn a living, how livelihood systems vary by area, and how economic activities are spatially interrelated and linked.

ecosphere *See* biosphere.

ecosystem A population of organisms existing together in a particular area, together with the energy, air, water, soil, and chemicals upon which it depends.

ecumene The permanently inhabited areas of the earth. *See also* nonecumene.

electoral geography The study of the delineation of voting districts and the spatial patterns of election results.

electromagnetic spectrum The entire range of radiation, including the shortest as well as the longer wavelengths.

El Niño The periodic (every 3 to 7 or 8 years) buildup of warm water along the west coast of South America; replacing the cold Humboldt current off the Peruvian coast, El Niño is associated with both a fall in plankton levels (and decreased fish supply) and short-term, widespread weather modification.

elongated state A state whose territory is long and narrow.

enclave A territory that is surrounded by, but is not part of, a state.

energy The ability to do work. *See also* kinetic energy, potential energy.

energy efficiency The ratio of the output of useful energy from a conversion process to the total energy inputs.

environment Surroundings; the totality of things that in any way may affect an organism, including both physical and cultural conditions; a region characterized by a certain set of physical conditions.

environmental determinism The theory that the physical environment, particularly climate, molds human behavior.

environmental perception The way people observe and interpret, and the ideas they have about, near or distant places.

environmental pollution *See* pollution.

epidemiologic transition The conversion of formerly fatal epidemic diseases into conditions continual within populations that develop partial immunity to them.

equal-area projection *See* equivalent projection.

equator An imaginary line that encircles the globe halfway between the North and South poles.

equidistant projection A map projection on which true distances in all directions can be measured from one or two central points.

equivalent projection A map projection on which the areas of regions are represented in correct or constant proportions to earth reality; also called equal-area.

erosion The result of processes that loosen, dissolve, wear away, and remove earth and rock material. Those processes include weathering, solution, abrasion, and transportation.

erosional agents The forces of wind, moving water, glaciers, waves, and ocean currents that carve, wear away, and remove rock and soil particles.

estuarine zone The relatively narrow area of wetlands along coastlines where salt water and fresh water mix.

estuary The lower course or mouth of a river where tides cause fresh water and salt water from the sea to mix.

ethnic cleansing The killing or forcible relocation of one traditional or ethnic group by a more powerful one.

ethnicity The social status afforded to, usually, a minority group within a national population. Recognition is based primarily upon culture traits such as religion, distinctive customs, or native or ancestral national origin.

ethnic religion A religion identified with a particular ethnic group and largely exclusive to it.

ethnocentrism The belief that one's own ethnic group is superior to all others.

ethnographic boundary *See* consequent boundary.

European Union (EU) An economic association established in 1957 of a number of Western European states that promotes free trade among member countries; often called the Common Market.

eutrophication The enrichment of a water body by the addition of nutrients received through erosion and runoff from the watershed. *See also* accelerated eutrophication.

exclave A portion of a state that is separated from the main territory and surrounded by another country.

exclusive economic zone (EEZ) As proposed in the Convention on the Law of the Sea, a zone of exploitation extending 200 nautical miles seaward from a coastal state that has exclusive mineral and fishing rights over it.

expansion diffusion The spread of ideas, behaviors, or articles from one culture to others through contact and exchange of information; the dispersion leaves the phenomenon intact or intensified in its area of origin. *See also* relocation diffusion.

extensive agriculture A crop or livestock system in which land quality or extent is more important than capital or labor inputs in determining output. May have either commercial or subsistence orientation.

extensive commerical agriculture *See* extensive agriculture.

extensive subsistence agriculture *See* extensive agriculture.

external economies *See* agglomeration economies.

extractive industries Primary activities involving the mining and quarrying of nonrenewable metallic and nonmetallic mineral resources.

extrusive rock Rock solidified from molten material that has issued out onto the earth's surface.

F

false-color image A remotely sensed image whose colors do not appear natural to the human eye.

fast breeder reactor A nuclear reactor that uses uranium-235 to release energy from the more abundant uranium-238.

fault A break or fracture in rock produced by stress or the movement of lithospheric plates.

fault escarpment A steep slope formed by the vertical movement of the earth along a fault.

filtering In urban geography, a process whereby individuals of one income group replace residents of a portion of an urban area who are of another income group.

fiord A glacial trough whose lower end is filled with seawater.

floodplain A valley area bordering a stream that is subject to inundation by flooding.

folds Rock layers that have buckled under pressure by the movement of lithospheric plates.

folk culture The body of institutions, customs, dress, artifacts, collective wisdoms, and traditions of a homogeneous, isolated, largely self-sufficient, and relatively static social group.

food chain A sequence of organisms through which energy and materials move within an ecosystem.

footloose A descriptive term applied to manufacturing activities for which the cost of transporting material or product is not important in determining location of production.

formal region An earth area throughout which a single feature or limited combination of features is of such uniformity that it can serve as the basis for an areal generalization and of contrast with adjacent areas.

form utility A value-increasing change in the form—and therefore in the utility—of a raw material or commodity.

forward-thrust capital A capital city deliberately sited in a state's frontier zone.

fossil fuels Hydrocarbon compounds of crude oil, natural gas, and coal that are derived from the accumulation of plant and animal remains in ancient sedimentary rocks.

fragmented state A state whose territory contains isolated parts, separated and discontinuous.

frictional effect In climatology, the slowing of wind movement due to the frictional drag of the earth's surface.

friction of distance A measurement indicating the effect of distance on the extent of interaction between two points. Generally, the greater the distance, the less the interaction or exchange or the greater the cost of achieving the exchange.

front The line or zone of separation between two air masses of different temperatures and humidities.

frontal precipitation *See* cyclonic precipitation.

frontier That portion of a country adjacent to its boundaries and fronting another political unit.

frontier zone A belt lying between two states or between settled and uninhabited or sparsely settled areas.

functional dispute A disagreement between neighboring states over policies to be applied to their common border; often induced by differing customs regulations, movement of nomadic groups, or illegal immigration or emigration.

functional region A region differentiated by what occurs within it rather than by a homogeneity of physical or cultural phenomena; an earth area recognized as an operational unit based on defined organizational criteria.

G

gated community An enclosed, master-planned residential development to which entry is restricted.

gathering industries Primary activities involving the harvesting of renewable natural resources of land or water; commercial gathering usually implies forestry and fishing industries.

gender The socially created, not biologically based, distinctions between femininity and masculinity.

gender empowerment measure (GEM) A statistic summarizing the extent of economic, political, and professional participation of women in the society of which they are members; a measure of relative gender equality.

gene flow The passage of genes characteristic of one breeding population into the gene pool of another by interbreeding.

genetic drift A chance modification of gene composition occurring in an isolated population and becoming accentuated through inbreeding.

gentrification The process by which middle- and high-income groups refurbish and rehabilitate housing in deteriorated inner-city areas, thereby displacing low-income populations.

geodetic control data The information specifying the horizontal and vertical positions of a place.

geographic data base In cartography, a digital record of geographic information.

geographic information system (GIS) A configuration of computer hardware and software for assembling, storing, manipulating, analyzing, and displaying geographically referenced information.

geometric boundary (*syn:* artificial boundary) A boundary without obvious physical geographic basis; often a section of a parallel of latitude or a meridian of longitude.

geomorphology The scientific study of landform origins, characteristics, and evolutions and their processes.

geopolitics The study of the economic, political, and military value of space.

geothermal energy Energy that is generated by harnessing the naturally occurring steam and hot water produced by contact with heated rocks in the earth's crust.

gerrymander To divide an area into voting districts in such a way as to give one political party an unfair advantage in elections, to fragment voting blocks, or to achieve other nondemocratic objectives.

glacial till The deposits of rocks, silt, and sand left by a glacier after it has receded.

glacial trough A deep, U-shaped valley or trench formed by glacial erosion.

glacier A huge mass of slowly moving land ice.

Global Positioning System (GPS) A method of using satellite observations for the determination of extremely accurate locational information.

global warming A rise in surface temperatures on earth, a process believed by some to be caused by human activities that increase the concentration of greenhouses gases in the atmosphere, magnifying the greenhouse effect.

globe properties The characteristics of the grid system of longitude and latitude on a globe.

GPS *See* Global Positioning System.

gradational processes The processes of weathering, gravity transfer, and erosion that are responsible for the reduction of the land surface.

grade (of coal) A classification of coals based on their content of waste materials.

graphic scale A graduated line included in a map legend by means of which distances on the map may be measured in terms of ground distances.

gravity model A mathematical prediction of the interaction between two bodies as a function of their size and of the distance separating them.

gravity transfer The downward movement of material at or near the earth's surface due to the gravitational attraction of the earth's mass.

great circle A circle formed by the intersection of the surface of a globe with a plane passing through the center of the globe. The equator is a great circle; meridians are one half of a great circle.

greenbelt A ring of parks, farmland, or undeveloped land around a community.

greenhouse effect The heating of the earth's surface as shortwave solar energy passes through the atmosphere, which is transparent to it but opaque to reradiated long wave terrestrial energy. Also refers to increasing the opacity of the atmosphere through the addition of increased amounts of carbon dioxide, nitrous oxides, methane, and chlorofluorocarbons.

greenhouse gases Carbon dioxide, chlorofluorocarbons, methane gas, and nitrous oxide.

Green Revolution The term suggesting the great increases in food production, primarily in subtropical areas, accomplished through the introduction of very high-yielding grain crops, particularly wheat and rice.

Greenwich Mean Time (GMT) Local time at the prime meridian (zero degrees longitude), which passes through the observatory at Greenwich, England.

grid system The set of imaginary lines of latitude and longitude that intersect at right angles to form a system of reference for locating points on the earth.

gross national product (GNP) The total value of all goods and services produced by a country per year.

groundwater Underground water that accumulates below the water table in the pores and cracks of rock and soil, supplying water to wells and springs.

H

half-life The time required for one-half of the atomic nuclei of an isotope to decay.

hazardous waste Discarded solid, liquid, or gaseous material that may pose a substantial threat to human health or the environment when it is improperly disposed of, stored, or transported.

heartland theory The belief of Halford Mackinder that the interior of Eurasia provided a likely base for world conquest.

herbicide A chemical that kills plants, especially weeds. *See also* biocide, pesticide.

hierarchical diffusion The process by which contacts between people and the resulting diffusion of things or ideas occurs first among those at the same level of a hierarchy and then among elements at a lower level of the hierarchy (e.g., small town residents acquire ideas or articles after they are common in large cities).

hierarchical migration The tendency for individuals to move from small places to larger ones.

hierarchy of central places The steplike series of urban units in classes differentiated by both size and function.

high-level waste Nuclear waste that can remain radioactive for thousands of years, produced principally by the generation of nuclear power and the manufacture of nuclear weapons.

hinterland An outlying region that furnishes raw materials or agricultural products to the heartland; the market area or region served by a town or city.

homeostatic plateau The equilibrium level of population that can be supported adequately by available resources; equivalent to carrying capacity.

humid continental climate A climate of east coast and continental interiors of midlatitudes, displaying large annual temperature ranges resulting from cold winters and hot summers; precipitation at all seasons.

humid subtropical climate A climate of the east coast of continents in lower middle latitudes, characterized by hot summers with convectional precipitation and cool winters with cyclonic precipitation.

humus Dark brown or black decomposed organic matter in soils.

hunting-gathering An economic and social system based primarily or exclusively on the hunting of wild animals and the gathering of food, fiber, and other materials from uncultivated plants.

hurricane A severe tropical cyclone with winds exceeding 120 kilometers per hour (75 mph) originating in the tropical region of the Atlantic Ocean, Caribbean Sea, or Gulf of Mexico.

hydrologic cycle The system by which water is continuously circulated through the biosphere.

hydropower The kinetic energy of moving water converted into electrical power by a power plant whose turbines are driven by flowing water.

hydrosphere All water at or near the earth's surface that is not chemically bound in rocks; includes the oceans, surface waters, groundwater, and water held in the atmosphere.

I

iconography In political geography, the study of symbols that unite a country.

ideological subsystem The complex of ideas, beliefs, knowledge, and means of their communication that characterize a culture.

igneous rock Rock that is formed as molten earth materials cool and harden either above or below the earth's surface.

incinerator A facility designed to burn waste.

inclination The tilt of the earth's axis about $23\,^1/_2°$ away from the perpendicular.

Industrial Revolution The term applied to the rapid economic and social changes in agriculture and manufacturing that followed the introduction of the factory system to the textile industry of England in the last quarter of the 18th century.

infant mortality rate A refinement of the death rate to specify the ratio of deaths of infants age one year or less per 1000 live births.

infrared Electromagnetic radiation having wavelengths greater than those of visible light.

infrastructure The basic structure of services, installations, and facilities needed to support industrial, agricultural, and other economic development.

innovation Introduction into an area of new ideas, practices, or objects; an alteration of custom or culture that originates within the social group itself.

insolation The solar radiation received at the earth's surface.

intensive agriculture The application of large amounts of capital and/or labor per unit of cultivated land to increase output; may have either commercial or subsistence orientation.

intensive commercial agriculture *See* intensive agriculture.

intensive subsistence agriculture *See* intensive agriculture.

interaction model *See* gravity model.

International Date Line By international agreement, the designated line where each new day begins; generally following the 180th meridian.

intrusive rock Rock resulting from the hardening of magma beneath the earth's surface.

irredentism The desire of a state to gain or regain territory inhabited by people who have historic or cultural links to the country but who now live in a neighboring state.

isochrone A line connecting points that are equidistant in travel time from a common origin.

isoline A map line connecting points of constant value, such as a contour line or an isobar.

isotropic plain A hypothetical portion of the earth's surface where, it is assumed, the land is everywhere the same and the characteristics of the inhabitants are everywhere similar.

J

J-curve A curve shaped like the letter J, depicting exponential or geometric growth (1, 2, 4, 8, 16 . . .).

jet stream A meandering belt of strong winds in the upper atmosphere; significant because it guides the movement of weather systems.

K

karst topography A limestone region marked by sinkholes, caverns, and underground streams.

kerogen A waxy, organic material occurring in oil shales that can be converted into crude oil by distillation.

kinetic energy The energy that results from the motion of a particle or body.

L

land breeze Airflow from the land toward the sea, resulting from a nighttime pressure gradient that moves winds from the cooler land surface to the warmer sea surface.

landform region A large section of the earth's surface characterized by a great deal of homogeneity among types of landforms.

landlocked state A state that lacks a seacoast.

Landsat satellite One of a series of continuously orbiting satellites that carry scanning instruments to measure reflected light in both the visible and near-infrared portions of the spectrum.

landscape A term referring to the appearance of an area and to the items comprising that appearance. A distinction is often made between "physical landscapes" confined to landforms, natural vegetation, soils, etc., and "cultural landscapes" (q.v.).

language An organized system of speech by which people communicate with each other with mutual comprehension.

language family A group of languages thought to have descended from a single, common ancestral tongue.

lapse rate The rate of change of temperature with altitude in the troposphere; the average lapse rate is about 6.4°C per 1000 meters (3.5°F per 1000 ft).

large-scale map A representation of a small land area, usually with a representative fraction of 1:75,000 or less.

latitude A measure of distance north or south of the equator, given in degrees.

lava Molten material that has emerged onto the earth's surface.

Law of the Sea Convention A code of sea law approved by the United Nations in 1982 that authorizes, among other provisions, territorial waters extending 12 nautical miles from shore and 200-nautical-mile-wide exclusive economic zones; generally referred to as UNCLOS.

leachate The contaminated liquid discharged from a sanitary landfill to either surface or subsurface land or water.

least cost theory (*syn:* Weberian analysis) The view that the optimum location of a manufacturing establishment is at the place where the costs of transport and labor and the advantages of agglomeration or dispersion are most favorable.

levee In agriculture, a continuous embankment surrounding areas to be flooded. *See also* natural levee.

lingua franca Any of the various auxiliary languages used as common tongues among people of an area where several languages are spoken.

liquefied natural gas (LNG) Methane gas that has been liquefied by refrigeration for storage or transportation.

lithosphere The solid shell of rocks resting on the asthenosphere.

loam Agriculturally productive soil containing roughly equal parts of sand, silt, and clay.

locational tradition One of the four traditions of geography; in this text, identified with economic, resource, and urban geography.

loess A deposit of wind-blown silt.

longitude A measure of distance east or west of the prime meridian, given in degrees.

longshore current A current that moves roughly parallel to the shore and transports the sand that forms beaches and sand spits.

low-level waste Hazardous material whose radioactivity will decay to safe levels in 100 years or less, produced principally by industries and nuclear power plants.

M

magma Underground molten material.

malnutrition Food intake insufficient in quantity or deficient in quality to sustain life at optimal conditions of health.

Malthus Thomas R. Malthus (1766–1834), English economist, demographer, and cleric, who suggested that unless checked by self-control, war, or natural disaster, population will inevitably increase faster than will the food supplies needed to sustain it.

map projection A method of transferring the grid system from the earth's curved surface to the flat surface of a map.

map scale *See* scale.

marine west coast climate A regional climate found on the west coast of continents in upper midlatitudes, rainy all seasons with relatively cool summers and relatively mild winters.

material culture The tangible, physical items produced and used by members of a specific culture group and reflective of their traditions, lifestyles, and technologies.

maximum sustainable yield The maximum rate at which a renewable resource can be exploited without impairing its ability to be renewed or replenished.

mechanical weathering The physical disintegration of earth materials, commonly by frost action, root action, or the development of salt crystals.

Mediterranean climate A climate of lower midlatitudes characterized by mild, wet winters and hot, dry, sunny summers.

megalopolis An extensive, heavily populated urban complex with contained open, nonurban land, created through the spread and merging of separate metropolitan areas; (*cap.*) the name applied to the continuous functionally urban area of the northeastern seaboard of the United States from Maine to Virginia.

megawatt A unit of power equal to 1 million watts (1000 kilowatts) of electricity.

mental map A map drawn to represent the mental image(s) a person has of an area.

mentifacts The central, enduring elements of a culture that express its values and beliefs, including language, religion, folklore, artistic traditions, and the like. Elements in the ideological subsystem of culture.

Mercator projection A true conformal cylindrical projection first published in 1569, useful for navigation.

meridian A north-south line of longitude; on the globe, all meridians are of equal length and converge at the poles.

mesa An extensive, flat-topped elevated tableland with horizontal strata, a resistant cap rock, and one or more steep sides; a large butte.

metamorphic rock Rock transformed from igneous and sedimentary rocks by earth forces that generate heat, pressure, or chemical reaction.

metropolitan area A large functional entity, perhaps containing several urbanized areas, discontinuously built up but operating as a coherent economic whole.

migration The movement of people or other organisms from one region to another.

migration field An area that sends major migration flows to or receives major flows from a given place.

mineral A natural inorganic substance that has a definite chemical composition and characteristic crystal structure, hardness, and density.

ministate An imprecise term for a state or territory small in both population and area. An informal definition accepted by the United Nations suggests a maximum of 1 million people combined with a territory of less than 700 square kilometers (270 sq mi).

monoculture An agricultural system dominated by a single crop.

monotheism The belief that there is only one God.

monsoon A wind system that reverses direction seasonally, producing wet and dry seasons; used especially to describe the wind system of South, Southeast, and East Asia.

moraine Any of several types of landforms composed of debris transported and deposited by a glacier.

mortality rate *See* death rate.

mountain breeze The downward flow of heavy, cool air at night from mountainsides to lower valley locations.

multiple-nuclei model The idea that large cities develop by peripheral spread not from one central business district but from several nodes of growth, each of specialized use.

multiplier effect The expected addition of nonbasic workers and dependents to a city's total employment and population that accompanies new basic employment.

N

nation A culturally distinctive group of people occupying a particular region and bound together by a sense of unity arising from shared ethnicity, beliefs, and customs.

nationalism A sense of unity binding the people of a state together; devotion to the interests of a particular nation; an identification with the state and an acceptance of national goals.

nation-state A state whose territory is identical to that occupied by a particular nation.

natural boundary (*syn:* physical boundary) A boundary line based on recognizable physiographic features, such as mountains, rivers, or deserts.

natural gas A mixture of hydrocarbons and small quantities of nonhydrocarbons existing in a gaseous state or in solution with crude oil in natural reservoirs.

natural hazard A process or event in the physical environment that has consequences harmful to humans.

natural increase The growth of a population through excess of births over deaths, excluding the effects of immigration or emigration.

natural levee An embankment on the sides of a meandering river formed by deposition of silt during floods.

natural resource A physically occurring item that a population perceives to be necessary and useful to its maintenance and well-being.

natural selection The process of survival and reproductive success of individuals or groups best adjusted to their environment, leading to the perpetuation of those genetic qualities most suited to that environment.

natural vegetation The plant life that would exist in an area if humans did not interfere with its development.

neo-Malthusianism The advocacy of population control programs to preserve and improve general national prosperity and well-being.

neritic zone The relatively shallow part of the sea that lies above the continental shelf.

net migration The difference between in-migration and out-migration of an area.

network cities Two or more nearby cities, potentially complementary in function, that cooperate by developing transportation links and communications infrastructure.

niche The place an organism or species occupies in an ecosystem.

nomadic herding The migratory but controlled movement of livestock solely dependent upon natural forage.

nonbasic sector Those economic activities of an urban unit that service the resident population.

nonecumene (*syn:* anecumene) The portion of the earth's surface that is uninhabited or only temporarily or intermittently inhabited. *See also* ecumene.

nonfuel mineral resource A mineral used for purposes other than providing a source of energy.

nonmaterial culture The oral traditions, songs, and stories of a culture group along with its beliefs and customary behaviors.

nonpoint source of pollution Pollution from a broad area, such as one of fertilizer or pesticide application, rather than from a discrete source.

nonrenewable resource A natural resource that is not replenished or replaced by natural processes or is used at a rate that exceeds its replacement rate.

North and South poles The end points of the axis about which the earth spins.

North Atlantic drift The massive movement of warm water in the Atlantic Ocean from the Caribbean Sea and Gulf of Mexico in a northeasterly direction to the British Isles and the Scandinavian peninsula.

nuclear fission The controlled splitting of an atom to release energy.

nuclear fusion The combining of two atoms of deuterium into a single atom of helium in order to release energy.

nuclear power Electricity generated by a power plant whose turbines are driven by steam produced by the fissioning of nuclear fuel in a reactor.

nutrient A mineral or other element an organism requires for normal growth and development.

oil shale Sedimentary rock containing solid organic material (kerogen) that can be extracted and converted into a crude oil by distillation.

organic Derived from living organisms; plant or animal life.

Organization of Petroleum Exporting Countries (OPEC) An international cartel composed of 11 countries that aims at pursuing common oil-marketing and pricing policies.

orographic precipitation The rain or snow caused when warm, moisture-laden air is forced to rise over hills or mountains in its path and is thereby cooled.

orthophotomap An aerial photograph to which a grid system and certain map symbols have been added.

out-sourcing Producing parts or products abroad for domestic use or sale.

outwash plain A gently sloping area in front of a glacier composed of neatly stratified glacial till carried out of the glacier by meltwater streams.

overburden Soil and rock of little or no value that overlies a deposit of economic value, such as coal.

overpopulation A value judgment that the resources of an area are insufficient to sustain adequately its present population numbers.

oxbow lake A crescent-shaped lake contained in an abandoned meander of a river.

ozone A gas molecule consisting of 3 atoms of oxygen (O_3) formed when diatomic oxygen (O_2) is exposed to ultraviolet radiation. As a damaging component of photochemical smog formed at the earth's surface, it is a faintly blue, poisonous agent with a pungent odor.

ozone layer A layer of ozone in the high atmosphere that protects life on earth by absorbing ultraviolet radiation from the sun.

P

Pangaea The name given to the supercontinent that is thought to have existed 200 million years ago.

parallel of latitude An east-west line indicating the distance north or south of the equator.

PCBs Polychlorinated biphenyls; compounds containing chlorine that can be biologically magnified in the food chain.

peak value intersection The most accessible and costly parcel of land in the central business district and, therefore, in the entire urbanized area.

perforated state A state whose territory is interrupted ("perforated") by a separate, independent state totally contained within its borders.

permafrost Permanently frozen subsoil.

perpetual resource A resource that comes from an inexhaustible source, such as the sun, wind, and tides.

pesticide A chemical that kills insects, rodents, fungi, weeds, and other pests. *See also* biocide, herbicide.

Peters projection An equal-area cylindrical projection developed by Arno Peters that purports to show developing countries in proper proportion to one another.

petroleum A general term applied to oil and oil products in all forms, such as crude oil and unfinished oils.

pH factor The measure of the acidity/alkalinity of soil or water, on a scale of 0 to 14, rising with increasing alkalinity.

photochemical smog A form of air pollution produced by the interaction of hydrocarbons and oxides of nitrogen in the presence of sunlight.

photovoltaic (PV) cell A device that converts solar energy directly into electrical energy. *See also* solar power.

physical boundary *See* natural boundary.

physiological density The number of persons per unit area of agricultural land. *See also* population density.

pidgin An auxiliary language derived, with reduction of vocabulary and simplification of structure, from other languages. Not a native tongue, it is employed to provide a mutually intelligible vehicle for limited transactions of trade or administration.

pixel An extremely small sensed unit of a digital image.

place utility 1. The perceived attractiveness of a place in its social, economic, or environmental attributes; 2. the value imparted to goods or services by tertiary activities that provide things needed in specific markets.

planar projection Any of several map projections based on the projection of the globe grid onto a plane.

planned economy The production of goods and services, usually consumed or distributed by a governmental agency, in quantities and at prices determined by governmental programs.

plantation A large agricultural holding, frequently foreign-owned, devoted to the production of a single export crop.

plate tectonics The theory that the lithosphere is divided into plates that slide or drift very slowly over the asthenosphere.

playa A temporary lake or lake bed found in a desert environment.

Pleistocene The geological epoch dating from 2 million to about 10,000 years ago during which four stages of continental glaciation occurred.

point source of pollution Pollution originating from a discrete source, such as a smokestack or the outflow from a pipe.

pollution The presence in the biosphere of substances that, because of their quantity, chemical nature, or temperature, have a negative impact on the ecosystem or that cannot be readily disposed of by natural recycling processes.

polychlorinated biphenyls *See* PCBs.

polytheism The belief in or worship of many gods.

popular culture The constantly changing mix of material and nonmaterial elements available through mass production and the mass media to an urbanized, heterogeneous, nontraditional society.

population density (*syn:* crude density) A measurement of the numbers of persons per unit area of land within predetermined limits, usually political or census boundaries. *See also* physiological density.

population geography That branch of human geography dealing with the number, composition, and distribution of humans in relation to variations in earth-space conditions.

population momentum (*syn:* demographic momentum) The tendency for population growth to continue despite stringent family planning programs because of a relatively high concentration of people in the childbearing years.

population projection A report of future size, age, and sex composition of a population based on assumptions applied to current data.

population pyramid A graphic depiction of the age and sex composition of a (usually national) population.

positional dispute In political geography, disagreement about the actual location of a boundary.

possibilism The philosophical viewpoint that the physical environment offers human beings a set of opportunities from which (within limits) people may choose according to their cultural needs and technological awareness.

potential energy The energy stored in a particle or body.

potentially renewable resource A resource that can last indefinitely if its natural replacement rate is not exceeded; examples include forests, groundwater, and soil.

precipitation All moisture, solid and liquid, that falls to the earth's surface from the atmosphere.

pressure gradient Differences in air pressure between areas that induce air to flow from areas of high pressure to areas of low pressure.

primary activities Those parts of the economy involved in making natural resources available for use or further processing; includes mining, agriculture, forestry, fishing or hunting, grazing.

primate city A country's leading city, much larger and functionally more complex than any other; usually the capital city and a center of wealth and power.

prime meridian An imaginary line passing through the Royal Observatory at Greenwich, England, serving by agreement as the zero degree line of longitude.

projection 1. *See* map projection; 2. an estimate of future conditions based on current trends.

prorupt state A state of basically compact form that has one or more narrow extensions of territory.

proto-language A recorded or assumed language ancestral to one or more contemporary languages or dialects.

proved reserves That portion of a natural resource that has been identified and can be extracted profitably with current technology.

psychological distance The way an individual perceives distance.

pull factor A characteristic of a region that acts as an attractive force, drawing migrants from other regions.

purchasing power parity (PPP) A monetary measurement that takes account of what money actually buys in each country.

push factor A characteristic of a region that contributes to the dissatisfaction of residents.

Q

quaternary activity That employment concerned with research, with the gathering or disseminating of information, and with administration, including administration of the other economic activity levels.

R

race A subset of human population whose members share certain distinctive, inherited biological characteristics.

radar A device for detecting distant objects by analysis of very high frequency radio waves beamed at and reflected from their surfaces.

rank (of coal) A classification of coals based on their age and energy content; those of higher rank are more mature and richer in energy.

rank-size rule An observed regularity in the city-size distribution of some countries. In a rank-size hierarchy, the population of any given town will be inversely proportional to its rank in the hierarchy; that is, the nth-ranked city will be 1/n the size of the largest city.

rate The frequency of occurrence of an event during a specified time period.

rate of natural increase The birth rate minus the death rate, suggesting the annual rate of population growth without considering net migration.

recycling The reuse of disposed materials after they have passed through some form of treatment (e.g., melting down glass bottles to produce new bottles).

reflection The process of returning to outer space some of the earth's received insolation.

region In geography, the term applied to an earth area that displays a distinctive grouping of physical or cultural phenomena or is functionally united as a single organizational unit.

regional autonomy A measure of self-governance for a subdivision of a country.

regional concept The view that physical and cultural phenomena on the surface of the earth are rationally arranged by complex but comprehensible spatial processes.

regionalism In political geography, minority-group identification with a particular region of a state rather than with the state as a whole.

relative direction (*syn:* relational direction) A culturally based locational reference, as the Far West, the Old South, or the Middle East.

relative humidity A measure of the moisture content of the air, expressed as the amount of water vapor present relative to the maximum that can exist at the current temperature.

relative location The position of a place or activity in relation to other places or activities.

relic boundary A former boundary line that is still discernible and marked by some cultural landscape feature.

religion A value system that involves formal or informal worship and faith in the sacred and divine.

relocation diffusion The transfer of ideas, behaviors, or articles from one place to another through the migration of those possessing the feature transported; also, spatial relocation in which a phenomenon leaves an area of origin as it is transported to a new location. *See also* expansion diffusion.

remote sensing Any of several techniques of obtaining images of an area without having the sensor in direct physical contact with it, as by air photography or satellite sensors.

renewable resource A naturally occurring material that is potentially inexhaustible, either because it flows continuously (such as solar radiation or wind) or is renewed within a short period of time (such as biomass). *See also* sustained yield.

replacement level The number of children per family just sufficient to keep total population constant. Depending on mortality conditions, replacement level is usually calculated to be between 2.1 and 2.5 children.

representative fraction (RF) The scale of a map expressed as a ratio of a unit of distance on the map to distance measured in the same unit on the ground (e.g., 1:250,000).

reradiation A process by which the earth returns solar energy to space; some of the shortwave solar energy that is absorbed into the land and water is returned to the atmosphere in the form of long wave terrestrial radiation.

resource *See* natural resource.

resource dispute A disagreement over the control or use of shared resources, such as boundary rivers or jointly claimed fishing grounds.

return migration The stream of migrants who subsequently decide to return to their point of origin.

rhumb line A line of constant compass bearing; it cuts all meridians at the same angle.

Richter scale A logarithmic scale used to express the magnitude of an earthquake.

rimland theory The belief of Nicholas Spykman that domination of the coastal fringes of Eurasia would provide a base for world conquest.

S

Sahel The semiarid zone between the Sahara desert and the savanna area to the south in West Africa; district of recurring drought, famine, and environmental degradation.

salinization The concentration of salts in the topsoil as a result of the evaporation of surface water; occurs in poorly drained soils in dry climates, often as a result of improper irrigation.

sandbar An offshore shoal of sand created by the backwash of waves.

sanitary landfill The disposal of solid wastes by spreading them in layers covered with enough soil or ashes to control odors, rats, and flies.

savanna A tropical grassland characterized by widely dispersed trees and experiencing pronounced yearly wet and dry seasons.

scale In cartography, the ratio between length or size of an area on a map and the actual length or size of that same area on the earth's surface; map scale may be represented verbally, graphically, or as a fraction. In more general terms, scale refers to the size of the area studied, from local to global.

S-curve The horizontal bending, or leveling, of an exponential J-curve.

sea breeze Airflow from the sea toward the land, resulting from a daytime pressure gradient that moves winds from the cooler sea surface onto the warmer land surface.

secondary activities Those parts of the economy involved in the processing of raw materials derived from primary activities; includes manufacturing, construction, power generation.

sector model A description of urban land uses as wedge-shaped sectors radiating outward from the central business district along transportation corridors. The radial access routes attract particular uses to certain sectors.

secularism An indifference to or rejection of religion and religious belief.

sedimentary rock Rock that is formed from particles of gravel, sand, silt, and clay that were eroded from already existing rocks.

seismic waves Vibrations within the earth set off by earthquakes.

self-determination The concept that nationalities have the right to govern themselves in their own state or territory, a right to self-rule.

shaded relief A method of representing the three-dimensional quality of an area by use of continuous graded tone to simulate the appearance of sunlight and shadows.

shale oil The crude oil resulting from the distillation of kerogen in oil shales.

shamanism A form of tribal religion based on belief in a hidden world of gods, ancestral spirits, and demons responsive only to a shaman, or interceding priest.

shifting cultivation (*syn:* slash and burn agriculture, swidden agriculture) Crop production of forest clearings kept in cultivation until their quickly declining fertility is lost. Cleared plots are then abandoned and new sites are prepared.

sinkhole A deep surface depression formed when ground collapses into a subterranean cavern.

site The place where something is located; the immediate surroundings and their attributes.

situation The location of something in relation to the physical and human characteristics of a larger region.

slash and burn agriculture *See* shifting cultivation.

small circle The line created by the intersection of a spherical surface with a plane that does not pass through its center.

small-scale map A representation of a large land area on which small features (e.g., highways, buildings) cannot be shown true to scale.

sociofacts The institutions and links between individuals and groups that unite a culture, including family structure and political, educational, and religious institutions; components of the sociological subsystem of culture.

sociological subsystem The totality of expected and accepted patterns of interpersonal relations common to a culture or subculture.

soil The thin layer of fine material that rests on bedrock and is capable of supporting plant life.

soil depletion The loss of some or all of the vital nutrients from soil.

soil erosion The wearing away and removal of soil particles from exposed surfaces by agents such as moving water, wind, or ice.

soil horizon A layer of soil distinguished from other soil zones by color, texture, and other characteristics resulting from soil-forming processes.

solar energy Radiation from the sun, which is transformed into heat primarily at the earth's surface and secondarily in the atmosphere.

solar power The radiant energy generated by the sun; sun's energy captured and directly converted for human use. *See also* photovoltaic cell.

solid waste The unwanted materials generated in production or consumption processes that are solid rather than liquid or gaseous in form.

source region In climatology, a large area of uniform surface and relatively consistent temperatures where an air mass forms.

southern oscillation The atmospheric conditions occurring periodically near Australia that create the El Niño condition off the coast of South America.

spatial diffusion The outward spread of a substance, concept, or population from its point of origin.

spatial distribution The arrangement of things on the earth's surface.

spatial interaction The movement (e.g., of people, goods, information) between different places.

spatial margin of profitability The set of points delimiting the area within which a firm's profitable operation is possible.

spatial search The process by which individuals evaluate the alternative locations to which they might move.

spine In urban geography, a continuation of the features of the central business district outward along the main wide boulevard characteristic of Latin American cities.

spring wheat Wheat sown in spring for ripening during the summer or autumn.

standard language A language substantially uniform with respect to spelling, grammar, pronunciation, and vocabulary and representing the approved community norm of the tongue.

standard parallel The tangent circle, usually a parallel of latitude, in a conic projection; along the standard line, the scale is as stated on the map.

state (*syn:* country) An independent political unit occupying a defined, permanently populated territory and having full sovereign control over its internal and foreign affairs.

steppe The name applied to treeless midlatitude grasslands.

strategic petroleum reserve (SPR) The petroleum stocks maintained by the federal government for use during periods of major supply interruption.

stratosphere The layer of the atmosphere that lies above the troposphere and extends outward to about 56 kilometers (35 mi).

stream load The eroded material carried by a stream in one of three ways, depending on the size and composition of the particles: 1. in dissolved form, 2. suspended by water, and 3. rolled along the stream bed.

subduction The process by which one lithospheric plate is forced down into the asthenosphere as a result of a collision with another plate.

subnationalism The feeling that one owes primary allegiance to a traditional group or nation rather than to the state.

subsequent boundary A boundary line that is established after the area in question has been settled, and that considers the cultural characteristics of the bounded area.

subsidence The settling or sinking of a portion of the land surface, sometimes as a result of the extraction of fluids such as oil or water from underground deposits.

subsistence agriculture Any of several farm economies in which most crops are grown for food, nearly exclusively for local consumption.

subsistence economy A system in which goods and services are created for the use of producers or their immediate families. Market exchanges are limited and of minor importance.

substitution principle In industry, the tendency to substitute one factor of production for another in order to achieve optimum plant location and profitability.

suburb A functionally specialized segment of a large urban complex located outside the boundaries of the central city.

superimposed boundary A boundary line placed over, and ignoring, an existing cultural pattern.

surface water Water that is on the earth's surface, such as in rivers, streams, reservoirs, lakes, and ponds.

sustained yield The practice of balancing harvesting with growth of new stocks in order to avoid depletion of the resource and ensure a perpetual supply.

swidden agriculture *See* shifting cultivation.

syncretism The development of a new form of, for example, religion or music, through the fusion of distinctive parental elements.

synfuel *See* synthetic fuel

syntax The way words are put together in phrases and sentences.

synthetic fuel (*syn*: synfuel) Crude oil and natural gas substitutes that can be synthesized from a variety of nonoil and nongas sources, including coal, tar sands, and wood.

systems analysis An approach to the study of large systems through
1. segregation of the entire system into its component parts,
2. investigation of the interactions between system elements, and
3. study of inputs, outputs, flows, interactions, and boundaries within the system.

T

talus slope A landform composed or rock particles that have accumulated at the base of a cliff, hill, or mountain.

tar sand Sand and sandstone impregnated with heavy oil.

technological subsystem The complex of material objects together with the techniques of their use by means of which people carry out their productive activities.

technology An integrated system of knowledge and skills developed within a culture to carry out successfully purposeful and productive tasks.

tectonic forces The processes that shape and reshape the earth's crust, the two main types being diastrophic and volcanic.

temperature inversion The condition caused by rapid reradiation in which air at lower altitudes is cooler than air aloft.

territorial dispute The rival claims of adjacent states to an ethnically homogeneous territory divided by a superimposed boundary.

territoriality The persistent attachment of most animals to a specific area; the behavior associated with the defense of the home territory.

territorial production complex In the economic planning of the former Soviet Union, a design for large regional industrial, mining, and agricultural development leading to regional self-sufficiency and the creation of specialized production for a larger national market.

tertiary activities Those parts of the economy that fulfill the exchange function and that provide market availability of commodities; includes wholesale and retail trade and associated transportation, government, and information services.

thermal pollution The introduction of heated water into the environment, with consequent adverse effects on aquatic life.

thermal scanner A remote sensing device that detects the energy (heat) radiated by objects on earth.

Third World Originally (in the 1950s), designating countries uncommitted to either the "First World" Western capitalist bloc or the Eastern "Second World" communist bloc; subsequently, a term applied to countries considered not yet fully developed or in a state of underdevelopment in economic and social terms.

threshold In economic geography, the minimum market needed to support the supply of a product or service.

topographic map One that portrays the surface features of a relatively small area, often in great detail.

toponym A place name.

toponymy The place names of a region or, especially, the study of place names.

tornado A small, violent storm characterized by a funnel-shaped cloud of whirling winds that can form beneath a cumulonimbus cloud in proximity to a cold front and that moves at speeds as high as 480 kilometers per hour (300 mph).

total fertility rate (TFR) The average number of children that would be born to each woman if, during her childbearing years, she bore children at the current year's rate for women that age.

town A nucleated settlement that contains a central business district but that is smaller and less functionally complex than a city.

traditional religion *See* tribal religion.

tragedy of the commons The observation that in the absence of collective control over the use of a resource available to all, it is to the advantage of all users to maximize their separate shares even though their collective pressures may diminish total yield or destroy the resource altogether.

transform fault A break in rocks that occurs when one lithospheric plate slips past another in a horizontal motion.

transnational corporation (TNC) A large business organization operating in at least two separate national economies.

tribal religion (*syn:* traditional religion) An ethnic religion specific to a small, localized, preindustrial culture group.

tropical rain forest The tree cover composed of tall, high-crowned evergreen deciduous species, associated with the continuously wet tropical lowlands.

tropical rain forest climate The continuously warm, frost-free climate of tropical and equatorial lowlands, with abundant moisture year-round.

troposphere The atmospheric layer closest to the earth, extending outward about 11 to 13 kilometers (7 to 8 mi) at the poles to about 26 kilometers (16 mi) at the equator.

truck farming The intensive production of fruits and vegetables for market rather than for processing or canning.

tsunami A seismic sea wave generated by an earthquake or volcanic eruption.

tundra The treeless area lying between the tree line of arctic regions and the permanently ice-covered zone.

typhoon The name given to hurricanes occurring in the western Pacific Ocean region.

U

ubiquitous industry A market-oriented industry whose establishments are distributed in direct proportion to the distribution of population (market).

underpopulation A value statement reflecting the view that an area has too few people in relation to its resources and population-supporting capacity.

unitary state A state in which the central government dictates the degree of local or regional autonomy and the nature of local governmental units; a country with few cultural conflicts and a strong sense of national identity.

United Nations Convention on the Law of the Sea (UNCLOS) *See* Law of the Sea Convention.

universalizing religion A religion that claims global truth and applicability and seeks the conversion of all humankind.

urban hierarchy The steplike series of urban units (e.g., hamlets, villages, towns, cities, metropolises) in classes differentiated by size and function.

urban influence zone An area outside of a city that is nevertheless affected by the city.

urbanization The transformation of a population from rural to urban status; the process of city formation and expansion.

urbanzied area A continuously built-up urban landscape defined by building and population densities with no reference to the political boundaries of the city; it may contain a central city and many contiguous towns, cities, suburbs, and unincorporated areas.

V

valley breeze The flow of air up mountain slopes during the day.

variable costs In economic geography, the costs of production inputs that change as the level of production changes. They differ from the costs incurred by agricultural or industrial firms that are fixed and do not change as the amount of production changes.

verbal scale A statement of the relationship between units of measure on a map and distance on the ground, as "one inch represents one mile."

vernacular 1. The nonstandard indigenous language or dialect of a locality; 2. of or related to indigenous arts and architecture, such as a vernacular house; 3. of or related to the perceptions and understandings of the general population, such as a vernacular region.

volcanism The earth force that transports subsurface materials (often heated, sometimes molten) to or toward the surface of the earth.

von Thünen model The model developed by Johann H. von Thünen (1783–1850) to explain the forces that control the prices of agricultural commodities and how those variable prices affect patterns of agricultural land utilization.

von Thünen rings The concentric zonal pattern of agricultural land use around a single market center proposed in the von Thünen model.

W

warping The bowing of a large region of the earth's surface due to the movement of continents or the melting of continental glaciers.

wash A dry, braided channel in the desert that remains after the rush of rainfall runoff water.

water table The upper limit of the saturated zone and therefore of groundwater; the top of the water within an aquifer.

weather The state of the atmosphere at a given time and place.

weathering The mechanical and chemical processes that fragment and decompose rock materials.

Weberian analysis *See* least cost theory.

Weber model The analytical model devised by Alfred Weber (1868–1958) to explain the principles governing the optimum location of industrial establishments.

wetland An area that is either occasionally or permanently saturated with moisture, such as a marsh or tidal flat.

wind power The kinetic energy of wind converted into mechanical energy by wind turbines that drive generators to produce electricity.

winter wheat Wheat planted in autumn for early summer harvesting.

world cities Cities that are control points for international production and marketing and for international finance.

Z

zero population growth (ZPG) A situation in which a population is not changing in size from year to year, as a result of the combination of births, deaths, and migration.

zoning Designating by ordinance areas in a municipality for particular types of land use.

INDEX

Page numbers in **boldface** indicate glossary terms. Page numbers followed by italic *f* and *t* indicate figures and tables, respectively.